International Federation of Automatic Control

# INTELLIGENT COMPONENTS
# AND INSTRUMENTS FOR
# CONTROL APPLICATIONS

IFAC Symposia Series, 1993. Number 3

# IFAC SYMPOSIA SERIES

Janos Gertler, *Editor-in-Chief*, George Mason University, School of Information Technology and Engineering, Fairfax, VA 22030-4444, USA

---

DHURJATI & STEPHANOPOULOS: On-line Fault Detection and Supervision in the Chemical Process Industries *(1993, No.1)*
BALCHEN *et al*: Dynamics and Control of Chemical Reactors, Distillation Columns and Batch Processes *(1993, No.2)*
OLLERO & CAMACHO: Intelligent Components and Instruments for Control Applications *(1993, No.3)*
ZAREMBA: Information Control Problems in Manufacturing Technology *(1993, No.4)*
STASSEN: Analysis, Design and Evaluation of Man-Machine Systems *(1993, No.5)*
RODD & VERBRUGGEN: Artificial Intelligence in Real-Time Control *(1993, No.6)*
FLIESS: Nonlinear Control Systems Design *(1993, No.7)*
DUGARD, M'SAAD & LANDAU: Adaptive Systems in Control and Signal Processing *(1993, No.8)*
TU XUYAN: Modelling and Control of National Economies *(1993, No.9)*
LIU, CHEN & ZHENG: Large Scale Systems: Theory and Applications *(1993, No.10)*
GU YAN & CHEN ZHEN-YU: Automation in Mining, Mineral and Metal Processing *(1993, No.11)*
DEBRA & GOTTZEIN: Automatic Control in Aerospace *(1993, No.12)*
ALBERTOS & KOPACEK: Low Cost Automation *(1993, No.13)*
HARVEY & EMSPAK: Automated Systems Based on Human Skill (and Intelligence) *(1993, No.14)*

---

BARKER: Computer Aided Design in Control Systems *(1992, No.1)*
KHEIR *et al*: Advances in Control Education *(1992, No.2)*
BANYASZ & KEVICZKY: Identification and System Parameter Estimation *(1992, No.3)*
LEVIS & STEPHANOU: Distributed Intelligence Systems *(1992, No.4)*
FRANKE & KRAUS: Design Methods of Control Systems *(1992, No.5)*
ISERMANN & FREYERMUTH: Fault Detection, Supervision and Safety for Technical Processes *(1992, No.6)*
TROCH *et al*: Robot Control *(1992, No.7)*
NAJIM & DUFOUR: Advanced Control of Chemical Processes *(1992, No.8)*
WELFONDER, LAUSTERER & WEBER: Control of Power Plants and Power Systems *(1992, No.9)*
KARIM & STEPHANOPOULOS: Modeling and Control of Biotechnical Processes *(1992, No.10)*
FREY: Safety of Computer Control Systems 1992

## NOTICE TO READERS

---

## *AUTOMATICA* and *CONTROL ENGINEERING PRACTICE*

The editors of the IFAC journals *Automatica* and *Control Engineering Practice* always welcome papers for publication. Manuscript requirements will be found in the journals. Manuscripts should be sent to:

*Automatica*

Professor H A Kwakernaak
Deputy Editor-in-Chief
AUTOMATICA
Department of Applied
 Mathematics
University of Twente
P O Box 217, 7500 AE Enschede
The Netherlands

*Control Engineering Practice*

Professor M G Rodd
Editor-in-Chief, CEP
Institute for Industrial
 Information Technology Ltd
Innovation Centre
Singleton Park
Swansea SA2 8PP
UK

*For a free sample copy of either journal please write to:*

Pergamon Press Ltd
Headington Hill Hall
Oxford OX3 0BW, UK

Pergamon Press Inc
660 White Plains Road
Tarrytown, NY 10591-5153, USA

*Full list of IFAC publications appears at the end of this volume*

# INTELLIGENT COMPONENTS AND INSTRUMENTS FOR CONTROL APPLICATIONS

*Selected Papers from the IFAC Symposium, Malaga, Spain,*
*20 - 22 May 1992*

Edited by

## A. OLLERO
*University of Malaga, Spain*

and

## E.F. CAMACHO
*University of Seville, Spain*

Published for the

INTERNATIONAL FEDERATION OF AUTOMATIC CONTROL

by

PERGAMON PRESS

OXFORD • NEW YORK • SEOUL • TOKYO

| UK | Pergamon Press Ltd, Headington Hill Hall, Oxford OX3 0BW, England |
|---|---|
| USA | Pergamon Press, Inc., 660 White Plains Road, Tarrytown, New York 10591-5153, USA |
| KOREA | Pergamon Press Korea, KPO Box 315, Seoul 110-603, Korea |
| JAPAN | Pergamon Press Japan, Tsunashima Building Annex, 3-20-12 Yushima, Bunkyo-ku, Tokyo 113, Japan |

First edition 1993

**Library of Congress Cataloging in Publication Data**

Intelligent components and instruments for control applications: selected papers from the IFAC symposium, Malaga, Spain, 20-22 May, 1992/edited by A. Ollero and E.F. Camacho. — (IFAC symposia series)
"Published for the International Federation of Automatic Control."
Includes index.
1. Intelligent control systems—Congresses. I. Ollero, A. II. Camacho, E.F.
III. International Federation of Automatic Control. IV. Series.
TJ217.5.I53  1993      629.8—dc20      92-44354

**British Library Cataloguing in Publication Data**

A catalogue record for this book is available from the British Library

ISBN: 9780080418995

*These proceedings were reproduced by means of the photo-offset process using the manuscripts supplied by the authors of the different papers. The manuscripts have been typed using different typewriters and typefaces. The lay-out, figures and tables of some papers did not agree completely with the standard requirements: consequently the reproduction does not display complete uniformity. To ensure rapid publication this discrepancy could not be changed: nor could the English be checked completely. Therefore, the readers are asked to excuse any deficiencies of this publication which may be due to the above mentioned reasons.*

*The Editors*

*Transferred to digital print 2009*

*Printed and bound in Great Britain by CPI Antony Rowe, Chippenham and Eastbourne*

# IFAC SYMPOSIUM ON INTELLIGENT COMPONENTS AND INSTRUMENTS FOR CONTROL APPLICATIONS

*Sponsored by*
International Federation of Automatic Control (IFAC)
- Technical Committee on Components and Instruments

*Co-sponsored by*
Technical Committees on Applications, Computers, Education and System Engineering

*Organized by*
Universidad de Málaga (Departamento de Ingeniería de Sistemas y Automática)
 *on behalf of the* Comité Español de la IFAC

*Supported by*
IDEA - Andalucia Technology Park
CICYT - Plan Nacional de I+D
Universidad de Málaga
Sociedad Estatal EXPO'92
Junta de Andalucia

*International Programme Committee*

P. Albertos (E) (Chairman)
J. Aracil (E)
K. Åstrom (S)
T. Boromisza (H)
E.F. Camacho (E)
T. Fukuda (J)
V. Haase (A)
R. Hanus (B)
R. Isermann (D)
U. Jaaksoo (ESTONIA)
E.K. Juuso (SF)
H. Leskiewicz (PL)

I. Loe (N)
Y.Z. Lu (PRC)
F. Morant (E)
J.L. Nevins (USA)
A. Ollero (E)
M.G. Rodd (UK)
M. Staroswiecki (F)
G. Suski (USA)
N. Tamura (J)
H.R. Traenkler (D)
H.B. Verbruggen (NL)
I.B. Yadikin (RUSSIA)

*National Organizing Committee*

A. Ollero (Chairman)
R. Aracil
L. Basañez
E.F. Camacho (Vice-Chairman)
J.F. Cañete (Secretary)
A.G. Cerezo
S. Dormido
J.A.G. Fortes
J. González
J.M. Martín
M. Silva
F. Triguero
A. Vergara
J. Vicente

# PREFACE

This volume contains a selection of papers presented at the IFAC Symposium on Intelligent Components and Instruments for Control Applications (SICICA´92) held between May 20-22, 1992 in Málaga (Andalucia), Spain.

The Symposium was the first of a series sponsored by the Working Group on Intelligent Components and Instruments of the IFAC Components and Instruments Technical Committee. The SICICA´92 was Co-Sponsored by the IFAC Technical Committees on: Applications, Computers, Education and System Engineering. The annual IFAC Council and Related Meetings were held in conjunction with the SICICA´92. The Symposium took place in the Hotel Alay, located in Benalmádena, one of the villages of the Costa del Sol by the Mediterranean sea in southern Spain, 12 km. west of Málaga.

The Symposium was funded by the Spanish National Research and Development Program, the Research Program of the regional government of Andalusia, the Andalusia Technology Park in Málaga, the Expo'92 Society in Seville, and the University of Málaga.

The objective of the Symposium was to bring together control systems specialists, equipment manufacturers, and end-users, to evaluate techniques, components, and instruments for intelligent control. Intelligent Control is an emergent field involving new control techniques as well as new smart controllers, sensors and actuators. Software and hardware components to implement advanced control strategies and intelligent functions such as reasoning, learning, and perception, are usually considered in the intelligent control domain. The development of these control techniques has a great impact on a number of applications including Robotics and Process Control.

About 200 people from 25 different countries attended the Symposium. This attendance included Universities, R&D Institutes, Industries, and Engineering Companies. The success of the SICICA'92 was due to the interesting Technical Program, the IFAC support, and the collaboration of many people and local institutions in supporting the organization.

The International Program Commitee was composed of 24 specialists from 17 countries, chaired by Prof. P. Albertos. IFAC was represented at the highest level at SICICA'92 by its president, Prof. B.D.O. Anderson, who participated in the opening session. Prof. G. Ferraté, president of the IFAC Spanish NMO, also participated in this session.

The National Organizing Committee was lead by Prof. A. Ollero (Chairman), University of Málaga, and Prof. E.F. Camacho (Vice-Chairman), University of Seville. They edited a 835 page volume with the Symposium Preprints. The Symposium was organized by the System Engineering and Automation Department of Málaga University with the enthusiastic work of about 25 people including professors, researchers, and students.

The final program of the Symposium included six Plenary Sessions, giving an overview of the Symposium scope, one Invited Session, and 20 Sessions with 132 papers selected by the International Program Committee from the 230 extended abstracts submitted. Authors from 30 different countries have contributed to the Symposium.

The Technical Sessions covered intelligent control techniques (fuzzy control, neural networks, qualitative methods, learning, expert systems for control), actuators, sensors, image processing, computer vision, software, and communications. About one third of the program was devoted to applications including robotics (arms and mobile robots), power and process control, manufacturing, aerospace, and traffic control.

Because of the large number of papers presented at the SICICA´92 and the limited budget for the Proceedings publications, it has not been possible to include in this volume a number of good papers. A very strict selection has been made based on Session Chairmen reports and IPC reviewers.

The editors would like to thank all who have devoted their work to this Symposium and this volume. Particularly we thank Prof. Albertos, chairman of the IFAC Committee on Components and Instruments, who has offered great support to the Symposium since its creation and helped with the editing of this volume.

The success of the SICICA´92 and the emergent field of Intelligent Components and Instruments for Control Applications encourages the continuation of this symposium series.

The editors.

# CONTENTS

# NEURAL NETWORKS

# EXPERT SYSTEMS

# QUALITATIVE METHODS

## *INTEGRATING QUALITATIVE AND QUANTITATIVE KNOWLEDGE IN SIMULATION AND CONTROL*

## *CONTROL METHODS*

## *ACTUATORS*

## *SENSORS*

## SENSORS AND MANUFACTURING

## IMAGE PROCESSING AND COMPUTER VISION

## *ROBOT ARM CONTROL*

## *MOBILE ROBOTS*

## *CONTROL APPLICATIONS*

## COMMUNICATIONS AND SOFTWARE

# INFORMATION PROCESSING IN SENSING DEVICES AND MICROSYSTEMS

**H.R. Traenkler**

*Institute of Measurement and Automation, Federal Armed Forces University Munich,
D-8014 Neubiberg near Munich, Germany*

**Abstract**. With the development of microsystems combining sensors and actuators with analog and digital electronic on one chip the benefits of digital information processing of the sensor signals become more and more evident. This paper will give a survey of the possibilities of digital processing of sensor signals. Examples of realized signal processing algorithms cover correction of cross sensitivities to ambient temperature, diverse sensor system structures, methods for differential sensors, hysteresis and creep error correction and object detection with ultrasonic sensors.

**Keywords**. Digital signal processing, microsystem, sensors, error compensation, modeling, linearization techniques, reliability, redundancy, differential sensors, hysteresis, ultrasonic object detection.

## INTRODUCTION

Nowadays the digital processing of sensor signals becomes more and more important especially in combination with the development of microsystems. Here the term microsystems names the combination of microelectronics with sensors and actuators, usually monolithically on one chip, as shown in Fig. 1.

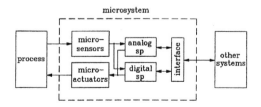

Fig. 1 Components of a microsystem

Both analog and digital signal processing (sp) units are implemented in such a microsystem, where the analog unit mainly is used for signal conditioning like amplification etc. . Additionally each system has a communication interface for the transmission of data to other systems or to a host computer. Because of that ability to communicate with each other, microsystems can be used as independent intelligent components within control systems.

Main advantages of microsystems are the reduction in size and weight combined with an increasing reliability of the whole system of sensor and signal electronic. Additionally sensors in microsystems will gain a higher cost-effectiveness than conventional sensors. The manufacturing of microsystems demands much technological experiences because of the necessary combination of analog and digital electronic on one chip. Microsystems offer the opportunity to develop powerful sensor systems, where the embedded microprocessor is used not only for the digital processing of the sensor signals but for the communication with other systems or a host computer, too.

The digital processing of sensor signals allows a more precise evaluation of sensor signals than analog methods. Especially the nonlinear correction of cross sensitivities, self-monitoring of sensors and the realization of complex evaluation algorithms like Fast Fourier Transformation (FFT), for instance, can be performed (Tränkler, 1988).

The general tasks of sensor signal processing in sensing devices can be summarized as follows:

- static sensor signal processing:
  - linearization
  - correction of cross sensitivities
  - hysteresis error correction
  - functional evaluation of signals
  - adaptive signal correction (self calibration)
- dynamic sensor signal processing:
  - filtering of interfering frequencies
  - dynamic correction
  - drift correction
  - control of dynamic behaviour at varying working conditions
  - creep correction
- complex sensor signal evaluation:
  e.g. evaluation of frequency or impulse responses

In microsystems, even more sophisticated tasks are to be performed by signal processing. In these employments, signal processing is inevitably a digital one because the evaluation of input signals and the generation of output signals of a complex sensor-actor-array can be done only digitally in an effective manner.

For multisensor arrays (either monolithically integrated or an accumulation of discrete sensors), for instance, different signal processing methods can be adequate, depending on the kind of array used. This array can consist of many sensors of the same kind or it can consist of different sensors, measuring all the same or different measuring quantities. Examples of tasks, which can be performed by such multisensor arrays are shown in TABLE 1.

TABLE 1 Examples for the Use of Multisensors

|  |  | one measuring quantity | several measuring quantities |
|---|---|---|---|
| same sensors | same position | reliability enhancement by redundancy | — |
| same sensors | different positions | image processing | image processing |
| different sensors | same position | reliability enhancement by diversity | • correction of cross sensitivities <br> • signal separation |
| different sensors | different positions | — | diagnosis systems |

An enhancement of the reliability of sensors, for instance, can be achieved both by a redundant array structure with identical sensor elements and also by a sensor array containing different sensor elements. In each case a certain strategy is necessary to decide which sensor signals can be relied on. Using several non-selective sensors within a sensor array even a selective measurement of measuring quantities is possible by signal separation if the used sensor elements show different cross sensitivities.

In general the development of algorithms for digital sensor signal processing is performed off-line during the manufacturing process by simulation methods using calibration data of the sensors. The following examples will give a survey of possible realizations of information processing in microsystems.

## DIGITAL CORRECTION OF CROSS SENSITIVITIES OF SENSORS TO AMBIENT TEMPERATURE

Nearly all sensors are sensitive not only to the measuring quantity, instead influences like the ambient temperature cause changes in their output signals, too. First step for the development of algorithms for a digital correction of these temperature influences to sensors has to be an analyzis of the sensor behaviour in its dependency on the measuring and the interfering quantity by calibration data. Using physical or mathematical models to describe the sensor characteristics, algorithms for the correction of the interfering quantity can be developed from this data. These algorithms then are valid for a whole kind of sensor elements, the adaptation to the single sensor specimen is done by calculating the parameters of the algorithm individually from calibration data.

Figures 2 and 3 give two typical examples of the temperature dependency of sensors. At the characteristic of the halleffect pressure sensor the temperature effect mainly causes a change in the gain and the offset of the sensor behaviour (Fig. 2), whereas at the gassensor a change in the basic gas sensitivity of the sensor results from the change of ambient temperature (Fig. 3). At the pressure sensor, the temperature has only an additional effect to the sensor output, therefore the pressure sensitivity of the sensor is not strongly affected by it. In the contrary at the gassensor the temperatur interferes dominantly with the principal gas sensing effect. This causes the enormous temperature effects in the sensitivity of the gassensor, which can be seen in Fig. 3. These different effects of ambient temperature to sensors have to be considered at the development of correction algorithms, too (Löschberger, 1991).

For the digital temperature correction of sensors it is advantageous to perform the correction by functional descriptions of the sensor behaviour in its dependency on the measuring quantity and the ambient temperature, where the temperature value

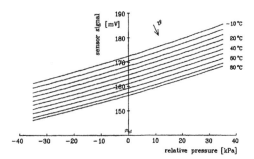

Fig. 2 Temperature characteristic of a halleffect pressure sensor

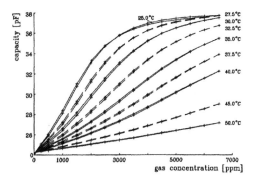

Fig. 3 Temperature characteristic of a capacitive gassensor

is obtained by a separate temperature sensor. In this case the corrected measuring value is easily obtained by only evaluating the describing equations of the sensor model using the measured signals from the both sensors, one for the original measuring value, the other for the ambient temperature. The use of look-up tables for sensor correction, which contain corrected output values corresponding to the temperature-affected original sensor signal, is memory-intensive. Moreover the calibration process for initializing these tables is very expensive in time and costs. If the actual temperature value is not obtained by a separate sensor but from a second signal of the sensor for the measuring quantity – e.g. by measuring the whole bridge resistance of an piezoresistive pressure sensor – then the two sensor signals cannot be separated and a coupled equation system has to be evaluated on-line to obtain the corrected measuring value.

Regarding the above mentioned differences in the temperature effects at different sensor principles, two ways of developing mathematical algorithms for digital temperature correction are possible. Physical model functions of the sensing effect are often not sufficient for the description of the sensor behaviour, because they do not consider additional effects coming from the manufacturing process of the sensor elements. If the temperature does not affect the principal sensing effect of a sensor, then the cor-

rection of the temperature influence to a sensor can be performed by a projection of the temperature-affected sensor output to the output of the sensor it would have at a certain reference temperature. Afterwards in a second step the measuring value is calculated from this corrected sensor signal. In the following this special kind of algorithms for temperature influence correction is called *base function method*. For a successful use of this method it is necessary, that the principal shape of the sensor characteristic will not be changed by temperature influences.

At the base function method the transformation of the temperature-affected sensor output $y(x, \vartheta)$ to the sensor output $y_N(x) = y(x, \vartheta_N)$ at the reference temperature $\vartheta_N$ is performed by the following exponential series:

$$
\begin{aligned}
y_N(x) = \; & \alpha_0(\vartheta) + [1 + \alpha_1(\vartheta)] \cdot y(x, \vartheta) + \\
& + \alpha_2(\vartheta) \cdot y^2(x, \vartheta) + \qquad (1) \\
& + \ldots + \alpha_n(\vartheta) \cdot y^n(x, \vartheta) \,,
\end{aligned}
$$

where the $\alpha_i(\vartheta)$ are polynomials of the temperature difference $\vartheta - \vartheta_N$:

$$
\begin{aligned}
\alpha_i(\vartheta) = \; & \beta_{i0} + \beta_{i1} \cdot (\vartheta - \vartheta_N) + \\
& + \beta_{i2} \cdot (\vartheta - \vartheta_N)^2 + \qquad (2) \\
& + \ldots + \beta_{im} \cdot (\vartheta - \vartheta_N)^m \,.
\end{aligned}
$$

Because of the necessary identity $y_N(x) = y(x, \vartheta = \vartheta_N)$ the parameters $\beta_{i0}$ in Eq. (2) have to be 0. The order of the exponential series and of the polynomial term of Eq. (2) can be determined from the shape of the sensor characteristics.

The linear equation system of Eq. (1) can be given in matrix notation $\mathbf{y}_N^* = \mathbf{U} \cdot \mathbf{b}$, where $\mathbf{y}_N^*$ names the vector of the differences $y_N(x_j) - y(x_j, \vartheta_k)$ at the calibration points $x_j$ at various temperatures $\vartheta_k$, $\mathbf{U}$ contains the appropriate products of $y(x_j, \vartheta_k)$ and $\vartheta_k$, and $\mathbf{b} = (\beta_{01}, \beta_{02}, \ldots \beta_{nm})^T$ is the vector of the parameters $\beta_{\nu\mu}$ of the correction model to be determined. The solution of this equation system is obtained by Multiple Linear Regression using the least squares method.

With this base function method the temperature effect of the pressure sensor shown in Fig. 2 is reduced to 0.75 % of the measuring range within an temperature range between -10 °C and 70 °C (see Fig. 4). This means an remaining temperature error of 0.09 %/10 K using the relatively simple correction model:

$$
y_N(p) = \beta_{01} \cdot \Delta\vartheta_k + [1 + \beta_{11} \cdot \Delta\vartheta_k] \cdot y(p, \vartheta_k) \,. \quad (3)
$$

For the correction of sensors whose principal sensing effect is greatly influenced by a change of ambient temperature, another method called *direct correction method* should be used. Here in a first step a mathematical description of the sensor behaviour at reference conditions is developed. Then in a second step this model will be enlarged to describe also

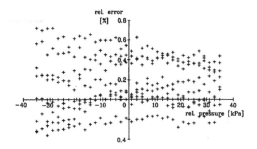

Fig. 4 Relative temperature error of the halleffect pressure sensor after digital temperature correction with the base function method

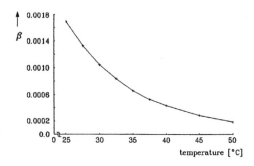

Fig. 5 Temperature dependency of parameter $\beta$

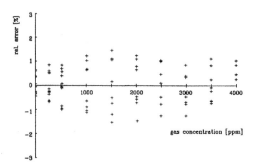

Fig. 6 Remaining relative temperature error of the gasssensor after temperature correction with the direct modelling method

the temperature behaviour of the sensor by examining the temperature dependencies of the particular model parameters.

The reference characteristic of the gassensor at a temperature of 25 °C shown in Fig. 3, for instance, can be described in its dependency on the gas concentration by the following model:

$$C(c) = C_0 + \frac{\Delta C}{1 + e^{(-\beta \cdot c + \gamma)}} . \qquad (4)$$

Here $C_0$ is the asymptotic minimum of the s-shaped capacitance characteristic of the sensor at a gas concentration $c$ of 0 ppm and $\Delta C = C_{max} - C_0$ names the capacity difference between the two asymptotic limits of the sensor capacity within the gas concentration range. The model parameter $\gamma$ can be determined from the basic sensor capacity $C(c = 0)$ and $\Delta C$, whereas the parameter $\beta$ is to be estimated by optimization methods.

The asymptotic values $C_0$ and $C_{max}$ are temperature independent as can be seen in Fig. 3. The parameter $\gamma$ of the model in Eq. (4) isn't a function of the temperature as well, because the basic sensor capacity $C(c = 0)$ is temperature independent. Therefore the whole temperature dependency of the model in Eq. (4) is to be described by the parameter $\beta$. The determination of $\beta$ at different temperatures gives its temperature dependence shown in Fig. 5. This dependency can be described by:

$$\beta = \beta(\vartheta) = \frac{1}{a_0 + a_1\vartheta + a_2\vartheta^2} . \qquad (5)$$

Hence the complete model to describe the behaviour of the capacitive gassensor in its dependency on the gas concentration and the ambient temperature is given by

$$
\begin{aligned}
C(c,\vartheta) &= C_0 + \qquad\qquad\qquad (6) \\
&+ \frac{\Delta C}{1 + e^{\left(-\frac{c}{a_0 + a_1\vartheta + a_2\vartheta^2} + \gamma\right)}} .
\end{aligned}
$$

With this functional description the temperature effect on the sensor behaviour is reduced to an remaining error less than 2 % of the signal span within a temperature range between 20 °C and 50 °C lying within the reproducability of the gassensor (see Fig. 6).

These both methods allow an systematic approach for the development of algorithms for digital temperature correction. The model finding and the parameter estimation has to be performed only once just after the manufacturing process of the sensor. Later on only the measured values of the temperature sensor and of the sensor for the measuring quantity have to be placed into the describing model function, whose simple evaluation gives the temperature corrected measuring value.

## ENHANCING THE RELIABILITY OF SENSORS BY A DIVERSE SENSOR SYSTEM STRUCTURE

In certain applications, e.g. gas-monitoring, it is necessary to have a high reliability of the measuring system used. Two different structures of sensor systems combined with a suitable digital signal processing will help to improve the safety of the sensor system compared to that of a single sensor. In a homogeneous redundant structure several sensors of the same kind are used. As output signal of the system the result of the majority of all sensors is taken. The diversity structure contains sensors with different physical principles for the same measuring value. The output signal of such system is derived from a combination of all output signals of the single sensors. The advantage of the second method is that there the influence of cross sensitivities of the single sensor elements can be reduced significantly.

Using these structures self-monitoring sensor systems for the early detection of sensor breakdown can be implemented. Figures 7 and 8 show the structure and a principal block diagram of such self-monitoring sensor system (Djamal, 1991). At

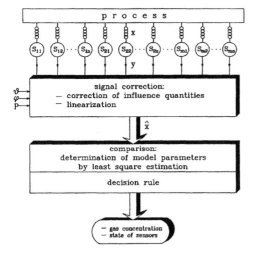

Fig. 7 Structure of a self monitoring sensor system

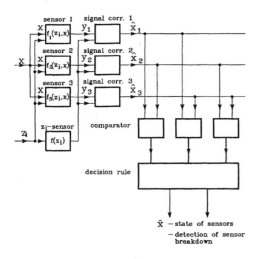

Fig. 8 Block diagram of a self monitoring sensor system

first for every sensor an individual signal correction, e.g. linearization and influence corrections (in Fig. 8 named $z_i$), is performed. Then the signal of the sensors are compared to each other regarding special decision rules to detect if one sensor is defective.

Assuming linear dependencies of the single sensors

to the measuring quantity, the estimated measuring value $\hat{x}_k$ is derived by evaluating the model

$$\hat{x}_k = x_{Nk}(b_{R0k} + b_{I0k}) + \qquad (7)$$
$$+ [1 + (b_{R1k} + b_{I1k})] \cdot K_{1k} \cdot x$$

for each sensor $k$. Here $x_{Nk}$ means the maximum measuring value of the sensor $k$, $K_{1k}$ the transmission factor of this sensor, while $b_{Rik}$ names the reversible and $b_{Iik}$ the irreversible components of influences to the sensor, which could not be eliminated by the preceding individual signal correction.

The comparison of the sensor signals is performed by regression methods. Other methods like Fourier transformation or cross correlation are not suitable here, because the signals are not ergodic and the needed spectral analysis had to be performed within limited time.

The comparison of the sensor signals to determine whether one sensor has broken down, should be demonstrated at two example sensors $p$ and $q$. The estimated values $\hat{x}_p$ and $\hat{x}_q$ of the measuring value of both sensors are determined by

$$\hat{x}_p = x_{Np}d_{0p} + (1 + d_{1p}) \cdot K_{1p} \cdot x , \qquad (8)$$
$$\hat{x}_q = x_{Nq}d_{0q} + (1 + d_{1q}) \cdot K_{1q} \cdot x , \qquad (9)$$

with $d_{ik} = b_{Rik} + b_{Iik}$. Putting Eq. (9) in Eq. (8) gives

$$\hat{x}_p = x_{Np}d_{0p} - x_{Nq}d_{0q} \cdot \frac{(1 + d_{1p}) \cdot K_{1p}}{(1 + d_{1q}) \cdot K_{1q}} +$$
$$+ \hat{x}_q \cdot \frac{(1 + d_{1p}) \cdot K_{1p}}{(1 + d_{1q}) \cdot K_{1q}} . \qquad (10)$$

By least squares approximation the interdependence between $x_{Np}d_{0p}$ and $x_{Nq}d_{0q}$ and the ratio $(1 + d_{1p}) \cdot K_{1p}/(1 + d_{1q}) \cdot K_{1q}$ is then estimated. With these values an estimated value $\hat{x}_p^* = \hat{\hat{x}}_p$ for $\hat{x}_p$ can be obtained.

For the comparison of the sensors two binary state variables $\chi_0$ and $\chi_1$ are defined, which are set in dependence of the following conditions:

$$\chi_0 = \begin{cases} 1 \text{ for } \left| x_{Nq}d_{0q} \cdot \frac{(1 + d_{1p}) \cdot K_{1p}}{(1 + d_{1q}) \cdot K_{1q}} \right. \\ \qquad \left. - x_{Np}d_{0p} \right| \le \delta_0 , \\ 0 \text{ for } \left| x_{Nq}d_{0q} \cdot \frac{(1 + d_{1p}) \cdot K_{1p}}{(1 + d_{1q}) \cdot K_{1q}} \right. \\ \qquad \left. - x_{Np}d_{0p} \right| > \delta_0 , \end{cases} \qquad (11)$$

$$\chi_1 = \begin{cases} 1 \text{ for } \left| 1 - \frac{(1 + d_{1p}) \cdot K_{1p}}{(1 + d_{1q}) \cdot K_{1q}} \right| \le \delta_1 , \\ 0 \text{ for } \left| 1 - \frac{(1 + d_{1p}) \cdot K_{1p}}{(1 + d_{1q}) \cdot K_{1q}} \right| > \delta_1 , \end{cases} \qquad (12)$$

where $\delta_0$ and $\delta_1$ are given tolerance limits.

The state of a sensor is determined by comparing it with all other sensors of the system by evaluating the appropriate values of $\chi_0$ and $\chi_1$. TABLE 2 gives an example for the comparison of four sensors. The rows and columns of the $\chi_k$-values of sensor 2 show all 0. This indicates that sensor 2 does not work correctly.

TABLE 2: Comparison of Sensors for Breakdown Detection

| Sens. | 1 | | 2 | | 3 | | 4 | |
|---|---|---|---|---|---|---|---|---|
| No. | $\chi_0$ | $\chi_1$ | $\chi_0$ | $\chi_1$ | $\chi_0$ | $\chi_1$ | $\chi_0$ | $\chi_1$ |
| 1 | – | – | 0 | 0 | 1 | 1 | 1 | 1 |
| 2 | 0 | 0 | – | – | 0 | 0 | 0 | 0 |
| 3 | 1 | 1 | 0 | 0 | – | – | 1 | 1 |
| 4 | 1 | 1 | 0 | 0 | 1 | 1 | – | – |

With this method an effective enhancement of the reliability of sensor systems can be obtained.

## SIGNAL PROCESSING FOR DIFFERENTIAL SENSORS

Main advantage of differential sensors is that due to the differential structure interfering quantities on the sensor elements and nonlinearities in the sensor signal are widely reduced. Two identical sensing elements of a displacement sensor, for example, can be positioned in such way that a displacement $x$ produces opposite responses of these two elements. Interfering quantities like temperature, however, will cause reactions of both sensor elements in the same direction. The dependence of the output signals $y_1$ and $y_2$ of the elements at a certain operating point $x_0$ can be expressed as functions of the sensor drive signals $x_1 = x_0 - x$ and $x_2 = x_0 + x$, respectively, and the interfering ambient temperature $\vartheta$. Assuming only linear and quadratic dependencies of the sensor elements to the displacement $x$ and a simple additive superposition of the temperature effects $f(\vartheta)$ the output signals can be expressed as (Tränkler, 1989):

$$y_1(x, \vartheta) = a_{01} + a_{11}(x_0 - x) + \\ + a_{21}(x_0 - x)^2 + f_1(\vartheta), \quad (13)$$

$$y_2(x, \vartheta) = a_{02} + a_{12}(x_0 + x) + \\ + a_{22}(x_0 + x)^2 + f_2(\vartheta). \quad (14)$$

With identical sensor parameters $a_i = a_{i1} = a_{i2}$ and identical temperature dependencies $f_1(\vartheta) = f_2(\vartheta)$ of both sensor elements, the signal difference $y_{res}(x, \vartheta)$ is given by

$$y_{res}(x, \vartheta) = y_2(x, \vartheta) - y_1(x, \vartheta) \\ = 2(a_1 + 2a_2 x_0)x. \quad (15)$$

It is exactly linear an independent from the ambient temperature $\vartheta$, but only if the above made assumptions of identical sensor parameters and temperature dependencies are true and if the additive superposition of the temperature effects and the displacement signals is possible. Otherwise errors in linearity and temperature effects will remain.

These errors can be reduced furthermore by implementing special algorithms for the digital processing of the output signals of differential sensors. One of those algorithms is called *differential structure algorithm* (DSA) (Richter, 1991). Here the output signal of one sensor element of the differential sensor is investigated in its dependence on the output signal of the other sensor element. The development of this algorithm is divided into 3 steps.

At first the sensor characteristic of the one sensor element is described at reference conditions of the interfering quantities (e.g. $\vartheta = \vartheta_N$) by a mathematical model function:

$$y_{2N} = y_2(y_1, \vartheta_N). \quad (16)$$

This is necessary to determine $\hat{y}_2(y_1, \vartheta_N) = y_2(y_1, \vartheta_N)$ at any given value of $y_1$.

In the next step the model for the temperature correction is developed. For that purpose the temperature dependent deviation between the measured value $y_2(y_1, \vartheta)$ and its temperature-corrected correspondant value $y_2(x, \vartheta_N)$ is described by a model function using the additional variable $h$ (see Fig. 9):

$$h = y_2(y_1, \vartheta) - \hat{y}_2(y_1, \vartheta_N), \quad (17)$$
$$\Delta y_2(x, \vartheta) = y_2(y_1, \vartheta) - y_2(x, \vartheta_N). \quad (18)$$

At the last step the measuring value $x$ is evalu-

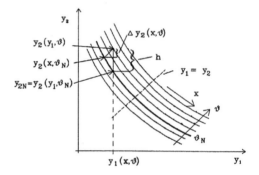

Fig. 9 Differential structure algorithm for signal correction at differential sensors

ated determining the linearization model function at reference conditions:

$$x = f(y_{2N}). \quad (19)$$

During measurement at first the temperature influence on the sensor signal $y_2$ is corrected using the variable $h$ for the determination of the appropriate value of $\Delta y_2(x, \vartheta)$. Then the measuring value $x$ is determined from the temperature-corrected sensor signal $y_{2N}$.

With this new algorithm especially designed for the signal correction of differential sensors improvements in the output characteristics of these sensors are obtained. Using this algorithm to correct the output signal of a differential magnetoresistor for displacement measurement, for instance, an improvement of the sensor characteristic of a factor 4–10 has been realized.

## ALGORITHMIC HYSTERE-SIS ERROR CORRECTION

Various sensor principles are characterized by non-unique, spread response curves, depending on the current direction of change in the input signal. Typical representants of sensors afflicted with hysteresis error due to the branching of the characteristic curve are capacitive humidity and gas sensors or magnetoelastic force sensors. In Fig. 10, the output capacity of a humidity sensor based on a hydrophil polymere coated thin film interdigital capacitor (IDC) is given in respect to the relative atmospheric humidity. For increasing humidity the lower branch is passed through. For descending values of the input humidity, the upper branch is valid. The branching results from the fact, that the sorption isotherm describing the equilibrium between the atmospheric humidity and the humidity of the sensing material differs for the absorption and the emission of moisture.

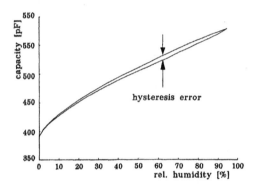

Fig. 10 Hysteresis error of a IDC-humidity sensor

The effect is observed below an excitation frequency given by the transient time for any quasi-static excitation and is therefore not by dynamic nature. This behaviour can be modelled as a static influence effect of previous values and states of the sensor input, i.e. as a memory effect of sensor history. Taking terms from magnetism, these branching phenomena are called hysteresis.

Previous approaches to reduce hysteresis phenomena were limited to constructive and technological measures. Sensor-specific signal processing gives not only an alternative to these cost-intensive efforts but renders the employment of several new sensor principles possible and feasible in applications demanding higher precision. (Haas, 1991)

A correction algorithmn for hysteresis errors has to take the history of the measuring quantity into account, i.e. the sequence of lokal extrema of the measuring quantity in time must be involved in the calculation of the corrected value. In the following, we propose a correction method based on the mathematical *Preisach model*, originally designed to simulate magnetic hysteresis phenomena, which has

been found to be suited best to model and correct hysteresis effects in measurement systems.

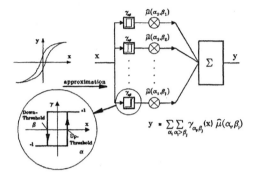

Fig. 11 Preisach model functionalism

In the Preisach model, a physical hysteresis-system is modelled by the aggregation of a infinite set of individually weighted, ideal hysteresis elements with two switching states +1 and −1, differing in their individual switching thresholds $\alpha$ (up) and $\beta$ (down). All hysteresis elements are exposed to the same input $x$ (the actual value of the measuring quantity, scaled in the standardization interval $[-1...+1]$) and the (scaled) output $y$ is given by the weighted sum of outputs of those hysteresis elements as shown in Fig. 11. The exact Preisach model

$$y(t) = \iint\limits_{\beta \leq \alpha} \mu(\alpha, \beta) \gamma_{\alpha\beta}(x(t)) \, d\alpha \, d\beta \,, \qquad (20)$$

with the hysteresis elements $\gamma_{\alpha\beta}(x(t))$ and the weighting function $\mu(\alpha, \beta)$ may be implemented numerically with an finite number of elements as the *Discrete Preisach model*

$$y(t) = \sum_{\alpha_i} \sum_{\beta_j \leq \alpha_i} \left( \hat{\mu}(\alpha_i, \beta_j) \gamma_{\alpha_i \beta_j}(x(t)) \right) \,. \qquad (21)$$

The characteristic shape of the yielding hysteresis loop is determined by the selected weighting function $\mu(\alpha, \beta)$. To model a given sensor response curve with hysteresis, extensive measurements are required to determine the corresponding weighting function.

The state of the hysteresis elements and the values of the weighting function may be presented advantageously in a triangle-shaped area of the $\beta - \alpha$-plane, where every point in the plane is related to a hysteresis element with its specific up-/down-thresholds (Fig. 12). The triangular contour is due to the finite (scaled) measuring range $[+1, -1]$ and to the assumption, that the down-threshold $\beta$ is less or equal the up-threshold $\alpha$.

Figure 13 shows the resulting state of the hysteresis elements when the modelling system is exposed to the given input signal $x(t)$. The elements belonging to the hatched area in the triangle are in

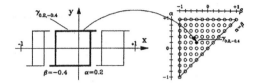

Fig. 12 Relation between hysteresis elements and points in the triangle in the $\beta - \alpha$-plane

state +1, the rest in state -1. The coordinates of the boundary line are given by a sequence of previous local input extrema declining in absolute value. This is, how the model stores the system history. The current system output is determined by the integral over the weights of the element-area in state +1 minus the integral over the weights of the area of elements in state $-1$.

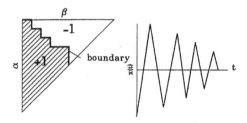

Fig. 13 State of the hysteresis elements for given input signal and resulting boundary in the triangle

To perform a sensor-specific hysteresis error correction using the Preisach model, two steps are necessary:

- Determination of model parameters
- On-line correction of the hysteresis error

First, the hysteresis of the sensing device – in terms of the Preisach model the weighting function $\mu(\alpha, \beta)$ – has to be determined by extensive measurements.

This may be done by taking the system's *first-order transition curves*. First order transition curves are received by increasing the input (measuring quantity) from the negative saturation to a point $x = \alpha'$ and returning again to the negative saturation as in Fig. 14. Points on this returning branch are uniquely identified by the transition point $\alpha'$ and the current value of input $x = \beta'$ and the output is therefore denoted $y_{\alpha',\beta'}$.

The weighting function is given by double differentiation:

$$\mu(\alpha', \beta') = \frac{1}{2} \frac{\partial^2 y_{\alpha'\beta'}}{\partial\alpha'\partial\beta'} . \tag{22}$$

It should be mentioned that the numeric implementation of this differentiation is critical because of the

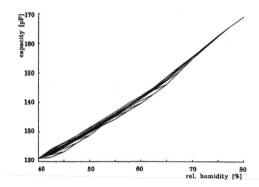

Fig. 14 Measured first-order transition curves for a capacitive humidity sensor

rugged character of the operation. That is one reason, why representations of the Preisach model expressing the input-/output-relationship directly using the measured values $y_{\alpha',\beta'}$ (Mayergoyz, 1991) have more practical relevance than Eq. (21).

Measurements for humidity sensors are especially time-consuming as setting times must be long enough to eliminate effects for dynamic response times of the sensor. The measurement in Fig. 14 with a setting time of 30 minutes took 2 weeks in all on a fully automatic calibrator. On an industrial scale, only production tolerances can be considered by an individual, short calibration. The global features of hysteresis including the basic characteristic of the weighting function however should not be determined individually.

Figure 15 shows the weighting function corresponding to the first-order transition curves of the humidity sensor in Fig. 14. The high weight of elements on the line $\alpha = \beta$ within the plane is explained by the relatively narrow hysteresis of the characteristic curve.

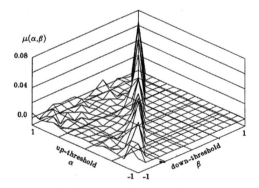

Fig. 15 Preisach weighting function $\mu(\alpha, \beta)$ for the capacitive humidity sensor in Fig. 14

Second, using the Preisach *simulation* model, a simple error correction algorithm may be performed

8

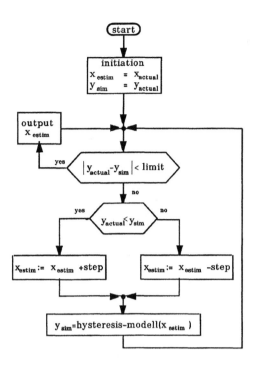

Fig. 16 Correction algorithmn using a Preisach simulation model

now. Hysteresis error correction may be attributed to the problem that seeks the inputs to the hysteresis-system from its output. The idea is, to stimulate the available model in a manner, that its output is identical with the current output of the real physical system. If at the starting point, the internal states were identical and the input of the simulation model is manipulated correctly to keep outputs on the same level, the inputs must be identical as well. The only problem is, that an iterative regulation algorithmn for the model input is not admissable, for it would play alternating and nonrelevant input signal changes to the modelling system. That is, why overshoots have to be strictly avoided using small input increments and a wide tolerance interval when comparing the system outputs.

An algorithm successfully applied to correct the hysteresis error of the introduced humidity sensor is given in the flow chart in Fig. 16.

Preliminary results indicate possible improvements in accuracy of a factor 3 - 7. Further examinations will have to take the significant effects of the ambient temperature into account.

## CREEP CORRECTION

Creep phenomena are, in contrast to the discussed hysteresis effects, of dynamical nature although the speed of changes by creeping is characteristically very slow. As a property known to many cheap and light materials (for instance plastics) employed in measurement systems, creeping causes systematic errors, especially in diminished components and systems.

Mechanical deformation processes can be distinguished by their (ir)reversiblity and by their dynamic characteristics. TABLE 3 gives a short survey of associated technical terms and $\varepsilon - \sigma$-diagrams.

TABLE 3 Mechanical Deformation Processes

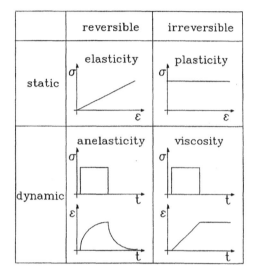

Creeping covers all dynamic deformation processes, i.e. anelasticity and viscosity.

Signal processing algorithmns to perform a creep correction have been developed and tested at a plastic weighing machine. In this machine, the displacement of a plastics (teflon, PTFE) leaf springs by the weight are detected by a distance sensor. (Kortendieck, 1991)

A theoretical approach to a description of material creeping can be derived from the theory of rate processes. Plastics consist of a three-dimensional network pattern of hooked macro molecules containing crystalline regions. Due to the external tensions and in combination with the thermal agitation, rate processes preferably take place in the direction of the external exitation. In the case of entropy- or engergy-elasticity the resulting creep processes are reversible. By irreversible creeping, links to neighbors are non-recoverably splitted.

Reversible creep phenomena with several (here: 4) relaxation mechansims, as observed at this weighing machine, are described by the *generalized Kelvin model* (Fig. 17).

The total displacement $x = x_0 + x_1 + x_2 + x_3 + x_4$ and the exciting force are related by the following

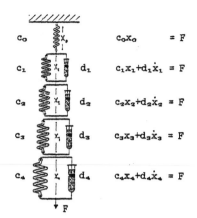

Fig. 17 Generalized Kelvin model

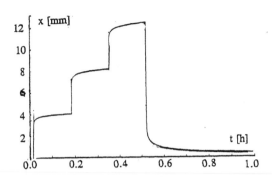

Fig. 18 Creep measurements at PTFE (temperature: $23^0$ C $\pm 0.7^0$ C). Test piece: 8 connected leaf springs, 130 mm × 15 mm × 5 mm. Netto loads (10 min. each): 4.90 N, 9.76 N and 14.6 N, discharged for 30 min. afterwards.

linear differential equation (Kortendieck, 1989)

$$\sum_{i=1}^{m} b_i \cdot x^{(i)} = \sum_{j=1}^{n} k_j \cdot F^{(i)} , \qquad (23)$$

with the coefficients $b_i$ and $k_j$ determined by the elongation by unit force $c_i$ and viscosity constants $d_i$. For calibration, an experiment with a constant load is performed. The solution of the generalized Kelvin differential equation is given in this case as

$$\varepsilon(t) = \sigma_0 \cdot \left[ \eta \cdot t + \frac{1}{E_0} + \right.$$
$$\left. + \sum_{i=1}^{m} \frac{1}{E_i} \left( 1 - e^{-\frac{t}{\tau_i}} \right) \right] . \qquad (24)$$

The parameters $\tau_i$ are fixed to be logarithmically equidistant in the period of the experiment with $\tau_1$ given as the sampling interval. Then the E-modules $E_i$ in Eq. (24), respectively the $c_i$, $d_i$ are received by a least-square estimation algorthmn from the time diagrams recorded in the constant load experment.

In Fig. 18, creep measurement time diagrams for PTFE (teflon: 2.15 g/cm$^3$, tensile strength 22 − 30 N/mm$^2$, elongation on fracture 250 − 400%, tensile E-module 750 N/mm$^2$, bending strength 7 − 8 N/mm$^2$) are given. The resulting parameters are summarized in TABLE 4.

An algorithm to perform creep correction can be deduced from Eq. (23). Converted to an integral equation, the loading force can be determined. Figure 19 shows creep correction results achieved for the exitations in Fig. 18.

## OBJECT DETECTION WITH ULTRASONIC SENSORS USING HOLOGRAPHICAL METHODS

As an example for complex sensor signal processing the object detection with ultrasonic sensors is

TABLE 4: Resulting Creep Model Data

| i | given: | estimated (least square): | |
|---|---|---|---|
| | relaxation time $\tau_i$ [s] | elongation by unit force $c_i$ [N/mm] | viscosity constant $d_i$ [N·s/mm] |
| 0 | | 1.7 | |
| 1 | 10.0 | 9.2 | $9.2 \cdot 10^1$ |
| 2 | 153.0 | 10.4 | $1.6 \cdot 10^3$ |
| 3 | 2.849 | 12.3 | $2.9 \cdot 10^4$ |
| 4 | 26.0 | 12.3 | $4.4 \cdot 10^5$ |

shown. The ultrasonic sensor element consists of a transmitter and a receiver for broadband wide-angle ultrasonic signals in a frequency range from 8 kHz to 280 kHz. Puls holographic methods using a synthetic aperture is used to get information about an object plane as shown in Fig. 20 (Löschberger, 1987). It gives the principal arrangement of the ultrasonic sensor and the objects to be detected. Here a combination of a little staircase and a small glass stick is chosen as an example. The sensor element is moved along a straight line. At a certain amount of points a short pulse is sent out and the received echo is stored digitally. Figure 21 shows the echo signals of 80 measuring points with a distance of 1 mm. To compute the intensity of one point in the object plane all echo signals with a time of flight relating to the distance of that point are accumulated. Because of the reflection mode the time of flight is calculated as $t = 2d/c$, with $d$ naming the distance between the ultrasonic sensor and a certain point in the object plane and $c$ the sound velocity.

The resolution in lateral direction (i.e. the direction of the x-axis in Fig. 20) depends mainly on geometrical relations and the center frequency of the transducers. Here the lateral resolution is about 1 mm.

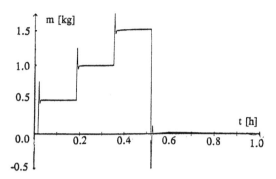

Fig. 19 Creep correcture results for PTFE

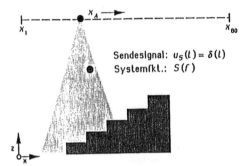

Fig. 20 Object detection with ultrasonic sensor

The resolution in axial direction (i.e. the direction of the z-axis in Fig. 20) depends on the point resolution of the single echo signal. To improve this resolution, the echo signals are deconvoluted with the system function of the transducers. Therefore, the echo signals are transformed to frequency domain via fast Fourier Transformation (FFT) and then divided by the system function. A filter has to be used because the echo signals are band-limited. After that operations the signals are transformed back to time domain:

$$r(t) = \mathcal{F}^{-1}\left\{\frac{U_E(f) \cdot N(f)}{U_S(f)}\right\}, \qquad (25)$$

where $U_E(f)$ is the spectrum of the echo signals, $N(f)$ the spectral transmission function of the filter and $U_S(f)$ the spectral function of the ultrasonic sensor.

The intensity of the real part and the imaginary part of those deconvoluted echo signals are computed seperately:

$$I_R(x,z) = I_R(x,z) + \text{real}\left(r\left(\frac{2d}{c}\right)\right), \quad (26)$$

$$I_I(x,z) = I_I(x,z) + \text{imag}\left(r\left(\frac{2d}{c}\right)\right). \quad (27)$$

The intensity of the reflections within the object plane is then calculated for each point of interest with:

$$|I(x,z)| = \sqrt{(I_R(x,z))^2 + (I_I(x,z))^2} \qquad (28)$$

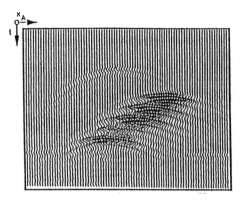

Fig. 21 Echo signals of the objects received by the ultrasonic sensor

Using this method a resolution of approximately 1 mm in axial direction can be achieved.

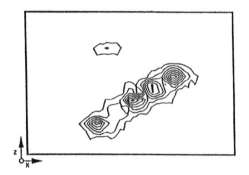

Fig. 22 Holographic picture of the objects

Figure 22 shows the intensity map of the reflection within the object plane as a contour plot. The picture of the staircase can be seen. The shadow of the small glass stick produces a smaller intensity at the second stair.

With this method it is possible to get information about the surrounding using a small, robust and cheap sensor. In combination with the movement of a robot gripper, information about an object can be sampled during the movement of the gripper towards the object. Key elements of this processing method are broadband ultrasonic transducers (Kleinschmid, Magori, 1985). Using special matched pulses, deconvolution is not longer necessary (Lschberger, 1988) and the computation can be done on a small microprocessor without a numeric coprocessor.

CONCLUSIONS

The above discussed algorithms give an overview over the possibilities of digital signal processing at sensors and sensor systems. Various other examples could be found. Common aim at sensor signal processing is to improve the accuracy and reliability of sensors or to generate a more sophisticated signal deducing the interesting parameters from the sensor outputs.

With the increasing significance of the microsystems, where sensors and actuators will be combined with digital signal electronic and microprocessors, the algorithms for digital sensor signal processing will gain a major role in the developing of those systems.

## ACKNOWLEDGEMENT

The following current and former scientists of the Institute of Measurement and Automation have contributed to this paper:

| | |
|---|---|
| Dipl.-Phys. | M. Djamal, |
| Dipl.-Phys. | M. Haas, |
| Dr.-Ing. habil | M. Horn, |
| Dipl.-Phys. | H. Kortendieck, |
| Dr.-Ing. | C. Löschberger, |
| Dr.-Ing. | J. Löschberger, |
| Dr.-Ing. | H. Richter, |
| Prof. Dr.-Ing. | H.-R. Tränkler, |
| Dipl.-Ing. | R. Ullmann. |

This work was supported by the *Deutsche Forschungsgemeinschaft (DFG)* by contract numbers TR 232/1-1, TR 232/4-1 and by the *Bundesministerium für Forschung und Technologie (BMFT)* by contract number 13 AS 010 89.

## REFERENCES

Djamal, M. (1991). Sensorausfall-Früherkennung durch homogene und diversitäre Redundanz. *Report for the BMFT-project "FET-gassensors"*, Institute of Measurement and Automation, Federal Armed Forces University, Munich.

Haas, M. (1991). Untersuchung und Erprobung von Algorithmen zur Kriech- und Hysteresiskorrektur an ausgewählten Sensoren. *Report for the DFG-project "Kriechkorrektur"*, Institute of Measurement and Automation, Federal Armed Forces University, Munich.

Kleinschmidt, P., and V. Magori (1985). Ultrasonic robotic sensors for exact short range distance measurement and object identification. *Proceedings of the 1985th IEEE Ultrasonic Symposium*, 457-462.

Kortendieck, H. (1989). Untersuchung und Erprobung sensorspezifischer Meßsignalverarbeitungs-Maßnahmen unter Berücksichtigung der Vorgeschichte der Meßgröße. *Report for the DFG-project "Meßsignalverarbeitung"*, Institute of Measurement and Automation, Federal Armed Forces University, Munich.

Kortendieck, H. (1991) Kriech- und Hysteresiskorrektur am Beispiel einer Kunststoff-Federwaage. *Proceedings of SENSOR 91 Congress, Nürnberg, Germany*, Vol. 2, 303-318.

Löschberger, C. (1991). Modelle zur digitalen Einflußgrößenkorrektur an Sensoren. *Ph.-D. thesis*, Federal Armed Forces University, Munich.

Löschberger, J. (1987). Ultraschall-Sensor-System zur Bestimmung axialer und lateraler Strukturen mit Hilfe bewegter Wandler zum Einsatz in der industriellen Automation. *Ph.-D. thesis* Federal Armed Forces University, Munich.

Löschberger, J., and V. Magori (1988). Luft-Ultraschall-Sensoren mit lateraler Auflösung für Robotik-Anwendungen. *Symposium Sensoren – Technologie und Anwendung 1988, Bad Nauheim, Germany, VDI-Berichte* No. 677, VDI-Verlag, Düsseldorf, pp. 21-22.

Mayergoyz, L. D. (1991). *Mathematical modells of hysteresis*. Springer-Verlag, New York.

Richter, H. (1991). Statische Meßeigenschaften von Differentialsensoren und deren Verbesserung durch spezifische Korrekturalgorithmen. *Ph.-D. thesis* Federal Armed Forces University, Munich.

Tränkler, H.-R. (1988). Signalvorverarbeitungskonzepte. *Technologietrends in der Sensorik*, study in comission of VDI/VDE-Technology Center of Information Technology, Berlin.

Tränkler, H.-R. (1989). Signal processing. In T. Grandke (Ed.), *Sensors*. Vol. 1. VCH-Verlag, Weinheim. Chap. 10, pp. 279-311.

# INTELLIGENT ACTUATORS - WAYS TO AUTONOMOUS ACTUATING SYSTEMS

### R. Isermann and U. Raab

*Technical University of Darmstadt, Institute of Automatic Control, Laboratory for Control Engineering and Process Automation, Landgraf Georg Str 4., D-6100 Darmstadt, Germany*

Abstract. The integration of microelectronics within the actuator allows to not only replace the analog position controller but to add several functions which give the actuator more intelligent functions. The actuator control is performed in different levels and includes adaptive nonlinear control, optimization of speed and precision, supervision and fault diagnosis. The actuator knowledge base comprises actuator models based on parameter estimation, controller design and a storage of the learned behavior. An inference mechanism makes decisions for control and fault diagnosis and a communication module operates internally and externally. As examples electromagnetic and pneumatic actuators are considered and it is shown how the control can be improved considerably by model based nonlinear control, taking into account time varying nonlinear characteristics and hysteresis effects. The supervision with fault detection indicates faults in the electrical and mechanical subsystems of the actuator. Several experimental results are shown including the implementation on a low-cost microcontroller.

Keywords. Actuators, intelligent control functions, nonlinear adaptive control, supervision, fault diagnosis

## 1. Introduction

From the beginning actuators and sensors play an important role in automatic control system. They must operate precisely and function reliably as they directly influence the right operation of the control system. In many cases actuators manipulate energy flows, mass flows or forces as a response to low energy input signals like electrical voltages or currents, pneumatic and hydraulic pressures or flows. Basic components are usually a power switch or a valve, an electrical, pneumatical or hydraulic amplifier or motor, sometimes with feedback to generate a specific static and dynamic behavior, and a sensor for the actuator output, like a position or force. Because of the continuous motion or changes and the power amplification, actuators usually underlie wear and aging. Hence their properties change at least gradually with time and the performance may diminish. Faults may appear and develop until a failure occurs.

Industrial sensors usually show a different behavior in dependence on life time than actuators. The influence of wear and aging may be less. However this very much depends on the environment and it is difficult to state something general. Sensor failures seem to occur randomly and suddenly, Halme and Selkäinako (1991), Henry and Clarke (1991).

Figure 1.1 shows the scheme of a classical actuator without and with analog position control. The analog command signal $U_R$ is the reference value for the position controller. Dependent on the actuator type one distinguishes

- proportional actuators (e.g. piezoelectric or electromagnetic actuator)
- integral actuators with varying speed (e. g. pneumatic/hydraulic cylinder or d.c. motor)
- integral actuators with constant speed (e. g. a.c. motor)
- actuators with quantization (e. g. stepper motor)

Isermann (1989), (1991a). Figure 1.2 shows the corresponding

position control loops in some more detail.

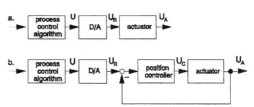

**Fig. 1.1.** Classical actuator control
a. feedforward position control
b. analog feedback position control

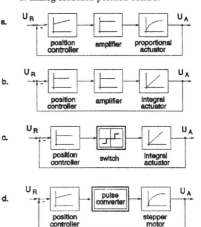

**Fig. 1.2.** Actuator position control dependent on
a. proportional actuator
b. integral actuator with varying speed
c. integral actuator with constant speed
d. quantizing actuator

The goal of the position controller is to obtain a precise positioning, independent on disturbances like power supply voltage, shaft and gear friction, backlash or reactive forces from the manipulated mechanism or medium. Because the analog position controllers are mostly linear P-, PI- or PID-Controllers the reachable control performance is not very high, as the actuators show frequently nonlinear behavior.

The position control covers also faults of the actuator up to a certain size. If the faults of an actuator are large enough, they may be detected indirectly by monitored variables like the power supply current or the control deviation of the position controller or of the superimposed controller or, of course, by inspection.

The further development of actuators will be determined by following requirements

- larger reliability and availability
- higher precision of positioning
- faster positioning without overshoot
- simpler and cheaper manufacturing

The design and the manufacturing of classical actuators has reached a very high standard. If the numbers of produced pieces are high, the prices are relatively low. Therefore no significant changes are to be expected from this side. However, new impacts can be expected from

- new actuator principles
- integration of microelectronics

New actuator principles are for example piezoceramic and magnetostrictive effects or electrochemical reactions, e.g. Raab, Isermann (1990a). A stronger influence may come from the <u>integration of microelectronics</u> on (classical) actuators, as the prices for the microcontrollers are now low enough. Then not only the analog position controller can be replaced but much more functions can be added. This may lead with time to actuators with more "intelligent" properties.

Saridis (1977) considers intelligent control as a next hierarchical level after adaptive and learning control to replace the human mind in making decisions, planing control strategies, and learning new functions by training. Merrill (1988) defines intelligent control systems as those which integrate traditional control concepts with real-time fault diagnostic and prognostic capabilities. According to Åström (1991) an <u>intelligent control system</u> possesses the ability to comprehend, reason and learn about processes.

Care should be taken in using the word "intelligence" for automatic control in order not to expect too much in comparison to a really intelligent human operator . Here, only a very low degree of "intelligence" as meant with "ability to model, reason and learn about the actuator and its control". Figure 1.3 shows the different modules of the information flow of a "<u>low-degree intelligent actuator</u>". They comprise:

- control in different levels
  - selftuning/adaptive (nonlinear) control
  - optimization of the dynamic performance (speed vs. precision)
  - supervision and fault diagnosis
- knowledge base
  analytical knowledge:
  - parameter and state estimation (actuator models)
  - controller design methods
  heuristic knowledge:
  - normal features (storage of learned behavior)
- interference mechanism
  - decision for (adaptive) control
  - decisions for fault-diagnosis
- communication
  - internal: connecting of modules, messages
  - external: with other actuators and the automation system

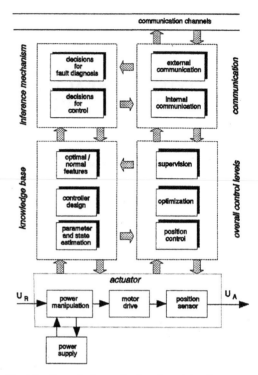

Fig. 1.3. General scheme of a (low-degree) intelligent actuator

Hence, the "intelligent" actuator adapts the controller to the mostly nonlinear behavior (adaptation) and stores its controller parameters in dependence on the position and load (learning), supervises all relevant elements and performs a fault diagnosis (supervision) to request for maintenance or if a failure occurs to fail safe (decisions on actions). In the case of multiple actuators supervision may help to switch off the faulty actuator and to perform a reconfiguration of the controlled process. Other words for the low-degree intelligent actuator would be "<u>smart actuators</u>" or "<u>autonomous actuators</u>".

In the following model based methods of selftuning and adaptive digital control and supervision with fault detection (diagnosis) are described. The models are based on the physics of the actuator and comprise as well nonlinear characterics as hysteresis effects. Their mostly unknown and time varying parameters are obtained by parameter estimation. Applications to different drives show practical results and also how the methods can be implemented on cheap standard microcontrollers.

## 2. Actuator Principles

The considered actuators transform electrical inputs in mechanical outputs such as position, force, angle or torque. The output energy level is much higher than the input signal, so the usage of a supporting energy like electricity, pneumatic or hydraulic pressure is required. With regard to the important actuator concepts, a classification and evaluation can be concentrated on three major groups

- electromechanical actuators
- fluid power actuators
- alternative actuator concepts

A further subdivision leads to different operating principles,

Fig. 2.1.

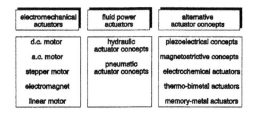

**Fig. 2.1.** Low power actuator principles ( < 10 kW)

Based on the power constraints of the supporting energy and on related constructive design properties, the actuator concepts show different characteristic features as well as limitations. Therefore an evaluation in terms of typical application areas, flexibility, robustness, safety and reliability is provided in Raab, Isermann (1990a).

In this paper we focus on a performance comparison of common actuator concepts. The presented results are obtained under the constraints of low power servo applications (< 5-10 kW). Each of the analyzed actuator concepts was designed only for the translational motion. The comparison starts with some graphical presentations, evaluating the generated force versus different characteristic dimensions.

In Fig. 2.2 logarithmical scaled force was plotted over the translational positioning speed. This choice represents the actuating power output (force•speed) for the evaluated concepts. A comparison of force versus typical positioning ranges is shown in Fig. 2.3. This figure also contains the positioning accuracy represented as left boundary of the x-axis. The widest range, starting from some μm up to m is covered by electromechanical concepts. An extremely high positioning accuracy is performed by piezoelectric actuators.

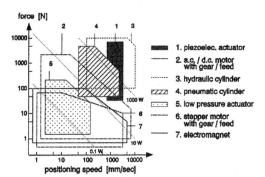

**Fig. 2.2.** Force vs. speed for common actuator principles

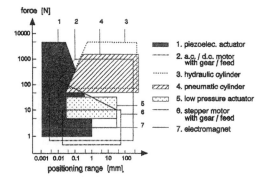

**Fig. 2.3.** Force vs. positioning range for common actuator principles

The power to weight ratio in [Watt/kg] is presented in Fig. 2.4. It underlines the leading position of fluid power systems as well as the restricted ratios of electromagnetic and -mechanical concepts (if the power supply is not considered).

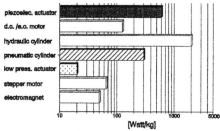

**Fig. 2.4.** Power to weight ratios for common actuator principles

Discussing actuator applications implies a specified closed loop performance in terms of accuracy, dynamics, positioning ranges e.g.. Therefore, system immanent characteristics such as the static I/O-behavior, nonlinear effects (friction, backlash, hysteresis) and (time) varying process parameters are important. They are obtained by an evaluation of the uncontrolled actuating device and presented in Table 2.1.

TABLE 2.1  Input/output Behavior of Important Actuator Types

| characteristic → <br><br> actuator typ ↓ | static linea-rity | nonlinear effects | | | varying process parameters | |
|---|---|---|---|---|---|---|
| | | fric-tion | back-lash | hyste-resis | internal 1 | external 2 |
| d.c./a.c. motor with feed | + | o - | o - | o | | o - |
| stepper motor with feed | + | o - | o - | o - | | o - |
| electromagnet | o - | - | | - | o | o - |
| pneumatic cylinder | | - | | | - | o |
| hydraulic cylinder | | - | | | - | o |
| piezo-stack actuator | o | | - | - | | - |

*Symbols:* + *good, neglectible*  o *average, common*  - *bad, significant*

*1: caused by internal physics (position dependent parameters etc.)*
*2: caused by external influence (varying supporting energy potential, thermal properties etc.)*

The evaluated terms as friction, nonlinear characteristics and varying process parameters are present in every actuator with a mechanical output. They limit or hinder the overall performance of position control in a closed loop, especially if their influence is large in case of low-cost actuator manufacturing. Hence, classical actuator control is not flexible enough to deal with these major restrictions. Therefore it is a challenging task to combine the given actuator hardware, a micro-controller and sophisticated control-software to improve the dynamics as well as the static characteristics of actuators and to add other more intelligent functions.

# 3.  Modelling and Identification of Actuators

Precise actuator models are of substantial importance for the design of model based control algorithms and the supervision with fault diagnosis. They are developed by theoretical modelling and describe the dynamic relations between the electrical input U and the mechanical output Y. For a more detailed approach considering electromechanical, pneumatic or hydraulic actuators see e.g. Pfaff (1982), Kuo (1982),

Backé (1988) or Isermann (1988).

For most actuator models there exist certain similarities. In the case of translational motions they have a SISO-structure as shown in Fig. 3.1.

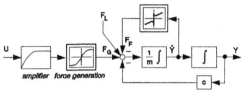

**Fig. 3.1.** Simplified actuator model for translational motions and proportional I/O-behavior

The actuator model describes the energy transducer, consisting of an amplifier with delayed dynamics and a following force generator. Because of energetic boundaries and material properties, the static behavior between the electrical input and the generated force $F_G$ is nonlinear to some extend. The dynamics can be simplified to a linear first order system, e.g. a closed loop controlled current subsystem, Freyermuth and Held (1990), or second order system, e.g. servo-valves, Backé (1990).

$F_G$ acts together with the sum of other forces on a mechanical spring-mass system. The resulting displacement is characterized by the position output Y and velocity dY/dt, and can be modelled by

$$m \ddot{Y}(t) = \sum F(t) = F_G(t) + F_L(t) - c\, Y(t) - F_F(\dot{Y}) \quad (1)$$

where $F_L$ denotes external loads and $F_F$ frictional forces. In the case of an integral I/O behavior the spring feedback, represented by spring constant c does not exist.

In general the motion of the mass is influenced by frictional forces. Once the motion begins, this effect can be modelled by a superposition of Coulomb friction $F_C$ and linear damping d ('viscous friction')

$$F_F(\dot{Y} \neq 0) = F_{C\pm}\, sign(\dot{Y}) + d_{\pm} \cdot \dot{Y} \quad (2)$$

where the index +/- denotes coefficients for positive (+) or negative (-) motion directions. Equation (2) delivers a simple but for practical purposes adequate friction model. A more sophisticated approach, which describes stip-slick instabilities during slow motion, requires Stribeck's extended friction model, Stribeck (1902), Maron (1991).

In reality actuators show a nonlinear behavior. If well known linear design methods are to be used, the model must first be linearized and the range of operation must be confined to a small range. A better approach is to use two differential equations for the actuators dynamics

$$y(t) = -\sum_{i=1}^{n} a_{i+}^{*} \frac{d^i}{dt^i} y(t) + \sum_{j=0}^{m} b_{j+}^{*} \frac{d^j}{dt^j} u(t) + c_{DC+}^{*} \quad (3)$$

for dy/dt > 0 (positive motion) and

$$y(t) = -\sum_{i=1}^{n} a_{i-}^{*} \frac{d^j}{dt^j} y(t) + \sum_{j=0}^{m} b_{j-}^{*} \frac{d^j}{dt^j} u(t) + c_{DC-}^{*} \quad (4)$$

for dy/dt < 0 (negative motion) with position dependent parameters. Note that standstill or the transients from one motion direction to the other are not modelled. $d^i/dt^i$ denotes thereby the derivatives of the input and output signal and the parameters $a_i^{*}$, $b_j^{*}$ contain expressions of physical coefficients as mass, spring constants, resistances etc.

The linear approach in Eq. (3)(4) models Coulomb friction $F_C$ as a simple d.c. value, which is approximately constant for all speeds, if the mass is moving only in one direction, Kofahl (1988). For symmetrical and position independent friction, the following correspondences are valid

$$a_{i+}^{*} = a_{i-}^{*} \; ; \;\; b_{j+}^{*} = b_{j-}^{*} \; ; \;\; |c_{DC+}^{*}| = |c_{DC-}^{*}| \quad (5)$$

Otherwise the parameter sets will differ in their values.

For process identification or design purposes we may use a discrete time representation of Eq. (3)(4). Assuming a zero-order hold, the adequate difference equations are

$$y(k) = -\sum_{i=1}^{n} a_{i+}\, y(k-i) + \sum_{j=1}^{n} b_{j+}\, u(k-j) + c_{DC+} \quad (6)$$

for y(k) > ... > y(k-n) (positive motion) and

$$y(k) = -\sum_{i=1}^{n} a_{i-}\, y(k-i) + \sum_{j=1}^{n} b_{j-}\, u(k-j) + c_{DC-} \quad (7)$$

for y(k) < ... < y(k-n) (negative motion), where k=t/$T_0$=0,1,2 .. represents the discrete time and $T_0$ the sampling time.

The parameters of the SISO-actuator model are usually unknown and must be computed by adequate estimation methods, e.g. Åström and Eykhoff (1970), Ljung (1988), Isermann (1988). Depending on the representation type (continuous time versus discrete time model), data processing (online versus offline operation) and process behavior (time invariant or time variant), different identification techniques are required.

In practical applications least squares estimators (LS) can deliver a sufficient identification result. Therefore, the unknown discrete time parameter vector

$$\underline{\Theta}^T = [\; a_{1\pm} \ldots a_{n\pm}\, , \, b_{0\pm}^{*} \ldots b_{m\pm}^{*}\, , \, c_{DC\pm} \;] \quad (8)$$

or continuous time parameter vector

$$\underline{\Theta}^T = [\; a_{1\pm} \ldots a_{n\pm}\, , \, b_{1\pm} \ldots b_{n\pm}\, , \, c_{DC\pm} \;] \quad (9)$$

is obtained by minimizing the loss function

$$V = \sum_{k=0}^{N} e^2(k) = \underline{e}^T \underline{e} \quad (10)$$

where

$$e(k) = y(k) - \underline{\psi}^T(k) \cdot \hat{\underline{\Theta}} \quad (11)$$

represents the equation error and $\hat{\underline{\Theta}}$ the estimated value of $\underline{\Theta}$. This results in the nonrecursive LS-estimation equation, which can be transformed to recursive algorithms. Modifications in the form of discrete square root filtering (DSFI) or adequate factorization methods show better numerical properties, e.g. Biermann (1977). In the case of a discrete time model, the data vector $\underline{\psi}$ directly involves the measured I/O-data of u and y. For the identification of continuous time models the unknown derivatives in the data vector must be determined by e.g. state variable filter techniques, see Young (1981), Peter, Isermann (1989).

If data processing can be performed offline, usually better identification results are obtained by nonlinear parameter estimation methods. Based on an output error approach

$$e(k, \underline{\Theta}) = y(k) - y_M(k, \underline{\Theta}) \quad (12)$$

the minimization of the quadratic cost function

$$\frac{\delta V}{\delta \hat{\underline{\Theta}}} = \min_{\underline{\Theta}} \left( \sum_{k=0}^{N} e^2(k, \hat{\underline{\Theta}}) \right) \quad (13)$$

with any hill climbing method offers an improved robustness as well as reduced estimation bias, see Kabaila (1983), Ljung (1978), Drewelow (1990). Using the discrete time representation, model output $y_M$ can be directly computed by Eq. (6) (7). For the continuous time representation an additional discretization of Eq. (3)(4) in each iteration steps is required. Although the computational effort therefore is high, the omission of state variable filter techniques may be an advantage, Raab (1992).

According to the proposed estimation methods, it is important to point out, that the identification procedure for the actuator is carried out seperatly for both motion directions. As the models are only valid for velocities unequal zero, only data vectors which fulfil the adequate conditions should be used. Therefore a sufficient excitation must be guaranteed, which can usually be obtained during a pre-identification period using special input signals. A suitable one, which delivers good identification results even in the case of high order actuator models is shown in Fig. 3.2, Raab (1992). Note that only I/O-datas within the shaded areas are used for the identification of the actuator model.

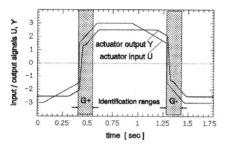

**Fig. 3.2.** I/O-signals and related identification ranges for a proportional actuator with friction (simulation)

# 4. Model Based Nonlinear Control of Actuators

In order to obtain a specified I/O-performance in terms of accuracy, dynamics and robustness, actuating systems require a closed loop position control. Assuming an approximately linear process behavior, the basic discrete time SISO-control algorithms are of type

$$u(k) = - \sum_{i=1}^{\mu} p_i \cdot u(k-i) + \sum_{j=0}^{\nu} q_j \cdot e_w(k-j) \qquad (14)$$

which include for example the P, PI or PID type.

$$e_w(k) = W(k) - Y(k) \qquad (15)$$

denotes thereby the control error and u the controller action. If a state control is used

$$u(k) = -[\ k_1\ k_2\ \dots\ k_n\ ] \cdot [\ x_1(k)\ x_2(k)\ \dots\ x_n(k)\ ]^T \qquad (16)$$

$\underline{x}$ are the measured or observed state variables and $\underline{k}$ is a constant gain vector.

The design and tuning of these algorithms is based on identified parametric actuator models (see section 3) and supported by appropriate software design packages. The used computer aided controller design and system analysis is described in e.g. Isermann (1984a), (1989).

Because actuator properties as friction, hysteresis, nonlinear characteristics and time varying process parameters are present (see section 2), well tuned linear control algorithms do not give satisfying results, Raab, Isermann (1990b). Therefore these nonlinear effects are taken into account for the design of the actuator position controller.

## Correction of Nonlinear Static Characteristics

Nonlinear static characteristics are present in most of the actuators, either in specific local areas or over the whole range. This leads to a loss of control performance or even closed loop instability.

The objective is to compensate the main static nonlinearity f by an approximate inverse function $f^{-1}$, which can be implemented in the microprocessor, Franz (1973), Lachmann (1983). According to Fig. 4.1, the regular actuator input U is then substituted by the 'corrected' value

$$U^* = f^{-1} (U, \underline{x}) \qquad (17)$$

such that the I/O-behavior U-Y becomes (approximately) linear. $G_{P1}$ usually represents here the dynamics of the energy transducer, $G_{P2}$ the mechanical system and $\underline{x}$ the involved process states.

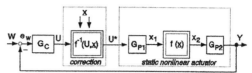

**Fig. 4.1.** General structure of a series correction (compensation) for nonlinear process statics

Assuming an actuator model as shown in Fig. 4.1, the nonlinear static relation $f^{-1}$ follows from

$$x_2 = U^* \cdot K_{P1} \cdot f ( \underline{x} ) \qquad (18)$$

and

$$x_2 = K \cdot U \qquad (19)$$

K describes the determined gain of the 'linearized' system and $K_{P1}$ the gain of the input system $G_{P1}$, Raab (1992).

In practical cases the dynamics of module $G_{P1}$ are often neglectible compared to the time constants of $G_{P2}$ (e.g. energy transducer versus mechanical system dynamics). If $f(\underline{x})$ offers then a precise approximation, good and robust compensation results are obtained.

## Friction Compensation

The main control problem with friction occurs when high positioning accuracy is required. If the process stops within the hysteresis before the set point is reached, only the integral part of the control algorithm can compensate for the offset. This leads to a significant loss of dynamics and accuracy, especially during small position changes.

The basic idea of friction compensation is to compensate the relay function of the Coulomb friction by adding an adequate compensation voltage $U_{comp}$ to the normal control action u, see Fig. 4.2. Different methods as dithering, feedforward compensation and adaptive friction compensation will be described.

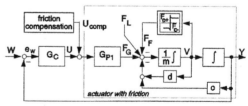

**Fig. 4.2.** General structure for friction compensation

In general, the success of each compensation depends on the quality of $U_{comp}$, but also on the frequency response of the energy transducer $G_{P1}$. However, an overcompensation may destabilize the position control loop, Maron (1991).

_Dithering._ Dynamic linearization or so called dithering is the classical way of analog and even digital friction compensation. By adding a high frequent, periodic signal to the control action U, the friction is compensated during half the period, whereas during the second half friction is undercompensated. The method is quite robust with regard to the amplitude and frequency of the dither signal. A little overcompensation results only in a small armature dither. However, if the amplitude is too large, the control performance is deteriorated.

Another disadvantage is a loss of dynamics, which is caused by the second half of the dither signal, stopping or even accelerating the mass in the wrong direction, Maron (1991).

Feedforward compensation. This approach is from the theoretical point of view the ideal control strategy for friction compensation, see e.g. Wallenborg (1987). By adding the compensation value

$$U_{comp}(k) = -\frac{F_{C\pm}}{K_{PI}} \cdot sign[e_w(k)] \qquad (20)$$

to the controller action U, an optimal inverse function of the Coulomb relay characteristic is obtained. Note that instead of the unknown velocity dY/dt the control error $e_w$ is used for the sign of $U_{comp}$.

In practical applications the accurate value of the Coulomb friction $F_{C+}$ and energy transducers gain $K_{P1}$ is not exactly known and has to be approximated by the measured/estimated static behavior (hysteresis characteristic), see e.g. Maron, Raab (1989) or the measured/estimated dynamic friction relation in Eq. (2), Maron (1991). To avoid an overcompensation, a safety factor $\alpha < 1$ can be introduced

$$U_{comp}(k) = -\alpha \cdot \frac{F_{C\pm}}{K_{PI}} \cdot sign[e_w(k)] \qquad (21)$$

allowing $100 \cdot (1-\alpha)$ % compensation of the effective friction value. The remaining offset is then controlled by the integral part of the position controller.

Adaptive Friction Compensation. In the preceding methods, the friction compensation was realized by a feedforward control strategy. Better results may be expected, if the actual friction value can be adapted in an additional feedback 'friction control loop'. Therefore an adaptive friction compensation was developed, which interprets the abbreviation

$$e_M(k) = y(k) - y_M(k) \qquad (22)$$

between the measured output Y(k) and a linear reference model $Y_M(k)$ as frictional effect. Using a nonlinear friction controller as described in Maron (1991), an inherent adaptation of the compensation value $U_{comp}$ to slowly time varying frictional forces is performed. Due to the fact, that external loads $F_L$ act in the same way as frictional forces, see Fig. 4.2, especially transient load changes may affect an transient overcompensation for several sampling instants.

Varying Process Parameters and Adaptive Position Control

During normal operation, most of the actuating systems change their parameters in a significant way. This is caused by several environmental conditions, wear or immanent physical principles as for example position dependent forces or dampings. Hence, fixed and robust algorithms are usually not suitable therefore. An improved control performance over the whole range of operation as well as lifetime may be obtained by fitted control techniques for the whole actuator.

According to actuating systems, we focus here on parameter scheduling and so called model identification adaptive control systems. Both concepts were extensively examined, see eg. Åström, Wittenmark (1989), Isermann, Lachmann, Matko (1992).

Parameter scheduling. Parameter scheduling based on the measurement of varying operation conditions is an effective method to deal with known and approximately time-invariant process nonlinearities. Supposing measurable auxiliary variables $\underline{V}$, that correlate well with the process changes, the adaptation of the controller parameters $\underline{\Gamma}$ is performed as functions of $\underline{V}$ (parameter schedule), see Fig. 4.3.

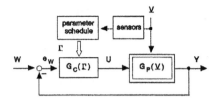

Fig. 4.3. Adaptive control with parameter scheduling

Parameter scheduling offers the specific advantages of a simple microcontroller implementation and a fast reaction to modelled process changes, providing an adaptation even during transient operations. Typical applications therefore are the feedforward adaptation to a varying supporting energy behavior as for example the electric potential in automotive applications, e.g. Raab (1992), or the compensation of position dependencies in pneumatic/hydraulic systems, e.g. Anders (1986).

Parameteradaptive control systems. Parameteradaptive control systems for the closed loop position control of actuators are characterized by using identification methods for parametric process models. The overall structure, performing online parameter estimation, controller design, supervision and coordination is shown in Fig. 4.4.

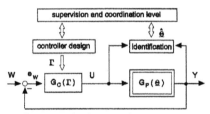

Fig. 4.4. Parameteradaptive control structure

According to typical sampling frequencies from 40 Hz up to 1 kHz and more, the implementation requires adequate microcontrollers/-computers or even digital signal processors. The practical application of parameter-adaptive control techniques is determined by the identification of the high dynamic actuating system in closed loop. The objective is to get good estimates of varying process parameters under the given constraints of transient load changes and several nonlinearities such as frictional forces or hysteresis. This can usually (only) be realized for the mentioned large sampling frequencies if the dominant process changes are described by low order actuator models, Raab (1992). Practical applications, considering also additional identification conditions as discussed in section 3 show that actuator models with integral or first order are good enough for this purpose, see Raab (1990), Köckemann (1991), Glotzbach (1991).

# 5. Experimental Results with Model Based Actuator Control

The proposed methodology for process identification and nonlinear model based control techniques was tested on different actuator types. Experimental results which show some disadvantages of the actuator behavior are presented in this chapter. Because the actuator design remains unchanged, the results show the development of high performance systems by using only a more sophisticated control software and intensified digital signal processing. An implementation of the presented algorithms has been tested on a standard 8-bit microcontroller (SIEMENS 80535). A transfer to similar actuating systems is possible.

## 5.1 Electromagnetic Actuators

Electromagnetic actuators play an important role as linear motion elements in e.g. hydraulic/pneumatic valves, Backé (1990) or fuel injection pumps, Häfner, Noreikat (1985). A precise position control is a challenging task as there are severe nonlinearities in the system. These include friction forces, magnetic hysteresis and nonlinear force-current characteristics, e.g. Lee (1981), Lu (1984), which limit the closed loop control performance in terms of accuracy and dynamics.

### Solenoid Drive

The specified d.c. solenoid drive, Fig. 5.1, has a positioning range of 25 mm and shows a nonlinear force characteristic as depicted in Fig. 5.2. The displacement of the armature is working against a spring and can be measured by an inductive position sensor. Process input is thereby a pulse width modulated (PWM) and amplified voltage U, which manipulates the coil current I.

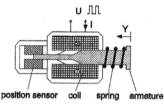

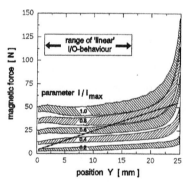

**Fig. 5.1.** Scheme of the investigated d.c. solenoid drive (BINDER MAGNET GmbH)

**Fig. 5.2.** Position dependent nonlinear force-current characteristic of the solenoid drive. The dotted line represents the linear spring characteristic.

The objective is to design a robust position control loop, which includes the correction of the nonlinear static characteristic in Fig. 5.2 and compensation of dominant frictional forces. The low cost solenoid, which usually performs simple mechanical switching tasks then offers similar features as a sophisticated magnet with proportional I/O-behavior.

Therefore the static force-current-position dependency has to be linearized by a nonlinear correction as shown in Fig. 4.1. An appropriate function, describing the nonlinear characteristic of Fig. 5.2 is obtained by a polynomial approximation

$$f\,(I,Y) = I \cdot \sum_{i=0}^{2} \frac{K_i}{(Y_0 - Y)^i} \quad \text{with } Y_0 = 26 \ mm \quad (23)$$

Raab (1992). The resulting statics of the linearized actuator are shown in Fig. 5.3, where a typical hysteresis characteristic becomes obvious. Its gradient represents the local gain $K_P$ of the actuator, which can be assumed now as constant. The position dependent width of the hysteresis characteristic is a measure for frictional forces and magnetic hysteresis, see e.g. Maron, Raab (1989).

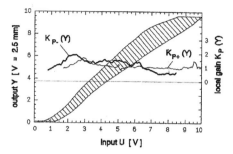

**Fig. 5.3.** Hysteresis characteristic and position dependent local gain of the 'linearized' solenoid drive

According to the 'linearized' system and the equations for the inner current loop

$$T_I \, \dot{I}(t) + I(t) = K_I \cdot U(t) \quad (24)$$

and the mechanical subsystem

$$m \, \ddot{Y}(t) + d \, \dot{Y}(t) + c \, Y(t)$$
$$= K_{Mag} \, I(t) - F_C \, sign(\dot{Y}) + F_L(t) \quad (25)$$

the I/O behavior of the actuator can be modelled as a third order system in form of Eq. (3)(4), e.g. Raab (1992). The unknown parameters are obtained during a pre-identification phase, exciting the actuator with the shown input signal (Fig. 3.2) and sampling with 400 Hz. Considering the effect of Coulomb friction as discussed in section 3, the nonlinear output error parameter estimation leads to the following "direction dependent transfer functions"

$$G_+(s) = \frac{Y(s)}{U(s)} = \frac{382400}{(s + 116.4)\,(s^2 + 40.4s + 3329.4)}\, e^{-0.0025\,s} \quad (26)$$

$$G_-(s) = \frac{Y(s)}{U(s)} = \frac{220100}{(s + 47.9)\,(s^2 + 47.9s + 3444.5)}\, e^{-0.0025\,s} \quad (27)$$

Index +/- denotes the direction of the armature motion and the additional deadtime describes the effect of an asynchronous PWM-generation.

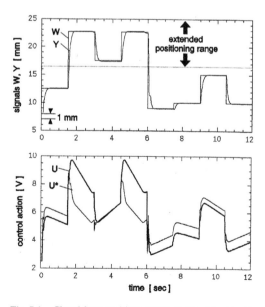

**Fig. 5.4.** Closed loop position control of the solenoid with correction of the nonlinear actuator characteristic, but without friction compensation, $T_0$=2.5 msec

Figure 5.4 shows the obtained control performance,

using a numerical optimized position controller

$$G_C(q^{-1}) = \frac{u(k)}{e_w(k)} = \frac{2.231 - 4.204q^{-1} + 2.000q^{-2}}{(1 - q^{-1})(1 - 0.616q^{-1})} \qquad (28)$$

(PIDT1-type, $T_0$=2.5 msec) where $q^{-1}$ is a shift operator for one sampling time ( $u(k)q^{-1} = u(k-1)$ ). Although there is a change in the actuators dynamic behavior, the controller designed for the slower negative motion (worst case) is robust enough for positive motions. The dynamic features are suitable and stability is obtained even in the extended positioning range (17 mm < Y < 25 mm).

Using the same linear control algorithm for small setpoint changes, the typical effects by the system-immanent friction and hysteresis occur. Fig. 5.5 shows the unsatisfying positioning and dynamics. In steady state an offset of about 110 .. 140 µm remains. The improved control performance with adaptive friction compensation is obvious as a positioning accuracy up to 25 - 50 µm could be achieved. At the same time an adaptation of the actual compensation value to the present hysteresis width is performed. The course of the control action U shows an automatic generated "dither signal", which adapts its amplitude and frequency with regard to the control performance.

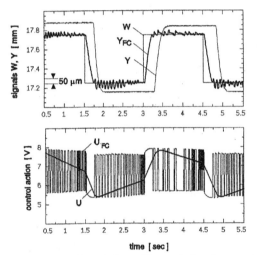

**Fig. 5.5.** Comparison of the position control performance for small setpoint changes without/with adaptive friction compensation (index $_{FC}$), $T_0$=2.5 msec

### Proportional Magnet (Fuel Injection Pump)

Similar experimental results could be obtained, using a proportional magnet drive in a diesel fuel injection pump, Fig. 5.6. The electromechanical subsystem, which performs continuous positioning over a range of 20 mm, is usually manipulated by a well tuned analog controller. According to position dependent frictional forces, the I/O behavior shows significant deviations in the hysteresis width and local gain characteristic, Fig. 5.7.

Because of the fast current loop, the modelling of the actuator dynamics can be concentrated here on the mechanical subsystem. The identification of the resulting second order model leads to the direction dependent transfer functions

$$G_+(s) = \frac{Y(s)}{U(s)} = \frac{16117.7}{(s^2 + 140.6\,s + 7683.3)} \cdot e^{-0.002\,s} \qquad (29)$$

$$G_-(s) = \frac{Y(s)}{U(s)} = \frac{10758.5}{(s^2 + 90.8\,s + 4005.3)} \cdot e^{-0.002\,s} \qquad (30)$$

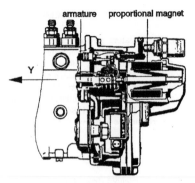

**Fig. 5.6.** Schematic of the electromechanical subsystem in a diesel fuel injection pump, BOSCH (1990). The armature is driven by an proportional magnet

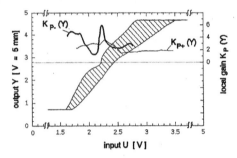

**Fig. 5.7.** Hysteresis characteristic and position dependent local gain of the actuating system

A position controller

$$G_C(q^{-1}) = \frac{U(k)}{e_w(k)} = \frac{2.911 - 5.579q^{-1} + 2.680q^{-2}}{(1 - q^{-1})(1 - 0.642q^{-1})} \qquad (31)$$

which is designed by the numerical minimization of a quadratic cost function (PIDT1-type, $T_0$=1 msec), shows the closed loop control performance in Fig. 5.8.

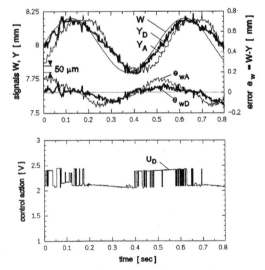

**Fig. 5.8.** Comparison of closed loop position control for small setpoint changes between the analog (index $_A$) and nonlinear digital controller (index $_D$), $T_0$=1 msec

20

Frictional and hysteresis effects are thereby compensated by a feed forward compensation algorithm with the safety factor $\alpha=0.9$ (see section 4). In comparison to the industrial analogue position controller, an improved positioning dynamics as well as accuracy during small setpoint changes is obtained, Raab (1992).

## 5.2 Pneumatic Actuator

The chosen pneumatic actuator is characterized by a rugged and temperature resistant design, which offers in general a high reliability, Fig. 5.9. Designed for actuating tasks in modern vehicle carburators, the membrane drive performs a positioning range of 20 mm, Baumgartner (1982), Schürfeld (1984). The supporting energy is low air pressure from the manifold, varying in automotive applications from 100 mbar to nearly atmosphere. A positioning control is obtained by manipulating the internal pressure potential with two pulse width modulated on/off valves.

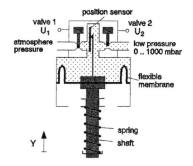

**Fig. 5.9.** Scheme of the low pressure membrane drive (PIERBURG GmbH)

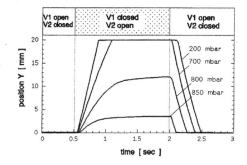

**Fig. 5.10.** Varying dynamics of the pneumatic actuator for different low pressure potentials within the manifold

The present system incorporates some typical drawbacks of low cost membrane drives. According to a changing low pressure support, the dynamics of the system are varying in wide ranges, Fig. 5.10. In addition nonlinear process dynamics are coupled with the shaft and membrane position. They depend on the motion direction as well as on the external load.

Hence a fixed, robust designed closed loop position controller cannot give a good control performance. Therefore it was tried to improve the control performance of the actuator by using adaptive control techniques. The developed methods are based on a discrete time actuator model, which is obtained by theoretical modelling. Under the given constraints, the dynamics for each direction can be described by an integral behavior

$$Y(k) = Y(k-1) + \frac{K_{I1}}{T_0} \cdot U_1(k-1) \qquad \text{für } Y(k) < Y(k-1) \ (down \downarrow)$$
$$Y(k) = Y(k-1) + \frac{K_{I2}}{T_0} \cdot U_2(k-1) \qquad \text{für } Y(k) > Y(k-1) \ (up \uparrow)$$
(32)

where $K_{I1/I2}$ represent integration constants with the physical dimension of a velocity [mm/sec], Raab (1992). In comparison to the sampling time $T_0=20$ msec, the model parameters are supposed to be constant or slowly time varying.

For closed loop position control a digital P-algorithm with motion dependent parameters

$$U_1(k) = q_{01} \ e_W(k) \ , \qquad e_W(k) < 0$$
$$U_2(k) = q_{02} \ e_W(k) \ , \qquad e_W(k) > 0$$
(33)

is used. According to an estimated process parameter range from $K_I=10 .. 80$ [mm/sec], the design of the controller gain $q_0$ by a numerical minimization of a quadratic cost function delivers the adequate values presented in Fig. 5.11.

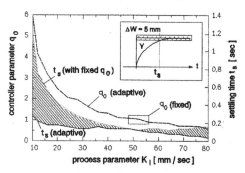

**Fig. 5.11.** Parameters and closed loop settling times for a fixed and scheduled adaptive position controller

Due to the fact, that the process parameters $K_{I1}$ and especially $K_{I2}$ is highly dependent on the low pressure $P_L$, an additional measurement (and sensor) of $P_L$ enables the design of a parameter scheduled control strategy

$$\Gamma = [ \ q_{01}, q_{02} \ ] = f [ \ K_{I1} \ (P_L) \ , \ K_{I2} \ (P_L) \ ] \qquad (34)$$

for both operation directions. The obtained control performance, evaluated by the settling time $t_S$ for a closed loop step response is presented in Fig. 5.11. Compared to a fixed, robust tuned position controller, which is designed for the worst case of the integration constant $K_I$, a considerably improvement in the range of low pressures is achieved (shaded area in Fig. 5.11).

A more sophisticated approach is the application of a parameteradaptive control structure, which performs the tasks shown in Fig. 4.4. For online/realtime identification of the unknown process parameters $K_{I1}$ and $K_{I2}$, the recursive DSFI-algorithm with the forgetting factor $\lambda=0.975$ is applied. Estimation is started by the supervision level, if appropriate exciting conditions are fulfilled. A valid estimate is already obtained after 3 .. 4 sampling instants. Hence the actual system parameters are obtained during small setpoint changes.

Process identification and controller tuning have to be coordinated during real time application. The used algorithm depends on monitored process I/O signals, an online comparison with a reference model and an evaluation of time varying eigenvalue of the recursive estimator

$$z_n = 1 - \underline{\psi}^T \cdot \underline{\chi}(k) \qquad (35)$$

, see Kofahl (1988), Isermann, Lachmann, Matko (1992).

The time history of the parameteradaptive position control is shown in Fig. 5.12 for different supporting pressures and external load changes. Adaptation is performed during appropriate process excitation phases. This guarantees a good estimation result and leads to a well-tuned control algorithm. An accuracy up to 15-20 $\mu$m even by using the maximal system dynamics can be obtained over the whole range of supporting energy changes and external loads.

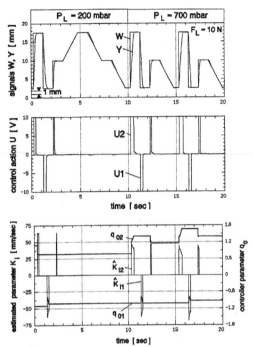

**Fig. 5.12.** Time history of the parameteradaptive position control for different pressures of the air supply and load $F_L$, $T_0$=20 msec

Compared to the positioning accuracy and dynamics of the fixed controller, an improved control performance is obtained. An evaluation of closed loop performance for sinusoidal reference values is shown in Fig. 5.13. The bandwidth with the adaptive controller is increased to about 6 Hz for all cases, compared to 1 .. 6 Hz with the fixed controller. Process parameter $K_I$ varies thereby between 12 and 50 mm/sec, which represents a low air pressure range from $P_L$=800 .. 400 mbar.

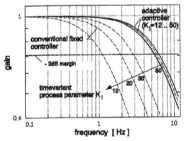

**Fig. 5.13.** Measured frequency response of the closed loop for the parameteradaptive and conventional fixed controller

# 6. Model Based Fault Detection, Diagnosis and Supervision of Actuators

An important feature of an intelligent actuator is the automatic supervision and fault diagnosis of its components. Fig. 6.1 shows an actuator influenced by faults. External faults are for example caused by the power supply, contamination or collision, internal faults by wear, missing lubrication, sensor faults or other malfunctions of components like springs, bearings or gears.

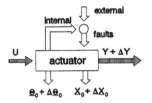

**Fig. 6.1** Scheme of an actuator influenced by faults

If the faults influence direct measurable output variables they may be detected by an appropriate signal evaluation. The corresponding functions are called <u>monitoring</u>, if the measured variables are checked with regard to a certain tolerance of the normal values and alarms are triggered if the tolerances are exceeded. For actuators e.g. the current of the input supply or the control deviation can be monitored. In the cases where the limit value violation signifies a dangerous state an appropriate action can be indicated automatically. This is called <u>automatic protection</u>. An example is the actuator switch-off at the end of the positioning range.

The classical ways of limit value checking of some few important measurable variables are appropriate for the overall supervision. However, developing actuator faults are only detected at rather late state and the available information does not allow an in-depth fault diagnosis. Research efforts have shown, that the use of process models allows an early fault detection in connection with normal measured variables, Isermann (1984b)(1991b). Then nonmeasurable quantities like state variables and parameters may be estimated. With this improved knowledge a <u>supervision with fault diagnosis</u> becomes possible.

Fig. 6.2 shows the scheme of a knowledge based fault diagnosis with process models. The knowledge base consists of an automatic and a heuristic part. The analytic problem solution includes parameter or state estimation and generates symptoms as changes from the normal behaviour. An inference mechanism then performs the fault diagnosis.

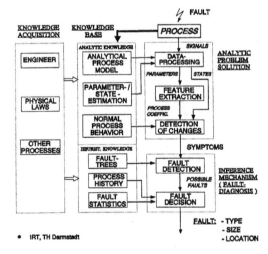

**Fig. 6.2** Scheme of the knowledge based fault diagnosis

The application of the model based fault detection will now be shown for an electromagnetic actuator.

## 6.1 Fault Diagnosis with Parameter Estimation

The electromagnetic actuator shown in Fig. 5.1 is now considered for the 'linearized' operation range from 0 .. 25 mm. From the equations for the current circuit Eq. (24) and the mechanical subsystem Eq. (25) a third order differential equation follows

$$Y^{(3)}(t) + a_2^* \dot{Y}(t) + a_1^* \dot{Y}(t) + a_0^* Y(t) = b_0^* U(t) + c_{DC}^*(t) \qquad (36)$$

The parameters of the continuous time representation

$$\underline{\Theta}^T = [\, a_2^* \,,\, a_1^* \,,\, a_0^* \,,\, b_0^* \,,\, c_{DC}^* \,] \qquad (37)$$

depend thereby on the physical process coefficients

$$\underline{P}^T = [\, T_I \,,\, D \,,\, \omega_0 \,,\, K_P \,,\, c_{DC}^* \,] \qquad (38)$$

with e.g.

$$D = \frac{d}{2\sqrt{m\,c}} \,,\quad \omega_0 = \sqrt{\frac{c}{m}} \qquad (39)$$

These process coefficients can be expressed in terms of the parameter estimates $\hat{\underline{\Theta}}$, Raab (1992). Hence after estimation of the model parameters $\underline{\Theta}$ by measuring the voltage U and the position Y all process coefficients $\underline{P}$ can be calculated.

In the following some experimental results are shown for artificially generated actuator faults:

F1: too large pretension of the spring
F2: decrease of the spring constant (by break or aging spring change from c=1650 → 1200 N/m)
F3: increase of friction (increase of surface roughness and jamming)
F4: fault in the current circuit (weak controller gain)

The parameters were estimated by the proposed output error minimization using specific excitation signals, see section 3. Sampling time was $T_0$=2.5 msec. Figure 6.3 and Table 6.1 show the results for different faults. Based on the deviations (symptoms) all faults can be identified. This can be performed by a pattern recognition or a systematic treatment of fault-symptom-trees, Freyermuth (1991). In all cases different patterns of coefficient changes result. This enables an unique diagnosis of the 4 faults.

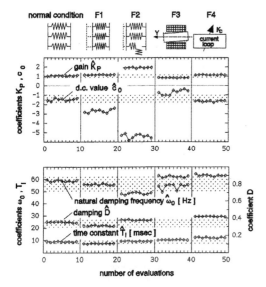

**Fig. 6.3.** Parameter estimates for an electromechanical drive with different faults (positive motion direction)

TABLE 6.1 Changes of Process Coefficients for an Electromechanical Drive in Dependence on Different Faults

| fault type ↓ | static coefficients | | dynamic coefficients | | |
|---|---|---|---|---|---|
| | $K_{P+}$ | $c_{0+}$ | $\omega_{0+}$ | $D_+$ | $T_{I+}$ |
| F1 | o | -- | o | o | o |
| F2 | ++ | -- | - | o | o |
| F3 | o | + | o | ++ | o |
| F4 | o | o | + | + | ++ |

*(estimates for positive motion direction, $c_0 = c_{DC}^*/a_0^*$)*
*Symbols: + increasement - decreasement o no changes*

## 6.2 Fault Detection with State Estimation

With the basic equations (24)(25), the continuous time state representation of the electromechanical actuator

$$\dot{x}(t) = \underline{A}^* \, \underline{x}(t) + \underline{b}^* \, U(t)$$
$$Y(t) = \underline{c}^{*T} \, \underline{x}(t) \qquad (40)$$

can be obtained with

$$\underline{x}(t) = [\, \dot{x}_1(t)\ \dot{x}_2(t)\ \dot{x}_3(t)\ \dot{x}_4(t)\,] \equiv [\, Y(t)\ \dot{Y}(t)\ I(t)\ c_{DC}^*(t)\,]$$

$$\underline{A}^* = \begin{pmatrix} 0 & 1 & 0 & 0 \\ -\dfrac{c}{m} & -\dfrac{d}{m} & \dfrac{K_{Mag}}{m} & \dfrac{1}{m} \\ 0 & 0 & -\dfrac{1}{T_I} & 0 \\ 0 & 0 & 0 & 0 \end{pmatrix} \quad \underline{b}^* = \begin{pmatrix} 0 \\ 0 \\ \dfrac{K_I}{T_I} \\ 0 \end{pmatrix} \quad \underline{c}^* = \begin{pmatrix} 1 \\ 0 \\ 0 \\ 0 \end{pmatrix} \qquad (41)$$

If the process coefficients are known, the state variables $\underline{x}(t)$ can be estimated by a disturbance observer based on the measurement of U(t) and Y(t), see e.g. Bakri (1988).

Now only the static behavior is considered, resulting in

$$Y(t) = K_P \, U(t) + Y_0 + x_4(t) \qquad (42)$$

This equation is depicted for the nominal state in Fig. 6.4. It describes the left and right hysteresis characteristics with two different steady state values $x_{4n+}$ and $x_{4n-}$ in relation to the nominal actuator statics without hysteresis $K_P U + Y_0$.

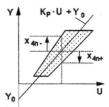

**Fig. 6.4** Simplified hysteresis curve of the electromagnetic actuator in the nominal state (index $_n$)

Now a discrete time state observer is used

$$\underline{x}_B(k+1) = \underline{A}\,\underline{x}_B(k) + \underline{b}\,u(k) + \underline{h}\,e_B(k) \qquad (43)$$

with the residual

$$e_B(k) = y(k) - \underline{c}^T \underline{x}_B(k) \qquad (44)$$

The representation is obtained by discretization of Eq. (40) for the sampling time $T_0$. $\underline{x}_B(k)$ describes thereby the continuous time state variables $\underline{x}(t)$ at the sampling instants $kT_0$, k=0,1,2.. . The observer feedback is designed by pole placement and resulting in

$$\underline{h}^T = [\, 0.5459\ \ 47.485\ \ 7.374\cdot 10^{-9}\ \ 1.017\,] \qquad (45)$$

Figure 6.5 shows the measured signals of the residuals and

two observed state variables of the actuator without faults.

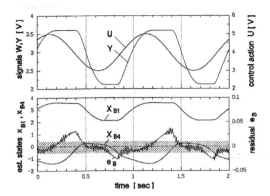

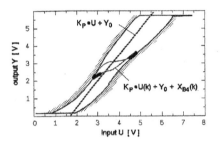

**Fig. 6.5.** Measured actuator signals and disturbance observer signals for the actuator without faults, $T_0 = 2.5$ msec

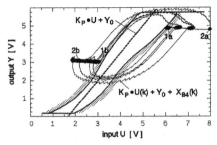

**Fig. 6.8.** Measured characteristic hysteresis for the actuator without faults and trajectories of Eq. (42) for increasing friction (with fault)

A comparison of both ways of fault detection shows:

- parameter estimation:
  - less a priori knowledge of the process model required
  - parameter deviations allow an in depth fault diagnosis of different faults, especially for multiplicative faults
  - extensive computations required

- state estimation:
  - more a priori knowledge required as process model parameters must be known
  - state estimates show fast response to sudden faults, especially for additive faults
  - no in depth fault diagnosis
  - less computations required

Hence parameter estimation gives more information on the type of faults which may develop slowly and the used state estimation approach gives a fast information for suddenly appearing faults with less computations.

**Fig. 6.6.** Measured hysteresis characteristic for the actuator without faults and trajectories of Eq. (42)

Now $x_{B4}(k)$ is monitored for the case that the observer is adapted for the motion in one direction, indicated by a small residual

$$|e_B(k)| = e_B \qquad (46)$$

As indicated in Fig. 6.6 by black areas, Eq. (42) then gives values of the nominal hysteresis characteristic.

Figure 6.7 shows the corresponding measured and observed signals if the actuator friction increases continuously from 1a, 1b to 2a, 2b. The position control variable Y(k) does not show change, only U(k) shows larger values. This shows that a closed loop compensates this fault. The observed state $x_{B4}(k)$, however, indicates significant changes from the nominal values (shaded ranges) for the adapted observer, Eq. (46) such that $x_{B4+} < x_{4n+}$ and $x_{B4-} > x_{4n-}$. Compare also Fig. 6.8 with 6.6. These deviations are now the symptoms for the fault detection with state estimation.

# 7. Implementation on Microcontrollers

Due to the rapidly advancing microcomputer technology the implementation of sophisticated digital control algorithms is possible even on low-cost hardware devices, see e.g Färber (1989). In the field of actuator control the microcontroller is therefore the dedicated processor type, which performs the embedded control including process interfacing and signal processing.

Before discussing the implementation, the proposed algorithms should be evaluated in terms of their arithmetic properties and related hardware requirements. As considered in e.g. Raab (1992) the most applied type of algorithm is the discrete time controller as formulated by Eq. (14). Because only limited ranges of variables and limited quantization of coefficients and variables are required, fixed point numbers and arithmetics are usually sufficient, if dealt with appropriately, Hanselmann (1985). Common 8 or 16 bit microcontrollers offer then an efficient realtime computation, even for nonlinear model based control algorithms. Problems usually arise, when state controllers Eq. (15), parameter and state estimators have to be implemented on a microcontroller, performing fast and precise computing. With respect to the wide number ranges and required precision, only floating point numbers and related arithmetics are considered as appropriate, Hanselmann (1985), Kofahl (1985). Although the floating point environment is easy to achieve through subroutine libraries, the increasing computational effort usually limits realtime applications to 'low order' algorithms or a more sophisticated hardware has to be used (for example 32-bit microcontrollers with custom VLSI designs).

For the presented algorithms in sections 3, 4 and 6 we have focused on the implementation using only a low-cost hardware. The chosen microcontroller SIEMENS 80535, 12

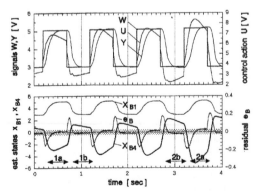

**Fig. 6.7.** Measured actuator signals and disturbance observer signals for the actuator with increasing friction from "1" to "2" with fault, $T_0 = 2.5$ msec

MHz system clock, performs 8-bit arithmetics under the given constraints of storage and register capacity, see e.g. Feger (1987). Using self defined data formats and a speed-optimized assembler code, computation times and sampling frequencies were obtained as presented in Table 7.1. The values are given for the tested and embedded realtime software including process interfacing (AD-conversion, PWM-generation).

TABLE 7.1 Computing Time and Obtained Sampling Frequencies with a Low-cost 8-bit Microcontroller

| algorithm | computation time [msec] | max. sampling frequency [Hz] |
|---|---|---|
| **electromagnet** | | |
| • closed loop position control | | |
| - PIDT1 | 0.48 (0.53) [1] | 1200 (1000) |
| - PIDT1 + feed forward friction control | 0.55 (0.60) [1] | 820 (600) |
| - PIDT1 + adaptive friction control | 2.11 (2.16) [2] | 430 (400) |
| • fault diagnosis and supervision | | |
| - via state estimation | 2.73 [2] | |
| - via parameter estimation | -- [3] | |
| **membrane drive** | | |
| • closed loop position control | | |
| - parameter scheduling | 0.42 [2] | 1200 [2] |
| - parameteradaptive control | 7.68 [2] | 120 [2] |

*(the values in brackets denote the position control algorithm with non-linear correction)*

1: *16 bit fixed-point data format and related arithmetics*
2: *24 bit floating-point data format and related arithmetics*
3: *implementation requires a sophisticated floating-point hardware*

Nonlinear closed loop position control can be performed with sampling frequencies from 400 Hz up to 1 kHz. The maximal value is limited by the computational and interfacing delay. In the case of adaptive control the implementation requires 24-bit floating point numbers, which result in decreased sampling rates. For fault diagnosis using parameter estimation a more sophisticated microcontroller hardware including an extended storage capacity has to be used, which is no problem for e.g. quality control after actuator manufacturing or maintenance computers.

## 8. Conclusions and further Development

The contribution has tried to show how the actuators can perform more intelligent functions if they are governed by microcontrollers. As pointed out "low degree intelligence" is meant which makes the actuator to a more autonomous system. One important basis is the theoretically derived mathematical model which includes the nonlinear behavior with friction and hysteresis. Another basis are parameter and state estimators which are driven by few measurable signals. It could be shown theoretically and experimentally for three actuator types that the positioning accuracy and dynamics could considerably be improved by adaptive control techniques by which the controllers "learn" about the process. Model based methods are then used for supervision and fault diagnosis of the actuator. Hence the microcomputer controlled actuators observe their own faults and "reason", close to realtime, about the causes. They may also make decisions with regard to the reached control performance and especially for the diagnosed faults. The decision may be a fail-safe-operation or a reconfiguration if other actuators can take over the task.

In the future the design of an actuator may be not limited to the mechanical side and a simple position controller, because the added microelectronics influence the static and dynamic behavior considerably and allows to perform new tasks. Hence an integrated design takes place, both on the mechanical and the electronic side, on the hardware and the software side. These are typical features of a mechatronic design.

Acknowledgements.  The research on actuators was supported by the Forschungsvereinigung Verbrennungskraftmaschine e.V (FVV) and the Bundesministerium für Forschung und Technologie (BMFT). The authors are grateful for the financial support and the discussions with the sponsoring committee under the chair of Dipl.-Ing. J.Gloger, VW AG, Wolfsburg.

References

Anders P. (1986). **Auswirkungen der Mikroelektronik auf die Regelungskonzepte fluidtechnischer Systeme**, Dissertation, RWTH Aachen

Åström K.J., Eykhoff P. (1970). **System Identification - A Survey,** Automatica Vol. 7

Åström K.J., Wittenmark B. (1989). **Adaptive Control**, Addison Wesley, New York

Åström K.J. (1991). **Intelligent control**, European Control Conference, Grenoble

Backé W. (1988). **Fluidtechnik im Wandel als Folge der Mikroelektronik**, VDI/VDE-Tagung ACTUATOR 88, Bremen 9.- 10. Juni

Backé W. (1990). **New developments in valve technology**, VDI/VDE-Tagung ACTUATOR 90, Bremen Juni 21.- 22.

Bakri N., Becker N., Ostertag E. (1988). **Anwendung von Kontroll-Störgrößenbeobachtern zur Regelung und Kompensation trockener Reibung**, AT 26 (1988) 2

Baumgartner H. (1982). **Elektropneumatische Stellglieder für Kraftfahrzeuge**, 5. Aachener Fluidtechnisches Kolloquium, Fachgebiet Pneumatik 23-25

Biermann G.J. (1977). **Factorization methods for discrete sequential estimation**, Academic Press, New York

BOSCH (1990). **Technische Unterrichtung - Dieseleinspritztechnik im Überblick**, Robert Bosch GmbH Stuttgart

Drewelow W. (1990). **Parameterschätzung nach der Ausgangsfehlermethode**, msr, Berlin, 33 (1990) 1

Färber G. (1989). **Entwicklungstrends der Mikroelektronik und der Informationstechnik**, ATP 31 (1989) 9

Feger O. (1987). **Die 8051 Mikrocontroller Familie**, Markt und Technik Verlag, München

Franz K.P. (1973). **Untersuchungen zur Kompensation unerwünschter Nichtlinearitäten in Steuerungs- und Regelungssystemen bei direkter Ansteuerung der Stelleinrichtung durch Prozeßrechner**, Dissertation A, TH Magdeburg

Freyermuth B., Held V. (1990). **Ein Vergleich von Methoden zur Identifikation von Übertragungsfunktionen elastischer Antriebssysteme**, at 38 (1990) 12

Freyermuth B. (1991). **Knowledge based incipient fault diagnosis of industrial robot mechanics** ,IFAC-Symp. SAFEPROCESS, Baden-Baden, Sept. 1991

Glotzbach J. (1991). **Adaptive Regelung eines hydraulischen Drehantriebs**, 36. Internationales wissenschaftliches Kolloquium, TH Ilmenau

Halme A.H., Selkäinako J. (1991). **Advanced fault detection for sensors and actuators**, IFAC-Symp. SAFEPROCESS, Baden-Baden, Sept. 1991

Häfner G., Noreikat K. (1985). **Stellglieder, Bausteine der Mikroelektronik**, VDI-Berichte 553, VDI-Verlag Düsseldorf

Henry M.P., Clarke D. W. (1991). **A standard interface for self-validating sensors**, IFAC-Symp. SAFEPROCESS, Baden-Baden, Sept. 1991

Hanselmann H. (1987). **Implementation of digital controllers - a survey**, Automatica 23 (1987) 1

Isermann R. (1984a). **Computer aided design and selftuning of digital control systems**, American Control Conference, San Diego 1984

Isermann R. (1984b). **Process fault detection based on modeling and estimation methods - a survey**, Automatica 20, 387 - 404

Isermann R. (1988). **Identifikation dynamischer Systeme**, Band I+II, Springer Verlag, Berlin

Isermann R. (1989, 1991a). **Digital control systems**, Vol.I (1989), Vol. II (1991a), Springer, Berlin

Isermann R. (1991b). **Fault diagnosis of machines via parameter estimation and knowledge processing**, IFAC-Symp. SAFEPRO-CESS, Baden-Baden Sept. 1991

Isermann R., Lachmann K.H., Matko D. (1992). **Adaptive control systems**, Prentice Hall, New York

Kabaila P. (1983). **On output-error methods for system identification**, IEEE Transactions on Automatic Control 28 (1983) 1

Kofahl R. (1986). **Verfahren zur Vermeidung numerischer Fehler bei Parameterschätzung und Optimalfilterung**, Automatisie-rungstechnik, 34.Jahrgang, Heft 11/1986

Kofahl R. (1988). **Parameteradaptive Regelungen mit robusten Eigenschaften**, Fachberichte Messen Steuern Regeln, Nr.39, Springer Verlag

Köckemann A., Konertz J., Lausch H. (1991). **Regelung elektro-hydraulischer Antriebe unter Berücksichtigung industrieller Randbedingungen**, Automatisierungstechnik 39 (1991) 6

Kuo B.C. (1980). **Digital control systems**, Holt, Rinehardt and Winston, Tokyo

Lachmann K.H. (1983). **Parameteradaptive Regelalgorithmen für bestimmte Klassen nichtlinearer Prozesse mit eindeutigen Nichtlinearitäten**, Dissertation TH Darmstadt, VDI-Fort-schrittberichte Reihe 8, Nr. 66, Düsseldorf

Lee C. (1981). **Untersuchungen an einem Proportionalmagneten**, Ölhydraulik und Pneumatik 25 (1981) Nr.6

Ljung L. (1978). **Convergence analysis of parametric identification methods**, IEEE Transactions on Automatic Control 23 (1978) 5

Ljung L. (1988). **System identification - theory for the user**, Prentice Hall Inc., New Jersey

Lu Y.H. (1984). **Neue Entwicklung von Proportionalmagneten**, Ölhydraulik und Pneumatik 28 (1984) Nr.7

Maron C., Raab U. (1989). **Identifikation und Kompensation von Reibung in mechanischen Prozessen**, VDI-Tagung "Mecha-tronik im Maschinen- und Fahrzeugbau", Bad Homburg

Maron C. (1991). **Methoden zur Identifikation und Lageregelung mechanischer Prozesse mit Reibung**, Dissertation TH Darm-stadt, VDI-Fortschrittberichte Reihe 8, Nr. 246, VDI Verlag Düsseldorf

Merrill W. C., Lorenzo C.F. (1988). **A reusable rocket engine intelligent control**, Joint Prepulsion Conference, Boston 1988, July 11-13

Peter K.H., Isermann R. (1989). **Parameter adaptive PID-control based on continuous-time process models**, IFAC Symposium on Adaptive Control and Signal Processing, Glasgow

Pfaff G. (1982). **Regelung elektrischer Antriebe Band I+II**, Oldenbourg Verlag, München

Raab U., Isermann R. (1990a). **Low power actuator principles**, VDI/VDE-Tagung Actuator 90, Bremen (1990)

Raab U., Isermann R. (1990b). **Application of digital control techniques for the design of actuators**, VDI/VDE-Tagung Ac-tuator 90, Bremen (1990)

Raab U. (1992). **Internal Report: Modellgestützte digitale Regelung und Überwachung von Kraftfahrzeugaktoren**, TH Darmstadt, Institut für Regelungstechnik

Saridis G.N. (1977). **Self-organizing control of stochastic systems**, Marcel Dekker, New York

Schürfeld A. (1984). **ECOTRONIC- Ein elektrisches Gemisch-bildungssystem für Ottomotoren**, MTZ 45 (1984) 2

Stribeck E. (1902). **Die wesentlichen Eigenschaften der Gleit- und Rollenlager**, Zeitschrift des VDI, Band 46, VDI Verlag Düssel-dorf

Wallenborg A. (1987). **Control of flexible servo systems**, LUTFD2/(TFRT-3188)/1-104/(1987), Department of Auto-matic Control, Lund Institute of Technology, Sweden

Young P. (1981). **Parameter estimation for continuous-time models - a survey**, Automatica, Vol 17, No. 1.23-39

# INTELLIGENT CONTROLLERS ISSUES

**P. Albertos\*, A. Crespo, F. Morant and J.L. Navarro**

*Departamento de Ingenieria de Sistemas, Computadores y Automatica, Universidad Politecnica de Valencia, Spain*
*\*on leave at CICS, University of Newcastle, Australia*

**Abstract.** Intelligent Controllers, that is, controllers being designed and implemented following some of the existing Artificial Intelligence techniques, are becoming the next generation of industrial controllers. In this paper an overview of the most interesting issues in these controllers, as new control devices, is presented. The controller structure and features, its connection to the process, the challenging software and hardware implementation issues, and the available design tools, are reviewed. An industrial application illustrates the feasibility of some of these new concepts.

**Keywords.** Direct Expert Control, Supervision, Distributed Control, Intelligent Control, Real Time Expert Systems, Neural Networks, Intelligent Regulators.

## 1. Introduction

One of the main advantages of the digital control over the continuous time control is that, implemented on a microprocessor or computer based device, it is able to handle many different control problems, such as regulation, tracking, monitoring, and supervision among others. Still some decision based activities are supported by the operator, sometimes following the indications of the controller. In that case we are in the fuzzy frontier of Intelligent Controllers (IC). In this paper we reserve this name to those controllers involving the use of any of the Artificial Intelligence (AI) techniques. In particular, we will consider controllers dealing with reasoning and learning.

The controller device may be approached from two different points of view. The control engineer as well as the process engineer, look at this device as a part of the global controlled system. Its off-line features are not as relevant as those related with the on-line closed loop behavior. On the other hand, from an instrumentation engineer point of view, the controller is a computer based device (the digital controllers, of course) providing some inputs to the process, computed from both process measurements and operator commands. The way each one handles the controller representation is significant. In the first case it is represented as a component of the closed loop controlled system and, in the second case, it is a device

with some inputs and outputs.

Of course, the first approach is more appropriated to design and analyze the controller effectiveness, whereas the second one allows the device maintenance, fitting and development. Typically, the controller engineer focuses his attention to the effect of the controller in the global system. When dealing with IC we must combine both approaches. In fact, sometimes, the IC just tries to implement the way the human operator performs the control. The plant knowledge is behind that but we are implementing the operations as previously defined by the human controller.

Among the functionalities we expect from IC are: handling of approximated knowledge, dealing with complex, non linear and hierarchical processes, allowance of decision based control actions, friendly interfacing with the different users, from the plant operator to the top management, and fault detection and basic maintenance, as well as ability to cover most of the operating conditions, including normal, abnormal, emergency, start-up or shut-down. As we see, the control algorithm under a given mode of operation, if there is one, is a very small part of the controller role.

It seems that IC are the base of the next generation of industrial controllers. Let us summarize the main

This work has been partially supported by the CICYT
project no. ROB-89 0442, and IMPIVA project n. 4639

features of IC supporting these ideas:

- AI techniques provide powerful representation, including numerical and heuristic information.
- They also provide an easy and understandable problem statement.
- Rule-based controllers: Fuzzy Logic Controllers (FLC) or Real Time Expert Systems (RTES) are a new version of decision-based controller implementation.
- Artificial Neural Networks (ANN) approach allows to deal with non-linear control problems.
- Learning is a common feature of AI systems.
- Knowledge structure improves understanding, updating, and modularity, allowing an easy interaction with the user.
- Advanced computers architectures, such as parallel or systolic processors, and transputers lead to faster and more reliable control systems.

In the last years, many commercial devices and technical papers show the heading of IC, (i.e.,[Saridis 1983], [Astrom 1991], [Navarro 1990], [Pang 1991]), dealing with some options and decision-based facilities, sometimes just dealing with a particular controller design, [Astrom 1992]. As mentioned above, our concept of IC is a more general one, trying to cover most of those items.

The paper is outlined as follows. In the next section, a review of the IC structures as well as the problems they are suitable to deal with are presented. In section 3, the practical implementation issues from both points of view, hardware and software, and the tuning methodologies, are discussed. Section 4 deals with the available design tools for IC, focusing the attention in expert systems environments and ANN design software. Finally, a typical application is presented. Although the proposed solution is original, and it is a particular one, it shows up most of the previously discussed concepts.

## 2. Controller Structure and Performances

According to the discussion above, the IC will be considered as a device to be connected to a real-time (RT) process that takes data, analyses them, computes control actions, interacts to the various users, and does some of the following activities: reasoning on data, self-organizing, learning from past experiences, auto-maintenance, fault detection, and operators guidance. These activities must be supported by a complex and powerful controller structure. It is clear that any IC will not perform all these activities.

Let us consider what is needed for such a job. Two items should be considered in the controller structure: i) The internal structure and ii) the interface with both, the process and the users. For any activity, some requirements must be fulfilled.

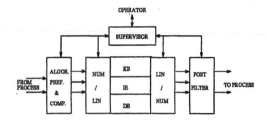

Fig 1

## Direct Control

In this case, the IC directly controls the process, [Broeders 1989]. The kind of controlled processes are those involving qualitative knowledge or goals, or non linear behavior. The controller scheme is as depicted in fig.1.

It must be able to deal with both, process measurements and control variables, and qualitative information and statements. The easiest way to merge all this information is to convert it to linguistic. In that way, it may be processed by an inference engine, and a control action may be computed from the reasoning results. Some algorithmic treatments may be implemented in order to evaluate trends, integrals, filtered values, and so on. It is also possible to postprocess the numeric output.

Examples of these IC are the **fuzzy controllers**, well established [Lee 1990] since the first published reports [Mandami 1974], [Kickert 1976]; the very simple **rule-based controllers**, [Ollero 1991], the **linguistic PID**, [Anderson et al 1988], and so on. The following structural issues may be pointed out:

- Directly connected to the process I/O. This implies A/D and D/A conversion, and extreme timing constraints in the controller response. The inference engine must be simple and the response time should be predictable.
- MMI involves the setting of control goals (set points) and parameters, under normal operating conditions. Explanatory facilities and graphic capabilities must be available to allow the control monitoring.
- Even under direct control, supervision is a natural option. The IC should be able to evaluate some performance indices and to reason both on them and on the current measurements, in order to tune some parameters, to select among basic rule-based controllers,such as linguistic PI or PD, to include or disregard some variables, to modify their universe of discourse, and so on.

The controller may be encapsulated, appearing to the operator as a classical controller, with simplified tuning parameters, such as the variables range (universe of discourse), the gains (input or output scaling), or the mode of operation. In this case, we just use the IC capabilities to initially tune up the controller, from a weak process knowledge. Also, this controller may be considered as an experimentally tuned **non linear controller**, [Ollero 1991]. For instant, if a control action is computed from the measurement error, and some linguistic variables (small negative, big positive,..) are attached to it, assigning experimentally to them the membership functions of the error current value, the controller I/O relationship may be expressed by means of a nonlinear function. But, of course, we can have the freedom to add many of the issues mentioned above.

As we will see later on, IC are also able to implement algorithmic control laws, combining the power of both classical and reasoning-based control.

Another option is the use of ANN. As it is well known an ANN is a system internally composed by many simple processing elements, functioning in a similar way to that of a biological neuron, [Hetch-Nielsen 1990]. These processing elements are interconnected by channels where information is weighted. The main issue is to select the weights of the network by means of a learning process in such a way that the desired input/output relation is matched. In control systems, this feature can be successfully applied where the conventional methods frequently tend to fail, specially under extreme conditions such as : no-knowledge of the plant, strong non-linearities, and/or changes in the parameters. The ANN offer an alternative in order to be used in non linear control algorithms. This affirmation is justified by the following properties: 1) Capability to model non-linear functions [Funakashi 1989] [Cybenko 1988], and [Hornik 1989], 2) Faster computation since it can be done in a parallel way 3) Capability to identify situations. The third property allow us to use the ANN as direct controller in such a way that, recognizing the state at every instant provides the more appropriated action. In this case it is necessary to perform an adequate training of the process to be controlled, by means of data taken at all possible situations [Anderson 1989].

## Supervision

Most of the available industrial control packages involve many logical and decision based actions. In fact, it is remarked [Astrom 1984] that control algorithms are a very small part of the code of a digital control program. IC provide the tools to take care of these activities in a structured and modular way.

In the supervisory level, the operator use to do some of the following tasks:

- to select the most appropriated controller and control structure
- to tune the controller parameters
- to change the controllers set points

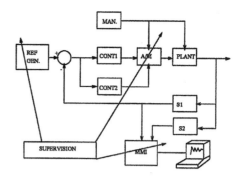

Fig. 2. Supervisory control

- to evaluate the performances of the controlled process
- to start-up and shut-down the process
- to perform some extra activities, such as addition of external excitation, computation of indicators, exchange of information between control loops, and so on
- to log incidence and alarms.

Most of these activities are based on approximated knowledge, being suitable to be implemented in IC. In that case the IC is connected to the lower level controllers and interacts with the operator. The external structure will be as in fig.2.

- It receives process I/O, but the actions are changes in control parameters and/or structure, i. e., mainly logical commands.
- MMI involves graphical capabilities, reporting facilities and querying.

The internal structure is similar to that of the expert systems for planning and monitoring. Those situations requiring a fast response, such as the switch on of a key alarm, may be handled directly, out of the ES shell.

Supervisory IC controllers include the new characteristics of **control monitors, logical and decision-based algorithms** and some advanced controllers, like the **self-tuning PID** [Krauss 1984].

## Mode selection

In process control, the process model use to be quite different according to the operating conditions. Adaptation or parameters change is not enough. Just different control problem and constraints are established and the control strategy may be totally different. The decision may be based on a huge amount of data, each one meaningless by itself.

Usually, it is the operator who selects the mode of operation, being typically the case in the starting up or

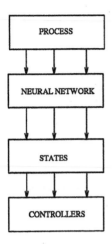

Fig. 3. ANN approach to mode detection

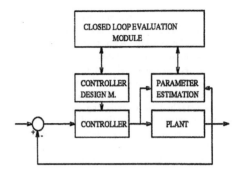

Fig. .4 Adaptive control modules

- perform different control activities (local control, adaptive control, hierarchical control, supervision)
- implement multiloop and large processes control
- split data and control actions at different users level
- handle complex control requirements leading to different control schemes

Let us consider in more detail the structure of an IC performing supervised adaptive control, as well as enabling a distributed control strategy.

shutting down of the plant, for instance. Under emergency or abnormal operating conditions, the control approach may also be quite different to that under almost steady-state functioning.

As before, it may happen that these operating conditions are easily related to a set of logical variables and the decision is quite simple. But in complex industrial processes, only the experienced operator can detect in advance the undesirable evolution of the process.

The use of **pattern recognition** approaches will be helpful. The non linear processing of large amount of data may be performed by ANN, leading to a fast and reliable situation detection.

This control activity will be located in the upper layer, being implemented as a separated module and/or device exchanging information to the supervisory level and providing it the information to select the most appropriated control strategy option.

The ANN output are the pre-scheduled modes of operation, dealing with almost any information from the process. This ANN should learn from historic records and update its weights according to the current experience.

### Multilayer controllers

Most of the previous activities involve a lot of information exchange with lower level controllers. But all this information, if the local controller is also implemented in a IC approach, may be handled by the same device. Usually, a control scheme involves data handling at different levels. These levels may be established in order to:

**Adaptive Control** In process control, techniques for controlling systems with temporal variable parameters have been applied. Adaptive control is a well studied topic, and a number of available products offer this possibility. They use two basic approaches, the **self-tuning regulator**, [Wittenmark 1984], and the **model reference adaptive control** [Landau 1979]. These techniques are applied for the case of linear or quasi-linear systems. A structure of the plant to be controlled must be *a priori* assumed and both techniques operate trying to on line adjust the parameters of the controller. It has been shown that both approaches are in some sense equivalent. But, as it is well known, they present some problems for critical processes, i.e., processes with unknown delay, non minimum phase, changing noisy environment and so on. Some of them still are useful but a lot of decision taken activity is required. Most of this activity is again based on approximated knowledge and IC techniques are well suited for. The most appropriate procedure is to have a higher control level taking care of the behavior of the system, [Isermann 1985], [Martinez 1991].

In an adaptive control scheme, fig. 4, the following modules must be considered:

- Parameter estimation module. It looks for the most appropriated parameters, in a family of models, to fit the process dynamic behavior.

- Controller design module. According to the stated objectives and the process model previously obtained, it offers at least one control law. Usually, the control solution is not unique and additional constraints, goals or requirements, will decide for the most suitable one.

- Closed-loop evaluation module. It verifies the fulfillment of the control specifications, looking for abnormal situations, such as bumbling, hidden oscillations, long term drift, and so on.

At the lower levels, the problems can be handled algorithmically. A number of indicators, [Martinez 1991], such as prediction error, trace of estimated covariance matrix, poles location, cost indices, among many others, may be evaluated. But these indicators together with some experimental knowledge, should be used at the top level. The use of heuristic will lead to the better option at the œrright time.

In this case, the controller device will support many algorithmic tasks. The interaction with the user is at the level of process knowledge, control goals and monitoring.

Again, another option is the use of ANN. For the case of the multilayer ANN, it has been shown that it is possible to control poorly modeled systems, owing to their **learning** capabilities to model non-linear mappings. In [Psaltis 1988], a classification of the ANN learning architectures applied to control are summarized. The most usual ones are: the indirect learning architecture (ILA), the general learning architecture (GLA) and the specialized learning architecture (SLA), as depicted in fig 5.

Both, the ILA and the GLA, are based upon the conventional supervised learning approach, using the error between the output of the network and a goal vector as information. The GLA requires a previous training of the neural controller (off line). Afterwards, this adjusted network is placed before the plant in order to control it. The approach is different for SLA. The training is done by means of a modified learning algorithm, taking as information the error between the plant output signal and its desired value; nevertheless, in order to apply this procedure, the Jacobian of the plant must be known beforehand which, in some cases, is not directly available. Again, the controller device will involve at least two ANN, arranged as shown in the fig.6.

On the other hand, once the plant model has been obtained, it can be used in predictive control. It is known that the goal of this approach is to match the process and the reference model output, [Ydstric 1990]. This tool also can be used in predictive algorithms based on the plant model such as the DMC, (Dynamic Control Matrix) [Cutler 1983] and the MAC, (Model algorithm Control), [Richalet 1978],[Martin 1981]. In [Bath 1990] and [Thibault 1990] examples where the ANN are used for non-linear modeling in control applications are shown.

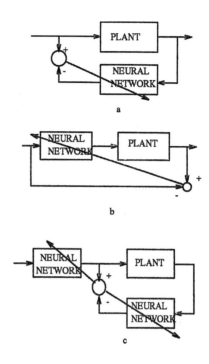

a General Learning Architecture

b Indirect Learning Architecture

c Specialized Learning Architecture

Fig 5. ANN learning structures

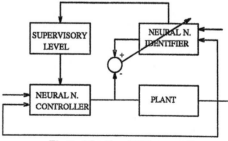

Fig 6. Adaptive ANN approach

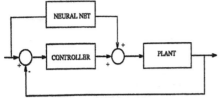

Fig. 7. Simple control scheme

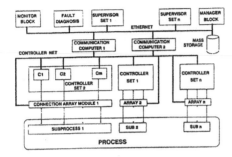

Fig. 8. Distributed control

In [Narendra 1989, 1990] a control scheme based on MRC (Model Reference Control) for the control of non-linear complex systems is proposed. These authors propose the use of two ANN. The first one acts as a controller and the second one is used to model the plant. The control parameters are tuning in such a way that the plant output follows a reference model. When the plant inverse model is available, the control scheme proposed by [Pesaltis 1988] is the simplest one. The control scheme is showed in figure 7.

In the scheme two controllers are used: a traditional controller (closed loop) and a controller implemented by ANN which models the plant inverse. The traditional control is used in order to counteract those disturbances no modeled by the ANN.

**Distributed control**  Usually, a large process is decomposed into subprocesses for control purposes. Industrial control reliability also requires backup controllers in order to substitute a controller that is improperly working. A set of controllers connected to a subprocess through a connection array module permits to change the connection between any subprocess and the controllers associated to it allowing the reduction to, say, one backup controller per subprocess.

A set of controllers takes care of the control of each subprocess, mainly using local information and fixed structure. On the other hand, some coordination between controllers is needed and a powerful MMI is required. This claims for a more structured control scheme, suitable for control of large processes. The control structure can be seen in figure 8.

Three levels may be distinguished: controller level, communication level and coordinator level.

- Controller Level

  This level consist of a set of controllers and a connection array module.

  Each controller is composed by three modules: the control algorithm itself, the adaptive module, with fixed estimation and controller design algorithms, and the local communication module. The control algorithm may be able to also include filtering and alarm modules. If required the signal from the process is filtered, and if a variable trespasses one of its limits, the attached alarm is set on. At this level, all the block parameters are fixed by either the user or another program (adaptive control, supervisory control, etc.). Furthermore all signal values can be transmitted to a computer through the communication module.

  An adaptive control strategy will permit to change the controller parameters according to the estimated process model parameters, but under a fixed algorithm scheme. This module will provide the data the upper level requires to decide about the most convenient control structure.

  Finally, the communication module will allow the connection between the controller and a computer. This module can access to all the controller internal parameters, process variables, control variables and alarms. Furthermore, the adaptive control program can be loaded from a program file stored in the computer.

- Communication Level

  To allow a good modularity and task oriented processors design, a communication level is desirable. This level will link both the controller level and the coordinator level. It must include some kind of redundancy for reliability purposes. The communication computer preprocesses the data from the controllers. Moreover alarm signals are processed and sent to the computer that manages them. On the other hand, adaptive programs, set points and parameters are sent to the controllers.

- Coordinator level

  The global system should be coordinated and it must friendly interact with the end-user. Although they may be implemented in just one computer, at least the following facilities must be available: Monitoring, Supervision, Fault diagnosis, and Control management.

    - The monitor block is the program that interacts whit the operator. It permits the whole control of the structure. The functions implemented by the monitor should include: to display the process status, to fix the control goals, to track the process variables, to allow the transfer to manual control accessing to the individual controller set point, to display the system structure configuration, and to introduce direct command actions. The interface must be written in a graphic symbolic form making easy the process status recognition and the input of operator commands. If implemented under AI techniques, it will provide the user some explanation and reasoning about the taken and/or suggested actions.

– Supervision module

Performs the activities as described in the section above.

– Fault Diagnosis module.

The fault diagnosis block checks all the elements in the control structure to verify if they are properly working. Once the fault is recognized, a set of recovery actions must be done, activating backup devices, if available. If the fault recovery is not possible, an emergency shut-down routine must be fired. If some level of sensor and actuator redundancy is provided, the array structure proposed above will allow to switch off the faulty devices to be repaired.

Clearly these functions are also suitable to be implemented by an expert system.

– Manager module

This module coordinates the different subprocess controllers to reach the control process goals. It fixes the partial subprocess goals which are depending on the process operating point. Among others, the following modes are always present: Normal operating point, Overloading, Emergency, Starting up, and Shutting down.

Each operating point can be decomposed into others, but this decomposition is process dependent.

Based upon these considerations, the manager block must first detect the current operating point and then plan the control subgoals for each subprocess. Again, this task is suitable to be implemented by an expert system approach.

## 3. IC Implementation

IC involves the use of AI in a RT environment. ANN do not pose any special problem, their main feature being the multiprocessor structure and the short computation time. But, rule-based systems were initially developed for off-line applications, and process control requires strong real time features: Short computation delay, to take into account the dynamic behavior of the process, the data obsolescence, leading to the need of checking the time consistency of knowledge with respect to the actions previously applied, and the time-defendant nature of data and conclusions. These features may be implemented either by software or hardware.

The basic hardware required to implement an IC is based on a PC or microprocessor-based device. The IC will present a software structure involving, at least, the following modules: i) the structured knowledge base, ii) the scheduler, iii) the event manager, and iv) the interface with the user and other software.

In this section, the main issues related to the implementation phase are reviewed. Although the problems are stated in a general framework, most of the proposed solutions refer to the IC application described in section 5.

## RT Constraints

When *real time knowledge-based control systems* are incorporated into traditional systems, problems come up to guarantee the deadline of activities. When several activities or tasks should be executed, *a priori* analysis should be performed in order to know its schedulability. Real time tasks may have a *periodic* or *sporadic* behavior as well as a set of timing constraints. When these tasks have an *intelligent* behavior and the predictability of their computation can not be guaranteed, timing constraints should be directly associated to the reasoning process. When this is not possible, the reasoning process can be defined as a separate task with explicit timing constraints. The scheduler will be responsible of starting the tasks at the appropriated time.

Some architectures have been defined for real time expert systems based on blackboard systems [Jagannathan 1990],[Hayes 1990], and [Erickson 1991]. Other approaches are more closely related to control systems, [Kligjsman 1991]. Most of them are defined improving some features of traditional expert systems in order to have fast response, continuous operation, access to external devices, etc.

Real time expert systems knowledge bases use the object oriented approach to represent the knowledge. It allows an easy understanding, redundancy avoidance, and modularity, among other interesting issues.

The features above stated require:

- Guaranteed response time

When we combine process control with expert systems techniques and we try to apply them under real time constraints, some criteria have to be applied in order to guarantee the response time of the system.

Actions may be taken at different levels:

1. Scheduling: establishing a plan to execute the appropriated tasks. Here, the term **appropriated** stands for the deadline specification and the resources to be used.

2. Knowledge Base: defining rules at different levels of knowledge (deep knowledge),

and allowing several sets of rules to cooperate in the solution of a problem. The inference engine should know which rules are appropriated and the time they require to be executed.

3. Inference engine matching algorithms: using RETE or TREAT pattern matching algorithms to improve the inference process [Forgy82] [Miranker87].

4. Data Base: providing concurrent and real time methods to access to object instances.

- Temporal representation and reasoning

IC need to handle some kind of temporal information and to reason about time. So, an efficient model to represent past, current and future facts and reasoning about these facts, should be considered.

Several temporal models, such as [Allen 1984], [Dean 1987], and [Barber 1990], have been proposed, based on time intervals or time points. The last option offers many possibilities to handle temporal knowledge, but, in order to have a simple and efficient model of the reasoning process when applied to real-time systems, a simplification of the model, as defined in [Barber 1990], would improve the response time.

Once a model has been selected, methods to reason about qualitative and quantitative temporal facts must to be provided.

- Continuous operation: The system has to operate in a loop, based on: i) data acquisition, ii) output calculation (using IC and other classical regulators), and iii) actuation.

- Focus of attention: when a concrete situation requiring a specific treatment is recognized, the system should change its mode and concentrate its attention over a reduced number of variables or processes. In this case, the system should try to keep the process under control, and to correct the situation putting all the available resources on it.

## Software Architecture

The IC is a part of an application software. Two different approaches may be followed.

- Direct IC. The controller controls the global system. It is the control kernel and should take care of the interaction with other Data Processing Software related with the application.

- The IC is part of a Digital control package, being treated as an intelligent resource in the system.

In both cases, a real time system with intelligent controllers could be structured as: i) a set of critical tasks, ii) a set of non critical tasks, iii) an expert task, and iv) a set of maintenance tasks. In fig. 9 the global architecture is shown.

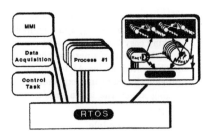

Fig 9.

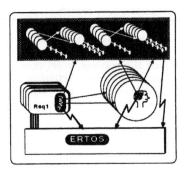

Fig 10.

This architecture has been proposed in [Crespo 1991], and [Paul 1991]. It provides mechanisms to split, structure, and isolate the search space, and to facilitate its integration in real time operating systems.

Expert tasks can model an independent problem by means of AI techniques. The internal structure of an expert system is shown in fig. 10

The main components are:

1. Tasks: define the activities to be carry out from external or internal events (periodic and sporadic behavior)

2. Agents: represent the knowledge base related to a concrete regulator concept.

3. Data objects: store the application variables in a common, concurrent and with real time performances blackboard.

4. Execution support: real time support to plan and execute internal actions.

The encapsulation of intelligent tasks in the expert server is a way to consider all these activities as a global task and it permits to apply the same policy (rate monotonic) to all real time tasks, in order to determine the priority attached to each task. Each server task will have a priority and a computation time that must be guaranteed by the internal scheduler of the expert task.

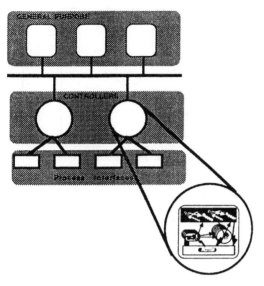

Fig 11. Distributed system architecture

## Hardware Issues

As previously discussed, some IC solutions may be implemented by the usual microprocessor-based digital controllers. Even IC controllers involving low number of reasoning levels, rough I/O discretization resolution, and simple control performances, such as the fuzzy controllers, can be implemented in a single chip, [Togai 1986]. The control of complex systems may require the use of intelligent resources at different parts/levels of the process. The computation time demands are, in most cases, very large, and it is difficult to assume that they can be fulfilled with small machines. As previously discussed, new trends in process control use distributed systems linked through a predictable and reliable network, and data bases with real time features.

Each node of this distributed system can be implemented by either, mono or multi computer architectures. When a concrete node has to perform a large activity, a multicomputer architecture can provide support to perform the following activities in each computer:

1. Complex scheduling policies of an intelligent server

2. Expert server activities

3. Real time critical tasks

4. Concurrent and temporal blackboard

5. Other tasks

In this case, non predictable activities are separated and computed with all the available computation resources.

Today it is possible to buy coprocessor cards where the neural network designed can be executed (SAIC Delta II, HNC ANZA Plus). These expensive cards provide software allowing implementation of several neural networks paradigms. For instance, the second one uses Weitex XL chip set with 10 Mb of fully addressable on board RAM. The coprocessor has a capacity of 2.5 million processing elements and can handle up to 10 million interconnections per second. Owing to the cost of this kind of boards, the transputer can used as a tool for implement neural networks since they are designed for parallel processing.

## IC Tuning

As it is always true, there is nothing for nothing. If you are able to design an IC, from a weak process knowledge and not well defined control specifications, and a good flexibility to update the controller is required, in some stage of the design procedure, a great design effort should be involved. And this happens in the IC training phase.

As previously mentioned, one desirable feature of the IC is learning. But learning is not yet a well established issue in IC, and it is the operator or designer job to train and tune the controller via trial and error approach. Once again we are faced to another AI typical activity. Much effort is also devoted to this issue in classical control theory. Recently, in [Anderson 1991], a new control design strategy, the so called windsurfing adaptive control, involving a learning capability, has been proposed. According to the controlled process knowledge you have, the controller design technique may be more or less ambitious. With poor process knowledge, a conservative control design must be implemented to avoid unexpected bad behavior.

The purpose of this paper is neither to deal with algorithms, nor to describe the technical staff behind the different approaches, fuzzy logic, ANN, RTES, and so on. There are many references related to these topics. Our main purpose is to present the IC as new devices and to discuss the main issues when dealing with their implementation. And the tuning is one of the more relevant topics.

We can say that the tuning approach complexity depends on both, the process knowledge and the precise control objectives statement. If the plant is well modeled and the goals are easily understood, probably the best is to forget the IC techniques and control the process by some classical control algorithm. The weaker is the knowledge the more appropriated the IC is.

The methodology we propose is in the way of the classical experimental tuning approaches.

- Start with the simplest control, such as a proportional-like controller, with

- reduced gain.

- more relevant variables

- low number of linguistic variables

- non-discriminant membership functions

- low certitude coefficients

Increase the gain until some undesirable response appears.

• Add knowledge according to the experience

- Looking for new linguistic or physical variables, to cover with the new control difficulties

- Tuning the weights, membership functions, controllers parameters

- Adding new components or changing the control structure

• Check the validity and coherence of the incorporated new knowledge under operating conditions others than those initially tried.

If some past experience over the controlled process is available, in the form of variables and events records, and a controller structure has been decided as convenient, try to estimate the controller parameters better fitting these records. This is the classical approach used in the learning phase of an ANN. It is also the approach Takagi and Sugeno [Takagi 1985] proposed to design a fuzzy controller.

As discussed in paragraph 2.1, supervision is a natural feature of IC. It may include some learning options, making easier the global controller tuning.

Anyway, in control of complex processes, the controlled process is frequently changing and control updating and tuning is a must. What that requires is a powerful MMI, allowing an easy operator interaction. The MMI must include specific options, such as editors for rules, requirements, objects, and procedures, simulators to foresee new controllers improvement, graphical capabilities, and a good explanatory subsystem, to show the involved reasoning in getting a result. [RIGAS 1992]

**An Intelligent Regulator**

Based on the previous considerations, a structure able to support different types of control algorithms is presented in this section.

The knowledge-based components may be defined as **Intelligent Regulators (IR)**, with reference to the classical regulators in process control.

Each **IR**, as shown in figure 12, defines a set of problem solvers (rule-based or algorithmic) in order to provide solution to a specific subsystem.

We can consider various components in an **IR**:

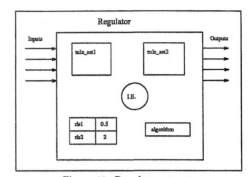

Figure 12. Regulator structure

1. Fast rule-based solver:
   Use of reduced set of variables, and simple functions in the rules. A predicted time of execution is related to this solver.

2. Complete rule-based solver:
   Use of all the variables and relations between them, as well as complex functions in rules (temporal reasoning). As in the above solver, a predicted time is associated to the solver.

3. Algorithmic solver:
   Classical regulator (PID, dead-beat, minimum variance, etc.) could be defined in order to supply alternative results when no solution is reached by the above solvers.

As many levels of rule-based solvers as the designer decide could be defined in the **IR**. Each level will add depth in the reasoning process with respect to the previous one. Moreover, each **IR** may have defined its own *inference engine* in order to evaluate the solution using the appropriated solver, taking into account its response time. After each evaluation, the average time attached to a rule set is updated.

Each regulator has information about the objects used as input or output in the rule sets. The object lists are read by the scheduler in order to know the resources needed by the **IR**.

A possible scheme of the **IR** module activity is the following: A message for starting a reasoning process arrives with a maximum time to give an answer. The **IR** evaluate the control algorithm, if it exists. The **IR**, taking into account the deadline, starts with the first (faster) level. When the reasoning process finish, the **IR** decides if the next level must be fired, considering the remaining time. The process follows until all level are considered or a deadline message is received. If the timeout message arrives before an answer is reached as a result of the application of any rule set, the algorithmic result or nil is answered.

The intelligent regulator has been described for the rule-base solver because it is the most complex module. The algorithmic one responds to the same messages or events, but its implementation is quite different. The regulator program can be built inside the regulator or can be resident in a different equipment (a local controller). In that case, the regulator must manage the communication with the external controller. The main idea behind this proposal is that the different regulators can be used with no changes in the existing control systems structure, and it allows the use of the most efficient regulator depending on the process characteristics.

## 4. IC Design Tools

Let us briefly review the current trends in RTES design environments and ANN design software. Tools related to the controller hardware design are mainly focussed to redundant and fault tolerant architectures issues and, for the sake of time constraints, will not be discussed here.

### RTES design environments

Real time expert systems has been applied to the control of different processes [Laffey 1988] [Foxvog 1991]: Monitoring in nuclear power plant (REALM and REACTOR) developed in the KEE environment, monitor and operator advisory system for batch manufacturing (COOKER) using a Lisp machine, rotary cement kiln process control (RIGAS), Autonomous vehicle guidance developed in CLIPS and Pascal, collision avoidance system developed in CROPS5, etc.

The main development environment are based on high level languages (Lisp, Smalltalk, Pascal) with extensions to real time management in the prototype versions, being translated to efficient languages as C, C++ or Pascal when all functionalities are obtained.

Some commercial environments for real time expert system development are:

- **G2**: provide the main features to real time environments as: fast inference method (zero order), access to external devices, continuous operation, focus of attention, past values management, simulator and man machine interface appropriated to process control. It is implemented in Lisp.
- **PICON**: high speed rule activation, access to external devices, man-machine interface, efficient garbage methods, focus of attention, and parallel inference supported by several machines. It has been written in Lisp for TI Explorer.
- **CHRONOS**: allows facts with timestamp and temporal reasoning, first order real-time inference engine, forward chaining, data acquisition, and access to external devices.

Other commercial environment are: Knowledge Craft, IPCS, FALCON, etc.

The main limitations of these RTES to be applied in RT environments are:

- Complete temporal model and fast methods for reasoning
- Deep reasoning
- Reason maintenance system
- Multi-agent and multi tasking
- Integration in embedded system
- Inference speed

New environments, such as DVMT [Decker 1990], Guardian [Hayes 1990], REAKT [REAKT 1990], RIGAS [Crespo 1991] etc., add some of these features in the framework of the application they are addressed.

### ANN design Software

Actually, there are a lot of tools to model, specify and test ANN. These tools can be run on PC's or workstations. In this sections we are going to present some of them. The environments we are going to present here do not appoint for an exhaustive list but, up to the best of our knowledge, they are the most used and known.

NeuralWorks Professional II, is perhaps the most well known ANN environment. This package is a powerful problem solving neural network development tool. This tool has a lot of flexibility to create networks and the user can choose from a large set of predefined networks (25 network paradigms, 22 learning rules, 16 transfer functions and 18 summation functions. It has two expansion software packages. The first one (User-Defined Neuro-Dynamics) enables the user to add new functions. The second one ( Neuralworks Designer Pack.) is a compiler that converts into C source code a network developed using NeuralWorks Professional II.

Quite similar to that one is Plexi. It also is a powerful environment that allows to experiment with new neural networks structures or to use predefined networks. The Plexi environment has a useful graphical interface in order to edit the new structures, as well as to present the result of each experience.

The Exploretnet 5000, is a low cost software package for developing neural networks applications on workstations. Its modular architecture enables the users to build applications from menus, without code writing. A program developed using the Exploretnet can be recompiled to run on the ANZA board, which is a neurocomputing coprocessor.

The Netset II, runs on the PC (Microsoft Windows). The system is graphically modeled by drawing icons such as pipes, data transformers, neural nets, database and graph display devices. The users connect these objects according to the desired structure.

In order to learning about, designing, developing, and training neural networks, we can use the Poplog-Neural environment. This tool has graphical facilities to on line show the adjustments of the weights between nodes. It is integrated with ISL'S multilanguage development tool, POPLOG.

The Neuroshell environment, is an easy to use package including utilities for showing graphically the results. Neuroshell incorporates a built-in back-propagation model that allow us to carry out the test of the designed networks in an easy way.

Finally, Nestor INc., provides a development system environment using the Nestor Learning System which is a neural network that learns how to organize solutions to pattern recognition applications from experience.

## 5. Application

An environment for building intelligent controllers is being developed at our Department. At this moment, it includes most of the characteristics proposed in this paper and is being used to control a cement kiln. In this section, the control architecture using intelligent regulators is shown, and the functioning modes of this process are discussed.

### Process description

The developed system has been used to control a rotary cement kiln in a cement factory, [Umbers 1980] [Holmblad 1981] [Morant 1992]. The kiln process heats the inlet flow of raw material (*lime*, *sand*, *pyrite*, *clime*) until the clinker is obtained. Three subprocesses may be distinguished: the **preheater**, where the material is heated using the residual heat of the combustion fumes; the **kiln**, a rotary large cylinder where the clinker is formed; and the **cooler**, where the clinker interchanges its heat with the combustion air.

Some of the characteristics of this process are:

- Distributed process
- The system can operate in several different situations.
- High number of variables (between 40 and 50): pressures(9), temperatures(14), gas concentration(6), laboratory analysis, etc.
- Large delay in some variables, due to the kiln condition.
- Absence of a model due to the process complexity.
- Only few people know how the system works (**expert operators**).

### Control goals

The proposed control goals, and in this order, are: Maximum quality, Maximum production, and Minimum production costs.

The first one is the most important and difficult to achieve because of the delay between the clinker formation and its measurement, obtained by a X-Ray analyzer, is about 20 minutes. Another difficulty is due to the impossibility of the clinker formation temperature measurement, which is directly related to the final quality. This question force to estimate the temperature using other variables (kiln torque, kiln temperatures, preheater temperatures, etc.).

Maximum production is achieved maintaining the control variables at their nominal values whenever it is possible.

Minimum production costs are obtained minimizing $O_2$ concentration in the fumes, CO concentration between two limits, and avoiding abnormal situations.

The variables used to control the kiln are: *raw material*, *fuel*, and *fumes* flows. Thus the IC has to conclude a value for each variable to achieve the control goals.

### Control modes

The kiln process can operate in different situations, some of them can be considered exceptional situations and, under them, actions to correct the process have to be taken. Some of these situations are: *normal, abnormal, repairing ring, inlet coolers obstruction, cyclone obstruction, raw cut off*, etc. For each situation a different control policy must be implemented.

The system is organized in three boxes modeling the **secure operation**, integrated by the normal and abnormal situations; **exception operation**, situations which require a set of specific actions outside of normal functioning; **repairing operation**, contains the set of actions in time to be taken to reach the secure operation. In the repairing operation, if the situation is out of control, the system informs the operator that manual control has to be taken. In the same case, if new exceptions are detected, the system evaluates which is the most important one and performs the actions to repair it.

This scheme is implemented by means of an attribute in the requirements, the **mode** attribute. It defines in which situation the system is. The set of requirements associated to a mode specifies the set of actions to be taken when the defined events are emitted. The **guard** attribute is another useful one. Attached to each requirement, it defines under which conditions the requirement is executed. At any time, an instance of the set of **regulator** objects is the agent

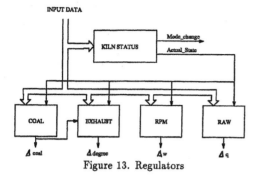

INPUT DATA

KILN STATUS — Mode_change — Actual_State

COAL  EXHAUST  RPM  RAW

$\Delta$ coal  $\Delta$ degree  $\Delta$w  $\Delta$q

Figure 13. Regulators

that evaluates the information of the kiln, in order to determine the operation mode.

## Normal Operating Mode

The **Normal** mode is going to be analyzed in order to show how to implement all the situations. In this mode the kiln state is divided in three static states (hot, normal, cold), and three dynamic ones (heating, stable, cooling). A first regulator determines from the measured values, which one is the current state, with an associated confidence coefficient.

This information is used by the control regulators in order to conclude numeric values associated to the output variables. A regulator can use information concluded by other regulator, this implies a synchronization through events.

Figure 13 shows a scheme of the regulators associated to the normal mode.

## Results

The control system of the aforementioned cement kiln has been working since march of 1991 with excellent results. With this closed loop control, we have increased the production with an important energy saving.

The application and development of this intelligent control has allowed both a better process knowledge of technician and a validation of real time artificial intelligent techniques in industrial environments.

## CONCLUSIONS

In this paper, a review of the main issues related to the implementation and design of IC has been done. It appears that intelligence is needed at different levels of the control strategy, and AI techniques provide different tools to efficiently implement it. RT constraints pose a conflict between depth of reasoning and guarantee of the response time. The solution must be different if the purpose is to get direct control or preventive maintenance or supervision.

The concept of Intelligent Regulator, in accordance to the classical point of view of a controller, has been introduced. It allows the user, to think about the control actions in the same way the operator does.

The main drawback of IC use to be the tuning phase. Learning algorithms and structures have to be developed to overcome this crucial IC implementation problem.

## References

[Allen 1984] Allen J.F. *Towards a general Theory of Action and Time.* Artificial Intelligence, 23, 1984.

[Anderson 1988] K.L. Anderson, G.L. Blankenship, and L.G. Lebow. "A Rule-based PID controller" *Proc. IEEE Conference on Decision and Control,* Austin, Texas, Dec 1988.

[Anderson 1989] C.W. Anderson, 1989. "Learning to Control an Inverted Pendulum Using Neural Networks". *IEEE Control Syst. Mag.* Abril 1989. pp 31-36.

[Anderson 1991] B.D.O. Anderson, and R.L. Kosut. "Adaptive Robust Control. On-line Learning" *Proc. IEEE Conference on Decision and Control,* Brighton. England. 1991.

[Astrom 1984] K.J. Astrom, and B. Wittenmark. *Computer Process Control* Ed. Prentice-Hall, 1984.

[Astrom 1991] K.J. Astrom "Intelligent Control" *ECC91 European Control Conference* ,Grenoble, France pp 2328-2339. 1991.

[Astrom 1992] K.J. Astrom, C.C Hang, P. Persson, and W.K. Ho. "Towards Intelligent PID Control" *Automatica* , Vol 28 , Num 1, pp 1-9. 1992.

[Barber 1990] F. Barber, V. Botti, A. Crespo *Temporal Expert System: Application to a Planning Environment.* IASTED Symposium. USA. December 1990. 78-82.

[Bath 1990] M. Bath and T. Mc Avoy "Use of Neural Nets for Dynamic Modeling and Control of Chemical Process Systems". *Proc 1989 A.C.C* Pittsburg. Vol 2, pp. 1342-1347. 1990.

[Broeders 1989] H.M.T. Broeders, P.M. Bruijn, and H.B. Verbruggen. "Real Time Direct Expert Control" *Engineering Applications of AI* , Vol 2 , Num 2, 1989.

[Crespo 1991] A. Crespo, J.L. Navarro, R. Vivó. A. Espinosa, A. García. "A Real-Time Expert System for Process Control". *9th IFAC Intl.Workshop on Artificial Inteligence in Real Time Control.* September 1991. California. U.S.A.

[Cutler 1983] J. Cutler. "Dynamic Matrix Control, An Optimal Multivariable Control Algortihm with constraints." Phd Thesis, University of Houston Tx. 1983.

[Cybenko 1988] G, Cybenko, 1988. "Aprproximation by superpositions of a sigmodial function". *Tech Rep N 850,* Dept. Elec. and Comp. Eng. University of Illinois at Urbana Champaign. 1988.

[Dean 1987] Dean T. and McDermott D. *Temporal Data Base Management* Articial Intelligence, 32, 1987

[Decker 1990] K.S. Decker, V.R. Lesser, R.C. Whitehair. "Extending a Blackboard Architecture for Approximate Processing". *Journal of Real Time Systems.* Vol 2, 1,2.

[Erickson 1991] W. Erickson, L. Braum. "Real-Time Erasmus". *Ninth National Conference on Artificial Intelligence. AAAI-91.* Jul, California, 1991.

[Forgy 1982] C. Forgy. "RETE: A fast algorithm for the many pattern/many object pattern match problem".*Artificial Intelligence.* 19, North-Holland. 1982

[Foxvog 1991] D. Foxvog, M. Kurki. "Survey of Real-Time and On-Line Diagnostic Expert Systems".*Proceedings Euromicro'91 Workshop on Real Time.* 1991

[Funakashi 1989] K, Funakashi. "On the approximate Realization of Continuos Maping by Neuronal Networks". *Neural Networks,* Vol 2 pp 183-192. 1989.

[Hayes 1990] B. Hayes-Roth *Architectural Foundations for Real-Time Performance in Intelligent Agents. Real Time Systems.* Vol 2, 1,2,

[Hecht-Nielsen 1990] R. Hecht-Nielsen. *Neurocomputing* Addison-Wesley, 1990.

[Hornik 1989] k, Hornik "Multilayer Feedforward Networks are Universal Approximation".*Neural Networks* Vol 2 pp 359-366. 1989.

[Holmblad 1981] L.P. Holmblad, J.J. Ostergaard. "Control of a cement kiln by fuzzy logic " *F.L. Smith Review,* n. 67, 1981.

[Isermann 1985] R. Isermann, and K.H. Lachmann "Parameter-adaptive Control with Confugrations Aids and Supervision Functions". *Automatica* ,Vol 21, N 6 pp 625-638.1985.

[Jagannathan 1990] V. Jagannathan, R. Dodhiawala and L. Braun. *Blackboard Architectures and Applications.* Academic Press. 1990.

[Kickert 1976] W.J.M. Kickert, and H.R. Van Nauta Lemke. "Application of a Fuzzy Controller in a Warm Water Plant " *Automatica* , Num 12, pp 301-306. 1976.

[Klijgsman 1991] A.J. Klijgsman, R. Jager, H.B. Verbruggen, P.M. Bruijn. "DICE: A Framework for Real Time Intelligent Control", *3rd IFAC Intl.Workshop on Artificial Inteligence in Real Time Control.* September 1991. California. U.S.A.

[Kraus 1984] T.W. Kraus, and T.J. Myron. "Self-tuning PID controller using pattern recognition approach " *Control Engineering,* Vol June 1984, pp 106-111.

[Laffey 1988] T.J. Laffey, P.A. Cox, J.L. Schmidt, S.M. Kao and J.Y. Read *Real-Time Knowledge-Based Systems. AI Magazine.* Spring. 1988.

[Landau 1979] I.D. Landau. *Adaptive Control-The Model Reference Approach.* Marcel Dekker, New York. 1979.

[Lee 1990] C.C. Lee. "Fuzzy Logic in Control Systems: Fuzzy Logic Controller" *IEEE Trans on Systems, Man, and Cybernetics,* Vol SMC-20, Num 2. pp 404-435. 1990.

[Mandami 1974] E.H. Mandami. "Application of Fuzzy Algorithms for Control of Simple Dynamic Plant" *Proc IEE,* Num 121, pp 1585-1588. 1974.

[Martin 1981] J. Martin, 1981 "Long Range Predictive Control". *Am. Inst. Chem. Journal.* Vol 27. Num 5, pp. 748-743. 1981.

[Martinez 1991]
Martinez M.A.,1991."Indicadores para la Supervisión del Control Adaptativo. Implementación mediante metodología de Sistema Experto".Doctoral Dissertation. DISCA. Univ. Politécnica Valencia,Spain. ( in spanish).

[Miranker 1987] D. Miranker. "TREAT: A better much algorithm for AI production systems".*Proc. AAAI.* July 1987.

[Morant 1992] F. Morant, P. Albertos, M. Martinez, A. Crespo, J.L. Navarro "RIGAS: An intelligent controller for cement kiln control" *Symp. on AIRTC* Delft. 1992.

[Narendra 1989] K.S. Narendra and K. Parthasarathy. "Adaptive Identification and Control of Dynamical Systems Using Neural Networks". *Proc. IEEE Confrence on Decision and Control.* Tampa. Florida. 1989.

[Narendra 1990] K.S. Narendra and K. Parthasarathy. "Identification and Control of Dynamical Systems Using Neural Networks". *IEEE Trans. on Neural Networks* Vol 1 Num 1 pp 4-27. 1990

[Navarro 1990] J.L. Navarro, P. Albertos, M. Martinez, and F. Morant. "Intelligent Industrial Control". *IFAC World Congress* , Tallin, Estonia. 1990.

[Ollero 1991] A. Ollero, A Garcia-Cerezo, and J. Aracil. "Design of Rule-based Expert Controllers". *ECC91 European Control Conference* ,Grenoble, France pp 578-583. 1991.

[Pang 1991] G.K.H. Pang. "A Framework for Intelligent Control " *Journal of Intelligent and Robotic Systemss,* Vol 4 , Num 2, 1991.

[Paul 1991] C.J. Paul, A. Acharya, B. Black, J.K. Strosnider. "Reducing Problem-Solving Variance to Improve Predictability". *Communications of the ACM.* Aug. 1991, Vol. 34, No.8.

[Psaltis 1988] Psaltis and co-workers. " A multilayer Neural Networks Controller".*I EEE Cont. Sys. Mag.* Vol 8, Num 2, pp. 17-21. 1988.

[REAKT 1990] Thomson, Syseca, Crin, GMV, UPV, Marconi, Etnoteam. *REAKT: Environment and Methodology for Real-time Knowledge Based Systems.* ESPRIT II 4651. 1990-93

[RIGAS 1992] RIGAS Reference manual. Internal Report DISCA-UPV.Enero 1992 (in Spanish).

[Richalet 1978] Richalet " Model Predictive Heuristic Control: Application to Industrial Processes". *Automatica* Vol.14 pp 413-428. 1978.

[Saridis 1983] G.N. Saridis. "Intelligent Robotic Control" *IEEE Trans on Auto. Control* , Vol AC-28, pp547-557. 1983.

[Takagi 1985] T. Takagi and M. Sugeno. "Fuzzy Identification of Systems and its Application to Modeling and Control" *IEEE Trans on Systems, Man, and Cybernetics,* Vol SMC 15, Num 1, pp 116-132. 1985.

[Thibault 1991] J.Thibault, and V. Van Breusegen "Modeling, Prediction and Control of Fermentation Process Via Neural Networks". *ECC91 First Europ. Control Conference* Grenoble. France. Vol 1. pp 224-229.

[Togai 1986] M. Togai, and H. Watanabe. "Expert Systems on a chip" *IEEE Expert* , Vol 1, pp 55-62. 1986.

[Umbers 1980] I.J. Umbers, P.G. King. "An analysis of humen decision making in cement kiln control and the implications" *Int. J. Man-Machine Std,* Vol 12 1980, pp 11-23.

[Wittenmark 1984] B. Wittenmark, and K.J. Astrom "Practical Issuses in the implementation of Self-Tuning Control".*Automatica,* Vol 20, pp 625-638. 1984.

[Ydstric 1990] B.E. Ydstric. "Forecasting and Control Using Adaptive Connectionist Networks". *Comp. Chem. Eng* Vol. 14 Num 4/5 pp 583-599. 1990

[Zadeh 1988] L. A. Zadeh *Fuzzy Logic Computer,* Vol 12, Num 4, April 1988.

# FIELD ROBOTS FOR THE NEXT CENTURY

W.L. Whittaker

*Field Robotics Center, The Robotics Institute, Carnegie Mellon University, Pittsburgh, PA 15213, USA*

*Abstract.* This paper takes a broad look at the evolution of Field Robotic Technology, beginning with the need to develop robotic systems for use in unstructured environments, their current performance niches, and their future opportunities to impact the world in construction, subsea, space, nuclear, mining, and military applications. It concludes that the evolution of field robotic technology is inevitable: new robotic forms will emerge with the capability and the strategic competence to construct, maintain, and demolish, but the evolution of field robotics will not culminate in a single form. Rather, classes of robots will emerge for classes of work within classes of constraints.

*Keywords.* Field Robotics, Unstructured Environments, Autonomous Robots, Teleoperation.

## Introduction

Robotics to date has produced undelying capabilities that enable robots to respond to a variety of task challenges. Robotics is maturing as a discipline, and the investment in prior research has yielded a wealth of technologies for a new generation of competent robots. It is no longer necessary to restrict research to work on testbed robots, since systems that meet performance specifications of end-users can now be developed. Given the existing technology base, robots that were unachievable five years ago now are within reach, provided that performance goals are established and development efforts in the near term are directed to meet them.

With seminal groundwork laid, robotics technology is now evolutionary, not revolutionary. Evolutionary technologies are born of knowledge-based research: efforts aimed at developing a new and better understanding of the application of scientific principles. From failures in early development come the insights that lead to successful future implementations, which show increasing utility and relevance as the technology evolves. Robots that meet new task challenges and exhibit

proficiency are feasible, since the knowledge we have gained is allowing us to cross the threshold from pure knowledge-based research to performance-based research.

## The Nature of Field Robots

Structured environments, like those found in factory settings, do not challenge robots with the dynamism and uncertainty of unstructured environments. Active and forceful manipulation of objects in unstructured environments requires much more than current industrial robotics can deliver. To work in a field site -say, digging up a gas pipe- a field robot must be able to recognize unknowns and respond to unplanned difficulties. It is paramount that the robot sense events and take responsive actions. Needs of the open work site, like robot intelligence and robustness, drive the agenda for the field robotics research.

Robots, in general, fall into three classes, each distinguished by the control procedures available to the robot and its relationship to human supervisors. The first of these classes, programmed robots perform predictable, invariant tasks according to pre-programmed instructions. Teleoperated robots, the second of these classes, includes machines where all planning, perception, and manipulation is controlled by humans. Cognitive robots, the third class, sense, model, plan, and act to achieve goals without intervention by human supervisors.

Programmed machines are the backbone of manufacturing; preprogramming is extensible to an important class of field work tasks (mostly on the periphery of the work mainstream, and mostly unenvisioned and untried at this time). Preprogramming commands actions through scripts that are played back by rote with branching of the script occurring at specified times or in response to anticipated events. Such scripts are only useful for predictable and invariant tasks, limiting the general use of preprogrammed robots for field work.

Teleoperated machines, servoed in real-time by human operators who close the strategic control loop, amplify and project the human. Because all perception, planning, high-level control, and liability rest with the human, teleoperation circumvents the most difficult issues that face other robot control modes, including the liability of passing control between machine and human and coping with unanticipated scenarios.

Teleoperation is proven where man does not tread, where demands are superhuman, where tasks are unstructured (by current measure), where liability is high, and where action is inevitable. A downside of teleoperation is that much is lost in translation across this man-machine interface. Robot bodies and senses are not optimal for coupling to man. Similarly, human minds are not optimal for the control of robots because of limitations in input/output bandwidth, memory structures, and numerical processing. The prospect exists for field robots to outperform their human counterparts in many ways.

Cognitive robots sense, model, plan, and act to achieve working goals. Cognitive robots servo themselves to real-time goals and conditions in the maner of teleoperators but without human controllers; they are their own supervisors. Cognitive robots pursue goals rather than play out scripts; they move toward goals and notions rather than to prescriptions and recipes. Although software driven, they are not programmed in the classical sense. Cognitive robots are perceptive and their actions are deliberate; they operate in the face of the vagaries and contingencies of the world. Task performance by a cognitive robot is responsive to the state of the environment and the robot itself.

Hybrid forms of teleoperated and programmed machines are becoming increasingly attractive as robots. For example, because factory processes are becoming more sophisticated as they integrate preprogramming and sensing, supervisory controllers and sensory feedback with teach/playback are becoming new research goals. Hybrid, supervisory, and programmable robots are also evolving from the roots of teleoperations in the nuclear service and decommissioning industries.

However they are classified, the most striking observation of present-day robots is that, with few exceptions,  robots lack the ability to perform with any generality, which is the goal of truly capable systems. Even when task directives and methods of procedure are explicit, unforeseen difficulties arise that impede or halt the robot's progress. Autonomous navigation systems, for example, lack the capacity to negotiate traffic or move quickly across and explore rough terrain. These robots are often debilitated by uncontrollable circumstances, such as bad lighting and inclement weater. Nor can they always cope with conficting data to resolve ambiguities. Only now are driving robots beginning to distinguish shadows from roads and separate real obstacles from the phantoms caused by spurious sensor readings.

## The Use of Field Robots

Factory robots bring the repeatibility, productivity, and quality control of automated mechanisms to manufacturing industries. The other historical motivation for using robots is to relieve humans of duty in hazardous environments. The nuclear industry was quick to adopt telerobotics so that human presence can be projected into places where the need for radiation protection hinders manual work or precludes it altogether. Teleoperated manipulators are presently saving thousands of man-rems of exposure in the routine servicing of reactors and associated steam generation equipment; recovery from the Three Mile Island and Chernobyl accidents would not have been possible without robotic worksystems specially commissioned to operate in those scenarios. For their specialized agenda, these nuclear-qualified robots exhibit high competence, owing to the fact that they were built to meet explicit performance goals and desing criteria.

The world is now positioned to apply robotic technologies in other commonplace scenarios. Non-factory work sites are ripe, virtually untouched, and inevitable arenas for robotic applications. Labor efficiency on field sites is alarmingly low and the need for improved productivity is evident. Worker time spent idle or doing ineffective work may exceed half the work week, and productivity has generally been in decline for two decades. Thus, industry size, economics, existing inefficiencies, and competition motivate the introduction of robotics to field work. Other motivations include quality assurance and the prospect for better control over the field work site of the future. Further, because field work is often hazardous, concerns for health and safety provide additional impetus for robotic implementations.

In addition to all these motives, certain applications are inevitable because man is not perfectly suited for field work; machines are often better equipped for many applications. Man, for example, is vulnerable to hostilities such as weather, dust, vacuum, submersion, and cave-ins, and limited by a lack of scale or power for activities such as mining, material handling and construction. Man lacks certain sensing modalities, memory structures, and computational abilities that will allow the robots of the future to precisely sense and execute tasks in scaled or measured environments, and optimize automatic material distribution throughout a site. The needs of the field industries drive the development of unstructured robotics just as manufacturing and assembly drove structured robotics and hazardous environments drove teleoperation.

Early applications of robotic arms in manufacturing leveraged on their accuracy, consistency, and repeatability to achieve productivity, performance quantified on the basis of speed and the efficiency of resource investment, particularly the human resource.

Similar increases in productivity are realizable in applications outside the factory. For example, proper characterization of a hazardous waste site requires an enormous amount of data to be taken over a large land area. There are current efforts to automate this process by replacing manual data collection with mobile robots that can acquire and spatially correlate site information. Orders of magnitude increases in the amount of site data, as well as higher precision position estimation, will enable more complete assessments and ultimately reduce the cost of the investigation process.

Excavation is another excellent application to further the evolution of robotics because of its significance in scale and economic importance. It operates on a universal and generic material (soil), and excavation's goal and state can be described adequately by models of geometry and kinetics. Further, escavation is tolerant of imprecision, well-understood as a human driven process, and prototypical of a host of spin-off applications. One motivation for robotic excavation is the hazard in such tasks as blind digging of gas lines, retrieval of unexploded ordnance or removal of hazardous waste from a landfill. Another motivation is the productivity and process control that could be realized in mass earth moving operations. Unmanned excavation will reduce the human injuries and property losses attributed to explosions, decrease operation costs, and increase productivity by lengthening the work day.

Automation of surface mining has the potential to increase safety, decrease cost, and revolutionize control of surface mining operations. Elimination of human operators could circumvent current variables of operator quality and availability and monotony of the task. Further, automation of surface mining is seen as a building block toward general work site automation. Surface mining lends itself well to automation. Driving and haulage are simple actions in comparison with the richness of other robotic tasks like manipulation. Off-road navigation can also be extended to the applications of agriculture and timber harvesting. The environment can be known in advance and rigged to an appropiate level. Because the task is repetitive (the same paths are traversed for years), explicit plans alleviate the need for the robot to explore or learn about its environment. Although it must be able to handle a range of contingencies such as obstacles, an autonomous haulage system is primarily a performer of preplanned actions relegating perceptive sensing to a mechanism of self-survival.

A new generation of robots, grounded in existing robotic technologies, is on the horizon and will find widespread utility.

Robotic navigators are one class of systems that have several applications, including haulage, material delivery, and waste site characterization. Through automation of off-road driving, these tasks can be performed with less direct human involvement, thereby increasing a worker's productivity through simultaneous control of several vehicles and removing his exposure to potential hazards.

Ground vehicles realizable in the near term will navigate under general lighting and weather conditions at productive rates of travel. Some will drive on streets and highways; others will negotiate rough terrain with variable geometry and natural surface characteristics. They will employ multiple sensory modes for guidance; use maps from several sources and of various resolution; detect, recognize and avoid obstacles; and be cognizant of their own dynamics. Future generations of robotic off-road navigators will focus on the design of robust navigational schemes. Obstacle detection and recognition will be extended to accommodate dynamic obstacles like other vehicles so that these robots will ultimately be capable of driving in traffic.

By coupling manipulation to locomotion, a robotic vehicle that can navigate off-road can be complemented with the ability to perform useful work. A terrestrial robot worksystem can be used in construction applications, such girder emplacement, excavation, and brick laying, and hazardous applications like handling of radiological material, waste packaging, and decontamination and decommissioning of nuclear facilities. These tasks share the common denominators that the robot physically engages and manipulates its environment and that the setting for these operations is often very hazardous.

These steps to enhance teleoperation of the worksystem provide the foundation for enhaced performance through increased task autonomy. The worksystem will evolve incrementally, as operations performed under human control in one generation are automated in the next. Interaction between man and machine will become simpler as the robot becomes able to acept higher level commands, and the human´s role will transition to supervisor.

Next generation worksystems will perform certain subtasks on their own, while the operator exercises direct control for the more difficult operations, monitors subtask execution, and intervenes as needed. In the case of excavation, subtasks might be the scooping and unloading phases of the digging cycle; for building construction, subtasks might include grasping an I-beam and carrying it to location where a building foundation is being established. These capabilities will develop from the basics of manipulator control and

geometric model building of the enhaced teleoperator by adding the capacity to recognize objects and the ability to reason on perceived geometry and force. Future generation worksystems will combine subtasks, automate more difficult aspects of the tasks, and add execution monitoring to achieve a higher degree of autonomy. Alternatively, it might be desirable to pursue execution of a variety of tasks using one worksystem with multiple tools and operating modes to achieve higher utility.

## The Evolution of Robotic Technology

Robotics research has reached a threshold where technologies are beginning to find performance niches in which their implementations show comparative advantages over older technology or allow the performance of tasks previously unperformable. We are also witnessing a shift in implementation process form ad hoc integrations to disciplined development of complete systems.

Robotic technology has gained competence in the key areas of sensing, cognition and control, to the point where new applications are feasible. Early robots had only mechatronic sensing with which they measured directly observable external variables, such as displacement and force, and could perform only simple operations, such as inspection, loading, and other positioning tasks. Increased understanding of vision and other sensory processes has made it possible for robots to make interpretations of their environment. Advanced robots extract and recognize certain features in data, often from multiple sensors, on the basis of pre-stored symbolic representations. This makes them capable of more challenging tasks, for example, manipulating irregularly shaped objects and assessing navigability of roads and paths. A very demanding task, like construction of a building, which requires not only the recognition of features and objects, but understanding of their semantic interplay, is presently beyond robotic technology.

Similarly, robots are able to undertake more challenging tasks as a result of advances in machine cognition. For early robots, planning was algorithmic and often no more than continuous state error correction, as in charting and following a trajectory. It is now feasible for robots to perform tasks like shaping soil and walking over rough terrain, which require automatic planning of significantly greater scope and depth: plans must be decomposed from goal specification into executable actions, and plan formulation has to be done in the face of uncertainty, requiring execution monitoring and use of contingencies. Coordination of multiple, potentially conflicting subgoals to fulfill a single, high-level directive, such as "clear obstacles from the road", remain too ambitious for existing robots.

The evolution of robotic technology is also evident in the increasing physical challenges met by robots. The first robots had kinematic control only, and their tasks domain spanned only operations that could be expressed by prescriptions of robot position. Better understanding of robot mechanisms and the application of more advanced control theory has enabled tasks that involve dynamic interaction of the robot with its environment, like stable walking and excavation. We are now implementing control at the task level, which goes beyond control theory and includes cognitive functions, such as error detection and fault recovery, so that occurrences, such as an unexpected obstacle, a sudden loss of traction, or a dropped payload, do not prevent completion of a task.

## The Erebus Project

The Erebus Project will explore an active volcano. Mount Erebus, Antarctica at year-end, 1992, using robots. The mission will advance scientific knowledge of this important volcano, and the research will create unprecedented technologies for unmanned exploration. This explorer will be a remote geologist, a forebear of polar robots, and a step to the planets.

The technical objectives of this program are to achieve extended autonomy, environmental survival, and selfsustainable mission performance in the harsh Antarctic climate, which demands as much of a robotic explorer as any location on Earth. This mission will set an important precedent in mobile autonomous operation and accomplish a necessary step in planetary exploration by achieving goals that are part of the joint NASA/NSF program to use Antarctic analogs in support of the space Exploration Initiative.

During the mission, two mobile robots, a *transporter* and *rappeller*, will work as a team. The transporter is a crawling robot whose principal function is to serve the rappeller. It will carry the rappeller to the volcano crater rim, navigating the mountain with a combination of a stored map and local terrain

sensing. The transporter will generate power, act as a communications relay between the rappeller and the base station, and provide a mechanical anchoring point during the rappeller's descent into the Mt. Erebus crater.

The rappeller is a legged mobile robot that carries a payload of scientific instruments into the volcano. The rappeller will be computationally self-sufficient, able to chart its own course, react to perceived terrain, and acquire data from mission payload sensors.

The robots will ascend Mount Erebus from a base station established about 2km from the crater rim. Once at the summit, the transporter will anchor itself and the rappeller will begin its descent into the crater. It will negotiate the crater wall by sensing the local terrain, groping with its legs and using its tether as a climbing rope. Within the crater the robot will measure lava temperature, collect samples, analyze gases, and make photographic records. It will then climb back to the transporter and return with it to the base camp.

Humans will interact with the exploration robots from a camp on the mountainside. This base station will house joysticks, displays and computers to process user commands and robot data, providing a flexible user interface in modes ranging from teleoperation to human supervision. A second command point will be established in Pittsburgh and linked to the Mt. Erebus base station via satellite.

Deployin robots to Antarctica will be logistically challenging. Ultimately, semi-autonomous robotic explorers will be commonplace in Antarctica but the first deployments will undoubtedly be difficult.

The Erebus project is generating innovative ideas in robot mechanisms, perception, planing, and task control - new ways of thinking to address the needs of NASA, NSF and the robotics community.

The physical robots will provide fundamental insights into the construction and control of environmentally survivable, capable systems. The perception component of our research will advance the state-of-art in mapping rugged terrain covered with difficult materials like ice and snow. The approaches used for modeling the sparsely-featured terrain and objects encountered in the Mt. Erebus mission are applicable to a wide range of navigation and manipulation tasks.

The problems of systems integration, systems architecture, and coordination of multiple, interacting software components are faced by every robotics program. In particular, issues of extended autonomy, long-distance and long-delay telemetry, and high-reliability software must necessarily be addressed to deploy robots in Antarctica.

This program is developing core robotics technologies, research robots and expertise that are essential for future missions and US leadership in space exploration and on Earth. Although the immediate objetive is an Antarctic mission, we intend to port results directly to planetary exploration and to expanded operations in polar regions.

The Erebus Project, as past and current NASA and NSF programs at Carnegie Mellon, encourages the exchange of technologies. We are actively involving industry and laboratory participation. Publication of articles, conference papers and technical reports spread our ideas and technology. In addition, the program is developing engineers, scientists and students that are in critical demand by the space community and the nation.

## Conclusions and Future Directions

Despite evident need and apparent promise, the evolution of field robotics has not been straightforward. Ancient crafts have been historically slow to embrace new technology. Research investment levels have been insignificant. No precedents in field work industries for development programs of the requisite magnitude exist. Because field problems are difficult, quick fixes or one-time solutions are few, running counter to historical insistence on short-term payoff for investment. Obstacles to the growth of field robotics are compounded by the lack of common ground between the field industries and the robotics research community. The industry cannot yet visualize a programmatic course of action for integrating the growing robotic technology with its own.

At this time, construction, subsea, space, nuclear, mining, and military applications are driving and pacing many field robotics developments. Subsea and space applications, in particular, present unique technical challenges to robots, specialized motivations for field work, and constraints and regulations that discourage the use of human workers. However, the formative integration and drive for field robotics must ultimately come from the field work industry itself. The inevitability of field robotics will drive its evolution despite the immediate immaturity and impotence of the field.

It is likely that all three classes of robots and their hybrids will find sustaining relevance. Experiences are too few and it is too soon to resolve the relative importance of these forms or to discount the potencial of any form. Though it now appears that attributes of intelligence, particularly the ability to deliberate performance of tasks, will eventually dominate field robotics, nonetheless, teleoperators and programmed machines have both short- and long-term relevance.

If robots eventually prove themselves infeasible for unstructured environments, our views on what constitutes structure must change. Robots other than teleoperators may be irrevocably synonymous with structure. Our judgement in this matter should not be too clouded by current measures of structure and machine perception. It is common to mistake or overestimate chaos in a task environment simply because form and understanding are not apparent. There is a great prospect for structuring the apparently unstructured either by discovering structure of by imposing it.

The evolution fo field robots will distill unique attributes for robots with working goals in unstructured environments. New robotic forms will emerge with the capability and the strategic competence to construct, maintain, and demolish. The evolution of field robotics will no more culminate in a single, ultimate form than did its biological counterpart. Rather, classes of robots will emerge for classes of work within classes of constraints.

Even the robot genus/species formed and proven in other application domains remains untested by field work. No doubt most of the forms evolved for other purposes will find relevance somewhere in field work, if only because field works umbrella is so broad. The discipline of field robotics is embryonic. Its maturation is inevitable, but its mature form is not apparent. Given the uncertainty of what robotic forms may be relevant to field work, we argue that the field should remain open to all possible.

The discipline must persevere to distill the unique identity and intellectual content of field robotics. The uniqueness of field robotics appears to lie in the cognitive skills and goals specific to the synthesis of an end product. Much research and many goals in field robotics, however, are generic to unstructured robotics, so field work can benefit from parallel developments in related fields. Little applicability would be lost by changing the domain specificity from field work to nuclear, mining, timbering, or military. It seems that field work will be dragged reluctantly to the opportunities of robotics. Nuclear, military, space , and offshore interests are embracing and driving the ideas now. It is essential that field robotics identify and drive the developments that will distinguish it as a discipline of its own.

## References

Bares, J., E. Krotkov, M. Hebert, T. Kanade, T. Mitchell, R. Simmons, and W. Whittaker, " An Autonomous Rover for Exploring Mars", Special Issue on Autonomous Intelligent Machines, Computer Magazine, June 1989.

Everett, H.R., " Robotics Technology Areas of Needed Research and Development", White Paper 90G/119, Office of Navy Research, September 1985.

Kanade,T., Thorpe,C., and Whittaker, W., " Autonomous Land Vehicle Project at CMU", ACM Computer Conference, February 1986.

Martin, H. and Kuban, D., " Teleoperated Robotics in Hostile Environments" , Dearborn : Robotics International of the Society of Mechanical Engineers,1985

Moavenzadeh,F., "Construction's High-Technology Revolution", TechnologyReview, October 1985.

Motazed, B. " Interpretation of Magnetic Sensing for Construction Inspection", Proceedings of the Second International Conference on Robotics in Construction, Pittsburgh, June 1985.

Motazed, B. and W. Whittaker, " Interpretation of Pipe Networks by Magnetic Sensing", Proceedings of the First International Conference on Applications of Artificial intelligence in Engineering Problems, Southampton, U.K., April 1986.

Osborn, J. D. Phanos, T. Stenz, C. Thorpe, and W. Whittaker, " Field Robots: The Next Generation", White Paper, The Robotics Institute, Carnegie Mellon University, January, 1990.

Paulson, B. " Automation of Robots for Construction", ASCE Journal of Construction Engineering and Management, Vol. 111, No3, September 1985, pp190-205.

Sagawa, Y. and Nakahara, Y., " Robots for the Japanese Construction Industry", IABSE Proceedings, No. P-86/85, May 1985.

Suzuki, S. " Construction Robotics in Japan", Third International Conference on Tall Buildings, Chicago, 1986.

Warzawski, A. " Application of Robotics to Building Construction", First International Conference on Robotics in Construction, Carnegie Mellon University, 1984.

Warzawski, A. and Sangrey D. "Robotics in Building Construction". ASCE Journal Construction Engineering and Management, Vol. 111, No. 3, September 1985.

Whittaker, W., " A Remote Work Vehicle for the Nuclear Environment ", Proceedings of the First Regional Meeting of the American Nuclear Society, Pittsbutgh, September, 1986.

Whittaker, W., " Construction Robotics: A Perspective", International Joint Conference on CAD & Robotics in Architecture and Construction, Marseilles, June 1986.

Whittaker, W., " Design Rationale for a Remote Work Vehicle", Proceedings of the 34th Conference on Remote Systems Technology, American Nuclear Society, Washington,D.C., November, 1986.

Whittaker, W., " Teleoperated Transporters for RERR", Proceedings of the Workshop on Requirements of Mobile Teleoperators for Radiological Emergency Response and Response and Recovery, Dallas, June 1985.

Whittaker, W. and Bandari, E., " A Framework for Integrating Multiple Construction Robots", International Joint Conference on CAD and Robotics in Architecture and Construction, Marseilles, France, June 1986.

Whittaker, W. and Motazed, B. " Evolution of a Robotics Excavator", International Joint Conference on CAD and Robotics in Architecture and Construction, Marseille, France, June, 1986.

Whittaker, W. and L. Champeny, " Capabilities of a Remote Work Vehicle", Topical Meeting on

Robotics and Remote Handling in Hostile Environments, American Nuclear Society, Seattle, March 1987.

Whittaker, W., G. Turkiyyah, and M. Hebert, " An Architecture and Two Cases in Range Based Modeling and Planning", Proceedings of the IEEE International Conference on Robotics and Automation, Raleigh, April 1987.

Whittaker, W., J. Bares, and L. Champeny, " Three Remote Systems for TMI-2 Basement Recovery", Proceedings of the 33rd Conference on Remote Systems Technology, San Francisco, November 1985.

Whittaker, W. et al., "First Result in Automated Pipe Excavation", Proceedings of the Second International Conference on Robotics in Construction. Pittsburgh, May 1985.

Whittaker, W. et al., " Mine Mapping by a Robot with Acoustic Sensors", Proceedings of the Second International Conference on Robotics in Construction, Pittsburgh, May 1985.

Whittaker, W., " Cognitive Robots for Construction":, Annual Research Review, The Robotics Institute, Carnegie Mellon University, 1986.

Whittaker, W., K. Dowling, J. Osborn, and S. Singh" Robots for Unstructured Environments", Unmanned Systems, 8, Winter 1990.

Yamada, B., " Developments of Robots for General Construction and Related Problems", Research Conference, Material and Construction Commitee, Architectural Institute of Japan, October 1984.

Yoshida,T and Ueno, T., " Development of a Spray Robot for Fireproof Tratment", Shimizu Technical Research Bulletin, No. 4, March 1985.

# INTELLIGENT ESTIMATION AND PREDICTION FOR SYSTEMS CONTROL AND DECISION - METHODOLOGY AND APPLICATIONS

## Y.Z. Lu[1]

*Research Institute of Industrial Control, Zhejiang University, Hangzhou, PRC*

**ABSTRACT:** This paper presents the concept , architecture and methodology of intelligent estimation and prediction (IEP) in systems control and decision, particularly the applications of both supervised and unsupervised learning in IEP. The IEP as described is mathematical model free estimation and prediction technique, and mainly based on knowledge base, fuzzy logic, artificial neural network and their combination. The major characteristics of the intelligent" in the proposed IEP system are learning and self-organizing in order to provide robust and adaptive system behavior. A few working examples and potential applications of IEP are also addressed in this paper.

**Key Words**: Estimation and Prediction, Intelligent Systems, Knowledge Base, Fuzzy Logic, Neural Networks, Supervised Learning, Unsupervised Learning, Neural Network Controller, Fluidized Catalytic Cracking Unit.

## INTRODUCTION

The estimation and prediction have been playing important roles in information gathering and forecast for system control and decision.

As is well known that the conventional mathematical model based estimation and prediction, for instance, observer and filtering [1,2] are mainly applicable to continuous function estimation for linear and some nonlinear deterministic or stochastic systems with specified mathematical model, at least, with given model structure.

During last decade, control scientists and experts have made great efforts to explore the future direction of control theory and technology for complex systems control and decision.The system complexity in the real-world problems may be characterized by the features of highly nonlinear, seriously inter-connected and large scale, time varying , system uncertainty,incomplete sensory information, high quality control, discrete-event driven and human factor involvement, etc..

It is clear that the existing conventionel mathematical model based estimation and prediction techniques obviously can not be applied to solving problems in many complex systems with the features as noted above. The main reasons are as follows:

* It is hardly possible to develop a mathematical model for complex system with the features as described above.

------
[1] The author currently is on leave with Bethlehem Steel Co. as Control System Control Consultant.

* Some system exists qualitative ( symbolic ) or multivalued ( fuzzy ) input, output, state and/or their relations which can not be modeled by existing mathematical tools, as a result,the approaches to mathematical model based techniques are not applicable in some complex problems;
* If system environment, such as parameter, structure and measurement, changes with significant uncertainty, it is difficult to provide robust and adaptive results.

With rapid development of fuzzy logic, knowledge engineering and artificial neural network ( ANN ), the intelligent estimation and prediction ( IEP ) has become an attractive new direction in systems control and decision. This paper will address the concept, strategy, methodology of IEP and their applications. The focus will be mainly on the issues of what is "intelligent" and how to apply "intelligent" in IEP, particularly, the potential application of supervised and unsupervised learning in IEP with knowledge acquisition and self-organization.

## PROBLEM STATEMENT

The task of the traditional functional estimation and prediction is to provide system "quantitative" information based on the given mathematical model and observed input/output data. The major advantage of this technology is that design of estimator or predictor has strong theoretical background and generic algorithms, and can provide quantitative knowledge of system external or internal information.

However, as we mentioned before that with increase of systems complexity, the existing technology is unable to be applied to many real-world complex systems. In last few years, the development of intelligent related technology provides a new path to investigate the novel estimation and prediction for complex systems, so-called "intelligent estimation and prediction".

## What is Intelligent Estimation and Prediction

Suppose a system with parameter, structure and / or measurements uncertainties which is governed by the following mapping in hyperspace:

$$X(t+1)=F(X(t),Y(t),U(t),D(t))+v(t) \quad (1)$$

$$Y(t)=G(X(t),U(t),D(t))+w(t) \quad (2)$$

Where

| | |
|---|---|
| $F(*),G(*)$ | Nonlinear mapping with significant uncertainties; |
| $X(k), Y(k)$ | system state and output vectors; |
| $U(k), D(k)$ | system control input and disturbance vectors ; |
| $v(k), w(k)$ | random noises of system model and measurement. |

The vectors, $X(k)$, $Y(k)$, $U(k)$ and $D(k)$, could be different types of information. Since the system under consideration is too complicated to be modelled by conventional mathematical approaches, the model structure is not available.

The problem of intelligent estimation and prediction can be described in a number of ways. Kosko [3] defines an intelligent system as: " *Intelligent Systems adaptively estimate continuous functions from data without specifying mathematically how output depend on inputs*". In addition, [3] also indicates that " Learning ", "Generalization", and "Creativity" are the major properties of intelligent systems.

In this paper, the concept of "intelligent systems" is extended in more generic sense, namely, not only for continuous function estimation, but also for system estimation with symbolic and multivalued functions. The investigation in this study will be based on the definition of Intelligent Estimation and Prediction: *An intelligent estimation system adaptively provides robust estimate of system output and / or state ( with numerical, symbolic or mutivalued information ) based on real-time input / output data and other information through using hybrid knowledge and supervised or unsupervised learning.*

## Background of Development of IEP

Based on the definition of IEP, the intelligent estimation does not only involve regular function estimation, but also other information estimation. Mathematical model-free estimation related technologies are given in Fig.1 [3].

As we know that rule based expert system being a structured knowledge based system can only be used in estimation of symbolic oriented information, such as fault diagnosis, production sequencing and planning, and discrete-driven systems.

Fuzzy systems also belong to structured knowledge oriented family, but fuzzy estimation can provide multivalued information. However, the level of estimation accuracy is highly related to the size of fuzzy set which corresponds to the number of fuzzy rules.

Finally, ANN is unstructured knowledge form with numerical computation framework. The knowledge with analog, discrete or binary form is stored in the network with many nonlinear processing elements, so-called "neuron" , and the high dimensional synaptic connections. ANN related information can be converted from one form to others through " encoding " and " decoding ", and ANN has good mathematical properties, as a result, ANN can effectively be applied to IEP for complex systems.

## The Major Characteristics of IEP

Based on the problem as we described above, the IEP has the following intelligent characteristics:

* IEP is a mathematical model free (or partly free) estimator or predictor;
* Signals of system inputs, outputs or states could be numeric, symbolic or multivalued ( fuzzy ), and they work together through their conversion;
* System mapping will be established through the learning of historical or real-time experience, as a result, the estimate or predicate will have robustness and adaptivity in nature;
* If feedback signals for learning are not available, the system knowledge related to system behavior or cluster will be established through unsupervised learning, which can be considered as " generalization " and " creativity ".
* the system design and implementation are based on the combination of various knowledge, such as rule base, fuzzy model, ANN model as well as mathematical model.

## THE GENERAL ARCHITECTURE OF IEP

A general architecture of IEP can be schematically shown in Fig.2. It can be seen that the major functional components in the IEP can be divided into the following categories:

## * Knowledge Representative and Conversion

In order to let various knowledge with different forms work together, both knowledge representative and their conversion ( encoding and decoding ) become a critical issue. It is required to select the main frame of IEP system, for instance, ANN, fuzzy system or rule base, which will well suit the problem under study. For instance, if the task of estimation is mainly related to symbolic oriented decisions, an expert system might be the main frame associated with other knowledge forms. If the problem understudy involves quality control related estimation, ANN or fuzzy systems could be the frame.

Obviously, ANN and fuzzy systems can also be applied to dynamic system estimation. Nevertheless what serves the main frame, the conversion of knowledge forms is needed in system design and implementation.

## * Learning and Adaptation

The function of learning components in the system.is to establish and / or adaptively adjust system model which could be ANN, fuzzy or knowledge base. In general, the driving force in learning algorithm is the mean squares error between the estimated and real value of state or output, so-called supervised learning. However, if the learning feedback knowledge is not available, then unsupervised learning might be used to provide system behavior, so-called self-organization, which has higher level of intelligence will be addressed later.

## * Self-Knowledge Acquisition

The knowledge acquisition has been a major concern in establishing a traditional expert systems. However, nevertheless what kind of systems, generally speaking, we can gather plenty of data which are related to system behavior. The extraction of information from system data has been one of important issues in pattern recognition, data communication and imagine processing, etc.. It should be noted that the application of ANN has provided a path to generalize system knowledge, for instance regular rules or fuzzy rules, through unsupervised learning.

## KNOWLEDGE REPRESENTATIVE AND CONVERSION

The system behaviors of complex systems usually are described by different levels of knowledge, which are cooperatively be used in system control and decision. Now we will introduce the knowledge representative and their conversion.

## Mathematical Description

Traditionally, A system can be described by mathematical formulas, such as differential or difference equations. Even a distributed parameter system with moving boundary which is governed by partial differential equations and corresponding boundary conditions can also be converted into nonlinear state space model to describe system internal knowledge, which could not be on-line measured by sensors [4].

Mathematical model developed by the combination of first principle and system identification provides parametric quantitative knowledge between system input, output and state. Obviously, mathematical model based system description are still most fundamental and important for the systems with transparency physical principle and less uncertainty. it should be emphasized that, mathematical description could still be a part of IEP , if necessary.

## Knowledge Base

The knowledge base model with rule base , frame or semantic structure can also describe the system dynamic behavior, so-called " temporal knowledge ". The most popular form of the rules is "If A and (or) B ,......., Than C , with time factors". However, knowledge base model can not provide the precisely qualitative knowledge of system dynamics. In contrast, knowledge base model is good at in representing the logic or linguistic knowledge and reasoning with symbolic forms.

## Fuzzy Model

In addition to the regular rule base, the system behavior sometimes can also be described by fuzzy rules, for example, the fuzzy mapping:

$$X(t) * U(t) \overset{R}{=====>} X(t+T)$$

The relevant fuzzy rules:

Ri : If  Xi(t) and Ui(t)
       Then Xi(t+T), i=1, N

Where Xi(t), Ui(t) and Xi(t+1) are described by fuzzy sets with corresponding membership function $\mu_{Si(t)}$, $\mu_{Ui(t)}$, and $\mu_{Si(t+T)}$, and then membership function for rule Ri and R can be described as follows:

$$\mu_{Ri}(X(t),U(t),X(t+T)) =$$

$$Min\{\mu_{Xi(t)}(X(t),\mu_{Ui(t)}(U(t),\mu_{Xi(t)}(X(t+T)) \quad (3)$$

$$\mu_{R}(X(t),U(t),X(t+T)) =$$

$$Max\{\mu_{Ri}(X(t),U(t),X(t+T)\}=Max\{Min[\mu_{Xi(t)}$$

$$(X(t),\mu_{Ui(t)}(U(t),\mu_{Xi(t+T)}(X(t+T)]\} \quad (4)$$

The fuzzy model governed by fuzzy relational matrix R can be calculated through fuzzy identification based on input/output data [5]

$$\underset{R1,..,Ri}{Min} \sum_{t} ||Xi(t)-\hat{X}i(t)||^2 , i=1,n \quad (5)$$

Where $\{\hat{X}i(t),i=1,n\}$ are the estimated outputs given by the fuzzy model.

As a result, the fuzzy relational model can be written in the following form:

$$X1(t+1) = X1(t)*U1(t) ... Ur(t)*R1$$
$$\cdot$$
$$\cdot \quad (6)$$
$$Xn(t+1) = Xn(t)*U1(t) ... Ur(t)*Rn$$

Where " * " denotes the composition or Max-Min product operator.

It can be seen that in contrast to knowledge base model, fuzzy model can also provide the quantitative knowledge with given accuracy which is dependent upon the size of reference fuzzy set being determined by designer. The detailed procedure for fuzzy model identification as shown in Fig.3.

The major advantages of fuzzy modeling can be summarized as follows:

* Fuzzy modeling can be effectively applied to nonlinear systems without knowing system structure;
* Fuzzy rules can be converted into fuzzy relational model based on fuzzy identification and I / O data, which

can provide system quantitative relations between input, output and state; Similar to regular approaches, fuzzy model, R, can also be upgraded through real-time identification.
* In the other hand, fuzzy rules can be directly established through learning based on I/O data

## ANN Model

ANN modeling techniques have become an attractive technique in system modeling [6,7,20], particularly for nonlinear systems with unknown dynamics. The general structure of an ANN model is shown in [6] . In general, an ANN consists of a few layers (input, hidden and output) which are interconnected through the adjustable weights. Each layer consists of many nonlinear processing elements,so-called "neurons" . The working process of ANN model can be functionally divided into "learning" and "retrieval" phases. An ANN model, in fact, is a large scale nonlinear dynamic systems with high dimensional adjustable connected parameters. As a result, it can be used to model complex nonlinear system with strong robust and adaptivity. The major advantages of ANN can be summarized as follows:

* Provide system input-output numerical relationship without having system mathematical structure;
* Can be effectively applied to system modeling with hybrid knowledge and various types of I / O data; As a result, ANN provide a flexible frame for combining "deep-knowledge", "fuzzy knowledge" and/or "shallow knowledge";
* The powerful learning function of ANN modeling provides an effective tool for carrying out adaptive modeling;
* The application of unsupervised learning in ANN can implement self-knowledge acquisition for complex system.
* Finally, the most important point is that even in principle, the approach of ANN can be reduced to of stochastical approximation, however, in comparison with conventional nonlinear system modeling and estimation techniques, for instance, polynomial or linear approximation, ANN provides much better mathematical properties and structures, and huge number of self-adjustable parameters. This is one of the most important reasons why ANN can emulate any continuous linear or nonlinear function with desired accuracy.

## Conversion of Different Knowledge Representatives

Fig.4 gives the general architecture of knowledge conversion. In general, knowledge can be divided into structured and unstructured two categories, as shown in Fig.1, they are related to expert systems,fuzzy systems and neural systems correspondingly. We can not directly encode the rule, for instance, "If temperature is high, Then pressure will be high" into numerical forms. However, if we introduce the degree of both "temperature high" and "pressure high" with corresponding membership, then structured knowledge can be directly encoded into high dimensional fuzzy relational matrix with numerical

framework to describe multi-valued-quantitative relationship between input and output.

In addition, it should also be emphasized that in contrast with expert systems, ANN has numerical framework with efficient numerical algorithms and theoretical theorems. Even ANN can not directly encode structured knowledge, the conversion from expert systems, fuzzy systems with symbolic information to numerical frame by using ANN have been an attractive subject [3,8,9].

Inversely, numerical information can also be converted into structured knowledge with symbolic information form through using ANN learning and/or fuzzy identification.

## LEARNING, GENERALIZATION AND CREATIVITY

The early learning studies [13,14] mainly applied to supervised learning with stochastic approximation. The emphasis focused on learning principle and theory rather than computational techniques. Tsypkin [14] proposed the fundamental of learning as " *The need for learning arises whenever available a priori information in incomplete. The type of learning depends on the degree of completeness of this a priori information. In learning with supervision, it is assumed that at each instant of time we know in advance the desired response of the learning system and we use the difference between the desired and actual response, that is, the error of learning system, to correct its behavior. In learning without supervision, we do not know the desired response of learning system*".

In order to emulate the human behavior, machine intelligence usually includes "learning from experience", " Associate memory ", " deep reasoning " and " generalization or creativity ", etc..

Now we will discuss supervised learning , unsupervised learning and associate reinforcement learning and their potential applications in IEP.

## Supervised Learning

Supervised learning is instruction-oriented, namely, the desired output pattern is specified and the objective of learning is to minimize the error between desired pattern and actual pattern through upgrading model parameters and/or structure.

Suppose we have a unknown mapping function F: X ==> Y. The mapping with respect to the relevant parameters and structure in hyperspace can be constructed through training based on the given training examples, { Xi, Yi , i=1,K }. In an ANN the learning applies to number of nodes and / or connected weights changes in network which is similar to synaptic changes in nervous systems.

As soon as the training phase is completed, the training patterns have been stored in the frozen weights of ANN as long-term memory. The system's new output Yj can be produced through the "retrial phase" based on the new input Xj. However, it should be emphasized that the accuracy of resulting estimates is heavily dependent on the difference between the training patterns and the patterns in real applications.

Back Propagation ( BP ) learning algorithms [6,10] have been one of the most popular learning algorithms in the supervised learning. The general architecture of BP based supervised learning in ANN is shown in Fig.5. Although many efforts have been made to try increase the learning speed of BP learning, however, since the BP learning algorithms are derived from gradient-decent approach in operation research, which we can not prove is same as the learning principle in real biological systems. From this point, the current learning approaches are still on low level of intelligence.

In stead of off-line learning, some adaptive learning algorithms [11,12] have been developed for ANN in order to follow system environment changes. However, it is a vital important topic how to select or produce the training examples through data compression or feature abstraction techniques in order to keep the most important information for training and to avoid losing important information during adaptive learning.

The applications of ANN with supervised learning in estimation and prediction are illustratively shown in Fig.6.

Unsupervised Learning and Applications

Another important feature of " intelligent " is self-knowledge acquisition. This means that an intelligent system should have the ability to generalize or create some knowledge which we do not recognize.

Even we can not require the computer doing what human can do, however, studies of machine learning with low level of generalization and creativity are certainly critical to promote the further development of intelligent systems. The purpose of developing the second generation expert system is to reach the goal of knowledge creation. However, it will be very difficult to create new knowledge through symbolic oriented rule base and its reasoning.

The " unsupervised learning" [3,6,7] or so-called " competitive learning ", " self-organizing learning ", has been playing an important role in pattern recognition, signal processing, communication and data compressing, etc.. However, the investigation of potential applications of unsupervised learning in control related technology is only at very beginning stage. This paper will preliminarily discuss the possibility to apply unsupervised learning in IEP.

For example, suppose we have a production system with input / output data and unknown mathematical model, describing quantitative or classified relationship between the operating conditions and the quality of finishing product. If we start to work with an expert system for this problem, we have

to deal with many rules describing the system behaviors. However, in practice, it will be very difficult to establish such kind rule base and related reasoning for some complex systems.

In contrast to supervised learning, unsupervised learning algorithms use unlabeled training samples to discover the general features without receiving any learning feedback, which can be used to classify a set of patterns or capture regularities ( clusters ) which represent system behavior. The problem concerning product quality as mentioned matches the functions and features of unsupervised learning.

The general architecture of competitive learning mechanism [6] is shown in Fig. 7. The units in a given layers are divided into a number of nonoverlapping clusters. Each unit within a cluster inhibits every other unit within cluster. The clusters are winner-take-all, such that the unit receiving the largest input achieves its maximum value while all other units in the cluster are pushed to their minimum value. A unit learns if and only if it wins the competition with other units in its cluster.

As a result, competitive leaning can effectively carry out the automatic cluster via learning without teacher Based on the fundamental feature of unsupervised learning, we can see that some new knowledge related to symbolic oriented regular rules or fuzzy associate memory rules might be discovered through unsupervised learning.

[3] indicates that ANN can adaptively generate fuzzy associate memory rules in fuzzy system with new technique of unsupervised product-space clustering. Suppose a system contains n fuzzy variables, and each fuzzy variable can assume m fuzzy-set values. This defines $m^n$ fuzzy associate memory cells in input-output product space $R^n$. The resulting fuzzy associate memory system architecture is given in Fig.8.

It can be seen that the creation of fuzzy associate memory rule is different from fuzzy identification, since the previous one does not require priori knowledge to provide structure of fuzzy model.

Obviously, this technology will be a challenge which control scientists are facing. Some potential applications of unsupervised learning can be listed as follows:

* Self-knowledge acquisition: Development of rule base or fuzzy systems;
* Data reduction and feature extraction for ANN training or regular identification;
* Clustering on quality space for quality prediction and estimation;
* Properties analysis for unknown complex systems;
* Self-tuning through cluster in controller parameter and/or structure space for complex dynamic systems, particularly for nonlinear and / or variable structure systems;
* Property Analysis for multiinput-

multioutput system structure identification;

## Associate Reinforcement Learning

Reinforcement learning system receives learning feedback signal with a binary variable representing " success " or " failure " or multi-valued signal indicating the degrees of "success" or "failure"which are similar to fuzzy set in fuzzy systems, then system changes based on " reward " or " penalty " respectively. obviously, there are many potentials to apply reinforcement learning in modeling, optimization and reasoning which are related to IEP.

## WORKING EXAMPLES

### Application of Intelligent Prediction and Optimization in Fluidized Catalytic Cracking Unit

The fuzzy model with real-time learning has been developed to predict the cracking products distribution. In addition, an intelligent optimization control strategy with combination of fuzzy model, rule base, optimization and coordination knowledge has been also developed. The criteria for optimization is to maximize the rate of light oil products with satisfying production constraints. The intelligent prediction and expert control have been successfully implemented at a large production scale FCCU [19]. The system architecture is given in Fig. 9.

### An ANN Based Industrial Controller

An ANN Controller with remarkable adaptive performance has been developed and patented [15]. The adaptive back propagation learning is used to upgrade ANN weights, as a result, this ANN controller provides robust behaviors, even the dynamics of controlled plant significantly change with both parameter and structure. the simulation results are given in Fig 10.

## REMARKS

This paper investigates the application of intelligence in estimation and prediction. However, since this is a novel field in systems control and decision, many potential problems are valuable for future studies.

(1) How to design an intelligent IEP system based on hybrid knowledge (structured and unstructured). The key of this issue is to effectively apply system knowledge with numerical, symbolic and multivalued information and to select the main framework to cast the problem most effectively and realistic.

(2) How to apply unsupervised learning in system control and decision. As we have noted that unsupervised learning can explore system behavior through clustering or adaptive vector-quantization. Based on the features of unsupervised learning, this technology might be applied to estimation and control for nonlinear, variable structure systems with incomplete priori

knowledge, and planning, scheduling and computer-aided design, etc..

(3) Even BP learning algorithms can solve many practical problems, however, many problems, such as local minimum, extremely slow convergent and how to determine numbers of layers and nodes, have not been solved yet. Either from theoretical or from application aspects, it is necessary to develop new learning algorithms based on fundamental research of biological technology.

(4) How to select training set of the patterns with rich information and less number of data obviously is still critical in order to reduce training time and increase ANN model accuracy. In ANN adaptive learning for dynamic systems, how to select the training signal sequences which cover various operating areas is still a problem. For instance, if a rolling mill operates within hundreds of physical sizes, such as gage, speed and width, it will be extremely difficult to select a few test sequences in order to obtain the data for ANN training.

In addition to establishing ANN model ( or inverse model) through thousands of training, from application point of view, dynamic adaptive algorithms for ANN modeling and control will be more realistic and feasible in order to avoid off-line data gathering and training, which are a tough job in industry.

## REFERENCES

[1] R.E. Kalman and R. Bucy " New Results in Linear Filtering and Prediction Problems", *Trans. ASME J. Basic Eng. Series D*, Vol.83, 1961;

[2] B. D. O. Anderson and J. B. Moore *"Optimal Filtering"*, Prentice Hall, Englewood Cliffs, NJ, 1979;

[3] B. Kosko " *Neural Network and Fuzzy systems*", Prentice-Hall Inc. New Jersey, 1992;

[4] Y.Z.Lu and T.J. Williams " *Modeling , Estimation and Control of the Soaking Pit* ", ISA Publisher, Reaserch Triangle Park, NC, 1983;

[5] C.W.Xu and Y.Z. Lu " Fuzzy Model Identification and Self-Learning for Dynamic systems ", *IEEE Trans SMC*, vol.17,no.4,1987;

[6] D.E. Rumelhart et. al. " *Parallel Distributed Processing*", Vol.1, MIT Press, Combridge, 1986;

[7] R.P. Lippmann " An Introduction to Computing With Neural Nets", *IEEE ASSP Magazine*, April, 1987;

[8] R.C. Lacher et. al " Back-Propagation Learning in Expert Network", *IEEE Trans on Neural Network*, Vol.3, No.1, 1992;

[9] K.M. Passino et. al. " Neural Computing for Numeric to Symbolic Conversion in Control Systems", *IEEE Control System Magazine*, April, 1989;

[10] G. Weiss " *Artificial Neural Learning*", Report FKI-127-90, Technical University of Munich, FRG, Feb. 1990;

[11] B. Widrow and R. Winter " Neural

Nets for Adaptive Filtering and Adaptive Pattern Recognition", *IEEE Computer Magazine*, March, 1988;

[12] K. Narendra and K. Pathasarathy " Identification and Control for Dynamical Systems Using Neural Networks ", *IEEE Trans Neural Network*, Vol.1, No.1, 1990;

[13] T.M. Mendel and K.S. Fu " *Adaptive Learning and Pattern Systems and Applications*" Academic Press, Olendo, FL, 1970;

[14] Y.Z. Tsypkin " *Foundation of the Theory of Learning Systems*", Academic Press, Translated by Z.J. Nikolic, Olando, FL, 1973;

[15] Y.Z. Lu, G. Cheng and M. Mannof " Universal Process Control Using Artificial Neural Networks ", *U.S. Patten approved and to be publishsed*, 1991;

[16] R.J. Williams " Reinforcement Learning Algorithms as Function Optimizers", *Proc. Intel. Conf on Neural Networks* , Vol.2, 1989;

[17] S. Grossberg " *Studies of Mind and Brain*",Reidel, Boston, 1980;

[18] Y.Z. Lu " The New Generation of Advanced Process Control ", *Control Engineering*, March, 1992;

[19] Y.Z. Lu " An expert Control System for Fluidized Catalytic Cracking Unit ", *Proc. of 17th Annual Advanced Control Conference*,Purdue University, Sept. 1991;

[20] S.Z. Qin, H.T. Su and T.J. McAvoy " Comparison of Four Neural Net Learning Methods for Dynamic System Identification ",*IEEE Trans Neural Network*, Vol. 3, No.1, 1992.

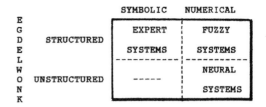

FIG.1 THE MATH MODEL FREE RELATED TECHNOLOGY

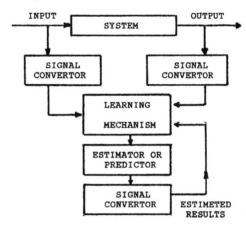

FIG.2 GENERAL ACHITECTURE OF IEP

(A) DETERMINE FUZZY MODEL STRUCTURE BASED ON THE GIVEN FUZZY RULES

(B) PREDICT OUTPUT BASED ON INITIAL FUZZY MATRIX R(t) AND REAL-TIME I/O DATA

(C) CALCULATE THE ERROR ,e(t), BETWEEN ACTUAL AND PREDICTED OUTPUT

(D) IF ( e(t) $\leqslant \varepsilon$ ) THEN
      R(t+1)=R(t) AND RETURN TO (B)
      ELSE CONTINUE

(E) CALCULATE POSSIBILITY DISTRIBUTION OF I/O ON CORRESPONDING FUZZY REFERENCE SETS AND FIND THE "CLOSET REFERENCE SETS", THEN DETERMINE THE RULES IN R(t) TO BE UPGRADED

(F) UPGRADE FUZZY MODEL AND RETURN TO (B)

FIG.3 PROCEDURES OF REAL-TIME FUZZY IDENTIFICATION

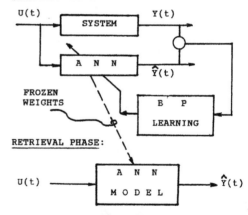

FIG. 5 BACK PROPAGATION LEARNING ARCHITECTURE

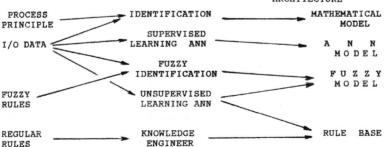

FIG. 4 THE CONVERSION OF DIFFERENT TYPE OF KNOWLEDGE

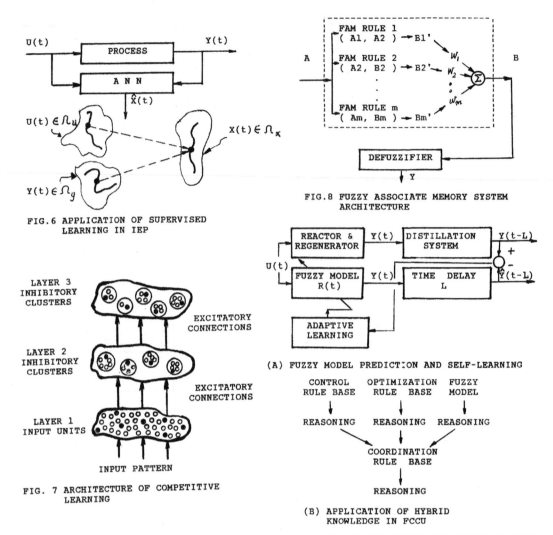

FIG.6 APPLICATION OF SUPERVISED
LEARNING IN IEP

FIG. 7 ARCHITECTURE OF COMPETITIVE
LEARNING

FIG.8 FUZZY ASSOCIATE MEMORY SYSTEM
ARCHITECTURE

(A) FUZZY MODEL PREDICTION AND SELF-LEARNING

(B) APPLICATION OF HYBRID
KNOWLEDGE IN FCCU

FIG.9 FUZZY MODEL PREDICTION AND EXPERT CONTROL OF FCCU

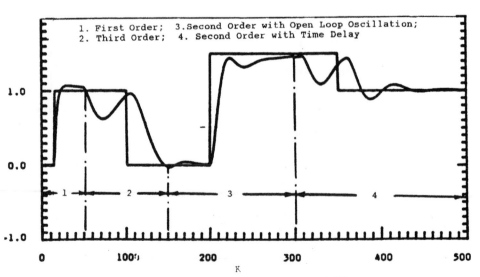

1. First Order;   3.Second Order with Open Loop Oscillation;
2. Third Order;   4. Second Order with Time Delay

FIG.10 THE DYNAMIC RESPONSE OF ANN CONTROLLER
FOR PLANT WITH VARIABLE STRUCTURES

# INTELLIGENT CONTROL: THEORY AND APPLICATIONS

**A.J. Krijgsman\*, H.B. Verbruggen\* and M.G. Rodd\*\***

*\*Delft University of Technology, Department of Electrical Engineering, Control Laboratory, P.O. Box 5031,*
*2600 GA Delft, The Netherlands*
*\*\*University of Wales, Swansea, Department of Electrical and Electronic Engineering, Singleton Park, Swansea, UK*

## 1   Introduction to Intelligent control

There is a vast, growing interest in the application of Artificial Intelligence methods in all fields of engineering. However, the real-time demands imposed by in-line control applications add an extra dimension to the problems arising when using AI methods in this field. This paper concentrates on in-line and on-line (supervisory) control applications and highlights the differences between the methods described, called "intelligent methods", and the methods previously applied in solving control problems.

A number of names have been introduced to describe control systems in which control strategies based on AI techniques have been introduced: intelligent control, autonomous control, heuristic control, etc. Why are these systems denoted by the controversial name: "intelligent"? What is the difference between these methods and conventional or more advanced and sophisticated controllers which have been applied in control engineering applications in the past? It seems rather presumptuous to call these systems intelligent after many decades have been spent on developing sophisticated control algorithms based on solid theoretical frameworks, such as linear systems, optimal and stochastic control systems and their extensions, such as adaptive and robust control systems.

There are many way to solve problems depending on the amount of knowledge available, the kind of knowledge available, the theoretical background, experience, personal preference, etc. After a pioneering phase in which simple process models, operator's experience and controllers with predetermined structure dominated the scene, control engineering was dominated for decades by the developments of linear (optimal) control theory. The need for mathematical models of the processes to be controlled stimulated research in fields such as system identification, parameter estimation and test signal generation. By formulating the requirements imposed on the controlled processes in a mathematical form (in terms of performance criteria or cost functions) it was possible to derive many control algorithms of whose parameters are analytical functions of the process model and the parameters of the cost function. Thus, a closely coupled design procedure was developed which automatically generated the desired control algorithm (structure, parameters). Knowledge about the process and intelligence of the designer woulde be completely captured in the off-line design procedure.

In a case where the process cannot be described by linear models and/or the requirements are not translated to simple criteria and quadratic cost functions, no analytcal solution can be found, and the design is translated to a numerical optimisation problem. There is a free choice of the structure and the number of parameters of the control algorithm to be determined. This procedure can be called a "loosely coupled" design procedure. There are more degrees of freedom in the design phase, and experiments on simulated processes support the designer in choosing the right controller structure and parameters. In this case mathematical models are also required, but knowledge and experience of the designer are necessary in the final stage of the design procedure as well based on extensive simulations.

A very natural approach, gaining more impetus in control engineering applications is the so-called "model-based predictive control method", in which the design procedure is based on:

- the prediction of the process output over a certain prediction horizon, based on a model of the process or on previous in- and output signals

- a desired process output behaviour, often called the "reference trajectory", over the same prediction horizon

- the minimisation of a cost function, yielding a control signal (process input), satisfying constraints on magnitude as well as on rate of change of the control signal and other signals related to the process. The control signal is calculated over a so-called "control horizon" which is smaller then the prediction horizon.

- only the control signal for the next sampling instant is applied, after which the whole procedure is reiterated.

The complete uncoupling of the different steps allows the application of all kinds of process models, reference trajectories and cost functions, and allows the inclusion of constraints. Due to these properties the application of some AI techniques such as artificial neural nets fits this approach quite naturally.

In the previous approaches, reasoning about the process behaviour and the desired overall behaviour is based on the formalism of mathematical descriptions. The level of the power of abstraction is rather high. In case where the mathematical model of the process is not available or can only be obtained with great effort and cost, the control strategy should be based on a completely different approach. Instead of a mathematical, a behavorial description of the process is needed, based on qualitative expressions and experience of people working with the process. Actions can be performed either as the result of evaluating rules (reasoning) or as unconscious actions based on presented process behaviour after a learning phase. Intelligence comes in as the capability to reason about facts and rules and to learn about presented behaviour. It opens up the possibility of applying the experience gathered by operators and process and control engineers. Uncertainty about the knowledge can handled, as well as ignorance about the structure of the system.

Most interesting is the case in which during operation new facts and rules are discovered and implemented, and knowledge is adapted to changing circumstances.
We are talking about a completely different approach in which the attainments of control theory (stability analysis, observability, controllability, etc) do not apply. That is the reason that many people in the control theory community are not convinced of the possible applications of these methods. Other people, confronted with difficult to describe or partly unknown processes, are willing to apply these methods, because experience based on real process behaviour can be implemented, regardless of whether the system is linear, non-linear, time-variant, etc. In those cases control theory can hardly provide them with adequate solutions.

In this paper a motivation is given to show why the methods based on control theory developed thusfar are not able to handle all real-life problems, even when they are enhanced by adaptation of the controller, or when the design is based on robust control methods.

In the next section it will be demonstrated that in an increasing number of applications we are forced to introduce alternative methods to control these systems. The alternative methods, however, are quite common in the daily lives of human beings.

The following methods are introduced, with emphasis on real-time applications:

- Knowledge-based Systems (KBS), based on expert systems (ES). This is a symbolic reasoning approach, commonly used for advisory systems in which the knowledge of experts in a small field of expertise is made available to the user.

- Artificial Neural Networks (ANN), which are learning systems based on subsymbolic reasoning.

- Cerebellar Model Articulation Control (CMAC) methods which are quite simular to ANNs, and also learn the system behaviour from aplied inputs and measured outputs

- Fuzzy Control (FC), which is very well suited handling heuristic knowledge to control a system

In many cases a combination of methods is most convenient, leading for instance to Fuzzy Expert Systems and Neural-Net-controlled systems supervised by a knowledge-based system. Intelligent control components can be used in various ways. In this paper the following configurations are distinguished:

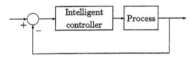

Figure 1: Direct intelligent control

- Direct control configurations (see Fig. 1), in which one of the above-mentioned intelligent control methods is included in the control loop. As in many conventional control configurations, the controller can be built up of different parts, i.e. a control and a process identification part.

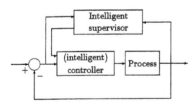

Figure 2: Indirect intelligent control

- Supervisory control configurations (see Fig. 2) in which the controller is either a conventional or an intelligent controller which is supervised by an intelligent component, for example a KBS.

Various examples of systems controlled by different intelligent components will be given. It will be shown that a combination of intelligent components is most usefull, because the methods support each other in many ways and are complementary in various aspects. However, some problems are still unsolved, for instance the stability problem in using the proposed intelligent methods.

Generally, control algorithms use a very fine quantization in order to approach the continuous case. This is most usefull when the system is well known and the requirements are very strict.

However, why should such a fine quantization be used when there is much uncertainty and lack of information about the process to be controlled? Why not using a knowledge-based model with a rough quantization? The rough quantization can easily be translated into symbolic variables and offers an opportunity to incorporate other symbolic knowledge in the controller(for instance, from operators). Besides, accurate control is not always necessary, but is only required in the neighbourhood of the desired behaviour. We are no longer bound to fixed controller strategies for all circumstances in which the process occurs.

What is interesting is that a more interesting "landscape of control" will come into being. How do we define the landscape of control? This landscape can be depicted only for very simple algoritms, but it illustrates very well the many additional possibilities introduced by intelligent controllers.

Assume a control strategy which produces a control output u as a function of e (the error signal between desired and measured process output). In a classical proportional controller every $u$ is calculated with a certainty of 100 %, given a measured or calculated value $e$. With a very fine quantization a thin wall illustrating the

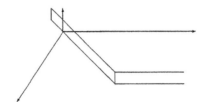

Figure 3: Landscape of control: classical proportional

relationship 1 (elswhere the relationship is zero) between $e$ and $u$ is given (see Fig. 3). Using a nonlinear relation between $e$ and $u$, a curved thin wall will be generated. By the introduction of a KBS, the influence of quantization and precision becomes visible. The relationship between $e$ and $u$ is illustrated by a number of

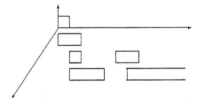

Figure 4: Landscape of control: rule-based

linked blocks with different bases but with the same height (see Fig. 4). Introducing uncertainty by fuzzy sets yields a more interesting landscape: pyramids or topped pyramids which are partly overlapping dominate the scene. Because a number of rules are fired at the same time with different influence on the output signal a nearly continuously changing $u$-signal is obtained despite a roughly quantized $e$-signal. The $u$-signal depends on the defuzzification method used, the shape of the pyramids and the logic being used. More interesting landscapes are created when for different areas of the $e, u$-plane different KBS are introduced, or when the relationship is assumed to be uncertain and depending on gained knowledge. In that case the landscape will change dynamically: some structures in some areas will be emphasized and change their shape, while others will fade away, and new structures will even come into existence!

## 2 The need for Intelligent Control

In the past control theory forced the application of control engineering to adopt a Procrustean approach, where the application is made to conform to the requirements of the control system.

In many cases real life is different, and so are the real life applications in the process and manufacturing industries. It can be stated that modern production and manufacturing methods introduce a number of interesting problems which cannot easily be solved by conventional control methods. Examples are:

- frequent changes in product throughput

- frequent changes in product mix and individualisation of products

- introduction of more advanced and highly complicated production methods

- increases in production and manufacturing speed, using fewer product buffers and flexible, lightweight mechanical constructions

- the introduction of plant-wide control systems which integrate in one system tactical, managerial, scheduling, operational, monitoring, supervision and control tasks.

As a result control problems will be more difficult to solve, because of:

- changing conditions and operating points

- highly nonlinear and time-varying behaviour

- increase of required speeds in relation to the dynamics of the process

- increase of the complexity of the process models (interactions, influence of higher-frequency dynamics)

- the mixture of qualitative and quantitative knowledge

Therefore new methods based on an entirely different approach were required to handle these problems. Artificial Intelligence methods emerged at the same time, and seemed to be able to solve some of the problems mentioned above. A number of trends tendencies caused the growing interest of control engineering people in intelligent control systems based upon AI techniques.

- The restrictions of conventional control methods based on mathematical models and the high emphasis on linearized systems

- The success of operators in controlling a number of practical control problems which are difficult to model, by applying a heuristic approach guided by a long-term experience. This approach yields satisfactory results in most cases results, but is very sensitive to changing conditions, and the know-how of a small number of operators, and has to be build up gradually when new problems arise or new installations are being set up.

- A multi-layered hierarchical information processing system has been defined for plant-wide control, based on various levels of decision making. On each level people are responsible for their part of the automation level. However, their decision should be based, not only on the informatiom available on the level concerned, but also on information provided by the adjacent levels. There is a growing need for concentrated and qualitative information, which is adapted to the people responsible for the decisions on each subsequent level. Intelligent systems, such as expert systems, should provide means to overcome the communication barriers which always seem to exist between different disciplines.

- The field of system analysis, design and simulation has developed considerably over the past decades, and has been concentrated in software packages and their accompanying toolboxes. Two types of knowledge are required from the user: specific knowledge related to a particular software package, and specific knowledge related to a certain analysis or design method. Expert systems can advise people to use these packages in a more-convenient and proper way.

- In practical control algorithms the code for the actual control algorithm plays only a minor role. There is much logic built around the algorithm, taking care of such activities as: switching between manual and automatic operation, bumpless transfer, supervision of automatic tuning and adaptation, etc. Continuous control and sequencing control are merging. A knowledge-based system can be used as an alternative to logic, sequencing and supervision code.

- The selection of the right control configuration and the right controller parameters can be implemented by a knowledge-based system. In addition, it can be advantageous to reconsider the resulting control configuration when the assumptions concerning the process operation no longer hold. Automatic strategy switching can be performed by an intelligent supervisory system.

- The successful proliferation of AI techniques in many engineering applications by the introduction of expert systems and artificial neural nets. Many hard- and software tools are available to facilitate the application of these methods. There are, however, still many problems to be solved when these tools are incorporated in real-time applications ([Rodd 1991]). Fuzzy sets have been successfully introduced to manage uncertainty.

However, as in many other developments taking place over a short time period and introduced as a panacea for solving all types of problems, the control community should also be warned against a too-rash application of these methods. The use and abuse of these methods should be carefully considered, keeping in mind that many of the implicit attainments of control theory are not available in intelligent control. Finally, a well-considered combination of conventional and intelligent methods, should be selected.

# 3  AI contributions to intelligent control

There are two main schools in AI research area, each focusing on a specific element of intelligence: *reasoning* and *learning*.

In the 1960s much research was done to model the human brain using a neuron-based structure. This subsymbolic processing is based on a model of biological neural systems in which a large number of simple processors (neurons) are connected in a highly interconnected network. Information is distributed among these connections. Systems based on these techniques are called "artificial neural systems" or symply "neural networks" and exhibit many of the properties of intelligent behaviour. In control engineering, expertise is based on both symbolic and subsymbolic processing. One of the first results of this research was the Perceptron, introduced by Rosenblatt [Rosenblatt 1961] and the ADALINE introduced by Widrow [Widrow 1960]. In such a control system the controller learns from experience during its operation. Research has been done into several learning schemes. In fact these systems were early attempts towards adaptive control, which has been a main research topic during the last 20 years.

The rule-based approach in control has been introduced by Mamdani [Mamdani 1975], using fuzzy logic in combination with rules for controlling cement kilns. The area of fuzzy control is based upon the work of Zadeh [Zadeh 1965]. In spite of the work of Zadeh not much attention was paid to this application of AI in the 1970s. This situation lasted till the beginning of the 1980s. Modern computer technology led to a renewed revival of AI research, and to the introduction of expert systems. These expert systems stimulated research into the application of these systems, including control applications.

## 3.1  Knowledge

The kernel of all intelligent or AI based systems is that they are designed to handle **knowledge**. Knowledge can be seen as a collection of facts, data, theory, beliefs etc. Knowledge-based development systems are developed by the symbolic AI community to offer the user built-in reasoning and explanation techniques to handle the knowledge provided by the user. This knowledge can be described in various ways. Over the past 25 years, numerous representational schemes have been proposed and implemented, each with its own strengths and weaknesses. They can be classified into four categories:

1. **Logical representation schemes.** This class of representations uses expressions in formal logic to represent a knowledge base. Inference rules and proof procedures apply this knowledge to problem instances. First-order predicate calculus is the most widely used logical representation scheme, but it is only one of a number of logical representations. PROLOG is an ideal programming language for implementing logical representation schemes.

2. **Procedural representation schemes.** Procedural schemes represent knowledge as a set of instructions for solving a problem. This contrasts with the declarative representations provided by logic and semantic networks. In a rule-based system, for example, an IF ... THEN rule may be interpreted as a procedure for solving a goal in a problem domain: to solve the conclusion, solve the premises in order. Production systems are examples procedural representation schemes.

3. **Network representation schemes.** Network representations capture knowledge as a graph in which the nodes represent objects or concepts in the problem domain and the links represent relations or associations between them. Examples of network representations are *semantic networks*, *conceptual dependencies*, and *conceptual graphs*.

4. **Structured representation schemes.** Structured representation languages extend networks by allowing each node to be a complex data structure consisting of named slots with attached values. These values may be simple numeric or symbolic data, pointers to other frames, or even procedures for performing a particular task. Examples of structured representations include *scripts*, *frames* and *objects*.

The choice for a specific knowledge description is highly-application dependent.

A very serious problem in the development and use of knowledge-based systems is the problem of knowledge acquistion. how to capture the knowledge for the system in an easy, fast and reliable way. Many techniques have been proposed in the last decade (interviewing techniques, structured knowledge analysis, etc.) These methods are mainly for expert systems. A very serious problem in the application of knowledge-based and expert systems exists if no specific expert is available to provide the knowledge. What happens when the system has to gather, refine or tune its own knowledge? This problem has been one of the key issues of AI research in subsymbolic reasoning: to develop learning systems.

## 3.2  Learning systems

The ability to learn must be a part of any system that exhibits intelligence. Feigenbaum (1983) has called the "knowledge engineering bottleneck" the major obstacle to the widespread use of expert systems, or knowledge-based systems in general. It refers to the cost and difficulty of building expert systems by knowledge engineers and domain experts. The solution to this problem would be to develop programs which start with a minimal amount of knowledge and learn from their own experience, from human advice, or from planned experiments. [Simon 1983] defines learning as: *any change in a system that allows it to perform better the second time on repetition of the same task or on another task drawn from the same population.*

Simon's definition of learning is very general. It covers various approaches to learning such as adaptive systems and artificial

neural networks, which will be described in the next section. The process of (structured) knowledge acquisition is also a part of learning. This kind of learning is done by human beings, in contrast to *machine learning*.

Memory plays an important role in this area of learning. The learning of a human being is strongly correlated to memorizing history. Experiences from previous events and actions can (and must) be used to improve control actions in the future. Learning from experience is therefore one of the main reasons for research into learning components like neural networks and associative memories.

## 3.3  Certainty and uncertainty

Expert-system technology has evolved from the logic-based theories in mathematics. The most important theory behind this is the Boolean algebra which uses two values for a variable in the system (true or false). This approach requires a strict and complete description of the system and its knowledge. In a practical situation, however, we are not only dealing with exact and precise knowledge, but also with inexact and inprecise knowledge. Many attempt have been made to deal with a kind of uncertainty management in knowledge-based systems. The following methods have been introduced:

- Bayes. This approach is based on the statistical approach, in which conditional probabilities are used to deduce the probablity of the related knowledge. To use this method all conditional probabilities must be known in advance. In a practical expert system this is impossible.

- Certainty factors. This approach was introduced in the Mycin expert system. These so-called $CF$ factors are values between $-1$ (false) and 1 (true). These factors are used to deduce the certainty factor of dependent knowledge. This certainty facter theory has no real mathematical background, but has proved to be useful in the MYCIN system. The main problem in using this approach is to the deduction of the certainty factors, and the fact that the certainty factors of deduced facts tend towards very small values for deep evaluations in a decision tree. Therefore, not many applications are found using this methodology nowadays.

- Fuzzy logic. This approach has been introduced as an extension to 'normal' conventional set theory. The degree of membership of a specific set is defined by a membership function $\mu$. Such a function has values in the range $[0, 1]$. Every variable is coupled to a fuzzy value via such membership functions. The fuzzy reasoning for the functions AND and OR can be implemented in various ways. In table 1 several

| | AND | OR |
|---|---|---|
| Lukasiewicz | $\max(p + q - 1, 0)$ | $\min(p + q, 1)$ |
| Probabilistic | $pq$ | $p + q - pq$ |
| Zadeh | $\min(p, q)$ | $\max(p, q)$ |

Table 1: *Possible implementations of AND- and OR-operation.*

of these implementations are summarized.

In the AI community there is no consensus on which method is the best to express uncertainty in the knowledge used for reasoning, but the trend nowadays is towards the use of fuzzy logic. Many applications have been developed in which fuzzy logic is successfully used to express and handle the uncertainty in a system.

## 3.4  Stability and convergence

Stability in a control system is one of the most widely accepted requirements, besides requirements for the performance. The stability issue is one of the most difficult problems to solve in intelligent control. Because expert systems and subsymbolic reasoning systems do not fit into the conventional (mathematical) description of a controller, it is seldom possible to guarantee stability in the conventional way.

## 3.5 Real-time issues

Artificial Intelligence methods have been tested and applied in many areas in which time was not the most important factor. For control systems however time plays a very important role. In those systems predictable and guaranteed response times are of crucial importance. In a dynamic system, data is not infinitely valid, or is only valid during a very short period. Intelligent systems functioning in a real-time control environment need special features to cope with this real-time problem. This issue is extensively treated in section 5.

## 4 Intelligent control components

In the application of intelligent control it is important to develop standard components for use in multilayered control systems using AI techniques. Components, based on different AI techniques are therefore investigated for their applicability for control purposes.

### 4.1 Expert systems

Expert-system technology is one of the fascinating possibilities to extend normal control theory. All kind of heuristic knowledge can be put into a knowledge base and mixed with more conventional decision and calculation mechanisms. The possibilities at the implementation level are very close to those of the subject of the next section on fuzzy control, except that in expert-system technology a wider range of knowledge representation is available (frames, networks etc.). These representation techniques are very well suited to expressing knowledge at a very high level like the supervision of advanced control mechanism. In [Krijgsman 1991a] an example is given of such a technique used to supervise a predictive control algorithm in an adequate way using advanced knowledge representation methods.

### 4.2 Fuzzy Control

Application of fuzzy set theory in control is gaining more and more attention from industry. An application of fuzzy control in a cement kiln was probably the first application of fuzzy control.

The theoretical approach of fuzzy inferencing is more based on the idea of relations rather than on the idea of implications. For example, a fuzzy controller is in theory a function of more than one variable, which is a combination (union) of the functions representing the individual fuzzy rules. The "direction" of inference is defined by the form of the fuzzy rules. A typical rule $'i'$ can be stated as:

$$\text{IF } x_1 \text{ is } X_1^i \text{ AND } \ldots \text{AND } x_N \text{ is } X_N^i \text{ THEN } y \text{ is } Y^i$$

in which $x_i$ is the input variable and $X_i$ is the fuzzy set the variable is mapped on. The result, however, describes a relation represented by a multi-dimensional function (a function with range $[0,1]$ of $(N+1)$ variables):

$$R(x_1, \ldots, x_N, y) = \bigvee_i R_i(x_1, \ldots, x_N, y) \qquad (1)$$

where

$$R_i(x_1, \ldots, x_N, y) = \bigwedge(\bigwedge(\mu_{X_1^i}(x_1), \ldots, \mu_{X_N^i}(x_N)), \mu_{Y^i}(y)) \quad (2)$$

Figure 5: *Fuzzy inference.*

and as can be seen from (1) and (2) the resulting fuzzy system no longer has a 'direction': it is a relation. This is why, for example, fuzzy models can be used for predictions as well as "inverse" control [Brown 1991].

By applying the "compositional rule of inference" using a fuzzification of the inputs, which is in fact also a relation presented by a function of $(N - 1)$ variables, a fuzzy output is obtained. In figure 5 this fuzzy inference is shown schematically. In practical applications however, a more efficient method is usually used [Harris 1989]. This method is based on the assumption that the fuzzification of a input is represented by a singleton. The application of the compositional rule of inference to relation, using the fuzzified input, results in a fuzzy output. The fuzzified input is represented by the union of all the singletons describing the individual (numerical) inputs. The same fuzzy output can be obtained by determining for every fuzzy rule $'i'$ the individual fuzzy output

$$\mu_Y^i(y) = \bigwedge(\bigwedge(\mu_{X_1^i}(x_1(t)), \ldots, \mu_{X_N^i}(x_N(t))), \mu_{Y^i}(y)). \qquad (3)$$

Combination of those individual fuzzy outputs by applying a union results in the complete fuzzy output of the system

$$\mu_Y(y) = \bigvee_i \mu_Y^i(y). \qquad (4)$$

Note that the inputs in (3) are represented in a numerical way $(\mu_{X_j^i}(x_j(t)))$ and the output is represented in a fuzzy way $(\mu_{Y^i}(y))$. Several defuzzification methods are available to translate the fuzzy output into a crisp, numerical output value. Based on this practical method of applying fuzzy rules and inference, a generalization of the inference mechanism can be made allowing fuzzy rules and inference to be applied in the way that rules and inference mechanisms are used in 'conventional' expert systems.

### Example of fuzzy expert control

Two control strategies are used within the rule-based controller: *direct expert control* (d.e.c.) and *direct reference expert control* (d.r.e.c.)

In **d.e.c.** the control signal depends directly on the classification of the measured state into one or more symbolic states. This state classification is a translation of a measured numerical state (error $e[kT_s]$ and error change $\Delta e[kT_s]$) into a symbolic state, which can be used within a rule base. Using a fuzzy state classification, one, two or at most four symbolic states can possibly be (partly) 'true', assuming that no more than two fuzzy sets per input (error or error change) overlap. Rules in this strategyrt of the controller are of the form:

> **IF** *error is big positive*
> **AND** *error change is small negative*
> **THEN** *change control signal medium positive*

In **d.r.e.c,** the control signal depends on a symbolic classification of a *predicted* behaviour, compared with a desired reference behaviour. Rules in this strategy are of the form:

> **IF** *predicted behaviour is too fast*
> **THEN** *change control signal big negative*

The inputs for this **d.r.e.c.** module are the angle between the reference and the predicted curve in the phase plane and a region around the present state in which both the desired and predicted next state are confined or not (alertness).

For improvement of the performance of the controller two types of adaptation are available: *in-line adaptation* and *on-line adaptation*:

**in-line adaptations** can become active during each sampling interval. They consist of a zoom mechanism, to improve steady-state control, a mechanism to adapt the membership function used for classification of the predicted behaviour, and a mechanism to achieve a constant steady-state value for the control signal. To achieve this constant value a scaling of the range of the control signal change is performed: every time the set point is crossed the scaling is decreased. Using this mechanism a limit cycle will be damped.

**on-line adaptations** in fact fulfils a supervisory function within the rule-based controller. Decisions made by rules in this section are based upon the overall behaviour of the system and should therefore extend over a time related to several times the slowest time constant of the system. The on-line adaptations are necessary to improve the overall performance of the controller, and detections and classifications are made of such performance indicators as the overshoot, undershoot, rise time and settling time.

The complete rule-based controller consists of several fuzzy sub-controllers, as shown schematically in figure 6. These controllers

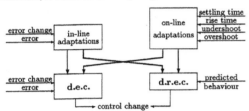

Figure 6: *Schematic representation of rule-based controller.*

perform on a time scale equal to the sampling time (d.e.c., d.r.e.c. and in-line adaptations) or on a larger time scale (on-line adaptations). The inputs as well as the outputs can differ for each sub-controller. The rule-based controller presented is able to control various processes even though no a priori knowledge about the actual process model is required. No restrictions are put on linearity, order, non-minimum phase character, delay time, etc. More details can be found in [Jager 1991].

**Simulation results**
To test the robustness of the rule-based controllers, several simulations were performed. In figure 7 the basic process used for the

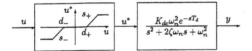

Figure 7: *Second-order process used in simulations.*

simulations is shown. In figure 8 the results are given of two of the many simulations that were performed to test the difference in performance between a Boolean rule-based controller and the fuzzy one. One can see that before adaptation of the controller(s) the fuzzy version (b) performs better than the Boolean version (a). After adaptation both versions perform satisfactorily.

### 4.3 Neural networks

Artificial neural networks are very appropriate tools for modeling and control purposes. The ability to represent an arbitrary nonlinear mapping is an especially interesting feature for complex nonlinear control problems. ([Cybenko 1989], [Cotter 1990],

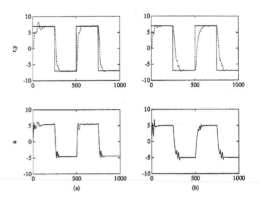

(a)                              (b)

Figure 8: *Simulation results with boolean (a) and fuzzy (b) version of rule-based controller. (Process: $d_- = -1$, $d_+ = 2$, $s_- = 5$, $s_+ = -4$, $K_{dc} = 2$, $\zeta = 0.7$, $\omega n = 0.14$ and $T_d = 0s$. Controller: $T_s = 1s$, $r_{min} = -10$, $r_{max} = 10$, $u_{min} = -10$ and $u_{max} = 10$*

[Billings 1992]).
The most relevant features for control concerning neural networks are:

(i) arbitrary nonlinear relations can be represented

(i) learning and adaptation in unknown and uncertain systems can be provided by weight adaptation

(iii) information is transformed into an internal (weight) representation allowing for data fusion of both quantitative and qualitative signals

(iv) because of the parallel architecture for data processing these methods can be used in large-scale dynamic systems.

(v) the system is fault-tolerant so it provides a degree of robustness.

**Modelling using neural networks**
Modelling dynamic systems using neural networks needs a special configuration in which information about the history of the input and output of the plant is used. In figure 9 the general identi-

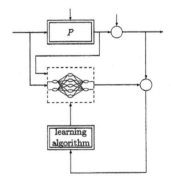

Figure 9: Plant identification

fication structure using neural networks is depicted. Of course many types of neural networks are suitable for use this configuration. The general objective of this configuration is to describe a general analytic system $\Sigma$ with a nonlinear difference equation:

$$\Sigma : y(k+1) = f(y(k), \cdots, y(k-n); u(k), \cdots, u(k-m)) \quad (5)$$

in which $n + 1$ is the order of the system, with $m \leq n$. If insufficient time-lagged values of $u(t)$ and $y(t)$ are assigned as inputs to the network, it cannot generate the necessary dynamic terms and this results in a poor model fit, especially in those cases where the system to be modelled is completely unknown. Therefore we have to use model validation techniques to be sure that the model developed is acceptable and accurate enough to be used in a control situation in which the control action depend on the model. Generally it is 'easy' (although time consuming) to train a neural network which predicts efficiently over the estimation set, but this does not mean that the network provides us with a correct and adequate description of the system. It is difficult to get analytical results but in [Billings 1992] some model validity tests are introduced which were proved to be useful experimentally.

### Neural control

Using a neural network as a general 'black box' control component is extremely interesting. There are several possibilities for neural control, which can all be thought of as the implementation of internal model control (figure 10), in which a model $M$ is used to control plant $P$ using a controller $C$. The structure is completed with a filter $F$.

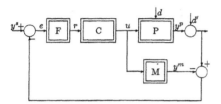

Figure 10: Internal Model Control

- **inverse control**: the model of the plant to be controlled is used to train the controller in a so-called "specialized learning architecture". The model, obtained by observation, is used to generate error corrections for the controller. Under special conditions this kind of control can be guaranteed to be stable ([Billings 1992]): the system has to be invertible and monotonic. In the next section on CMAC an example of this kind of control using the inverse of a system is shown.

- **model-based control**: the model of the plant to be controlled is used in a conventional (e.g. predictive) control scheme to calculate the proper control actions. One of the possibilities is to implement an iterative method to calculate the inverse. Other possibilities are the use of a neural model as the basis of an on-line optimization routine to calculate the optimal control action.
  A special situation exists when the controller is also a neural network, while the neural model (emulator) is used to generate the Jacobian of the process. This gradient information is then used to adapt the controller. The method is known as the "error backpropagation" method [Narendra 1990], [Narendra 1991].

- **reinforcement control**: no neural model is derived and used but the controller is adapted according to a criterion function which evaluates the desired behaviour.

The effect of this kind of control is that limitations concerning the linearity of the system to be controlled are released. The choice of the best approach, and determining when the training procedure is completed are the main issues of concern.
In the section about CMAC an example is given of neural-based control using an inverse model of the system to be controlled. Examples of control based on error backpropagation can be found in [Hertog 1991].

## 4.4 CMAC

The CMAC (Cerebellar Model Articulation Controller) [Albus 1975a] algorithm is based on the functioning of the small brain (cerebellum), which is responsible for the coordination of our limbs. Albus originally designed the algorithm for controlling robot systems. It is essentially based on lookup table techniques and has the ability to learn vector functions by storing data from the past in a smart way. The basic form of CMAC is directly useful to model static functions. However in control engineering one is mainly interested in controlling and modelling dynamic functions. To give CMAC dynamic characteristics the basic form has to be extended, either by adding additional input information related to previous in- and outputs, or by inherently feeding back the output.

CMAC is a perceptron-like lookup table technique. Normal lookup table techniques transform an input vector into a pointer, which is used to index a table. With this pointer the input vector is uniquely coupled to a weight (table location). A function of any shape can be learned because the function is stored numerically, so linearity is out of the question.

### The construction of the CMAC algorithm

The CMAC algorithm can be characterized by three important features, which are extensions to the normal table lookup methods. These important extensions are: distributed storage, the generalization algorithm and a random mapping.

### Distributed storage and the generalization algorithm

In control engineering many functions are continuous and rather smooth. Therefore, correlated input vectors will produce similar outputs. Distributed storage combined with the generalization algorithm will give CMAC the property called generalization, which implies that output values of correlated input vectors are correlated. Distributed storage means that the output value of

Figure 11: Illustration of (a) distributed storage and (b) generalization in CMAC

CMAC is stored as the sum of a number of weights. In this case the indexing mechanism has to generate the same number of pointers for each input. Figure 11 shows the effect of distributed storage. The input vector $S_1$ results in the selection of four weights.

The generalization algorithm uses the distributed storage feature in a smart way: correlated input vectors address many weights in common. Figure 11 reflects the effect of generalization by using distributed storage. An arbitrary input vector $S_2$ in the neighbourhood of $S_1$ shares three out of four weights with $S_1$. Thus, the output values generated by $S_1$ and $S_2$ are strongly correlated. A nice feature of the generalization algorithm is that it will work for arbitrary dimensions of the input vector.

### Random mapping

Because of an unknown range of a component of the input vector, a complex shape of the input space or a sparse quantized input space, it is very difficult to construct an appropriate indexing mechanism. These problems often result in very sparsely filled tables. The CMAC algorithm deals with this problem in a very elegant way. The generalization algorithm generates for every quantized input set vector a unique set of pointers. These pointers are n-dimensional vectors pointing in a virtual table space. CMAC uses a random mapping to map these vectors into the real linear table space. The essence of random mapping is similar to a hash function in efficient symbol table addressing techniques.

Using a random mapping implies the occurrence of mapping collisions, which gives rise to undesired generalization. However, because distributed storage is used, a collision results only in a minor error in the output value. The major advantage of using random mapping is its flexibility, because irrespective of signal ranges and the shape and sparseness of the input space, the size of the table can be chosen related to the number of reachable input vectors. In figure 12 all calculation steps have been depicted.

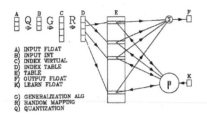

A) INPUT FLOAT
B) INPUT INT
C) INDEX VIRTUAL
D) INDEX TABLE
E) TABLE
F) OUTPUT FLOAT
K) LEARN FLOAT

G) GENERALIZATION ALG
R) RANDOM MAPPING
Q) QUANTIZATION

Figure 12: CMAC calculation steps

First, the weights must be addressed by pointers. In figure 12, vector $A$ is the 3-dimensional floating point input vector, which is supplied by the user. This vector is quantized into an integer vector $B$ by operation $Q$ (quantization). Vector $B$ is then processed by the generalization algorithm $G$ which generates a pointer $C$ in the virtual table space. Next, pointer vector $C$ is scrambled by the random mapping $R$ into vector $D$, pointing to the real linear table space of table $E$. Finally, the weights can be accessed in two ways:

- a retrieve operation calculates an output value $F$ by simply summing the selected weights.

- a storage operation is possible which adapts the weights according to the desired output value in vector $K$. This adaptation (learning) is done in a cautious way using a Widrow-Hoff type of adaptation rule (the delta learning rule [Widrow 1960]).

The main properties of CMAC are:

- Input quantization which adds quantizing noise to the inputs. To reduce this noise, the quantization interval must be chosen to be as small as possible. Because of the limited size of the table, the quantization interval should not be made too small.

- Generalization which correlates outputs of nearby input vectors. Generalization is strongly related to interpolation and smoothing. It reduces the required table size and, in contrast with the lookup table, it is not necessary that all possible input vectors in the input space have to be learned.

- By random mapping an arbitrary vector in the input space is mapped into the linear table space without problems.

- Using the storing concept, CMAC is able to learn. However, because function values are simply stored, CMAC can only generate a reasonable output in an area where learning has been performed.

To use CMAC a (small) number of parameters have to be set. Some important parameters are:

- Input resolution. The input resolution (= quantization interval) affects the amount of quantization noise, the table size and the generalization distance in the input space. The generalization distance is the distance between two input vectors where correlation vanishes.

- Generalization number. The generalization number is the number of weights which are addressed simultaneously. It affects the generalization distance and the speed of the CMAC algorithm.

- The size of the CMAC table. The number of weights in the CMAC table can not be chosen arbitrary. The size of the table is directly related to the number of reachable vectors in the input space. The physical size of the table is determined by the hashing scheme and the number of collisions that is still acceptable. Therefore, a more practical lower bound of the table size can be obtained.

- The learning factor. This affects the speed at which the weights are changed to meet the desired output. The choice for this parameter is a trade-off between distortion of the surrounding output values and speed of learning (adaptation, initialisation).

**Example using CMAC as a neural controller**

In this example CMAC is shown as a neural controller device for a nonlinear model. The configuration in which the controller is trained is based on the internal model control structure presented above. The system to be controlled is a nonlinear model of the pH value in a fluid. First of all a model is derived via an identification procedure. After that an inverse control configuration is set up. In figure 13 the result of this nonlinear inverse

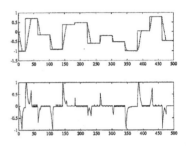

Figure 13: CMAC pH level control

controller is depicted. The behaviour of the controller depends on the settings of the CMAC module, especially the quantization interval.

## 5  Real-time issues

In a time-critical situation, extra demands are necessary when using intelligent systems in the lower levels of automation: the supervisory and control levels. In these layers the expertise of process operators and control system designers should be integrated with the time-varying information obtained directly from the process by measurements.

Real-time behaviour is often easier to recognize than to define. As discussed by [O'Reilly 1986] many definitions of real time exist, mostly related to "fast"; meaning that a system processes data quickly. In [O'Reilly 1986] a formal definition of real time is offered:

- *a hard real-time system is defined as a system in which correctness of the system not only depends on the logical results of a computation, but also on the time at which the results are produced.*

The most important item is the response time, When events are not handled in a timely fashion, the process can get out of control. Thus, the feature that defines a real-time system is the system's ability to guarantee a response before a certain time has elapsed, where that time is related to the dynamic behaviour of the system. The system is said to be real time if, given an arbitrary event or state of the system, the system always produces a response by the time it is needed.

## Real-time issues in expert systems

In a time-critical system the problem must be solved in real time, including the knowledge-based parts. Because the data is non static and changing as a function of time, interesting problems arise:

- **Nonmonotonicity:** incoming data does not remain constant during the complete run of the system. The data is either not durable or loses its validity during the run (because of external events). All dependent conclusions would then have to be removed.

- **Temporal reasoning:** time is a very important variable in real-time domains. A real-time system needs to reason about past, present and future events within a certain time-slot.

- **Interfacing to external software:** a real-time system must be integrated with conventional software for signal processing and application-specific I/O.

- **Asynchronous events:** a real-time system must be capable of being interrupted to accept inputs from unscheduled or asynchronous events.

- **Focus of attention:** when something serious happens the system must be able to change its focus of attention. The system will be divided into several knowledge sources, each with its own area of attention. The system must be able to focus very quickly on a specific knowledge source without adversely affecting the behaviour of the interrupted tasks.

In general expert-system shells available for diagnosis and advisory purposes, no features are available for temporal reasoning or the other issues mentioned above.

In real-time control an important parameter is the sampling period which has to be chosen. The minimum sampling period which can be used is defined by the amount of time needed by the system to make the necessary calculations. *Progressive reasoning* is an appropriate tool in a real-time expert system [Lattimer 1986]. The idea behind progressive reasoning is that the reasoning is divided into several subsequent sections go into more detail about the problem.

The progressive reasoning principle is very well suited to process-control purposes. With the application of a direct digital controller using AI techniques, a problem arises when the sampling period is chosen smaller than the time needed to evaluate the rule base. This will cause a situation where the inference engine cannot provide a hypotheses. Since each hypothesis is indirectly coupled to a control signal the expert system will not be able to control the process. To solve this problem, the rule base is divided into a set of different knowledge layers, each with its own set of hypotheses and therefore its own set of control signals. The minimum length of the sampling period which can be used is now defined by the time needed to evaluate the first knowledge layer. This will generate a fairly rough conclusion about the control signal. The next layers are used to produce better control signals using more information about the process to be controlled. In a typical example using this progressive reasoning principle the lowest (fastest) layers of the system are used to implement a direct expert controller, while the next layers are used for supervisory and adaptive control.

Several attempts have been made to develop expert-system shells which are able to deal with the real-time issues. G2[1] is probably the best known commercially available real-time expert system shell, to be used at the level at which response times of seconds are required. For higher speeds more dedicated solutions have to be used like DICE (**D**elft **I**ntelligent **C**ontrol **E**nvironment) [Krijgsman 1991b].

DICE is a real-time environment which consists of the following major parts: three expert system kernels ($ES_1$, $ES_2$ and $ES_3$), a

[1]G2 is trademark of the Gensym Corporation

blackboard for data storage and other global information, and a server which controls the information stream between programs running on other computers in the network. In figure 14 the functional scheme of DICE is given. The DICE environment has

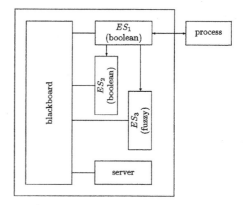

Figure 14: *Overall structure of DICE.*

been completed with an operator task to give commands to the expert-system kernels. A display task can provide the user with graphical output. DICE provides the user with the capability to describe the problem he wants to implement in a so-called "task file". This file describes which rules have to be compiled into a specific expert system kernel, also the rules which will activate a specific user-written routine, the variables which have to be displayed and the variables which are to be depicted by the display task. This task file compilation results in the creation of three expert-system tasks, an operator task and a display task. The rule-base compilation is implemented for reasons of speed and efficiency.

## Real-time issues in neural computing

Neural networks are very simple processing units. As a result of the highly parallel structure, several hardware implementations of neural network architectures have been developed and are currently the subject of research. The big advantage of these methods is the relatively simple calculations which have to be performed. The calculation effort in a feedforward and recall situation is rather low and high speeds are possible. Training takes a lot of time in evaluating the examples to be processed, but in many cases this can be done in an off-line situation.

Nowadays, a number of commercially available tools for neural computing are available. These tools are mainly used in an off-line situation, providing the user with high-level graphical interfaces (NWorks for example). Some hardware boards are available, in which neural-net configurations can be run at very high speeds. These hardware boards are currently very expensive and not widespread.

The CMAC algorithm shown in the previous section is a not particulary computationally intensive algorithm, but memory consumption is rather high. With the current move towards very cheap memories this methods becomes more and more appropriate. Hardware implementations of the algorithm could considerably speed up the calculation times.

## Fuzzy control hardware

The use of fuzzy control modules in real time gives rise to the same problems as in an expert-system environment. For fuzzy control purposes many calculations using fuzzy membership functions have to be carried out. Some dedicated hardware and software (e.g. Togai) is currently available to speed up calculation times.

# 6  Comparison

In the previous sections, intelligent control configurations and intelligent control components are introduced and described. Each component has its own strengths and weaknesses. According to the available knowledge (theoretical, empirical, experimental...) a proper choice for the use of a specific intelligent control module must be made.

In general there are a number of criteria which influence the choice of an intelligent control component:

- knowledge description
- ability to adapt
- explanation facilities
- quantization desired
- ...

According to these items a comparison can be made which is summarized in table 2 Expert systems and fuzzy-control compo-

| Features | Symbolic AI | Neural networks |
|---|---|---|
| variables | logic | analog |
| processing | sequential | parallel |
| | discrete | continuous |
| knowledge | programmation | adaptation |
| level | high | low |
| knowledge | explicit formulation | no formulation |
| learning | learning deals with rules | numerical learning by synaptic weights |
| explanation | explanation facilities | no explanation at all |

Table 2: Comparison of AI methods

nents are summarized as symbolic AI, while the other intelligent control component are indicated as neural networks. Of course we must be aware that this is a very strict separation wheras many overlapping situations may occur in practice.

In cases where we are dealing with problems for which experts are available, their knowledge (e.g. heuristics) can be implemented using symbolic tools like expert systems. But in a case where no real expert is available and knowledge about the system to be controlled has to be collected by the system itself, other tools are more appropriate: neural nets or fuzzy components.

# 7  Concluding remarks - future research

In the introduction to this paper it has already been mentioned that the need for intelligent control arises from plant-wide control approaches with varying operating conditions. Plant-wide control mechanisms require numerous control loops at several levels, which are interconnected. A decision at one level affects the behaviour of another level. To handle the information flow between several levels of automation and to reason about the effects of decisions several AI technologies can be used. In particular, expert-system technology has been developed for the purpose of symbolic data handling.

At the lower levels of automation a higher precision of the mechanisms used is required. Fuzzy control gives an attempt to achieve this higher level of precision, while neural networks are based on precise numerical data. Ther is therefore a tendency to use neural networks and memory elements at the lowest levels of control, while at higher levels symbolic AI techniques become more appropriate.

The development of intelligent control systems requires considerable research into suitable methods for use at a specific level. In [Wang 1990] it is postulated that the general conformation of an intelligent control scheme should meet the IPDI principle: Increased Precision with Decreased Intelligence. This principle says that at the higher levels in an intelligent control scheme the precision is not as high as at the lowest levels of control, but the intelligence level is higher.

# References

[Albus 1975a]  Albus J.S. (1975). *A New Approach to Manipulator Control: The Cerebellar Model Articulation Controller (CMAC)*. Transactions of the ASME, pp. 220-227, September.

[Billings 1992]  Billings S.A., H.B. Jamaluddin and S. Chen (1992). *Properties of neural networks with applications to modelling nonlinear dynamical systems*. International Journal of Control, Vol. 55, No. 1, pp. 193-224.

[Brown 1991]  M. Brown, R. Fraser, C.J. Harris, C.G. Moore. *Intelligent self-organising controllers for autonomous guided vehicles*. Proceedings of IEE Control 91, pp. 134-139, Edinburgh, U.K., March 1991.

[Cotter 1990]  Cotter N.E. (1990). *The Stone-Weierstrass Theorem and Its Application to Neural Networks*. IEEE Transactions on Neural Networks, Vol. 1, No. 4, pp.290-295, December.

[Cybenko 1989]  Cybenko G. (1989). *Approximations by superpositions of a sigmoidal function*. Mathematical Control Signal Systems, Vol. 2, pp. 303-314.

[Jager 1991]  Jager, R., H.B. Verbruggen, P.M. Bruijn and A.J. Krijgsman (1991). Real-time fuzzy expert control. *Proceedings IEE Conference Control 91*, Edinburgh U.K.

[Harris 1989]  C.J. Harris and C.G. Moore. *Intelligent identification and control for autonomous guided vehicles using adaptive fuzzy-based algorithms*. Engineering Applications of Artificial Intelligence, vol. 2, pp. 267-285, December 1989.

[Hertog 1991]  Hertog, P.A. den (1991). *Neural Networks in Control*. Report A91.003 (552), Control Laboratory, Fac. of El. Eng., Delft University of Technology.

[Krijgsman 1991a]  Krijgsman, A.J., H.B. Verbruggen, P.M. Bruijn and M. Wijnhorst (1991). *Knowledge-Based Tuning and Control*. Proceedings of the IFAC Symposium on Intelligent Tuning and Adaptive Control (ITAC 91), Singapore.

[Krijgsman 1991b]  Krijgsman A.J., P.M. Bruijn and H.B. Verbruggen (1991). *DICE: A framework for real-time intelligent control*. 3rd IFAC Workshop on Artificial Intelligence in Real-Time Control, Napa/Sonoma, USA.

[Lattimer 1986]  Lattimer Wright M. and co-workers (1986). An expert system for real-time control. *IEEE Software*, pp. 16-24, March.

[Mamdani 1975]  Mamdani E.H. and S. Assilian (1975). *A fuzzy logic controller for a dynamic plant*. Int. Journal of Man-Machine Studies 7, 1-13.

[Narendra 1990]  Narendra K.S. and K. Parthasarathy (1990). *Identification and Control of Dynamical Systems Using Neural Networks*, IEEE Transactions on Neural Networks, Vol.1, No. 1, March, pp.4-27.

[Narendra 1991]  Narendra K.S. and K. Parthasaraty (1991). *Gradient Methods for the Optimization of Dynmical Systems Containing Neural Networks.* IEEE Transactions on Neural Networks, Vol. 2, No. 2, pp. 252-262, March.

[O'Reilly 1986]  O'Reilly C.A. and A.S. Cromarty (1986). Fast is not "Real Time". in *Designing Effective Real-Time AI Systems.* Applications of Artificial Intelligence II 548, pp. 249-257. Bellingham.

[Rodd 1991]  Rodd M.G. and H.B. Verbruggen (1991). Expert systems in advanced control - Myths, legends and realities. *Proceedings 17th Advanced Control Conference*, Purdue, USA.

[Rosenblatt 1961]  Rosenblatt F. (1961). *Principles of Neurodynamics: Perceptrons and the Theory of Brain Mechanisms.* Spartan Books, Washington DC.

[Wang 1990]  Wang F. and G.N. Saridis (1990) *A Coordination Theory for Intelligent Machines.* Automatica, Vol. 26, No. 5, pp. 833-844.

[Simon 1983]  Simon, H.A. (1983) *Why should machines learn?* In: Michalski et al., Machine learning: An Artificial Intelligence Approach, Palo Alto, CA: Tioga.

[Widrow 1960]  Widrow B. and M.E. Hoff (1960), *Adaptive Switching Circuits,* In *Convention Record, Part 4.,* IRE Wescon Connection Record, New York, pp. 96-104.

[Zadeh 1965]  Zadeh L.A. (1965). *Fuzzy sets.* Inform. Control 8, pp. 338-353.

# PERFORMANCE EVALUATION OF FUZZY CONTROLLERS

**S. Boverie\*\'\*\*, B. Demaya\*\*, R. Ketata\*\*\*\'† and A. Titli\*\*\'\*\*\*\'†**

*\*Siemens Automotive SA, Avenue du Mirail, B.P. 1149, 31036 Toulouse Cedex, France*
*\*\*Mirgas, Avenue du Mirail, B.P. 1149, 31036 Toulouse Cedex, France*
*\*\*\*INSA de Toulouse, Avenue de Rangueil, 31077 Toulouse, France*
*†LAAS-CNRS, 7 Avenue du Colonel Roche, 31077 Toulouse Cedex, France*

Abstract

Fuzzy control is a much debated question in the automatic control community, and needs obviously an unbiased performance evaluation. This paper is an attempt in this direction, using an international and challenging benchmark found in the community of specialists in adaptive control. On problems with different levels of complexity, for systems of order 1,2 and 3, we then compare the performance of P.I.D controllers (very much used in the industry), adaptive controllers (advanced control methods) and fuzzy controllers (rule based control more depending on human expertise than on mathematical models).

Keywords : process control, P.I.D contollers, adaptive control, fuzzy control, benchmark examples.

## I. Introduction

The control of industrial processes needs more and more fast response times, as well as design times as short as possible.
Moreover, the environmental conditions of the processes create important constraints, for example :

  – external disturbances,
  – saturation on the control,
  – changes of the system parameters (ageing,...),
  – etc ...

Due to these different constraints, the methods of the "classical" automatic control are put in check all the more as the behavior is often better described in reality by non–linear models, varying with time, for which the automatic control techniques are even more at a loss.

Therefore, the use of fuzzy logic control, based on the expertise of the human operator, appears to be an interesting alternative.

This paper presents a comparative study on performance and robustness of different controllers, on a showcase used in the international adaptive control community.
Three examples have been defined, and, for each one, we consider different degrees of complexity.
Then we compare the results obtained with the different controllers (P.I.D Controller, Adaptive Supervised Controller, Fuzzy Logic Controller).

### II. Description of the process used :

See [MASTEN and COHEN, [1] & [2]]

The overall diagram of the system is represented by the following figure :

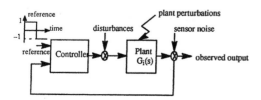

## II.1 Description of the plant

Three different linear plants are considered. They are described by the following models in the table below:

| | |
|---|---|
| First system | First order plant :<br><br>$G_1(s) = \dfrac{K}{s + a}$ |
| Second system | Second order plant :<br><br>$G_2(s) = \dfrac{K}{s^2 + a_1 s + a_2}$ |
| Third system | Third order plant :<br><br>$G_3(s) = \dfrac{K}{s^3 + a_1 s^2 + a_2 s + a_3}$ |

Nominal values of the $a_i$ terms and the gain K are :
$K = 1$ ; $a = 1$
$a_1 = 1.4$ , $a_2 = 1$
$a_1 = 1.75$, $a_2 = 2.15$, $a_3 = 1$

### II.2. Description of the perturbations applied to the process

Following MASTEN and COHEN ([1] & [2]) , three kinds of disturbances are applied to the process:

External disturbances : Bias on the control variable ( nominal value = 2 ), White noise on the measurement.

Change on the model parameters : poles, static gain.

Neglected dynamics : Time delay, Unmodelled poles, unmodelled zero.

We have, then, defined four levels of complexity :

plant perturbations

| level of complexity | Disturbance level in % of the nominal va- lue | sensor noise (R M S) | unmodelled dynamic. (sec) | time delay (sec) | unmodelled plant zero (sec) (only for second and third order plants) | pole variation max value | Gain variation % of the nominal gain |
|---|---|---|---|---|---|---|---|
| 0 | 100 % | 0 | 0 | 0 | 0 | $\Delta a=\pm 2$ for first order $\Delta a=\pm 3$ for 2nd, 3rd order | −50 to 200 for first order −50 to 300 for 2nd to 3rd order |
| 1 | 10 % | 0.02 | 0.10 | 0.05 | 0.10 | idem | idem |
| 2 | 60 % | 0.16 | 0.25 | 0.10 | 0.20 | idem | idem |
| 3 | 100 % | 0.2 | 0.33 | 0.33 | 0.30 | idem | idem |

### II.3 Requirements:

Following MASTEN and COHEN ( [1], [2]) we define three ideal process behaviors in response to a reference step input, with respect to the order of the plants .

They are introduced in the simulation as a reference model :

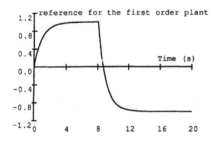

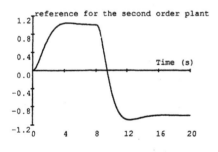

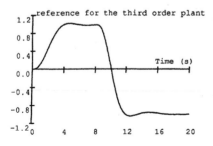

### III. Synthesis on the controllers :

### III.1. P.I.D controller :

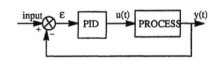

Determination of the P.I.D controller parameters is obtained by the pole compensation method. The P.I.D controller is tuned off line for the nominal plants.

### III.2. Adaptive controller :

A schematic diagram of the supervised adaptive controller is shown in the figure below:

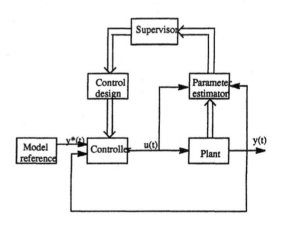

The results concerning this part are extracted from M'SAAD, LANDAU, SAMAAN ([3]), [4]).

## III.4. Fuzzy logic controller :

The operation of a Fuzzy controller can be described by means of the following sequence
(see L.A. ZADEH [6]) :

Input variables ... Control ouput

Human expertise is translated in the form of rules of the following type:

**If** < Fuzzy conditions on input variables > **Then** < Fuzzy control actions >

These rules are considered by the inference engine.

The fuzzyfication action enables to define a grade of membership to fuzzy sets characterizing input variables.

The defuzzyfication action enables to calculate the effective control from all the rules applicable for a given input set.

In our example the fuzzy logic controller uses in real time the deviation $<\varepsilon>$ between the set point and the process output , as well as the change of the error : $<\Delta\varepsilon>$ and the variation of $\Delta\varepsilon$ : $<\Delta^2\varepsilon>$ , according to the order of the plant (see BOVERIE, DEMAYA, TITLI [5] ).

From these quantized informations and after defining the membership grade of the variables, we determine a set of rules which allows us to build the control variable.

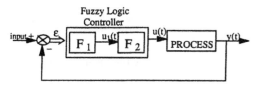

with $u_1 = f_1 ( \varepsilon, \Delta\varepsilon, \Delta^2\varepsilon)$ and $u_2 = f_2 ( \int u_1, u_1)$
For each case , the fuzzy controller is designed for the nominal plant.

## IV. Simulation results :

The simulation results obtained for the proposed problems are presented in this section.

### IV.1 Simulation for the first order system :

The more representative models for the first order example are defined by the following sets of parameters:

$K = 0.5$  $a \in \{-1 ; 1 ; 3\}$
$K = 1$  $a \in \{-1 ; 1 ; 3\}$
$K = 3$  $a \in \{-1 ; 1 ; 3\}$

Different simulations have been carried out according to the choice of different parameters for the same plant model.

All the proposed experiments have been performed successfully with adaptive controllers.

With Fuzzy controller all the proposed experiments have been performed successfully for the complexity levels: 0,1,2.
At the third complexity level, Fuzzy controllers present some fair

results as long as the pole plant is negative, when the pole plant becomes positive the process becomes unstable.
In general, the P.I.D controllers presents some bad quality results. Robustness, with respect to parameter modifications is ensured with P.I.D controller as long as the pole of the plant remains negative. Otherwise the closed loop process becomes unstable

The corresponding output performance results are shown in figure:

1(a) , 2(a) , 3(a), for adaptive control
( respectively for complexity levels, 0,1,3 ).
1(b) , 3(b), for P.I.D control
( respectively for complexity levels, 0,3 ).
1(c) , 2(c) , 3(c), for Fuzzy control
( respectively for complexity levels, 0,1,3 ).

*Synthesis and classification:*
The different controllers can be classified according to these performances in the following table (1, standing for the best controller for a given level of complexity)

| level | P.I.D | Adaptive | Fuzzy |
|-------|-------|----------|-------|
| 0 | 3 | 1 | 1 |
| 1 | 3 | 1 | 1 |
| 2 | 3 | 1 | 1 |
| 3 | 3 | 1 | 2 |

### IV.2 Simulation for the second order system :

The more representative models for the second order example are defined by the following sets of parameters:

$K = 0.5$  $a_1, a_2 \in \{ (1.4 , 1), (4.4 , 3), (-1.6 , 1) \}$
$K = 1$  $a_1, a_2 \in \{ (1.4 , 1), (4.4 , 3), (-1.6 , 1) \}$
$K = 3$  $a_1, a_2 \in \{ (1.4 , 1), (4.4 , 3), (-1.6 , 1) \}$

The same comments as for the 1st example can be made here on the performances of P.I.D and adaptive controllers.

With Fuzzy controllers all the proposed experiments have been performed successfully for the complexity levels: 0,1.
From second to third complexity level, Fuzzy controllers present some fair results as long as the pole plant is negative, when the pole plant becomes positive the process becomes unstable.

The corresponding output performance results are shown in figures:

4(a) , 5(a) , for adaptive control
( respectively for complexity levels, 0,2 ).
5(b) , for P.I.D control
( for complexity levels 2 ).
4(c) , 5(c) , for Fuzzy control
( respectively for complexity levels, 0,2 ).

*Synthesis and classification:*
The classification table becomes:

| level | P.I.D | Adaptive | Fuzzy |
|-------|-------|----------|-------|
| 0 | 3 | 1 | 1 |
| 1 | 3 | 1 | 1 |
| 2 | 3 | 1 | 2 |
| 3 | 3 | 1 | 2 |

### IV.3 Simulation for the third order system:

We have chosen as a representative set of parameters :

For the complexity level 0:

$K \in \{1,3,0.5\}$;
$a_1, a_2, a_3 \in \{(-0.25,0.15,-1), (1.75,2.15,1), (4.75,5.15,4)\}$

For the complexity level 1:

$K = 1$ ;
$a_1, a_2, a_3 \in \{(1.75,2.15,1), (4.75,5.15,4)\}$

In this case, results for adaptive control are not available.

No experiment has been performed in this case with P.I.D controller given the order of the plant to be controlled.

Fuzzy controllers have performed successfully the 0 complexity level proposed experiments.
For complexity level 1 the process becomes unstable when the poles of the plant are positive.

The corresponding output performance results are shown in figures:

6 , 7 , ( respectively for complexity levels, 0,1 )

### V. Conclusion :

In this paper, in order to evaluate in an unbiased way the real performance of fuzzy control, we have chosen to work on a recognized and challenging benchmark used by the international community of specialists in adaptive control.
After this intensive simulation work, it appears that fuzzy control has positively surprising properties of robustness with respect to both internal and external severe disturbances and can compete with advanced control methods such as supervised adaptive control, with probably a lower cost design!
This robustness can be explained by the nonlinearity of the control and some specific procedures like the agregation of rules and the defuzzyfication procedures.

The results obtained encourage us to work now in two directions:
– Investigating more in details the dynamic behaviour of the fuzzy controllers in order to improve the design of such controllers.
– Dealing with actual applications for which mathematical models are difficult to obtain but where an human expertise is available.

#### Acknowledgement

All those works have been done with the participation of the AFME (Agence Française pour la Maîtrise de l'énergie : French Agency for the Control of Energy).

#### References

[1] – MASTEN, M. K., COHEN, H. E., A Showcase of Adaptive Controller Designs, International Journal of Adaptive Control and Signal Processing, 1989, Vol. 3, pp. 95–101

[2] – MASTEN, M. K., COHEN, H. E., An advanced Showcase of Adaptive Controller Designs, International Journal of Adaptive Control and Signal Processing, 1990, Vol. 4, pp. 89–98

[3] – M'SAAD, I.D. LANDAU, M. SAMAAN, Further Evaluation of Partial State Model Reference Adaptive Design, International Journal of Adaptive Control and Signal Processing, 1990, Vol. 4. pp. 133–148

[4] – M'SAAD, I.D. LANDAU, M. DUQUE, M. SAMAAN, Example applications of the partial state model reference adaptive design technique, International Journal of Adaptive Control and Signal Processing, 1989, Vol. 3. pp. 155–165

[5] – S.BOVERIE, B.DEMAYA, A.TITLI, Fuzzy logic control compared with other automatic control approaches, Proceedings of the 30th IEEE conference on decision control, december 1991, pp. 1212–1216.

[6] – L.A. ZADEH, Fuzzy logic, IEEE computer, 1988, pp. 83–93.

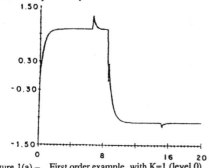

figure 1(a) –   First order example, with K=1 (level 0)
                Adaptive control

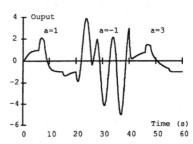

figure 1(b) First order example, with K=1 (level 0)
            P.I.D control

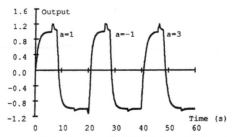

figure 1(c)  –First order example with K=1 (level 0) – Fuzzy control

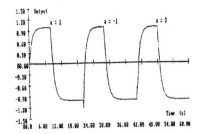

figure 2(a) –  First order example, with K=1 (level 1)
Adaptive control

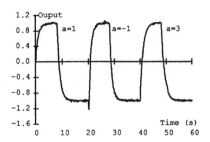

figure 2(b) –  First order example, with K=1 (level 1)
Fuzzy control

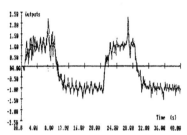

figure 3(a) –  First order example, with K=1, a=3 (level 3)
Adaptive control

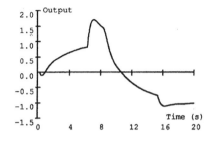

figure 3(b) –  First order example, with K=1, a=3 (level 3)
P.I.D control

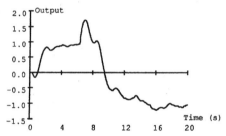

figure 3(c) –  First order example, with K=1, a=3 (level 3)
Fuzzy control

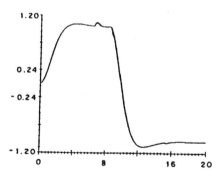

figure 4(a) –  Second order example, with K=1, (level 0)
Adaptive control

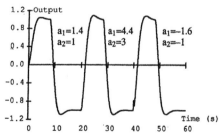

figure 4(b) –  Second order example, with K=1, (level 0)
Fuzzy control

73

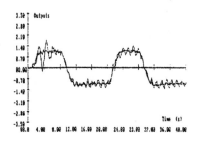

figure 5(a) – Second order example, with K=1, $a_1$=4.4, $a_2$=3
Adaptive control (level 2)

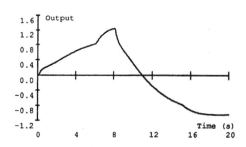

figure 5(b) – Second order example, with K=1, $a_1$=4.4, $a_2$=3
P.I.D control (level 2)

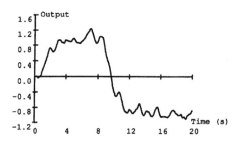

figure 5(c) – Second order example, with K=1, $a_1$=4.4, $a_2$=3
Fuzzy control (level 2)

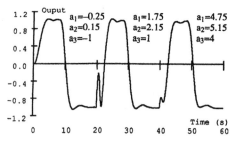

figure 6 – Third order example, with K=1,3,0.5
Fuzzy control (level 0)

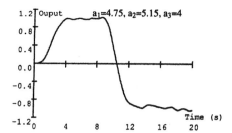

figure 7 – Third order example, with K=1,
Fuzzy control (level 1)

74

# THE ROLE OF DEFUZZIFICATION METHODS IN THE APPLICATION OF FUZZY CONTROL

R. Jager, H.B. Verbruggen and P.M. Bruijn

*Delft University of Technology, Department of Electrical Engineering, Control Laboratory,*
*P.O. Box 5031, 2600 GA Delft, The Netherlands*

**Abstract.** In this article the role of defuzzification methods in the application of fuzzy controllers is adressed. In most applications described in literature the concepts of the fuzzy controllers used are presented without motivation. The defuzzifier of a fuzzy controller can deteriorate a great deal of the results achieved by the inference of fuzzy rules. Therefor a closer look at the influence of this specific part of the fuzzy controller is worthwile. It appears that the application of a specific defuzzification method can reduce completely the fuzziness of the controller. Otherwise, it is possible to use crisp representations of fuzzy sets and still being able to realize a fuzzy character of the controller. Examples are given which show the (dis)advantages of several defuzzification methods.

**Keywords.** Defuzzification; Fuzzy Control; Fuzzy Systems.

## INTRODUCTION

The number of applications of fuzzy set theory in the field of process control is growing. Applications have been reported in the cement industry [4], waste water treatment [13] and glass melting [1]. Other applications can be found in elevator control, subway control, camcorders, televisions, air conditioners, washing machines, vacuum cleaners, and so on. Fuzzy controllers appear to be a satisfying alternative for the control of non-linear processes. In case of time-variant processes self-organising fuzzy controllers are a competetive alternative for 'conventional' adaptive controllers.

It appears that in many control applications the need of a mathematical model of the process to be controlled is not necessary to achieve satisfactory control. A model/controller consisting of heuristic rules of thumb (according to the human way of modelling/control) will be sufficient in many cases [7].

The basic structure of all fuzzy controllers is the same: subsequently a fuzzification-, an inference- and a defuzzification-part. We developed a real-time fuzzy expert system shell to replace the (more or less standard) inference-part [6, 7] of fuzzy controllers in order to be able to perform more complex reasoning with fuzzy rules. However, still a defuzzification method is needed to translate the fuzzy conclusions provided by the fuzzy inference engine.

This paper is concerned with the defuzzification-part of fuzzy (expert) controllers. An example of an application where more than one defuzzification method was used for comparison can be found in [9]. In this paper a comparison is made between different alternatives and the benefits of several approaches are illustrated on a number of experiments.

Note that fuzzy controllers using quantisations of in- and outputs and which are reduced to (simple) look-up tables are not considered in this paper. These controllers are no longer fuzzy, although they are designed using fuzzy concepts.

Before determining the influence of defuzzification methods, the conditions, which the discussed fuzzy controllers should met, are considered. Analysing the defuzzification methods the following properties of the fuzzy controller are assumed:

- the controller is a M.I.S.O.-system;

- the membership functions are convex;

- the maximum value of the membership functions is equal to one;

- no more than two overlapping membership functions exist;

- the sum of the (overlapping) membership functions is equal to one.

In applications of fuzzy control described in literature these conditions are almost always fullfilled. Fuzzy controllers in which the fuzzy rules can have more than one, weighted conclusions are not considered, although those fuzzy controllers are less restricted because of the extra degree of freedom introduced [8].

From literature several defuzzification methods are known. The two most commonly used are [3, 10]: Centre-of-Area (CoA) and Mean-of-Maxima (MoM) defuzzification method. Although the term 'gravity' is more appropriate than 'area', because it refers to a possible multi-dimensional fuzzy output to be defuzzified, we use the term 'area', because only 1-dimensional outputs are considered in this paper. The following two sections will discuss those two defuzzification methods.

## CENTRE-OF-AREA

The Centre-of-Gravity (CoA) of a fuzzy output $Y_o$, represented by the membership function $\mu_{Y_o}$, is in continious form defined

$$\text{CoA}_c(Y_o) = \frac{\int y \cdot \mu_{Y_o}(y) \cdot dy}{\int \mu_{Y_o}(y) \cdot dy} \quad (1)$$

and in the discrete form, using $N_q$ quantisations, represented by

$$\text{CoA}_d(Y_o) = \frac{\sum_{i=1}^{N_q} y_i \cdot \mu_{Y_o}(y_i)}{\sum_{i=1}^{N_q} \mu_{Y_o}(y_i)} \quad (2)$$

The CoA-method also has an modified version which is called the Indexed-Centre-of-Area and is defined by

$$\text{ICoA}(Y_o, \alpha) = \text{CoA}(\mu_{Y_o}(y) \wedge \mu_{Y_{o,\alpha}}(y)) \quad (3)$$

in which $\mu_{Y_{o,\alpha}}(y)$ represent the $\alpha$-cut [2] of fuzzy set $Y_o$. This is a classical (crisp) set, defined by

$$Y_{o,\alpha} = \{y \in Y | \mu_{Y_o}(y) \geq \alpha\} \quad (4)$$

with membership function

$$\mu_{Y_{o,\alpha}}(y) = \begin{cases} 1 & \text{iff } y \in Y_{o,\alpha} \\ 0 & \text{otherwise} \end{cases} \quad (5)$$

The ICoA-method in fact ignores those parts of fuzzy set $Y_o$ which have a membership grade below threshold $\alpha$.

In applications it is impractical to use the continous form of the CoA-method according to equation 1. Therefor the discrete version, which can be interpreted as the Riemann-sum, is used in practical situations. In the most simple representation of the CoA-method the number of quantisations $N_q$ is used which is equal to $N_y$, the number of fuzzy sets defined for the output. Each quantisation is chosen equal to the crisp representation (defuzzification of initial fuzzy set) of one of those $N_y$ fuzzy sets. This is sometimes called the Fuzzy-Mean (FM) defuzzification method (as will be done in the rest of the paper) [11], and can be calculated very easy by

$$\text{FM}(Y_o) = \frac{\sum_{k=1}^{N_y} \gamma_k \cdot y_k}{\sum_{k=1}^{N_y} \gamma_k} \quad (6)$$

where $\gamma_k$ represents the degree of truth for fuzzy set $k$. This method ignores the shape of the fuzzy sets chosen for the output and reduces the fuzzy controller to one with a crisp classification of the output instead of a fuzzy one. That is why emphasizing certain conclusions, by choosing fuzzy sets of different sizes/shapes, is not possible. Because of the easy and fast calculations use of the FM-method is prefered. In case emphasizing or prefering of certain conclusions is desired it seems more appropriate to use a weighted version of the FM-method (from now on called Weighted-Fuzzy-Mean (WFM) defuzzification method)

$$\text{WFM}(Y_o) = \frac{\sum\limits_{k=1}^{N_y} w_k \cdot \gamma_k \cdot y_k}{\sum\limits_{k=1}^{N_y} w_k \cdot \gamma_k} \qquad (7)$$

The weight-factors are an indication of the 'importance' of the corresponding fuzzy sets.

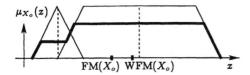

Figure 1: *Effect of using WFM- instead of FM-method.*

In figure 1 one can see the possible advantage of using a weighted version of the FM-method: 'emphasized' fuzzy conclusions are dealt with in a more 'natural' way. Actually, we apply the continious version of the CoA-method using block-shaped representations of the initial fuzzy sets defined for the output, in case the weights are chosen as

$$w_k = \int \mu_{Y_k}(y) \cdot dy \qquad (8)$$

and the block-shaped representations have a base of length $w_k$. In case the shape and magnitude of all fuzzy sets $Y_k$ defined for the output is the same, this method represents the Fuzzy-Mean defuzzification method as in 6.

| method | $*, /$ | $+, -$ |
|--------|--------|--------|
| CoA | $N_q + 1$ | $2(N_q - 1)$ |
| FM | $N_y + 1$ | $2(N_y - 1)$ |
| WFM | $3N_y + 1$ | $2(N_y - 1)$ |

Table 1: *Number and types of operations CoA-related defuzzification methods ($N_q \geq N_y$).*

In table 1 number and types of numerical operations necessary to calculate the defuzzified output are shown for various methods.

## MEAN-OF-MAXIMA

In the Mean-of-Maxima defuzzification method (MoM) only the maximum values of the membership function are used for the determination of a crisp representation. In fact the MoM of the fuzzy output is the Indexed-Centre-of-Gravity

(ICoA) of the fuzzy output, in case the level $\alpha$ is chosen equal to the height (Hgt) of the membership function of the fuzzy output

$$\text{MoM}(Y_o) = \text{ICoA}(Y_o, \text{Hgt}(Y_o)) \qquad (9)$$

in which $\text{Hgt}(Y_o)$ is the supremum of fuzzy set $Y_o$

$$\text{Hgt}(Y_o) = \sup_{y \in Y} \mu_{Y_o}(y) \qquad (10)$$

The discrete version of the MoM-method can also be represented by

$$\text{MoM}(Y_o) = \sum_{j=1}^{J} \frac{y_j}{J} \qquad (11)$$

with $\mu_{Y_o}(y_j) = \text{Hgt}(Y_o)$ and $J = |\{y_j\}|$. On the implementation level this discrete version of the MoM-method can be coded using only $2J$ summation operations and 1 divide operation in case symmetrical fuzzy sets are defined for the output.

Although the MoM-method is an obvious method for defuzzification [2], applying it in fuzzy control reduces the 'fuzziness' of the controller completely in many cases. In case the MoM-method is used for defuzzification of the fuzzy output, the membership functions describing the inputs can even be chosen as crisp (classical) sets, whitout having any effect on the numerical output.

Suppose the fuzzy controller satisfied the assumptions introduced in the introduction. When also is assumed that none of the variables has a value which results in membership values equal to $\frac{1}{2}$ and the T-norm chosen (representing the intersection) satisfies besides the conditions [12]

T-1   $T(a, 1) = a$

T-2   $T(a, b) \leq T(c, d)$
      whenever $a \leq c$, $b \leq d$

T-3   $T(a, b) = T(b, a)$

T-4   $T(T(a, b), c) = T(a, T(b, c))$

also a stronger condition for T-2, namely

T'-2   $T(a, b) < T(c, d)$
       whenever $a < \frac{1}{2} < c$, $b < \frac{1}{2} < d$

then there exists only one dominant fuzzy rule in the rule base. This dominant rule will completely determine the numerical output of the fuzzy controller in case the chosen S-norm (or T-conorm) for combining the individual fuzzy conclusions of the fuzzy rules (union) satisfies besides the conditions T-2, T-3 and T4 and

S-1 $\quad S(a, 0) = a$

instead of T-1 [5], also a stronger condition for S-2, namely

S'-2 $\quad S(a, b) < S(c, d)$
$\quad\quad$ whenever $a < c$, $b < d$

So every combination of T- and S-norms satisfying the above listed conditions will result in the same numerical output of the fuzzy controller. This means, for example, that redefinition of the shape of the fuzzy sets defined for the inputs in a way as is done in figure 2 has no effect for the numerical output at all.

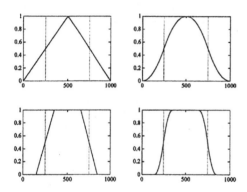

Figure 2: *Fuzzy sets with different shapes without having effect on numerical output in case of MoM-method.*

In case we consider also the cases in which more than one fuzzy rule are equally dominant, than the (finite) number of possible numerical values for the output is still restricted by

$$\sum_{n=1}^{N_y} \binom{N_y}{n} \qquad (12)$$

Another disadvantage of the MoM-method is the fact that non-symmetrical fuzzy sets defined for the output can result in undesired behaviour. In figure 3 two situations are depicted which point out that a shift of the fuzzy output possibly results in a shift in the opposite direction of the numerical output.

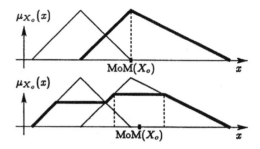

Figure 3: *Undesired result of MoM-method in case of non-symmetrical fuzzy sets: shift of numerical output opposite to shift in fuzzy output.*

A third main disadvantage of the MoM-method will be shown in section 4 where experiments with a fuzzy controller show the difference in perfomance in comparison with the CoA-related defuzzification methods.

## EXPERIMENTS

Comparing the defuzzification methods, it appears that the choice of the defuzzification method determines to a large extend the 'quality' of control as well as the computational performance of the controller. Comparing the results based on control and computational performance, one can choose the appropriate defuzzification method for a specific application.

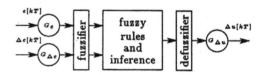

Figure 4: *Typical fuzzy controller.*

First of all we have to choose a simple fuzzy controller, as in figure 4, which represents most of the fuzzy controllers described in literature. For the inference of the fuzzy rules the max-min-operation is used as the compositional rule of inference.

In table 2 the used rule base is given. For every input and output the fuzzy sets are of triangular shape and are equidistant distributed. The process chosen in the experiments was a second-order process with the following parameters: $K_{dc} = 2$, $\tau_1 = 10$ and $\tau_2 = 20$.

In figure 5 on can see the main disadvantage of application of the MoM-method for defuzzification. Because of the fact that there is each sample only one dominant rule which determines the

| PB | AZ | PS | PM | PB | PB | PB | PB |
|----|----|----|----|----|----|----|----|
| PM | NS | AZ | PS | PM | PB | PB | PB |
| PS | NM | NS | AZ | PS | PM | PB | PB |
| AZ | NB | NM | NS | AZ | PS | PM | PB |
| NS | NB | NB | NM | NS | AZ | PS | PM |
| NM | NB | NB | NB | NM | NS | AZ | PS |
| NB | NB | NB | NB | NB | NM | NS | AZ |
|    | NB | NM | NS | AZ | PS | PM | PB |

Table 2: *Rulebase of fuzzy controller with conclusions for $\Delta u[kT]$ as 'function' of error $e[kT]$ (horizontal) and error change $\Delta e[kT]$ (vertical).*

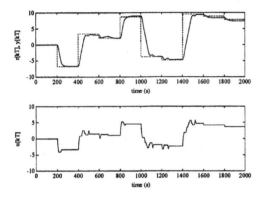

Figure 5: *Experiment with fuzzy controller using MoM-method.*

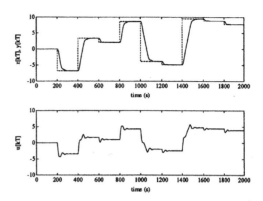

Figure 6: *Experiment with fuzzy controller using CoA-method.*

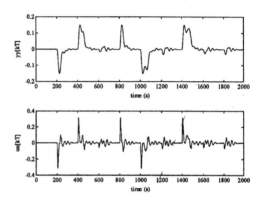

Figure 7: *Difference between experiments with FM- and CoA-method, with $yy[kT] = y_{FM}[kT] - y_{CoA}[kT]$ and $uu[kT] = u_{FM}[kT] - u_{CoA}[kT]$.*

output of the controller, a steady state error will occur. The maximum value of this steady state error is determined by the choice of the fuzzy sets representing $AZ$ for the error as well as the error change.

Trying to overcome this problem the fuzzy set $AZ$ defined for the error could be split into two fuzzy sets, for example $NZ$ and $PZ$. However, this will result in an oscillating control signal. These problems, the possible steady state error inequal zero and the possible oscillating control signal, are due to the discontinious behaviour of the MoM-method. Increasing the number of fuzzy sets for in- and outputs will reduce these problems, but will require more memroy and calculation time.

As can be seen in figure 6 the CoA-method does not suffer the problem of a steady state error as the MoM-method does.

Comparison of CoA- and FM-method (see figure 7) shows that the performance of the FM-method is comparable with the CoA-method. One could say that the FM-method also corresponds better with the way humans try to translate their fuzzy idea into a numerical quantity: making a kind of weighted average using some reference points.

What can be seen in figure 8 is that the number of quantisations used for the defuzzification with the CoA-method does not have much effect on the control performance. (In the experiments done: the less quantisations, the better the performance.)

Knowing the linear relation between the number of quantisations used for the CoA-method (see table 1) and looking at the control performance criteria as function of the number of quantisations used (see figure 8), one can conclude that the (W)FM-method is far more preferable than the CoA-method.

## CONCLUSIONS AND REMARKS

In the field of fuzzy control several defuzzification methods are available. In (real-time) applications a balance of computational requirements and control performance has to be found. To do

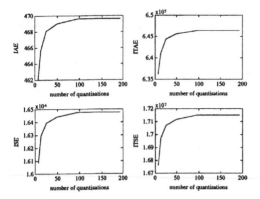

Figure 8: *Control performance criteria as function of the number of quantisations used for defuzzification with CoA-method.*

so several methods has been analysed and experimented with.

From analysing of and experimenting with the 'Mean-of-Maxima' defuzzification method it is seems that this method is not quite suitable for the application of fuzzy control. It reduces the fuzzy character of the fuzzy controller in many situations to nil. Possible steady state errors or oscillations in the control signal can occur. Trends in fuzzy conlusions can be 'opposite' to trends in numerical conclusions.

The 'Centre-of-Area' defuzzification method is in many cases better than the MoM-method, but is in terms of calculations not suitable for practical situations. For practical situations it is better to use the most simple estimation of this method: the 'Fuzzy-Mean' defuzzification method. The performance of both methods is comparable, because of the little influence of the number of quantisations used to calculate the CoA-method.

In case certain fuzzy conclusions are prefered and need to be emphasized a weighted version of the FM-method, the 'Weighted-Fuzzy-Mean' defuzzification method, can be used. Maybe the (W)FM-method is also the best approach to the way humans tend to combine 'fuzzy' conclusions and translate into a 'numerical' one.

## REFERENCES

[1] S. Aoki, S. Kawachi, M. Sugeno. *Application of fuzzy control logic for dead-time processes in a glass melting furnace.* Fuzzy Sets and Systems, no. 38, pp. 251-265, 1990.

[2] D. Dubois, H. Prade. *Fuzzy sets and systems: theory and applications.* Mathematics in science and engineering, vol. 144., Academic Press, 1980.

[3] C.J. Harris, C.G. Moore. *Intelligent identification and control for autonomous guided vehicles using adaptive fuzzy-based algorithms.* Engineering Applications of Artificial Intelligence, vol. 2, pp. 267-285, December 1989.

[4] T.J.M. Flintham. *Expert systems in control, why so few?* Control 91, Edinburgh, U.K., March 1991.

[5] J. Hellendoorn. *Reasoning with fuzzy logic.* PhD-thesis, Delft University of Technology, Delft, The Netherlands, 1990.

[6] R. Jager, H.B. Verbruggen, P.M. Bruijn, A.J. Krijgsman. *Real-time fuzzy expert control.* Control 91, Edinburgh, U.K., March 1991.

[7] R. Jager, A.J. Krijgsman, H.B. Verbruggen, P.M. Bruijn. *Rule-based controller using fuzzy logic.* EURISCON '91, Kanoni, Corfu, Greece, June 1991.

[8] R. Jager, H.B. Verbruggen, P.M. Bruijn. *Adaptive fuzzy control.* ESS '91, Ghent, Belgium, November 1991.

[9] L.I. Larkin. *A fuzzy logic controller for aircraft flight control.* Industrial Applications of Fuzzy Control, M. Sugeno (ed.), pp. 87-103, 1985.

[10] W. Pedrycz. *Fuzzy control and fuzzy systems.* Research Studies Press Ltd., Taunton, Somerset, England, 1989.

[11] B. Postlethwaite. *Basic theory and algorithms for fuzzy sets and logic.* Knowledge-based systems for industrial control, editors: J. McGhee, M.J. Grimble, P. Mowforth, IEE Control Engineering Series vol. 44, ch. 3, Peter Pereginus Ltd., 1990.

[12] B. Schweizer, A. Sklar. *Associative functions and abstract semi-groups.* Publicationes Mathematicae Debrecen, vol. 10, pp. 69-81, 1963.

[13] C. Yu, Z. Cao, A. Kandel. *Application of fuzzy reasoning to the control of an activated sludge plant.* Fuzzy Sets and Systems, no. 38, pp. 1-14, 1990.

# FUZZY CONTROLLER WITH MATRIX REPRESENTATION MODIFIED BY NEURAL NETWORKS

**M. Nakatsuyama and H. Kaminaga**

*Department of Electronic Engineering, Yamagata University, Yonezawa 992, Japan*

Abstract. Fuzzy control is supposed to be essentially nondeterministic, but fuzzy control
statements may be converted into a fixed matrix which is considered as deterministic. The
values of the matrix representation are not unique but even very changeable. We determine
the temporary values of the matrix and the weights of the neuron by 500 trainings. Then
we decide the most adequate value by addition or subtraction of the weights of the neuron at
each simulation or experiment with the simple heuristic algorithm. We show that the over
all delays of the vehicles is reduced by 22 % with this method.

Keywords. Fuzzy control; matrix representation; fuzzy control statement; traffic control; over
all delays.

## INTRODUCTION

Fuzzy controllers have been developed by a lot of
researchers. Some of them have a large number of
parallel processors and execute the computation for
control faster than even a super computer.
Yamakawa (1989) suggested that the membership
functions are chosen and fixed before computation in
his controller. Zadeh (1968) proposed fuzzy
algorithm and showed an example of fuzzy automatic
control of an automobile of which action is like a
human performance. Pappis (1977) controlled the
traffic signals by using fuzzy control statements.
In general, fuzzy control is supposed to be
nondeterministic . Fuzzy control program is
constituted of many fuzzy control statements which
are essentially the calculation of MINIMUM or fuzzy
AND operator. The final decision is made by
choosing the maximum values of these statements
by using the simple maximum value or the gravity
method (Mizumoto, 1991). Though fuzzy program
is based on the fuzzy sets or ambiguous concepts,
fuzzy control program must provide only one
determined value when some fixed inputs are applied
to a fuzzy control system. If plural results are to be
selected, then exact control can not be expected.
Nakatsuyama (1990a) showed that the fuzzy program
can be converted into matrices. At first, the value
of the matrix is not adequate, so the fine adjustment

of the value is required. We trained the neural
networks to get the adequate value of the matrix
representation. Combining the matrix
representation and the neural network, we get the
good result for controlling the traffics. The fuzzy
controller with matrix representation is simpler and
is faster than the fuzzy control statements.

## FUZZY PROGRAMS

Fuzzy control statements are assembly of statements
such as "if ... then ... else ... ". We show an
example of fuzzy program which controls the traffic
signals as follows.

.

> if    T = medium and A = mt(medium)
>           and Q = lt(small)    then E = medium
> else if   T = long    and A = mt(many)
>           and Q = lt(medium) then E = long

The terms mt and lt denote "more than" and "less
than" respectively. The symbols T, A, Q and E are
the fuzzy variables of time, arrived vehicles, queue of
vehicles and the signal duration respectively. In
fact, the real calculation is executed as follows.

$$\mu_c = \max( \ \mu_R(t,a,q,e), \ \ \mu_s(t,a,q,e)...) \quad (1)$$

The symbol $\mu$ denotes the membership function.
If $\mu_R = \mu_s = ....$, we are able to choice any

number as the most suitable value. The fuzzy control is nondeterministic in this sense. However, it is necessary to select only one value to guaranty the stable control. Then fuzzy control becomes deterministic. We get the following equation,

$$FP ; f(t,a,q,e) \rightarrow d \qquad (2)$$

In general, Eq.(1) can be written as follows.

$$FM : gm(k1,k2,k3,.....kn) \rightarrow dm \qquad (3)$$

The term ki is an integer. Let M be a matrix, then this equation will become

$$dm = M(k1,k2,k3,...kn). \qquad (4)$$

The fuzzy control statements are represented as a matrix which we call the matrix representation.

## MATRIX REPRESENTATION MODIFIED BY NEURAL NETWORK

At first, the value of the elements of the matrix representation are supposed not to be optimal, so the fine adjustment of the values should be done before using the matrix for control. Nakatsuyama (1990b) proposed the method for the matrix modification which modifies the values of the matrix by the repetition of the control and by using the simple self-tuning. We got the good result, but we found that the correction is not sufficient. To get the more precise amendment, we adopt the neural networks as shown in Fig. 1.

After several hundred training, we get the adequate values of weights. The output of the output layer is almost equivalent of the value of the matrix representation. The over all delay of the present and the past, the mean traffic, the mean value of the signal duration and the mean value of the queue are the input of the input layer . An input data set just corresponds one  element of the matrix, so we determine the values of $5 \times 5$ or $3 \times 3$ elements at each input data set. To get more precise value, we use the mask matrix shown in Fig. 2.

If a data set corresponds only one neuron, it does not need the mask matrix, while a large number of data sets are necessary. We use the function

$$out = 1 / ( 1 + exp ( - p)), \qquad (5)$$

which is called as a simple sigmoidal activation function in the reference (Wasserman, 1989). The term p is the value of the output layer. After 500 training, we get almost the same value as the matrix

representation by using 49 data sets which include the traffics 32, 64, 128, 256, 512,1024 and 1280 and cover all the necessary area. The original matrix representation obtained from the fuzzy program is shown in Fig. 3.

The term tr and tm are the traffics and the time duration of traffic signal respectively. The value of the matrix represents the queue q. If the mean value of tr, tm and q are 15.18, 13.15 and 4.38, then we get the trained data shown in Fig. 4 a. The calculated time delay of the vehicles is 56.08. Fig. 4b is the original data.

Both the trained data and the original data are normalized by 20.0. The value of the central element differs from the original data slightly, but does not have a bad influence upon the performance of the control.

## MATRIX IMPROVED BY HEURISTIC ALGORITHM

The trained data simulate only the original matrix. Therefore it does not improve the performance of the control. Since the neural networks itself has not the ability of the reasoning, it is necessary to get this ability by using the training algorithm. Generally speaking, it is supposed that there is no explicit goal in fuzzy control. It is required only to get the better result than before.

We adopted the so-called heuristic algorithm to get the adequate value of the neural networks. The outline of the heuristic algorithm is as follows.

STEP 1
    Determine the matrix representation and temporary goal. swt = 0. cv = 0.

STEP 2
    Simulation or experiment. If swt = i, then the result i is obtained.

STEP 3
    If cv is 0, then swt = 1 and go to STEP 5.

STEP 4
    If swt is 3 and  the result i is better than  result j, the temporary goal is result i  and the weights i of the neural networks  is selected and swt becomes 1 and
cv = cv +1.

STEP 5
    Calculate the output of the neural networks and the weight adjustment by using  the temporary goal and result i.

## STEP 7

If swt is 1, the weights 1 of the neural networks are to be subtracted by the weight adjustment. If swt is 2, the weights 2 of the neural networks are to be added by the weight adjustment.

## STEP 8

Determine the matrix representation by using the modified neural networks. swt = swt + 1. Go to STEP 2.

## FUZZY CONTROLLER

The fuzzy controller with parallel processors (Yamakawa, 1989) has an architecture shown in Fig. 5. This fuzzy controller is very fast in calculation and controls the inverted pendulum successfully.

We proposed the matrix representation instead of the fuzzy control statements. The matrix representation requires a lot of memories. Nowadays the price of the memory is not expensive, so we may use a large number of memory. Nakatsuyama (1990a) proposed the fuzzy controller with the matrix representation shown in Fig. 6 which is fast in computation.

Our new proposition is based on the matrix representation modified by the neural networks which are improved by the simulation or experimental data. Its architecture is shown in Fig. 7 and is very similar to the one shown in Fig. 6.

The values of the matrix representation will be modified by the repetition of the experiment or simulation. And the heuristic algorithm is very simple.

## FUZZY CONTROLLER FOR TRAFFICS

We adopt the fuzzy controller to the traffic control. We suppose there are 8 one-way roads and there are 16 traffic junctions shown in Fig. 8. Each traffic junction may be controlled by a simple fuzzy controller which provide a good result. In fact, these 16 traffic signals are controlled by a single 20 $\times$ 20 matrix which is shown in Fig. 3. If some one-way roads are designated to the artery road, the phase control is very effective. The time-chart of the traffic signal for phase control is also shown in Fig. 9. The term $\tau$ is the signal duration. The term L and L' are determined to control the phase. We showed that a personal computer can control 16 traffic junctions easily. We use the data sets shown in TABLE 1 for training the neural networks. It is not effective to improve the time delay if the traffics are less than the 1280 vehicles / hour, since the values have already been precisely adjusted. So we

adopt the traffics 1280 vehicles / hour.

At first, we calculated the over all delays of the vehicles shown in TABLE 1. All the traffic junctions are controlled by a simple fuzzy controller with the matrix representation.

### TABLE 1 Over All Delays

| tr1 | tr2 | delay | tr1 | tr2 | delay |
|-----|-----|-------|-----|-----|-------|
| 32 | 32 | 3.14 | 32 | 64 | 3.06 |
| 32 | 128 | 2.73 | 32 | 256 | 1.67 |
| 32 | 512 | 1.19 | 32 | 1024 | 1.67 |
| 32 | 128 | 2.21 | | | |
| 64 | 32 | 3.23 | 64 | 64 | 3.64 |
| 64 | 128 | 3.62 | 64 | 256 | 2.61 |
| 64 | 1280 | 5.27 | | | |
| 128 | 32 | 2.44 | 128 | 64 | 3.26 |
| 128 | 128 | 4.15 | 128 | 256 | 4.46 |
| 128 | 512 | 4.45 | 128 | 1024 | 6.32 |
| 128 | 1280 | 8.32 | | | |
| 256 | 32 | 1.68 | 256 | 64 | 3.01 |
| 256 | 128 | 4.37 | 256 | 256 | 5.25 |
| 256 | 512 | 5.90 | 256 | 1024 | 8.36 |
| 256 | 1280 | 10.51 | | | |
| 512 | 32 | 4.01 | 512 | 256 | 5.90 |
| 512 | 512 | 6.96 | 512 | 1024 | 9.49 |
| 512 | 1280 | 12.44 | | | |
| 1024 | 32 | 1.87 | 1024 | 64 | 4.32 |
| 1024 | 128 | 6.22 | 1024 | 256 | 8.41 |
| 1024 | 512 | 9.45 | 1024 | 1024 | 11.48 |
| 1024 | 1280 | 27.22 | | | |
| 1280 | 32 | 2.21 | 1280 | 64 | 5.21 |
| 1280 | 128 | 8.50 | 1280 | 256 | 10.78 |
| 1280 | 512 | 12.58 | 1280 | 1024 | 40.33 |
| 1280 | 1280 | 56.08 | | | |

The term tr1 and tr2 are the traffics of the direction east-west and the one of the direction north-south

### TABLE 2 Error

| tr1 | tr2 | error |
|-----|-----|-------|
| 1280 | 32 | 0.00000644183 |
| 1280 | 64 | 0.00000143224 |
| 1280 | 128 | 0.00000130032 |
| 1280 | 256 | 0.00000029831 |
| 1280 | 512 | 0.00000067673 |
| 1280 | 1024 | 0.00000218982 |
| 1280 | 1280 | 0.00000441928 |

respectively. It requires 500 training to get the adequate weights of the neural networks. We show a part of the error of the training data in TABLE 2. The output p of the neural networks is calculated by Eq. 6.

$$p(i,j) = \sum_k wk(i,j,k) \times ( \sec(k,0) \times v + \sec(k,1)$$
$$\times w + \sec(k,2) \times x + \sec(k,3) \times y + \sec(k,4) \times z ) \qquad (6)$$

The terms wk and sec are both the weights of the neural networks. The term v, w, x, y, and z are the present delay time, the past delay time, the traffics, the queue, the signal duration respectively. The weight adjustment del and the error is calculated by the following equations.

$$del(i,j) = \sum_k h \times ( temp - mean) \times ( \sec(k,0) \times v$$
$$+ \sec(k,1) \times w + \sec(k,2) \times x + \sec(k,3) \times y$$
$$+ \sec(k,4) \times z ) \qquad (7)$$

$$error = \sum_{i=cr-1}^{cr+1} \sum_{j=cm-1}^{cm+1} del(i,j) \times del(i,j) \qquad (8)$$

The terms cr and cm are the mean traffics and the mean signal duration respectively. The term h is 0.005. The term temp is the temporal goal of the fuzzy control. The term mean is the present delay time. In this case, the term of the dither is not effective. By using the matrix representation modified by the neural networks, we get the delay time 43.63 sec / vehicle at the traffics tr1 = tr2 = 1280, while the original delay time is 56.08. The delay time is improved by about 22%.

## CONCLUSION

The performance of the fuzzy controller is improved by using the neuron networks, though it is necessary to use the simple heuristic algorithm. We showed the example of the control of the traffic signals and we got the good result. This method may be easily applied to any other control system. The term of the dither is not effective in the traffic control, but it might be effective for the other control.

We acknowledge Mr. Mizunuma and Ms Sugimoto for their help.

## REFERENCES

Mizumoto, M. (1991). Min-max-gravity method versus product-sum-gravity method for fuzzy controls, *IFSA'91 Brussels*, (Engineering) pp.127-130.

Nakatsuyama, M., H.Nagahashi, N. Nishizuka and K. Watanabe (1990a). Matrix representation for fuzzy program and its application to traffic control, *11th IFAC WORLD Congress in Tallinn*, 7, pp.83-88.

Nakatsuyama , M., H. Nagahashi and K. Watanabe (1990b). Matrix representation for fuzzy programming and reconstitution of fuzzy program from modified matrix, *Proc. of Sino-Japan Joint Meeting on Fuzzy Sets and Systems*, Beijing.

C.P.Pappis, C.P. and E.H. Mamdani (1977). A fuzzy logic controller for a traffic junction, *IEEE Trans. Syst. Man Cybern.*, SMC-7,10,pp.701-717

Wasserman, P.D. (1989). *Neural Computing Theory and Practice*, VAN NOSTRAND REINOLD, New York.

Yamakawa, T. (1989). Stabilization of an inverted pendulum by a high-speed fuzzy logic controller hardware system, *Fuzzy Sets and Systems*, 32, 2, pp.161-180.

Zadeh,L.A. (1968). Fuzzy algorithm, *Information and Control*, 12,pp.91-102.

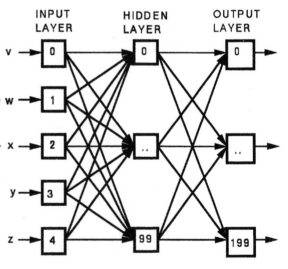

Fig.1 Neural networks.

0.6 0.7 0.8 0.7 0.6
0.7 0.8 0.9 0.8 0.7
0.8 0.9 1.0 0.9 0.8
0.7 0.8 0.9 0.8 0.7
0.6 0.7 0.8 0.7 0.6

Fig. 2. The mask matrix.

tm

| tr | 0 | 1 | 2 | 3 | 4 | 5 | 6 | 7 | 8 | 9 | 10 | 11 | 12 | 13 | 14 | 15 | 16 | 17 | 18 | 19 |
|---|---|---|---|---|---|---|---|---|---|---|---|---|---|---|---|---|---|---|---|---|
| 0 | 1 | 1 | 1 | 1 | 1 | 1 | 1 | 1 | 1 | 1 | 1 | 1 | 1 | 1 | 1 | 1 | 1 | 1 | 1 | 1 |
| 1 | 1 | 1 | 1 | 1 | 1 | 1 | 1 | 1 | 1 | 1 | 1 | 1 | 1 | 1 | 1 | 1 | 1 | 1 | 1 | 1 |
| 2 | 2 | 1 | 1 | 1 | 1 | 1 | 1 | 1 | 1 | 1 | 1 | 1 | 1 | 1 | 1 | 1 | 1 | 1 | 1 | 1 |
| 3 | 2 | 1 | 1 | 1 | 1 | 1 | 1 | 1 | 1 | 1 | 1 | 1 | 1 | 1 | 1 | 1 | 1 | 1 | 1 | 1 |
| 4 | 2 | 1 | 1 | 1 | 1 | 1 | 1 | 1 | 1 | 1 | 1 | 1 | 1 | 1 | 1 | 1 | 1 | 1 | 1 | 1 |
| 5 | 3 | 1 | 1 | 1 | 1 | 1 | 1 | 1 | 1 | 1 | 1 | 1 | 1 | 1 | 1 | 1 | 1 | 1 | 1 | 1 |
| 6 | 3 | 1 | 1 | 1 | 1 | 1 | 1 | 1 | 1 | 1 | 1 | 1 | 1 | 1 | 1 | 1 | 1 | 1 | 1 | 1 |
| 7 | 4 | 2 | 1 | 1 | 1 | 1 | 1 | 1 | 1 | 1 | 1 | 1 | 1 | 1 | 1 | 1 | 1 | 1 | 1 | 1 |
| 8 | 6 | 3 | 2 | 1 | 1 | 1 | 1 | 1 | 1 | 1 | 1 | 1 | 1 | 1 | 1 | 1 | 1 | 1 | 1 | 1 |
| 9 | 8 | 4 | 2 | 1 | 1 | 1 | 1 | 1 | 1 | 1 | 1 | 1 | 1 | 1 | 1 | 1 | 1 | 1 | 1 | 1 |
| 10 | 10 | 5 | 3 | 2 | 1 | 1 | 1 | 1 | 1 | 1 | 1 | 1 | 1 | 1 | 1 | 1 | 1 | 1 | 1 | 1 |
| 11 | 12 | 6 | 4 | 2 | 1 | 1 | 1 | 1 | 1 | 1 | 1 | 1 | 1 | 1 | 1 | 1 | 1 | 1 | 1 | 1 |
| 12 | 14 | 7 | 5 | 3 | 2 | 1 | 1 | 1 | 1 | 1 | 1 | 1 | 1 | 1 | 1 | 1 | 1 | 1 | 1 | 1 |
| 13 | 16 | 11 | 9 | 7 | 5 | 4 | 3 | 2 | 1 | 1 | 1 | 1 | 1 | 1 | 1 | 1 | 1 | 1 | 1 | 1 |
| 14 | 17 | 15 | 13 | 10 | 8 | 7 | 6 | 4 | 4 | 4 | 4 | 3 | 2 | 1 | 1 | 1 | 1 | 1 | 1 | 1 |
| 15 | 18 | 16 | 15 | 14 | 12 | 12 | 8 | 7 | 6 | 6 | 6 | 6 | 5 | 4 | 3 | 2 | 1 | 1 | 1 | 1 |
| 16 | 19 | 18 | 17 | 16 | 16 | 16 | 14 | 9 | 8 | 8 | 8 | 8 | 7 | 6 | 5 | 3 | 4 | 2 | 1 | 1 |
| 17 | 19 | 19 | 19 | 19 | 18 | 18 | 17 | 17 | 16 | 16 | 14 | 14 | 12 | 12 | 11 | 10 | 8 | 6 | 4 | 3 |
| 18 | 19 | 19 | 19 | 19 | 19 | 19 | 19 | 19 | 18 | 18 | 16 | 16 | 14 | 14 | 13 | 12 | 10 | 10 | 8 | 6 |
| 19 | 19 | 19 | 19 | 19 | 19 | 19 | 19 | 19 | 19 | 19 | 19 | 19 | 19 | 19 | 18 | 17 | 16 | 15 | 14 | 12 |

Fig. 3. The original matrix representation.

0.43 0.35 0.24
0.66 0.56 0.67
0.90 0.78 0.81

(a)

0.50 0.40 0.35
0.70 0.60 0.60
0.80 0.80 0.80

(b)

Fig. 4. The trained data (a) and the original data (b)

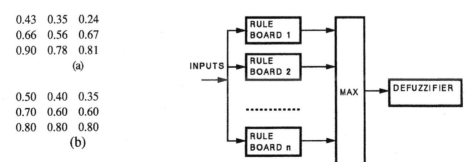

Fig. 5. Architecture of fuzzy controller.

85

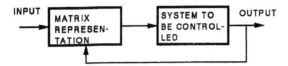

Fig. 6. Fuzzy controller with matrix representation.

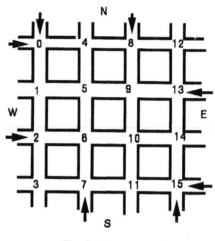

Fig. 8 One-way roads.

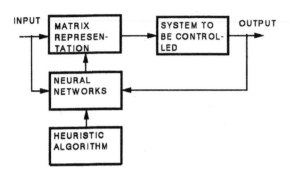

Fig. 7. Fuzzy controller with matrix representation
modified by neural networks.

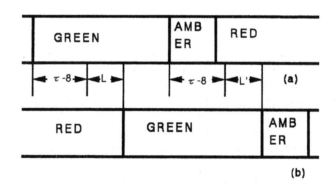

Fig. 9  (a) Traffic signal of junction i.
(b) Traffic signal of junction j.

# SYSTOLIC IMPLEMENTATION OF DBT-TRANSFORMED FUZZY CONTROL SYSTEMS

## A. Bugarin*, S. Barro*, R. Ruiz** and J. Presedo*

*Departamento de Electronica, Facultad de Fisica, Universidad de Santiago de Compostela,
15706 Santiago de Compostela, Spain
**Departamento de Ingenieria Electromecanica, E.T.S.I.I. Cartagena, Universidad de Murcia,
30203 Cartagena, Spain

ABSTRACT

A systolic architecture for the evaluation of Fuzzy Control Systems is presented. The homomorphism existing between certain strategies for the evaluation of these systems and the classical matrix-vector product is used. This permits the use of techniques for partitioning and projection onto systolic architectures described for this type of product (the DBT: Dense to Band matrix transformation by Triangular-block partitioning).

KEYWORDS: Fuzzy Control, Fuzzy Systems, Fuzzy Set Theory, Parallel Processing, Intelligent Process Control.

## INTRODUCTION

The control systems based on fuzzy logic (Zadeh, 1965, 1971b) try to model the knowledge and experience accumulated by the human expert about a specific process and not the process itself. Therefore, this approach to the design of control systems is particularly adequate in complex applications, for which there are no well established mathematical models. The design of the rule base which describes the fuzzy controller can follow two paths: the projection of human experience by means of the translation of natural language into Fuzzy Conditional Statements (FCSs) or a more methodical process which analyzes the operating mode of the expert (specially interesting when the operator does not have well defined models about the control process in his mind).

The FCSs are formulated as an implication relation of the type IF-THEN between fuzzy linguistic propositions, which produce an estimation of the value of one or more linguistic variables of the form

$$\text{"}X_j \text{ is } A_j\text{"} \qquad (1)$$

where $A_j$ is one of the possible linguistic values of the variable $X_j$, to which we associate a probability distribution $\mu_{A_j}$ defined over a discourse universe $U_j$ (which we will consider discrete)

$$\mu_{A_j}: \quad U_j \quad \rightarrow [0,1] \qquad (2)$$
$$u_{j_{p_j}} \quad \mu_{A_j}(u_{j_{p_j}}) = a_{j_{p_j}}$$

which encompasses the meaning of the term $A_j$.

A fuzzy control system is configured as a set of unlinked FCSs of the type

$$\text{IF} \quad \text{"}X_1\text{" is "}A_1^r\text{" AND ... AND "}X_J\text{" is "}A_J^r\text{"}$$
$$\text{THEN "}Y_1\text{" is "}B_1^r\text{" AND ... AND "}Y_K\text{" is "}B_K^r\text{"}$$
$$r=1,...,R \qquad (3)$$

where R represents the number of rules, J the number of antecedents, K the number of consequents, $A_j^r = \{a_{j_{p_j}}^r\}$, $p_j=1,...,P_j$, and $B_k^r = \{b_{k_{q_k}}^r\}$, $q_k=1,...,Q_k$, are linguistic values associated with the input (state) and output (control) variables of the process, defined over the discourse universes $U_j$ and $V_k$, respectively.

The evaluation of a fuzzy control system as the one we have presented has been contemplated in the literature mainly from two perspectives. In one of them (Manzoul, 1987; Togai and Watanabe, 1986; Zadeh, 1988), a degree of fulfillment ($\sigma$), which will modulate the consequent part of each FCS, will be derived from the evaluation of the antecedent part for some specific values of the state variables. . An option which is often used although it has some problems (Takefuji and Lim, 1989), is to obtain $\sigma$ from the possibility of each one of the linguistic values associated in each rule to each input variable, with respect to the particular values taken in each situation for evaluation by the fuzzy control system. Finally, the evaluation of the whole rule base will end with the combination of the action values which derive from the particular evaluation of each rule.

The other perspective (Mamdani and Assilian, 1981), under which we will approach our proposal, consists in the construction of a Global Implication Matrix (GIM) which links, in a global manner, the condition and action parts of the rules of the FCS. This will imply the reduction of the set of R rules to a multidimensional matrix with a size independent from R, which seems speculatively interesting for high values of R. A study of the viability in calculation time and memory required with respect to other evaluation methods was carried out in Ruiz and co-workers (1991).

As our proposal is based on this method for the evaluation of FCSs, we will now present a more detailed description adapting it to the specific solution of the problem we propose.

## METHODOLOGY

Each one of the rules of a Fuzzy control system such as the one we presented before can be decomposed into $K$ subrules with a single consequent, so that once the global system has been reordered it is described by expression

IF $X_1$ is $A_1^r$ AND ... AND $X_J$ is $A_J^r$ THEN $Y_k$ is $B_k^r$

$$r=1,\ldots,R \qquad k=1,\ldots,K \qquad (4)$$

Therefore, the problem of evaluating the output variables in a multiple-antecedent multiple-consequent fuzzy control system can be solved by means of the parallel evaluation of $K$ multiple-antecedent single-consequent subsystems by simply performing a doubling and regrouping of the set of rules.

For each of the $K$ subsystems we can define a Global Implication Matrix (GIM)

$$R_k(p_1,\ldots,p_J,q_k) = \bigvee_{r=1,\ldots,R} \left[ a_{1p_1}^r \wedge \ldots \wedge a_{Jp_J}^r \wedge b_{kq_k}^r \right]$$
$$\forall\, p_j \in U_j, j=1,\ldots,J; \quad \forall\, q_k \in V_k$$
$$(5)$$

where $x \wedge y = \min\{x,y\}$.

This way, assuming we want to evaluate a situation in which each variable $X_j$ has the value $A_j' = \{a_{jp_j}'\}$ (distribution of possibility over $U_j$) associated with it, the calculation of $B_k' = \{b_{kq_k}'\}$ (distribution of possibility over $V_k$) can be obtained through the compositional rule of inference

$$B_k' = R_k \circ A_1' \circ \ldots \circ A_J' \qquad (6)$$

Using the sup-star composition rule (Zadeh, 1971a), and the intersection operator (Pedrycz, 1989) as T-norm, we have

$$b_{kq_k}' = \bigvee_{\substack{p_j=1,\ldots,P_j \\ j=1,\ldots,J}} \left[ R_k(p_1,\ldots,p_J,q_k) \wedge a_{1p_1}' \wedge \ldots \wedge a_{Jp_J}' \right]$$
$$q_k=1,\ldots,Q_k$$
$$(7)$$

The calculation of this expression by means of a systolic array forces the reduction of the dimensionality of matrix $R_k$ (whose dimension is $J+1$). This process is carried out by means of the projection both of the components of $R_k$, and those of the cartesian product of the input linguistic values. We will associate to matrix $R_k(p_1,\ldots,p_J,q_k)$ a 2D matrix $Dk=\{Dk_{q_ks}\}$ such that

$$Dk_{q_ks} = R_k(p_1,\ldots,p_J,q_k)$$

where, for each $s=1,\ldots,S$, with

$$S = \prod_{j=1}^{J} P_j \qquad (8)$$

the set of $p_j$ is given by the solution of the diophantine equation (9) subject to the

$$s = p_J + \sum_{i=1}^{J-1}\left[ (p_i-1) \times \prod_{j=i+1}^{J} P_j \right] \qquad (9)$$

initial conditions $1 \le p_j \le P_j$, $\forall\, j=1,\ldots,J$; $p_j \in \mathbb{N}$, which guarantee that the solution to the problem is unique.

In an analogous manner we obtain the expression for the cartesian product of the input linguistic values as a vector with $S$ components:

$$a_s' = a_{1p_1}' \wedge \ldots \wedge a_{Jp_J}', \quad s=1,\ldots,S \qquad (10)$$

where the $p_j$ must equally verify equation (9).

This way, we can rewrite expression (7) as

$$b_{kq_k}' = \bigvee_{s=1}^{S} Dk_{q_ks} \wedge a_s', \quad q_k=1,\ldots,Q_k \qquad (11)$$

Formally, the operation indicated in expression (11) coincides with the classical product of a matrix with $q_kS$ elements times a vector with $S$ components, by simply substituting operator "$\wedge$" by "$\cdot$" and operator "$\vee$" by "$\sum$". This way, through a simple operator exchange, we can carry out the evaluation of fuzzy control systems adopting architectures which are specifically designed for the product of matrices, and projection partition techniques for algorithms adequate for this type of operations. This last aspect is specially interesting as it permits the independence of the evaluation architecture with respect to the size of the problem.

## SYSTOLIC ARCHITECTURE FOR THE EVALUATION OF FUZZY CONTROL SYSTEMS.

A systolic array is made up of a network of elementary processors (PEs) which transfer data among themselves in a rhythmic and regular manner, performing in each stage a series of brief operations and which maintain a regular flux throughout the network. Its characteristics of modularity, regularity and local interconnection make them very attractive for their implementation with VLSI techniques.

Different approaches to systolic architectures aimed at the evaluation of FCSs from the two perspectives outlined in the Introduction have been presented in the literature. However, in all of them, the number of PEs required depends on the dimensions of the problem (number of elements in the discourse universes where the linguistic variables are defined and their number). Therefore, it is interesting from a practical point of view to develop a systolic architecture which can approach the evaluation of Fuzzy Control Systems independently from their dimensions, giving way to a highly versatile hardware structure with respect to its application possibilities.

Making use of the parallelism between the evaluation of Fuzzy Control Systems and the classical matrix-vector product (as can be deduced from expression (11)), we will take as a base a technique for projection onto systolic architectures using matrix calculus operations (Navarro and others, 1987), and adapting it to the specific needs of the evaluation of this type of systems.

## DBT Projection Technique.

Fig. 1 illustrates with an example the theoretical base for the DBT partitioning and projection method (Dense to Band matrix transformation by Triangular-block partitioning) proposed by Navarro and co-workers. In Fig. 1a we perform the product according to the traditional technique. In Fig. 1b and 1c we describe an alternative method for the computation of the product: in Fig. 1b we present a problem transformed from the previous one and in Fig. 1c we establish the procedure for the composition of the solutions of the transformed problem to produce the solution of the original problem.

Even though in both methods the number of operations to be performed is identical, the second one presents an advantage that we could call "of structural type". As the number of operandi which intervene in the computation of the components of this vector is constant, (depends exclusively on the number of non zero elements in a column or row of the transformed matrix, that is, its bandwidth) we will be able to perform the global calculation in a structure which is completely independent from the dimensions of the original problem. This fact, added to the simplicity and reiteration of the operations implied make this procedure very adequate for its implementation on a systolic architecture.

The key point of this method is the production of the transformed matrix from the original matrix. In order to do this we can use several options, one of which we will now describe applied to the evaluation of FCSs.

## Application of the DBT Method to the Evaluation of FCSs

Following a method which is analogous to the one we have already presented we can approach the evaluation of fuzzy control systems from expression (11). Once the two-dimensional equivalent matrix of the GIM for each rule subsystem has been obtained, the solution of the transformed problem, whose structure is formally identical to that of the original problem, can be approached:

$$B_T' = Dk_T \bullet A_T' \qquad (12)$$

where $Dk_T$ is a band matrix calculated from Dk. The process for obtaining the transformed matrix is briefly described below:

First we divide the matrix Dk into square submatrices $C_{i,j}$ of dimension $\omega$, being $\omega$ the number of PEs of the architecture onto which we are going to project the FCS; we add rows and/or columns of zeroes when necessary, choosing $N'$, $M' \in \mathbf{N}$ so that the following equalities are verified:

$$N \Delta \prod_{j=1}^{J} P_j = \omega N' \qquad (13)$$

$$M \Delta Q_k = \omega M' \qquad (14)$$

Each square submatrix is divided into its lower triangular, diagonal and higher triangular components

$$C_{i,j} = C_{i,j}^L + C_{i,j}^D + C_{i,j}^U = C_{i,j}^L + C_{i,j}^{DU} \qquad (15)$$

After this, a band matrix $Dk_T$ is constructed

by means of the algorithm we describe now:

```
FOR m=1 TO m=M'
DO
   {
   FOR n=(m-1)N'+1 TO n=N'+m-1
   DO
      {                              (16)
      Dk_Tn,n  = C_m,n-(m-1)N'^DU
      Dk_Tn,n+1 = C_m,n+1-(m-1)N'^L
      }
   Dk_TmN',mN'  = C_m,N'^DU
   Dk_TmN',mN'+1 = C_m,1^L
   }
```

As we have already pointed out, the cartesian product of inputs must also be transformed. In order to do this we employ the following expression:

$$a_{Ts}' = a_{s\text{-}mod\text{-}N}', \quad s=1,..,(N'M'+1)\omega-1 \quad (17)$$

It is evident, therefore, that the main advantage of approaching the solution of expression (11) with this technique is that the calculation is divided into $N'$ stages of $\omega$ operations each, and this permits the use of a cyclic cumulative hardware structure for the solution of the problem, which does not depend on the size of the FCS to be evaluated. This will be the main advantage of the systolic architecture we propose and the description of which we are now going to present.

## Description of the Systolic Architecture

In Fig. 2 we show the scheme of a basic hardware cell (PE) which permits the evaluation of a fuzzy control system. Each of the PEs carries out the following operations:

$$Y_{out}(t+1) = Y_{in}(t) \vee (X_{in}(t) \wedge A(t))$$
$$X_{out}(t+1) = X_{in}(t) \qquad (18)$$

considering discrete instants of time. We must point out that with an adequate coding of the data, the operation MIN can be implemented using AND gates and the operation MAX using OR gates. This evidently produces a great simplicity in the PE and, therefore, in the systolic array.

The interconnections between PEs, as well as the global configuration of the array, is shown, with a specific example, in Fig. 3. In a generic case, the use of an array with $\omega$ PEs requires a transformation of matrix Dk of expression (11) into a band matrix with a bandwidth equal to $\omega$. This way, each processor will carry out one of the $\omega$ calculations necessary in each of the $N'$ stages into which the computation of each component of the fuzzy control variable is divided, as was previously established. This dependence between the number of PEs and the dimension of the matrices into which Dk is divided during the process of approaching the transformed problem, conditions, as we will see, important parameters in the functioning of the array such as the data output frequency and the total computation time.

A fundamental point in the description of the systolic is the adequate control of the input sequence and, in particular, the flux in the feedback of partial results.

In this sense, the dumping of the components of the transformed input vector follows, as can be seen in Fig. 3, an extremely simple sequence, with an alternation between waiting cycles and data input cycles. The flux of the elements of the transformed matrix also follows a simple sequence, dumping into each PE one of the diagonals of this matrix, after the initial waiting and synchronization states.

With respect to the control of the partial result feedback loop, as the global computation of each component of the action variable of the controller is carried out cumulatively in several stages, it will be necessary to feedback, at the beginning of a new computation stage, the partial result obtained until then. In a general case, the first partial result is obtained in the cycle $2\omega-1$. From this moment, it begins to generate every two clock cycles the partial results of the first $\omega$ components of the output vector. As these partial results will not be used in the calculation until $\omega$ clock cycles have elapsed (instant in which the next computation stage begins) it will be necessary to endow the feedback circuit with a retention mechanism (FIFO memory with $\omega$ words). The first feedback must occur when element $a'_{\omega+1}$ of the transformed input vector has reached the input $X_{in}$ of the $\omega$-th PE. From that moment, the new partial results must be feedback every two clock cycles until the final results of the first $\omega$ components of the action variable are obtained (this happens in cycle $2N-1$ and is completed in cycle $2(N+\omega)-3$). From here on, the feedback sequence is repeated, as the input situations for each PE are analogous every $2N$ cycles. All this synchronism is achieved in an immediate way using the retention mechanism we have already described.

Apart from the output instant for the first definitive data (output vector components) item and the data output frequency, which we have already obtained, a important parameter when describing a control system is the cycle in which the last definitive data item is obtained, as it measures the minimum temporal separation between two inputs to the control system (successive observations of the process). This last data item will come out of the first PE after $2N(M'-1)$ cycles from the generation of the $\omega$-th element of the control variable, this means in cycle

$$2(N+\omega)-3 + 2N(M'-1) = 2(NM'+\omega)-3 \quad (19)$$

### ANALYSIS AND COMPARISON

Table 1 shows the comparison between the results obtained in the analysis of different parameters of interest for the systolic proposed in the present work. Apart from the temporal parameters, analyzed in the previous section, we have included an indicative evaluation of the number of memory locations necessary for the correct operation of the array, including the necessary registers in the feedback control.

The clock cycle in which the last output data item of the array is obtained is equal to the one obtained in other solutions which do not present partition (Kung, 1988; Manzoul, 1988) (Kung's was proposed for matrix-vector multiplication processes, although, as we have said, it is a problem which is formally identical to this one) when the following condition is verified:

$$M'=1 \text{ or } N'=1 \quad (20)$$

or in an equivalent way,

$$\omega=M \text{ or } \omega=N \quad (21)$$

As, in general, a Fuzzy Control System will verify $N \gg M$, the option $M'=1$ makes the proposed solution imply a lower number of PEs with respect to those commented, which require $N+M-1$ PEs. However, the equality with respect to the clock cycle in which the last result is obtained is only maintained if, besides,

$$\exists \ \alpha \in N \ / \ N = \alpha M \quad (22)$$

In any other case, the application of the projection algorithm to the array, needs the addition of a number of columns ($M-1$ in the most unfavorable case) to the equivalent two-dimensional matrix, thus making the value of N slightly higher. Despite this, and as the condition $N \gg M$ is still valid, the following approximation is possible

$$N+M-1 \approx N \quad (23)$$

Also, in most cases, the number of elements of the discourse universes over which the linguistic values of the variables of the Fuzzy Control System are defined, will be constant. That is,

$$P_j = Q_k = p, \quad \forall \ j=1,\ldots,J; \quad \forall \ k=1,\ldots,K \quad (24)$$

And therefore, substituting in expressions (13) and (14), we have

$$N=p^J \qquad M=p \quad (25)$$

where it is trivial to proof that expression (22) is verified.

The same thing does not happen with respect to the memory requirements of both arrays, whose analysis shows a clearly unfavorable result for those proposals which depend on the size of the problem, this comes from the large number of processors they require.

We could finally comment on the possibility of grouping into one physical PE (each one of those really implemented in the systolic) the functions which are theoretically carried out by several PEs. This way we could reach a degree of use of PEs close to one, achieving a noticeable increase with respect to the limit of 0.5 imposed by the DBT partition.

### CONCLUSIONS

We have presented a systolic architecture for the evaluation of Fuzzy control systems, based on a projection/partitioning technique applied to the expressions obtained for describing this evaluation. We achieve this way, complete independence between the size of the problem and the dimensions of the array required (number of PEs). The comparison of the efficiency of the systolic we propose with other solutions which do not present partitions produces similar results with respect to temporal parameters (throughput rate) and highly favorable results from the point of view of hardware savings (number of PEs and memory).

## REFERENCES

Kung, S.Y. (1988). In T. Kailath (Ed.), *VLSI array processors*. Prentice Hall, Englewood Cliffs, N. Jersey.

Mamdani, E.H. and S. Assilian (1981). Advances in the linguistic synthesis of fuzzy controllers. In *Fuzzy reasoning and its applications*. Academic Press.

Manzoul, M.A. (1987). Fuzzy inference on a systolic array. *Proc. 18th Modelling and Simulation Conference*, 1103-1108.

Manzoul, M.A. and H.A. Serrate (1988). Fuzzy systolic arrays. *Proc. 18th Int. Symp. Multiple-Valued Logic*.

Navarro, J.J., J.M. Llabería and M. Valero (1987). Partitioning: An Essential Step in Mapping Algorithms into Systolic Array Processors. *IEEE Computer*, July, 77-89.

Pedrycz, W. (1989). In M.J.H. Sterling (Ed.), *Fuzzy control and fuzzy systems*. Research Studies Press, Taunton, Somerset, England.

Ruiz, R., S. Barro, A. Bugarín and J. Presedo (1991). Fuzzy production systems on a VLSI architecture. *Proc. 1991 IFSA Congress*, Brussels, 175-178.

Takefuji, Y. and M.H. Lim (1989). Computation scheme for the general purpose VLSI fuzzy inference engine as expert system. *Knowledge-based systems, 2*, 109-116.

Togai, M. and H. Watanabe (1986). Expert system on a chip: an engine for real-time approximate reasoning. *IEEE Expert*, Fall, 55-62.

Zadeh, L.A. (1965). Fuzzy sets. *Information & Control*, 8, 338-353.

Zadeh, L.A. (1971a). Similarity relations anf fuzzy orderings. *Informat. Sci.*, 3, 177-200.

Zadeh, L.A. (1971b). Towards a theory of fuzzy systems. In R.E. Kalman and N. DeClaris (Eds.), *Aspects of network and systems theory*. Holt, Rinehart, Winston.

Zadeh, L.A. (1988). Fuzzy logic. *IEEE Computer*, April, 83-93.

#### TABLE 1 Parameters of interest for the Systolic we propose.

General case

First result (cycles)
$2N-1$

Last result (cycles)
$2(NM'+\omega)-3$

Requested Memory (words)
$\omega[2\omega(M'N'+1)-3]$

$$\begin{bmatrix} y_1 \\ y_2 \\ y_3 \\ y_4 \end{bmatrix} = \begin{bmatrix} a_{11} & a_{12} & a_{13} & a_{14} \\ a_{21} & a_{22} & a_{23} & a_{24} \\ a_{31} & a_{32} & a_{33} & a_{34} \\ a_{41} & a_{42} & a_{43} & a_{44} \end{bmatrix} \times \begin{bmatrix} x_1 \\ x_2 \\ x_3 \\ x_4 \end{bmatrix}$$

$$y_1 = a_{11}x_1 + a_{12}x_2 + a_{13}x_3 + a_{14}x_4$$
$$y_2 = a_{21}x_1 + a_{22}x_2 + a_{23}x_3 + a_{24}x_4$$
$$y_3 = a_{31}x_1 + a_{32}x_2 + a_{33}x_3 + a_{34}x_4$$
$$y_4 = a_{41}x_1 + a_{42}x_2 + a_{43}x_3 + a_{44}x_4$$

a)

$$\begin{bmatrix} y_1^T \\ y_2^T \\ y_3^T \\ y_4^T \\ y_5^T \\ y_6^T \\ y_7^T \\ y_8^T \end{bmatrix} = \begin{bmatrix} a_{11} & a_{12} & 0 & 0 & 0 & 0 & 0 & 0 \\ 0 & a_{22} & a_{23} & 0 & 0 & 0 & 0 & 0 \\ 0 & 0 & a_{13} & a_{14} & 0 & 0 & 0 & 0 \\ 0 & 0 & 0 & a_{24} & a_{21} & 0 & 0 & 0 \\ 0 & 0 & 0 & 0 & a_{31} & a_{32} & 0 & 0 \\ 0 & 0 & 0 & 0 & 0 & a_{42} & a_{43} & 0 \\ 0 & 0 & 0 & 0 & 0 & 0 & a_{33} & a_{34} \\ 0 & 0 & 0 & 0 & 0 & 0 & 0 & a_{44} & a_{41} \end{bmatrix} \times \begin{bmatrix} x_1 \\ x_2 \\ x_3 \\ x_4 \\ x_1 \\ x_2 \\ x_3 \\ x_4 \\ x_1 \end{bmatrix}$$

$$y_1^T = a_{11}x_1 + a_{12}x_2$$
$$y_2^T = a_{22}x_2 + a_{23}x_3$$
$$y_3^T = a_{13}x_3 + a_{14}x_4$$
$$y_4^T = a_{24}x_4 + a_{21}x_1$$
$$y_5^T = a_{31}x_1 + a_{32}x_2$$
$$y_6^T = a_{42}x_2 + a_{43}x_3$$
$$y_7^T = a_{33}x_3 + a_{34}x_4$$
$$y_8^T = a_{44}x_4 + a_{41}x_1$$

b)

$$y_1 = y_1^T + y_3^T$$
$$y_2 = y_2^T + y_4^T$$
$$y_3 = y_5^T + y_7^T$$
$$y_4 = y_6^T + y_8^T$$

c)

Figure 1. Computation of a Matrix Vector Product. a) Dense matrix. b) and c) Band Matrix.

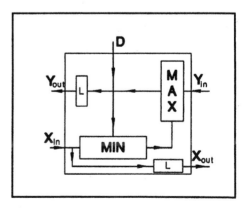

Figure 2. Internal Structure of the PE.

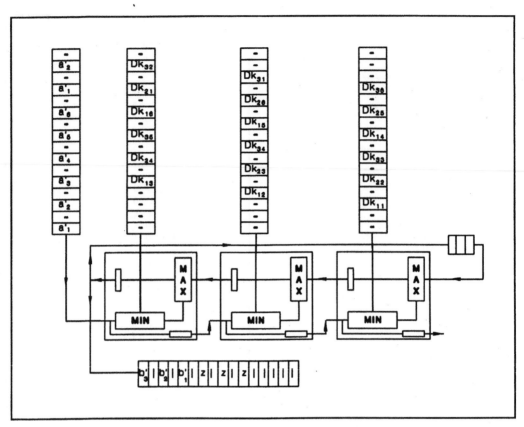

Figure 3. Data I/O Sequence in the Evaluation of a Fuzzy Control System (N=6, M=3) on an Array of 3 PEs. (In the data Outputs, the Values "z" represent partial Results).

# A HIGHLY NONLINEAR FUZZY CONTROL ALGORITHM
# FOR SERVO SYSTEMS POSITIONING

R. Herrero, J. Landaluze, C.F. Nicolas and R. Reyero

*Department of Control Engineering, Ikerlan, P.O. Box 146, E-20500, Mondragon (Guipuzcoa), Spain*

Abstract. This paper proposes to apply new fuzzy control techniques for servo systems positioning, in order to improve performance of traditional Proportional and Proportional-Derivative position controllers. The proposed knowledge base generates better highly nonlinear controllers compared with other recent results. The behaviour of this controller includes nonlinear actions that correct operation under a wide range of set points.

Keywords. Fuzzy control; nonlinear control systems; motor control; d. c. motors.

## INTRODUCTION.

During the last years many authors have been working to solve the problem of positioning in motion control by means of fuzzy control algorithms (Li, 1989; Agüero, 1990). The main objective of these works was to improve the performance of the fuzzy positioning systems versus Proportional (P) and Proportional plus Derivative (PD) controllers. Most of these works just attain the performance of the best tuned PD controllers.

In this paper, we propose the use of innovative fuzzy control techniques as a way of obtaining oustanding performances over conventional controllers. These algorithms suit the sampling period to the time constant of the motor, and the number of fuzzy sets and relevant rules to the required system dynamics.

## OBJECTIVES AND SYSTEM DESCRIPTION.

The objective of this paper is to suggest a set of fuzzy control rules after comparing four different positioning algorithms when testing its transient and frequency responses. Two of these controllers are conventional P and PD, while the other two controllers belong to the set of fuzzy algorithms. The knowledge base of the first fuzzy controller reproduces the behaviour of an optimal PD controller. The knowledge base of the second fuzzy algorithm generates a highly nonlinear fuzzy controller, whose performance is far better than the previous algorithms and will be the main objective of this paper.

In order to evaluate this controller, we have worked with an unloaded DC motor, the model TT-2952 from Inland, and a 5,000 pulses per revolution encoder. The fuzzy control algorithms were implemented with an i486/33 based PC/AT. Two ISA cards, an encoder input card and a D/A output card, were used to close the position loop as described in figure 1.

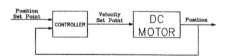

*Fig. 1. Position control loop.*

## DESCRIPTION OF CONTROL ALGORITHMS.

*Conventional Linear Controllers.*

The control law that characterizes P and PD controllers is the well known

$$u(k) = K_p [e(k) + X_d de(k)] \qquad (1)$$

where $u(k)$ is the dynamic velocity set point fed into the motor at kth sampling period, the signal error $e(k)$ is the difference between the desired position set point and actual position of the motor, and the change of error $de(k)$ is $de(k) = e(k) - e(k-1)$. In particular, when $X_d$ is zero we have a P-type controller. As (1) is a linear equation, the characteristic surface of this controller is a plane (graphical representation of control signals $u$ in relation to $e$ and $de$). This surface is represented in figure 2, when using unity parameters as $K_p = X_d = 1$.

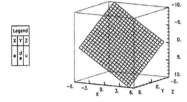

*Fig. 2. PD characteristic surface.*

## *Fuzzy Controllers.*

A fuzzy set A of a universe of discourse $X$ is characterized by a membership function $\mu_A : X \longrightarrow [0,1]$, which associates with each element $x$ of $X$ a number in the interval [0,1]. As the membership function is not restricted to a discrete set $\{0,1\}$ as occurs with the classical set, the fuzzy sets offer treatment of vagueness and qualitative concepts.

Fuzzy controllers are performed with the objective of replacing the experience of a human operator, and are implemented by an algorithm programmed into a computer. The human operator infers a control action of a universe of discourse $U$ in relation to data examined in previous universes, $E$ and $DE$ when there are two. His experience, which will have vague concepts such as "big", "small", "around -2", etc., could be summarized as a knowledge base. This is a set of N rules as

$$
\begin{array}{l}
\text{IF } e \text{ is } E_1 \text{ AND } de \text{ is } DE_1 \text{ THEN } u \text{ is } U_1 \\
\text{IF } e \text{ is } E_2 \text{ AND } de \text{ is } DE_2 \text{ THEN } u \text{ is } U_2 \\
\qquad\qquad\qquad \vdots \\
\text{IF } e \text{ is } E_N \text{ AND } de \text{ is } DE_N \text{ THEN } u \text{ is } U_N
\end{array}
$$

$E_i$ and $DE_i$ are fuzzy sets defined in the object spaces $E$ and $DE$ (antecedents), but $U_i$ are fuzzy sets in the space image $U$ (consequents). Every set includes vague notions as mentioned before.

If we consider the objects $e_o$ and $de_o$, our fuzzy algorithm will get the degree of suitability for each rule as a minimum of the antecedents membership function, namely

$$
\sigma_i = \mu_{\text{Si AND DEi}}(e_o, de_o) = \min ( \mu_{Ei}(e_o), \mu_{DEi}(de_o) ) \quad (2)
$$

Based on the observed antecedents or inputs, the fuzzy controller deduces a fuzzy consequent U whose membership funtion is

$$
\mu_U(u) \qquad = \max [ \min( \sigma_i, \mu_{Ui}(u) ) ] \qquad (3)
$$
$$
1 \leq i \leq N
$$

Depending on the way the ith rule obtains a better degree of occurrence, the inferred consequent U will be closer to the consequent $U_i$. This corresponds to the sup-min inference procedure. The relation (3) can be expressed as

$$
\mu_U(u) \qquad = \max [ \min ( \alpha_j, \mu_{Uj}(u) ) ] \qquad (4)
$$
$$
1 \leq j \leq n
$$

where n is the different number of fuzzy sets defined in the space U and $\alpha_j$ is the maximum degree of membership of those rules whose consequent is $U_j$.

The control output sent to different processes is the centre of gravity of the figure described by the membership function of the inferred fuzzy control signal U. The equation is

$$
u_o = \frac{\int_U u\, \mu_U(u)\, du}{\int_U \mu_U(u)\, du} \qquad (5)
$$

Every procedure referred to before is described in the references (Agüero, 1990; Pedrycz, 1989; Lee, 1990; Sugeno, 1985; Huang, 1990; Mizumoto, 1988).

Similar to the PD controller, our fuzzy controllers will have as inputs the position error $e$ and the derivative position error $de$, and as output the velocity set point $u$. Each variable will be changed by a scale factor GE, GDE and GU, which will be the only adjustable controller parameters to be changed and tuned. Then, the fuzzy controller inputs will be $e_o = GE \cdot e_o^*$ and $de_o = GDE \cdot de_o^*$ instead of the actual error $(e_o^*)$ and derivative error $(de_o^*)$. In the same way, the real control fed into the system will be $u_o^* = GU \cdot u_o$ instead of the control signal $u_o$ inferred by the fuzzy controller.

The spaces of errors and derivative errors are divided into eleven fuzzy sets, so we will obtain 121 rules. These sets contain notions such as "around -5", "around -4", etc. The membership functions have a triangular distribution with equidistant vertices, as is shown in figure 3.

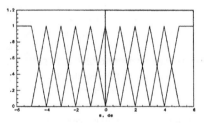

*Fig. 3. Membership functions in E and DE spaces.*

The overlapping between sets is the minimum one that assures the stationary error suppression, as is described in Mizumoto(1988).

The control signal space will be divided into eleven fuzzy sets (see figure 4).

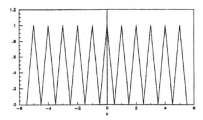

*Fig. 4. Membership functions in U space.*

In this case, in order to simplify the defuzzyfying mechanism, the overlapping between sets become zero. The inferred fuzzy control will have the shape shown in figure 5.

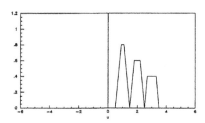

*Fig. 5. Inferred Control without overlapping.*

The centre of gravity of this figure can be computed by means of the following analytical procedure

$$u_O = \frac{\Sigma \, u_i \, A_i}{\Sigma \, A_i} \qquad (6)$$

where $u_i$ is the centre of gravity of each isosceles trapezium, whose value coincides with the central value of maximum membership, and $A_i$ is the area of each trapezium. These areas can be computed as a function of their height, which is the degree of likelihood $\alpha_i$ that corresponds to the control signal $u_i$. Then, it is very easy to conclude that the final control law is:

$$u_O = \frac{\Sigma \, u_i \, (2 - \alpha_i) \, \alpha_i}{\Sigma \, (2 - \alpha_i) \, \alpha_i} \qquad (7)$$

Two fuzzy controllers will be implemented, with two different knowledge bases. The first knowledge base is described in table 1.

| e \ de | -5 | -4 | -3 | -2 | -1 | +0 | +1 | +2 | +3 | +4 | +5 |
|---|---|---|---|---|---|---|---|---|---|---|---|
| -5 | -5 | -5 | -5 | -5 | -5 | -5 | -4 | -3 | -2 | -1 | +0 |
| -4 | -5 | -5 | -5 | -5 | -5 | -4 | -3 | -2 | -1 | +0 | +1 |
| -3 | -5 | -5 | -5 | -5 | -4 | -3 | -2 | -1 | +0 | +1 | +2 |
| -2 | -5 | -5 | -5 | -4 | -3 | -2 | -1 | +0 | +1 | +2 | +3 |
| -1 | -5 | -5 | -4 | -3 | -2 | -1 | +0 | +1 | +2 | +3 | +4 |
| +0 | -5 | -4 | -3 | -2 | -1 | +0 | +1 | +2 | +3 | +4 | +5 |
| +1 | -4 | -3 | -2 | -1 | +0 | +1 | +2 | +3 | +4 | +5 | +5 |
| +2 | -3 | -2 | -1 | +0 | +1 | +2 | +3 | +4 | +5 | +5 | +5 |
| +3 | -2 | -1 | +0 | +1 | +2 | +3 | +4 | +5 | +5 | +5 | +5 |
| +4 | -1 | +0 | +1 | +2 | +3 | +4 | +5 | +5 | +5 | +5 | +5 |
| +5 | +0 | +1 | +2 | +3 | +4 | +5 | +5 | +5 | +5 | +5 | +5 |

*TABLE 1  LFC Knowledge Base.*

The inputs and outputs of this table must not be considered as parameters, on the contrary these values are linguistic labels "around -5", "around -4", etc. This knowledge base contains a linear distribution of rules. It means that if one of the inputs "increases" one fuzzy category (for example $e$ increases "around +2" to "around +3"), while the other one remains constant (for example $de$ = "around +1") the output also increases one fuzzy category (from "around +3" to "around +4"). That is why the obtained controller behaviour looks like a PD controller. We will refer to this controller as "Linear Fuzzy Controller" (LFC). This controller is not a linear one because the composition rule sup-min is an intrinsic non linear process. The characteristic surface of this controller can be seen in figure 6, when using unity parameters GE=GDE=GU=1. As we expressed before, this surface is similar to a PD one, but shows saturations by the corners because we use a finite number of fuzzy sets.

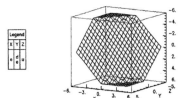

*Fig. 6. LFC characteristic surface.*

The second knowledge base is founded on the previous one, but includes a certain amount of modifications based on heuristic reasoning in order to improve transient response to a step change in the motor set point. The controller obtained with this knowledge base will be highly nonlinear, and we will refer to it as "Nonlinear Fuzzy Controller" (NFC). The mentioned modifications have as a main objective to reduce the derivative effect when the response starts or stops, so that we obtain a faster acceleration both when the response begins and when the actual position of the motor is close to set point and deceleration is neccessary. The NFC knowledge base is that described in table 2.

| de<br>e | -5 | -4 | -3 | -2 | -1 | +0 | +1 | +2 | +3 | +4 | +5 |
|---|---|---|---|---|---|---|---|---|---|---|---|
| -5 | -5 | -5 | -5 | -5 | -5 | -5 | -5 | -5 | -5 | -5 | -5 |
| -4 | -5 | -5 | -5 | -4 | -4 | -4 | -4 | -4 | -3 | -3 | -3 |
| -3 | -5 | -5 | -4 | -4 | -3 | -3 | -3 | -2 | -2 | -1 | -1 |
| -2 | -5 | -5 | -4 | -3 | -2 | -2 | -2 | -1 | +0 | +1 | +1 |
| -1 | -5 | -4 | -3 | -2 | -1 | -1 | +0 | +1 | +2 | +3 | +3 |
| +0 | -5 | -4 | -3 | -2 | -1 | +0 | +1 | +2 | +3 | +4 | +5 |
| +1 | -3 | -3 | -2 | -1 | +0 | +1 | +1 | +2 | +3 | +4 | +5 |
| +2 | -1 | -1 | +0 | +1 | +2 | +2 | +2 | +3 | +4 | +5 | +5 |
| +3 | +1 | +1 | +2 | +2 | +3 | +3 | +3 | +4 | +4 | +5 | +5 |
| +4 | +3 | +3 | +3 | +4 | +4 | +4 | +4 | +4 | +5 | +5 | +5 |
| +5 | +5 | +5 | +5 | +5 | +5 | +5 | +5 | +5 | +5 | +5 | +5 |

*TABLE 2  NFC Knowledge Base.*

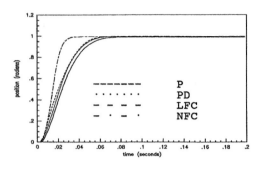

*Fig. 8. Step responses.*

The most important difference related to the LFC is the rule "IF $e$ is "around +5" AND $de$ is "around -5"", whose output is now "around +5" instead of "around +0". With this procedure we ignore the derivative effect when response begins. The rest of the NFC knowledge base is performed while smoothly changing the control signal between that rule and the central rule, " IF $e$ is "around +0" AND $de$ is "around +0" THEN $u$ is "around +0" ", in order to eliminate the stationary errors. The characteristic surface of the NFC is shown in figure 7, once more using unity parameters.

We see that the P, PD and LFC controllers produce a similar response, as were previously supposed. The NFC algorithm has a faster response, as we intended on introducing as a new heuristic behaviour of this controller. In this case, the rise time is half the time obtained with conventional linear controllers and the LFC. The stationary error is even smaller because of the Coulomb friction.

The excellence of the NFC can be tested looking at figure 9, which also shows control signals.

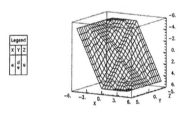

*Fig. 7. NFC characteristic surface.*

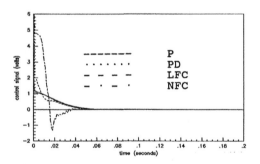

*Fig. 9. Control signals.*

## EXPERIMENTAL RESULTS

The four controllers were previously tuned in order to obtain a critically damped response when changing position set points in 1 radian. The optimal parameters obtained using a sampling period of 1 millisecond are

P:    $K_p=0.94$  $X_d=0$
PD:   $K_p=1.10$  $X_d=1.15$
LFC:  GU=2.00 GE=1.10 GDE=6.00
NFC:  GU=2.00 GE=2.40 GDE=20.0

The obtained transient responses are shown in figure 8.

P, PD and LFC algorithms give small and smooth control outputs. NFC, however, initially gives a larger control signal in order to accelerate the motor. When the position is near to the set point, the control signal is sharply modified, decelerating the motor. The effect as a whole is a faster response.

Gains and phases of the output signals were analyzed, compared with sinusoidal inputs of 1 radian of amplitude. The related Bode diagrams are those described in figure 10, which shows that the bandwidths obtained with fuzzy controllers are far better (15 Hz.) than those obtained with P and PD algorithms (9 Hz.). The phase losses are also improved using fuzzy control algorithms, because they reach half the losses of classical P and PD controllers inside the bandwidths of LFC and NFC controllers.

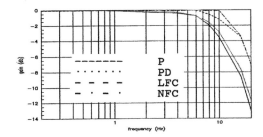

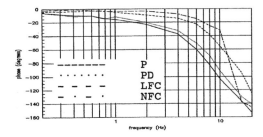

*Fig. 10. Bode plots.*

Two different characteristics are pointed out to show efficiency and correct functioning of the NFC controller: the number of rules and selected sampling period. The number of rules must be big enough to introduce additional heuristic behaviours, and to obtain the characteristic surface described in figure 7. The sampling period must be small enough to permit the controller to generate sharply controlled responses as described in figure 9. Several tests performed both by simulation and experiments carried out in the laboratory, tell us that the minimum appropriate sampling period is half the time for the NFC controllers than those used in classical P and PD controllers. For instance, if T is the settling time of a critical damped classical controller, a sampling period of T/80 will make a NFC controller to have a settling time of T/2 or less.

## MODIFIED NFC ALGORITHM: VARIABLE PARAMETERS CONTROLLER

Nevertheless, a controller like NFC could have some disadvantages, such as offering different dynamic responses to different set points. The only condition in which the behaviour could be considered appropriate is near the set point in which the controller was tuned. Because of that, the Bode Diagram is very much dependent on the amplitude of the sinusoidal input signal. That is because set points (W) different to the tuned set point ($W_{sint}$) make different trajectories in the transformed phase space (we refer to the space $e_o$ - $de_o$ and not to the real space $e_o{}^*$ - $de_o{}^*$) . It means that the evaluated rules also become different. Then, changing the initial scale factors ($GE_{sint}$, $GDE_{sint}$ and $GU_{sint}$) with the set point evolution, this problem could be avoided properly.

For example, let us consider a set point W double the one used in the tuning process. If we expect the transient response

to be similar, every sampling period the variables $e_o{}^*$ and $de_o{}^*$ must be double those obtained when the set point was $W_{sint}$. If we want $e_o$ and $de_o$ to have the same value in both transient responses, in order to verify similar rules, it is obvious that the initial scale factors $GE_{sint}$ and $GDE_{sint}$ must be reduced by half. On the other hand, the control signals $u_o{}^*$ fed into process must be double, that is why the scale factor GU must be double. A summary of the variations that must be contained in the algorithm are those described by the equations (8), (9) and (10).

$$GE \qquad = GE_{sint} / (W / W_{sint}) \qquad (8)$$

$$GDE \qquad = GDE_{sint} / (W / W_{sint}) \qquad (9)$$

$$GU \qquad = GU_{sint} \cdot (W / W_{sint}) \qquad (10)$$

We name the obtained controller as "Nonlinear Fuzzy Controller of Variable Parameters" (NFCVP). As was mentioned before, when the process is linear and the initial conditions are zero, the behaviour of this controller is similar to different set points. If the system has no offset, the frequency response of the system becomes independent of the sinusoidal input signal. Nevertheless, the frequency response of the system could be different to the original NFC because the parameters change continuously with sinusoidal set points.

The Bode diagram of the NFCVP appears in figure 11, compared with those of the NFC and LFC algorithms.

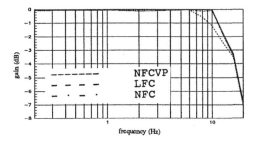

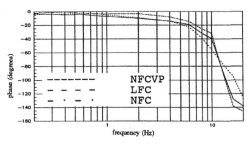

*Fig. 11. Bode plots.*

We can see that the NFCVP gain is similar to the NFC, whereas the phase is lightly worse than described by LFC. The changes introduced into the original algorithm produce nearly the same dynamic effect, and obtain similar frequency responses.

## CONCLUSIONS

An optimal control of positioning in DC motors can be obtained using fuzzy control algorithms, and subsequently improves performance reached with classical position controllers. In order to reach these objectives we need a number of suitable rules, a minimum sampling period and, of course, and efficient knowledge base. The irregular behaviour of the controller in different set points, due to the fact that is nonlinear, can be solved by simply redefining scale factors, as this does not change frequency response significantly.

## REFERENCES

E. Agüero,E. and others (1990). *"Estudio de la Aplicabilidad de Técnicas de Control Difuso en la Optimización del Control de una Máquina Herramienta"*. VIII Congreso INVEMA. October, 1990. San Sebastián.

Li,Y. F. and C. C. Lau (1989). *"Development of Fuzzy Algorithms for Servo Systems"*. IEEE Control Systems Magazine, pp. 65-71.

Pedrycz, W. (1989). *"Fuzzy Control and Fuzzy Systems"*. Research. Studies Press Ltd.

Lee C. C. (1990). *"Fuzzy Logic in Control Systems: Fuzzy Logic Controller"*. IEEE Transactions on Systems, Man, and Cybernetics, vol. 20, no. 2, pp. 404-435.

Andersen,T. R. and S. B. Nielsen (1985). *"An Efficient Single Output Fuzzy Control Algorithm for Adaptive Applications"*. Automatica, vol. 21, no. 5, pp. 539-545.

Sugeno, M. (1985). *"An Introductory Survey of Fuzzy Control"*. Information Sciences, no. 36, pp. 59-83.

Ying,H., W. Siler and J. J. Buckley (1990). *"Fuzzy Control Theory: A Nonlinear Case"*. Automatica, vol. 26, no. 3, pp. 513-520.

Huang,L. J. and M. Tomizuka (1990). *"A Self-Paced Fuzzy Tracking Controller for Two-Dimensional Motion Control"*. IEEE Transactions on Systems, Man, and Cybernetics, vol. 20, no. 5, pp. 1115-1124.

García,M. C. (1991). *"Inteligencia Artificial en el Control de Procesos: Controladores Borrosos"*. Mundo Electrónico, Feb. 1991, pp. 42-49.

Mizumoto,M. (1988). *"Fuzzy Controls Under Various Fuzzy Reasoning Methods"*. Information Sciences, no. 45, pp. 129-151.

# FUZZY CONTROLLER FOR GENERALLY LOADED
# DC ELECTRIC MOTOR

### A. Boscolo*, C. Mangiavacchi**, F. Drius*** and M. Golak***

*Zeltron S.P.A., V. Principe di Udine 114-30330 Campoformido, Italy
**D.E.E.I. Department of Electrical and Electronics Engineering and Computer Science, University of Trieste,
V.A. Valerio 10-34100 Trieste, Italy
***LASA Laboratory, D.E.E.I. University of Trieste, Italy, V.A. Valerio 10-34100 Trieste, Italy

Abstract. In the paper an application of the fuzzy logic to the control of an electric motor assigned to a generic activity, and therefore loaded in a unknouwn way, will be presented, pointing out with particular attention the tuning and optimization methods, based on neural networks. The developed controller, even if showing a particular interest in many applicative areas both industrial or not, will only represent an applicative example of the possibilities given by the proposed approach.

Keywords. Fuzzy Systems; Fuzzy Controller; Neural Network; Tuning.

## INTRODUCTION

This paper shows, by use of an applicative example, how a control system could be realized starting from simple considerations and avoiding the need of an explicit identification of the process to be controlled. Such control system shall be able to tolerate large variations affecting the characteristics of the process under control, granting at the same time a global behaviour allowing many different applications.

The possibility of developing a control system able to drive a general process to a defined dynamic behaviour, is strictly bound to the availability of dedicated procedures allowing to automatically tune the values of the controller parameters.

More generally, such procedures shall show their capability to modify the regulator behaviour according to the characteristics of the process under control and how they evolve.

$$G(s) = K \left( 1 + s\, Td + \frac{1}{s\, Ti} \right) \qquad (1)$$

The PID controllers Eq.(1), are presently the most extensively used and their performances are generally reported to be satisfactory in many application fields. Nevertheless, when dealing with processes showing a time variant or a strongly non linear behaviour, the use of fixed structure and parameters regulators (like the PIDs) results in severe limitations. Particularly, the tuning of this type of regulator, aiming to reach the required time domain specifications in a closed loop, seems to ask the use of skills that could be regarded more as a sort of art than as a scientific activity.

When different reasons don't allow the process under control to be modelled, the tuning process is performed by a skilled operator whose process knowledge is strong enough to optimize in a subjective form the control performances. The operator is not requested to know the theoretical ground on which the process is based, he is simply supposed to have a good experience on it and to intuitively perceive its behaviour.

In these circumstances it is obviously impossible to refer to the optimisation in terms of mathematics. A different operator would certainly use different options which would nevertheless lead to a similar global behaviour of the system. Different approaches are used to perform a time domain tuning of regulators. Beside the Ziegler-Nichols method based upon the stability limits of the system (Ziegler 1942.), the most known are those addressing the optimization of a performance index (Lopez 1967.) those based upon the pattern recognition (Kraus 1984) and those based upon the use of expert-systems (Litt 1991.).

The latter include the fuzzy logic based expert control systems to which a growing interest is currently devoted both for speculation and application purposes (Mamdani 1974, 1984.).

The fuzzy control systems can be basically divided in two main classes:

1) Control systems in which the the fuzzy logic informs the realization of the regulator to the extent that this one is fully described by the fuzzy rules (Mamdani 1974, 1984.).

2) Control systems in which the principles of the fuzzy logic inform the expert tuning of regulator parameters in accordance with the behaviour of the process under control (Litt 1991, Tzafestas 1990.).

This paper addresses a system belonging to the second class and will provide a description of a method whose use allows to improve the controller performances by means of an aided tuning procedure.

## THE CONTROLLER

The controller we are introducing is a Fuzzy controller adapting the parameters of a generic PID regulator. This could be as well of fuzzy type, even if it could be looked as an adaptive control system working on error function and its derivative (Ying 1990.).

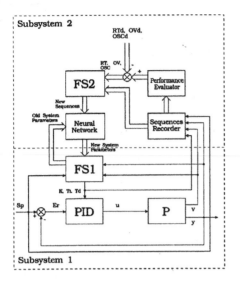

Fig. 1 System Architeture.

The system architecture is shown in Fig. 1 where two different sub-systems can be recognised, the first properly devoted to the process control, the second to be used to optimize, when required, the performances of the first in stationary situations.

A structure, very similar to those used for gain scheduling adaptive systems (Scattolini 1990), has been selected for the first sub-system, the real process controller. The aim is to reduce the complexity of the regulator granting at the same time satisfactory results in terms of global performances even for processes whose modelling would not be available or would result not affordable.

The main difference between the controller we are introducing and a gain scheduling adaptive system is to be basically found in the mechanism driving the instantaneous selection of the control parameters. These, instead of being drawn from a given matrix, are supplied by a fuzzy system (FS1) built upon a set of rules deriving from the knowledge and intuitive perception of a generic skilled operator.

The fuzzy system 1 [FS1] operates over the PID parameters in order to minimize some of the inherent characteristics (overshoot, dumping and response time) of the error related to the system step response.

The extended experience deriving from a reaction analysis of the generic process under the PID action performed by the regulator, could be used to define the rules for the parameters variation.

The following heuristic consideration can be derived from the observation of the process behaviour :

1) When the overshoot is mainly caused by the integral term, a slight decrease of it, at the moment in which the system answer exceeds the set-point, can result in a significant reduction of the overshoot. On the other side, a slight increase during the leading edge can allow a reduction of the leading time.

2) When the flatness of the step response is mainly caused by the derivative term, a slight increase of its value during the leading edge and the steady state could speed-up the system response and dampen the

oscillations that often could be generated under these conditions.

3) Increasing the proportional term effect, its would be possible to reduce the leading time but the oscillations would increase.

An example has been selected to support a clear and efficient description of the different sub-systems. It will be seen how, although related to a limiting application, this approach allows to discuss the method features avoiding to loose the needed character of total applicability of results. A speed regulator for a permanent magnet electrical motor has been selected. The motor will drive a time varying load and, under the above assumptions, the FS1 fuzzy controller will exclusively perform the control upon the integral action of the PID.

The FS1 controller input variables are the regulator input (set point - Sp) the process output (motor speed - Y) and an auxiliary signal (V). The FS1 internal variables are the error (Er), the error first derivative (dEr) and the V variable. V allows to monitor the existence of non-linearities potentially resulting, in the selected example, from the saturation of the power supply or from physical limits of the motor. In these circumstances V will be defined as:

$$V = K - Ia \qquad (2)$$

where K is a constant and Ia the current value of the motor armature.

In this case the FS1 controller provides only one output PI representing the weight of the integral action of the PID regulator. Three fuzzy sets N - Z - P have been defined for the values range of input variables, the abbreviation stand for *negative , zero , positive* . As well three fuzzy sets Z - PN - PB have been defined for the values range of the output variables, the abbreviations stand for *zero, positive normal, positive big*.

The membership of fuzzy sets used in the fuzzy expert controller are shown in fig. 2.

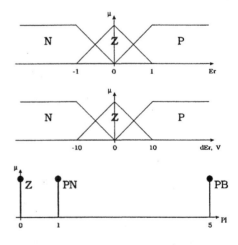

Fig.2 Input and Output Memebership Function of FS1

The rules supporting FS1, deriving from the above declared criteria, are expressed in this example under the following form:

TABLE 1  Rules of Fuzzy System 1 (FS1)

| Er | dEr | V | PI |
|----|-----|---|-----|
| N | / | N | Z |
| N | / | P | PB |
| P | / | N | PB |
| P | / | P | Z |
| Z | / | / | PN |
| / | / | Z | PN |
| / | Z | / | PN |
| N | N | Z | PB |
| P | P | Z | PB |

The classical min-max inference mechanism suggested by Zadeh (Zadeh 1965.) is used and the defuzzify operations are based upon the determination of the center of gravity of the envelope of the output set.

The fuzzy system FS1 is merely heuristic. It follows that, the process expert is supposed to find himself the input and output fuzzy sets. Many times, in spite of the soundness of its skills, the process operator reveals unable to transpose his experience in the correct definition of input and output membership functions, which is strictly needed to properly realize an expert control system. Because these reasons, the availability of controllers based upon a limited number of rules could be of greatest interest when dealing with system realization. Beside the quantity limitation, the rules shall be properly set and able to consistently cope with the phenomena involved in the process to be controlled.

Controllers developed accordingly to the above reveal to be particularly suitable for integration in control systems whose HW platforms are base on general purposes micro-controllers.

It follows the need to optimize and validate the system global behaviour using suitable criteria, i.e. minimizing the error over a known input-output sequence. These goals could be reached without changing the structure of the fuzzy model set by the expert through a global modification of the input and output fuzzy sets (Boscolo 1992.).

This will be achieved by use of the second sub-system as indicated in Fig. 1 which will perform, when needed, a procedure of aided tuning for the above described controller. This second sub-system stay in an upper layer in respect to the controller which will be tuned through the minimization of a performance index (i.e. the output standard deviation over an I/O sequence of known examples).

The definition process of the FS1 parameters (new membership functions) is performed through the learning phase of expressly developed neural network.

Interest is devoted to neural networks in this field because these structures are deemed able to autonomously "spot" even the rules informing the expert operator natural behaviour from which a given input-output sequence has been originated (Kosko 1992.).

In this application the use of the neural network is aimed only to tune, for a defined sequence of examples, the parameters of a generic fuzzy system.

## THE TUNING PROCESS

The tuning process identifies the input and output membership functions able to find a global minimum in the output variance over a known sequence of input-output examples (Boscolo 1992.).

A key element to run a successful process, beside a proper system description and initial definition of fuzzy sets, is to select a sequence of examples able to fully represent the expected system behaviour.

The required sequence could be originated in different ways, the most used methods are the following:
a) the sequence is predefined by the expert
b) the sequence is derived from dedicated laboratory experimentations
c) the sequence is iteratively and heuristically derived from subsequent adjustments operated on to the plant.

In the present example all the above methods will be used in different forms and times to automatically generate a sequence of examples. In respect of the former, this final sequence shall result in a global improvement of the system behaviour which would approach a given reference behaviour.

The tuning procedures are usually performed as follows (see Fig. 1):

the generic step response of the system is recorded together with the corresponding evolution of the auxiliary variable and PID parameters. With reference to the above conditions, the global system performances are determined by evaluating at the same time how much far they are from the required performances set by the expert operator (OVd, DMPd, RTd).

A sequence of weights is generated by the above information which infer into the FS2 together with those relating to the instantaneous error and to the auxiliary variable.

The new weights act upon the actual sequence of PID parameters, by which the response under examination was originated, leading to the generation of the new sequence of examples allowing, in the expert opinion, the system to approach in a closer way the reference response.

Next step in the tuning procedure is the determination of FS1 membership functions which allow to globally minimize the error on this new sequence. To this end FS1 is represented by use of a neural network whose inputs are the error, its derivative, and the auxiliary variable relating to the new examples sequence and whose output is the PID parameters. The network representing FS1 is forced, during its learning procedure, performed by use of a modified back-propagation mechanism, to minimize the global error upon the examples sequence through a modification of input and output membership function. The new parameters enabling the optimization of the FS1 behaviour in respect to the given sequence, will be available at the end of the learning procedure.

A short description of the second sub-system composition is provided below.

## I/O Sequence Recorder

The main function of this block is to record, during a consistent interval of time, the patterns of input variables of the performances evaluator (Y, V, Sp) and the output variables (K, Ti, Td) of the FS1 fuzzy system controlling the PID regulator. By this way the recorder is able to develop the actual I/O sequence on which the FS2 controller will operate.

## Performances Evaluator

The existing literature provides different suggestions to evaluate the dynamic performances of a system (Maeda 1990.).

In the above explained example, as well as in many applications, the overshoot level OV, the dumping DMP and the time response RT (expressed by the parameters shown in Fig. 3) have been deemed consistent enough to evaluate system performances.

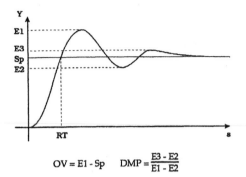

$$OV = E1 - Sp \qquad DMP = \frac{E3 - E2}{E1 - E2}$$

Fig. 3 Systems Performance Measurement

#### Fuzzy system 2

The main function of the fuzzy system FS2 is to generate a new input-output sequence on which the tuning of FS1 parameters will be performed. The FS2 system is based as well upon the above described criteria.

The inputs of the Fuzzy system 2 can be divided in two classes: the first includes global parameters (see Fig. 3) while the second includes the errors and the auxiliary variable.

Peculiar global parameters are represented by the following differences that show how much the actual response is far from the required response expressed by the RTd, OVd, DMPd terms:

$$\Delta RT_k = RTd - RT_k \qquad (3)$$
$$\Delta OV_k = OVd - OV_k \qquad (4)$$
$$\Delta DMP_k = DMPd - DMP_k \qquad (5)$$

In this example FS2 provides only one output which represents the weight to be used to transform the corresponding actual sequence of PID parameters (FS1 output) into a new sequence to which a new tuning interactions shall be applied.

The Input and Output fuzzy sets of FS2 are show in Fig. 4.

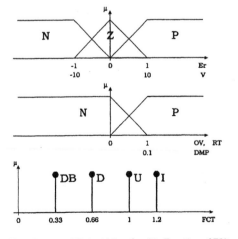

Fig.4 Input and Output Memebership Function of FS2

To derive the rules describing FS2, criteria similar to those above explained have been used. These rules can be expressed as show in Table 2:

TABLE 2 Rules of Fuzzy System 2 (FS2)

| ΔRT | ΔOV | ΔDMP | E | V | FCT |
|-----|-----|------|---|---|-----|
| N | N | N | / | / | U |
| P | N | N | Z | Z | I |
| P | N | N | Z | N | U |
| P | N | N | Z | P | U |
| / | N | P | Z | / | U |
| / | N | P | N | / | D |
| / | N | P | P | / | D |
| / | P | N | Z | / | D |
| / | P | N | N | / | U |
| / | P | N | P | / | U |
| / | P | P | Z | / | DB |
| / | P | P | N | / | U |
| / | P | P | P | / | U |

The new I/O sequence will result by multiplying the sequence of the PID actual parameters, (from which the last step response has been provided) by the corresponding factor defined by FS2.

#### The neural network

The following paragraph illustrates the topology of the neural network expressly developed for the system. During the definition of its structure focus has been devoted in holding an explicit correspondance between the parameters defining the fuzzy input and output fuzzy sets of the system and the weights of the neural network. The global configuration of the resulting network is shown in Fig. 5 from which clearly appears how nodes performing the operations of sum, minimum, maximum and divisions have been used (Kosko 1992, Masuoka 1990.).

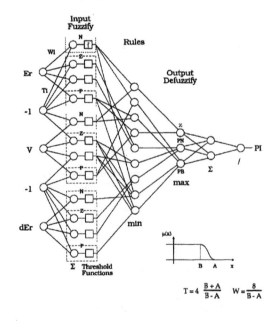

Fig. 5 Neural Network Devoted to Tune FS1

The threshold functions, restricted to the first layer of the network, cover a basic role into the realization of the input membership functions (these can be both saturation non-linearities both sigmoids). The use of a threshold function alone allows to get decreasing and increasing type of membership functions, while two threshold functions shall be combined with a minimum operation when compound type membership functions are required. The rules are built using two minimum nodes (one each) allocated in the second layer. The output membership functions are realized using the maximum nodes (third layer) and are identified by two parameters: the area, corresponding to the fuzzy sets, and the center of gravity abscissa. The last two layers of the network are used to perform the defuzzify operation and to implement the calculation of the barycenter abscissa of the total envelope. The solution is reached by combining the output fuzzy sets following the min-max criteria.

The definition of the neural network structure follows the calculation methodology which is used to derive in a fuzzy system the output value starting from the input values. Under this condition, the network structure reflects the structure of the fuzzy system that must be tuned (the FS1 in the present application). A simple back-propagation algorithm, properly modified to allow the use of nodes performing the minimum-maximum operations, is used in the learning phase of the network.

### OBTAINED RESULTS

The results arising from the application example are illustrated below. It must be noted that, to allow the method to be described as simply as possible, only the weight of the PID integral term has been modified. In spite of this limit, the resulting improvements reveal to be consistent even in presence of variations in the load driven by the motor.

A reference system has been selected to evaluate how the process-regulator system behaves when changes occur in characteristics of the load driven by the motor. The system, whose parameters are shown below, is composed by a PID regulator tuned according Ziegler-Nichols, by a motor and by a reference load.

#### Motor Parameters

$Ra$ $= 0.6\ W$
$La$ $= 6\ mH$
$KF$ $= 0.55\ \dfrac{N\,m}{A}$
$VMax$ $= 110\ V$

#### Reference Load

$B = 0.08\ \dfrac{N\,m\,s}{rad}$

$J = 0.09\ Kg\,m^2$

#### PID Parameters

$K$ $= 30;$
$Ti$ $= 0.0175\ s$
$Td$ $= 0.004375\ s$

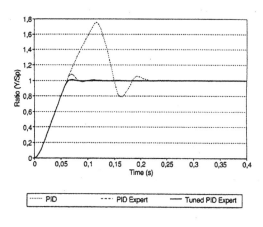

Fig. 6 Step Response of the System With the Reference Load

Fig. 6 shows the typical patterns of the system step response with the motor driving the reference load and
- being controlled by the PID regulator tuned according to Ziegler Nicholts or
- provided by a control system whose tuning was performed according only to the operator experience in which FS1 operates only on the integral part leaving unchanged the original P & D values or
- being controlled by the regulator used in this application whose tuning was performed according the present suggested method after four interactions.

Fig. 7 shows the step responses of the system whose tuning has been left unchanged under a load which has been consistently modified. The load changes have been a 300 % increase in the original inertia moment and a further friction effect, resulting in a constant term and in a term varying as square function of the motor speed. It clearly appears that, under the above conditions, the stability limits are nearly reached when the traditional PID regulator is used. The system here introduced, although not specifically tuned for this new load configuration, holds nevertheless an step response that is fully consistent with the needs of many real applications.

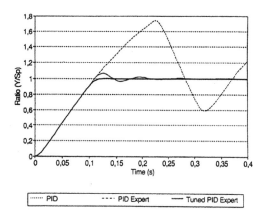

Fig. 7 Step Response of the System With Modified Load

## REFERENCE

Boscolo A.,Drius F. (1992).
   Computer aided tuning and validation of fuzzy system.
   *IEEE international conference on fuzzy system March*
   *1992 San Diego.*
Kosko B. (1992).
   *Neural networks and fuzzy system.*
   ed.Prentice Hall,New Jersey
Kraus T.W., Myron T.J. (1984).
   Self tuning PID controller uses pattern recognition
   approach.
   *Control Eng.* June
Litt J. (1991).
   An expert system to perform on line controller tuning.
   *IEEE tran. Control System* April
Lopez A.M., Miller J.A., Smith C.L., Murrill P.W. (1967).
   Tuning controllers with error integral criteria.
   *Instrumentation Technol.* November
Maeda M., Sato T., Mirakami S. (1990).
   Desing of the self tounig fuzzy controller.
   *Proceeding of the international conference on fuzzy*
   *logic and neural network June*
Mamdani E.H., Ortergaard J.J., Lembessis E. (1984).
   Use of fuzzy logic for implementation rule based
   control of industrial processes.
   *Fuzzy set and decision analysis .* Nort Holland New
   York
Mamdani E.H. (1974).
   Applications of fuzzy algorithms for control of simple
   dynamics plant.
   *IEE proc N.21*
Masuoka R., Watanabe N., Kawamura A., Owada Y.,
Asakawa K. (1990).
   Neurofuzzy system fuzzy inference using a structured
   neural network.
   *Proceeding of international confernce on fuzzy logic*
   *and neural networks Lsuka Japan July 1990*
Scattolini R., Schiavoni N. (1990).
   Introduzione al controllo adattativo.
   *Automazione e Strumentazione* October
Tzafestas S., Papanikolopoulos N.P. (1990).
   Incremental fuzzy expert PID control.
   *IEEE Trans. Indus. Electron. V.37* October
Ying H., Siler W., Buckley J.J. (1990).
   Fuzzy control theory: a nonlinear case.
   *Automatica V.26 N.3*
Takagi T., Sugeno M. (1983).
   Derivation of fuzzy control rules from human
   operator`s control actions.
   *Proc. of IFAC Symp. on Fuzzy information,knowledge*
   *rapresentation and decision analysis   Marseilles*
   *(France)*
Zadeh L.A. (1965).
   Fuzzy sets.
   *Inform. Contr. V.8*
Ziegler J.G., Nichols N.B. (1942).
   Optimun setting for automatic controllers.
   *Trans. A.S.M.E. V.64*

# A CONTROL ARCHITECTURE FOR OPTIMAL OPERATION
# WITH INDUCTIVE LEARNING

L. Magdalena, J.R. Velasco, G. Fernandez and F. Monasterio

*E.T.S.I. Telecomunicacion, Ciudad Universitaria S/N, E-28040 Madrid, Spain*

**ABSTRACT**

This paper describes an architecture for expert process control for those processes where learning is needed. Machine learning let the system to discover new rules to act over new circumstances and improve process performance in known areas. The architecture is open enough to allow developing such different control systems as fuzzy or classical logic based ones. The real application described in section III shows the capability of two different process control systems designed under this scheme.

The rule evaluation mechanism let the system to modify the credibility of every rule which is able to be used. This algorithm makes possible the existence of a *limbo*: a place where induced new rules are tested, before to be included into the knowledge base.

**Keywords:** Intelligent control systems, Genetic Algorithms, Fuzzy Control, Machine Learning, Credit Assignment

## I.- INTRODUCTION. STATE OF THE ART.

Expert control idea was suggested some years ago [Åström et al, 86]: rule based expert systems might be used to tune a PID controller, to help operators when any alarm is blinking, etc... Expert systems may be used like a conventional control system where classical control method are not adequate (the problem is sufficiently complex, there is no time to call for an expert, there is not a complete mathematical model of the process) and we have an expert who knows how to control the process [Shirley, 87]. Nevertheless, there are some problems that expert control is not able to solve: expert knowledge elicitation (which is the typical bottleneck) and the limited domain of the expert system (which is not able to work fine on situations which are not known) are the main barrier. Machine learning let us to override this barrier, by generating new knowledge to control unexplored regions of the state-space of the process.

Some rule learning systems have been developed to control small processes. Fogarty applies genetic algorithms to control the combustion in multiple burner installations ([Fogarty, 88] and [Fogarty, 89]), and Grefenstette talks about an architecture that uses production rules to control a simple process and generates new rules with genetic algorithms [Grefenstette, 89]. The main difference between Grefenstette architecture and the work we have developed is on the inference engine, the ability to use fuzzy logic, and an special area of the knowledge base called *limbo* were new rules are tested prior to be used ([Velasco, 91a] and [Velasco, 91b]).

The main application of this architecture has been to build an expert system capable of help operators to optimize the heat rate in a fossil power plant (as described in section III), suggesting control actions, refining the rules weight using the results of the operator actions, and learning new rules.

## II.- NEW CONTRIBUTION

The architecture of the control system (fig. 1) is conventional system based. A group of data is received from the process each T seconds. These groups of data, called *context vectors*, build the facts base. This context vector groups some variables of the process that may be divided in three subvectors:

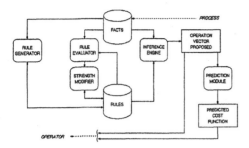

Fig. 1 - System architecture

*Operation vector*: With all the variables which can be read from the process and modified by the operator. The output of the control system is a new operation vector, where new values for this variables are suggested to the operator.

*Exploitation vector*: With all the variables that define the objective and the particular working conditions for the process. Most of these variables are determined by economic and political criteria.

*Installation vector*: With all the variables measured from the external world (environment) and those that represent the internal structure of the process.

Rules of the knowledge base owns a weight or strength which is used to measure its credibility. This weight is modified growing when the system suggest good actions to reach the goal, and shrinking when they are bad.

The proposed architecture has four added modules to conventional expert systems: The **rule evaluator** is used to estimate the payoff for any rule used at the inference process. It is able to evaluate those rules that could have been used too. This evaluator let the **strength modifier** to alter the rule's weight. A **rule generator** let the system to enlarge the knowledge base, in order to learn new actions both for known and new situations. This allow the system to expand the domain of the expert system (which was one of the conventional expert system limitations), and to increase the knowledge elicited to the experts (which was the other one). When the new suggested operation vector is generated, a **prediction module** is used to estimate the cost function value, in order to know how good or bad are going to be these actions.

These learning systems are able to work with random initial knowledge. Good rules will increase their strength and bad ones will decrease it. We only need wait until the rule generator make new good rules. The prediction module helps the operator to select between good and bad

suggestions. Nevertheless, it is better to have some good initial knowledge which allow the system to work properly since its installation. We can get initial knowledge from experts, if their are accessible. Anyway, experts are not the unique source of knowledge: historical databases, with sequences of context vectors, may be used to select viable actions to reach a good performance from a selected situation. Of course, if we are able to use a process simulator, we can learn control rules from observation of the simulated evolution of the process, which may be introduced in the initial knowledge base too.

Next paragraphs will describe in detail the main aspects of the shown architecture: the evaluator and the strength functions, two different approaches used to build the inference engine and the induction module which have been used to develop two different systems (described at section III), and the prediction module, used for both system to show the operator the predicted goodness of the suggested actions.

**II.1.- Evaluator and strength modifier**
The rule evaluation is made by combining two different values, both of them in the [-1,1] interval. The first one is related with the objective of the process and it will be the same value for all rules evaluation. The second one measure how nearly is the suggestion of the rule from the action made by the process operator (we must remember that our control system just suggest some actions to the operator, but it is him who has to do them).

The first value, the objective evaluation -OE-, is calculated assigning the maximum value (1) if the process is working at the best condition. In other case, if the process is working inadequately, the objective evaluation is set to -1. On those processes where the objective is defined as the optimization of an specified value, we should use a function like the fig. 2.

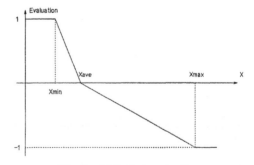

Fig. 2 - Objective evaluation

The abscise coordinate represents the variable to optimize (in this case to minimize), X, and the ordinate the objective evaluation. Xmax, Xmin and Xave are the maximum, minimum and average value for the variable over the last $n$ inputs. If X have a value lower than Xmin, evaluation is set to 1 (it is the best situation that the system knows). When X is getting higher, evaluation is decreasing, until X is higher than Xmax, when evaluation is set to -1.

In the other hand, the nearly evaluation -NE- will be different for any of the rules involved in the inference process, as it is set on base to how nearly the suggested values is from the real value in the next vector received. Function used is shown in the fig. 3. The abscise coordinate symbolize the difference between the suggested and the real values, where the minimum evaluation is set to those rules which suggest the most dissimilar values, and the maximum to those ones which suggest exactly the real read value. he point where evaluation is 0 is set as a percent of the maximum difference.

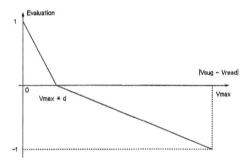

Fig. 3 - Nearly evaluation

| RE | OE + | OE - |
|------|------|------|
| NE + | + | - |
| NE - | - | + |

Fig. 4 - Rule evaluation

Rule evaluation is set as *RE = OE · NE*. As we can see in the fig. 4, rule evaluation will be positive when:

- Process is working well (OE +) and the rule suggests a value near the read one (NE +): this rule is a good one, because we get good results with it.
- Process is working bad (OE -) and our rule suggests a value far away from the read one (NE -): this rule may be a good one, because we got bad results if we don't apply it.

In the other hand, evaluation will be negative when:

- Process is working well (OE +) and the rule suggests a value far different of the real one (NE -): the suggested value is not the good one (the real value is).
- Process is working bad (OE -) and the rule suggest a value near the real one (NE +): we have used the rule, and the results were not good.

The mechanism described above let us to evaluate all the rules we would be able to apply at any time: The objective evaluation is a common value for all the rules, an the nearly evaluation may be applied to any rule with true premise. This is why we can create the *limbo*, where young rules are tested, as we are able to evaluate all the rules that would have been used, both applied or not. The function used to modify rules strength is:

$$S(r,t+1) = \begin{cases} S(r,t)+K \cdot E \cdot (1-S(r,t)) & E>0 \\ S(r,t) & E=0 \\ S(r,t)+K \cdot E \cdot S(r,t) & E<0 \end{cases}$$

$$E \equiv Rule\ Evaluation$$

We are able to let the control system to learn in *dumb mode*, getting context vectors from the process, and throwing away the suggested values. System can learn by looking how the process operator works.

### II.2.- Fuzzy Control Based System

The fuzzy control based system has a two level architecture with three different modules. Discrimination module constitutes the first level, Optimization and Prediction modules are contained in the second one. This second level is divided in a family of rule sets, where each set is adapted to a particular situation. Discrimination module defines the actual working situation as a function of some exploitation variables and selects a set of rules (from Optimization and Prediction modules) to be used in decision making.

Fuzzy rules used have a structure as follows:

*If ($x_i$ is $C_{ij}$ and ... and $x_k$ is $C_{kl}$) then A*

where $x_i$ is the actual value of variable $i$ and $C_{ij}$ is a fuzzy set belonging to this variable, and A is the consequence of the rule. When translating knowledge from human operators to fuzzy rules, each fuzzy rule has attached a linguistic label.

Different consequences are defined for each module of optimization system:
- At Discrimination level A:
  Use the second (optimization) level labeled n.
- At Optimization level A:

$y_i$ is $C_{ij}$. Where $y_i$ is an output variable.

The Inference engine works as a fuzzy controller [Pedrycz, 89]. The union of both levels, discrimination and optimization can be viewed as an unique rule system. In fact if considering a discrimination rule and an optimization rule belonging to level selected by previous rule, a whole system rule is obtained, as shown in the following example taken from the system described on section III:

1st level rule:

| | |
|---|---|
| IF | Generated Power is LOW |
| AND | Cos φ is MEDIUM-HIGH |
| AND | %Fuel is VERY HIGH |
| THEN | Use second level no. 7 |

2nd level (no. 7) rule:

| | |
|---|---|
| IF | Live Steam Temperature is HIGH |
| AND | Live Steam Spray is LOW |
| THEN | Let Live Steam Spray to VERY LOW |

Whole system rule:

| | |
|---|---|
| IF | Generated Power is LOW |
| AND | Cos φ is MEDIUM-HIGH |
| AND | %Fuel is VERY HIGH |
| AND | Live Steam Temperature is HIGH |
| AND | Live Steam Spray is LOW |
| THEN | Let Live Steam Spray to VERY LOW |

**II.3.- Classic system**

The second approach uses classical logic, with a non usual operator that is *near of*. Rules are written like *IF X ≈0.01 ≈0.4 AND ... THEN Y = 0.2* (every variable is normalized in [0, 1]), where X may be any variable from the context vector, and Y, any variable from the operation vector. Rule has to be read as *IF X is between 0.39 and 0.41 AND ... THEN let Y = 0.23*. Full syntax let us to write expressions and references to past vectors, which are stored in the facts base. Our syntax is similar to such used by Grefenstette with a little difference: we specify both center and radius for the interval while Grefenstette uses the upper and lower limits [Grefenstette, 89]. We believe that center is the important value, and it must be explicitly specified. We have preferred classical logic rather than other knowledge representation mechanism (like classifier systems proposed in [Holland, 86] because of its better understandingly, and the ability to represent real numbers. An example of typical rule for the system described in section III could be:

| | |
|---|---|
| IF | LSP ≈0.05 ≈0.456 |
| AND | ShST ≈0.013 ≈ 0.951 |
| THEN | LST = 0.984 |

*(IF Live Steam Pressure IS NEAR OF 0.456 AND Superheated Steam Temperature IS NEAR OF 0.951 THEN SET Live Steam Temperature TO 0.984).*

In order to provide some memory to the control system, knowledge base must be split into small files, each one adapted to a particular state of the process. The variables we have to use to select the adequate knowledge base must be specified in the exploitation vector, as it is where the main aspects of how the process is working are stored.

Each time, when a new context vector is received, our system select the actual state file, and inference engine selects, from all possible rules which may be fired (true premise), the rules with highest strength for each operation variable suggested. We prefer to select the strongest rule rather than to use a random procedure (like Grefenstette system) because in a real control system we have to select the best action known: it's not time to try new ideas. Random procedure is needed when only used rules are evaluated, but our system, as has been shown in section II.1 is able to evaluate every possible rules.

More possible inference engines, with suggested value interpolation among the different available rules are proposed in [Velasco, 91a].

Rules are generated by genetic algorithms ([Goldberg, 89]) with three main operators: crossover, inversion and mutation. Two rules, selected because of their high strength (with a pseudo-random procedure) generate two new ones, applying the crossover operator. Over these two new rules, both, inversion and mutation, are applied with a low probability (we have used 0.1 for inversion and 0.01 for mutation). Next paragraphs show some examples about how operators work:

**CROSSOVER**

| | |
|---|---|
| Parents: | IF X AND Y AND Z THEN A |
| | IF R AND S THEN B |
| Children: | IF X AND S THEN B |
| | IF R AND Y AND Z THEN A |

**INVERSION**

| | |
|---|---|
| Parent: | IF R AND Y AND Z THEN A |
| Child: | IF R AND Z AND Y THEN A |

**MUTATION**

| | |
|---|---|
| Parent: | IF R AND Y AND Z THEN A |
| Child: | IF R AND H AND Z THEN A |

In fact, inversion does not generate new knowledge, but allows to join some antecedents and let the system to generate new rules with these antecedents together. Mutation can act over every part on the antecedent: the variable, the interval, the value, etc.. may be mutated.

New rules are installed in the *limbo*. When they increase their strength, they will replace weak rules from the knowledge base.

### II.4.- Prediction module
This module is a set of plant descriptions by means of fuzzy implications with premises similar to those showed above, and assertions like *Cost function* = $a_0 + a_1 x_1 + ... + a_n x_n$, where $x_i$ is the actual value from the input variable *i*. Fuzzy implication are divided into sets corresponding to those of Optimization level. The parameters $a_0...a_n$ are adapted using a parameter identification method.

### III.- APPLICATION RESULTS

Fossil power plants usually work when all other power plants (nuclear, water fall, etc...) are insufficient to generate demanded energy. They are switched on at last position because price of fossil fuel (gas, fuel-oil, etc...) make this energy more expensive than the others. Expert systems may be good tools to improve operator's work in a complex process like this. In addition, a fossil power plant has a typical cost measure, that is heat rate, to be used as optimization function. Having in mind all this ideas, fossil power plant control optimization seems to be a good test problem for the control architecture described above.

Both of systems proposed (Genetic and Fuzzy) have been implemented and installed in a real power plant, assisting operator to decide about the best action in order to obtain a heat rate reduction.

### III.1.- Application description
The expert system must receive the values concerned with some variables of the plant, and on the base of this values, it will propose some actions to operate the plant. The variables considered for the decision are a subset of the set of plant variables, which is named context vector. This subset is divided into the three different groups refereed above:

*Operation Vector* contains those parameters that are under direct control of the plant operators (i.e. *Main-steam temperature*).

*Exploitation Vector* contains those one that are fixed by central dispatching (i.e. *%fuel used*).

*Installation Vector* contains parameters that concerns with plant environment situation and cannot be changed (i.e. *sea water temperature*).

System receives each two minutes a vector with 40 parameters (16 from OV, 6 from EV and 18 from IV) and propose new values for the 16 parameters contained in Operation Vector. These new values suggested are presented to the plant operator in addiction with some other information and on the basis of this global output, the operator acts over the plant.

### III.2.- Test and results
In a first phase, the very large amount of parameters of the system, (genetic algorithms, evaluation margins, reinforcement constants, ...) where set to initial values, using very simple plant models before the final implementation.

In a second phase, each system where put to work in the real plant for two days: the first day was dedicated to let the system learn getting context vectors from the observation system and suggesting actions, that were not showed to the plant operator. During this phase both systems adjusted their initial rules. The second day, different for each system, the actions suggested where applied as near as possible, to see how the system really worked.

The goal of these tests were more qualitative than quantitative, because it was very difficult to compare the two days results. The different conditions and some external events made it impossible. Nevertheless some good actuation were appreciated. Both systems were able to learn some good rules that were used to optimize plant performance, and some of which were a surprise for operators. The next paragraphs remarks some examples of them. Systems generated bad rules too, which were killed by genetic algorithms after some tests in the first system, and faded its influence in the second one. At last, evaluation algorithm were tested, showing that the functions used were fine adjusted.

Fuzzy system discovered a new rule saying that for low power, it was better to work with a lower number of fuel burners, finding a (more or less) linear relation between power and burners when power was lower that 200 Mw, from this point, all burners were needed. This was surprising to operators that only worked usually with maximum burners, or minimum when plant worked at minimum power maintenance.

Some aspects of genetic system tests must be remarked too.

The reinforcement function, and the evaluation mechanism were tested by doing some suggested actions that operators known to produce a higher heat rate. The system noted the error setting down the rule strength, and didn't suggest the same action anymore.

The system memory was proved when at the evening, the system received a context vector very similar to other one received in the morning. The action realized by operators in the morning was remembered, suggesting to eliminate the injection sprays. Five minutes after, operators (who didn't know suggestion) realized the action putting sprays at 0, decreasing the eat rate. Rules applied were rewarded and stay suggested until a status changed have been produced.

Rule generator produced good and bad rules, the goods were rewarded, and the bad were killed after some cycles but they realized some bad suggestion before to be killed. To avoid this kind of suggestion, the idea of "limbo" have been added to the architecture but is still not implemented in working versions.

The ideas obtained from these tests have made possible to change some aspects of the programs to setup the last version that is working at this moment.

Described test have been realized in Dec. 90. Since these days both systems have been tested in real work with acceptable results for plant operator, and they are still working now.

## IV.- ACKNOWLEDGEMENTS

This project has been supported by the Spanish electric companies, and coordinated by OCIDE. It has been made in association with UITESA, and it has been tested and installed in the second group of Térmicas del Besós, S.A., which is a company of Enher and Hidroeléctrica de Cataluña (50% - 50%), at Sant Adriá del Besós, (Barcelona, Spain).

We would like to thank P. Hernán, J. García from UITESA and J. Riera, E. Villagut and J.M. Oliva from Térmicas del Besós, S.A. and all plant operators who have contributed with their knowledge and interest to tests both systems.

## V.- REFERENCES

[Åström et al, 86] "Expert Control". Åström, K.J., Anton, J.J. y Årzen, K.E. *Automática*, Vol. 22, pp. 277-286, 1.986.

[Fogarty, 88] "Adapting to Noise". Fogarty, T.C. *Proc. of the IFAC Conference on Artificial Intelligence in Real-Time Control*, UK, 1.988.

[Fogarty, 89] "The machine learning of rules for combustion control in multiple burner installations". Fogarty, T.C. *Proc. of the 5th Conference on Artificial Intelligence Applications*, IEEE Computer Society Press, 1.989.

[Goldberg, 89] Genetics Algorithms in Search, Optimization and Machine Learning. Goldberg, D.E. Addison-Wesley, Mass, 1.989.

[Grefenstette, 89] "A System for Learning Control Strategies with Genetic Algoritms". Grefenstette, J.J. *Proc. of the Third International Conference on Genetic Algorithms*. Schaffer, J.D. ed. Morgan Kaufmann, San Mateo, CA, 1.989.

[Holland, 86] "Escaping Brittleness: The posibilities of general-purpose learning algorithms applied to parallel rule-based systems" Holland, J.H. en *Machine Learning: An Artificial Intelligence Approach, Vol II*. Michalski, R.S, Carbonell, J.G. y Mitchell, T.M. eds. Cap 20, pp 593-623. Morgan Kaufmann Publishers, 1.986.

[Pedrycz, 89] Fuzzy Control and Fuzzy Systems. Pedryzc, W. John Wiley and Sons, Inc., 1.989.

[Shirley, 87] "Some Lessons Learned using Expert Systems for Process Control". Shirley, R.S. *IEEE Control Systems Magazine*, Dec, 1.987, pp. 11-15.

[Velasco, 91a] Arquitectura para Sistemas de Control Inteligentes. Velasco, Juan R. PhD. Thesis Departamento de Ingeniería de Sistemas Telemáticos. Universidad Politécnica de Madrid, 1.991.

[Velasco, 91b] "Algoritmo para evaluación de de reglas en un sistema de control con aprendizaje". Velasco, Juan. R.. Asociación Española Para la Inteligencia Artificial, IV Reunión Técnica AEPIA'91 Madrid, 1.991.

# A NEURAL NETWORK-BASED INVERSE SYSTEM CONTROLLER FOR A KIND OF SYSTEM WITH UNKNOWN DYNAMICS

**F.G. Zhu, Y.Y. Yang and Y.Z. Lu**

*Research Institute of Industrial Control, Zhejiang University, Hangzhou 310027, PRC*

Abstract. In this paper, we propose a novel neural network-based finite frequency band inverse system and corresponding controller for a kind of system with unknown dynamics. Their some properties such as restructure and existence in terms of neural networks are discussed. The two numerical simulations and a pilot liquid level control experiment are given to demonstrate that the proposed approach possesses strong adaptability and robustness.

Keywords. Nonlinear systems; inverse systems; neural networks; feedforword control; feedback control; level control.

## INTRODUCTION

In original systems, the observed state of a system $y(t)$ is assumed to be a function of current action $u(t)$ and prior state of the system, i.e. $y(t)=F(u(t),...,u(t-n_u), y(t-1),...,y(t-n_y))$. An inverse system is one that recover the input $u(t)$ by observing its output data $y(t)$. Studies of inverse systems are essentially related to the analysis of their input-output structure $F(\cdot)$. Most of the inverse systems studied so far is limited to the systems with exact mathematical model [1,2,3] and have a large number of applications in various fields such as decoupling, model-matching, control. However, in practical systems with unknown dynamics, we cannot recover the real value of its input from its output by above methods.

Currently there are a wide variety of neural networks being studied or used in control areas. Of them, the multilayer network is most popular and has proved sucessful in pattern recognition as well as in associative learning [4,5,6]. It has been proved [1] that a multilayer network with hidden layers and any fixed continuous sigmoidal nonlinearity can approximate any continuous function arbitrarily well on a conpact set. In this paper, an attempt is made to demonstrate that the multilayer networks will find applications in inverse system control for a kind of systems with unknown dynamics. We don't need to know what $F(\cdot)$ is, however, we must assume that $F(\cdot)$ is invertible as a function of $u(t)$. Given examples of $y(t)$ which resulted from actual input $u(t)$ which were tried out in some experiment. One may use supervised learning to adapt a multilayer network H such that $u(t)=H(u(t-1),...,u(t-n_u), y(t),...,y(t-n_y))$. In practice, it is impossible to implement $F(\cdot)$ -inverse when there exists the integral action in dynamical characteristics of the plant. This paper proposed a novel finite frequency band inverse system for the general type system. In this inverse system, only desired frequency component of the input is reconstructed by the pre-processor and the multilayer network whereas the high-frequency part is neglected. This inverse system, has several desirable characteristics from the application viewpoint: (1) it is dynamical (2) it reproduced the approximate value of the input (3) it can be realized by the multilayer networks. The use of multilayer network-based inverse systems raises many theoretical questions such as the existence

of the inverse, its properties, its reconstruction in terms of the multilayer networks, the latter two questions of which are discussed in this paper. Furthermore, the multilayer network-based inverse system controller as shown in Fig.2 is proposed for the nonlinear systems with unknown dynamics. Given a desire output $y_d$, the inverse system can output a feedforword control vector $\ddot{u}(t)$, according to the system states $y(t-1),...,y(t-n_y)$ and prior controls $u(t-1),...,u(t-n_u)$. The self-tuning PID control $\Delta u(t)$ is used as feedback control to eliminate the high-frequency error of the system such that the summation of $\ddot{u}(t)$ and $\Delta u(t)$ leads the output of the system $y(t)$ to match the desired trajectory $y_d(t)$. Finally various numerical simulations are carried out and a pilot liquid level control are given to test the performance of the proposed controller.

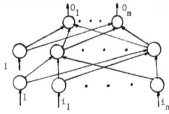

Fig.1  The Multilayer Feedforward Network

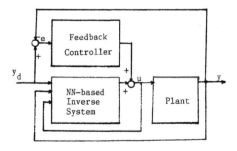

Fig.2  Feedforward Feedback Control Structure

## A MULTILAYER NETWORK-BASED INVERSE SYSTEM

Consider the non-linear system with unknown dynamics described by the following equation

$$y(k)=f(y(k-1),\ldots,y(k-n_y),u(k),\ldots,u(k-n_u)) \quad (1)$$

where $f(\cdot)$ is unknown non-linear vector function, $y(k)=[y_1(k),\ldots,y_m(k)]^T \epsilon R^m$ is the output vector of the system, and $u(k)=[u_1(k),\ldots,u_r(k)]^T \epsilon R^r$ is the vector.

Assumption 1: the non-linear system with unknown dynamics (1) is invertible and minimum-phased system.

The inverse system of the system (1) may be represented by

$$u(k)=g(y(k),y(k-1),\ldots,y(k-n_y),u(k-1),\ldots, u(k-n_u)) \quad (2)$$

where $g(\cdot)$ describes the inverse characteristics of the system (1) and is unknown. It is known that a multilayer network with hidden layers and any fixed continuous sigmoidal nonlinearity can learn to approximate any continuous function $y=f(x)$ from examples $(x_i,y_i)$. It is natural that the multilayer network is an ideal candidate for approximating the inverse system (2). The multilayer network shown in Fig.1 consists of three layers, where the number of input nodes $N_1=m(n_y+1)+r\cdot n_u+1$, the number of hidden nodes $N_2$ is given by genetic algorithm or experience[9], the number of output nodes $N_3=r$, the node function at the input and output layer is linear function $f(x)=x$, the hidden node function is sigmoidal function $f(x)=1/(1+\exp(-x))$. In the learning phase, $[y^T(k),\ldots,y^T(k-n_y), u^T(k-1),\ldots, u^T(k-n_u)]$ is used as the input of the multilayer network and $u^T(k)$ as the desired output. Fast recursive learning algorithm [8] of the multilayer network is adopted to adapt the network. When the number of examples is large enough, the multilayer network can approximate the inverse characteristics $g(\cdot)$ well enough. In the action phase, if the desired output of the system $y_d(k)$ is appropriately chosen, $[y_d^T(k), y^T(k-1),\ldots,y^T(k-n_y), u^T(k-1),\ldots, u^T(k-n_u)]$ is used as the input of the multilayer network and the output of the network $\ddot{u}(k)$ is used as the control of the system to force the output of the system approach the desired output $y_d(k)$. In order to track the change of the inverse characteristics of the system, the learning phase alternates with the action phase, Thus the multilayer network-based inverse system have strong adaptability. In practical problems, there may exist some integral action in dynamical characteristics of the system (1). It is known that the inverse characteristics $g(\cdot)$ will contain some differential action. From theory and experiments, we know that it is impossible for the multilayer network to implement the complete differential action, but it can implement the partial differential action. This stimulates us to propose the concept about the neural network-based finite frequency band inverse system which only require the multilayer network to learn the low frequency compoment of the inverse system (2) whereas the high-frequency part is neglected, i.e. A low frequency filter is added in front of the multilayer net H and the output of the filter $\bar{y}$ instead of the output of the system y is used as the input of the net. Thus the multilayer net learns the approximate inverse system in low frequency band from the examples $(u,\bar{y})$.

## A NEURAL NETWORK-BASED INVERSE SYSTEM CONTROLLER

Generally speaking, the neural network-based inverse system controllers can be classified as three kinds of structures.

### Open-loop Structure

The inverse system cascades with the original system, i.e. its control strategy $u(k)=g(y_d(k), y(k-1),\ldots,y(k-n_y), u(k-1),\ldots,u(k-n_u))$ is adopted, where $y_d(k)$ is desired output. Its control performance completely depends on the accuracy of the multilayer network-based inverse system. There always exist some error and disturbance in practical inverse system and the system stability isn't guaranteed so that this structure isn't used.

### Internal Model Structure

The inverse characteristics and model of the original system may be realized by two multilayer networks and be modified through on line learning novel information. If the inverse characteristics and model are accurate enough, the disturbance in the system can be eliminated and stability is guaranteed.

### Feedforward-Feedback Structure

As shown in Fig.2, the traditional self-tuning PID controller may be adopted as feedback controller and the neural network-based inverse system is used as feedforward controller, where T,H,P are operators of PID controller, inverse system, and original system. The control system may be described by the following operator equation:

$$(I+PT)Y=(PH+PT)y_d \quad (3)$$

If there doesn't exist integral action in the original system P, the neural network-based inverse system H can approximate the inverse system (2) well, i.e. $H=P^{-1}$, where $P^{-1}$ is the inverse system of the original system. From the expression (3), we know that the output of the system y approaches the desired output $y_d$. If there exists some integral action in the original system. The neural network-based finite frequency band inverse system can approximate the inverse system in the low frequency band $W\leq W_o$, where $W_o$ is upper cut-off frequency. From the expression (3), we obtain:

$$y = \begin{cases} y_d & W\leq W_o \\ (I+PT)^{-1}PTy_d & W>W_o \end{cases} \quad (4)$$

From the above expression, it is known that the high-frequency component in the system is partially complemented by the feedback controller T. If the parameters of the feedback controller T is appropriately chosen to make the system stable, the feedforward-feedback controller can force the system output track the desired output $y_d$. Thus this control structure is adopted in the following simulations and experiement.

## SIMULATIONS AND EXPERIMENTL

### Simulation 1.

Consider the original system

$$y(s)/u(s)=Ke^{-\tau}dS/(S_2+c_1S+c_2) \quad (5)$$

where $K_p=3$, $c_1=2$, $c_2=1$, $\tau_d=10$, and the ratio of signal to noise is 8. The parameters of the multilayer network $N_1=9$, $N_2=8$, $N_3=1$ and the sampling interval $\Delta T=1$ second. The fixed proportion 1 is adopted as the feedback controller and the original system (5) is viewed as unknown. The results of the simulation are shown in Fig.3.

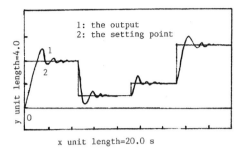

Fig.3  The Output of Simulation 1

Simulation 2.

Consider the original system

$$y(s)/u(s)=K_p(y)/S(S+c_1)$$

where $K_p=3+0.1y$, $c_1=5$. the ratio of signal to noise is 8. The parameters of the multilayer network and the feedback controller are the same as simulation 1. The results of the simulation are shown in Fig.4.

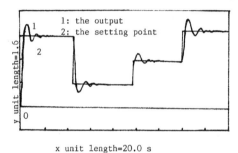

Fig.4  The Output of Simulation 2

A pilot liquid level control experiment

The liquid level control system is shown in Fig.5. The action u controls the valve 7 in order to make liquid level $H_3$ at setting point. When only the valve 3 is switched on, the plant is first-order nonlinear system. The valve 4, the valve 5 and the valve 6 are the disturbance valves which adds disturbance to liquid level $H_3$. D/Q is an electricity-gas transducer, PDT is a pressure difference transducer which transform the liquid level signal into electrical signal. The microcomputer is Apple II. The programs of the neural network and feedback controller are written in BASIC language. The parameters of the neural network $N_1=6$, $N_2=5$, $N_3=1$ and the same feedback controller as simulation 1 is adopted. The output of the system, i.e. liquid level $H_3$, is recorded as the curve.

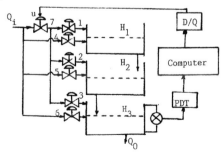

Fig.5  The Liquid Level Control System

When the seting point is 8.0, the output $H_3$ is shown in the curve AB in Fig.6 and its steady error is 0.01. When the setting point is 4.5, the output $H_3$ is shown in the curve BC and its steady error is 0.02. When the valve 6 is switch on abruptly. The output $H_3$ is recorded in the curve CD.

The results shows that the neural network-based inverse system control system can eliminate the disturbance and track the desired output well.

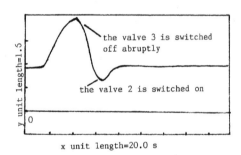

Fig.6  The Output of the System when the Setting Point is Changed and Disturbance is Added

In order to test the adaptability and robustness of the system, the valve 3 is abruptly switched off and the valve 2 is switched on, i.e., the dynamical characteristics of the system changes from first-order into second-order. The output of the system is recorded as the curve in Fig.7. The valve 3 is abruptly switched off and the valve 1 is switched on, i.e. the dynamical characteristics of the system changes from first-order into third-order. The output $H_3$ is recorded in the curve in Fig.8.

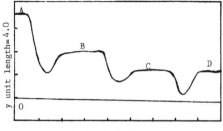

Fig.7  The Output of the System when the Plant Changes Abruptly from First-order into Second-order

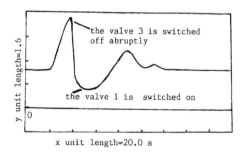

Fig.8  The Output of the System when the Plant Changes Abruptly from First-order into Third-order

## CONCLUSION

From the above discussion, it is known that the neural network-based inverse system controller can deal with a kind of systems with unknown dynamics. The results of the various numerical simulations and a pilot level control show that proposed controller possess strong adaptability and robustness. Although the assumption of invertibility of the system is sometimes restricted, the simplicity and efficiency of the approach are strong points for many practical applications.

## REFERENCES

Tsuneo Yoshikawa, Toshiharu Sugie (1986). Filtered inverse systems. INT.J. Control, 43, 1661-1671

Chu-Wen Li, and Yuan-Kun Feng (1987). Functional reproducibility of general multivariable analytic non-linear systems, INT.J. Control, 45, 255-268

R.M. Hrischron (1979). Invertiblity of multivariable nonlinear control systems, IEEE Trans. Auto. Control, AC-24, 855-865

Allon Guez, James L. Eilbert, and Moshe Kam (1988). Neural network architecture for control, IEEE Control Systems Magazine, 2, 22-25

Charles W. Anderson (1989). Learning to control on inverted pendulum using neural network, IEEE Control Systems Magazine, 2, 31-35

Naveen V. Bhat, et al. (1990). Modeling chemical process systems via neural computation, IEEE Control Systems Magazine, 2, 24-29

Robert Hecht-Nielsen (1989). Theory of the back-Propagation neural network, The Proceding of IJCNN, 1, 593-611

Fu-Gen Zhu and Yong-Zai Lu (1992). A new recursive learning algorithm of BP network, to be published in the proceding Singapre International Conference on Intelligent Control and Instrumentation, Marina Mandarin Singapre

David B. Fosel-Lawrence J. Fogel, V. William Porto (1990). Evolutionary programming for training neural networks, IEEE IJCNN, 1, 601-605

# ITERATIVE LEARNING CONTROL OF FLEXIBLE-JOINT ROBOTS USING NEURAL NETWORKS

### J. Fu and N.K. Sinha

*Department of Electrical and Computer Engineering, McMaster University,
Hamilton, Ontario, Canada L8S 4L7*

**Abstract.** Due to the increased complexity of the dynamics of robots with joint flexibility, many conventional robot control strategies are incapable to solve this problem effectively. To overcome this difficulty, a neural controller based on iterative learning has been proposed for the control of a flexible joint robot in this paper. The controller design is based the fact that most industrial robots perform repetitive tasks. The neural network for this purpose is trained as an inverse dynamic model of a flexible joint robot and is implemented as a feedforward controller. Our modifications of the back-propagation neural network [11] are employed in this study and result in a much better learning efficiency. A case study of a two-link flexible joint robot demonstrated that the proposed iterative neural learning controller can reduce the tracking errors after a few learning steps. Also, the proposed neural learning has advantages of easy implementation and requiring very little on-line computation.

Keywords. Learning systems; neural nets; robots; nonlinear control systems.

## 1. INTRODUCTION

In this paper, we consider the motion control of robotic manipulators with flexible joints. Compared with the large volume of literature available on the control of rigid robot, relatively little has been published on the control of flexible-joint robots. On the other hand, experiments indicate that joint flexibility should be considered in both the modelling and control of robots if high performance is to be achieved. Due to the increased complexity of the dynamics of flexible joint robots, nice features, associated with rigid robots, such as independent control input for each degree of freedom and the passivity, have been lost. To overcome these problems, various control strategies have been proposed. One approach is based on feedback linearization. The drawbacks of this approach require measurement of link acceleration and jerk. Another approach is based on the concept of integral manifold. In this case, the flexible-joint robot dynamics are restricted to a suitable integral manifold in state space and a reduced-order model can be derived. This reduced model can be linearized by non-linear feedback using only position and velocity information. The main drawback to this approach is its lack of robustness to parametric uncertainties.

The recent resurgence of neural networks has inspired the interest of the control community because of its high potential to solve control problems of complicated nonlinear systems, such as robotic manipulators. Some researchers have made considerable progress in applying neural networks to the control of robotic manipulators. In particular, reference [16] has applied neural networks to the control of flexible-joint robots. However, like other neural learning controllers, they are configured on-line in a feedback loop. This is difficult to implement in real time since the neural controller is composed of a huge number of neurons.

To overcome this problem, we proposed a neural controller based on iterative learning for control of flexible-joint robots in this paper. The proposed learning scheme requires little on-line computation and achieves satisfactory control precision after a few learning steps.

Adaptive controllers and robust controllers try to accomplish their control objectives in one operation. Intuitively, it will make a controller more complicated and more expensive in order to fulfil control objectives in one operation. On the other hand, most industrial robots perform repetitive tasks within a given time period. If experience of the previous operation is used, the controller will become simpler and require less knowledge of the system. Motivated by this idea, an iterative learning control scheme for robotic manipulators has been studied in recent years. Iterative learning control can be defined as "a control scheme that improves the performance of the device being controlled as actions are repeated, and to do so without the necessity of a parametric model of the system" [6]. Basically, the iterative learning controller is a kind of adaptive feedforward controller, which modifies the control inputs by making use of the previous operational data, such as input signals, position errors, velocity errors and force errors. Also, iterative learning control is implemented off-line while adaptive control is realized on-line. Thus, iterative learning control saves valuable on-line computation time. There is a great deal of research is being done on this topic. The main idea of iterative leaning control is to improve the performance of the system as control actions are repeated, and to do so without the necessity of a parametric model of systems. As a matter of fact, convergence of an iterative learning scheme may not guarantee satisfactory performance and the controller may have a very slow convergence rate, as reported in reference [5]. Intuitively, ignoring whatever knowledge we may have of systems parameters in favour of general learning schemes seems a bad idea. We should model what can be modeled and use learning for what cannot be modeled. Thus, a practical iterative learning control should utilize the approximate model of the controlled system in order to achieve better performance and faster rate of convergence. One way to derive this approximate model is to use neural network as shown in this paper.

This paper is organized as follows: In section 2, we introduce a modification of back-propagation neural network. In section 3, an iterative learning scheme using neural network is studied. Simulation results will be presented in section 4. Finally, some concluding remarks are given.

## 2. MODIFIED ALGORITHM OF BACK-PROPAGATION NEURAL NETWORK

Neural networks are systems which have been derived through models of neurophysiology. In general, they consist of a collection of simple nonlinear computing elements whose inputs and outputs are tied together to form a network. The main advantages of neural network for robotics is their ability to adaptively learn nonlinear function whose analytical forms are difficulty to derive and whose solutions are hard to compute. The most dominant forms of neural networks used in robotics are the multi-layer perceptron, such as back-propagation neural network.

The basic computational element in a multi-layer neural network is the perceptron. A perceptron receives a number of inputs from external source or other perceptrons and performs a nonlinear mapping as well as forms an output of the perceptron. The following figure represents a typical perceptron in a back-propagation neural network.

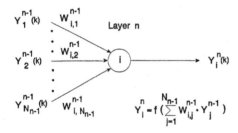

**Figure 1**

The operation of the perception in figure 1 is expressed by

$$X_i^n(k) = \sum_{j=1}^{N_{n-1}} W_{i,j}^{n-1}(k) \times Y_j^{n-1}(k) \qquad (1)$$

$$Y_i^n(k) = f[x_i^n(k)] \qquad (2)$$

In this paper, the function $f$ associate with hidden layers is assumed a sigmoid function
and the function associated with output layer is just a

$$f(x) = \frac{1}{1 + e^{-x}} \qquad (3)$$

linear function: $f(x) = x$.

Since the functions associated with neurons in the output layer are linear, thus the Linear Least Squares Learning Algorithm can be applied to the training of the output layer. This is expected to have much faster rate of convergence than the gradient method. The training of the hidden layers still utilizes the gradient method. However, the gradient used here is with respect to the weights based on the whole training samples rather than a single training sample. Iterative computation of the gradient is studied in order to save computational effort and time in the presence of large number of training samples.

### 2.1 Training of the hidden layers - gradient method

Assume that there are $L$ training data available in all. The training process of neural network is to minimize the following performance index.

$$J(L) = \sum_{l=1}^{L} [Y_s(l) - Y(l)]^T [Y_s(l) - Y(l)] = \sum_{l=1}^{L} e^T(l)e(l) \qquad (4)$$

where

$L$:    the number of total training data
$l$:    denoting the sequential number of the training data, $l=1,2,...,L$.
$Y(l)$:    $N_m \times 1$ vector, outputs of the neural network, as show in figure 1
      $Y(l) = [Y_1^m(l) \quad Y_2^m(l) \quad ......Y_{Nm}^m(l)]$;
$Y_s(l)$:    $N_m \times 1$ vector, outputs of the real systems.
$N_m$:    number of the output nodes
$e(l)$:    $N_m \times 1$ vector, representing the error between the outputs of neural network and the outputs of real systems.

Note that the index $J(L)$ is different from the one used in the back-propagation algorithm described in reference [11]. Here, the index is the sum of the square of the errors over the total training samples. Direct computation of the gradient is quite time consuming when there exist tens of thousands of training data. Therefore, it is important to find an efficient computing technique to derive the gradient. The ideal approach is to develop an iterative method to obtain the gradient rather than to compute it directly.
Let $1 < k \le L$, from eq.(4) we have

$$J(k) = \sum_{l=1}^{k} e^T(l)e(l) \qquad (5)$$

Let the vector $W_i$ denote the weights connecting the neurons in the lower hidden layer or the input layer to neuron $i$ in the upper hidden layer, as shown in figure 1, where

$$W_i = [W_{i,1}^n \quad W_{i,2}^n \quad ......W_{i,Nn-1}^n]$$

Using the original back-propagation algorithm described in reference [11], we have

116

$$W_i(k+1) = W_i(k) + \eta \frac{\partial J(k)}{\partial W_i}\Big|_{W_i = W_i(k)} \qquad (6)$$

where $\eta$ is the learning rate. From eq.(5), we obtain

$$J(k) = \sum_{l=1}^{k} e^T(l)e(l) = e^T(k)e(k) + J(k-1) \qquad (7)$$

Let

$$g(k) = \frac{\partial[e^T(k)e(k)]}{\partial W_i}\Big|_{W_i = W_i(k)} \qquad (8)$$

Since the computing method for $g(k)$ is described in detail in reference [11], we shall omit it here. From equation (7), we can obtain

$$\frac{\partial J(k)}{\partial W_i}\Big|_{W_i = W_i(k)} = g(k) + \frac{\partial J(k-1)}{\partial W_i}\Big|_{W_i = W_i(k)} \qquad (9)$$

Although the partial differential term on the right hand side of the equation is unknown at the $k$th iteration step, by using Taylor's formula, we have the following approximate expression:

$$\frac{\partial J(k-1)}{\partial W_i}\Big|_{W_i = W_i(k)} = \frac{\partial J(k-1)}{\partial W_i}\Big|_{W_i = W_i(k-1)} + \frac{\partial^2 J(k-1)}{\partial^2 W_i}\Big|_{W_i = W_i(k-1)} [ \qquad (10)$$

and the second-order partial derivative can be estimated by following equation:

$$\frac{\partial^2 J(k-1)}{\partial^2 W_i} = \frac{\dfrac{\partial J(k-1)}{\partial W_i} - \dfrac{\partial J(k-2)}{\partial W_i}}{W_i(k-1) - W_i(k-2)} \qquad (11)$$

Substituting the eq.(9) into eq.(6), we have

$$W_i(k+1) = W_i(k) + \eta g(k) + \eta \frac{\partial J(k-1)}{\partial W_i}\Big|_{W_i = W_i(k)} \qquad (12)$$

Comparing with the back-propagation algorithm with additional momentum term described in reference [11], which is expressed by

$$W_i(k+1) = W_i(k) + \eta g(k) + \alpha[W_i(k) - W_i(k-1)] \qquad (13)$$

where $g(k)$ is defined by equation (8), we find that the additional momentum term in equation (13) is trying to accomplish the same tasks that equation (12) has achieved. From this point of view, it is believed that the proposed training algorithm in this paper is better than that of reference [11]. Our later simulations will confirm this.

Now, we discuss the selection of an important parameter $\eta$, the rate of learning. The smaller we make the learning rate, the smaller will the changes to the weights in the network be, and therefore the better will be the approximation. This improvement, however, is obtained at the cost of a slower rate of learning. If, on the other hand, we make the learning rate too large so as to speed up the training process, the resulting large changes in the weights may assume such a form that the network becomes unstable(i.e.,oscillatory). A simple method of increasing the learning rate and yet avoiding the danger of instability is to select the learning rate as:

$$\eta = \frac{\alpha/k}{\left\| \dfrac{\partial J(k)}{\partial W_i} \right\|}$$

where $\left\| \dfrac{\partial J(k)}{\partial W_i} \right\|$ is the Euclidean norm

$$(14)$$

where $\alpha$ is a very small positive number.

## 2.2 Summary of the training procedures for the modified back-propagation neural network

Combining the training procedures of the hidden layers and the output layer, we derive the modified training algorithm for backpropagation neural network.

Step 1: Set the initial values;
Step 2: Present the network with input vectors and output response vectors of systems for iteration $k=1,2,...,L$;
Step 3: Train the output layer using the iterative Linear Least Squares Algorithm;
Step 4: Train the hidden layers using the modified gradient algorithm;
Step 5: Repeat the computation by going back to step 2 until satisfactory results are achieved.

## 3. A NEURAL LEARNING CONTROLLER FOR FLEXIBLE-JOINT ROBOTS

Consider a robotic manipulator with flexible joints expressed by the following nonlinear dynamic model [14]

$$M(q_1)\ddot{q}_1 + C(q_1,\dot{q}_1)\dot{q}_1 + g(q_1) + K(q_1 - q_2) = 0$$
$$J\ddot{q}_2 + B\dot{q}_2 - K(q_1 - q_2) = 0 \qquad (15)$$

where $M(q_1)$ is the $n \times n$ symmetric and positive definite matrix ( also called the generalized inertia matrix), $C(q_1, \dot{q}_1)$ is the $n \times 1$ vector due to Coriolis and centripetal forces, $g(q_1)$ is the $n \times 1$ vector due to gravitational forces. $u(t)$ is the $n \times 1$ vector of joint torques supplied by actuator. The vectors $q_1(t)$ and $q_2(t)$ represent the link angels and motor angles. K is a diagonal matrix consisting of stiffness constants.

Experimental investigations of industrial manipulators indicate that joint flexibility contributes significantly to the overall dynamics of the systems. The increased complexity of dynamics of flexible joint manipulator makes it much more difficult to control. Here, we use the idea of the iterative learning controller described in reference [2] and the learning capability of neural networks to design an iterative learning controller, which is described by figure 2

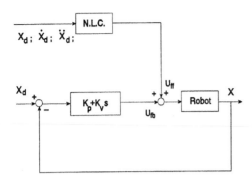

Figure 2

where     N.L.C.       the neural learning controller;
             $K_p$:             position feedback gains;
             $K_v$:             velocity feedback gains.

The controller consists of two major parts: a feedback P.D. controller and a feedforward neural learning controller (N.L.C.), which is an approximate inverse dynamic model of robot in the neighbourhood of the desired trajectory. The difference between true model of the robot and the approximate neural network model are regarded as disturbances. At the initial control stage, the feedback controller plays a role in making the whole system stable and contributes a relatively larger portion of control inputs to the robot. However, as the learning process continues, the dominant control input to robot will be shifted to the feed-forward neural controller, because the neural network model of the robot is more and more close to the true model of the robot. We should point out that the weights of the neural controller are fixed during the control period. After each operation of the robot, the neural network will be retrained with the operational data obtained from last trial. Basically, the neural network will be trained as an inverse dynamic model of the robot. The training scheme of the neural network can be described by figure 3.

During the training process, the output of the robot, i.e. measurement of link angles, velocity angels of both link and motor as well as desired acceleration, will be fed into the input layer of the neural network. Notice that no measured acceleration is needed in the training of the neural network. The weights of the neural network are adjusted based on the errors between the input torque of manipulator U(t) and the output of neural network Y(t) by using modified back-propagation algorithm of the neural network. We point out here that the actual trajectory from the previous operation is fed into the input layer of the neural network during training process while the desired trajectory is presented in the control process.

When we apply neural network to the control of robots, two important things should be taken into consideration. One is that most industrial robots execute tasks along a fixed trajectory. Secondly, a trajectory close to the desired is easily obtained by using a simple controller. Thus, our approach is to train the neural network as an inverse dynamic model of the robot in the neighbourhood of the desired trajectory based on the experience from the realized trajectory. After each operation of the robot, the learning process is repeated and as the number of repeated operations increases, the trajectory tracking errors can be reduced to almost zero.

As we know, a teacher is needed to start the training of a back-propagation neural network. Here, we choose a control scheme introduced in reference [7] as the teacher, which is a modified version of inverse dynamic scheme for rigid robot. The controller is described as

$$u(t) = u^r(t) + K_v(\dot{q}_1 - \dot{q}_2) \qquad (16)$$

where $u^r(t)$ is control input derived based on the inverse dynamic controller for rigid robot and $K_v$ is a constant diagonal matrix.

Since parameters of robot dynamics are partially unknown, the above modified inverse dynamic control law cannot result in a linearized system in practice. However, this scheme is still a good teacher for the neural controller. It is expected that even the approximate inverse dynamic control can bring the robot to the neighbourhood of the desired trajectory. This is good enough to start the learning of the neural controller. The iterative learning controller with neural network can improve the tracking precision by itself as the number of learning steps increases. Our simulation studies confirm it.

## 4.   RESULTS OF SIMULATION

In this section, we apply the proposed iterative learning scheme to the control of a two-link planar manipulator with joint flexibility. In our simulation, we assume that the two links of the manipulator have the same length, with mass $m_1 = 2.0$ kg and $m_2 = 2.5$ kg respectively. The stiffness constant $K = 1000$. The joints of simulated manipulator are required to move along the following desired trajectories

Joint one:    $q^1(t) = 3t^2 - 2t^3 + 0.375t^4$
Joint two:    $q^2(t) = 4.5t^2 - 3t^3 + 0.5625t^4$    (17)

The running time is two seconds.

The neural network employed in the simulation consists of an input layer with eight neuron nodes , the first

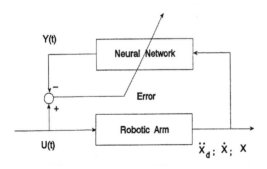

**Figure 3**

hidden layer with 20 neuron nodes, the second hidden layer with 40 neuron nodes and an output layer with 2 neuron nodes. Thus it is symbolized as $N_{8,20,40,2}$.

Since the exact model of the robot is not available, in our simulation , we assume that the parameters of the model are different from the true values, while $m_2$ equals 3.0 kg.

As discussed before, we employ a modified inverse dynamic control scheme in joint space based on an approximate model of the flexible joint robot to bring the robot into the neighbourhood of desired trajectory. The resulting trajectory are shown by figure 4. In the following figures, the solid lines represent desired trajectories and the dotted lines represent actual trajectories. We observe that the performance is not good because there is a significant difference between the modelled parameters of the robot and the true parameters of the robot. However, this is sufficient to start training for learning process, in which the neural learning controllers are able to improve the control performance by themselves and to achieve satisfactory performance after a few learning steps. Using the operational data obtained from the results of the teacher, we train the neural network as an approximate model of the robot as in figure 3. Figure 5 is the training results of neural network after three lessons. Figure 6 shows the curves of learning performance vs training times. These figures demonstrate that the modified back-propagation algorithm has a very fast rate of convergence and many simulations conducted by the authors have shown that the training is always convergent. After finishing the training of the neural network, we implement the neural learning controller as described in figure 2. During the control stage, all weights of the neural network are fixed. The feedforward control signals are calculated beforehand, where the inputs are desired position, velocity, acceleration. Only the PD controller requires the on-line computation. Figure 7 represents the control results after the second trial. Figure 8 is the result of the eighth trial. Figure 9 shows the convergence rate of the learning controller. We find the proposed learning controller does not have the property of asymptotic convergence and the tracking errors fall within a specified bound. However, the satisfactory control precision can be achieved after a few learning steps. This shows that the proposed neural learning scheme is an effective approach for control of flexible joint robotic manipulators.

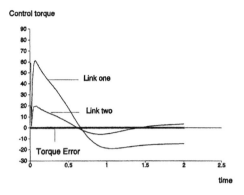

**Figure 5** joint control torque history

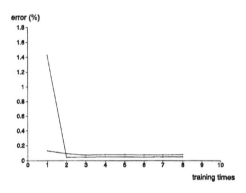

**Figure 6** training results

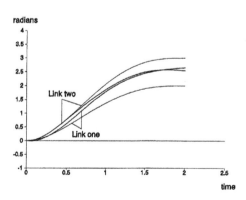

**Figure 4** control results of the teacher

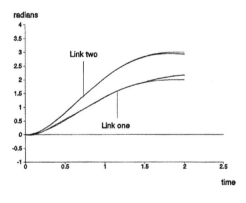

**Figure 7** results after two learning steps

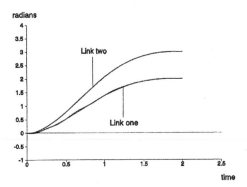

**Figure 8** results after eight learning steps

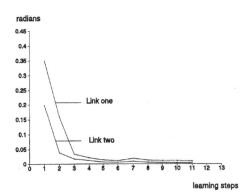

**Figure 9** tracking errors vs learning steps

## 5. CONCLUSIONS

In this paper, we have presented an iterative neural learning scheme for control of flexible joint robot by using authors' modified backpropagation algorithm. A case study on a two-link robot has demonstrated that the proposed neural learning controller is very promising and achieved satisfactory performance after a few trials. Generally, the proposed iterative neural learning controller has a faster rate of convergence and better robustness than the common iterative learning controller. The learning process occurs between two consecutive operations of the robot and the neural learning control is implemented as a feedforward controller. The required on-line computation is only for the PD feedback controller. Thus, this avoids the problem of heavy on-line computation required by adaptive controllers and robust controllers. In practice, we can develop an approximate model of the robot. Then we simulate this robot on a computer by using the inverse dynamic control scheme. The neural network is trained by using the simulation data.

After this, the learning scheme described in figure 3 will put into the actual robot. It is expected that more precise the model of the robot, the faster will be the learning process. However, the neural learning controller needs retraining when the robotic trajectory has been changed. This is one drawback of the proposed approach compared with adaptive controllers and robust controllers. Our future work will be directed into experiments of real robotic manipulators using the learning controller proposed in this paper and more strict theoretical analysis.

## References

[1] Aicardi M., (1989), Combined Learning and Identification Control Techniques in the Control Manipulator, in Proceedings of the 28th Conference on Control and Decision, 1651-1656

[2] Arimoto S., (1984), Bettering Operation of Dynamic Systems by Learning: A New Control Theory for Servomechanism or Mechanics Systems, in Proceedings of Conference on Decision and Control, 1064-1069

[3] Arimoto S.,(1990), Robustness of Learning Control for Robotic Manipulators, in International Conference on Robotics and Automation, 1528-1533

[4] Bondi Paola, (1988), On the Iterative Learning Control Theory for Robotic Manipulator, in IEEE Journal of Robotics and Automation, Vol.4, 14-21

[5] Chae H. An, (1988), Model-based Control of A Robot manipulator. The MIT Press,

[6] Craig J.J., 1988, Adaptive Control of Mechanical Manipulators, Adison-Wesley, MA,

[7] F. Ghorbel, (1989), Adaptive Control of Flexible-Joint Manipulators, IEEE Control Systems Magazine, December, 9-13

[8] You-Liang Gu, (1990), On Nonlinear Systems Invertibility and Learning Approaches by Neural Networks, 1990 American Control Conference, 3013-3017

[9] B. Horne, (1990), Neural Networks in Robotics: A Survey. Journal of Intelligent and Robotic Systems 3, 51-66.

[10] Kawato M., (1988), Hierarchical Neural Network Model for Voluntary Movement with Application to Robotics, IEEE Control Systems Magazine, 9-15

[11] D.F. Rumelhart, (1986), Parallel Distributed Processing, Vol.1, Foundations, MIT Press.

[12] N.K. Sinha and B. Kuszta, (1983), Modelling and Identification of Dynamic Systems, Van Nostrand Reinhold,

[13] M.W. Spong and M. Vidyasagar, (1989), Robot Dynamics and Control, John Wiley & Sons.

[14] M.W. Spong, (1987), Modeling and Control of Elastic Joint Robots, Journal of Dynamic Systems, Measurement, and Control, Vol. 109, 310-311.

[15] T.Yabuta, (1990), Possibility of Neural Networks Controller for Robot Manipulator, in 1990 IEEE International Conference on Robotics And Automation, 1686 -1691

[16] V. Zeman, (1989), A Neural Network Based Control Strategy for Flexible-Joint Manipulators, in 1989 IEEE 28th Conference on Decision and Control, 1759-1764

# NEURAL NETWORK MODELING OF AN AEROSUSPENDED VEHICLE

**J.M. Fuertes, B. Morcego, J. Codina and A. Catala**

*Departamento de Ingenieria. de Sistemas, Automatica e Informatica Industrial, Universidad Politecnica de Cataluña,
Pau Gargallo 5, 08071 Barcelona, Spain*

Abstract. This paper presents the neural network modelization of an aerosuspended platform prototype. The modeled prototype, a single motor-fan system which is confined to move only vertically, is a nonlinear, time varying system. The neural model is trained using experimental data of the platform system and, after the training, it will be used for the design of a neural network controller of the platform. Along the paper it is discussed the methodologies followed to evaluate the results obtained from different neural network internal configuration, and it finish with some simulation results.

Keywords. Autonomous Aerosuspended Vehicles, Neural Network Modeling, Machine Learning, Advanced Control.

## INTRODUCTION

Aerosuspended vehicles, or hovering mechanisms, are platforms which, using some kind of motorized wind propellers, are able to keep their position or move in the desired direction and orientation at an arbitrary distance of the ground. An example of those vehicles are the helicopters, but they are not well suited for lateral approximation to scarped walls or for travelling close to ground because of their large rotors and because ground and wall proximity effects in their aerodynamics.

There is a project at the Polytechnic University of Catalonia which is designing and developing a scaled vehicle module with multiple aerosustentation system. That system can be used, for example, as an alternative to helicopters in applications in which they can not exhibit adequate characteristics.

Aerodynamic influences from diverse perturbation sources cause instability effects to aerosuspended platforms, while the multivariable character of those systems and their non linear and time varying characteristics impose the utilization of advanced modeling and design methods.

The methodology followed in the research project is the construction and experimentation of scaled prototypes and the use of classical and advanced techniques to obtain their models and design their control systems. Two prototypes are currently in operation: a single fan, one degree of freedom platform, and a four fan free flight platform.

## THE PLATFORM PROTOTYPE

Designing a multifan hovering mechanism has had a long run and has attracted the engineering efforts for long time (Stimson, 1957) while some prototypes have been built since then (Nakamura, 1986). Similar mechanisms with mobility and small size have been developed, but they have not been put for practical use. Among the reasons for that, it is counted the difficulties encountered for handling the instabilities, and the inefficiencies of the powering system. The development of new engines with a better power-weight relation and the search for new control paradigms can change the forecasting for practical application of those systems.

Fig.1. The free flight aerosuspended prototype.

In order to study this kind of vehicles, our project research team has constructed a free flight prototype with six degrees of freedom which is now in the basic operation testing phase (Fig. 1). That prototype is a circular platform of about 80 centimeter in diameter; it uses four independent thermic, internal combustion engines which move 28 centimeter fans. The control commands drive the fuel feeding into de carburetor, modifying the sustentation force. It is called IVOLA, (Investigation Vehicle Operated by a Levitation Automatism). The present system uses three accelerometers to measure pitch, yaw and roll. The initial tests are based on radio control of the prototype just with proportional feedback of accelerometer data. We have prepared a free force arm to confine the operation of the vehicle into a reduced area and to provide an easy path for communicating sensor data and actuator commands to/from a digital microcontroller.

Fig. 2. The one degree of freedom platform.

But for data gathering of the response and the operating characteristics of the propulsion engines, a reduced operation prototype has been developed. This prototype, which is called IVOLA-I, is a single engine platform, confined by a four track rail system to move only vertically so it has only one degree of freedom (Fig. 2). The control of fuel feeding gives the propulsion-suspension force, and the elevation of the platform is measured through a vertical distance sensor. The experiments carried out with this prototype were the mathematical model identification based on input-output data, and the design of a classical three term discrete time controller (Quevedo, 1991), while here it is presented the development of an abstract open loop model of the prototype using a neural network system.

The experimental results obtained from the platform tests show a highly non linear response, so that a linear mathematical representation gives a poor description of the system. The approximate model obtained in (Quevedo, 1991) had transfer function,

$$G(s) = \frac{K}{s(1+0.2s)} e^{-0.2s} \tag{1}$$

where the gain K had two different identification values depending on the direction of the movement, $K_{up}=2$, $K_{down}=0.8$. Although the model was useful

for the analytical design of a pole-placement controller, it was not enough precise because the nonlinearities of the platform and the time variant dependencies were not taken into account. In particular, the weight of the platform varies with time due to the fuel consumption of the motor (1.250 / 0.750 kg. platform weight with a full/empty fuel tank).

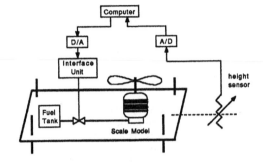

Fig. 3. Diagram of the prototype IVOLA-I

## NEURAL NETWORK MODELING

Recently, there has been a great resurgence of interest in the application of neural networks to the modeling of dynamical systems and real-time process control. They have been used successfully in numerous areas including train control, robot control, consumer electronics (Narendra, 1990b), in vehicle guidance (Nguyen, 1990), plant time series modelization (Griño, 1991), chemical plant modeling and control (Bhat, 1990), and simulation (Aynsley, 1991).

Traditional approaches to the modeling of a dynamic system involve the construction of an explicit logical or analytical model. Obtaining such models are hard tasks because there are some difficulties that have to be solved: the model has to be well defined for all circumstances and operating conditions in which the system has to function, including the changing environment conditions and the unmodelizable perturbations. Also, the model must be obtained in an explicit, computerizable form.

Neural networks have the capability to "learn" by adjusting internal parameters through experience. They have good robustness properties with respect to noise and incomplete information. Therefore, neural networks offer an attractively simple approach for parallel computation in the field of process and plant modeling. It allows the creation of an abstract model (black box model) which can be quickly completed, whose quality can be improved over time (experience), can adapt itself to changes of environment, and as long as it can be implemented on a parallel hardware, can be processed at high speed. The neural network does not provide a mathematical model of the system, although it is a reliable method to obtain the output response given an input signal.

## Network structure, inputs and outputs

The platform is a dynamical system where it is assumed that the output depends on a bounded number of previous inputs, outputs, and time. The system can be described (Narendra, 1990a) by the following nonlinear and time varying difference equation:

$$x(k+1) = f \ [x(k), x(k-1),...x(k-n+1);$$
$$u(k), u(k-1), u(k-m+1); t] \qquad (2)$$

where $\{u(k), x(k)\}$ represents the input/output pair sequences of the plant at time k, and t is an incremental variable representing time. (Fig. 4).

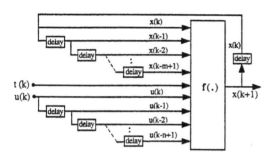

Fig. 4. Plant dynamical model

The objective of such modelization with a neural network is to generate a good approximation to the platform system, from input/output measurements, (training phase of the model), and after that, use the model as a feedforward information to obtain the appropriate neural controller by back-propagating the error through the model (training phase of the controller), (Fig. 5).

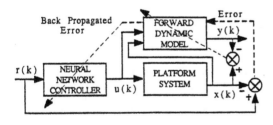

Fig. 5: Controlled system diagram.

The plant modelization is obtained by an appropriate training strategy: the training set includes platform input commands and output heights as input patterns while the output patterns are the actual responses of the plant. Such training data has been obtained from an experimental plant data recorded during a successful operation.

Initial attempts where done using a network with one single output (the platform height). The inputs were n plant present and past command data (fuel injection into the carburetor) and m plant state data (previous heights) (Fig. 6). The number n was selected from the time response characteristics of the platform system. After training the network, some apparently good results were obtained. But it was observed that there was delay between the output of the network and the desired output. Some of the trained networks had learnt that a good approximation of the next output was the present output. So it was necessary to evaluate the output with a better error parameter than the mean square error.

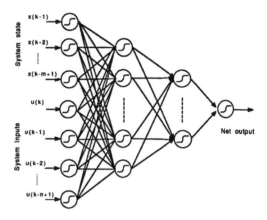

Fig. 6: Network initial structure.

## Evaluation of the model.

The chosen method to evaluate the goodness of the model consists in calculating the cross correlation of the desired and obtained outputs:

$$r_{xy}(k) = \sum_{j=0}^{N} x(j) \ y(k-j) \ ; \qquad k=0, k=\pm 1, k=\pm 2, ... \qquad (3)$$

where $x(k)$ is the desired output and $y(k)$ is the obtained output at time k.

If both signals are exactly equal, then their cross correlation should be equal to the autocorrelation of the desired output, and the difference of both correlations should be equal to zero for every k.

$$R_{xy}(k) = r_{xx}(k) - r_{xy}(k) = 0 \qquad (4)$$

The cross correlation also indicates the delay between signals x and y. If both signals are not shifted in time, the cross correlation maximum is at k=0. If it is different from zero, then the network is considered inappropriate. This is a very reliable method to evaluate time shifts between signals but does not say much about how similar they really are.

It was possible to reduce the number of network models which were giving apparent good results using the previous criterion with the initial configuration described in Fig. 6. In particular, it was seen that including m (state inputs) approximately equal to n (fuel inputs) was not appropriate. Since this result, some new configuration features were included into the neural network structure.

## Resulting neural network configuration

Three improvements were made to the previous configurations to obtain the resulting approach. First was to reduce the inputs from about 2n to n+m (where m is a small number, i.e. 1,2,...). This was done to prevent the network from considering height inputs as the only inputs to be taken into account for minimizing mean square errors at the output. On the other side, the network collects the information that it strictly needs: a sequence of fuel injection data, which determine the trend and quantity of height variance, and few previous heights, which must be added to that quantity.

The second change was to obtain two outputs instead of only one. The outputs were the next height and the actual velocity of the platform. When a network is trained it adjusts its ponderation weights to create an internal representation of the input-output relationship and, as noted in (Suddarth, 1988), when two different outputs are correlated, their internal representation can be similar and the learning of one will reinforce the learning of the other. In the present case, adding the actual velocity as an output helped the network to train in less cycles and with better accuracy.

The third change was the addition as a new input a decreasing signal which represents the weight decrement of the system (decreasing with time because the fuel tank is part of the platform). The major advantage of adding this signal is that the network, apart from the weight, was able to extract information on time varying parameters.

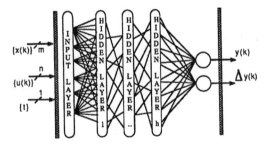

Fig. 7. Resulting neural network configuration.

Those previous changes, when introduced into the initial neural network structure, led to some different configurations which gave good results, as a first approximation, and discarded configurations which did not conform to the phase delay criterion. Each configuration was defined by the number of inputs and outputs, number of hidden layers and number of

neurons into each hidden layer (Fig. 7). Next step was to classify and evaluate those configurations by finding the resemblance among the platform height responses and the model outputs.

## Configuration evaluation.

The goal of this evaluation was to check if a given configuration, which was already without delay, provides statistically correct results.

The network output is a random variable, as well as the mean of the difference between the plant output and the network output (x-y). The mean of the difference will behave as a normal random variable (due to the central limit theorem), with mean $\mu$ and variance $\sigma^2$. Since the number N of data values is large enough and they are independent and identically distributed, the central limit theorem stands and the variance can be approximated by the standard deviation of the measurements.

In order to apply the central limit theorem, it was assumed that the differences between the plant output and the network output were independent. The independency assumption is being checked, although we believe it is approximately true.

Schematically the argument to get a confidence interval for $\mu$, and thus, a criterion to decide upon the correct configuration is the following:

Due to the central limit theorem,

$$\frac{\overline{x-y} - \mu}{s / \sqrt{N}} \approx N(0,1) \tag{5}$$

hence,

$$\text{Prob}\left[ -1.96 \leq \frac{\overline{x-y} - \mu}{s / \sqrt{N}} \leq 1.96 \right] = 0.95 \tag{6}$$

and therefore a confidence interval of 95% is given by

$$\left[ \overline{x-y} - 1.96\frac{s}{\sqrt{N}} \ , \ \overline{x-y} + 1.96\frac{s}{\sqrt{N}} \right] \tag{7}$$

The criterion, now, is to take as correct configurations those with a confidence interval that contains the value 0. This means that 0 is a likely value for the mean of the differences between the plant output and the network output, so the neural network configuration gives a response almost equal to the platform response.

## SIMULATION RESULTS

There were trained more than fifty different neural network configurations, going from the initially described networks to the more conditioned configurations. The modelizations were obtained using a neural network coprocessor while a mathematical package was used for statistical manipulations.

The experimental data set was a sequence of about 3500 fuel and height patterns which correspond to a experimental realization of the system (Fig. 8). Training set was a subsequence of 1000 pattern pairs (patterns 500 through 1500), using the resting data for testing the trained network.

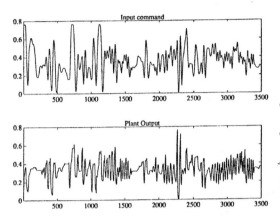

Fig. 8: Experimental input-output data set.

Among the resulting configurations, Table 1 shows a sample of some correct phase network results. Input column corresponds to the number of fuel data, previous height number values and time respectively. Hidden column shows the number of neurons on each hidden layer (there are cases with two and three hidden layers). Epochs column is the number of trained cycles needed to reach a certain limit on the error function (mse and stdn). Confidence interval column is the application of (8) on the results. Only the first and second networks are statistically correct according to previous analysis, because their confidence interval includes the mean μ=0.

TABLE 1. Sample of configurations.

| NN | Input | Hidden | Epochs | Conf. Int. (E-3) |
|----|-------|--------|--------|------------------|
| 1 | 20/1/1 | 20 | 1000 | -0.37, +0.37 |
| 2 | 20/1/1 | 25 | 1400 | -0.59, +0.39 |
| 3 | 22/1/1 | 8 | 1100 | +0.15, +1.84 |
| 4 | 20/2/1 | 20/12 | 500 | +1.85, +3.54 |
| 5 | 20/1/1 | 10/3 | 1900 | +7.09, +8.32 |
| 6 | 22/1/1 | 17/9/4 | 500 | -9.39, -6.80 |
| 7 | 24/1/1 | 18/10/5 | 1000 | +19.5, +22.0 |
| 8 | 15/1/1 | 35 | 1000 | +29.6, +34.2 |

The best results correspond to networks with 20 inputs and one hidden layer with similar number of elements. Increasing the number of hidden layers accelerates the trainnig, but does not improve results.

Error plots for both networks, NN 1 and NN 2 (Fig. 9) and (Fig. 10), reveal that the largest errors occur in three zones. The longer is for data 600 through 1500, which belongs to the data used to train the network. This means that the behavior of the system during

this period of time was probably the most unpredictable and the one that had more difficulties, which is exactly the reason why it was chosen to be the training period (although it was surprising obtaining such large error).

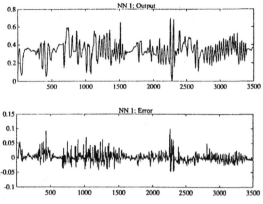

Fig. 9: Output and error plots for network NN 1.

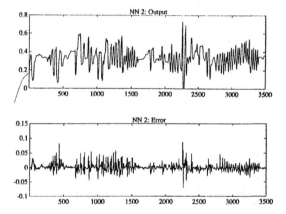

Fig. 10: Output and error plots for network NN 2

The other two zones correspond to around data 450 and 2300, respectively. The explanation for such errors is, as can be seen on the output plots, that they were produced because the engine got stuck on a certain height for a moment. This is a random hazard of the system which is not modelizable and so it is predictable to obtain errors of this kind.

CONCLUSIONS

In this paper it has been shown that the neural network approach to model the dynamics of a complex system is a fruitful task. The example presented here was not an easy modeling problem, where classical approaches did not give good enough results. It has been shown that, even the large number of expected possibilities for the configuration of the neural network, it seems that they confluence to some particular configurations for a good model of the system.

The trial and error method was used for selecting the different configurations, but this is a rather discouraging method because two radically different configurations (for example, one with a big hidden layer and another with two small hidden layers) can have very similar results after being evaluated, while adding a single neuron to any of the hidden layers could result in a useless network.

Genetic Algorithms (Chalmers, 1990) is one of the most encouraging methods to solve the previous problem. It consists basically to have a population of networks and, according to their goodness factor and some genetic rules, let them evolutione until they reach a state in which all members of the population become adapted (i.e. their output response is very similar to the desired output). The major advantage of this kind of algorithms is that the final configuration of the network can be a complete surprise because all kinds of random mutations are allowed. Its disadvantages are obvious: too many networks are trained and tested, and it means wasted time.

Acknowledgment. This work has been funded by the Comisión Interministerial de Ciencia y Tecnología, project ref. ROB89-0278. We want to thank Carles Soler, for constructing the prototypes, and to CERCA for its support. The authors are members of CERCA (Collective of Study and Research in Control and Automation) based in Terrassa (Barcelona).

## REFERENCES

Aynsley, M., Hofland, D., Montague, G.A., Morris, A.J. (1991). Real-Time Expert Systems in the Control of a Fermentation Plant. Proc. First European Control Conf., Grenoble, 236-241

Barto, A.G. (1989). Connectionist Learning for Control: an Overview. COINS Technical Report 89-89 and in Miller,W.T., Sutton, R.S., Werbos, P.J. (Ed.) Neural Networks for Control, MIT Press, Cambridge, pp. 5-58, 1990.

Bhat, N., McAvoy, T.J. (1990). Use of Neural Nets for Dynamic Modeling and Control of Chemical Process Systems. Computers in Chemical Engineering, 14, 4, 573-583.

Chalmers, D.J. (1990). The Evolution of Learning: An Experiment in Genetic Connectionism. Proc. of the 1990 Connectionist Models Summer School, San Mateo, CA. Touretzky, D.S. (Ed.) Morgan Kaufman

Griñó, R. (1991). Neural Networks for Water Demand Time Series Forecasting, in A. Prieto (Ed.), Artificial Neural Networks, Proc. Int. Workshop IWANN'91, Granada, September, Springer Verlag, Berlin. pp. 453-460.

Miller, T., Sutton, R.S., Werbos P.J. ed. (1990). Neural Networks for Control. MIT Press, Cambridge.

Nakamura, T. (1986). Control Characteristics of a Multifan Hovering Mechanism. IEEE Control Systems Magazine, 15-19, October

Narendra, K.S. (1990a). Adaptive Control Using Neural Networks in Miller,W.T., Sutton, R.S., Werbos, P.J. (ed) Neural Networks for Control, MIT Press, Cambridge, pp. 115-143

Narendra, K.S. (1990b). Identification and Control of Dynamic Systems Using Neural Networks. IEEE Trans. Neural Networks, 1, 1, 4-27.

Nguyen, D.H., Widrow, B. (1990). Neural Networks for Self-Learning Control Systems. IEEE Control Sys. Mag. April, pp. 18-23

Quevedo, J., Argelaguet, R., Escobet, T., Vehí, J., (1991). Height Control of a Hovering System with a Degree of Freedom. Proc. First European Control Conf., Grenoble, France, pp. 2447-2450

Stimson, T.E. (1957). Habrá Sedanes Aéreos en 1967. Mecánica Popular, 26-30, Septiembre.

Suddarth, S.C., Sutton, S.A., Holden, A.D. (1988). A Symbolic-Neural Method for Solving Control Problems. Proc. Int. Conf. Neural Networks, San Diego, pp. I-516-I-523

# THE FUZZY NEURAL NETWORK:
# THE EMERGING PARADIGMS

### M.M. Gupta

*Intelligent Systems Research Laboratory, College of Engineering, University of Saskatchewan,
Saskatoon, Saskatchewan S7N 0W0, Canada*

### Abstract

In recent years, an increasing number of researchers have become involved in the subject of fuzzy neural networks in the hope of combining the reasoning strength of fuzzy logic and the learning and adaptation power of neural networks. This provides a more powerful tool for fuzzy information processing and for exploring the functioning of human brains. In this paper, an attempt has been made to establish some basic models for fuzzy neurons. First, several possible fuzzy neuron models are proposed. Second, synaptic and somatic learning and adaptation mechanisms are proposed. Finally, the possibility of applying non-fuzzy neural networks approaches to fuzzy systems is also described.

## 1. Introduction

A typical neural network has multiple inputs and outputs which are connected by many neurons via weights to form a parallel structure for information processing. The potential benefits of such a structure are as follows: First, the neural network models have many neurons or computational units linked via the adaptive weights arranged in a massively parallel structure. This structure is believed to be essential for building systems with a faster response and a higher performance than the modern sequentially arranged digital computers. In fact, the structure is built after biological neural systems in the hope of emulating and taking advantage of the capabilities of human brains. Second, because of its high parallelism, problems with a few neurons do not cause significant effects on the overall system performance. This characteristic is also called fault-tolerance. Third, probably, the biggest attraction of the neural network models is their adaptive and learning ability. Adaptation and learning are achieved by constantly monitoring the system performance and modifying the weights and other elements. The ability to learn and adapt from the environment means that only a minimum amount of apriori information about the environment is needed and that the changes in system characteristics can be compensated.

Neural network models mainly deal with imprecise data and ill-defined activities. The subjective phenomena such as reasoning and perceptions are often regarded as the targets of neural network modeling. It is interesting to note that fuzzy logic is another powerful tool for modeling phenomena associated with human thinking and perception. In fact, the neural network approach fuses well with fuzzy logic [1] and some research endeavors have given birth to the so-called 'fuzzy neural networks' which are believed to have considerable potential in the areas of expert systems, medical diagnosis, control systems, pattern recognition and system modeling.

The term 'fuzzy neural networks' (FNN) has existed for more than a decade now. However, the recent resurgence of interest in this area is motivated by the increasing recognition of the potential of fuzzy logic, some successful examples, and the belief that both fuzzy logic and neural networks are two of the most promising approaches for exploring the functioning of human brains. Recently, an increasing number of researchers have become involved in the area of fuzzy neural networks. Yamakawa and Tomoda [2] described a FNN model and applied it successfully to a pattern recognition problem. Kuncicky and Kandel [3] proposed a fuzzy neuron model in which the output of one neuron is represented by a fuzzy level of confidence and the firing process is regarded as an attempt to find a typical value among the inputs. Kiszka and Gupta [4] studied a FNN described by the logic equations. However, no specific learning algorithms are given in these three cases. Gupta and Knopf [5] proposed a fuzzy neuron model which is similar to the first two cases except that a specific modification scheme was proposed for weights adaptation during learning. Nakanishi et al [6] and Hayashi et al [7] used the non-fuzzy neural networks approach for the design of fuzzy logic controllers with adaptive and learning abilities. Similarly, Cohen et al [1] used non-fuzzy neural network learning techniques to determine the weights of antecedents for use in fuzzy expert systems. However, no fuzzy neuron models were used in their work.

Some new fuzzy neuron models are proposed in this study which could overcome the limitations of the models described above. Improvements have been made by extending non-fuzzy neural models to fuzzy ones and, more importantly, by adding learning algorithms to fuzzy neural models. A class of weighting functions and aggregation operators are proposed which are also called 'synaptic' operators and 'somatic' operators, respectively. Basically, two kinds of fuzzy neuron models are

discussed. One is the 'fuzzification' of non-fuzzy neural models. The other is where the input-output relations are described by 'If-then' rules. Learning algorithms are proposed for both types of fuzzy neurons.

## 2. Basic Models of A Fuzzy Neuron

The most popular non-fuzzy neural model is proposed by McCulloch and Pitts [8] more than 40 years ago. The neuron has N inputs $x_i$ which are weighted by weights $w_i$ and then passed on to the node i. The node sums the weighted inputs and then transfers the results to a nonlinearity. A neuron has an internal threshold level, a predetermined value $\theta$, and the neurons fires when the sum of the weighted inputs exceeds the value $\theta$. A mathematical representation of such a neuron is given by:

$$y = f(\sum_{i=0}^{N} x_i w_i - \theta) \qquad (1)$$

where y is the output of the neuron, f represents the nonlinear function, $x_i$ and $w_i$ are the i-th input and its corresponding weighting factor, respectively.

A fuzzy neuron is designed to function in much the same way as a non-fuzzy neuron, except that it reflects the fuzzy nature of a neuron and has the ability to cope with fuzzy information. The inputs to the fuzzy neuron are fuzzy sets $X_1$, $X_2$, ..., $X_N$ in the universes of discourse $U_1$, $U_2$, ..., $U_N$, respectively. These fuzzy sets may be labelled by such linguistic terms as *high, large, warm*, etc. The inputs are then 'weighted' in ways much different from those used in the non-fuzzy case. The 'weighted' inputs are then aggregated not by the summation but by the fuzzy aggregation operations. The fuzzy output Y may stay with or without further operations depending upon specific circumstances. It is also noted that the procedure from the input to the output may not necessarily always be the same.

In the following, detailed discussions of three types of fuzzy neural models are given. Some possible learning (training) schemes are also proposed.

## 2.1. A Fuzzy Neuron Described by Logical Equations: Model I

In knowledge based systems, one often uses a set of conditional statements, 'If-then' rules to represent human knowledge extracted from human experts. Very often this knowledge is associated with *uncertain* and *fuzzy* terms. Therefore, antecedents and consequents in the 'If-then' rules are treated as fuzzy sets. The first fuzzy neuron model we discuss here is described by such rules. A fuzzy neuron with N inputs and one output with input-output relations are represented by 'If-then' rules:

$$R_i: \qquad If X_{1i} \ and \ X_{2i} \ and \dots X_{Ni} \ then \ Y_i \qquad (2)$$

where $X_1$, $X_2$, ..., $X_N$ are the current inputs, and $Y_i$ the current output of the i-th neuron which is described by the i-th rule $R_i$ of the overall M rules. This means that each neuron represents one of the M 'If-then' rules.

According to the fuzzy logic theory, the i-th fuzzy neuron can be described by a fuzzy relation $R_i$. For example,

$$R_i = X_{1i} \times X_{2i} \times ... \times X_{Ni} \times Y_i \qquad (3)$$

or in the general case:

$$R_i = F(X_{1i}, X_{2i}, ..., X_{Ni}, Y_i) \qquad (4)$$

where F represents an implication function.

Given the current inputs (fuzzy or non-fuzzy) $X_1, X_2, ... X_N$, according to the compositional rule of inference, the i-th rule gives an output as

$$Y_i = X_1 \circ (X_2 \circ (... \circ (X_N \circ R_i)...)) \qquad (5)$$

where o represents any composition operation, such as sup-T-norm.

Now, we propose a fuzzy neural model whose outputs are related to its outputs by a fuzzy conditional statement or a *'If-then'* rule. The experience of the neuron is stored in a fuzzy relation $R_i$, and its output is composed from the current inputs and the past experiences $R_i$. Therefore, it seems that this artificial fuzzy neuron behaves in much the same way as a biological neuron does.

It should also be noted that inputs to the neuron can be either fuzzy or non-fuzzy (crisp). Crisp values are special cases of the fuzzy ones.

The learning algorithms for this fuzzy neuron may vary depending on the real-world problems. Here some basic considerations are given about how a fuzzy neuron changes its rules and weights during learning and adaptation. This goal may be achieved by *'synaptic '* and *'somatic '* adaptation. *'Synaptic '* adaptation implies that all the inputs are constantly modified according to the synaptic weights and then forwarded to the neuron's body, the soma. The *'somatic '* adaptation implies modifying the past experience stored in the soma of the neuron. More detailed discussions are given in the next section.

## 2.2 A Fuzzy Neuron Given by Direct 'Fuzzification' of its Non-Fuzzy Counterpart.

Unlike the above model, the fuzzy neuron proposed in this section is not described by a *'If-then'* rule, rather it is obtained by a direct 'fuzzification' or extension of a non-fuzzy neuron model. Similar to a non-fuzzy neuron, first, all the inputs (fuzzy or crisp) of a fuzzy neuron are modified by the synaptic weightings and then these weighted inputs are aggregated through a somatic operation. In the following section, we will discuss two fuzzy neural models as well as their learning schemes.

### 2.2.1. A Fuzzy Neuron with Crisp Inputs: Model II

This fuzzy neuron has N non-fuzzy inputs, and the weighting operations are replaced by membership functions. The result of each weighting operation is the membership value of the corresponding input in a fuzzy set. All these membership values are aggregated together to give a single output in the interval of [0, 1], which may be considered as the 'level of confidence'. The aggregation process represented by $\otimes$ may use any aggregation by a 'somatic' operator, such as, MIN, MAX and any other operation based on T-norm and T-conorm [9, 10]. A mathematical representation of such a fuzzy neuron is described by

$$\mu(x_1, x_2, ..., x_N) = \mu_1(x_1) \otimes \mu_2(x_2) \otimes ... \otimes \mu_i(x_i) \otimes ... \otimes \mu_N(x_N) \qquad (6)$$

where $x_i$ is the i-th input to the neuron, $\mu_i(.)$ the membership function of the i-th weight, $\mu(.)$ the output of the neuron, and $\otimes$ an aggregation operator.

### 2.2.2. A Fuzzy Neuron with Fuzzy Inputs: Model III

Another fuzzy neuron which seems to be very much similar to a non-fuzzy neuron except that all the inputs and the output are fuzzy sets rather than crisp values. Each fuzzy input undergoes a 'synaptic' operation which results in another fuzzy set. All the modified inputs are aggregated to produce a N-dimensional fuzzy set. Because the output is rather complicated, it may go through further operations. It must be noted, however, that the weighting operation, unlike the one in the above, is not a membership function. Instead, it is a modifier to each fuzzy input. The fuzzy set $X_i$ is modified into another fuzzy set $X_i'$. The aggression operator $\otimes$ may be the same as the one mentioned in Eqn.(6). This fuzzy neuron is mathematically described as

$$X_i' = G_i(X_i) \qquad i = 1, 2, ..., N \qquad (7a)$$

$$Y = X_1' \otimes X_2' \otimes ... \otimes X_i' \otimes ... \otimes X_N' \qquad (7b)$$

where $X_i$ and $X_i'$ are the i-th inputs before and after the synaptic

weighting operation, respectively, $G_i$ is the weighting operation on the i-th synaptic connection, and Y is the fuzzy set representing the output of the fuzzy neuron.

## 2.3. Learning and Adaptation in Fuzzy Neural Systems

An adaptive neuron must go through the learning and adaptation processes in order to improve its performance. In conventional neural networks, this goal is achieved through the adaptation of synaptic weights. In the fuzzy neural systems, in addition to 'synaptic' adaptation, the 'somatic' adaptation may also be used. The learning and adaptive mechanisms discussed here are applicable to all the neuron models discussed in the previous sections.

### 2.3.1. 'Synaptic' Adaptation

During a learning phase, a neuron constantly changes itself to adapt and improve its performance. 'Synaptic' adaptation is one of the scenarios to realize this purpose. In a neuron (fuzzy or not), all inputs are modified by weighting or 'synaptic' operations. In the fuzzy case, however, the weighting or 'synaptic' operations are rather complex.

In the case of the fuzzy neuron Model I, one may also introduce the weighting operations. If so, the fuzzy neurons Models I and III are considered as the same as far as the weights adaptation is concerned. In both cases, all the weights simply serve as mapping functions which transform or modify each fuzzy input into another fuzzy set, this adaptation process continues until the training results are satisfactory and depends on the problem of concern. In synaptic learning and adaptation the following may occur: (a), the 'synaptic' operation may be a shifting process, (b), the width of the fuzzy number changes, and (c), the shape of the fuzzy number changes. In the case of fuzzy neuron Model II, the weights are membership functions which transforms numerical inputs into their corresponding membership values. However, the above three adaptation schemes can also be used for the adaptation of the membership functions of the weights.

### 2.3.2. 'Somatic' Adaptation

During a learning or 'training' process, a fuzzy neuron may also change its somatic structure rather than modifying only the input weightings. In the case of fuzzy neuron Model I, this means changing or updating the past experience, which includes

(i) changing the rules,
(ii) changing the membership functions assigned to the fuzzy terms in the rules, or
(iii) changing the way of representing the rules, for example, various implication functions and aggregation operations may be considered [9, 10].

In the cases of fuzzy neurons Models II and III, many options are available for the aggregation operator $\otimes$, such as T-norms, T-conorms, etc. [9, 10].

## 2.4. Non-Fuzzy Neural Network Approaches to Fuzzy Systems: Fuzzy Logic Controllers and Fuzzy Expert Systems

In the above sections, several models of a fuzzy neuron are proposed in the hope of combining the strengths of fuzzy logic and neural networks and achieving a more powerful fuzzy information processing tool. However, it is also possible to use the well-established non-fuzzy neural networks approach to fuzzy system modeling and construction where the neural networks' ability to learn and to adapt is particularly attractive. For example, in fuzzy logic controllers and expert systems, one often represents expert knowledge by the *'If-then '* rules. A membership function is assigned to each fuzzy term in the rules. By using the backpropagation learning algorithm of conventional neural networks, the membership functions for fuzzy terms are learned based on the training data. More recently, there is a subject starting to attract researchers, it is fuzzy systems modeling using non-fuzzy neural networks approach where membership functions involved in input and output models are learned either from the *' If-then '* rules extracted from experienced human operators, or from the numerical data obtained by data logging experiments.

## 3. Conclusions

Compared with the conventional neural networks theory which has been successfully applied to various areas concerning information processing, the fuzzy neural networks approach, to fuzzy information processing is still in its infancy. This paper is an attempt to contribute to the further theoretical development of fuzzy neural network theory.

In this paper, three types of fuzzy neural models are proposed. The neuron Model I is described by logical equations or the *' If-then '* rules, its inputs are either fuzzy sets or crisp values. The neuron Model II, with numerical inputs, and the neurons Model III, with fuzzy inputs, are considered to be a simple extension of non-fuzzy neurons. A few methods of how these neurons change themselves during learning to improve their performance are also given. The notion of ' *synaptic* ' and ' *somatic* ' learning and adaptation is also introduced, which seems to be a powerful approach for developing a new class of fuzzy neural networks. Such an approach may have powerful applications especially in the processing of fuzzy information and designing of new expert systems with learning and adaptation abilities.

## References

[1]    M.E. Cohen and D.L. Hudson, "An Expert System Based on Neural Network Techniques" in the Proceedings of NAFIP's 90 (I.B. Turksen, Ed.), Toronto, June 6-8, 1990, pp.117-120.

[2]    T. Yamakawa and S. Tomoda, "A Fuzzy Neuron and its Application to Pattern Recognition" in the Proceedings of the 3rd IFSA Congress (J.C. Bezdek, Ed.), Seattle, Washington, Aug. 6-11, 1989, pp.30-38.

[3]    D.C. Kuncicky and A. Kandel, "A Fuzzy Interpretation of Neural Networks" in the Proceedings of the 3rd IFSA Congress (J.C. Bezdek, Ed.), Seattle, Washington, Aug. 6-11, 1989, pp. 113-116.

[4]    J.B. Kiszka and M.M. Gupta, "Fuzzy Logic Neural Network", BUSEFAL, No.4,1990.

[5]    M.M. Gupta and G.K. Knopf, "Fuzzy Neural Network Approach to Control Systems", InternationalSymposium on Uncertainty Modeling and Analysis, University of Maryland, College Park, Dec. 3-5, IEEE Computer Society Press,1990, pp. 483-488.

[6]    S. Nakanishi, T. Takagi, K. Uehara and Y. Gotoh, "Self-Organising Fuzzy Controllers by Neural Networks" in the Proceedings of International Conference on Fuzzy Logic & Neural Networks (Vol. 1), IIZUKA'90, Fukuoka, Japan, July 22-24, 1990, pp. 187-192.

[7]    I. Hayashi, H. Nomura and N. Wakami, "Artificial Neural Network Driven Fuzzy Control and its Application to the Learning of Inverted Pendulum System" in the Proceedings of the 3rd IFSA Congress J.C. Bezdek, Ed.), Seattle, Washington, Aug. 6-11, 1989, pp. 610-613.

[8]    R.P. Lippmann, "An Introduction to Computing with Neural Nets", IEEE ASSP Magazine, pp. 4-22, 1987.

[9]    M.M. Gupta and J. Qi, "Theory of T-norms and Fuzzy Inference methods", Fuzzy Sets and Systems Vol. 40, No. 3, 1991, pp. 473-489.

[10]   J. Qi, "Design of Fuzzy Logic Controllers Based on T-Operators", M.Sc Thesis, Intelligent Systems Research Laboratory, University of Saskatchewan, 1991.

# NEUROFUZZY COMPONENTS BASED ON THRESHOLD

F.A. Gomide and A.F. Rocha

*UNICAMP/FEE/DCA, Caixa Postal 6101, 13081 Campinas, Brazil*

**Abstract.** A new mathematical model which generalizes the processing capabilities of the usual artificial neuron, is presented. Also, a new class of neurons, called Recurrent Controlled Threshold Neuron is introduced, which seems particularly usefull for control system purposes. These devices includes Max and Min neurons as well as Adaptive Sensors. A scheme for implementation of a Neurofuzzy Controller (NFLC) using these components is presented. This NFLC seems particularly appropriate to implement adaptive control, since the control strategy is coded into the network weights.

**Keywords.** Fuzzy Control, Neural Nets, Adaptive Sensor, Adaptive Control

## INTRODUCTION

The use of fuzzy logic in control applications has shown to be a very convenient way to handle control problems when the system is not well defined, and when qualitative information processing is required. A number of successful applications of fuzzy logic controllers (FLC) have bee reported (Lee, 1990). The rebirth of interest in neural networks (NN) appeared in the 80's as a consequence of some technical improvements of the learnign algorithms and network structures (Rumelhart et alii, 1986). The massive parallelism and the learning capabilities of neural nets offer the promise of better solutions for a class of control, identification and system optmization problems (Narendra, 1990; Francelin and Gomide, 1991). During the last years, the relationships between fuzzy logic and neural netwroks have been taken into serious consideration, providing an alternative mathematical modelling and analysis approach which seems to be particularly useful for control and decision making problems in general (Gomide and Rocha, 1992; Rocha, 1993; Takagi, 1990). The success of this approach is demonstrated by several applications of the neuro-fuzzy technology in daily use machines in countries like Japan, Korea, etc.

Besides this success, this emergent neuro-fuzzy technology has not yet exploited all the potentiality of the neural fuzzy devices. The aim of this paper is to present some neuro-fuzzy components obtained by generalizing the structure of the artificial neuron and to show how these devices may be used to build neuro-fuzzy controllers (NFLC).

## A MODEL OF FUZZY NEURON

The artificial neuron is usually taken as a computational device that power averages its $n$ inputs $ai$ as

$$v = \sum_{i=1}^{n} ai \ wi \qquad (1)$$

where $wi$ are the weigths of the synapsis linking the input (pre-synaptic) neurons $ni$ to the post-synaptic neuron $np$. The neuron $np$ recodes the aggregated value $v$ into the axonic activation $ap$

$$ap = \begin{cases} f(v) & \textit{if } v \geq \alpha \\ 0 & \textit{otherwise} \end{cases} \qquad (2)$$

where $\alpha$ is the axonic threshold and f is the encoding function. If

$$f(v) = 1 \qquad (3)$$

the neuron is called a crisp neuron. If

$$f : V \to [0,1] \qquad (4)$$

it is named fuzzy neuron. The encoding function of the fuzzy neuron are in general of the type

$$ap = \begin{cases} 1 & \text{if } v \geq \alpha2 \\ 0 & \text{if } v \leq \alpha1 \\ f(v) & \text{otherwise} \end{cases} \qquad (5)$$

The aggregation function in Eq. 1 may be generalized with a t-norm (Gomide and Rocha, 1992)

$$v = \sum_{i=1}^{n} ai \lceil vi \qquad (6)$$

where $\lceil$ enjoys the following properties (Dubois and Prade, 1980):

$$\begin{array}{l} \lceil(a,1) = \lceil(1,a) = a \; ; \; \lceil(0,0) = 0 \; ; \\ \lceil(a,b) \leq \lceil(c,b) \quad \text{if } a \leq c \; ; \; b \leq d \; ; \; (7) \\ \lceil(a,b) = \lceil(b,a) \quad , \text{ and} \\ \lceil(a,\lceil(b,c)) = \lceil(\lceil(a,b),c) \end{array}$$

The recurrent synapsis is established if the axon of the neuron $np$ makes contacts with dendrites or the cell body of $np$ itself (Fig. 1). If the recurrent synapsis is located near the axon, then it may control the axonic threshold as a function of the $np$'s activity itself.

Let the following types of neuron to be defined:

● high threshold neuron (HTN): the bias of this type of neuron is set at a very high value, such that its output is maintained zero no matter the value of its inputs. From time to time, this threshold is lowered under the control of another neuron, called here setting neuron. At these specific moments, the actual value of the $v$ in Eq. 6 is encoded into ai at the axon of $ni$ and transmitted to the HTN post-synaptic cell. In this condition the neuron is said to fire at $t$;

● low threshold neuron (LTN): the bias of this type of neuron is set at a low value, such that its output changes in time according with the temporal modification of its inputs. The neurons commonly used in artificial neural nets are LTN neurons.

In this condition, the LTN axon thresholds may be controlled by a recurrent synapsis, such that

where gi: $[0,1] \times [0,1] \to [0,1]$ is an appropriate

$$\begin{array}{l} \alpha1(t) = g1(ap(t-1), v(t-1)) \\ \alpha2(t) = g2(ap(t-1), v(t-1)) \end{array} \qquad (8)$$

threshold controlling function. Also the output $ap(t)$ is defined as, for instance,

$$ap(t) = \begin{cases} 1 & \text{if } v \geq \alpha2(t) \\ 0 & \text{if } v \leq \alpha1(t) \\ f(v(t)) & \text{otherwise} \end{cases} \qquad (9)$$

with $v$ being obtained according to Eq.(6). The neuron defined by Eq.(9) will be called here Recurrent Threshold Controlled Neurons (RCTN), and it will be used to build special neuro-fuzzy devices. Note that the inequality signs in Eq.(9) may be reversed, depending of the application.

### MAX and MIN NEURONS

Let the axonic threshold $\alpha1(t)$ of the RTCN $np$ in Fig.1 be set equal to 0 at the time $t = 0$ by the setting neuron $ns$. Assume that at the instant $t$ the same $\alpha1(t)$ is set equal to the firing level $ap(t-1)$ at the time $t-1$, that is

$$\alpha1(t) = ap(t-1) \qquad (10)$$

Also, the output $ap(t)$ is supposed to be

$$ap(t) = \begin{cases} \alpha1(t) & \text{if } v(t) \leq \alpha1(t) \\ v(t) & \text{otherwise} \end{cases} \qquad (11)$$

where $v(t)$ is the post-synaptic activation of $np$ at time $t$

$$v(t) = ai(t) \lceil wi \qquad (12)$$

Hence, the output $ap(t)$ of the neuron $np$ at time $t$ encodes (see Fig. 2)

$$ap(t) = \overset{t}{\underset{k=1}{V}} v(k) \qquad (13)$$

where $V$ is the maximum operator. Now, if the firing of the pre-synaptic neurons $ni$ are timely ordered then

$$ap(t) = \overset{t}{\underset{k=1}{V}} ak(k) \lceil wk \qquad (14)$$

Finally, if $wk = 1$ for all $k$

$$ap(t) = \overset{t}{\underset{k=1}{V}} \, ak(k) \qquad (15)$$

Note that, the neuron $np$ computes in this case the maximum of the $t$ inputs. Because of this, the neuron defined by Eq.(11) will be called here max-neuron.

Analogously, the min neuron is obtained if the axonic threshold $\alpha2(t)$ of the RTCN $np$ in Fig.1 is set equal to 1 at the time $t = 0$ by the setting neuron $ns$. Assume that at the instant $t$ the same $\alpha2(t)$ is set equal to the firing level $ap(t-1)$ at the time $t-1$, that is

$$\alpha2(t) = ap(t-1) \qquad (16)$$

Also, the output $ap(t)$ is supposed to be

$$ap(t) = \begin{bmatrix} \alpha(t) & \text{if } v(t) \geq \alpha2(t) \\ v(t) & \text{otherwise} \end{bmatrix} \qquad (17)$$

where $v(t)$ is the post-synaptic activation of $np$ at time $t$

$$v(t) = ai(t) \lceil wi \qquad (18)$$

Hence, the output $ap(t)$ of the neuron $np$ at time $t$ encodes (see Fig. 2)

$$ap(t) = \overset{t}{\underset{k=1}{\cap}} v(t) \qquad (19)$$

where $\cap$ is the minimum operator. Now, if the firing of the pre-synaptic neurons $ni$ are timely ordered then

$$ap(t) = \overset{t}{\underset{k=1}{\cap}} ak(k) \lceil wk \qquad (20)$$

Finally, if $wk = 1$ for all $k$

$$ap(t) = \overset{t}{\underset{k=1}{\cap}} a(k) \qquad (21)$$

Note that, the neuron $np$ computes in this case the minimum of the $t$ inputs. Because of this, the neuron defined by Eq.(17) will be called here min-neuron.

## AN ADAPTIVE SENSOR DEVICE

Let be considered the measurement device defined by the characteristic curve depicted in Fig. 3. It is clear that the values $U^{n-1}$ and $U^n$ of the universe of discourse $U$ are within the measuring range defined by $U_m$ and $U_M$. In this case, the sensor device provides a correct evaluation of both $U^n$ and $U^{n-1}$. However, for $U^{n+1} \geq U_M$ the sensor saturates. The adequate behavior of the sensor may

be restored if its characteristic curve is conveniently shifted (see dashed line in Fig. 3) to put $U^{n+1}$ inside the measuring range $U_{m+1}, U_{M+1}$. A sensor device being able to automatically adjusting its measuring range is called an adaptive sensor device.

Let be considered three neurons $nL$, $nS$ and $nH$ having their correspondent encoding functions fL , fS and fH defined as in Fig. 4. Now, it is possible to assume for sake of simplicity and without loss of generality that:
for $nL$

$$\begin{aligned} \alpha L1(t) &= \alpha L1(n) = V_m^n \\ \alpha L2(t) &= \alpha L2(n) = v^{n-1} \\ \text{fL} &: VL \to [0,1] \\ VL &= \{ v : \alpha L1 \leq v \leq \alpha L2 \} \end{aligned} \qquad (22)$$

for $nS$

$$\begin{aligned} \alpha S1(t) &= \alpha S1(n) = V_m^n \\ \alpha S2(t) &= \alpha S2(n) = V_M^n \\ \text{fS} &: VS \to [0,1] \\ VS &= \{ v : \alpha S1 \leq v \leq \alpha S2 \} \end{aligned} \qquad (23)$$

for $nH$

$$\begin{aligned} \alpha H1(t) &= \alpha H1(n) = v^{n-1} \\ \alpha H2(t) &= \alpha H2(n) = V_M^n \\ \text{fH} &: VH \to [0,1] \\ VH &= \{ v : \alpha H1 \leq v \leq \alpha H2 \} \end{aligned} \qquad (24)$$

where $v^{n-1}$ is the measurement or the unity weighted input for neurons $nL$, $nS$ and $nH$. If the next measurement is taken as $v^n$, to achieve the desired adaptive behavior, the neurons thresholds should be adjusted accordingly. For example, for upper threshold adjustment (see figure 4) assuming that the new measurement should be at a point where both, fL and fH are null ( for scale symetry purposes ), we may set

$$V_M^{n+1} = v^n + k \cdot ( v^n - v^{n-1}) \qquad (25)$$

$$k = [ \text{fH}(v^n) ]^{-1} \qquad (26)$$

## A NEUROFUZZY CONTROLLER

A tipical FLC has a base of rules of the type:

$$\textit{If X is A \quad then \quad Y is B} \qquad (27)$$

where $A$ and $B$ are fuzzy sets defined in the universes of discourse $U$ and $V$, respectively. The following compatibility functions are associated with these fuzzy sets:

$$\mu A : U \to [0,1]$$
$$\mu B : V \to [0,1] \qquad (28)$$

Let $A$ to be described by a vector $X$ of size $n$, such that

$$xi = \mu Ai \quad if\ \tau i < u \le \tau i+1 \qquad (29)$$

where $i = 1, ..., n-1$. Thus, the fuzzy set $A$ is:

$$A = \begin{bmatrix} x1 \\ \cdot \\ \cdot \\ \cdot \\ xn \end{bmatrix} \qquad (30)$$

Similarly, let $B$ to be described by a vector $Y$ of size $m$ such that

$$yi = \mu Bi \quad if\ \beta i < v \le \beta i+1 \qquad (31)$$

where $i = 1, ..., m - 1$. Thus, the fuzzy set $B$ is:

$$B = \begin{bmatrix} y1 \\ \cdot \\ \cdot \\ \cdot \\ ym \end{bmatrix} \qquad (32)$$

The implication $A \to B$ supporting the fuzzy proposition *If X is A Then Y is B* is defined by the following fuzzy relation (Zadeh, 1983):

$$R : A \times B \to [0,1] \qquad (33)$$

$$\mu R(x,y) = \mu A(u) \lceil \mu B(v) \qquad (34)$$

where $\lceil$ is a t-norm.
The reasoning in a fuzzy data base composed by rules like that in Eq.(27) is supported by the Generalized Modus Ponens (GMP) (Zadeh, 1983):

$$\begin{array}{c} \textit{If X is A then Y is B} \\ \underline{\textit{X is A}'} \\ \textit{Y is B}' \end{array} \qquad (35)$$

where $A'$ is defined by $\mu A'$

$$\mu A' = U \to [0,1] \qquad (36)$$

Hence, using the same vector notation as above to describe $A'$:

$$x'i = \mu A'i \quad if\ \tau i < u \le \tau i+1 \qquad (37)$$

where $i = 1, ..., n-1$, the fuzzy set $A'$ becomes:

$$A' = \begin{bmatrix} x'1 \\ \cdot \\ \cdot \\ \cdot \\ x'n \end{bmatrix} \qquad (38)$$

Similarly, $B'$ becomes

$$B' = \begin{bmatrix} y'1 \\ \cdot \\ \cdot \\ \cdot \\ y'm \end{bmatrix} \qquad (39)$$

The most popular way to implement the GMP in Eq.(35) is to use the compositional rule of inference introduced by Zadeh,1983:

$$B' = A' \circ R \qquad (40)$$

with $R$ as in Eq.(34), (50).

## MAPPING FUZZY RULES IN FUZZY NEURAL NETS

The fuzzy neurons defined before may be used to assemble fuzzy neural nets (FNN) to compute Eq. (40). Let initially be discussed the simplest case for which the dimensions $n$ and $m$ of the vectors $X$ and $Y$ are equal to 2. In this condition, let be given:

$$A = \begin{bmatrix} x1 \\ x2 \end{bmatrix} \qquad (41)$$

$$B = \begin{bmatrix} y1 \\ y2 \end{bmatrix} \qquad (42)$$

Thus:

$$R = \begin{bmatrix} x1 \lceil y1 & x1 \lceil y2 \\ x2 \lceil y1 & x2 \lceil y2 \end{bmatrix} \qquad (43)$$

Now let be the FNN displayed in Fig. 5, where:
- the HTN input neurons $n1$, $n2$ are in charge of measuring $A'$, such that their activities $x1$ and $x2$ are calculated as:

$$x'i = \mu A'i \quad if\ \tau i < u \le \tau i+1 \qquad (44)$$

Hence, neurons $n1$ and $n2$ may be considered as sensory neurons in charge of matching the actual output variables with the desired set points of the FLC;

- the output neurons $p1$, $p2$ are max neurons defined according to Eq. (9),

- the weight $wij$ of the linkages of the input neurons $ni$ with the output cells $pj$ is set as:

$$wij = xi \lceil yj \qquad (45)$$

- and the recoding $vij$ at the synapsis between $ni$ and $pj$ becomes

$$vij = x'i \lceil wij \qquad (46)$$

Now, if the firing of $n1$ and $n2$ is synchronized by the setting neuron such that $n1$ fires always before $n2$, the output $apj(t)$ ( $j = 1, 2$ ) represented by the activity of the neurons $p1$ and $p2$ for $t = 2$ is

$$B^l = A^l \circ R \qquad (47)$$

because

$$apj(2) = v1j \ V \ v2j \qquad (48)$$

that is

$$apj(2) = (x'1 \lceil (x1 \lceil y2)) \ V \ (x'2 \lceil (x2 \lceil y2)) \qquad (49)$$

The synchronization required may be provided by the circuit of setting neurons $si$ in Fig. 5; which may control the sensory neurons besides the output neurons. The role played by these cells $si$ is to control the initial value of the threshold of the post-synaptic max neurons $pi$ and the actual value of the threshold of the HTN neurons $ni$.

The same kind of structure may be used to construct the FNN in Fig. 6, which generalizes the above processing when the size of the vectors $X$ and $Y$ are $n$ and $m$, respectively. The synchronizing system is now composed by $n+1$ setting neurons disposed in a serial chain. The circuit is omitted in Fig. 6 to simplify the drawing. Now let be considered the following RB of a FLC:

*If X is A1 then Y is B1 else*

$$\cdot$$
$$\cdot \qquad (50)$$
$$\cdot$$

*If X is An then Y is Bn*

where the connective *else* is either a conjunction or a disjunction. In this condition, the entire RB may be considered a relation $R$ obtained from the composition of the individual relations $Rk$ associated with each rule $k$ in RB:

$$R = \bigoplus_{k=1}^{n} Rk \qquad (51)$$

where $\oplus$ is a t-norm if RB is conjunctive or a s-norm if RB is disjunctive. The most popular approach is to assume as the max-operator. This implies, that the same FNN in Fig. 6 may compute

$$B^l = A^l \circ R \qquad (52)$$

if $wij$ is assumed to be

$$wij = \bigoplus_{k=1}^{n} (xki \lceil ykj) \qquad (53)$$

The complexity of the FNN implementing the RB in 50 is mostly dependent on the degree of precision required by the input/output discretization.

## CONCLUSION

A new mathematical model for artificial neurons has been presented in this paper. The model generalizes the processing capabilities of the usual artificial neuron. Also, a new class of neurons, called Recurrent Controlled Threshold Neuron is introduced. This generalized artificial neuron is used to construct a class or neuro-fuzzy components, which seems particularly usefull for control system purposes. These devices includes Max and Min neurons as well as Adaptive Sensors. A scheme for implementation of a neurofuzzy controller (NFLC) using these components was presented. This NFLC seems particularly appropiate to implement adaptive control, since the control strategy is coded into the network weights. This is the subject of the present research in our group. The results will be presented in a near future.

## REFERENCES

Dubois, D, H. Prade (1982). A class of fuzzy measures based on triangular norms. *Int. J. General Systems*, vol.8; pp 43-61.

Francelin, R; F. Gomide, K. Loparo (1991). System Optimization with Artificial Neural Networks. Proc. IJCNN'91, Singapore.

Gomide, F, A. F. Rocha (1992). Neurofuzzy Controllers (submitted).

Lee, C. C. (1990). Fuzzy Logic Control Systems *IEEE Trans. SMC*, vol. 20, pp. 419-439.

Rumellhart, D.E., J.L. McClelland and the PDP group (1986). *Parallel Distributed Processing*, MIT Press.

Narendra, K.S., K. Parthasarathy (1990). Identification and Control of dynamical systems using neural networks. *IEEE Trans Neural Networks*. vol. 1, pp. 4-27.

Rocha, A.(1982). Basic Properties of Neural Circuits, *Fuzzy Sets and Systems*, vol. 7, pp. 109 - 121

Rocha, A. (1983) *Neural Nets: A theory for brains and machines*. Lecture Notes in Artificial Intelligence, Springer Verlag.

Takagi, H. (1990) Fuzion Technolgy of Fuzzy Theory and Neural Networks. *Proc. Int. Conf. On Fuzzy Logic and Neural Networks*, Iizuka, Japan, pp. 13 - 26.

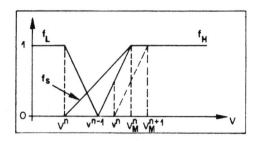

Fig. 4 - Adaptive Sensor

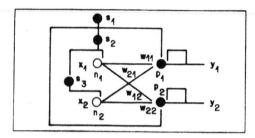

Fig. 5 - Simple Neurofuzzy Controller

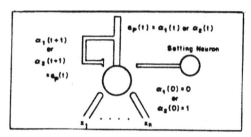

Fig. 1 - The RCTN

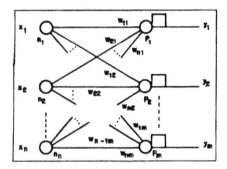

Fig. 6 - Neurofuzzy Controller

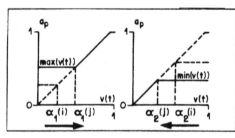

Fig. 2 - Max, Min Neurons

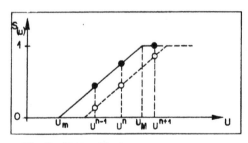

Fig. 3 - Sensor Device

Acknowledgements: The first author acknowledges the CNPq for grant # 300729/86-3. The authors also thank Prof. José Raimundo de Oliveira for his help in preparing the final version of the paper.

# SENSOR FAULT ISOLATION IN DYNAMIC SYSTEMS BY A NEURAL NETWORK HIERARCHY

**A. Bulsari†, A. Medvedev\* and H. Saxen**

*Kemisk-Tekniska Fakulteten, Abo Akademi, SF 20600 Turku/Abo, Finland*

Abstract. A sensor fault detection system comprising a finite–memory observer for residuals generation and a hierarchical bank of the feed–forward neural neworks for fault diagnosis is discussed. Instead of the ubiquitous back–propagation, Levenberg–Marquardt method is used to train the neural networks to isolate the sensors faults based on the observer's residuals. By simulation, the system is shown to permit timely detection and isolation of sensor faults in a biochemical system.

Keywords. Failure detection; observers; neural networks; dynamic systems

## STATE OF THE ART

A fault detection/isolation (FDI) procedure can be implemented using physical and/or analytical redundancy. Physical redundancy exists when an identical plant variable is measured by two or more sensors. In this case, a fault may be identified by simple majority voting logic. Obviously, physical redundancy leads to a rise in hardware expenses, but it solves the fault tolerance problem for a control system in a straightforward manner and does not need any kind of plant model.

Analytical redundancy takes two forms (Frank, 1990). The algebraic relationship among instantaneous outputs of sensors can be refered to as a direct analytical redundancy while the relationship among the histories of sensor outputs and actuator inputs is known as temporal analytical redundancy. Both types of analytical redundancy may be implemented without any additional hardware.

State vector estimators, like Kalman filter in the stochastic framework or Luenberger observer in deterministic one, have seen a lot of applications in the FDI system area, detecting abrupt sensor or/and actuator faults. Such methods as the parity space approach (Chow and Willsky, 1984), the dedicated observer approach (Tylee, 1982) or fault detection filter, also known as Beard–Jones filter (Gertler and Luo, 1989), implicitly or explicitly use the state vector estimation technique to generate the fault accentuated functions, so–called residuals.

Subsequent analysis of residuals may provide an information about a fault component localization but techniques for multiple fault isolation are rather obscure.

The potentially harmful drawback inherited by the observer–based FDI systems from the state estimator structure is the infinite memory, accompanied by the model errors, which can cause the observer divergence. Moreover, the infinite memory inflicts observer insensitivity to recent measurements. The latter property makes the observer response to a sensor fault (known also as a fault signature) rather sluggish. This shortcoming brings us to the conclusion that an observer structure with finite memory would be more appropriate to the FDI system's area.

More recently, neural networks have been studied as a tool for pattern recognition and classification. Promising results were obtained in a neural network application for sensor fault detection in a control system (Naidu et al, 1989) and a multilayer feedforward neural network for incipient fault diagnosis of chemical processes (Watanabe et al., 1989). Once trained, a neural network contains the knowledge about possible process faults and their effects on plant outputs. It should be pointed out that up to now a combination of model–based and knowledge–based methods has not been addressed, though it offers reasonable decomposition of the initial task by implementing a state vector estimator for the residuals generation and a neural network for decision making in the fault isolation procedure.

---

† on leave from Lappeenranta University of Technology

\* now with Division of Automatic Control, Luleå University of Technology, Sweden

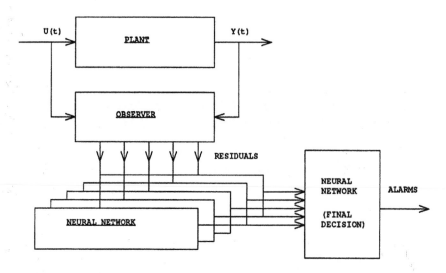

Figure 1. A fault detection and isolation system structure

## SENSOR FAULT ISOLATION SYSTEM

This paper presents an approach to sensor fault detection and isolation in dynamic systems implementing a combination of a state vector observer and feed–forward neural networks (Fig. 1). The FDI system that includes a finite memory observer (Medvedev and Toivonen, 1991) for the fault accentuated residuals generation and a bank of feed–forward neural networks for the fault signature recognition is shown in Fig. 1. The observer yields a number of residuals which correspond to the different fault detection window widths. They are fed to separate neural networks trained to detect and isolate sensor faults from the residuals of particular window widths. The lower layer neural network outputs are sensor fault indicators that switch to one if the sensor is faulty and remain zero under the normal conditions. Note that indicator value may be also any real number between zero and one if observer residuals do not belong to a training data set. The upper layer network is responsible for the final decision making based on the fault indicators of the lower layer.

The same decision making procedure can be implemented by a single neural network, but such a network would be rather complicated and most likely, hard to train. As considered, all neural networks may be trained separately as totally independent parts of a network structure.

### Observer structure

Let us consider the nonhomogeneous equa-

tion

$$\dot{x}(t) = Ax(t) + Bu(t)$$
$$y(t) = Cx(t) \tag{1}$$

where $x(t) \in R^n$ is the state vector, $u(t) \in R^m$ is the control vector, $y(t) \in R^\ell$ is the observation vector $A, B, C$ are real matrices of appropriate dimensions.. Provided the knowledge of the control signal, we introduce measurement vectors as

$$y_i \triangleq y(t - \tau_i)$$
$$+ C \int_{t-\tau_i}^{t} \exp(A(t - \delta - \tau_i))Bu(\delta)d\delta$$
$$\tau_i \geq 0, i = 0, 1, \dots, k$$

Theorem 1: Provided that the matrix

$$\mathcal{W}_k = \sum_{i=0}^{k} \exp(-A^T \tau_i)C^T R_i^{-1} C \exp(-A\tau_i)$$
$$R_i > 0, \ i = 0, \dots, k$$

is positive definite, then the observer

$$\hat{x}_k(t) = \mathcal{W}_k^{-1} \sum_{i=0}^{k} \exp(-A^T \tau_i)C^T R_i^{-1} y_i \tag{2}$$

has the following properties
(*i*) finite memory, limited by the largest time–delay $\tau_k$;
(*ii*) dead–beat performance, i.e $e(t) = x(t) - \hat{x}_k(t) = 0$; $t > \tau_k$ for any initial function $\phi_0 = y(t), t = [-\tau_k, 0]$.

*Proof:* For proof see Medvedev and Toivonen (1991).

<u>Observer residuals</u>

Consider the application of the state estimation algorithm (2) to sensor fault detection, using temporal redundancy. The main idea is comparison of two estimates: one based on process output history, excluding the last $r$ measurements, and the other based on the same data plus the last $r$ output measurements. Actually, the former estimate gives the predicted value of the state vector, but utilizing only reliable data, while the latter one uses the whole data available. The number $r$ will be refered to as fault detection window width while the fault detection window lies between the moments $t - \tau_{k-r}$ and $t - \tau_k$.

Assume that there are no sensor failures before the moment $t - \tau_{k-r}$. The aim of residual generation is to provide information for the decision making procedure which completely characterizes a fault and the faulty sensor(s).

Consider the real matrices $W_k \in R^{(k+1)\ell \times n}$, $Y_k \in R^{(k+1)\ell \times 1}$ of the following block structure

$$W_k = \begin{pmatrix} C \exp(-A\tau_0) \\ \vdots \\ C \exp(-A\tau_k) \end{pmatrix} ; Y_k = \begin{pmatrix} y_0 \\ \vdots \\ y_k \end{pmatrix}$$

If $\text{rank}(Q_k W_k) = \text{rank}(\mathcal{W}_k) = n$ and $R$ is a block–diagonal matrix $R = \text{diag}(R_0, \ldots, R_k)$, then from (1.2) we have

$$\hat{x}_k(t) = (W_k^T R^{-1} W_k)^{-1} W_k^T R^{-1} Y_k \qquad (3)$$

The observer (3) performs state vector estimation from the $\tau_k$–process history. At each time instant $t$, $k+1$ output vector measurements are available.

Let the weighting matrix $R$ be block–diagonal positive definite and of the form

$$R^{-1} = Q_k^T Q_k = \text{diag}(q_0^T q_0, q_1^T q_1, \ldots, q_k^T q_k)$$

where $q_i^T q_i = R_i^{-1}$.

Moreover, let

$$Q_{k+1} W_{k+1} = \begin{pmatrix} Q_k W_k \\ q_{k+1} w_{k+1} \end{pmatrix}$$

$$Q_{k+1} Y_{k+1} = \begin{pmatrix} Q_k Y_k \\ q_{k+1} y_{k+1} \end{pmatrix}$$

be partitioned matrices for the least-squares estimates $\hat{x}_k = (Q_k W_k)^+ Q_k Y_k$ and $\hat{x}_{k+1} = (Q_{k+1} W_{k+1})^+ Q_{k+1} Y_{k+1}$.

Define $\hat{x}_k(t)$ as the estimate computed from $k + 1$ sample output values $i = 0, \ldots, k$ and $\hat{x}_{k-r}(t)$ the estimate computed from $k$ sample output values $i = 1, \ldots, k - r$. It can be shown that the observer residual $\varepsilon_r$ has the form

$$\varepsilon_r = \hat{x}_{k-r} - \hat{x}_k = T_r'(W_r \hat{x}_{k-r} - Y_r)$$
$$T_r' = \mathcal{W}_{k-r}^{-1} W_r^T (R_r + W_r \mathcal{W}_{k-r}^{-1} W_r^T)^{-1} \qquad (4)$$

It is easy to see that the $\varepsilon_r$ vector shows to which extent newly measured data, $Y_r$, suit $(\tau_k - \tau_r)$ process history expressed in $\hat{x}_{k-r}$. Now, if all sensors are operating properly the estimate $\hat{x}_{k-r}$ and the estimate $\hat{x}_k$ will be nearly identical. If, however, a sensor fault occurs at the moment inside the fault detection window, the $\varepsilon_r$ value would be of sufficient scale when the output change caused by this sensor fault is noticeable. To provide more information for decision making under fault detection/isolation procedure a number of $\varepsilon_i$ can be generated. Since all matrix coefficients $T_i', i = 1, \ldots, r$ are independent from the current measurements and do not use any additional information in comparison with the estimate $\hat{x}_k$, the set of $\varepsilon_i, i = 1, \ldots, r$ seems to be reasonable choice.

The essential point to be addressed is whether the $\varepsilon(t)$ provides uniqueness of its trajectory for each faulty sensor, and if so, under which conditions a faulty sensor can be pinpointed. To simplify consideration, let us assume that the weighting matrix $R$ in (4) is a unit matrix.

<u>Theorem 2</u> If all components of the plant (1) output vector $y = C_1 x(t), y = (y_1, \ldots, y_\ell)^T$ satisfy the condition $y_i \not\equiv 0, i = 1, \ldots, \ell$ for the time interval $t_0 < t < t^*$, and $\mathcal{W}_{k-1}$ is positive definite, then the faulty sensor is recognizable from the $\varepsilon(t)$ vector.

*Proof:* For proof see Bulsari *et. al* (1991)

<u>The feed–forward neural network</u>

Artificial neural networks (ANNs), also called connectionist models, parallel distributed models, neuromorphic systems or simply neural nets consist of many simple computational elements called nodes, or neurons each of which collects by weighted addition the signals from various other nodes which are connected to it directionally. This sum, the net input to the node, is processed by a function (usually a sigmoid or a step) resulting in the output or the activation of that node. In multilayer feed–forward networks of the type considered in this paper , there are a few layers (the input layer, the output layer, and possibly some hidden layers) across which all the nodes of each layer are connected to all the nodes in the layer above it, but there are no connections within the layer. A constant term, a bias, is usually added to each of the nodes.

The feed–forward neural network has an input layer which simply transmits the input variables without any processing to the next layer. The bias is a weighted unit input to each node, thus adding a constant term to the net input of the node. Two, one or no hidden layers are considered. Each node in the upper layers (hidden or output layers) receives weighted inputs from each of the nodes in the layer below it. These weighted inputs are added to get the net input of the node, and the output $x_i$ (or the activation) is calculated by an activation function of the net input $a_i$ as

$$a_i = \sum_{j=0}^{N} w_{ij} z_j; \;\; z_0 = 1$$

$$z_i = \frac{1}{1 + e^{-\beta a_i}}$$

where N is the number of nodes in a hidden layer or the number of input nodes and $w_{i0}$ is the bias of node $i$. $\beta$ is referred to as the gain term and is usually set to 1.

Levenberg–Marquardt method was used to determine the weights in the neural networks. Network training aims at minimizing the sum of squares of errors, the errors measured as the difference between the calculated output and the desired output. Back propagation by the generalised delta rule, a kind of a gradient descent method is one popular method for training feed–forward neural networks. Many algorithms for least–squares optimization use Taylor–series models. The Levenberg–Marquardt method uses an interpolation between the approaches based on the maximum neighbourhood (a "trust region") in which the truncated Taylor series gives an adequate representation of the non–linear model. The method has been found to be quite efficient.

APPLICATION

Biochemical processes have highly non-linear characteristics, and have operability in limited domains. In the process considered here, *Saccharomyces cerevisiae*, a yeast, is grown on glucose substrate in a chemostat (a biochemical continuous stirred tank reactor) producing ethanol as a product of primary energy metabolism. There are three state variables : microbial concentration, $X$; substrate concentration, $S$; and product concentration, $P$. The kinetic and stoichiometric parameters

were taken from a recent study on kinetics of this system (Warren *et al.*, 1990) The feed to the chemostat is sterile, *i.e.* there are no microorganisms in the feed. The feed concentration of substrate, glucose, is $S_0$, and $D$ is the dilution rate (volumetric flow rate per volume of the chemostat. ) The dynamics of this system can be described by the following equations.

$$\frac{dX}{dt} = (\mu - D)X$$
$$\frac{dS}{dt} = D(S_0 - S) - Y_{S/X} \mu X$$
$$\frac{dP}{dt} = -DP + Y_{P/X} \mu X$$

where the growth rate, $\mu$ and the yield coefficients, $Y_{S/X}$ and $Y_{P/X}$ are given by

$$\mu = \frac{0.427S}{0.245 + S}(1 - (P/101.6)^{1.95})$$
$$Y_{P/X} = 3.436, \;\; Y_{X/P} = 0.291$$
$$Y_{X/S} = 0.152(1 - P/302.3), \;\; Y_{S/X} = 1/Y_{X/P}$$

The growth rate, $\mu$ cannot exceed 0.427, and that is also the upper limit for the dilution rate, $D$. If the dilution rate is higher than the growth rate, microbial concentration in the bioreactor decreases to zero. Table 1 shows typical steady state values for the process parameters.

TABLE 1   Steady State Values

| | |
|---|---|
| $S_0$ | 101.416 gm/lit |
| $D$ | 0.345 hr$^{-1}$ |
| $X$ | 12.0 gm/lit |
| $S$ | 10.0 gm/lit |
| $P$ | 41.232 gm/lit |
| $\mu$ | 0.345 hr$^{-1}$ |
| $Y_{X/S}$ | 0.1313 |
| $Y_{X/P}$ | 0.291 |

In the neighbourhood of the steady state, the biochemical process can be described by the linearized model (1), were the matrices are

$$A = \begin{pmatrix} 0 & -0.041 & 0.010 \\ 1.185 & -0.485 & 0.034 \\ -2.628 & 0.190 & -0.420 \end{pmatrix}$$

$$B = \begin{pmatrix} 0 & -12. \\ 0 & -41.232 \\ 0.345 & 91.416 \end{pmatrix}$$

and the state and control vectors are $x^T = (x\ s\ p)$, $u^T = (s_0\ d)$. Variables in small letters are the deviation from their corresponding operating parameters.

### Training the lower layer neural networks

The lower layer networks were trained for 0, 1, or 2 faults with 1206 training instances. Neural networks with one hidden layer, three inputs and three outputs were trained for recognising the sensor faults. Two hidden layers were also considered, but they did not significantly improve the results, as has been experienced with several other systems, partly due to the difficulty in training (running into local minima often) and partly due to reduced flexibility of these networks. The number of nodes in the hidden layer was varied between 12 and 18. 16 seemed to be a suitable number of nodes in the hidden layer based on various training results, and was therefore used for all the three lower layer networks.

For a window width of one, the sum of squares of errors (SSQ) was $0.4 \times 10^{-10}$. The SSQ for a window width of 2 was 3.64 for the (3,16,3) network, and was 5.25 for a window width of 3. These SSQ values are quite low, when one takes into consideration that the number of training instances is 1206.

### Training the upper layer neural network

To train the upper layer neural network, 726 instances were generated with known sensor faults. The lower layer networks produced 3 outputs each, which were not always correct, and not always the same as the final output of the upper layer, although this was usually the case. Therefore, the upper layer network had to decide only when there was some discrepancy among the three sets of outputs from the lower layer.

Neural network with a configuration (9,3) without hidden layers was found to be adequate, and hidden layers were almost redundant. On training the upper layer starting with random initial weights, the optimisation procedure converged quickly resulting in a near-zero SSQ. However, the weights showed strong interactions between the various inputs to the network, which was counter–intuitive. When the first output of the upper layer (sensor fault for variable $X$) was allowed to be a function only of the variable $X$

(first) outputs of the lower layer networks, and the second output of the upper layer, a function of the second outputs of the lower layer networks, and similarly the third, the SSQ was found to be 2.997, contributed exclusively by the second variable $S$.

### Testing the complete FDI structure

To test the complete FDI structure, simulations were carried out with a fault in sensor 3 in the first 1.2 hours, starting at 0.9 hours and ending at 1.2 hours; faults in sensors 1 and 3 in the next 1.2 hours, starting at 2.1 hours and ending at 2.4 hours; and with faults in sensors 2 and 3 in the last 1.2 hours, starting at 3.3 hours and ending at 3.6 hours.

Fig. 2b shows the outputs of the three lower layer (3,16,3) networks. The upper three plots are for a window width of one, and the lower ones are for two and three respectively. The plots on the left show faults in $X$, the middle ones, $S$, and the ones on the right show faults in $P$. The results for a window width of one are quite accurate in this case, and in general, were found to be more accurate than the wider windows.

These outputs were fed to the upper layer network (9,3). The outputs from the upper layer, shown in Fig. 2a, can be seen to be accurate.

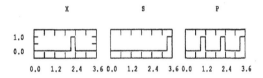

Figure 2a. Outputs of the upper neural network (9,3) which takes inputs from the 3 lower (3,16,3) networks

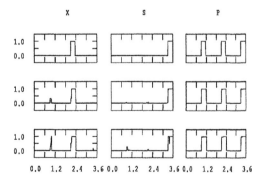

Figure 2b. Outputs of the 3 lower neural networks (3,16,3)
(Upper 3 plots show results with window width 1)

## CONCLUSIONS

A new approach combining state vector estimation and artificial neural networks for sensor fault detection and isolation was studied in the course of this work, and was found to work successfully. The residuals generated by windows containing and excluding the last instances of measurement incorporate the information necessary to detect and isolate single or multiple sensor faults.

Neural networks were trained using the Levenberg–Marquard method, which was found to be a good choice compared to the conventional back–propagation or its variants. Lower layer networks of structure (3,16,3) and upper layer network of structure (9,3) permitted timely detection and isolation of all single and double sensor faults in this biochemical system. Linearized dynamics were found to be adequate for the purpose.

## REFERENCES

Frank, P. M. (1990). Fault diagnosis in dynamic systems using analytical and knowledge–based redundancy— a survey and some new results", Automatica, 26, 459–474.

Chow, E. Y., and A. S. Willsky (1984). Analytical redundancy and the design of robust failure detection dystems, IEEE Trans. Aut. Control., 29, 603–614.

Tylee, J. L. (1982). On-line failure detection in nuclear power plant instrumentation, Kalman Filtering: Theory and Application, IEEE Press, New York.

Gertler, J. and Q. Luo (1989). Robust isolable models for failure diagnosis", AIChE Journal, 35, 1856–1868.

Naidu, S. R , E. Zafiriou, and T. J. McAvoy (1989). Use of neural networks for sensor failure detection in a control system", 1989 American Control Conference.

Watanabe, K. et al. (1989). Incipient fault diagnosis of chemical processes via artificial neural networks, AIChE Journal 35, 1803–1812.

Medvedev, A. V., and H. T. Toivonen (1991). Investigation of a finite memory observer structure, Report 91-6, Process Control Laboratory, Åbo Akademi, Åbo, Finland.

Warren, R. K., G. A. Hill, and D. G. Macdonald, (1990) Improved bioreaction kinetics for the simulation of continuous ethanol fermentation by Saccharomyces cerevisiae, Biotechnology Progress, 6, 319-325.

Marquardt, D. W. (1963), An algorithm for least-squares estimation of nonlinear parameters, J. Soc. Indust. Appl. Math., 11, 431-441.

Levenberg, K. (1944), A method for the solution of certain nonlinear problems in least squares, Quart. Appl. Math., 2, 164-168.

Bulsari, A., A. V. Medvedev, and H. Saxén (1991). On sensor fault detection and isolation in dynamic systems, Report 91-9, Process Control Laboratory, Åbo Akademi, Åbo, Finland.

# BINARY PATTERN DESCRIPTION BASED ON UNL-FOURIER FEATURES - A NEURAL NETWORK IDENTIFIER WITH LEARNING CAPABILITY

**T.W. Rauber and A. Steiger-Garçao**

*Universidade Nova de Lisboa, Faculdade de Ciencias e Tecnologia, Departamento de Informatica,*
*Quinta de Torre, 2825 Monte de Caparica, Portugal*

Abstract. The paper presents a general purpose method to describe the form of binary patterns. Features based on a Fourier analysis of transformed parametric curves are introduced which are insensitive to translation, rotation and scaling. If patterns are different they can be distinguished without explicitly specifying the differences due to the information preserving qualities of the transform. The calculus of the features is inexpensive compared to the potentiality to characterize a arbitrary formed pattern. An identifier using a backpropagation neural network simulator is constructed on top of the feature model. In order to prove the general purpose capability of the technique, we present results originating from three different pattern types: Object contours from vision system images, 2-D projections of CAD modelled solids and handwritten numerals.

Keywords. Pattern recognition, shape features, Fourier analysis, computer vision, backpropagation neural network.

## INTRODUCTION

In many pattern recognition applications only the form of an object is the matter of interest. Change in scale, translation and rotation should not affect the characterizing features. Other qualities of the features should be easy representability as numerical values and low cost to obtain them computationally. Also the complexity of the shape should be no barrier to the description method.

Fourier descriptors can be used to describe patterns (Zahn and Roskies, 1972), (Persoon and King-Sun Fu, 1977). The patterns however are limited to single plane closed curves. Our method distinguishes itself clearly from the Log-polar transform of Massone et al. (1985) which was also investigated by Wechsler and Zimmerman (1988) and Reber and Lyman (1987). It is not possible to apply that transform to thin isolated curves.

In this text a pattern transformation is introduced which is information conserving and creates an optimal input for a 2-dimensional Fourier analysis. It is based on a centroid oriented polar coordinate transform of curves and will be denoted as the UNL Transform. Patterns are composed by a set of thin, derivable parametric curves. The curves are transformed from the Cartesian coordinate system to the polar coordinate system with the centroid as origin.

The theory is introduced in the analytical case and then extended to the discrete case.

As we deal always with single pixel thin curves, a thinning algorithm must always ensure this condition in both coordinate representations.

An identifier is build around the feature model. The Fourier analysis yields a huge amount of features because the features are sampled values of a 2-dimensional continuous function. Their number increases with the resolution of the pixel matrix.

A simple feature selection algorithm divides the feature pool into "good" and "bad". This preselection limits the number of usable features to a manageable amount.

A subset of the good features makes up the input for a neural network simulator based on the backpropagation algorithm. We choose this knowledge representation schema because of its easy "programmability". It was experimentally proved (Burr, 1988) that the classification potentiality of backpropagation neural networks are at least as high as conventional techniques, like nearest neighbor classifiers.

## THE UNL TRANSFORM

### Basic Ideas

The UNL Transform creates an ideal input for Fourier analysis by transforming any curve pointset given in Cartesian coordinates to a 2-D function which is $2\pi$-periodic in one of its parameters and invariant in the other parameter. No information is gained or lost during the transform because there is a bijection between the pattern and its transform.

The important qualities as a shape descriptor are respected by the process of the transformation. The Fourier analyzed function of the transformed pattern yields features that are not affected by translation, scaling and rotation.

### Description

A finite set of thin smooth curves can assemble any shape, either as a closed boundary of an object or isolated e.g. handwritten letters. The curves can be represented in a parametric form. As an example consider the number "83"; it is a pattern which consists of 3 different smooth curves. The condition of smoothness ensures that the curve is differentiable in all of its points.

The centroid of all curves becomes the origin of a polar coordinate system. Each curve can now be transformed from its Cartesian coordinate system to a normalized polar representation without loss of information. It is important to note that we transform the curve as a whole, not single points.

The resulting 2-D function with values between 1.0 (pixel on) and 0.0 (pixel off) is invariant to translation and scaling of the original pattern. Rotations of the original result in translations in the polar coordinate space.

The translation theorem of the Fourier analysis states that the amplitudes of a Fourier transformed function do not vary when the function is translated. Hence the Fourier amplitudes of the center oriented polar transform of the original pattern are features which are insensitive to the 3 linear transformations translation, scaling and rotation. They are shape descriptors which only represent the _form_ of the pattern.

## Definition

### Definition 1: UNL Transform

Let

$$C = \{C_i\}_1^n, \quad C_i(t_i) = (x(t_i), y(t_i)),$$

$$t_i \in [a_i, b_i] , \quad a_i, b_i \in \mathbb{R} \tag{1}$$

be a finite number of smooth curves in a parametric form. C represents a pattern in the Cartesian $\mathbb{R}^2$ plane.

Let

$$f(x,y) = \begin{cases} 1 & \text{if } \exists\, i,j: C_i(t_{ij}) = (x,y) \\ 0 & \text{else} \end{cases} \tag{2}$$

be the intensity function in point $C_i(t)$

Let $O \in \mathbb{R}^2$ be the centroid of all curves, $O = (O_x, O_y)$

$$O_x = \frac{1}{L} * \sum_{i=1}^{n} \int_{a_i}^{b_i} x(t_i) \sqrt{x'^2(t_i) + y'^2(t_i)}\, dt_i$$

$$O_y = \frac{1}{L} * \sum_{i=1}^{n} \int_{a_i}^{b_i} y(t_i) \sqrt{x'^2(t_i) + y'^2(t_i)}\, dt_i$$

$$\tag{3}$$

where L is the total length of all curves.

$$L = \sum_{i=1}^{n} \int_{a_i}^{b_i} \sqrt{x'^2(t_i) + y'^2(t_i)}\, dt_i \tag{4}$$

Let $d \in \mathbb{R}$, $d = \max_{i=1}^{n} \{d_i\}$, $\tag{5}$

be the maximum distance between any curve point and O, where

$$d_i = \max_{t_{ij}} \left\{ \ | \ \vec{r}_{ij}(t_{ij}) \ | \ \right\} ,$$

$$t_{ij} \in [a_i, b_i],\ \vec{r}_{ij}(t_{ij}) = \vec{C}_i(t_{ij}) - \vec{O} \tag{6}$$

is the maximum length of any vector from the geometric center to a pattern point. Fig. 1 outlines the situation.

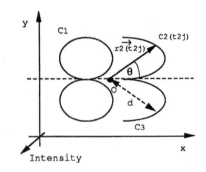

Fig. 1. Parameter transformation of the UNL Transform.

Then there exists a bijective transform U which transforms the Cartesian coordinates $(x,y)$ of function f to the normalized polar coordinates $(R, \theta)$,

$$U: \mathbb{R}^2 \rightarrow [0,1] \times [-\pi, \pi] , \quad (x,y) \mapsto (R, \theta)$$

$$f(R(x,y),\theta(x,y)) = f(U(x,y)) = f(x,y)$$

$$R_{ij}(t_{ij}) = \frac{1}{d}\sqrt{r_x^2 + r_y^2},$$

$$\theta_{ij}(t_{ij}) = \text{Arctan}(r_y/r_x) \qquad (7)$$

where $r_x = r_{ij}(t_{ij})_x = x(t_{ij}) - O_x$
and $r_y = r_{ij}(t_{ij})_y = y(t_{ij}) - O_y$ (8)

are the abscissa and ordinates of the vector that connects the centroid O to a pattern point given by a curve $C_i$ and its respective parameter value $t_{ij}$. (The "Arctan" function is quadrant respective).

Fig. 2 shows a pattern in the Cartesian coordinate system and its UNL Transform.

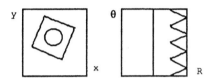

Fig. 2. Pattern and its UNL Transform

## UNL Fourier Features

Definition 2: UNL Fourier Features

Let $f(x,y)$ be the intensity function of a pattern in the Cartesian coordinate plane and let $f(R,\theta)$ be its its UNL transform. Then the spectrum of the Fourier transform of $f(R,\theta)$

$$|F(p,q)| = |\mathcal{F}(f(R,\theta))| = |\mathcal{F}(f(U(x,y)))| \qquad (9)$$

is the function of the UNL Fourier Features.

## THE UNL TRANSFORM - DISCRETE CASE

In the discrete case we have to deal with a number of additional problems that are senseless in their analytical form. The digital nature of the data conditions errors caused by noise. A scaling or rotation operation on a pixel matrix normally changes the number of points or destroys fine details. How do we estimate the analytical equations of the parametric curves if only a bitmap pattern is known?

### Definition

A first approach to estimate the analytical form of the curve is to use linear interpolation between neighboring pixels. For every point pair of neighbors in the original (and thinned) pattern we define the equation of the line segment that connects them. It is then possible to generate analytically the corresponding curve in the polar coordinate system. It is only necessary to divide the line segments between two Cartesian pixels into sufficiently fine interval steps.

The center of mass of the curves becomes approximately the mean of all point coordinates. This can be proved analytically by considering the whole pattern as a composition of tiny line segments between neighboring pixels.

Definition 3: Discrete UNL Transform

Let $P = \{p_i\}_1^n$,

$$p_i = (x_i, y_i), \quad x_i, y_i \in \mathbb{N} \qquad (10)$$

be a pointset that defines a 2-dimensional curve pattern in the the Cartesian $\mathbb{R}^2$ coordinate system, and let N be the resolution of the image with

$$0 \le x_i \le N-1, \ 0 \le y_i \le N-1.$$

Let $f(x,y) = \begin{cases} 1 & \text{if } p_i \in P \\ 0 & \text{else} \end{cases} \qquad (11)$
be the intensity function in point $p_i$.

Let $O \in \mathbb{R}^2$ be the centroid of all curves, $O = (O_x, O_y)$

$$O_x = \frac{1}{n} * \sum_{i=1}^n x_i, \quad O_y = \frac{1}{n} * \sum_{i=1}^n y_i.$$

Let $d \in \mathbb{R}$, $d = \max_{i=1}^n \{ |\vec{r_i}| \}$, $\qquad (12)$
be the maximum distance between any pattern point and O, where

$$\vec{r_i} = \vec{p_i} - \vec{O} \qquad (13)$$
is the vector from the geometric center O to a pattern point $p_i$.

Then there is a transform U which transform the Cartesian coordinates $(x,y)$ of the intensity function f to normalized polar coordinates $(R,\theta)$,

$$U: \mathbb{R}^2 \to [0,1] \times [-\pi,\pi], \ f(x,y) \mapsto f(R,\theta)$$

$$R_i = \frac{1}{d}\sqrt{x_i^2 + y_i^2},$$

$$\theta_i = \text{Arctan}(r_y/r_x) \qquad (14)$$

where $r_x = r_{ix} = x_i - O_x$
and $r_y = r_{iy} = y_i - O_y \qquad (15)$

are the abscissa and ordinates of the vector that connects the geometric center O to a pattern point $p_i$.

### Particular Problems of the Discrete UNL Transform

Analytically a pattern consists of a set of curves which have an infinite small thickness. In practice however we analyze the pattern as a signal of pixels that have finite area inside the image matrix.

Curve thickness. We must ensure that the discrete curves have only the thickness of one pixel because a pattern is composed of thin curves by definition. Therefore we apply a thinning algorithm (Zhang and Suen, 1984) to i.) the original pattern as well as ii.) to the UNL transformed pattern. This operation introduces a tolerable degree of information loss.

Furthermore we allow only m-connectivity (Gonzalez and Wintz, 1988) between neighbors which are no special points (links, joints, endpoints). This guarantees that between two normal pixels (no endpoints etc.) there is a unique connection. This allows an estimation of the analytic equations of pattern curves.

## The UNL-Fourier Features - Discrete Case

Signal normalization. The DC component of the pixel matrix represents the basic signal energy. This is the Fourier magnitude for both polar parameters equal to zero: $F(0,0)$. This magnitude has always the maximum values of all magnitudes. Consequently we normalize all other magnitudes by dividing them by $F(0,0)$. This has further the advantage that normalized features represent an ideal input for the neural network identifier.

Conjugate Symmetry and Normalization Redundancy. Since $f(R,\theta)$ is real, the property of conjugate symmetry is valid which means a redundancy in the feature values. We have:

$$|F(p,q)| = |F(-p,-q)| \qquad (16)$$

For this reason we only have to calculate half the form feature values. Furthermore the normalization implies that the frequency $UFF(0,0)$ is always the same and has no more discriminatory information.

We are now able to define the set of features with which we will characterize the shape of the pattern:

Definition 4: Discrete UNL-Fourier Features

Let $f(x,y)$ be the intensity function of a pattern in the Cartesian coordinate plane and let $f(R,\theta)$ be its UNL transform. Then the normalized spectrum of the Fourier transform of $f(R,\theta)$

$$|F(p',q')| = |\mathcal{F}(f(R,\theta))| = |\mathcal{F}(f(U(x,y))|$$

$$UFF(p,q) = (|F(p,q)| \, / \, |F(0,0)|) \subset (|F(p',q')| \, / \, |F(0,0)|) \qquad (17)$$

are the UNL Fourier Features (UFF).

## Calculus of the Fourier Transform

A pattern identifier normally consists of several working phases: training, training data analysis and classifier. Sometimes it is necessary to calculate all UFFs. In this case we use of course a Fast-Fourier-Transform algorithm (Gonzalez and Wintz, 1988) to analyze the UNL transformed pattern. In the classification phase we only have to calculate a very small portion of all existing UFFs.

We have one further great advantage when calculating not all UFF: Since the majority of the pixel intensity values in the $(R,\theta)$ -plane are 0, we can skip them in the explicit Fourier transform. This makes the calculus of one UFF extremely fast which is a great advantage in the identification phase.

## FEATURE SELECTION

The proposed method to calculate the shape descriptors yields an enormous amount of features for each sample of each class. As the features are sampled values of a 2-D function their number even grows with a higher resolution of the pixel matrices. We are obliged to perform a feature selection.

We use a statistical approach to store the knowledge about the feature values. A Gaussian probability density is assumed for the feature values and the parameters mean and standard deviation are estimated.

### Interclass Distance Measure

Heuristic and Exacter Interclass Distance Measure. For two classes $\omega_1$ and $\omega_2$ , we use a measure which characterizes the separability potential for a certain feature. Based on the Gaussian probability distribution we calculate a percentage of interclass distance.

First a quick heuristic is used which needs only the mean and standard deviation of the two classes. Given two classes $\omega_1, \omega_2$ and their mean and standard deviation $(\mu_1, \sigma_1)$, $(\mu_2, \sigma_2)$, we use a normalized probability distance measure

$$d_i(x) = |x - \mu_i|/\sigma_i \;\; ; \; i = 1,2 \qquad (18)$$

for an arbitrary feature value x.

For $d_1(x) = d_2(x) = d$ one can verify that $d = (\mu_1 - \mu_2)/(\sigma_1 + \sigma_2)$ if $x$ is supposed to be located between $\mu_1$ and $\mu_2$ and $\mu_1 < \mu_2$.

For decreasing $\sigma_i$; $i = 1,2$ and a growing distance between $\mu_1$ and $\mu_2$ we notice that $d$ is growing. Logically big values of $d$ express a great interclass distance between the two classes. They are a fast thumbnail guess for "good" or "bad" features.

Chebychev's inequality

$$P\{|x - \mu_i| \geq \varepsilon\} \leq \sigma_i^2/\varepsilon^2 \; ; \; i = 1,2 \quad (19)$$

applied to the normalized distance $d_1(x)$ and for that $x$ where the distances $d_1(x)$ and $d_2(x)$ are equal, we obtain

$P\{d_i \geq d\} \leq 1/d^2$. This probability should be as small as possible. A smaller probability is equivalent to the fact that a value of the feature lies nearer to its own class than to the other class.

Hence $1/d^2$ should also be minimal for a good feature. This allows to establish a quantitative power of class separability measure

$$pd_{heu}(\omega_1, \omega_2) = 1 - \left(\frac{\sigma_1 + \sigma_2}{\mu_1 - \mu_2}\right)^2 \quad (20)$$

This function states a percentage how well a certain feature can distinguish two classes. (Negative values can occur for very high class interference.)

Secondly another more exact probability distance measure calculates the minimum error probability for the two Gaussian distributions (Duda and Hart, 1973).

$$pd_{exact}(\omega_1, \omega_2) = 1 - P\{minerr\} \quad \text{with}$$

$$P\{minerr\} = P(x \in \mathfrak{R}_2, \omega_1) + P(x \in \mathfrak{R}_1, \omega_2) \quad (21)$$

which for two equiprobable classes is

$$\int_{\mathfrak{R}_2} p(x|\omega_1) \, dx + \int_{\mathfrak{R}_1} p(x|\omega_2) \, dx \quad (22)$$

This integral calculus can be approximated by numerical methods.

We decide now to proceed the following way to obtain an interclass distance measure for each class. i) calculate first heuristically; if the distance lies above a certain threshold then ii) calculate more exactly otherwise set distance as 0.

The Global Power of Discrimination for one UNL-Fourier Feature is defined as the average value of all interclass distance measures compared by pairs:

$$GPD(f) = \frac{2}{c(c-1)} \sum_{i=1}^{c-1} \sum_{j=i+1}^{c} pd_{exact}(\omega_i, \omega_j) \quad (23)$$

with $c$ being the number of all classes and $f$ being one particular UFF. This methodology permits to create a ranking for all features in percent of discriminatory power. Fig. 3 gives an overview over the two probabilistic distance measures.

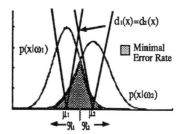

Fig. 3. Probabilistic Distances : heuristic and exact measure.

## Selecting the Feature Set for the Neural Network Classifier

Our ad hoc feature selection technique provided a limited number of "good" features, ordered following the interclass separability criterion. We have to go a step further and restrict even more the number of features which will feed the identifier.

The correlation problem. The fact that the UNL Fourier features are sampled function values raises the problem of highly correlated features. If we detect UFF(1,2) and UFF(1,3) as good feature, then it is probable that this neighboring pair is highly correlated, thus bringing only apparently more classification information. On the other hand, it can be experimentally shown that backpropagation neural network architecture for an identifier is robust in relation to redundant features.

## BACKPROPAGATION NEURAL NETWORK CLASSIFIER

Having reduced the features to a dominable amount, we train a 3-layer feedforward neural network simulator with these features. The architecture is based on a mapping structure which tries to associate different feature values to different classes. The learning procedure bases itself on the error backpropagation algorithm of Rummelhart et al. (Rumelhart and McClelland, 1986), (Rumelhart, Hinton, and McClelland, 1986). The method compares the desired to the actual values of the output layer and adapts internal weights to reduce the error. The theory of backpropagation can be consulted in more detail in (Burr, 1988) and (Pao, 1989).

## Motivation

We choose this identifier model for several reasons.

1) Interface to Features and Classes: We use feature vectors of dimensionality F and map them to C pattern classes. We can directly connect the normalized output of our feature generation method to the layer of the network. The output layer has one neuron for each class.
2) Ease and Speed of Training: For a small number of features, the time for learning cycles is reasonable to achieve the optimal weight state.

<u>Architecture</u>

A 3-layer fully interconnected feedforward -network is used. The <u>Kolmogorov mapping neural network existence theorem</u> (Hecht-Nielsen, 1987) suggests 2F+1 neurons for the middle (hidden) layer.
Obeying this fact we obtain a F — 2F+1 — C network architecture where F is the dimension of the feature vector and C the number of the classes (fig. 4).

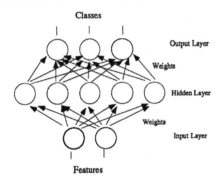

Fig. 4. A 3-layer feedforward neural network. The example has 2 features and 3 classes.

RESULTS

We present the outcome of experiments based on the proposed method for three different pattern classes. It will be proved that the pattern form descriptor is robust in relation to patterns of quite distinct nature.

* The first set of examples stems from a vision system. A photograph of an object is taken in the usual way, digitized, grey level preprocessed and finally segmented which yields the contours of an object. The resulting boundary curves are then processed as the input for the pattern classification unit. The used vision system hardware is the Magiscan 2 image analyzer from Joyce-Loebl.

* Secondly polygons are used which are generated by a polygonal oriented solid modeler. Corresponding to a video camera in reality, the viewpoint for the solids is placed along the working platform's normal vector in an orthogonal

perspective. Linear transformations can be simulated well by applying them to the object. Also the simulated distortion of a real camera is tunable in the CAD model.

* Handwritten characters are a great challenge in the pattern recognition world. The results of an experiment with the ten numerals is presented. The data source is a bitmap-tablet which registers the coordinated where the device's pen touches the surface of the tablet. The benchmark comes from a single person.

Fig. 5 shows one representative of each class for the different universes.

*Scissor  Tacker  Vulcan_14   Vulcan_22  Vulcan_2m*

Fig. 5a. Object Universe: *Vision*.

*Box_1  Box_2  Screw   Wheel   Wheel_1  Wheel_2*

Fig. 5b. Object Universe: *CAD*.

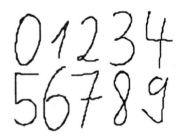

Fig. 5c. Object Universe: *Numerals*.

<u>General Classification Results</u>

a.) Vision
Good classification performance. Classes seem to be easily distinguishable.

b.) Numerals
Data set from one writer. Numbers '6' and '9' can be distinguished because they are not rotation symmetric for this particular writer. Encouraging results.

c.) CAD
The three "wheel" classes are too similar for the identifier. The feature selection yields only one feature that has a heuristic value above 0.0.

## CONCLUSIONS AND FUTURE WORK

A complete system was presented which generates a classifier for an universe of arbitrary shapes. No special feature set has to be designed patterns by conventional means. The emphasis of the work lies on the feature model. A neural network identifier completes the knowledge representation and inference parts of the identifier.

Future ambitions will focus on the analytic representation of curves which are given as mere bitmaps. It will be tried to estimate better the mathematical descriptions of the curves in order to perform the UNL transform with even better results.

## REFERENCES

Burr D. J. (1988). Experiments on Neural Net Recognition of Spoken and Written Text. IEEE Trans. on Acoust., Speech, Signal Processing, 36, NO. 7, 1162-1168.

Duda R. O. and Hart P. E. (1973). Pattern Classification and Scene Analysis. Wiley, New York.

Gonzalez R. C. and Wintz P. (1987). Digital Image Processing, 2.Ed. Addison-Wesley Publishing Corporation.

Hecht-Nielsen R. (1987). Kolmogorov's mapping neural network existence theorem. Proc. IEEE First Int. Conf. on Neural Networks, San Diego, 3, 431-439.

Massone L., Sandini G. and Tagliasco V. (1985). "Form-Invariant" Topological Mapping Strategy for 2D Shape Recognition. Computer Vision, Graphics and Image Proc., 30, No. 2, 169-188.

Pao Y.-H. (1989). Adaptive Pattern Recognition and Neural Networks. Addison-Wesley Publishing Corporation.

Persoon E. and King-Sun Fu (1977). Shape Discrimination using Fourier Descriptors. IEEE Trans. Systems, Man, and Cybernetics, SMC-7, No. 3, 170-179.

Reber W. L. and Lyman J. (1987). An Artificial Neural System Design for Rotation and Scale Invariant Pattern Recognition. Proc. IEEE First Int. Conf. on Neural Networks. San Diego, CA, IV, 277-283.

Rumelhart D. E., Hinton G. E., and McClelland J. L. (1986). A general framework for parallel distributed processing. In Rumelhart and McClelland (eds.), Parallel Distributed Processing, MIT Press, Cambridge, MA, 45-76.

Rumelhart D. E. and McClelland J. L. (1986). Parallel Distributed Processing: Explorations in the Microstructure of Cognition. 1, MIT Press, Cambridge, MA.

Wechsler H. and Zimmerman G. L. (1988). 2-D Invariant Object Recognition Using Distributed Associative Memory. IEEE Trans. Pattern Analysis and Machine Intelligence, 10, No. 6, 811-821.

Zahn C. T. and Roskies R. Z. (1972). Fourier Descriptors for plane closed curves. IEEE Trans. Computers C-21, 269-281.

Zhang T. Y. and Suen C. Y. (1984). A Fast Parallel Algorithm for Thinning Digital Patterns. Comm. ACM. 27, NO. 3, 236-239.

## TABLE 1  CLASSIFIER RESULTS

|  | VISION | NUMERALS | CAD |
|---|---|---|---|
| **General** | | | |
| # Classes | 5 | 10 | 6 |
| # Training samples | 122 | 950 | 230 |
| # Test samples | 50 | 195 | 60 |
| **Feature Selection** | | | |
| Heuristic threshold | 0.0 | 0.0 | 0.0 |
| # Features | 30 | 30 | 30 |
| **Neural Network** | | | |
| Learning rate eta | 0.5 | 0.5 | 0.5 |
| Smoothing factor alpha | 0.2 | 0.2 | 0.2 |
| Rejection threshold | 0.1 | 0.3 | 0.1 |
| **Learning & Classification** | | | |
| Patterns learned (Y/N) | Y | Y | N |
| Learning iterations | 122 | 323 | – |
| Correct # (%) | 121(99.17) | 192 (98.5) | 20(40) |
| Reject # (%) | 1(0.82) | 3 (1.5) | 0(0) |
| Error # (%) | 0(0) | 0 (0) | 30(60) |
| Global correct class. (%) | 99.17 | 98.5 | 40 |

# EQUILIBRIA OF THE HOPFIELD'S NEURON: EXISTENCE, ISOLATION AND EXPONENTIAL STABILITY[1]

R. Kelly and G. Davila

*Instituto Tecnologico de Monterrey, Centro de Sistemas de Manufactura, Aptdo. Postal 2103,
Sucursal de Correos "J", Monterrey, N.L., 64841, Mexico*

## Abstract

In this paper we address the equilibria study of the continuous–time dynamic Hopfield's neuron. We focus our attention on sufficient conditions for equilibria existence, isolation, exponential stability, uniqueness and global exponential stability.

Keywords. Neural nets, stability, Lyapunov methods.

## 1 Introduction

In this paper we address the equilibria study of the continuous–time neuron described by:

$$\dot{x} = -ax + wg(x) + b \qquad x(0) \in \Re \qquad (1)$$

where $x : \Re_+ \to \Re$ is the state variable, $a > 0$ is a constant, $w$ is the weight or interconnection constant, $b$ is a real constant, $x(0) \in \Re$ is the initial condition, and $g(\cdot) : \Re \to \Re$ is the sigmoidal function:

$$g(x) = \tanh(x). \qquad (2)$$

The description given by (1) include networks of the Hopfield's type (Hopfield, 1984) with:

$$
\begin{aligned}
a &= \frac{1}{RC} \\
w &= \frac{T}{C} \\
b &= \frac{I}{C}
\end{aligned}
$$

where $T$ is the conductance connecting the neuron with another one, $I$ is the input bias current, $R$ is the input "resistence", and $C$ is the capacitance of the neuron.

An important property of sigmoidal function $g(x) = \tanh(x)$ is that it is a continuous bounded function:

$$-1 < g(x) < 1 \quad \text{for all } x \in \Re$$

and it is a Lipschitz one, i.e. there is a constant $l > 0$ such that, for all $z, y \in \Re$:

$$|g(z) - g(y)| \le l\,|z - y|. \qquad (3)$$

It is straightforward to show that (3) holds for $l = 1$.

Hopfield's neuron (1) is used to design associative memories. The fundamental philosophy of associative memories can be stated as follows (see e.g. Li and co-workers (1988), Michel and Farell (1990), Sudharsanan and Sundareshan (1991) and Hecht-Nielsen (1991)). Given a set of $m$ $n$-dimensional constant vectors $x_1^*, x_2^*, \cdots, x_m^*$, and an initial condition $x(0) \in \Re^n$, the state vector solution $x(t)$ must tends asymptotically to "the nearest", in some sense, vector $x_i^*$ of $x(0)$.

The associative memory models have a rich dynamic behavior. Instead to propose a memory synthesis methodology, we believe that a first step is to understand the equilibria nature of the simplest element of an associative memory: the Hopfield's neuron. In this paper we focus our attention on sufficient conditions for equilibria existence, isolation, exponential stability, uniqueness, and global exponential stability of the Hopfield's neuron (1) and (2).

In spite of these subjects have been previously considered in the literature, we analyze them by using different approaches. The equilibria existence is studied via the Schauder's fixed point theorem, while for equilibria isolation we use the inverse function theorem. Exponential stability of equilibria is performed by proposing a new Lyapunov function and invoking the Lyapunov's direct method. Finally, equilibrium uniqueness and its global exponential stability are analyzed via the contraction mapping theorem and Lyapunov's direct method respectively.

Stability and asymptotic stability of equilibria in associative memories, which includes as a particular case the neuron described by (1), have attracted attention in the recent past. Hopfield (1984) uses an energy function approach to show Lyapunov stability. This approach is revisited in details in the excellent book by Khalil (1991). Li and co-workers (1988) present a qualitative analysis including equilibria asymptotic stability, and a synthesis procedure for associative memories. More recently, Sudharsanan and Sundareshan (1991) present, under equilibrium isolation assumption, simple conditions for the exponential stability and inestability of the equilibrium points of the network, and for the former case, they provide explicit estimation of the rate of decay of the trajectories and region of attraction to these equilibrium points.

---

[1] Work partially supported by CONACyT–México

The paper is organized as follows. In Section 2, we present a sufficient condition for equilibria existence of the neuron model (1) and (2). Section 3 is devoted to the equilibria isolation analysis. By using a simple, new Lyapunov function, we give sufficient conditions for exponential stability of equilibria in Section 4. The equilibrium uniqueness and its global exponential stability is addressed in Section 5. Finally, we offer brief concluding remarks in Section 6.

## 2 Equilibria existence

A vector $x^* \in \Re^n$ is said to be an equilibrium of the autonomous differential equation $\dot{x} = f(x)$, if it has the property that whenever the initial condition $x(0)$ starts at $x^*$, the equation solution $x(t)$ will remain at $x^*$ for all future time. The equilibria of $\dot{x} = f(x)$ are the roots of the algebraic equation:

$$f(x) = 0.$$

The equilibria of the neuron differential equation (1) are the roots of equation:

$$-ax + wg(x) + b = 0.$$

In this section, we are interested in showing that memory model (1) has equilibria. Since $a$ is a positive constant, then there exist equilibria for neuron model (1). In the case when interconnection constant $w$ is zero, the unique equilibrium is $x^* = a^{-1}b$. On the other hand, if $w$ is non zero, then there exists at least one equilibrium. To show this, we can employ the following lemma whose proof is straightforward from the Schauder's fixed point theorem (see e.g. Hale, 1980).

*Lemma 1.* Consider the continuous vector–valued function $s(\cdot) : \Re^n \to \Re^n$. If there exists $k > 0$ such that $\|s(x)\| \leq k$ for all $x \in \Re^n$, then there exists at least a solution $x^*$ of $s(x) - x = 0$, where $x^*$ satisfy $\|x^*\| \leq k$.

$$\nabla\nabla\nabla$$

Define $f(x)$ and $s(x)$ as

$$f(x) = -ax + wg(x) + b \qquad (4)$$

and

$$s(x) = a^{-1}wg(x) + a^{-1}b$$

Hence, the roots of $f(x) = 0$ are the roots of $s(x) - x = 0$. Afterwards, we show, by invoking straightforward *lemma 1*, existence of roots for $s(x) - x = 0$. This roots are the equilibria of the neuron equation (1).

The term $|s(x)|$ can be upper bounded by:

$$
\begin{aligned}
|s(x)| &\leq |a^{-1}| \, [|wg(x)| + |b|] \\
&\leq k
\end{aligned}
$$

where $k = \frac{1}{a}[|w| + |b|]$. Finally, by using *lemma 1*, we conclude existence of roots for $s(x) - x = 0$, hence, existence of equilibria for neuron model (1). This is summarized in the following

*Proposition 1.* Consider the neuron model (1) and (2). As $a$ is a positive constant, then there exists at least one equilibrium $x^* \in \Re$ of neuron model.

$$\nabla\nabla\nabla$$

## 3 Equilibria isolation

A well–known necesary condition for asymptotic stability for equilibria is that they must be isolated equilibria. An equilibrium $x^* \in \Re^n$ is an isolated equilibrium if there exists a neighborhood $N$ of $x^*$ in $\Re^n$ such that $N$ contains no equilibrum states other than $x^*$.

Now, we present a lemma whose proof is shown in Vidyasagar (1978) and follows immediately from the inverse function theorem (see e.g. Nijmeijer and Van der Schaft, 1990).

*Lemma 2.* Consider the continuous (smooth) vector–valued function $f(\cdot) : \Re^n \to \Re^n$. Suppose that $x^* \in \Re^n$ is a root of $f(x) = 0$, i.e. $f(x^*) = 0$. If the Jacobian matrix $P$:

$$P = \left. \frac{\partial f(x)}{\partial x} \right|_{x = x^*}. \qquad (5)$$

is nonsingular, then there exists a neighborhood $N$ of $x^*$ such that $N$ contains no roots of $f(x) = 0$ other than $x^*$.

$$\nabla\nabla\nabla$$

Afterwards, we present a sufficient condition for equilibrium isolation of neuron (1).

Assume $x^* \in \Re$ is an equilibrium state of (1). Considering the definition of $f(x)$ in equation (4), $P$ is given by:

$$
\begin{aligned}
P &= \left. \frac{\partial f(x)}{\partial x} \right|_{x = x^*} \\
&= \left. \frac{\partial \left( -ax + wg(x) + b \right)}{\partial x} \right|_{x = x^*} \\
&= -a + wG(x^*)
\end{aligned}
$$

where $G(x^*) \in \Re$ is defined by:

$$G(x^*) = \operatorname{sech}^2(x^*).$$

Invoking *lemma 2*, $x^*$ is an isolated equilibrium of neuron model (1) provided that $P = -a + wG(x^*)$ is non zero. Notice that if $wG(x^*) - a$ or $a - wG(x^*)$ is a positive constant, then $P$ is non zero.

Since we have $\left| \operatorname{sech}^2(z) - \operatorname{sech}^2(y) \right| \leq 2\,|z - y|$ for all $z, y \in \Re$, thus

$$|G(\xi + x^*) - G(x^*)| \leq 2\,|\xi| \qquad \forall\, x^* \in \Re \qquad (6)$$

which will be used in next section.

The main results of this section are summarized in the following:

*Proposition 2.* Consider neuron model (1) and (2). Suppose $x^* \in \Re$ is an equilibrium. If $a - wG(x^*)$ is non zero, then $x^*$ is an isolated equilibrium.

$$\nabla\nabla\nabla$$

## 4 Equilibria stability

In this section we provide sufficient conditions to guarantee exponential stability for an isolated equilibrium $x^*$ of the neuron (1).

We digress momentarily to present some definitions and theorems on exponetial stability shown in Vidyasagar (1986).

**Definition 1.** The origin $x = 0 \in \Re^n$ is an exponentially stable equilibrium of $\dot{x} = f(x)$ if there exist positive constants $\alpha$, $\beta$, $\delta$ such that:

$$\|x(t)\| \leq \alpha \|x(0)\| e^{-\beta t} \quad \forall \, t \geq 0 \qquad (7)$$

whenever $\|x(0)\| < \delta$. The origin is a globally exponentially stable equilibrium if (7) holds for all initial condition $x(0) \in \Re^n$.

$$\nabla\nabla\nabla$$

**Theorem 1.** The origin $x = 0 \in \Re^n$ is an exponentially stable equilibrium of $\dot{x} = f(x)$ if there exists a continuously differentiable function $v : \Re^n \to \Re$ and positive constants $\alpha$, $\beta$, $\gamma$ such that:

$$\alpha\|x\|^2 \leq v(x) \leq \beta\|x\|^2 \qquad (8)$$

$$\dot{v}(x) \leq -\gamma\|x\|^2 \qquad (9)$$

for all $x$ in a neighborhood of the origin in $\Re^n$ ( $\alpha$ and $\beta$ are not the same throughout).

$$\nabla\nabla\nabla$$

**Theorem 2.** The origin $x = 0 \in \Re^n$ is a globally exponentially stable equilibrium of $\dot{x} = f(x)$ if there exists a continuously differentiable function $v : \Re^n \to \Re$ and positive constants $\alpha$, $\beta$, $\gamma$ such that (8) and (9) hold for all $x \in \Re^n$.

$$\nabla\nabla\nabla$$

Assume that $x^* \in \Re$ is an isolated equilibrium of (1). As shown in Section 3, a sufficient condition for that is to select $w$ such that $a - wG(x^*)$ is a positive constant.

Since $x^*$ is an equilibrium, we have:

$$-ax^* + wg(x^*) + b = 0. \qquad (10)$$

Define $\xi \in \Re$ as:

$$\xi = x - x^* \qquad (11)$$

Considering (1), (10) and (11), the time derivative $\dot{\xi}$ yields:

$$\dot{\xi} = -a\xi + w\left[g(\xi + x^*) - g(x^*)\right] \qquad (12)$$

whose origin $\xi = 0 \in \Re$ is an equilibrium. From $\xi$ definition in (11), equilibrium $x^*$ of neuron equation (1) will be exponentially stable provided that $\xi = 0$ is also an exponentially stable equilibrium of (12).

To show $\xi = 0$ is an exponentially stable equilibrium of (12), we use theorem 1.

Let us define function $h(\xi)$ as the right hand side of equation (12):

$$h(\xi) = -a\xi + w\left[g(\xi + x^*) - g(x^*)\right] \qquad (13)$$

Consider the following Lyapunov function candidate:

$$v(\xi) = \frac{1}{2}h(\xi)^2. \qquad (14)$$

Since we assume $x^*$ to be an isolated equilibrium of (1), hence $\xi = 0$ is also an isolated one of (12), i.e. $h(0) = 0$, and there exists a neighborhood $N$ in $\Re$ such that $h(\xi)^2 > 0$ for all $\xi \neq 0$ in $N$. In other words, Lyapunov function candidate (14) is a locally positive definite function and there exist positive constants $\alpha$ and $\beta$ such that (8) holds (Slotine and Li, 1991).

The time derivative of $v(\xi)$ along the trajectories of (12) is given by:

$$\dot{v}(\xi) = h(\xi)\left[\frac{\partial h(\xi)}{\partial \xi}\right]\dot{\xi}$$

$$= -h(\xi)^2\left[a - wG(\xi + x^*)\right] \qquad (15)$$

In agreement with the *Theorem 1*, if $\dot{v}(\xi) \leq -\gamma|\xi|^2$ for some $\gamma > 0$ in a neighborhood $N$ of $\xi = 0 \in \Re$, then the origin $\xi = 0 \in \Re$ is an exponentially stable equilibrium of (12). Last condition is satisfied provided that $a - wG(\xi + x^*)$, at least locally, is a positive definite function, because this implies there exists $r > 0$ such that, for all $|\xi| \leq r$:

$$h(\xi)^2\left[a - wG(\xi + x^*)\right] \geq \varepsilon\, h(\xi)^2$$

with:

$$\varepsilon = \min_{|\xi| \leq r}\left[a - wG(\xi + x^*)\right] > 0$$

and $\dot{v}(\xi)$ holds $\dot{v}(\xi) \leq -\gamma|\xi|^2$ with $\gamma = 2\alpha\varepsilon$.

As it will be shown afterwards, $a - wG(x^*) > 0$ implies $a - wG(\xi + x^*) > 0$ for all $|\xi| \leq r$ with some $r > 0$. To show this, notice that

$$
\begin{aligned}
&|(a - wG(\xi + x^*)) - (a - wG(x^*))| \\
\leq\ & |wG(\xi + x^*) - wG(x^*)| \\
\leq\ & |w|\,|G(\xi + x^*) - G(x^*)| \\
\leq\ & 2|w|\,|\xi| \qquad\qquad\qquad\qquad (16)
\end{aligned}
$$

where we have used (6).

Assume that $|\xi|$ satisfies:

$$|\xi| < r = \frac{a - wG(x^*)}{2\,|w|} \qquad (17)$$

where the right side hand is positive because $a - wG(x^*)$ was assumed to be positive. Notice from (16) and (17) that:

$$|(a - wG(\xi + x^*)) - (a - wG(x^*))| < a - wG(x^*).$$

Now, using the fact that if $y > 0$ and $|x - y| < y$, then $x > 0$, we conclude from the above inequality that:

$$a - wG(\xi + x^*) > 0$$

(for $\xi \in \Re$ satisfying (17)).

The sufficient condition for equilibrium exponential stability can be summarized as follows. If $a - wG(x^*) > 0$ , i.e. $x^*$ is an isolated equilibrium of (1), then there exists $r > 0$ such that:

$$a - wG(\xi + x^*) > 0 \qquad \forall\,\xi : |\xi| \leq r \qquad (18)$$

and the equilibrium $\xi = 0$ of (12) is exponentially stable.

The main conclusions of this section are abridged in the following:

*Proposition 3.* Consider the neuron model (1) and (2). Assume $x^* \in \Re$ is an equilibrium. If $a - wG(x^*) > 0$, then $x^*$ is an isolated equilibrium and it is exponentially stable.

$$\nabla\nabla\nabla$$

Finally, by using well known instability theorems (see e.g. Vidyasagar, 1978; Sánchez, 1968), it is straightforward to show that a sufficient condition for equilibria inestability is $wG(x^*) - a > 0$.

# 5 Equilibrium uniqueness and global exponential stability

In this section we present sufficient conditions for uniqueness equilibrium for the neuron modeled by (1).

Let us denote by $x^* \in \Re$ an equilibrium for the differential equation (1). We will show that $x^*$ is the unique equilibrium provided that $a - |w| > 0$.

Since $x^*$ is an equilibrium, we have:

$$-ax^* + wg(x^*) + b = 0. \tag{19}$$

Define $\xi \in \Re$ as:

$$\xi = x - x^* \tag{20}$$

Considering (1), (19) and (20), the time derivative $\dot{\xi}$ yields:

$$\dot{\xi} = -a\xi + w[g(\xi + x^*) - g(x^*)] \tag{21}$$

whose origin $\xi = 0 \in \Re^n$ is an equilibrium. From $\xi$ definition in (20), equilibrium $x^*$ of neuron equation (1) will be the unique one provided that $\xi = 0$ is also the unique equilibrium of (21).

As we will show, $a - |w| > 0$ is a sufficient condition for the existence and uniqueness of equilibrium. This can be proven by invoking the *contraction mapping theorem* (see e.g. Kolmogorov and Fomin, 1970; Khalil, 1991).

*Theorem 4 (Contraction mapping).* Let $T$ be a mapping that maps $\Re^n$ into $\Re^n$. Suppose that:

$$\|T(z) - T(y)\| \leq k\|z - y\| \quad \forall z, y \in \Re^n, \quad 0 \leq k < 1$$

then, there exists a unique vector $x^* \in \Re^n$ satisfying $T(x^*) - x^* = 0$.

$$\nabla\nabla\nabla$$

Let us define $T(x)$ as $T(x) = a^{-1}[wg(x) + b]$. Hence, if there exists a unique root of $T(x) - x = 0$, then $-ax + wg(x) + b = 0$ has always a unique root too, which means neuron model (1) has a unique equilibrium. Notice that:

$$
\begin{aligned}
|T(z) - T(y)| &= |a^{-1}w[g(z) - g(y)]| \\
&\leq a^{-1}|w||z - y|
\end{aligned}
$$

hence, in agreement with contraction mapping theorem, if:

$$a^{-1}|w| < 1 \tag{22}$$

then there exists a unique root of $-ax + wg(x) + b = 0$. It is interesting to remark that inequality (22) is equivalent to:

$$a - |w| > 0.$$

## Global exponential stability

An other interesting stability result is concerned with a sufficient condition for global exponential stability. In this case, we need to guarantee that the equilibrium is the only one. This is a well–know necessary condition for global stability for it.

Assume that $w$ is selected in such a way that $a - |w| > 0$. Hence, we can assure existence and

uniqueness of an equilibrium $x^*$ of (1), or equivalently, that $\xi = 0 \in \Re$ is the unique equilibrium of:

$$
\begin{aligned}
\dot{\xi} &= h(\xi) \tag{23} \\
&= -a\xi + w[g(\xi + x^*) - g(x^*)]. \tag{24}
\end{aligned}
$$

This implies in turn that $h(0) = 0$, $h(\xi)^2 > 0$ for all $\xi \neq 0 \in \Re$, and $h(\xi)^2 \to \infty$ as $|\xi| \to \infty$.

To show exponential stability of $\xi = 0 \in \Re$, we propose again the Lyapunov function candidate $v(\xi) = \frac{1}{2}h(\xi)^2$ which holds (8) for all $\xi \in \Re$. The time derivative of Lyapunov function candidate along the trajectories of the equation (24) is given by (see (15) for details):

$$\dot{v}(\xi) \leq -h(\xi)^2 [a - wG(\xi + x^*)] \tag{25}$$

We digress momentarily to show that $a - |w| > 0$ implies $a - wG(\xi + x^*) > 0$ for all $\xi, x^* \in \Re$.

Because $G(\xi + x^*)$ is a positive function and it has an upper bound less or equal than one for all $\xi, x^* \in \Re$, we have

$$a - |G(\xi + x^*)||w| > 0 \tag{26}$$

On the other hand, we have for all $\xi, x^* \in \Re$ that:

$$
\begin{aligned}
|G(\xi + x^*)||w| &\geq |wG(\xi + x^*)| \\
&\geq wG(\xi + x^*).
\end{aligned}
$$

By using last inequality and (26), we have finally:

$$a - wG(\xi + x^*) > 0$$

for all $\xi, x^* \in \Re$.

Using this result, from (25) we have:

$$\dot{v}(\xi) \leq -\varepsilon h(\xi)^2$$

where:

$$\varepsilon = \min_{\xi, x^* \in \Re}[a - wG(\xi + x^*)]$$

which is a positive constant because $w$ was assumed to be selected so that $a - |w| > 0$. Thus, by invoking theorem 2, we conclude that origin $\xi = 0 \in \Re$ is a globally exponentially stable equilibrium of (24).

In brief, the main results of this section are presented in the following:

*Proposition 4.* Consider the neuron model (1) and (2). If $a - |w| > 0$, then there exists an unique equilibrium $x^*$, which is isolated, and it is globally exponentially stable.

$$\nabla\nabla\nabla$$

# 6 Conclusions

In this paper, we consider the neuron model proposed by Hopfield. For this neuron model, we analyze several equilibria properties, namely, existence, isolation, exponential stability, uniqueness and global exponential stability. The main results obtained in this paper are summarized in Table 1. Some of these results have been previously reported in the literature, but we have employed, in this paper, different approaches to prove them. We believe that the main

**Table 1: Main results**

| Property | Condition |
|---|---|
| Equilibria existence | $a > 0$ |
| Equilibria isolation | $a - wG(x^*)$ non zero |
| Equilibria exponential stability | $a - wG(x^*) > 0$ |
| Equilibrium uniqueness | $a - |w| > 0$ |
| Equilibrium global exponential stability | $a - |w| > 0$ |

contribution is the exponential stability result for isolated equilibria of the neuron by using an original Lyapunov function together with Lyapunov's direct method.

It is interesting to remark that the approches presented in this paper can be straightforward used to analyze the equilibria of associative neural network memories.

## Acknowledgment

The authors are grateful to H. Sira-Ramirez for his insightful comments.

## References

1. Hale, J. K. (1980). *Ordinary Differential Equations*, Krieger Pub. Co.

2. Hecht–Nielsen, R. (1991). *Neurocomputing*, Addison–Wesley.

3. Hopfield, J. J. (1984)."Neurons with graded response have collective computational properties like those of two–state neurons", *Proceedings of the National Academy of Sciences 81*, May.

4. Khalil, H. K. (1991). *Nonlinear Systems*, Macmillan Pub. Co.

5. Kolmogorov A. N. and S. V. Fomin (1970). *Introductory Real Analysis*, Dover Pub. Inc.

6. Li, J. H., A. N. Michel and W. Porod (1988). "Qualitative analysis and synthesis of a class of neural networks", *IEEE Transactions on Circuits and Systems*, Vol. 35, No.8, August.

7. Michel, A. N. and J. A. Farell (1990). "Associative memories via artificial neural networks", *IEEE Control Systems Magazine*, Vol. 10, No. 3, April.

8. Nijmeijer, H. and A. Van der Schaft (1990). *Nonlinear Dynamical Control Systems*, Springer–Verlag.

9. Sánchez, D. A. (1968). *Ordinary Differential Equations and Stability Theory: An Introduction*, Dover.

10. Slotine, J. J. E. and W. Li (1991). *Applied Nonlinear Control*, Prentice Hall.

11. Sudharsanan, S. I. and M. K. Sundareshan (1991). "Equilibrium characterization of dynamical neural networks and a systematic synthesis procedure for associative memories", *IEEE Transactions on Neural Networks*, Vol. 2, No. 5, September.

12. Vidyasagar, M. (1978). *Nonlinear Systems Analysis*, Prentice Hall.

13. Vidyasagar, M. (1986). "New directions of research in nonlinear system theory", *Proceedings of the IEEE*, Vol. 74, No. 8, August.

# KNOWLEDGE-BASED ADAPTIVE GENERALIZED PREDICTIVE CONTROL FOR A GLASS FOREHEARTH

Q. Wang*, G. Chalaye**, G. Thomas* and G. Gilles*

*LAGEP, URA CNRS D.1328, Ecole Centrale de Lyon, Labo. Mathematique, Informatique et Systeme,
B.P. 163, 69131 Ecully Cedex, France
**BSN emballage, Ecole Centrale de Lyon, Labo. Mathematique, Informatique et Systeme,
B.P. 163, 69131 Ecully Cedex, France

Abstract. This paper describes the development of a control strategy for the surface temperature of glass in a forehearth. The surface temperature set point sequence is provided by un  expert system obtained with heuristic method. An adaptive Generalized Predictive Control(GPC) is installed for each zone of the forehearth to maintain the glass surface temperature. The algorithm parameters of  GPC are tuned with help of a intelligent unit which is established for the specific forehearth model. The experimental results have  shown that the knowledge-based adaptive GPC controller is applicable to the glass forming process.

Keywords. Glass forehearth; Knowledge-based Control; GPC; industrial control; Adaptive Control.

## I. Introduction

The Forehearth are used to condition the molten glass as it flows from the working end of the furnace to the glass forming machine. This conditioning process attempts to provide the required homogeneous properties and hence uniform viscosity prior to glass forming by controlling the glass surface temperature down the length of the forehearth. The structure of the forehearth is given in Fig.1

The conditionning of the glass melt as it flows down the forehearth channel is achived by surface heating in all four zones and surface cooling in zones 2 and 3. A series of burners is mounted laterally on both sides of the forehearth channel to control the glass surface temperature in the mannual manner or the automatic manner according to the different zones. The cooling action is provided by ducts in the forehearth superstructure through which cool air is forced over the glass surface. The efficacy of the cooling action can be adjusted by positionning the dampers in the superstructure.

The actual control strategy used is composed of four industrial PID controllers which accomplish independently the temperature regulation in the different zones of forehearth as illustrated in Fig.2. It can be seen that each zone has an independent feedback control loop. The control variables in zones 3 and 2 are the ventilation valve position, for the zone 1 and the condition zone the control variables are the combustion air valve  position. The thermocouple sensors monted at the end of each zone monitor the temperature of the glass melt and provide the feedback signal for the controllers. A signal within the range 0 to 10 volts from the thermocouples is compared with the set point for the surface temperature.

This type of control strategy exists three principal inconveniences. Firstly, it is restricted within the process which possess a small dead-time; secondly it is difficult to eliminate the mesurable disturbance by introducing directly a feedforward control; last the lack of a supervisor makes the incoordination between the four controllers, so the global optimisation does not exists.

A major problem for the glass forehearth process is that when a forehearth is stopped, it can introduce a large variation in the working end, so another forehearth which shares the same working end will be perturbed, because the feedback corrective action will not be taken until this disturbance reaches the end of zone 3. This situation could be improved by introducing a feedforward control.

We have developed a knowledge-based adaptive generalized predictive control strategy which consists of a adaptive GPC for each zone of the forehearth and a supervisor for all zones. The structure of the system is illustrated in Fig.3.

The GPC has many advantages(ref.1, 2), it can provide a high robustness and a stability in the cas of variable deat-time and non-minimum phase process, especially the closed loop performance can be specified by uniquely choosing the algorithm parameters, such as minimum, maximum predictive horizons $N_1$, $N_2$, control horizon $N_u$ and control weighting factor $\lambda$. But the lack of a analytical relation between the parameters and the performances of system impeds a wide application of GPC in industry process.

Certain researches have been carried out with respect to the parameters tuning of GPC, some preliminary results on the first order plant have been obtained(ref.4) . We will adapt some results to the glass forehearth process in this paper.

## 2. Control strategy for each zone

### 2-1. Zone control algorithm

The forehearth zone model represents the relationship between the control variable(valve position) and the glass surface temperature at the centre exit from the zone, it can be expressed as:

$$A(q^{-1})y(t)=B(q^{-1})u(t-1)$$
$$+ D(q^{-1})p(t-1) +C(q^{-1})e(t)/\Delta \quad ......(1)$$

where:

$A(q^{-1})=1+a_1q^{-1}+ \ldots\ldots +a_{na}q^{-na}$.
$B(q^{-1})=b_0+b_1q^{-1}+\ldots\ldots+b_{nb}q^{-nb}$.
$D(q^{-1})=d_0+d_1q^{-1}+\ldots\ldots+d_{nd}q^{-nd}$.

$C(q^{-1})=1+c_1q^{-1}+ \ldots\ldots +c_{nc}q^{-nc}$.
$\Delta = 1-q^{-1}$.
y(t): output(surface temperature)
u(t): input(valve position)
p(t): mesurable disturbance
e(t): uncorrelated random sequence
the polynomial $D(q^{-1})$ represents the mesurable disturbance model.

The predictive output can be formulated as:

$$y(t+j)=G_j\Delta u(t+j-1) + S_j\Delta u_f(t-1)+$$
$$M_j\Delta p(t+j-1)+N_j\Delta p_f(t-1)+F_jy_f(t)$$
$$\ldots\ldots\ldots (2)$$

The matrix $G_j$ is composed of the coefficients of the step response of the open loop process, it can be obtained by an iteration calculation of Diophantine equations proposed by CLARKE(ref.1,2).

$$C=E_j\Delta A+q^{-j}F_j \quad \ldots\ldots\ldots\ldots\ldots\ldots (3)$$
$$E_jB=G_jC+q^{-j}S_j \quad \ldots\ldots\ldots\ldots\ldots\ldots (4)$$

The polynomial $M_j$ and $N_j$ are calculated as the polynomials $G_j$ and $S_j$.

Rewriting the equation (2) in the vector form:

$$\mathbf{Y=G\Delta U + f} \quad \ldots\ldots\ldots\ldots\ldots (5)$$

with:
$$f_j=S_j\Delta u_f(t-1)+M_j\Delta p(t+j-1)+N_j\Delta p_f(t-1)$$
$$+F_jy_f(t). \quad \ldots\ldots\ldots\ldots\ldots\ldots (6)$$

By minimising the criterion J:

$$J = \sum_{j=N1}^{N_2} (y(t+j)-w(t+j))^2 + \sum_{j=1}^{N_2} \lambda(j)\Delta u^2(t+j-1)$$
$$\ldots\ldots\ldots\ldots\ldots\ldots (7)$$

where:

$\lambda$: control weighting factor.
$N_1$: minimum costing horizon.
$N_2$: maximum costing horizon.

w(t+j): reference sequence, it can be obtained from a first-order model
$w(t+j) =\alpha w(t+j-1)+(1-\alpha)y_r$
with j=1,N2; $\alpha \in (0,1)$.
$y_r$: set point.

We have the incremental control vector as:

$$\Delta U = (G_1^t G_1 + \lambda I)^{-1} G_1^t (W - f) \quad \dots \dots \quad (8)$$

where: $G_1$ is a matrix $(N_2 - N_1 + 1, N_u)$.
$N_u$ is the control horizon.

## 2-2. Influence of the algorithm parameters on the closed-loop system

The zone model of forehearth is estimated by L.S method, both cooling model and heating model can be expressed as a first order CARIMA model:

$$(1 + a_1 q^{-1}) Y(t) = b_0 u(t-1) + C(q^{-1}) e(t) \quad \dots \dots \quad (9)$$

For this type plant model, it is not difficult to find a analytical relation between the parameters of GPC and the closed-loop performance in the cas of $N_u = 1$ (ref.4). We shall analyse here the influence of the control parameters on the behaviour of forehearth process.

The charaterestic equation of closed-loop system can be formulated as(ref.6):

$$A_m = C(A\Delta + \sum_{i=N_1}^{N_2} k_i (B - A\Delta G_i) q^{i-1}) \quad \dots \quad (10)$$

where:

$$(k_{N1} \dots k_i \dots k_{N2}) = (1 \quad 0 \dots 0)(G_1 G_1^t + \lambda I) G_1^t \quad \dots \dots \quad (11)$$

For the reason of simplicity, the minimum output horizon $N_1$ is chosen as 1, so in order to have the inverse of the matrix $(G_1 G_1^t + \lambda I)$ for certain situations, the factor $\lambda$ is restricted superior to the value zero.

A exemple of real zone model is used during the analyse:

$$(1 - 0.927 q^{-1}) y(t) = -0.131 u(t-1) + Ce(t) \quad \dots \dots \quad (12)$$

(a). Influence of maximum output horizon $N_2$

In the Fig.4, the root locus of closed-loop system is drawn with $N_1 = 1$, $N_u = 1$, $\lambda = 0.001$, $N_2$ varing from 1 to 40. We can see that the root locus asymptotically approaches the open loop pole as $N_2 \to \infty$. The forehearth is an open loop stable and a relatively simple

process, with the parameters chosen the roots of the closed-loop system are all inside of the unit circle.

(b). Control weighting factor $\lambda$

For the forehearth process studied here, both the cooling action and heating action are restricted inside 10% of the actual valve position. So it is important to introduce the weighting factor $\lambda$. In the Fig.5, we can see that the root locus are inside the unit cycle for the values $N_1 = 1$, $N_u = 1$, $N_2 = 10$. The closed loop system can be satabilized with any value of $\lambda > 0$.

(c). Control horizon $N_u$

Generally it is sufficient for forehearth control with $N_u = 1$, because the temperature system possess a slow response. In the Fig.6 the root locus are drawn with $\lambda = 0.001$, $N_1 = 1$, $N_2 = 10$ and $N_u$ varing from 1 to 10. We can see that after $N_u = 2$(degree of $A\Delta$) the root locus of the system are the same value, so it is proposed that $N_u$ does not exceed the value of 2 during the calculation of controller.

(d). Time delay of the process

For a process possessing the deat-time d, its model can be expressed as:

$$A(q^{-1}) y(t) = B(q^{-1}) u(t-d) \quad \dots \dots \quad (13)$$

The controller can be calculated with the model as

$$A(q^{-1}) y(t) = B(q^{-1}) q^{-d} u(t) \quad \dots \dots \quad (14)$$

and the parameters as $N_1 = N_1'$, $N_u = N_u'$ $\lambda = \lambda'$ $N_2 = N_2' + d$, if we want to obtain the same poles as the process without time delay, here $N_1'$, $N_u'$, $N_2'$ and $\lambda'$ sont the parameters used for the model without time delay.

(e). Reference sequence model and noise polynomial $C(q^{-1})$

We have given a firsrt order model for the reference sequence calculation, but if the transition of this model is more slower than the closed-loop system, certainly the response of the closed-loop system will be filered by reference model. In the like manner the noise polynomial will filter the response for a step disturbance in the output.

## 2-3. Parameters tuning rules

In order to help the user of the control strategy, a knowledge-based system on the GPC conception can provide the algorithm parameters by the iterative researche programme. For exemple, the performances of system can be principally represented by the values as overshoot, maximum amplitude of incremental control and rapidity of disturbance rejection. The control parameters($N_1, N_2, N_u, \lambda$) are tuned until the performance specifications are satisfied. Some rules of deduction for researche of parameters can be formulated as follows:

(Defining $\Delta Tmax$ and $\Delta Pmax$ as the overshoot of the surface temperature for a step input and a step disturbance in the output respectively, $\Delta Vmax$ as maximum position of the control valve )

IF.  The zone model of the
     forehearth is given,
$\Delta Tmax$ and $\Delta Vmax$ are
specified.
Then. Start the GPC algorithm
     with a small value of $\lambda$
     (0.0001), a large value of
     $N_2$(10),$N_u$=1 and $N_1$=1.

IF.  The process possess a
     time delay d.
THEN. $N_2 = N_2 + d$

IF.  The control input is
larger than $\Delta Vmax$.
THEN. Increase $\lambda$ until the
control input is lower
than $\Delta Vmax$.

IF.  The overshoot exceeds
     $\Delta Tmax$.
THEN. Increase the $N_2$ and
continue the rules          above.

IF.  The performance is not
     satisfying.
THEN. Recommence the
     researche from the
beginning with
$N_u = N_u + 1$.

## 3. Supervisor for the forehearth control

The forehearth is divided in four zones, a coordination between the ragulation of each zone is necessary in order to minimize the energy consumed. A intelligent unit on the global characteristics of the forehearth is created, it achive the following tasks:

* Providing the zones models of forehearth according to the results of on-line identification and the global status of process.

* Providing the surface temperature set point sequence, it is obtained with heuristic method on the forehearth conducting, the possible interaction between the different zones and current operation environment.

A global optimisation criterion is formulated in this unit during the establishment of set point in each zone:

$$J_t = Min\{ \sum_{i=1}^{4} [\Delta T_i^t Q \Delta T_i +$$

$$(T_i - T_{i-1})^t R (T_i - T_{i-1})] \}$$
$$\dots\dots\dots (15)$$

where: $T_i$ is the surface temperature of the $i^{th}$ zone.

## 4. Experimental results

Some results of the experiments are given in the Fig.7 and Fig.8. The conditions of the experiments are given in Table.1.

Table.1

|  | zone 3 | zone 2 | zone 1 | zone condition |
|---|---|---|---|---|
| temperature (°c) | 1155 | 1100 | 1100 | 1122 |
| valve | 91% | 79% | 12% | 18% |

The Fig.7 shows that the rejection of a mesurable disturbance with the feedforword control. In the Fig.8 we have shown that after a changment of the glass type the old controller adapts badly to the process, the new controller is introduced with some step inputs about the set point in order to excite the process. It can be seen that the surface temperature becomes more stable with the new controller.

## 5. Conclusion

This paper analyses the problems existing in the old controller for the glass forehearth. We have shown that the knowledge-based adaptive GPC controller is applicable to a industrial process. The idea combing heuristic and theoritical knowledge for tuning GPC is feasible concerning the forehearth control. Notwithstanding it is limited within the special cases. Last the introduction of a supervisor makes the regulation of glass forehearth more systematic and practical.

## References:

(1) Clarke.D.W, C.Mohtadi and P.S.Tuffs "Generalized predictive Control-Part I. The Basic Algorithm" Automatica.Vol.23, No.2,PP.137_148.1987

(2) Clarke.D.W, C.Mohtadi and P.S.Tuffs "Generalized predictive Control-Part II. Extension and Interpretation" Automatica.Vol.23, No.2,PP.149_160.1987

(3) Clarke.D.W, C.Mohtadi and P.S.Tuffs "Properties of Generalized predictive Control" Automatica.Vol.25, No.6,PP.859_875.1989

(4) S.NUNGAM and T.H.LEE "Investigative studies for a knowledge-based intelligent adaptive generalized predictive controller" ITAC 91,15-17 Jan.1991,SINGAPORE

(5) KARL-ERIC ARZEN "An Architecture for Expert System Based Feedback Control" Automatica,Vol.25, No.6,PP.816-875. 1989

(6) ROBERT R.BITMEAD, M.GEVERS and V.WERTZ "Adaptive Optimal Control " Prentice Hall,1990

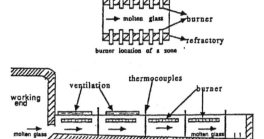

Fig.1 cross-section view of the forehearth

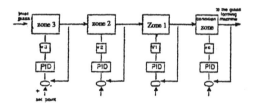

Fig.2 former glass forehearth control

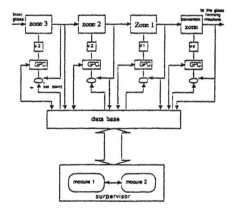

Fig.3 New glass forehearth control

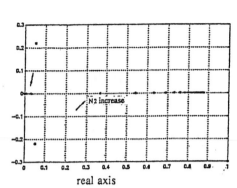

real axis

Fig.4 root locus with $N_2$ varing

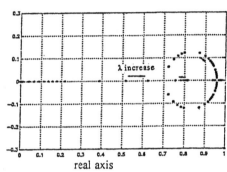

real axis

Fig.5 root locus with $\lambda$ varing

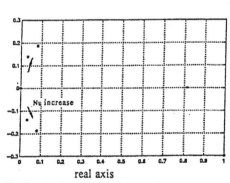

real axis

Fig.6 root locus with $N_u$ varing

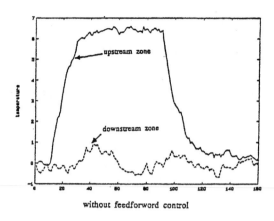

without feedforword control

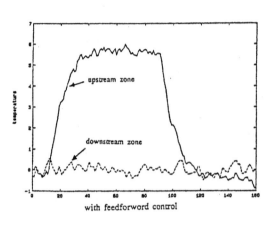

with feedforword control

Fig.7 Introduction of feedforword control

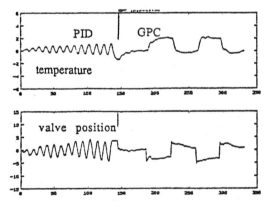

Fig.8 Comparison of the controllers

# EXPERT SYSTEM FOR A BIOREACTOR CONTROL

## I. Dumitrache and G. Galcev

*Department of Automatic Control, Polytechnical Insitute of Bucharest, Bucharest, Romania*

## ABSTRACT

The paper describes an expert system for control of fed-batch process for single cell protein obtained from methylotroph yeast Hansenula Polymorpha CBS 4732 in a CSTR bioreactor. The complex dynamics of biosystem are very sensitive to environment and make it a system with many uncertainties. For cover this difficulties we propose an expert system which is integrated in a hierarchical structure were the low level is conventional implemented.

## INTRODUCTION

Most biotechnological processes utilize a fermentation process which consists of oxidation and/or reduction a substrate by microorganisms such as yeast or bacteria. These processes may conviently classified , according to the mode chosen for process operation , either batch , fed-batch or continuous. During batch operation of a process neither substrate is added to the initial reactor contents nor product is removed until the end of process. In fedbatch mode , the feed rate may be changed during the process but generally no product is removed until the end. In continuous operation substrate is continuous added and product continuous removed.

In a biological reactor several biological reactions occur simultaneously in the liquid medium. These reactions can be classified in two classes : microbial grows reactions , enzyme catalyzed reactions [1,2].

A general biological process is defined as a set of biological reactions such as :

$$S + O ---> X + P$$
autocatalyses reaction

$$S + E ---> P + E$$
enzyme catalysed reaction
or $\quad$ $S + X ---> P + X$

where S represents the carbon source , O is the dissolved oxygen , X the biomass , P is product and E enzyme.

Process plants in the biotechnological industry are highly interactive , interconnected systems . These process involve complex dynamics which are normally incompletely know. The associated models are varying in time and from operation to operation , nonlinear and contain structured and parametric uncertainties.

Many techniques has been applied to control these complex process like adaptive control , long range predictive techniques , linear quadratic gaussian control (LQG) [3,5,7].

The efficiency of these control techniques is determined by accuracy of mathematical models of process (3).

### MATHEMATICAL MODELING OF BIOLOGICAL REACTORS

The reactions which occur in the reactor are assumed to be encoded by m reactions involving n components (n)m) . The model express the mass balance of various components inside the reactor. The mass balance dynamics of each components in the reactor resultsfrom two mechanisms : reaction kinetics and material exchange with the environment , and may be expressed as follows :

$$x_i = \sum_{i=j} c_{ij} \, r_j(x) - Dx_i + F_i - Q_i \qquad (1)$$

$x_i$ is the state of the model

$F_i$ represents the rate of supply of component x to reactor per unit of volume

$Q_i$ represents the rate of removed of component x from reactor in gas form per unit of volume

$Dx_i$ represents the rate of removal of component x from the reactor in liquid form in the input stream in continuous reactors and represents the dilution of component x in the tank due to the increase in volume in fed-batch reactors

$$\sum_{i=j} c_{ij} r_j(x) \qquad \text{represents the}$$

rate of consumption and / or production of component x

The coefficients $c_{ij}$ are dimensionless yield coefficients and the reaction rate

$r_j(x)$ is associated with each others reaction. The vector reaction rates denoted :

$$r(x) = [r_1(x) \ r_2(x) \ ... \ r_m(x)]$$

A general dynamical model of biological reactors can be represented if we introduce :

$x(t) = [x_1 ... x_n]^T$ — the state of model

$A = [a_{ij}]$ — matrix of yield coefficients c

$F = [F_1 ... F_n]^T$ — input flows

$Q = [Q_1 ... Q_n]^T$ — output flows

$$x = A r(x) - Dx + F - Q \qquad (3)$$

The reaction kinetics r(x) are generally represented by rational functions of the state x [1,2]
For a reaction network such as :

$$S + D \longrightarrow X \qquad (4)$$
$$S + X \longrightarrow P + X$$

the associated dynamical model is written

$$\begin{bmatrix} S \\ X \\ P \\ O \end{bmatrix} = \begin{bmatrix} -C_1 & -C_2 \\ 1 & 0 \\ 0 & 1 \\ -C_3 & 0 \end{bmatrix} \begin{bmatrix} r_1 \\ r_2 \end{bmatrix} - D \begin{bmatrix} S \\ X \\ P \\ O \end{bmatrix} + \begin{bmatrix} F_1 \\ 0 \\ 0 \\ F_2 \end{bmatrix} \qquad (5)$$

or

$$x = A \ r(x) - Dx + F \qquad (6)$$

If we introduce the following additional assumptions:
- dissolved oxygen is not limiting
- product P has no inhibiting action on the process
- the reaction rates r1(S,X,P,O) and r2(x,S,P,O) depend only on X , S as follows:

$$r_1 = \mu(s) x \qquad r_2 = v(s) x \qquad (7)$$

where $\mu(s)$ and $v(s)$ are specific growth and production rates , respectively:

$$\mu(s) \triangleq \frac{\mu_{max} \ s}{a_0 + s} \quad , \qquad v(s) \triangleq \frac{v_0 \ s}{a_1 + s + a_2 s^2}$$

$$\mu_{max} \ , \ v_0 \ , \ a_0 \ , \ a_1 \ , \ a_2$$

( represent the kinetic coefficients)

- the substrate feed rate F1 is represented as F1=DS0 , where S0 is the input substrate concentration
- the time period T of the fedbatch

operation , the initial and final volumes V(0) and V(T) , and the dilution D(t),t ∈ [0,T] are fixed by operation planning.
With this assumptions , the associated dynamical model is written :

$$\begin{aligned} \dot{X} &= \mu(s) \ X - D \ X \\ \dot{S} &= -c_1 \mu(s) \ X - c_2 \ v(s) \ X - D \ (S-S_0) \qquad (8) \\ \dot{P} &= v(s) \ X - D \ P \\ \dot{pO} &= -c_3 \mu(s) \ X + F \end{aligned}$$

In this model we can use as a control inputs $S_0$ or D as an output function for the system , the biomass/substrate rate.
The general form of mathematical model of the bioreactors can be written as nonlinear, variant and multivariable system.

$$x = A(x)x + B(x)u$$
$$y = C(x)x$$

where
$$x = [X \ S \ P \ O]$$
$$u = [D \ F]$$
$$y = [X \ P]$$

The matrix A(x), B(x) , C(x) have the variable forms for different type of biological reactors and for different operation modes.

## CONTROL ALGORITHMS OF BIOLOGICAL REACTORS

The bioreactor control are one of the most difficile problem in automatic control due to chemical and biological constraints with nonliniar and stochastic characteristics.
Since the heuristic specific of many decisions in control of bioreactors , the most efficiently structures are with adaptive control either selftuning or with reference model.
Figure 2 show a adaptive optimal control structure.
In this structure we consider the dilution rate as control input , the influent substrate as external measurable disturbance input and the substrate concentration S and the product P are measurable outputs.
The uk control is generated in each of selected time periods when the model parameters are estimated and the process output Yk+1 is predicted as an optimal piecewise constant.
This structure may be easily modified for the substrate concentration control or for the production rate control [ ]
If we consider an optimal trajectory S(t),the objective of adaptive optimal controller is to realize an adaptive feedback control law of substrate concentration S(t) to track the optimal trajectory S(t) by using the influent substrate concentration or dilution rate as control action.
The reduced information , time variance of parameters and model for each stage of process,complex nonliniarities explains the importance of heuristic techniques in control of this kind of process.
Fuzzy based controller and linguistic

based controller are today used in bioreactors control.

## EXPERT SYSTEMS IN BIOREACTORS CONTROL

The expert systems are an efficient alternative in bioreactor control .Combining heuristics and classical control they embed operator experience The complexities of mathematical relations which describes microorganisms grows in the biochemical environment made very difficult to obtain analytical relations between biological parameters (biomass , growth rate , product concentration ) and physic and chemical parameters (temperature , pressure, oxygen concentration , acidity, speed of stir ) . The conection of the biological and phisico-chimical aspects are determined by microorganism , bioreactor geometry , substrate compositions and they are formulate in empirical observations , after many experiments. This observations are made by the biologs and bichemests engineers which survey the process and appear like:

IF conditions THEN actions

An other characteristics of this processes is the correlations between dispersion of contol signals and process behavior , so is very indicated a fuzzy model.
In order to control a biotechnological process we can distingue more developement stages:
-lag phase
-exponential growth phase
-stationary phase
-death phase
For each phase exists an optimal criterion , and an optimal model so the control strategies are specific . The evaluation of current stage may be proved by process data and experience of operator wich is empiric and euristic and is formulated by linquistic statements .
This paper presents a hierachical control structure which use in the high level AI technics. The low level asures the biological enviroment for microorganisms evolution . The high level use the data from sensors and knowledge base and in this level is a knowledge based system implemented.
The main problems solved by expert system in high level implemented are:
-control and optimisation
-supervision
-man-machine interface
-learning
In order to accomplish this problems the system must realise many tasks :
To control:
-state evaluation based on data from proces and rules from knowledge base
-optimal model selection in relation with current state-behavior prediction and choice of optimal command
-the set points prescription for low level
To supervise:
-sensor supervision by means of measurments tolerances , alarm levels thresholds , constraints , all contained in the rule base
-actuators supervision — wind-up detection, normality of device

-qualitative supervision of normality evolution of process (health of population),system stability
- statistical control and indication
- the conection with learning module
To man machine interface:
- display all essentials information in user friendly form
- asures the rules edit posibilities
- show the explanations forthe decisions
To learn:
- the learnig process is a supervised learning , and this module detect how frequently a rule is used and their effect

## IMPLEMENTATION

Thesystem control was implemented in hte PC architecture , in C language and is used for a CSTR bioreactor control (12 1), for yeasts grows (Hansuella Polymorpha).
In the low level system is the convetional control of:
-pressure
-temperature
- pH
The oxigen concetration is controlled by means of aeration and agitation speed.
On-line the system measure pH,oxigen concentration , temperature , pressure , aeration flow and speed agitation .
Off-line in the laboratory analisys determine in time
-substrate concentration
-viscosity
in the numerical form and in the qualitative
-biomass concentration
The expert system is in forward chaining strategy implemented with the maintask : the process achieving of desired behavior.
The knowledge base is in the rules base form . The precondition partin the rules contains deterministic and fuzzy decisions and also time related decisions.There are 186 rules and can be classified:
-configuration rules
-the starting process rules : for the set-points fixation in orderto establish phisico chimic enviroment (aeration , temperature , oxigen disolved concetration , acid or base inputs.
-to inoculate rules : depending liquid volume in bioreactor , inhibition type , residual concentration of methanol
-control rules : depending inhibition type , healthy of microorganisms , current state and current dinamics , can modify the set-points , analisys rate
-advertising rules for human operator
- learning rules in order to change the old existing rules from both successful and unsuccesful actions

## CONCLUSIONS

In this paper we present an hierarchical expert system for a bioreactor control. The program was tested on IBM-PC , and a bioreactor CSTR (12 1).
The comparative analysis of results between conventional control and intelligent control show the possibility ofthe improvement of biomass rate growth and of the evolution of the process.

REFERENCES

[1] G. Bastin , " Nonlinear and Adaptive
Control in Biotechnology : A Tutorial ",
EEC 91, Grenoble , pp 2001-2012

[2] I. Chen ,G. Bastin ," On the Model
Identifiability of Stirred Tank
Bioreactors " , EEC 91 , Grenoble , pp
242-247

[3] T. Proell , "Comparison of Different
Optimisation and Control Schemes in an
Industrial Scale Microalgae Fermentation
", ACC 91 ,pp 1323-1328

[4] X.F.Ni , "Expert control for a
papermaking process " , Eng. Appli. of
AI , 1988, Vol. 1, June, pp. 119-125

[5] I. Queinnec , "Theoretical and
Experimental Results in Optimisation of
Fedbatch Fermentation Processes ",J.
Syst. Eng. (1991),pp 31-40

[6] K.H. Bellgart ," Optimum Quality
Control of a Baker's Yeast Production
Process ", EEC 91 , Grenoble ,
pp.230-234

[7] I. Dumitrache " Computer Based
System for a Bioreactor Supervision ",
1987 , Buletin IPB

[8] I. Dumitrache "On -line Control of
biosynthesis process ", 1986 , Buletin
IPB

[9] Aarne Halme, Arto Visala ,"
Combining Symbolic and Numerical
Information in Modelling the State of
Biotechnological Process",EEC91
, Grenoble, pp.219-223

[10] M. Marcos ," An Intelligent
Rule-based Compensator for Control
System Input Nonlinearities , ACC 1990,
pp 1467-1473

[11] Kang. Shin," Design of a Knowledge
- Based Controller for Intelligent
Control Systems", ACC 1990, pp.
1461-1466

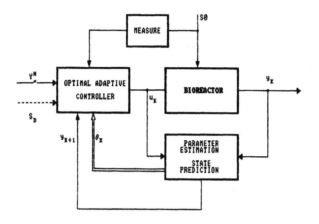

FIG. 1

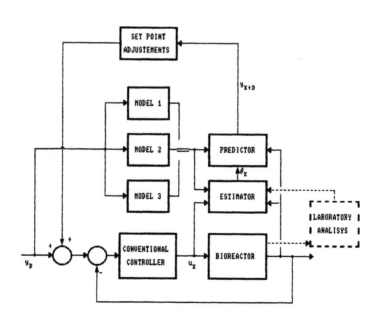

FIG. 2

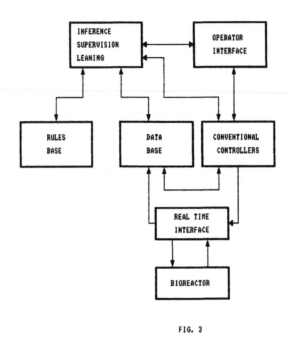

FIG. 3

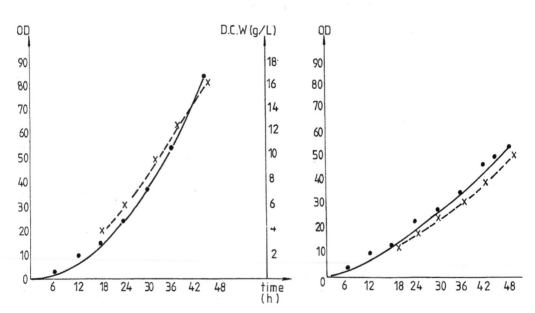

Fig. 4   OD AND  DRIED CELLULAR  WEIGHT
         HANSENULA  POLYMORPHA   CBS 4732 ●●OD -×-×-DCW

# COMBINED METHODS SIGNAL PROCESSING MANAGEMENT

**N.B. Jones\*, A.S. Sehmi\*\* and S. Kabay\*\*\***

*\*Department of Engineering, University of Leicester, UK*
*\*\*Sharp Laboratories of Europe Ltd., Abingdon, UK*
*\*\*\*British Telecom, Martlesham Heath, Ipswich, UK*

Abstract. Work is presented describing a scheme to bring
together procedural, knowledge based and neural network
techniques to tackle large signal processing problems. The
scheme, which is still under development,is based on the
blackboard model and allows efficient exploitation of data
gathered from complex signal sources for data compression,
record keeping, control and for the diagnosis of malfunction.

Individual techniques within the three broad classes above
and some combining more than one method are treated as knowledge sources about the system under consideration. Simulations have a special role in this context and qualitative
simulations based on linguistic descriptions can offer advantages over numerical simulations. Ways of prioritising the
information generated by the knowledge sources and of integrating the whole process are discussed.

Examples are presented from electrophysiology.

Keywords. Artificial intelligence; database management
systems; electrophysiology; expert systems; knowledge
engineering; pattern recognition

## INTRODUCTION

Procedural methods of signal
processing have become established
as powerful tools for data compression, for feature extraction
and for the analysis of system
dynamics. Such methods are powerful because they make use of a
detailed numerical description of
the signals being studied. In
practice, however, signals and
other information extracted from
systems often do not lend themselves to such detailed quantitative methods and the understanding of their origins is often
obscured if such methods are used
in an analysis or simulation situation which does not warrant it.

Recently methods based on expert
systems and on neural networks have
been shown to have advantages in
some circumstances, particularly
where the information is of an imprecise nature or where the numerical value of the signals are of
less importance than a qualitative
understanding of the signal features. There are also situations
where groups of signals or other
information sources about a process
being studied are best treated by a
combination of methods in order to
extract the relevant information in
a satisfactory time scale.

This paper uses as examples methods
developed for use in cardiology and
neurology where the three classes
of processing methods address three

rather different types of problem which need to be solved if data analysis and interpretation is to be made largely automatic and easily understood by users.

At present the task of data evaluation and interpretation, in either a medical or industrial context, is usually based on pattern recognition techniques and human knowledge about the domain of interest. Decisions about data interpretation are usually made using incomplete and uncertain information. Decisions about the possible states of the system under consideration are made by a human using a deep understanding of the subject arising partly from past experience. The combination of the three signal processing methods (i.e. numeric procedural algorithms, symbolic heuristic methods and neural computing structures) can simulate decisions normally made by a combination of automatic and human procedures.

In the examples described here, the signals are either noise-like or are a series of transients from different sources which are embedded in noise. These types of signals can arise in the monitoring of industrial machinery and in radar and sonar systems as well as in the medical areas studied. It is the transients which contain the information of interest. Some of the transients from different sources overlap in time and the signal is sometimes quasi-periodic in nature as it would be, for example, if it arose from rotating machinery or from cardiac activity.

## Knowledge Sources

The state of the system under study can be observed indirectly by the use of transducers which generate signals and by individual tests to determine values for parameters of interest.The information gathered can be analysed using many schemes, and incorporating any one, or a combination of more than one of the classes of techniques referred to. Each will produce results which can range from a new signal, such as a spectrum, to a single number, or a statement of probability or even a qualitative natural language state-

ment. Each of these procedures for changing the data and presenting it in a new way is called a *knowledge source* (KS).

One rather special class of knowledge source is a *simulation*. Such a knowledge source may attempt to simulate part or all of the activity of the real system and may be numerical or non-numerical. It is special because in addition to its role as a generator of synthetic data for test purposes it can be seen as a structure to hold the present state of understanding about the real system. It can in its fullest form be seen to occupy the place of the *hypothesis* in the classical *scientific method*. This view can be taken as either short term, by the use of simulations which adapt during the diagnostic process, or, perhaps additionally, in the long term by regarding the simulation as an encapsulation of present knowledge of the system.

Knowledge sources, of whatever type, can be developed by researchers of various professions, in a variety of computer languages and over a considerable time span. Thus an efficient means of incorporating diperate knowledge sources and of managing them in order to achieve the required goals, needs to be established.

## Knowledge Source Management

After reviewing the alternatives available it was considered most appropriate to implement a blackboard system (BBS) to achieve the objectives defined. A blackboard architecture provides an opportunistic problem solving environment calling on multiple knowledge sources to reach a solution.

The main benefits of using the BBS architecture include:

- Problem-solving may be partitioned into discrete modules corresponding to levels in a hierarchy

- Data-driven strategies may be employed

- Goal-driven strategies may be employed

- Incremental system evolution can take place

- Autonomous problem-solving modules that use strategies which are most appropriate to their particular task can be used

- The system may respond opportunistically to changes in the environment

- Decisions may be retracted in the light of new evidence

Some of the problem-solving applications of blackboard architectures include control (Hayes-Roth, 1985), plan recognition (Carver et al., 1984), signals and data interpretation (Engelmore & Terry, 1979; Nii et al., 1982) and vision (Williams et al., 1977). Consideration has also been given to its use as a general problem-solving frame-work (Hayes-Roth, 1983). Several reviews and applications of the blackboard model may be found in Engelmore & Morgan (1989).

A typical BBS system has three defining features:

- The *blackboard* which is a database of facts and deductions

- The *knowledge sources* which embody the specialist knowledge

- The *scheduler* which is the principal mechanism for activating knowledge sources

The main idea behind this structure is to provide a framework within which to solve a complex problem in (possibly several) hierarchical steps. The problem-solving strategy attempts to solve lower-level sub-problems, until a higher hierarchical level is reached. Using the information generated at this level an attempt is made to reach the next higher level, and so on until the final overall solution is found. For each of the sub-problems an appropriate but different knowledge source is activated according to the particular sub-problem knowledge domain.

The *blackboard* is a global data structure that contains, at any given time, all known facts and hypotheses made by the system. The objects of the solution space are input data, partial solutions, alternatives, final solutions, and possibly control data. The purpose of the blackboard is to hold the computational and solution data required and produced by the KSs.

The knowledge needed to solve a problem is partitioned into the KSs referred to which are kept separate and independent. Using the information currently contained on a particular layer of the blackboard, the KSs calculate or deduce new information and hypotheses and place their findings on another layer above or below. Interactions between the KSs take place solely through changes on the blackboard.

The KSs modify only the blackboard or control data structures, and the blackboard is only modified by them. Each KS is responsible for knowing the conditions under which it can contribute to a solution and each has an associated set of pre-conditions indicating the conditions on the blackboard that must be satisfied before its body is activated. When a KS is activated, it will modify or amend the blackboard, independently of other KSs.

The *scheduler* is a control process which orchestrates problem-solving in the BBS architecture. The BB model assumes KSs respond opportunistically to changes in the blackboard. However, in implementation a scheduling mechanism is used to simulate opportunistic reasoning by building an agenda of potentially active KSs based on the status of the blackboard and the KS trigger conditions. The criteria used for selecting a KS on this agenda may be based on many factors, including:

- The reliability of the KS

- The credibility of the KS triggering information

- The potential for generating important information

- Urgency/priority.

The scheduler must monitor changes in the blackboard and use various kinds of information to determine a focus of attention, i.e. which KS to process next, which (partial) solution on the blackboard to pursue, or which KS to apply on which blackboard objects. As a result one or more of several reasoning types (forward-chaining, etc.) can be applied at any step of the search, and so the sequence of KS triggering becomes dynamic rather than pre-programmed.

It should be noted that the basis of the schedulers decisions can itself be amenable to reasoning in the blackboard.It is possible to customise *intelligent* schedulers which can manipulate the agenda as required (Hayes-Routh 1985).

Blackboard architecture allows the use of diverse, indepedent, potentially inaccurate sources of knowledge. In the presence of uncertainty, both in the KSs and the data, a solution is found (if at all) by converging on the highest rated candidate explanation

The prototype BBS adopts a simple scheduling strategy where new agenda entries are created in response to changes in the black-board in conjunction with the existing status of a shadow black-board. Each task on the agenda has associated with it various user defined measures of the reliabil-ity, usefulness and correctness of the task to be performed. The sche-duler uses this information to sel-ect the best task to perform next.

The effect of activating a KS can be to add an entry or amend an entry already present. The addition or amendment of a blackboard entry poses many problems in terms of truth and consistency maintenance.

The interpretation of what it means to amend an entry is based on the assumption that in the proceeding inference process the amended version of the entry should be used in preference to the original. The problem is in determining precisely under what conditions it is appropriate for the new entry to supplant the old.

The knowledge representation scheme adopted is rule based and includes a frame memory which acts as a repository for structured object representations.The use of a frame-based knowledge representation scheme allows for constructing an object-oriented rulebase with inheritance of data and classes between parent and child frames. Such an approach means that knowledge about a particular domain can evolve with time, is easy to maintain and provides flexibility.

## Examples of Knowledge Sources

The processing of neurological signals for medical diagnosis provides a good vehicle for exploring the ideas described. Several knowledge sources have been developed which analyse the electromyogram (EMG). Some of these are numerical procedures for analysing the macroscopic statics of the interference EMG (Jones, Parekh and Lago 1990). Others (Loudon, Sehmi and Jones 1990) are mixed procedural and knowledge based methods aimed at extracting transients from noise. The first group produces outputs which indicate the distance of the present data from three clusters indicating broad disease classi-fications. The second provides waveforms and a list of waveform features which have diagnostic significance when combined with medical knowledge. Both of these types of output can be handled by the blackboard system by assoc-iation with them a measure of their relevance to the decision making process. The interference EMG is more relevant at high levels of contraction and the transient extraction process at lower levels.

Features generated by one knowledge source, of the type described above for example, have been used to pro-vide classification via a three layer neural network and have been shown to give consistent classifi-cations in the restricted number of tests tried so far. This has allow-ed the conclusions of a knowledge source which extracts waveform features to be compared with those of a knowledge source using statistical pattern recognition methods.

Combinations of these types of knowledge sources with others providing simple numerical data such as conduction velocities can be brought together to set up a numerical simulation of the type described by Jones, Lago and Parekh (1987) or a quantitative simulation as described by Jones and others in 1992 all or part of the process under investigation. An iterative procedure of adjustment can then be initiated which allows the most likely status of the system to be estimated.

## RESULTS

This work is at an intermediate stage and only an intermediate set of results is available.

Eight knowledge sources have been prepared in a format to be integrated into the blackboard system; these are Zero-crossing Analysis; Turning Point Analysis; Spectral Analysis; Decomposition; Principal Component Analysis (PCA); PCA Residual Analysis; Simulation; and Doctor's Inspection. Many results have been generated from each of these knowledge sources. The full set would be too extensive for inclusion here. One example showing the progression of a neurogenic disorder as revealed by knowledge source PCA described by Jones and others (1990) is given in Fig.1.

The simulation programme is the most important knowledge source for the reasons outlined earlier. This is capable of simulating up to 20 motor units. The inter-spike interval is a Gaussian renewal process and the number, size and shape of the action potentials can be set by the user and will in the future be auto-matically adjustable via the blackboard. Random variation simulating measurement noise and movement artifacts is incor-porated. Examples of simulated EMG is shown in Fig.2.

## SUMMARY

A method of integrating various forms of procedural, knowledge based and neural network signal analyis methods via a blackboard system is proposed. Several knowledge sources have been written for problems in neurology and cardiology which are being integrated into a decision making system based on this method.

Experience so far indicates that the problems of bringing together disparate knowledge sources to provide interpretations of signals eminating from complex sys-tems can be solved. The use of an internal model, using either one or more numerical or qualitative simulations, provides a means of testing hypotheses.

## REFERENCES

Carver, N.F.,Lesser, V.R. and McCue, D.L. (1984). Focusing in plan recognition, AAAI-84, pp.42-48.

Englemore, R. and Terry, A. (1979) Structure and function of the CRYSALIS system, IJCAL, pp.250-256.

Englemore, R. and Morgan, T. (1989). Blackboard systems. Addison-Wesley.

Hayes-Roth, B. (1983). The blackboard architecture: a gen-eral framework for problem solving, HPP-83-30,Stanford University.

Hayes-Roth, B. (1985) The blackboard architecture for control, AI, 26, pp.251-321.

Jones, N.B., Lago, P.J. and Parekh, A. (1987). Simulations of the electromyogram and their use in assessment of new diagnostic tests.Proc. of UKSC Conf. (Zokel (Ed)), pp.120-125.

Jones, N.B., Wang, J.T., Sehmi, A.S. and de Bono, D.P. (1992). Knowledge based qualitative simulation of related signals for hypoth-esis testing. Submitted to IMAC 1992.

Jones, N.B., Parekh, A.K. and Lago, P.J. (1990). Classification of point processes using principal component analysis. Mathematics in Signal Processing. (McWinter (Ed)). OUP. pp.369-384.

Loudon, G., Sehmi, A. and Jones, N.B. (1990). Intelligent classification in EMG composition. Condition monitoring and diagnostic management. (Rao, Au and Griffiths (Eds)). Chapman and Hall. pp.317-325.

Nii, H.P. Feigenbaum, E.A., Anton. J.J. and Rockmore, A.J. (1982). Signal-to-symbol transformation. HASP/SIAP case study. AI Magazine, pp. 23-25.

Williams, T, Lowrance, J., Hanson, A. and Riseman, E. (1977) Model-building in the VISIONS system, IJCAI, pp.644-645.

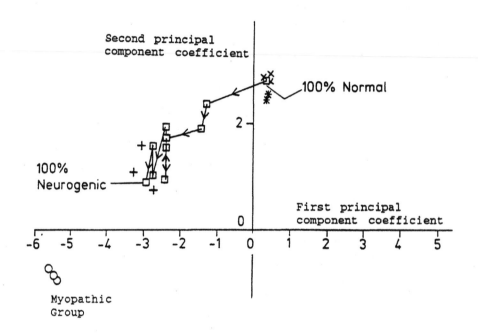

Fig.1. Example of the knowledge source PCA

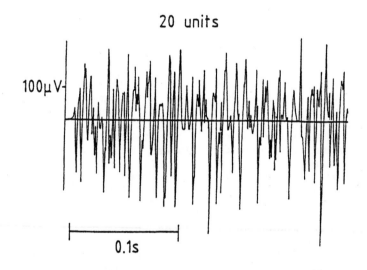

Fig.2. Example of a simulated EMG

# FORMALIZATION APPROACHES OF ERGONOMIC KNOWLEDGE FOR "INTELLIGENT" DESIGN, EVALUATION AND MANAGEMENT OF MAN-MACHINE INTERFACE IN PROCESS CONTROL

## C. Kolski

*Laboratoire d'Automatique Industrielle et Humaine, URA CNRS 1118,*
*Université de Valenciennes et du Hainaut Cambresis, Le Mont Houy,*
*59326 Valenciennes Cedex, France*

Abstract. This paper aims at describing some original works relating to the formalization of ergonomic knowledge in the field of process control. For this purpose, explicit or implicit concepts, advice and methods issued from Ergonomics, Cognitive Psychology and Human Automation have been taken into account. The medium and long-term objective is to develop engineering tools using ergonomic knowledge base. The concepts and principles used by three rule-based approaches for the "intelligent" ergonomic evaluation, design and management of man-machine interface illustrate this paper.

Key words. Process control, man-machine interface, management, evaluation, design, ergonomic knowledge, engineering tools

## INTRODUCTION

The design and the evaluation of man-machine interface for process control are actually subjected to a lot of researches, leading to methodologies, techniques, tools and models already operational or currently to being validated (see for instance Helander, 1988; Gilmore, Gertman and Blackman, 1989; Millot, 1990; Tendjaoui, Kolski and Millot, 1991b). Nevertheless, according to Scapin, Reynard and Pollier (1988) about Man-Machine interaction, the present ergonomic models are often too general, too informal and lack - for the greatest part - organization to be directly usable by the designers. There are difficulties to computer scientists, automatists and even to ergonomists for the ergonomic recommendation utilization. The interface's designer is not always in a position to allow the necessary time - which is often very long - to read completely a manual of recommendations. These recommendations are often very general, and they are problems of access for a beginner. Interactions between the ergonomic criteria lead to compromises that are difficult to estimate for a beginner. The ergonomic data do not always converge according to the different design criteria (use performance, learning performance, etc ). So several necessities appear : (i) to collect and structure the design ergonomical principles used, (ii) to try to bring about formal solutions in the field of man-machine interaction, (iii) to integrate these formalized ergonomic solutions into the design, evaluation and management tools.

This paper aims at describing some original works relating to the formalization of ergonomic knowledge in the field of process control. For this purpose, explicit or implicit concepts, advice and methods issued from Ergonomics, Cognitive Psychology and Human Automation have been taken into account. The medium and long-term objective is to develop engineering tools using ergonomic knowledge base. Different approaches of tools will be the subject of this paper. Their realization is based on advanced tools and techniques such as : knowledge-based systems, objects, hypermedias, graphical toolboxes and/or machine learning. The first part of this paper describes an approach of knowledge-based system for ergonomic evaluation of man-machine interface. A knowledge-based approach for ergonomic design is detailed in the second part. Finally, the last part of this paper presents a knowledge-based approach for ergonomic management of man-machine interface. Of course it is not possible (for lack of place) in this paper to describe in detail these different formalization approaches. Our main goal consists in presenting the main concepts and principles used. For more details please see the references given.

## AN APPROACH TOWARDS "INTELLIGENT" ERGONOMIC EVALUATION

SYNOP is an expert system used for the static ergonomic evaluation of graphic industrial displays. This system can improve displays automatically and give advice to the designer, through production rules centralized on knowledge bases (Kolski and colleagues, 1988; Kolski, 1989; Kolski, Millot, 1991). The general structure of the system is described in the following part.

### The Structure of the Expert System SYNOP

Developed in LISP, the system uses a first order inference engine (Grzesiak, 1987) and notions of frame (Minsky, 1975) and semantic network for knowledge representation. For the time being, it is interfaced with a graphic editor using a G.K.S. package for creating displays. The structure of the expert system is described in figure 1.

SYNOP includes a module for the static evaluation of graphic control displays. The role of this module is to evaluate and improve the displays which have been initially created by the designer with a graphic editor, and automatically stocked into initial files. An interface between SYNOP and the graphic editor interprets these files to create a semantic network of structured objects describing the display. The ergonomic rules contained in the knowledge base are able to modify the attributes of the objects; for instance the height of a character or the colour of a curve, according to ergonomic rules concerning presentation on a graphic screen.

The improved semantic network is then reconstructed into final files by the interface between SYNOP and the graphic editor. These files correspond to an ergonomic display which can be displayed on the screen by the designer. SYNOP has two man-machine interfaces managed by a supervisor: the user interface is used by the designer to start and control the module of evaluation; the expert interface is used by the knowledge engineer to introduce and update the ergonomic rules of the knowledge base. This knowledge base is shared in "knowledge sub-bases" each related to an ergonomic theme, and into "meta-knowledge sub-bases" which contain selection criteria for the "knowledge sub-bases". The module of static ergonomic evaluation is presented below.

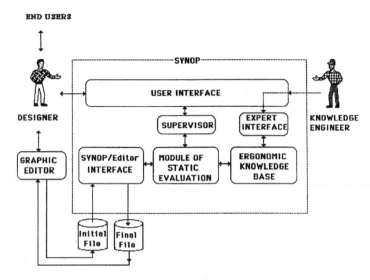

Fig. 1. Structure of the expert system SYNOP.

## Module of Static Ergonomic Evaluation

Three phases define the ergonomic assessment and improvement process of a graphic display. The first phase consists in interpreting the graphic files related to the display which has been processed. This helps create a semantic network of LISP structured objects. The second phase deals with the ergonomic evaluation of the network's objects. Rules contained in the selected "sub-bases" are therefore activated to improve the presentation of the graphic display. The third and final phase helps rebuild the graphic files corresponding to the improved display. A comprehensive description of this process is given in the following part.

### Initial phase of interpretation of the graphic files into a semantic network.
The first phase consists in interpreting the static and dynamic graphic functions with a view to creating a semantic network of LISP structured objects describing the display. This information is created and automatically centralized by the graphic editor into binary-coded files. These files are decoded by the displays interpreter integrated into the interface between SYNOP and the graphic editor (see Fig. 1). The principle of this interpreter consists first in reading the graphic files and in transfering their content in the LISP work environment. Then, structured objects are progressively created by the instantiation of predefined LISP object prototypes for each possible type of graphic object. The structure of these objects is derived from the notion of "frames" (Minsky, 1975). Following this creation phase, links between objects are analyzed and calculated to allocate a value to the attributes of structural relations. After this phase, the objects constitute a semantic network, ready for the second phase of evaluation and improvement.

### Phase of evaluation and improvement of the display.
The ergonomic evaluation is carried out by inference of the rules contained in the knowledge base of SYNOP. The ergonomic knowledge is represented by production rules operated by a first order inference engine (Grzesiak, 1987). The engine is based on the sequence : definition of the set of applicable rules, choice of a rule, inference of the rule and updating. The production rules formalism is of the following type :

IF (P1 * P2 * ...* Pn) THEN C

The condition of application of a rule is a logic arrangement of premises Pi. The operator "*" can take the value "AND" or "OR". To each premise is associated a procedure, written in LISP, which allows evaluation of the verity value by a direct test on the graphic attributes of the objects describing the display. In the same way, conclusion C is associated to a procedure which is able to improve with graphic attributes automatically. The evaluation of a display begins by a global assessment of the objects to select knowledge "sub-bases" contained in the knowledge base of SYNOP. Each of these "sub-bases" is related to the study of an ergonomic concept or theme, and contains a set of rules related to an ergonomic sub-problem. For instance, a

"sub-base" concerns the ergonomic presentation of bargraphs and regroups a set of rules related to this sub-problem. Another "sub-base" can concern the legibility of characters, coloured contrasts, and so on. "Sub-rules" are selected by the activation of selection criteria, in the form of meta-rules contained in meta-knowledge "sub-bases". This method helps restrict the work space, by limiting the step of applicable rules selection to a set of effectively applicable rules. Then it helps reducing the processing time, by avoiding testing useless rules. The display is then evaluated by inference of action rules contained in each selected "sub-base". Two simple examples are presented below. The first rule allows the separation of two parallel texts, x1 and x2, from a distance equal to 100% of the height of their characters, if this distance is initially lower to this reference value. The second rule allows the change in colour of an object x1 if its color is red and if it is not in relation with the alarm notion. At present, the system contains approximatively hundred rules. These rules have been validated by experts in the field of Man-Machine Communication.

| | | |
|---|---|---|
| IF | (parallel-texts | x1 x2) |
| AND IF | (distance-between-then<100%-height | x1 x2) |
| THEN | (set-distance- = 100 % height | x1 x2) |

| | | |
|---|---|---|
| IF | (color-red | x1) |
| AND IF | (don't-represent-an-alarm | x1) |
| THEN | (change-color) | x1) |

Modifying graphic attributes can sometimes be difficult or impossible. If so, the expert system provides the designer with general ergonomic advice concerning the processed display.

After this second phase of evaluation and improvement, the modified semantic network is coded in binary files which can be exploited by the graphic editor.

### Phase of reconstruction of the graphic files related to the improved display.
The third phase concerns the automatic reconstruction of the binary files issued from the expert treatment (according to the inverse principle of those described). These files can be used directly by the designer. The expert system provides the designer with more advice and explanation files which help him alter his display and also to understand the reasoning of SYNOP. These files have been progressively created during the previous phase and provide the designer with a written trace of the ergonomic process. The processing ends this phase. The designer can then leave the software environment of SYNOP in order to improve his displays with the given advice. He can also ask for another evaluation.

## Discussion

Potential interests and consequences of such an expert system are presented below with regards to the ergonomy of graphic information presentation. Four classes of applications should be considered.

Evaluation of man-machine graphic interface. In the first class, the system aims at ergonomically assisting the designer during the "static" evaluation step. SYNOP can thus help him optimize the means of presenting the graphic information, without direct help of an ergonom. SYNOP has thus been validated during the "static" evaluation of the graphic interface between the operators and a real-time expert system. This application has taken place in the French project ALLIANCE aiming at implementing a real-time expert system for alarm filtering, diagnosis and trouble shooting (Taborin and Millot, 1986, 1988, 1989). The expert system which has been developed, currently predicts (in a continuous way) the future evolution of the process along a prediction horizon of several minutes (up to 15 min) with a prediction period of 1 min. It compares the future evolution of the process with alarm thresholds, and if need be, provides the operator with preventive advice.

Our research work concerned the development of the interface between operators and the expert system. The interface has been designed according to the operator's informational needs in the different operation modes of the process : (1) monitoring of the normal operation and fault detection; (2) problem solving and trouble shooting in the abnormal operation mode. The ergonomic evaluation by SYNOP of this interface is detailed by Kolski and colleagues (1988) and Kolski, Millot (1991). Many ergonomic improvements have been automatically made during the processing of SYNOP. The designer has thus been discharged of correction tasks, which are often dull and long lasting. Following this static evaluation, the interface has been implemented in the Cadarache power station and is being dynamically tested and evaluated with the operators.

Saving and structuration of ergonomic knowledge. The second class is related to the saving and structuring of the experts' ergonomic knowledge through an "expert" interface meant to create new ergonomic rules in the knowledge bases of the expert system. This data can be general, like the use of colours or the density of information. It can also be specific to certain applications, like process control, consultation of data bases... Such a step will help making an exhaustive list of the existing ergonomic knowledge to develop other expert system tools for interface design or evaluation. The ideal during this knowledge acquisition would be to dispense with the knowledge engineer while permitting the expert to manage himself his knowledge in the bases of the system himself. This would avoid the skew involved by the difference between the real expert knowledge and its interpretation and implementation by the knowledge engineer. Such an approach is being studied in our laboratory.

Computer-aided training in the field of graphic information presentation. The third class concerns the use of SYNOP or of such an expert tool for Computer-Aided Training in the field of graphic information presentation. This tool would assess and correct graphic displays created by the operators still training to be specialists in information presentation, or operators trying to update their knowledge. This category is directly linked to the previous category. Such applications should be carefully considered, given the growing importance of ergonomy in interactive processing applications (Kolski and Millot, 1989).

Ergonomic design of man-machine graphic interface. The objective of the last class is to develop "intelligent" editors of industrial control displays, using rules for ergonomic design. See the second part of this paper concerning a tool called ERGO-CONCEPTOR.

AN APPROACH TOWARDS "INTELLIGENT" ERGONOMIC DESIGN

This field can be illustrated by the description of a tool called ERGO-CONCEPTOR. It aims at automatizing several stages of design of man-machine interface for process control (Moussa, Kolski, Millot, 1990). The ERGO-CONCEPTOR system relies on the idea that it is possible to automatically generate graphical views from an exhaustive description of the process. This description is made according to various abstraction levels while taking account of control objectives (in terms of tasks to be achieved by the operators).

ERGO-CONCEPTOR includes three main modules (see Fig. 2) : (i) the first one is used for describing a process.This description must be guided by the operator's informational needs, (ii) the second module allows the automatic generation of the file of specifications by exploiting the ergonomic knowledge stored in a knowledge base, (iii) finally, the third module gets back the file of specifications which was generated by the previous module and thus creates graphical displays.

Process Description according to the Human Operator Needs

With a goal of implementation in ERGO-CONCEPTOR, a method to describe the automated process has been proposed by Moussa, Ben Hassine, Kolski (1991) and Moussa, Kolski (to appear). The basic concepts on which the methodology is built on, are presented below.

The Simple Functional Group (SFG). It's a sub-system - which can not be broken down - and which is presented as a set of variables reflecting its state at any moment. It allows to control an irresolvable function. The SFG resources are the variables forming it and called intrinsic variables. The links between the different sub-systems are described by input-output links. The information transported by the inputs and outputs of a SFG are represented by variables called communication variables.

The Composed Functional Group (CFG). It's a grouping of sub-systems forming a sub-system of a higher level of abstraction. The elements of a CFG - which can be SFG or CFG - represent the resources of the current CFG which have the purpose to control a macro-function.

The causality network. Further to the structural description the process, the third main concept is the functional description. It is done from the representation of the interactions between different variables of a sub-system by influence links grouped in a causality network. The role of the link is to describe the impact of a variable perturbation of a FG on the rest of the system.

So, the process functional and structural descriptions are completed. It is then necessary to describe the system in terms of the hierarchy of abstraction "means-goals". At the abstraction level $N_i$, the goals of a given system $S_i$ can be reached by acting on the sub-systems of $N_{i-1}$ level forming it. At the abstraction level $N_{i+1}$, this sub-system $S_i$ is considered as being a means or a resource allowing to realize the sub-system $S_{i+1}$ which include it. This description allow to have an idea which is precise enough to deduce the specifications of the interface to be designed.

Automatic Generation of Interface Specifications

The second module includes three given sub-modules : (i) the sub-module of data extraction, (ii) the inference engine and (iii) the sub-module of specifications creation (see Fig. 2). The first sub-module is in charge of the extraction of the data describing the process from the data base "process". The second sub-module is focused on the ergonomic side of the system. To that end, several ergonomic rules have been stored in a knowledge base. A specific inference engine, working in a forward chaining, passes on the results to the next sub-module. The ergonomic rules have to make apparent : (i) for each functional group, the different types of displays which have to be foreseen, (ii) for each type of display, the different zones forming it and the different modes of representation of the sub-system state, (iii) for each mode, the ergonomic aspects relative to the representation of information which have to be implemented (for example : ergonomic rules about colours, characters size, and so on).

The ergonomic knowledge formalization consists on the one hand in stating some concepts and rules for the information graphical presentation and the display structuration and on the other hand in formalizing them in production rules. These production rules are exploited by an inference engine working in a forward chaining. The mechanism of the knowledge processing is based on the classic sequence : definition of all the applicable rules, execution of the rule and updating. The formalism of the production rules is : *If Condition Then Conclusion*. The knowledge base contains about 50 rules.

A sub-module of specification creation allows to generate a specification file while respecting a predefined syntax. The content of the specification file describes for each view : its type (supervision, control...), the concerned sub-system, the zones it includes (structural definition) and the contents of each zone (informational entities).

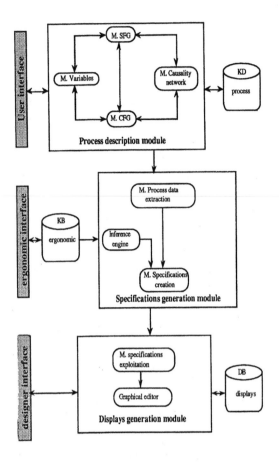

Fig. 2. Architecture of ERGO-CONCEPTOR

and that the operator tasks amount to the supervision of the different sub-systems constituting the nuclear power plant. In case of dysfunctionning, the operator has to make a diagnosis, to evaluate the tendency of variables and to correct by using directly the interface.

According to the methodology described, the nuclear power plant has been split up into an arborescence : 16 SFG and CFG, and 50 interconnected variables have been described. For the data description in the first module, about four hours were necessary. The automatic generation of the specifications has taken about 2 minutes. Then 55 graphical displays have been created. Using systematically the specification proposed by the expert system, the designer needs around 30 seconds for each display created. An example of view can be seen in (Kolski et Moussa, 1991). Another article is planned to describe entirely this first application of ERGO-CONCEPTOR.

The results obtained by this first model are encouraging insofar as it was quite easy to realize some first control displays according to this methodology. It remains to be realized the animation module of the generated displays. This module has to take into account other ergonomic rules for the dynamic presentation of the display graphical objects. Furthermore, ERGO-CONCEPTOR proposes to the designer several ergonomical representation modes adapted to the control tasks. But, of course, the control displays, issued from ERGO-CONCEPTOR, must be validated with the operators. Nevertheless, the first laboratory version of ERGO-CONCEPTOR permits to consider in a promising manner the "intelligent" ergonomic design of man-machine interface for the industrial process control. This system is being validated on several simulated industrial process (Moussa, Kolski, Millot, 1992).

## AN APPROACH TOWARDS "INTELLIGENT" ERGONOMIC MANAGEMENT

Our research works in the field of "intelligent" ergonomic management of man-machine interface concern particularly man-machine cooperation problems. These works aim at optimizing in real time the information communication between the human operator and a decision support system, using the "intelligent interface" notion (Kolski, Tendjaoui and Millot, 1990 ; Hancock and Chignell, 1989).

### The Concept of the "Intelligent" Interface

We call "intelligent interface" an independent interface able to adapt itself to the operator's informational needs, using expert knowledge related to the operational context of the problem to be solved, to the characteristics of the different users of the system and on to the tasks they have to perform. In brief, our approach consists of using an expert system to ensure the communication between the decision support system and the human operator (Tendjaoui, Kolski and Millot, 1990). This "intelligent" approach is tailored to the area of supervisory process control. It's goal is to design an intelligent imagery manager called "Decisional Module of Imagery" (D.M.I.). This approach can be integrated into the global model of the Man-Machine system in automated process control rooms to obtain an overall assistance tool (see Fig. 3) : the supervisory calculator centralizes the whole process scored data. These data are accessible by both the decision support expert system and the D.M.I. Using this data, the decision support expert system infers information such as predictive, diagnosis or recovery procedures. This set of information is transmitted to the D.M.I., which selects those that can be presented to the operator. This selection is based on a task model to be performed by the operator, and on "operator" model containing information about the operator (Tendjaoui, Kolski, Millot, 1991b).

### The Principles of the Task and Operator Models

The task model is currently restricted to problem solving tasks and results from a previous analysis of fixed tasks which have to be performed by the operator. This model is based on the qualitative general model of Rasmussen (1980). Whereby a task is built through four information processing steps : event detection, situation assessment, decision making and action. This task model contains a set of process significant variables used by the operator while performing his different tasks.

### Graphical Displays Generation

From the specifications file of the graphical displays, which results from the previous module, this module is charged to assist the designer during the effective generation of graphical displays. This semi-automatic generation of the graphical displays follows this step : (i) the system analyses the specifications file in order to deduce the set of information which is necessary to realize the graphical displays. This analysis is based on the lexicon defined in the previous module. (ii) The operator use the "intelligent" functionalities of the graphical editor which integrate the process knowledge. These functionalities exploit all the information describing the structure and the functionalities of the displays, and elementary graphical routines to generate the process graphical displays. (iii) The operator can use "classic" functionalities of a graphical editor according to two complementary aspects : on the one hand, graphical routines, like designing a rectangle, a circle, editing a text, and so on, are used as inputs -under the form of graphical objects- for the "intelligent" editor, on the other hand, this editor ensures a high degree of freedom for the displays designer. The latter can refuse some of the suggestions made by the "intelligent" editor. He can modify at any time a part of the display which is automatically generated by the intelligent functionalities, and even do completely without it so as create his personal displays. In order to do it, in addition to the basic graphical routines, the interface suggests to the designer -under the form of a text, the necessary specifications concerning the process to help him to design the displays.

### First Results

First tests have been made in laboratory. From the schemas of a nuclear power plant, we have realized a human-machine interface permitting to control it. Of course, we have started from the hypothesis that the human task analysis has been made,

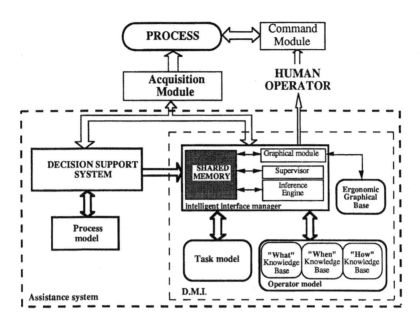

Fig. 3. Global Man-Machine system integrating the Decisional Module of Imagery

The operator model integrates a set of ergonomic data which is presently limited to : (i) three possible levels of expertise for the human operator (unskilled, experienced, expert), (ii) the type of displays associated to each of the operator's cognitive behavior, corresponding to Rasmussen's model, (iii) the representation mode associated to each type of display

### The Aims of the D.M.I.

The aims of the D.M.I are : (i) to select the data that can be displayed on the screen taking into account both the operational context of the process and the informational needs of the operator, in order to enable the operator to supervise the process and to define possible corrective actions when a failure appears; (ii) to define the ergonomic parameters associated with presentation of this information in order to make the human operator's understanding easier; (iii) to add to this supervisory imagery the corrective advice given by the decision support expert system in order to justify its reasoning and thus to prevent possible conflicts between the system and the human operator (Taborin and Millot, 1989).

### The "What", "When" and "How " Problems

In order to meet these goals, the D.M.I. has to be autonomous and able to adapt itself to the operators' needs. The D.M.I. must therefore integrate expert knowledge on : (i) the different operational contexts of the system to be supervised; (ii) the characteristics of the different users of the system; and (iii) the cognitive and sensorimotor tasks that these operators have to perform. These types of knowledge allow us to answer the three following questions related to ergonomics considerations : (i) **What** information is to be presented to the operator (we consider here, that the "what" includes the "why", by using justification scales of the information displayed); (ii) **When** shall we display it, (iii) **How** to display it.These ergonomics considerations are more detailed by Tendjaoui, Kolski and Millot (1991b).

The D.M.I. is implemented in an experimental platform that integrates a supervisory system of a simulated process. It is written in "C" language. The platform is based on a VS/3100 station. Graphical displays are created with a graphical tool called DATAVIEWS (V.I. Corporation, 1988). The software architecture of the platform is described by Tendjaoui, Kolski and Millot (1991a, 1991b).

### Discussion

In order to optimize man-machine communication, we use Artificial Intelligence techniques and it seems that Artificial Intelligence can contribute solutions to some of the problems encountered in the field of process control/man-machine interaction (Elzer and Johanssen, 1988 ; Millot, 1990).

The first version of the D.M.I., contains approximately 80 rules in the "WHAT" database, 70 rules in the "WHEN" database, and 110 rules in the "HOW" database. These rules are compiled in such a way as to be easily exploited by the inference module that controls the graphics display (for more details see Tendjaoui, Kolski and Millot, 1991a).

At present the D.M.I. attempts to fulfil the following objectives (Tendjaoui, Kolski and Millot, 1991b) : (i) It must adapt itself to the operator. This problem is solved in three steps. Firstly, the D.M.I. uses rules from the "WHAT" database in order to decide which information is to be presented to the operator. Secondly, armed with data on the severity of the problem and the skill level of the operator, the D.M.I. uses rules from the "WHEN" database to decide at which moment this information is to be presented. This temporal control over the display allows the D.M.I. to adapt itself to the different operational situations of the process. Finally, using data known about ergonomic and operator preferences, the D.M.I. uses rules from the "HOW" database to decide on the form of data presentation. (ii) it must help the reasoning of the operator : the D.M.I. identifies the operators task using information on the severity of the problem and the status of the process, and guides him with useful information.

The D.M.I.'s capacity to adapt and reason has an impact on : (i) The operators learning curve : the flow of information and its level of detail are managed by the D.M.I. according to the experience level of the operators. The D.M.I. can be used as a formative tool and, in a working situation, its characteristics can improve the operators experience. (ii) The operators workload : The operators workload will decrease because the D.M.I. displays only information to the operator that is useful and in a form that he can readily understand.

These points will be analyzed and evaluated at a later date and will almost certainly highlight some shortcomings of the D.M.I., but it is possible to list now some problems that are the object of our current research and that have allowed us to arrive at an evolved system : (i) consideration of the experience that the operator has of all the different sub-processes : he may, for example, be considered an expert in one sub-process and a novice at another. This experience level depends particularly on operator performance, operator error and on unknown situations that the operator may have already solved. (ii) optimizing the adaptation between the state of the process and the mental image of it that the operator has, using the D.M.I. (iii) integrating a human operator model into the D.M.I. in order to identify, in real time, his cognitive task and to provide appropriate support to his decision. (iv) dynamic evaluation of the operators mental workload. (v) Keeping records of uncommon situations and making them available to the operator via the D.M.I. when necessary. The results of the evaluation will be published in others papers.

## CONCLUSION

Throughout this paper, we have been able to present our current research works concerning the use of formalized ergonomic knowledge in three different fields : evaluation, design and management of man-machine interface used in the process control. For each field, three representative tools have been described and discussed : SYNOP, ERGO-CONCEPTOR and D.M.I.

These different rule-based approaches are not yet completely validated and it is still necessary to proceed to many rigorous experiments in different cases of evaluation, design and management of man-machine interface. But they offer new and interesting perspectives for man-machine interface designers in process control. Such concepts and principles can be considered in "intelligent" software engineering tools dedicated to process control.

## ACKNOWLEDGEMENTS

I would like to thank Professor Patrick Millot, and the future doctors Emmanuelle Le strugeon, Faouzi Moussa, Thierry Poulain and Mustapha Tendjaoui for their very strong contribution and personal works to these different projects.

## REFERENCES

Chignell , M.H and Hancock, P.A. (1988). Intelligent Interface Design. In : Handbook of Human Computer Interaction. M. Helander (ed), 1988.

Elzer, P., Siebert, H., Zinser, K. (1988). New possibilities for the presentation of process information in industrial control. 3rd IFAC congress on Analysis design and evaluation of man-machine systems, Oulu, Finland, June 1988.

Gilmore W.E., Gertman D.I., Blackman H.S. (1989). User-Computer Interface in Process Control. A Human Factors Engineering Handbook, Academic Press.

Grzesiak F. (1987). Représentation des connaissances et techniques d'inférence pour le maniement d'objets graphiques : application au système expert SYNOP. Thèse de Docteur-Ingénieur, Laboratoire d'Automatique Industrielle et Humaine, Université de Valenciennes, Mars 1987.

Hancock P.A., Chignell M.M. (1989). Intelligent interfaces : Theory, Research and Design. North-Holland.

Helander M. (1988). Handbook of Human Computer Interaction, M. Helander (ed.), Elsevier Science Publishers B.V. (North Holland).

Kolski C. (1989). Contribution à l'ergonomie de conception des interfaces graphiques homme-machine dans les procédés industriels : application au système expert SYNOP. Thèse de Doctorat, Université de Valenciennes, Janvier 1989.

Kolski C., Millot P. (1989). Démarches ergonomiques d'évaluation et de conception d'interfaces graphiques homme-machine à l'aide de techniques d'intelligence artificielle : évolutions vis-à-vis de l'ergonomie, XXVème Congrès de la SELF : Evolutions technologiques et Ergonomie, Lyon, 4-6 Octobre 1989.

Kolski C., Millot P. (1991). A rule-based approach for the ergonomic evaluation of man-machine graphic interface. International Journal of Man-Machine Studies, 35, pp. 657-674.

Kolski C., Moussa F. (1991). Une approche d'intégration de connaissances ergonomiques dans un atelier logiciel de création d'interfaces pour le contrôle de procédé. Quatrièmes Journées Internationales : Le génie logiciel et ses applications, Toulouse, 9-13 décembre 1991.

Kolski C., Tendjaoui M., Millot P. (1990). An "intelligent" interface approach. The second International Conference on "Human aspects of advanced manufacturing and hybrid automation", Honolulu, Hawaii, USA, August 12-16, 1990.

Kolski C., Van Daele A., Millot P., De Keyser V. (1988). Towards an intelligent editor of industrial control views, using rules for ergonomic design. IFAC Workshop "Artificial intelligence in real-time control", Clyne Castle, Swansea, Great Britain, 21-23 September 1988.

Millot P. (1990). Coopération Homme-Machine : Exemple de la téléopération, Journées du GR Automatique, 17-19 Octobre, Strasbourg, France.

Minsky M. (1975). A framework for representing knowledge, The Psychology of Computer Vision, Editions P.H. Winston, Mc Graw Hill, New-York, p. 211-280, 1975.

Moussa F., Ben Hassine T., Kolski C. (1991). Ergo-conceptor : Etat d'avancement des travaux. Research report, L.A.I.H., Université de Valenciennes, Mars 1991.

Moussa F., Kolski C. (to appear). Ergo-conceptor : système à base de connaissances ergonomiques pour la conception d'interface de contrôle de procédé industriel. In : Technologies Avancées.

Moussa F., Kolski C., Millot P. (1990). Artificial intelligence approach for the creation and the ergonomic design of man-machine interfaces in control room. Ningth European Annual conference on "Human decision making and manual control", Varese, Italy, September 10-12, 1990.

Moussa F., Kolski C., Millot P. (1992). A formal methodology for ergonomic design of Man-Machine interfaces. Proposition for : 5th IFAC/IFIP/IFOR/IEA Symposium on Analysis, Design and Evaluation of Man-Machine Systems, The Hague, The Netherlands, June 9-11, 1992.

Rasmussen, J. (1980). The human as a system component. Dans H.T. Smith and T.R.G. Green Editors , Human Interaction with Computer, London Academic Press, 1980.

Scapin D.L., Reynard P., Pollier A. (1988). La conception ergonomique d'interfaces : problèmes de méthode, Rapport de recherche n° 957, INRIA, Décembre 1988.

Taborin V., Millot P. (1986 and 1988). ALLIANCE : Système de gestion d'alarmes utilisant les techniques de l'intelligence artificielle, Rapports de contrat MRES-ADI en collaboration avec CEA, LAG, ITMI, IIRIAM, SHELL RECHERCHE, SGN, EDF, France, Université de Valenciennes et du Hainaut Cambrésis, Juillet et Novembre 1986, Mai 1988.

Taborin, V., Millot P. (1989). Cooperation Between Man and Decision Aid System in Supervisory Loop of Continuous Processes, 8'th European Annual Conference on "Human Decision Making and Manual Control", June 1989, Lyngby, Danemark.

Tendjaoui M., Kolski C., Millot P. (1990). Interaction between real-time aid expert system, intelligent interface and human operator. International Symposium Computational Intelligence 90 "Heterogeneous knowledge representation systems", September 24-28, 1990, Milano, Italy.

Tendjaoui M., Kolski C., Millot P. (1991a). Knowledge based interface approach for real-time aid expert system. IFAC/IMACS "Safeprocess'91" Symposium, 10-13 September, Baden-Baden, Germany.

Tendjaoui M., Kolski C., Millot P. (1991b). An approach towards the design of intelligent man-machine interfaces used in process control. International Journal of Industrial Ergonomics, 8, pp. 345-361, 1991.

# MODEL BASED CONTROL SYSTEMS: FUZZY AND QUALITATIVE REALIZATIONS

### R. Gorez*, M. de Neyert*,**, D. Galardini* and J. Barreto*,**

*Université Catholique de Louvain, Laboratoire d'Automatique, Dinamique et Analyse des Systemes,
Bat. Maxwell, Pl. du Levant, 3 B-1348 Louvain la Neuve, Belgium
**Université Catholique de Louvain, Laboratoire de Neurophysiologie, UCL 54.49 Avenue Hippocrate,
54 B-1200 Bruxelles, Belgium

*Abstract*: Some control systems based on fuzzy and qualitative models of the controlled process are introduced in a unifying presentation. The tracking capability and the disturbance rejection of the various control schemes are analyzed and some simulation results are presented.

*Keywords*: Fuzzy control, Qualitative control, Model based control, Process control, Motion control.

## INTRODUCTION

Fuzzy concepts have been introduced in the design of controllers for some process control applications approximately twenty years ago (Mamdani, 1974; Kickert and van Nauta-Lemke, 1976); nowadays they are gaining an increasing use in low cost control systems. Fuzzy control indeed can be seen as a clever way of interpolating between some control decisions, which is probably close to the way of reasoning of the human beings and may be useful if the controlled process is ill-defined or the control objectives are loose. Moreover fuzzy variables and control laws can be expressed in linguistic terms, which allows a straightforward translation of the control strategy of human operators into the rule-base of a controller.

Basically fuzzy control loops have the same properties as conventional quantitative control loops; for example, in steady-state operation, full disturbance rejection can be achieved if and only if there is a reset action in the controller. This may be one of the reasons why many fuzzy controllers proposed in the literature are discrete-time incremental controllers providing an update of the actual control variable at each sampling time. Again this is probably close to the way of reasoning of a human operator; unfortunately this implies the presence of an accumulator or an integrator in the control line. For 2 terms-controllers using only the error (between the reference signal and the process output) and its time-increment this can lead to closed-loop instability unless updating is relatively moderate; in other words, the controller gain should be relatively low and the sampling rate rather slow, which forbids fast control. Faster control can be achieved only with 3 terms-controllers using also the second time-increment of the error. But an appropriate a-priori weighting of three actions in the rule-base of a fuzzy controller may be more difficult, contrary to the case of 2 terms-fuzzy controllers where it is relatively easy to set up a decision table with two entries, through a kind of analogy with the phase-plane of a bang-bang control system.

On the other hand, various model based control systems have been proposed for process control applications since the fifties, because they have nice disturbance rejection capabilities and robustness properties. A model-based control scheme consists of three basic parts: the process which must be controlled, a model of this process and a controller or compensator. This control scheme can be set up independently of the realizations of the controller and the model of the process. Therefore either the controller or the model or both of them can be implemented in the process computer through fuzzy and qualitative control concepts. Hybrid implementation including fuzzy or qualitative

realization of some parts of the control system and numerical or quantitative realization of others is also possible. Besides different sampling rates may be used in different parts of the control system, for instance in the controller and in the model, resulting in dual-rate or multi-rate control systems. Again analogy with the human behaviour is straightforward. It can be reasonably thought that the control strategy of a human operator is based on a qualitative understanding of the process that he has to control. In other words, the operator would have in his mind a qualitative model of the process response, either static (gain) or even dynamic (stable or unstable process, oscillatory or damped transients, estimate of the settling time,...). Then the operator would correct the manipulated control variable according to the deviation of the actual process output from the output which had been forecast by his internal model. Thereafter the introduction of fuzzy or qualitative models in control systems is a natural extension of the use of fuzzy concepts in control. In the followings several model based control schemes are presented and fuzzy or hybrid realizations are investigated.

## MODEL BASED CONTROL

Basically there are three model based control schemes, which differ by the way the control system uses the "return-difference" signal (difference between output signals of the model and the actual controlled process; note that the output of the actual process is not necessarily the controlled output variable, it can be an auxiliary output variable which is measurable). First, this signal can be used for correcting the common control variable which is acting upon both the controlled process and the model. This is the "Internal Model Control" (I.M.C.) scheme shown in Fig. 1, probably the oldest and the best known model based control system; see e.g. Morari and Zafiriou (1990). Internal Model Control has good robustness properties and an appealing intuitive interpretation, but actually, it is an open-loop control system and its use must be restricted to the control of open-loop stable processes. In the two other schemes, the return-difference is used for correcting either the input signal of the model or that of the controlled process only. The first of these two schemes gives in fact nothing else as the well known Luenberger observer (Luenberger, 1971), where the return-difference signal is used for forcing the state of the model to track that of the actual process. The second scheme (Fig. 2) is the dual of the first one and as such, it has the same nice properties as shown by Landau (1974); here it can be said that the return-difference aims to forcing the actual output signal to track that of the model, the effect of unmodelled disturbances being

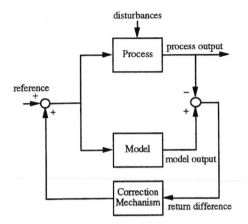

Fig. 1. Internal Model Control Scheme.

counteracted by the return-difference. This scheme is generally called "disturbance observer" because, under some general conditions, the correction of the control signal can be seen as an estimate of an equivalent disturbance signal acting upon the process input.

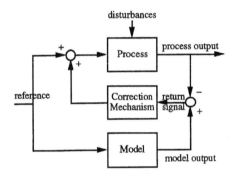

Fig. 2. Disturbance observer.

The use of a disturbance observer results in complete disturbance rejection in steady-state operation provided that the correction of the control signal is achieved through an integrator or that the controlled process itself includes an integrator; see e.g. Gorez et al (1991). For stability reasons the first case requires a slow correction, while the second case

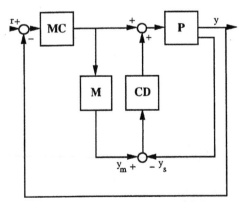

Fig. 3. Disturbance observer inside a main control loop.

| r | = reference signal | y | = controlled output |
|---|---|---|---|
| $y_m$ | = model output | $y_s$ | = auxiliary output |
| P | = process | M | = model |
| MC | = main controller | CD | = correction device |

implies that the disturbance observer is used via a secondary feedback loop inside the main control loop stabilizing the process and/or giving it the desired dynamics (Fig. 3). Another way to cope with integrating or unstable processes would be to equip them with a stabilizing feedback loop and to use internal model control (Fig. 4) or a disturbance observer (Fig. 5) for the closed-loop system. In fact, it has been shown by Gorez (1991) that all these schemes share the same basic properties: first, the tracking dynamics is the same if the return-difference is used or not in the control system; second, the regulation dynamics can be tailored to the designer's objectives, thanks to a suitable filter processing the return-difference signal; third, their robustness properties are equivalent. Therefore internal model control and disturbance observer can be thought as complementary schemes, the first being used if the controlled process is stable, the second if the process itself includes integrators.

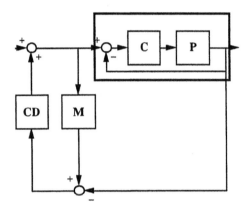

Fig. 4. Internal model control of a process stabilized by a secondary feedback loop.

| P | = process | M | = model |
|---|---|---|---|
| C | = secondary controller | CD | = correction device |

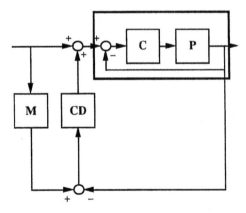

Fig. 5. Open-loop control with disturbance observer for a process stabilized by a secondary feedback loop.

| P | = process | M | = model |
|---|---|---|---|
| C | = secondary controller | CD | = correction device |

## CONTROL WITH QUALITATIVE INTERNAL MODEL

### Control schemes

Several applications of model based control systems using fuzzy and qualitative concepts in the implementation of the model and/or the controller have been considered by Gorez and De Neyer (Gorez , 1990, 1991; De Neyer , 1991a, 1991b, 1991c;...). The process to be controlled is a servomechanism for positional control. Let us assume that the driving torque can be taken as the manipulated control variable and that

182

friction and gravity torques, as well as Coriolis and centrifugal torques in multi-body mechanical systems, can be neglected or considered as internal disturbances. Then, if the servomechanism moment of inertia is compensated by the gain of the output stage of the control system, the model of the system is a double integrator, the first integrator relating velocity to the difference between a driving acceleration proportional to the control variable and an acceleration disturbance due to loading torques and the second one relating position to velocity. Then stabilization and the required tracking dynamics can be provided by a main controller using position and velocity feedbacks as shown in Fig. 6, disturbance rejection being achieved through a model based control structure.

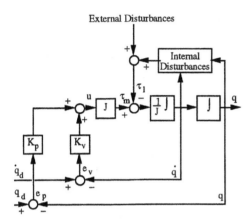

Fig. 6. P-D control of a double integrator plant (servomechanism)

Various implementations of the model based control structure have been tested. First, in the control scheme of Fig. 3 the main controller is a fuzzy controller using fuzzy interpretations of position and velocity errors. Inside the main control loop there is a disturbance observer based on the model relating the velocity to the driving torque: with the above assumptions this model is a simple integrator. Several implementations of this model have been tested: numerical, qualitative or hybrid (combination of the two previous ones).

In other control schemes, stabilization and the required dynamics are obtained via an auxiliary position + velocity feedback loop, through an analogous or a fuzzy controller. Outside this stabilizing loop there is a slow sampling qualitative model of the desired response, used either in the internal model control scheme of Fig. 4 or in a disturbance observer as in Fig. 5. The model can be viewed as a one-step ahead predictor which, at each sampling time, predicts the next value that the process output should reach given the current data, and uses the difference between predicted and measured values for correction of the control variable at the next sampling time. Due to the robustness of the model based control schemes modelling must not be accurate, an approximate model representing roughly the static gain and the settling-time of the controlled system being satisfactory in most cases. The simplest model is a one-step ahead shift operator, the sampling interval being at least equal to the settling-time of the closed-loop controlled system (it can be seen as a "wait and see" strategy leading to small updates of the control variable). This model can be imbedded either in the internal model control scheme of Fig. 4 or in the disturbance observer of Fig. 5; the disturbance rejection capabilities are the same in the two cases, but the errors when tracking time-varying signals are different. Therefore the use of such a model allows changes of the control strategy according to the control task. The second model is a first-order predictor, with a sampling interval approximately equal to half the rise-time of the closed-loop controlled system. For stability reasons and ease of implementation it has been used only in the internal model control scheme of Fig. 4.

## Implementation of fuzzy controllers and qualitative models

The above control schemes include fuzzy controllers and/or qualitative models. Here fuzzy controllers are Proportional Derivative (P-D) controllers using fuzzy values of the position and velocity errors. Model may be a first-order predictor or an accumulator, using values of two fuzzy input variables. Implementation of the control scheme also requires sums or differences of two fuzzy variables. This implies that the actual measurements of position and velocity errors, and other signals such as return difference signals and control variables, are represented as fuzzy values, the latter being expressed in linguistic terms. Interpretation of real data in terms of fuzzy values is achieved by "Real to Fuzzy Interpreters" where the range of real data ([-1,1]) is covered by seven intervals corresponding to seven fuzzy values with asymmetric triangular membership functions, as shown in Fig. 7.

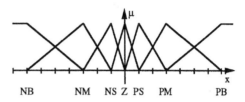

Fig. 7. Membership function of a fuzzy variable
N = Negative    P = Positive
Z = Zero    S = Small
M = Medium    B = Big

All the above fuzzy processors having two inputs and one output, each of them can be implemented as a set of 7 by 7 decision rules expressed in linguistic terms as follows:

*If input 1 has fuzzy value $A_1$ and input 2 has fuzzy value $A_2$ then give the output fuzzy value B .*

The rule-base of each of these processors can be summarized by a table with two entries: the output fuzzy value is given at the crossing of the row and the column corresponding to the two input fuzzy values, the latter being shown in the first row and the first column of the table. Table 1 summarizes the rule-base of the P-D fuzzy controller; it can be seen also as a kind of fuzzy phase-plane representation of the controlled system.

TABLE 1  Rule-Base of a P-D Fuzzy Controller

|  |  | $E_p$ | | | | | | |
|---|---|---|---|---|---|---|---|---|
|  | U | NB | NM | NS | Z | PS | PM | PB |
| $E_v$ | PB | Z | PS | PM | PB | PB | PB | PB |
|  | PM | NS | Z | PS | PM | PB | PB | PB |
|  | PS | NM | NS | Z | PS | PM | PB | PB |
|  | Z | NB | NM | NS | Z | PS | PM | PB |
|  | NS | NB | NB | NM | NS | Z | PS | PM |
|  | NM | NB | NB | NB | NM | NS | Z | PS |
|  | NB | NB | NB | NB | NB | NM | NS | Z |

Tables 2 and 3 give respectively the rule-bases of an accumulator and a first-order predictor, Y and $Y_u$ being respectively the last and the updated value of the output variable, U being the last value of the input variable (either the current value for the control scheme of Fig. 5 or the past value for the control schemes of Fig. 3 and 4). The rule-base of a fuzzy adder can also be represented by Table 2. The decision table for a fuzzy subtracter is the same provided that values of one of the input variables are replaced by the opposite values.

Fuzzy data must be converted into real values before being applied to the controlled system. This is achieved in a "Fuzzy to Real Converter": at any sampling time, the actual value of the control variable is a weighted sum of the real output values associated to the decision rules which are fired at that time, the weighting factor for a given rule being the smallest of the

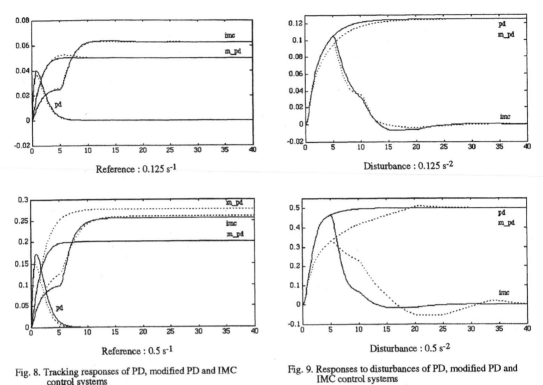

Reference : 0.125 s$^{-1}$

Reference : 0.5 s$^{-1}$

Fig. 8. Tracking responses of PD, modified PD and IMC control systems

Disturbance : 0.125 s$^{-2}$

Disturbance : 0.5 s$^{-2}$

Fig. 9. Responses to disturbances of PD, modified PD and IMC control systems

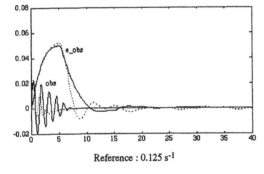

Reference : 0.125 s$^{-1}$

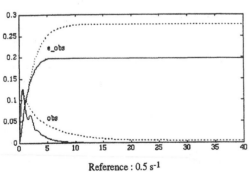

Reference : 0.5 s$^{-1}$

Fig. 10. Tracking responses of control systems with an external observer or with an observer inside the main control loop

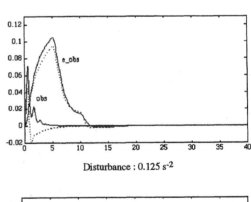

Disturbance : 0.125 s$^{-2}$

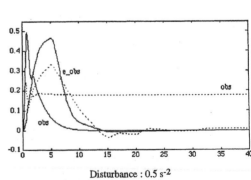

Disturbance : 0.5 s$^{-2}$

Fig. 11. Responses to disturbances of control systems with an external observer or with an oberver inside the main control loop

Solid lines: control systems using identical membership functions
Dotted lines: control systems using asymmetric membership functions

membership values of the two input variables firing that decision:

$$u = \frac{\sum\limits_i \mu_i \, u_i}{\sum\limits_i \mu_i},$$

where $\mu_i = \underset{j}{Max} \, Min\{\mu_{A_{1j}}, \mu_{A_{2j}}\}$ and $\mu_{A_{1j}}, \mu_{A_{2j}}$ are the membership values of the two input variables of the $j^{th}$ rule firing the $i^{th}$ decision.

TABLE 2 Rule-Base of a Fuzzy Accumulator

| | $Y_u$ | U | | | | | | |
|---|---|---|---|---|---|---|---|---|
| | | NB | NM | NS | Z | PS | PM | PB |
| | PB | Z | PS | PM | PB | PB | PB | PB |
| | PM | NS | Z | PS | PM | PB | PB | PB |
| | PS | NM | NS | Z | PS | PM | PB | PB |
| Y | Z | NB | NM | NS | Z | PS | PM | PB |
| | NS | NB | NB | NM | NS | Z | PS | PM |
| | NM | NB | NB | NB | NM | NS | Z | PS |
| | NB | NB | NB | NB | NB | NM | NS | Z |

TABLE 3 Rule-Base of a First-Order Fuzzy Predictor

| | $Y_u$ | U | | | | | | |
|---|---|---|---|---|---|---|---|---|
| | | NB | NM | NS | Z | PS | PM | PB |
| | PB | PS | PM | PM | PM | PM | PM | PB |
| | PM | Z | PS | PS | PS | PS | PM | PB |
| | PS | NS | Z | Z | Z | PS | PM | PM |
| Y | Z | NS | NS | NS | Z | PS | PS | PS |
| | NS | NM | NM | NS | Z | Z | Z | PS |
| | NM | NB | NM | NS | NS | NS | NS | Z |
| | NB | NB | NM | NM | NM | NM | NM | NS |

Simulation results

Several of the proposed control schemes had been tested through simulation by Gorez and De Neyer (Gorez , 1990, 1991, De Neyer , 1991a, 1991b, 1991c, ...). Here new results are presented for the double integrator plant of Fig. 6. Three model based control systems are considered: a control system with disturbance observer (Fig. 3), an internal model control system (Fig. 4) and an open-loop control system with disturbance observer (Fig. 5). Parameter values are the following: $J = 1$, $K_p = 1$, $K_v = 2$, resulting in a closed-loop frequency $\omega_n$ rad/s. Implementation of the control system is in discrete-time: for the main PD controller sampling interval is equal to 0.1s; it is 0.1s for the model used in the disturbance observer control scheme of Fig. 3, and 5s for the other control schemes. Fuzzy control systems being nonlinear, input signals of different amplitudes have been tried. Moreover two different types of membership functions have been used: asymmetric triangular shapes as shown in Fig. 7 or more conventional symmetric identical triangles (except for the extreme linguistic values). Tracking and disturbance rejection responses are presented in Fig. 8 to 11. Dotted lines refer to responses with asymmetric triangular membership functions and solid lines to responses with symmetric triangles. As expected, for the PD control alone (see pd curves on Fig. 9) there is a steady state error due to persistent disturbances. A variant of PD control scheme where there is only velocity feedback but no velocity feedforward is also considered (see m_pd curves on Fig. 8 and 9). Then there is also a steady state error during tracking; the magnitude of this error is obviously related to the gains of the PD controller. In this control structure, tracking of ramp is impossible for reference velocities bigger than a fixed value (0.5) determined by the PD gains and the saturations of the fuzzy controller.

Figures 8 and 9 show also performances of an internal model control scheme (Fig. 4) where stabilization of the plant is achieved using the modified PD control mentioned above (see imc curves). This IMC structure is quantitatively implemented because of the poor performances of its fuzzy counterpart which exhibits sustained oscillations during tracking. As expected for a first order model, the tracking error does not vanish asymptotically; it becomes constant (in fact, it is proportional to the model sampling period). On the other hand, disturbances are rejected as shown in Fig. 9. Figures 10 and 11 show that control schemes using disturbance observers have better performances. The steady state error is asymptotically null. Plots labelled 'e_obs' are the responses of the system with external disturbance observer (Fig. 5); they are slower than the responses of systems using a disturbance observer inside the main control loop (Fig. 3) which are labelled 'obs' in Fig. 10 and 11. This was expected because the latter control scheme has a smaller sampling period for the disturbance observer. It can be seen also that for reference velocities larger than 0.5, the system using an external observer has a steady state error in tracking; the latter can be attributed to the use of a modified PD controller.

Surprisingly, the influence of the membership function shape is different on the tracking and on the disturbance rejection performances. Performances are similar in tracking; on the contrary, the transient responses with asymmetric triangles are slower with smaller peak values in presence of disturbances. It appears also that saturation phenomena occur earlier with asymmetric triangles both in tracking and in disturbance rejection.

CONCLUSIONS

From an extensive simulation study results of which have been presented here and previous papers, it can be claimed that control based on a qualitative internal model or disturbance observer results into a closed-loop behaviour similar to that of quantitative control systems of the same kind. Robustness of such control systems versus external disturbances, parameter variations or unmodelled dynamics appears to be very good. The effect of factors such as the sampling rate of the model, the number of classes in the range of a fuzzy variable or the shape of the membership functions, has also been investigated. Due to the fact that this study has been done in cooperation with neurophysiologists (Barreto, 1991; Lefèvre, 1991), in the frame of a general study on the control of motions of eye, head and arms of living beings, a particular attention has been also given to the error while tracking ramp-wise time-varying reference signals; it may be influenced by the selected control scheme. So this characteristic and some problems of implementation are the main points which allow the selection of a particular scheme among all those which have been considered.

ACKNOWLEDGMENTS

*This research is supported by the Belgian National Incentive-Program for Fundamental Research in Artificial Intelligence, Prime Minister's Office - Science Policy Programming. The scientific responsibility is assumed by its authors.*

REFERENCES

Barreto J., M. De Neyer, Ph. Lefèvre & R. Gorez (1991). Qualitative physics versus fuzzy sets theory in modelling and control. *IECON'91: IEEE/SICE International Conference on Industrial Electronics, Control and Instrumentation*, Kobe, Japan, 2, 1651-1656.

De Neyer M., R. Gorez and J. Barreto (1991a). Disturbance rejection based on fuzzy models. In M.Singh and L. Travé-Massuyès (Ed.). *Decision Support Systems and Qualitative Reasoning*, North-Holland, Amsterdam. pp.215-220.

De Neyer M., R. Gorez and J. Barreto (1991b). Fuzzy controllers using internal models. *Proc. IMACS - IFAC Symposium on Modelling and Control of Technological Systems*, Lille, France, 1, 726-731.

De Neyer M, R. Gorez & J. Barreto (1991c). Simulation analysis of control systems with internal fuzzy models. *ESS'91: Intelligent process control design and scheduling*, Gent, Belgium, 145-150.

Gorez, R. and J. O'Shea (1990a). Robots positioning control revisited, *J. of Intelligent and Robotic Systems*, 3, 213-231.

Gorez, R., M. De Neyer and J. Barreto (1990b). Disturbance rejection in fuzzy control of manipulators. *Proc. ESS 90: European Simulation Symposium*, Ghent, Belgium, 38-43.

Gorez R., M. De Neyer & J. Barreto (1991a). Fuzzy internal model control. Presented at *EURISCON '91: European Robotics and Intelligent Systems Conference*, Corfu, Greece.

Gorez, R., D. Galardini and K. Y. Zhu (1991b). Model based control systems. *13 th IMACS World Congress on Computation and Applied Mathematics*, Dublin, Ireland, 781-783.

Kickert, W. and H. R. van Nauta-Lemke (1976). The application of fuzzy set theory to control a warm water process. *Automatica*, 12, 301-308.

Landau Y.(1979), Adaptive control: The model reference approach, J.M. Mendel (Ed.), M. Dekker Inc, New-York.

Lefèvre Ph., M. De Neyer, R. Gorez & J. Barreto (1991). Fuzzy internal models in vision systems modelling. *IROS'91: IEEE/RSJ International Workshop on Intelligent Robots and Systems*, Osaka, Japan, 1, 105-110.

Luenberger D.G.(1971). An introduction to observers, *IEEE Trans. on Automatic Control*, Vol.AC-16, 596-602.

Mamdani, E. H.(1974). Application of fuzzy algorithms for the control of a dynamic plant. In *Proc. I.E.E.*, 121, 1585-1588.

Morari M.and E. Zafiriou (1989). Robust process control. Prentice-Hall, Englewood Cliffs, N.J.

# FROM QUALITATIVE SUPERVISION TO QUANTITATIVE CONTROL OF WATER DISTRIBUTION SYSTEMS

## S. Sawadogo*, A.K. Achaibou*, J. Aguilar-Martin* and F. Mora-Camino**

*LAAS - CNRS, 7 Avenue du Colonel Roche, 31077 Toulouse, France
**ENAC, 7 Avenue Edouard Belin, 31055 Toulouse, France

<u>Abstract</u> In this paper, a management structure including supervision and control layers of large water distribution systems is investigated. The supervision layer is devoted to the efficient management on the long run of the upstream water stock and to the definition of users' priority. The control level is devoted to the regulation of inflows so that the water allocated to each user at the supervision level is efficiently distributed in a hourly basis.

<u>KeyWords:</u>; Supervision; Qualitative Control; Adaptive Control; Water Distribution Systems.

## The Problem Considered

Irrigation of agricultural lands is an important activity which results in increased productivity in temperate regions and turns possible cultivation in arid areas. Surface irrigation usually requires an extensive distribution system between storage and users implying in high investiment costs which must be carefully planned.

At the operations level, in the short term, demand for water varies stochastically with the changing weather conditions which influence also on the long run water storage levels. Regulation of water inflows is thus necessary to ensure an efficient operation of the distribution system. An efficient operation of such kind of system must avoid wastes and must guarantee the satisfaction of local demand in due time. Today no completely effective procedure exists for the design of automatic control systems for canals and the majority of them are operated manually.

Manual systems do not permit flexibility in operation: In normal operation the users must submit their requests for water 24 or 48 hours in advance and once the water inflow is sent, it must be taken, otherwise it will be wasted downstream. In general also, no means are available to take into account small perturbations (changes in water demand, rainfalls). This implies that the water diverted to canals is then often greater than the actual needed amount and thus is not efficiently used.

In the case of larger perturbations (longuer drought periodes than planned, new demands ) or failures of the system, its operation must be completely modified to fulfil basic priorities.

The basic system considered in this paper is a stretch of canal used for water distribution and composed of the following elements :

. an upstream reservoir of unlimited capacity,

. an entry section with a flow control gate,

. a sequence of pumping stations distributed all along the canal,

. a sequence of measurement points which in general coincide with the pumping stations,

. a final exit section equiped with a flow measuring device.

In normal operation, the main objective of the regulation of inflows is to satisfy, in spite of uncertainties, the water demand at each pumping station while guaranteeing a minimum water level all along the canal and spending a minimum water volume from the upsteam reservoir. A basic assumption is that it is always possible to get a "good" prediction on the short term of water demand all along the canal and at its exit section ( "short term" means a period of time two or three time-lags ahead, here up to 48 hours, while "medium term" means a period of some weeks and "long term" means a period of some months).

## The Solution Approach

The management of flows is based on two interdependant layers: a Supervision and a Control layer.

The structure of the proposed **control layer** is as follows: To get nominal inflows, an accurate but tractable model of flow dynamics is needed. We start from the Saint Venant equations of flows. Since the solution of general Saint-Venant equations is numerically prohibitive, simplifying assumptions must be made. It is supposed that no distributed lateral inflows are present in the system and that inertial terms are negligeable in front of gravity terms. So, instead of Saint-Venant equations a linear flow model is used. This model is numerically tractable since the sampling period for the controller system is slow (an hour in general) and demand either at pumping stations or at the exit section can be well represented by steps varying on this basis. To get nominal inflows from given outflows, this linear model must be inverted. This can be done simply as soon as the parameters of the model have known values. However these values must be updated through an autoregressive process from the current inflows to maintain the validity of the linearization. So, the linear model is inverted to get the next nominal control action ( a receding horizon technique is used here) at the entrance of the canal.

To the already described control structure is added a **supervision layer** which acts as a filter with respect to the nominal inflows obtained from the inverse adaptive approach. The main objectives for the supervision of this kind of system which are considered in this study are:

- the efficient management on the long run of the upstream water stock,

- the guarantee of equity between users along the canal.

The current situation is assessed every week through the comparison between the current water stock levels and the estimated future needs. Qualitative rules are used to detect this current condition of operation for the whole system ( nominal, low severity restriction, medium severity restriction, high severity restriction). This leads to the regulation of the nominal inflows obtained from the control layer to get the effective inflows for the system.

The second objective implies the use of distributed control devices so that local pumped flows can be bounded in accordance with an equity principle. According with the degree of severity of the water deficit the pumping restrictions take two forms:

- for low severity conditions, water is distributed proportionally to the predicted demand, so that its deficit is balanced equally between users,

- for high severity conditions, the priority is given to the satisfaction of basic needs. In this case, the inverse adaptive control approach may suspended.

## The Supervision Layer

Planned discharge curves have been used traditionnally to assess the current situation of water stokcs in dams devoted to irrigation. This approach relies heavily on statistics and supposes that deviations from mean values remain limited.

However, in many regions where irrigation is important, two characteristics of rain are relevant: the deficit in rain water and the irregulary of these rains.So if from one year to the next rains may change their pattern (volume and periods), the use of a mean discharge curve may be largely misleading.

Here we consider an approach which directly takes into consideration the current and future needs of the cultures. Depending of the seeding period, of the culture acreage, of the intermediate rains and volumes of drained water, at current time (a week in fact), needs for the whole irrigation sector can be estimated or their estimation can be updated for the whole remaining period of irrigation.

So an index such as $\nu(t)$ given by :

$$\nu(t) = \frac{V(t) - V_{min}}{\sum_{\tau=t}^{T} \hat{D}_\tau^t} \qquad t \leq T \qquad (1)$$

where T is such as $\hat{D}_{T+\tau}^t = 0 \qquad \forall \tau \geq 0$
can be estimated.
Here:

- V(t) is the current water stock in the barrage at the beginnig of week t,

- $V_{min}$ is the minimum water reserve allowed in the barrage,

- $\hat{D}_\tau^t$ is the estimated demand of water at time t for period $\tau$.

To compare two situations at different times such as $\nu_1$ at $t_1$ and $\nu_2$ at $t_2$, some additional hypothesis must be made. First we see that if $\nu(t) < 1$ there is a shortage in water for irrigation while if $\nu(t) \geq 1$ the water stock seems sufficient to meet the demand until the end of the irrigation campaign.

A prolongated high deficit of water will cause damages which can be irreversible and which can ruin the whole plantation, while if the deficit is maintained within some limits, only its yield will be diminished. In the last case, a prolongated small deficit situation will be equivalent to a severe deficit of short duration. For instance if the damage caused by water deficit grows according to weight function $g(t, T)$, then two equivalent deficits situations will be such as:

$$\sum_{\tau=t_1}^{T} (1 - \nu_1) \hat{D}_\tau^{t_1} g(\tau, T) = \sum_{\tau=t_2}^{T} (1 - \nu_2) \hat{D}_\tau^{t_2} g(\tau, T) \qquad (2)$$

or

$$\frac{1 - \nu_1}{1 - \nu_2} = \frac{\sum_{\tau=t_2}^{T} \hat{D}_\tau^{t_2} g(\tau, T)}{\sum_{\tau=t_1}^{T} \hat{D}_\tau^{t_1} g(\tau, T)} \qquad (3)$$

So, a deficit of $\nu_0$ over the whole period will be considered equivalent to a deficit of $\nu(t)$ over period [t,T]. So we get deficit curves which permit the definition of different levels for the satisfaction of demand for water (Fig 1). For instance we could define:

1. High Severity conditions (HS) when : $\nu_0 \leq 0.50$

2. Medium Severity conditions (MS) when : $0.50 \leq \nu_0 \leq 0.80$

3. Low Severity conditions (LS) when : $0.80 \leq \nu_0 \leq 0.95$

4. Normal Situations (NS) when : $0.95 \leq \nu_0$

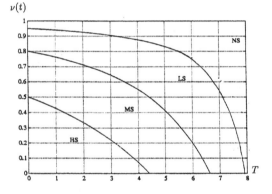

Figure 1 :

Cultures must also be classified to provide some priority ordering. We have for instance cultures classes such as:
Very essential (VE), medium essential (ME), lightly essential (LE) and secondary essential (SE). So in HS conditions irrigation water must be devoted with priority to VE cultures, and so on.

Applying a proportionnal-saturation rule, at the beginnig of a "week" the volume of water which is planned to be consumed during that week is given by:

$$Q^s(t) = min \{Q_{max}^s, \nu(t)\hat{D}_t^t\} \qquad (4)$$

where

$$\hat{D}_t^t = \sum_{\alpha \in C} \hat{D}_{t_\alpha}^t \qquad with \ \ C = \{VE, ME, LE, SE\}$$

where $\hat{D}_{t_\alpha}^t$ is, at time t, the predicted needs of water by cultures of class $\alpha$ for the next week.

Membership functions can be defined for classes:

- HS : high severity conditions

- MS : medium severity conditions

- LS : low severity situation

- NS : normal situation

They are displayed in fig. 2 for a given period t:

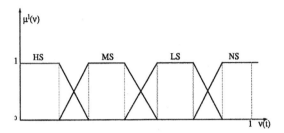

Figure 2 : membership functions for situation classes

However, membership functions can be considered for each category of water consumers( class of culture, industries, towns ...). Let j be one of these categories. We define the index $\lambda_j(t)$ by:

$$\lambda_j(t) = P_j n_j(t) \qquad (5)$$

with $0 \leq P_j \leq 1 \qquad 0 \leq n_j(t) \leq 1$
where:
$P_j$ are the arbitrary weights and thus reflects the a priori choices of the adopted irrigation policy,
$n_j(t)$ is an index which informs about the importance of the satisfaction of water demand during period t for category j of consumers.
A KBS( Knowledge Based System) system can be used to select current values for these indexes. For cultures this KBS must integrate informations about:
. nominal needs of Water per stage of their life cycle,
. past irrigation history (chronology and volumes),
. past rain history and present precipitations over cultures during the current campaign.
For categories characterized by a permanent demand, $n_j(t)$ can be chosen constant. For instance, for human use $n_h = 0.9$ and for industrial use $n_i(t) = 0.8$
So we obtain membership functions( here independant of time) such as:

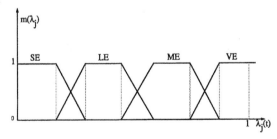

Figure 3 :

Cultures can be also classified to provide some priority ordering. The relative importance of each kind of culture may vary with the current and predicted irrigation balance(parameter $\nu$). Here the following cultures and water consumers categories are considered: VE: Very essential,

ME: Medium essential, LE: Lightly essential and SE: Secondary essential and priority weights are defined for each situation by matrix P:

|  | **HS** | **MS** | **LS** | **NS** |
|---|---|---|---|---|
| **VE** | $P_{VE}^{HS}$ | $P_{VE}^{MS}$ | $P_{VE}^{LS}$ | $P_{VE}^{NS}$ |
| **ME** | $P_{ME}^{HS}$ | $P_{ME}^{MS}$ | $P_{ME}^{LS}$ | $P_{ME}^{NS}$ |
| **LE** | $P_{LE}^{HS}$ | $P_{LE}^{MS}$ | $P_{LE}^{LS}$ | $P_{LE}^{NS}$ |
| **SE** | $P_{SE}^{HS}$ | $P_{SE}^{MS}$ | $P_{SE}^{LS}$ | $P_{SE}^{NS}$ |

with: $\forall$ i,j : $0 \leq P_i^j$ and $\sum_i P_i^j = 1$

Finally, we get a set of current weights for each consumer type given by the expression:

$$\Pi_j(t) = \sum_{l \in L} \sum_{k \in K} \mu^l(\nu) m^k(\lambda_j) P_k^l \qquad (6)$$

with L={HS,MS,LS,NS} and K={SE,LE,ME,VE}

Once these weights are available a linear quadratic programming problem can be formulated to allocate weekly water volume to each class of culture:

$$\min_n \sum_{j=1}^M \left( Q_{t_j}^s - \hat{D}_{t_j}^t \right)^2 \Pi_j(t) \qquad (7)$$

subject to

$$\sum_{j=1}^M Q_{t_j}^s \leq Q_t^s \qquad and \quad Q_{t_j}^s \geq 0 \;\; j = 1 \; to \; M$$

Using a Lagrangian it is easily shown that the solution of this problem is given by:

$$Q_{t_j}^s = \hat{D}_{t_j}^t - \frac{[1/\Pi_j(t)]}{\sum_{k=1}^M [1/\Pi_k(t)]} \left( \hat{D}_t^t - Q_t^s \right) \qquad (8)$$

i.e. the water deficit is allocated to each class of cultures inversely to the priority weight of that class. However if minimun volumes guaranteeing survival for some classes of cultures are taken into account, the above formulation must be completed with restrictions such as:

$$Q_{t_j}^s \geq Q_{j_{min}}^s \qquad j \in [1, \ldots, M]$$

In this case only a numerical solution will be available.
A <u>sensitivity study</u> can be performed in the following way:

1. If $\nu(t)$ is such that $\exists \, l \in L$:
   $0 < \mu^l(\nu(t)) < 1 \qquad and \qquad \mu^{l+1}(\nu(t)) = 1 - \mu^l(\nu(t))$
   then two extreme situations can be considered:
   a) $\mu^l = 1 \quad and \quad \mu^{l+1} = 0$
   b) $\mu^l = 0 \quad and \quad \mu^{l+1} = 1$

2. Analyse each category of users and for each case where $\exists \, k \in K$ such that:
   $0 < m^k(\lambda_j) < 1 \qquad and \qquad m^{k+1}(\lambda_j) = 1 - m^k(\lambda_j)$
   consider the two extreme situations:
   c) $m^k = 1 \quad and \quad m^{k+1} = 0$
   d) $m^k = 0 \quad and \quad m^{k+1} = 1$

So, we get $2^{(1+M)}$ possible situations to be considered. Since M is not very large, for each situation weights $\Pi_j(t)$ can be recomputed and the LQP Problem rerun to get a solution $Q_{t_j}^n$ . Let

$$Q_{t_j}^{smin} = \max_n \left[ Q_{t_j}^{s(n)} \mid Q_{t_j}^{s(n)} \leq Q_{t_j}^s \right] \qquad (9)$$

and

$$Q_{t_j}^{smax} = min\left[Q_{t_j}^{s(n)} \mid Q_{t_j}^{s(n)} \geq Q_{t_j}^s\right] \qquad (10)$$

So, we get for each user j a possible interval for the volume of water which should be available for the next week:

$$Q_j(t) = [Q_{t_j}^{smin}, Q_{t_j}^{smax}] \qquad (11)$$

Now considering the pattern of water demand during a week for each kind of culture we can get the nominal volume of water which will be allocated every day of the week to each culture.
Let $\lambda_k^j$, $k = 1 to J$, be this pattern. We have:

$$\hat{d}_{t_j}^k = \lambda_k^j(t)\hat{D}_{t_j}$$

where $\hat{D}_{t_j}$ is the demand of water for period t by user j. Then a possible water delivery interval is defined as:

$$d_{t_j}^{kmin} = \lambda_k^j(t)Q_{t_j}^{smin} \leq \tilde{d}_{t_j}^k \leq \lambda_k^j(t)Q^{smax} = d_{t_j}^{kmax} \qquad (12)$$

shown in figure 4 :

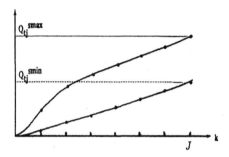

Figure 4 :

So the volume of water allocated to user j during day k of period t will be given by:

$$\begin{cases} d_{t_j}^k = \hat{d}_{t_j}^k & if \quad d_{t_j}^{kmin} \leq \hat{d}_{t_j}^k \leq d_{t_j}^{kmax} \\ d_{t_j}^k = d_{t_j}^{kmin} & if \quad \hat{d}_{t_j}^k < d_{t_j}^{kmin} \\ d_{t_j}^k = d_{t_j}^{kmax} & if \quad \hat{d}_{t_j}^k > d_{t_j}^{kmax} \end{cases}$$

At the tactical level this volumes may be modified to take into account recent precipitations, so that nominal daily volumes of water for each class of culture will be given by :

$$\tilde{d}_{t_j}^k = d_{t_j}^k - \delta d_{t_j}^k \qquad (13)$$

with

$$\delta d_{t_j}^k = r_j(\rho)d_{t_j}^k$$

where $\rho$ , $0 \leq \rho \leq 1$, indicates the importance of the precipitation and $r_j(\rho)$ is an increasing monotonic function.
And the total volume for the next day is : $\sum_k \tilde{d}_{t_j}^k = \hat{\delta}$.
Now to build the hourly pattern of water which will be allocated to each user, we have to consider they demand for the following day.
Let $\delta_i^t$ be the demand of water (in $m^3/s$) by user i for hour t. Supposing that each user grows a unique class of culture, let $S_j$ be the set of users which grow cultures of class j. The total demand for hour t is given by:

$$\delta^t = \sum_{j=1}^M \left(\sum_{l \in S_j} \delta_l^t\right) = \sum_{j=1}^M \mathcal{D}_j^t \qquad (14)$$

and the total demand for the day is then :

$$\delta = \sum_{t=1}^{24} \delta^t$$

The water allocated to user i during hour t will be given then by :

$$\hat{\delta}_i^t = min(\delta_i^t, \tilde{\delta}) \qquad i \in S_j \qquad (15)$$

with

$$\tilde{\delta} = \tilde{d}_{t_j}^k \frac{\delta_i^t}{\mathcal{D}_j^t}\frac{\delta^t}{\delta} \qquad (16)$$

## The Control Layer

Let T be the receding horizon for the control of inflows which coincides with the horizon of prediction of flow demand. Actions taken at the current time t at the entrance of the canal will not influence water flows at section i until time:

$$t^i = t + \sum_{j=1}^i L_j\Theta(t)$$

where

$$\Theta_{t+1} = \alpha\Theta_t + \beta(Q_t^I - \bar{Q}_t^t) + (1-\alpha)\bar{\Theta}(\bar{Q}_t^I) \qquad (17)$$

$0 < \alpha \leq 1 \qquad \beta < 0$
where $\bar{Q}_t^I$ is a mean flow value up to period n given by:
$\bar{Q}_t^I = \sum_{m=t-N+1}^t Q_m^E/N$

So that events at section i will be the consequence of past decisions and of perturbations taking place upstream to this section. This gives to the problem a degree of uncontrollability which can be only overcome by an accurate prediction of the different classes of demand and perturbations. The receding horizon of control for section i is then at instant t :

$$h_i(t) = \left[t + \sum_{j=1}^i L_j\Theta(t), \quad t + \sum_{j=1}^i L_j\Theta(t) + T\right] \qquad (18)$$

$i = 1, ...N$
The predicted flow at section i is given by

$$Q_i(t) = \hat{\delta}_i(t) + Q_{min} + H^{-1}(Q_{i+1}(t + L_i\Theta(t))) \qquad (19)$$

$i = N, N-1, ...1$
where $\hat{\delta}_i(t)$ given by relation (15) is the flow allocated to user i during period t:
$H^{-1}$ indicates that the linear flow model is inverted to get the inflow at section i.
So, the total flow at the entrance of the canal is given finally by:

$$Q(t) = H^{-1}(Q_1(t + L_1\Theta(t))) \qquad t \in h_1(t) \qquad (20)$$

The inversion of the flow model is as follows:
For a given section of canal of length L, the value of $\Theta_{t+1}$ is obtained from relation (17), so that an estimate of the transport delay for the inflows during period $t + 1$ is given by $d_{t+1} = L\Theta_{t+1}$. Then if $Q^S(t + 1 + d_{t+1})$ is the predicted flow at the exit section, inverting the modified Hayami's relation , we get the inflow:

$$Q^E(t+1) = max\left\{\beta_{t+1} + Q^E(t), \ 0\right\} \qquad (21)$$

with

$$\beta_{t+1} = \frac{[Q^S(t+1+d_{t+1}) - \sum_{t<t+1+d_{t+1}}\beta_t\Phi(t+1+d_{t+1},t)]}{\Phi(t+1+d_{t+1},t+1)} \qquad (22)$$

Since the flow at successive sections may vary abruptly, the inversion process leads to large variations of predicted inflows which can exceed the flow capacity of the canal.
So, at time $t + 1$, the effective inflow $Q^I(t + 1)$ is also submitted to capacity constraints such as:

$$Q_{min} \leq Q^I(t+1) \leq Q_{max}(t+1)$$

where $Q_{min}$ is a minimum inflow level related to other considerations than water distribution (protection of environment, animal life, navigation,...), while $Q_{max}(t+1)$ is related to the characteristics of the inflow control gate and to the level of water in the upstream reservoir.

## Simulation

A 29.5km long stretch of canal with two pumping stations is considered (Fig. 5).
We assume an initial steady state of $Q^I = Q^S = 0.5 m^3/s$. The desired final outflows are given equal to $Q^* = 0.5 m^3/s$. In Fig. 6 are displayed water demands ($\delta_1^t$, $\delta_2^t$) by users 1 and 2, as well as the water flows ($\hat{\delta}_1^t$, $\hat{\delta}_2^t$) allocated by the supervision level.
For this water allocated, the computed inflow is displayed in Fig.7, resulting in the outflows of Fig. 8.

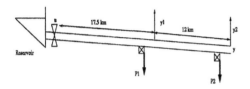

Figure 5 : The simulated system

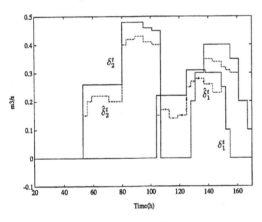

Figure 6 : The pumped flow sequence

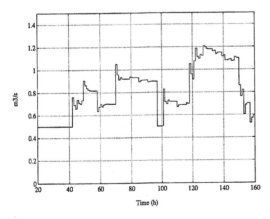

Figure 7 : Selected inflows for the controlled system

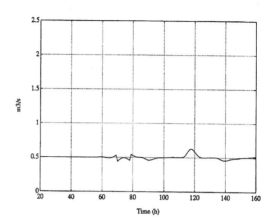

Figure 8 : Resulting outflows for the controlled system

## Conclusion

This paper describes an attempt to manage a large water distribution systems through the use of modern techniques with a two level approach. It is an example of cooperation between qualitative techniques at the supervision level and quantitative techniques at the control level.
The resulting approach, thanks to its simplicity, is compatible with normal conditions of operation.

### REFERENCES

J. Aguilar-Martin, (1991). Knowledge based real time supervision of dynamical processes, LAAS report No. 91148.

S. Gentil and al. (1987). Qualitative Modelling Process Supervision Systems, presented at the first european meeting on cognitive science approaches to process control, Marcoussis.

S. Sawadogo; A.K. Achaïbou; J. Aguilar-Martin and F. Mora-Camino, (1991). Output Tracking by Inverse Adaptive Control: Application to Water Distribution Systems, Proceeding of the IMACS Symposium, pp. 88-93.

S. Sawadogo; A.K. Achaïbou; J. Aguilar-Martin and F. Mora-Camino, (1991). An Application of Adaptative Predictive Control to Water Distribution Systems, Proceeding of the IFAC.ITAC Symposium.

# KNOWLEDGE-BASED COORDINATION OF QUALITATIVE ONLINE DIAGNOSTIC TESTS

G.G. Koch

*Department of Automatic Control, Swiss Federal Institute of Technology, CH-8091 Zurich, Switzerland*

ABSTRACT. It is shown, how active diagnostic tests and their coordination can be integrated within an existing prototype of a knowledge-based extension of a process management system. A generic coordination unit loads the specific tests from a library. Once loaded, a test activates, adapts, interprets the data and unloads itself. The structure of all the tests is the same, while the tests can be quantitative or qualitative. On the example of blocked dampers in a heating and ventilating plant, three qualitative diagnostic tests are described in order to illustrate how qualitative knowledge can be used to detect and diagnose faults. The explicit representation of the assumptions and constraints upon which a test is based, allows to make tests during special operation phases that seldom appear, but that are very instructive.

KEYWORDS. Process management systems, expert systems, qualitative models, on-line failure detection and diagnosis, heating and ventilating plant.

## INTRODUCTION

Humans as operators and service engineers of processes like for example heating and ventilating plants, are found to diagnose faults in a way different from the purely statistical as well as from the quantitative model-based approaches. Studies, as the ones conducted by Rasmussen (1985, 1986) indicate, that human operators make use of a hierarchically organized mental representation of the plant when diagnosing a fault that they cannot immediately and intuitively recognize by their experiences. An important characteristic of this mental representation is the incorporation of qualitative models to describe the behavior of the components, their relations to each other and the functionality of the overall plant.

Qualitative reasoning or qualitative physics is the new area of research that deals with the technical implementation of this human approach (Bobrow, 1984; Struss, 1989; Mavrovouniotis, 1990). While qualitative models are more feasible to obtain than the quantitative process and fault models required by the traditional model-based approaches, it seems that the qualitative models soon become too general, and that they prevent a meaningful reasoning due to their ambiguities. A human operator faced with the same problem usually runs simple tests in order to actively gather the additional quantitative or qualitative information necessesary to solve and clear the ambiguities. That is the background in front of which the work presented in this paper must be seen.

First a generic coordination of online diagnostic tests is described and then it is shown how such tests have been integrated within an existing prototype of a knowledge-based extension of a conventional process management system. Finally, the results of experiments on a physical model of a heating and ventilating plant are given.

## GENERIC COORDINATION UNIT

The task of the generic coordination unit within the knowledge-based extension of a process management system is to monitor the working space of the knowledge-based system and to respond to a hypothetic fault present. In such a case, the coordination unit searches the library of specific diagnostic tests to load the test most suited to diagnose the fault, given the current situation. Once loaded and initiated, each test runs on its own, entirely driven by the information stored in the library.

The generic coordination unit can be described by the following items:
- the *generic coordination* of the search for the diagnostic test and its instantiation.
- the *structure* common to all the diagnostic tests.
- the *library* of the diagnostic tests.

### Generic Coordination of Diagnostic Tests

The activation of any diagnostic test within for instance a process management system can depend on different reasons:
- the test to be performed has been chosen by an operator or service engineer.
- the system itself periodically activates the tests to ensure that there are no faults.
- a fault has been detected or suspected and a test is desired to verify and to find the cause.

The tests can be thought as being organized in a library from where they can be loaded. Out of security reasons and to avoid interferences, it is best to only keep one test loaded at a time. Since certain faults might be more important to be diagnosed than others, it must be possible to set priorities and to end a current test in order to start a more urgent and important one.

Figure 1 illustrates the points mentioned above in form of a finite state machine.

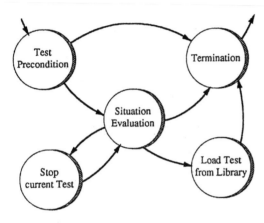

Fig. 1. Description of the generic coordination of diagnostic tests by a finite state machine.

The actions performed within each state are the following:

Test precondition: It is checked, whether the need or demand to carry out any diagnostic test is given. If no hypothetic fault is suspected either by the operator or by a fault detection unit within the knowledge-based system, the task of the generic coordination is terminated. If a hypothetic fault is present, the situation is evaluated.

Situation evaluation: The library of diagnostic tests is searched to find the set of tests that are suited to diagnose the given fault in the current situation of the plant. If there are several tests, the one with the lowest costs is chosen. If the chosen test is currently already loaded nothing is changed and the task of the generic coordination is terminated.

Stop current test: The test currently loaded is eliminated, but only if its priority is less than the one of the new test chosen.

Load test from the library: The chosen test is loaded from the library into the working space of the knowledge-based system and forced to instantiate itself. While the test remains loaded, the task of the generic coordination is being terminated.

Termination: The task of the coordination is completed and control is given back to the higher levels of the knowledge based system.

## The Structure of the Diagnostic Tests

Once a diagnostic test is loaded and instantiated, it is on its own, so to speak. As illustrated in Fig. 2, the test observes the situation of the plant and activates itself when its preconditions are fulfilled. After its activation, the test supervises itself, first to ensure that things are under control and mainly to decide when parameters need to be adapted and when the test is completed. The interpretation of the test-data will lead to the diagnostic decisions. Depending on the test, certain parameters need to be adapted before the test data can be interpreted. After each adaptation, the test needs to reactivate (reset, precondition, activation) itself. The test is completed successfully, when a diagnostic decision could be reached. The test then eliminates itself from the knowledge-based system.

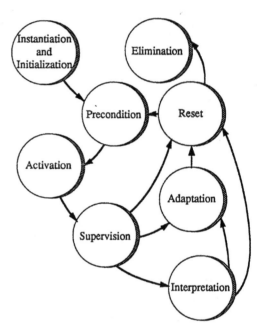

Fig. 2. The different states of the diagnostic tests in form of a finite state machine.

The actions performed within each state are the following:

Instantiation and Initialization: The test is forced by the coordination unit to instantiate itself before it is connected to an observer that supplies the necessary measurement values or estimates of the process variables.

Precondition: The test observes the behavior of the plant in regard to the constraints and assumptions which must be fulfilled that the test can be performed.

Activation: If the preconditions are fulfilled the test is being activated by sending commands to the process management system.

Supervision: After the activation of the test, the supervision is crucial. The test observes the behavior of the plant and decides when it becomes harmful, when its parameters need to be adapted and when it is completed.

**Interpretation:** The test data collected is being interpreted towards diagnostic decisions.

**Adaptation:** Given the information collected during supervision and/or interpretation, test parameters are being adapted so that the repeated activation of the test will eventually lead to the collection of useful data.

**Reset:** The test is stopped by sending commands to the process management system. The values are set back to the initial conditions (with the exception of the adapted variables).

**Elimination:** The test unloads itself from the working space of the knowledge based system, either because the test has been successful, or because the test is being stopped from the coordination unit (stop current test), or because the test has been in the system for a long time without leading to any diagnostic decisions.

### The Library of Specific Diagnostic Tests

A specific test is defined by the information that specifies the different states described above. It is this information that entirely determines the course of the test. The human expert only needs to provide this information that is stored in a library for easy access. When loaded into the working space of the knowledge-based system, the test is provided with the needed values of the process variables and estimates. As the name indicates, the tests themselves are very specific and aimed at diagnosing specific causes of faults. Although we have supposed until now that the tests are qualitative, they could be of any kind; quantitative or qualitative, heuristic or model-based. Such tests can be developed out of the need and be added to the library at any time.

Before examples of three specific qualitative tests are given, we take a look at the integration of the generic coordination unit within the knowledge-based system.

## INTEGRATION OF THE GENERIC COORDINATION UNIT

### Knowledge-Based Extension of the Process Management System

An important aim of this work was to show, that the generic coordination unit can be integrated into an existing prototype of a knowledge-based extension of a process management system. A description of this prototype applied to a heating and ventilating plant is given in (Koch and Grünenfelder, 1990). The prototype contains an object-oriented, hierarchically structured representation of the plant. Based on this representation of the plant, different reasoning strategies can be activated (such as monitoring, fault detection, fault diagnosis, documentation, report generation, etc.). The generic coordination unit can be seen as just another reasoning strategy in that list.

Each reasoning strategy within the knowledge-based system is implemented in a so named *reasoning-module*. The functionality of the knowledge-based system results from the set of such modules that are inserted into the so named *reasoning-loop*. The

reasoning-modules within that loop are cyclic evaluated and if their conditions are fulfilled, the modules are activated.

### Integration of the Coordination Unit

The integration of the coordination unit has been achieved in the following way (see Fig. 3): The coordination unit is implemented as one of the various reasoning modules (named "online diagnostic test module"). It depends on the "update-module", a reasoning module that sets up an observer which updates the quantitative and symbolic measurement values and estimates of the process variables. Since the diagnostic tests depend on those updated values, the specific test found in the library is instantiated as a part of the (extended) observer. This way, once the specific diagnostic test is being loaded, it is independend from the "online diagnostic test module". On the other hand, this reasoning module is part of the reasoning loop, where it is only activated if a possible fault has been detected or suspected (by some other reasoning module or by the human operator). The activated "online diagnostic test module" then looks for the suitable specific online test in the library and forces its instantiation.

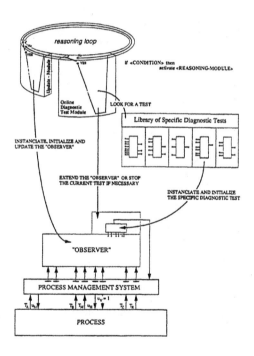

Fig. 3: The integration of the coordination unit as a reasoning module. (The indicated input/ output variables of the process relate to the example in the following section).

## EXAMPLES

### Description of the Plant

The plant considered is a physical model of a heating and ventilating plant, confined to the process and its management system shown in Fig. 4.

The process contains three dampers, two electrical air heaters (the one in the room is to simulate an internal heat source), two ventilators, five temperature sensors, some air pipes and a room. The two running ventilators cause a flow of fresh air entering from the outside through the inlet damper. This air is mixed with extracted air flowing through the damper in the bypass pipe before it is heated up by the heater and blown into the room. One part of the extracted air is flowing back through the bypass pipe while the rest leaves through the outlet damper. The three dampers are synchronously moved by the motor.

The control variables used by the process management system in the given set up are $u_M$ for the angle of the dampers, $u_H$ for the power of the air heater and $u_V = 1$ for the ventilators. A cascade control algorithm is used to maintain a constant set-point air temperature in the room. The measurement needed for the cascade control are the room inlet air temperature ($T_Z$) and the room temperature ($T_R$). The internal heat source ($q_i$) and the outside temperature ($T_A$) are disturbances.

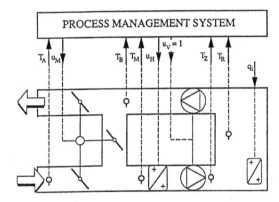

Fig. 4: The physical model of a heating and ventilating process and its process management system.

## Specific, Qualitative Diagnostic Tests

A realistic fault that can be simulated by this physical model, is the example of blocked dampers. In real plants, this fault is often hard to detect, when the dampers are not visible and the motor is still working but not turning the root of the dampers. Depending on the position of the blocked dampers, this fault will either cause discomfort in the rooms (additional heat can not be taken away fast enough) or lead to a waste of energy (too much cold air from the outside has to be heated up).

In the following Table 1, three simple and qualitative diagnostic tests are described by the information specifying their states as illustrated in Fig. 2.

Test 1: Opened Inlet Damper, Closed Bypass Damper.
In its stationary state, the value of the mixing temperature $T_M$ will drop to the value of the outside temperature $T_A$.

In order to avoid a decrease in the room temperature, the test may only be activated, when the outside temperature is not too cold for the heater to heat up the air to the desired temperature. Another possibility is to activate the test during a non-occupation phase, when a decrease in the room temperature is not critical.

The "costs" to run this test follow from the additional energy necessary to heat up the colder air.

Test 2: Closed Inlet Damper, Opened Bypass Damper.
Keeping the fresh, cold air from entering, the stationary mixing air temperature $T_M$ will become close to the temperature $T_B$ of the extract air.

This test can only be performed when not too many people are in the room, because the flow of fresh air is cut off.

The costs to activate this test are minimal. (Energy consumption is even decreasing).

Test 3: Step Response of the Dampers.
With this test, the dynamics of the mixing air temperature $T_M$ as a result to a step response of the inlet and bypass dampers is qualitatively analyzed. Two steps are performed in two test cycles. A positive step should lead to a decrease in the mixing air temperature and a negative step to an increase.

Because the test is performed around the working point of the dampers, the steps may not be too big. As a consequence, the test is very sensitive towards disturbances and set point changes of the room temperature and it can only detect and diagnose that both dampers are blocked or that the damper motor is turning the wrong way.

The costs to run the test are minimal. However, its sensitivity towards small disturbances and its diagnostic limits make it less effective.

## Experimental Results

Some of the experimental results are shown on the example of test 2: Closed Inlet Damper in Fig. 5 (both dampers working alright) and Fig. 6 (both dampers blocked in their working position). First, a non-occupation phase of the room is simulated by lowering the set-point of the room temperature from 20°C to 16°C. This is one of the prerequisits for the test to be loaded. As soon as the mixing air temperature is stationary, the test activates itself by forcing the damper motor to close the inlet damper and to open the bypass damper. The test then waites a minimal time interval (Fig. 6) or until the mixing air temperature is stationary again (Fig. 5). The results of the interpretations are given as diagnostic reports and shown in the figures. After the adaptation state, the test resets and unloads itself.

The figures show the graphs (recorded by the process management system) of the room temperature, the mixing air temperature and the inlet damper position. The messages generated by the knowledge based system concerning the different states of the test are given in their relation to the graphs.

TABLE 1 Definitions of the Specific Online Tests to Diagnose Blocked Dampers

| | Test 1: Opened Inlet Damper | Test 2: Closed Inlet Damper | Test 3: Damper Step Response |
|---|---|---|---|
| Tested Element | Bypass Damper | Inlet Damper | Damper Group (inlet & bypass) |
| Initialization | Process variables needed: outside temp., mixed air temp., extract air temp., room temp., control var. of damper motor, etc. Internal variables: cost, min. outside temp., min./max. test time, etc. | Process variables needed: outside temp., mixed air temp., extract air temp., room temp., control var. of damper motor, etc. Internal variables: cost, min./max. test time, etc. | Process variables needed: outside temp., mixed air temp., extract air temp., room temp., control var. of damper motor, etc. Internal variables: cost, min./max. test time, size of step, phi-limit, test cycle, etc. |
| Precondition | cascade control working, stationary state, damper motor not moving, either outside temp. > min. outside temp. or non-occupation phase of the room. | cascade control working, stationary state, damper motor not moving, non-occupation phase of the room. | cascade control working, stationary state, damper motor not moving. |
| Activation | disconnect the dampers from the cascade control, open the inlet damper and close the bypass damper, set the starting time, store the initial values. | disconnect the dampers from the cascade control, close the inlet damper and open the bypass damper, set the starting time, store the initial values. | disconnect the dampers from the cascade control, change the inlet damper by +step and the bypass damper by -step. If the position of the inlet damper < phi-limit then step is positive else negative, set the starting time, store the initial values. If test cycle = 2 reverse the step. |
| Supervision | Reset if: cascade control not working, or inlet damper not open, or outside temp. not constant, or set point changes of the room temp. Interpret if: max. test time is over, or min. test time is over and mixed air temp. is constant. Adapt parameters, if: room temperature is dropping far under its set point value. | Reset if: cascade control not working, or inlet damper not closed, or set point changes of the room temp. Interpret if: max. test time is over, or min. test time is over and mixed air temp. is constant. | Reset if: cascade control not working, or inlet damper changed , or set point changes of the room temp. Interpret if: max. test time is over, or min. test time is over and mixed air temp. is constant. Adapt parameters, if: number of qualitative states of mixed air temp. > max. |
| Interpretation | Bypass damper blocked: if mixing air temp. did not change. Bypass damper working alright: if mixing air temp. decreases to the outside temp. Wrong wiring of the damper motors: if mixing air temp. increases close to the extract air temp. Bypass damper not closing tight: if mixing air temp. decreases but not to the outside temp. | Inlet damper blocked: if mixing air temp. did not change. Inlet damper working alright: if mixing air temp. increases to the temp. of the extract air. Wrong wiring of the damper motors: if mixing air temp. decreases close to the outside air temp. Inlet damper not closing tight: if mixing air temp. increases but not to the temp. of the extract air. | Both dampers blocked: if mixing air temp. did not change in either test cycle. Wrong wiring of the damper motors: if positive step causes an increase and a negative step causes a decrease of the mixed air temp. |
| Adaptation | Increase the max. test time if the interpretation was not successful and test time = max. test time. Increase min. outside temp. if the room temp. is decreasing too far below its set point value. | Increase the max. test time if the interpretation was not successful and test time = max. test time. | Increase the max. test time if the interpretation was not successful and test time = max. test time. Increase the limits of the intervals defining the qualitative states, if the number of recorded states > max. |
| Reset | Delete the results. Send the command to connect the dampers to the control again. | Delete the results. Send the command to connect the dampers to the control again. | If test cycle = 1: Save the interpretation results, increase test cycle. If test cycle = 2: Delete the results. Send the command to connect the dampers to the control again. |
| Elimination | Unload the test. | Unload the test. | Unload the test. |

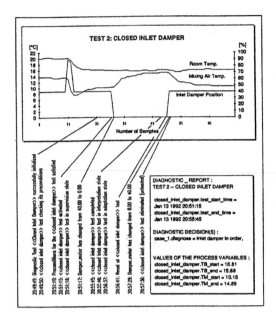

Fig. 5: Example of Diagnostic - Test 2: "Closed Inlet Damper", diagnosing that both dampers are working alright.

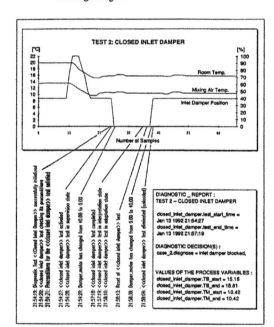

Fig. 6: Example of Diagnostic - Test 2: "Closed Inlet Damper", diagnosing that both dampers are blocked.

## CONCLUDING REMARKS

It has been shown, how active diagnostic tests and their coordination can be integrated within an existing prototype of a knowledge-based extension of a process management system. A generic coordination unit loads the specific tests from a library. Once loaded, a test activates, adapts, interprets the data and unloads itself.

The structure of all the tests is the same, while the tests can be quantitative or qualitative. On the example of blocked dampers in a heating and ventilating plant, three qualitative diagnostic tests are described in order to illustrate how qualitative knowledge can be used to detect and diagnose faults.

The explicit representation of the assumptions and constraints upon which a test is based, allows to make tests during special operation phases that seldom appear, but that are very instructive. Since a test runs on its own once it is loaded and instantiated, it can stay within the working space of the knowledge-based system and wait for its preconditions to be fulfilled without disturbing any of the reasoning strategies.

A specific diagnostic test is entirely determined by the information stored in the library. New test are added by only providing this information.

The implementation of the generic diagnostic unit within the existing prototype of the knowledge-based extension of a process management system is straightforward, owing to the modularity and flexibility of the knowledge-based system.

The principle illustrated on the generic coordination unit with diagnostic tests can be applied to other online information processing methods, such as online model identification or online adaptation of parameters.

## ACKNOWLEDGMENT

This wok is financed by a grant from the Swiss Research Foundation and Landis & Gyr Zug AG (Nr. 20-5430.87). The author expresses his gratitude to Dr. Jürg Tödtli, Landis & Gyr Zug AG, for his support and substantial contributions.

## REFERENCES

Bobrow, D.G. (Ed.) (1984). *Qualitative Reasoning about Physical Systems*. Reprinted from the journal Artificial intelligence volume 24, Elsevier Science, Amsterdam.

Koch, G. and W. Grünenfelder (1990). Extension of a Process Management System with a Knowledge Based Online Reasoning. *Proceedings of the 1990 European Simulation Symposium*, Ghent (Belgium), Nov. 8-10, Simulation Councils, 91-95.

Mavrovouniotis, M. L. (Ed.) (1990). *Artificial Intelligence in Process Engineering*. Academic Press, San Diego.

Rasmussen, J. (1985). The Role of Hierarchical Knowledge Representation in Decisionmaking and System Management. *IEEE Transactions on systems, man and cybernetics*, SMC- 15/2, 234-243.

Rasmussen, J. (1986). *Information Processing and Human-Machine Interaction*. Series Vol. 12, Elsevier Science Publishing, New York.

Struss, P. (1989). *Structuring of Models and Reasoning about Quantities in Qualitative Physics*. Ph.D. Thesis, University of Kaiserslautern, Germany.

# NATURAL LANGUAGE INTERFACE IN CONTROL.
# APPLICATION TO A PILOT PLANT

**J.M.G. Romano, J.A. Ternero and E.F. Camacho**

*Dpto., Ing. Electronica, de Sistemas y Automatica, Universidad de Sevilla, Spain*

**Abstract.** This paper presents a prototype of a friendly man–machine interface for control centers that uses a natural language (Spanish) processor. The vocabulary is adapted to the context of physical processes. The developed prototype is illustrated with an application to a pilot plant.

**Keywords.** Natural Language, Man–Machine Interfaces, Control Systems.

## INTRODUCTION

The design of friendly man–machine interfaces is a major concern in process control centers because in the majority of industrial processes most significant decisions are taken by control center operators. The reason for this is that completely automated systems are extremely problematic as pointed out by Martin and co-workers (1990) because *in situations where surprises can lead to critical situations and matters of life and death, responsible decisions cannot be left to such systems alone. Instead final decisions must be left to people, because people have the ability to master unprecedented situations.*

Although it can be said, in general, that man–machine interfaces of control centers are quite friendly, they have some drawbacks, such as the need of an extensive training period for operators and the great quantity of information, some of it irrelevant, they tend to present to the operator. Presenting the operators with too much information can be a problem as it tends to shift their attention from the important issues at that moment. In reality, the amount of information given to the operator can normally be specified when configurating the system, but as it is not precisely known at this stagewhat information the operator is going to need in every situation, practically every screen is configured presenting more information than necessary, just in case it is needed.

Natural Language interfaces have been used successfully for this purpose in other fields such as graphical management systems (Kasturi and co-workers, 1989),information retrieval in stock market systems (Rodríguez, 1989) and in robotic assembly systems (Selfridge and Vannoy, 1986). Natural language interfaces may be used in control systems to dialog with the operators, accepting commands and presenting information to them, in their natural languages. This may overcome some of the difficulties mentioned previously.

Another field of interest of natural language processing is the explanation generation for expert systems. An expert system is a computer program that acts as an expert, that is, offering advice in most cases. An example of such an application can be found in (Kosy, 1989) where explanations in natural language are applied to financial modelling. As expert systems are a growing field in control centers, see for example (Rich and Venkatasubramanian, 1987) and (Niemann and co-workers, 1990), and some of them act as monitors offering advice or explaining the causes of a given situation, natural language processing may be, in general, of great interest as a tool for man–machine interfaces in control rooms.

This paper presents a prototype of a friendly man–machine interface for control centers that uses a natural language (Spanish) processor. The presented prototype uses a grammar restricted to control center environments and the vocabulary is adapted to the context of physical processes. The prototype is an extension of the one described in (Romano and co-workers, 1991) and is illustrated by an application to a pilot plant.

The paper is organized as follows: Section 2 introduces the main concepts of natural language interfaces and the knowledge bases needed for natural language processing in control centers. The developed prototype is described in section 3, and section 4 is dedicated to describing the process and the specific knowledge base. Dialog examples are presented in section 5 and finally some conclusions are made.

## NATURAL LANGUAGE INTERFACE

This paper centers on the problem of understanding natural language in control center environments. The procedure for solving this problem can be decomposed into the following tasks:

- Lexical analysis: The sentence is broken into words and each of them is checked in a dictionary where necessary information about them is obtained for the next step.

- Syntactical analysis: The grammatical correctness of the sentence is checked.

- Semantical analysis: The meaning of the sentence is extracted. It may occur that a syntactically correct phrase has no meaning. In general, a phrase is considered to be semantically correct if the corresponding control command can be constructed.

The two fundamental elements for understanding NL are the dictionary and grammar. The dictionary contains all words that can be recognized by the interface and is necessary in all three phases of analysis. The grammar is used in syntactical and semantical analysis. In the latter, the compositionality principle of Frege is applied. This principle establishes that the meaning of the whole depends on the meaning of the parts, that can be the same as the ones obtained in the syntactical analysis. In the case of control centers with NLI, the operators can express their commands to the control system in their language (or close to it). These commands will be analyzed, understood (translated into elementary commands to the control system) and processed.

Operators do not have to know complicated Man-machine interfaces but will be able to use their natural languages referring to the process and to the related components without needing to remember precise names or labels for each of them.

The necessary knowledge for the analysis is, on one hand, a linguistic knowledge and on the other, knowledge about the process to be controlled. The first type of knowledge is general, whilst the second depends on the process at hand and has to be changed if the NLI is to be applied to another process.

**Linguistic Knowledge.** The linguistic knowledge includes the dictionary and grammar. The dictionary is composed of all the words that the operator can use to communicate with the control system and is adapted to the context of physical processes. Thus the process is considered to be composed of components (pumps, tanks, pipes, ...) and elements (water, fuel, coolants, ...). There are also physical and chemical magnitudes associated to components and elements (level, flow, temperature, ...).

Words are entries for the dictionary and the output is their syntactical category and their associated meaning. For example, the word *valve* has the syntactical category *<noun>* and the associated meaning *<component,type valve>*. The word *open* has the syntactical category *<verb>* and the associated meaning *<order, type action, open>*.

The grammar is composed of a set of rules describing the language, that is describing how the correct phrases can be formed. The grammar is restricted to the environment of a control center. Two types of commands are recognised, those of action and of information. The first of these allows actions to be

carried out on the process and the second one for information to be obtained about it. Gazdar and Mellish (1984) have shown that Recursive Transition Networks (RTN) have many advantages to process natural languages. The grammar used, expressed as a RTN, is shown in Fig. 1, where $S$ = *Sentence*, $V$ = *Verb*, $NP$ = *Noun Phrase*, $DET$ = *Determiner*, $PP$ = *Prepositional Phrase*, $PREP$ = *Preposition*, $R$ = *Relational Clause*, $REL$ = *Relative*, $CNJ$ = *Conjunction*, $ID$ = *Element Identification* and $ADJ$ = *Adjective*.

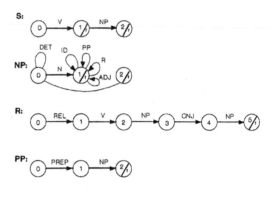

Fig. 1: Recursive Transition Network for the proposed grammar

This type of grammar simplifies the analysis, allowing real time processing, and can cope with most of the possible dialog between an operator and the control system.

**Knowledge about the process.** Knowledge about the process being controlled is necessary for extracting the meaning of the phrases introduced by the operators, that is, in order to perform a semantical analysis. The information about the process is composed of two types, fix and variable, of information. The first type corresponds to information such as component identification and type, component functional dependencies, inclusion relationships, process topology and actions that can be executed on components. The second type corresponds to information such as the state of the components and measurements of the different variables. The variable knowledge is kept fresh by the control system that reads the sensors located on the process as indicated in Fig. 2.

The dialog of an operator with the control system is composed of phrases telling the control system to give some information, or phrases telling the control system to perform some action on the process. These phrases can be a direct command to an element which is directly specified or specified by its relationship to other elements or systems, such as *open valve V4* or *open the inlet valve of the main deposit*. They can be direct commands to a set of elements specified by a condition such as being a part of a certain system such as *open all valves of the refrigeration system*. Some

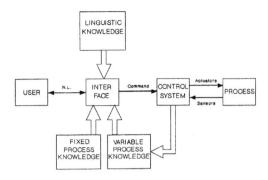

Fig. 2: Knowledge Bases diagram

times they may be imprecise commands (*slightly open the filling valve of the main deposit*) or goal oriented commands which require the execution of a series of elementary commands in order to be accomplished. All these orders are not, in general, completely specified. In some cases the phrases can be specified by using an anaphoric resolution method, in other cases this is not possible and a dialog with the operator is necessary.

## DEVELOPED PROTOTYPE

The ideas presented in the previous sections were implemented on a prototype of a NLI. The principal characteristics of the presented prototype are as follows:

- PROLOG (Clocksin and Mellish, 1984) was chosen as the programming language due to its being a declarative and recursive language, easy to use with symbols and structures and thus ideal for the development of natural language applications.

- The vocabulary is adapted to the context of physical processes. Thus the entries for the dictionary are words related to the process, such as components (pumps, tanks, pipes, ...), elements (water, fuel, coolants, ...), physical and chemical magnitudes associated to components and elements (level, flow, temperature, ...), and so forth.

- The grammar used corresponds to a subset of the Spanish Language, and has been implemented as a Recursive Transition Network (RTN), due to this parsing technique being clean, clear and efficient (Gazdar and Mellish, 1989).

- The grammar is restricted to the environment of a control center. Two types of commands are recognised, those of action and of information. The first of these allows actions to be carried out on the process and the second one for information to be obtained about it.

- The prototype uses a procedure for anaphoric resolution of phrases which are not completely specified.

The NLI works as follows: Once the operator iniciates the procedure by introducing a sentence a lexical analysis is first performed. That is, the sentence is broken into words and it is determined if all the words are in the dictionary and their syntactical category and meaning.

The second step is the syntactical analysis. A RTN parsing algorithm is used. Only phrases which are grammatically correct will pass this phase of the NLI. Part of the semantical analysis is done in this phase of the algorithm because, as has been was indicated before, it is assumed that the syntactical analysis breaks the sentence into parts which are semantically significant (this is known as the *principle of compositionality*).

The semantical analysis starts by instantiating the referred components to their respective values. This is accomplished in a bottom-up manner (Notice that syntactical analysis breaks the sentence in a top-down way whilst semantical analysis recomposes the phrase in the opposite direction).

Once the lexical, syntactical and semantical analysis have been performed, the operator's sentence has to be translated into a precise command to the system. The command may be executed, producing some changes upon the process, or may not be executed due to various reasons. In this last case the interface will show the operator a message explaining why the order cannot be executed.

The whole process can be generally described by the following algorithm:

```
begin
    Read operator's sentence
    while there is a sentence do
        Perform sentence lexical analysis
        Perform sentence syntactical analysis
        Perform sentence semantical analysis
        if the resulting command can be executed then
            execute the command upon the process
            write command result
        else
            write error message
        end if
    end while
end
```

## PROCESS DESCRIPTION

The developed prototype has been applied to the process shown in Fig. 3, which corresponds to a control laboratory pilot plant. The plant is composed of a heat exchanger and a deposit interconnected with a set of pipes. Different configurations may be obtained by changing the position of a set of manual valves. The main variables found in the process industry, temperatures, pressure, flows and levels may be controlled in the plant which is provided with a set of industrial sensors and actuators.

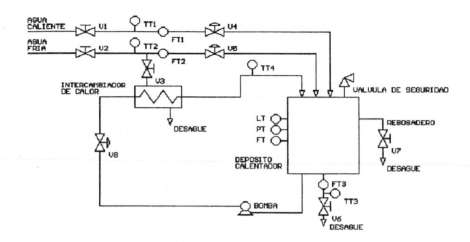

Fig. 3: Diagram of pilot plant

The elements composing the plant are: deposit, recirculation pump, heat exchanger, control valves, manual valves, heating element, pipes and sensors (pressure, level, flow and temperature). The process can be controlled by acting upon the controlled valves, the recirculation pump and the heating element.

The knowledge about the process is coded in a set of PROLOG predicates. Some of them are shown below:

- **component**: types of component existing in the process. For example, *component (PUMP)*.

- **system**: The whole process is composed of a series of systems. Each component is a part of one of these systems. For example, *system (RECIRCULATION_CIRCUIT)*.

- **is_a**: each component of the process has an associated identifier and belongs to a determined class of components. It also may be associated to other components. For example: *es_un (V4, VALVE, [ENTRY], [DC])*.

- **is_part_of**: indicates the system to which each element belongs: *is_part_of (BR, RECIRCULATION_CIRCUIT)*.

- **actions**: describe the possible actions on components that can be controlled by actuators: *actions (V4, [OPEN, CLOSE])*.

- **connect**: establishes the process topology, indicating the existing connections between the different components. In this case, the connections correspond to pipes. For example: *connect (T3, V4, DC)*.

- **measure**: this predicate describes the existing measures, indicating their identifier, the element to which it is associated and the corresponding type of magnitude. For example: *measure (LT, DC, LEVEL)*.

The predicates given above correspond to fixed information. Variable information is coded as shown by the following predicates:

- **state**: state of the elements that can be controlled. For example, *state (V4, OPEN)*.

- **failure**: A component may be in a faulty state. In this case it will not be possible to actuate upon it: *failure (V5)*.

- **measure_value**: values associated to the variables described with the measure predicate. For example, *measure_value (LT, 50)*.

The initial situation of the system is described by a determined combination of all these predicates. The action orders (OPEN, CLOSE, ACTIVATE, ...) may change the state while other orders (LOCK, UNLOCK) may change the failure conditions of the elements.

Commands accepted by the NLI prototype can be classified in one of the following types:

- Direct commands upon one element as:

    - open V4.
    - open valve V4.
    - open entry valve of main deposit.
    - open the valve connecting pipes T2 and T3.

Notice that these commands are associated with direct actions upon the process. Some commands may demand information retrieval:

    - show flow of T8.
    - show flow of pipe T8.
    - show flow of the entry pipe of the main deposit.
    - show flow of the pipe connecting the heat exchanger to the main deposit.

Notice that all these orders are referred to a single element that can be addressed in various ways: directly by its identifier, by describing its relationship to other elements, by indicating the system to which it belongs, or by indicating its topological situation. All these orders will originate an action once the element which it refers to has been precisely identified.

- Direct commands on a set of elements. Certain commands may refer to a determined set of elements of the same type. These elements can be determined by a relationship such as being members of the same system, being related to another element or connecting other components. As an example consider the following commands:

  - open all valves.
  - open all valves of the cold water circuit.
  - open all outlet valves of the main deposit.
  - open all valves connecting the main deposit to the heat exchanger.

  These types of commands may also be found when demanding information:

  - show flows in all pipes.
  - show flows in all pipes of the recirculation circuit.
  - show flows in all entry pipes of the main deposit.
  - show flows in all valves connecting the main deposit to the heat exchanger.

- Goal oriented commands. All commands described previously are direct, that is, they involve a simple action upon a determined element or set of elements. Goal oriented commands correspond to a set of actions upon different elements in order to achieve a goal. For example:

  - Heat the main deposit.
  - Activate recirculation circuit.

  Notice that goal oriented commands require a plan of simple actions to be executed. This plan of actions must be a part of the knowledge base.

All commands which have been described before refer to the NLI input. Regarding the NLI output, three cases can be distinguished:

1. The command is correct and can be carried out. In this case a message is displayed with the result of the action taken (describing the new state of the elements) or showing the information demanded.

2. The command is correct but cannot be executed because of different reasons:

   - The state of the element is incompatible with the command.
   - The command cannot be applied to the specified element.
   - The demanded information is not available.
   - The specified element is not operative.
   - There are no elements which fit the determining conditions.

- The element does not correspond to the specified type.
- The element is not completely specified.

3. The command is not correct. This is detected because the sentence fails in the lexical (containing unknown words), syntactical (the sentence cannot be generated with the grammar) or semantical (the sentence is syntactically correct but has no meaning) analysis. The appropiate message is generated in every case.

## APPLICATION EXAMPLES

In order to illustrate the types of commands that the system is capable of processing and understanding, an example of a dialog with the NLI is presented. The control system and process referred to are simulated. The original dialog in Spanish is given in italics with the responses of the NLI in capital letters.

- *cerrar la válvula de entrada del depósito principal*
  (close entry valve of main tank)

  *-> VALVULA V4 CERRADA*
  (VALVE V4 CLOSED)

- *abrir la válvula que comunica la bomba de recirculación con el intercambiador de calor*
  (open valve connecting recirculation pump to heat exchanger)

  *-> VALVULA V8 ABIERTA*
  (VALVE V8 OPEN)

- *activar la bomba de recirculación*
  (activate recirculation pump)

  *-> BOMBA BR ACTIVADA*
  (PUMP BR ACTIVATED)

- *mostrar las temperaturas de todas las tuberías del circuito de recirculación*
  (show the temperatures of all pipes in the recirculation circuit)

  *-> EL VALOR DE LA MEDIDA DE TEMPERATURA EN TUBERIA T10 ES 25°C*
  (THE VALUE OF THE TEMPERATURE MEASURE IN PIPE T10 IS 25°C)
  *-> NO EXISTE MEDIDA DE TEMPERATURA EN LA TUBERIA T11*
  (THERE IS NO FLOW MEASURE IN PIPE T11)
  *-> NO EXISTE MEDIDA DE TEMPERATURA EN LA TUBERIA T12*
  (THERE IS NO FLOW MEASURE IN PIPE T12)
  *-> NO EXISTE MEDIDA DE TEMPERATURA EN LA TUBERIA T13*
  (THERE IS NO FLOW MEASURE IN PIPE T13)

- *mostrar la temperatura del depósito principal*
  (show the temperature of main tank)

  *-> EL VALOR DE LA MEDIDA DE TEMPERATURA EN DEPOSITO DC ES 25°C*
  (THE VALUE OF THE TEMPERATURE MEASURE IN TANK DC IS 25°C)

- *mostrar el nivel*
  (show level)

-> *EL VALOR DE LA MEDIDA DE NIVEL EN DEPOSITO DC ES 50 %*
(THE VALUE OF THE LEVEL MEASURE IN TANK DC IS 50 %)

- *abrir la válvula de salida*
  (open exit valve)

  -> *LA ACCION ABRIR NO ES APLICABLE AL COMPONENTE V6*
  (OPEN ACTION CANNOT BE APPLIED TO COMPONENT V6)

- *cerrar todas las válvulas del circuito de recirculación*
  (open all valves in the recirculation circuit)

  -> *VALVULA V8 CERRADA*
  (VALVE V8 CLOSED)

- *desactivar todas las bombas*
  (deactivate all pumps)

  -> *BOMBA BR DESACTIVADA*
  (PUMP BR DEACTIVATED)

- *cerrar la válvula que conecta las tuberías T2 y T3*
  (open valve connecting pipes T2 and T3)

  -> *LA VALVULA V4 YA ESTA CERRADA*
  (VALVE V4 IS ALREADY CLOSED)

- *abrir la válvula V5*
  (open valve V5)

  -> *LA VALVULA V5 ESTA EN ESTADO DEFECTUOSO*
  (VALVE V5 IS IN A FAULTY STATE)

- *desbloquear V5*
  (unlock V5)

  -> *VALVULA V5 DESBLOQUEADA*
  (VALVE V5 UNLOCKED)

- *abrir la válvula V5*
  (open valve V5)

  -> *VALVULA V5 ABIERTA*
  (VALVE V5 OPEN)

- *mostrar el nivel del depósito principal del circuito de agua caliente*
  (show level of the hot water circuit tank)

  -> *EL VALOR DE LA MEDIDA DE NIVEL EN DEPOSITO DC ES 50 %*
  (THE VALUE OF THE LEVEL MEASURE IN TANK DC IS 50 %)

- *activar el depósito principal*
  (activate main tank)

  -> *LA ACCION ACTIVAR NO ES APLICABLE AL COMPONENTE DC*
  (ACTIVATE ACTION CANNOT BE APPLIED TO COMPONENT DC)

- *calentar el depósito principal*
  (heat main tank)

  -> *VALVULA V5 CERRADA*
  (VALVE V5 CLOSED)
  -> *LA VALVULA V8 YA ESTA CERRADA*
  (VALVE V8 IS ALREADY CLOSED)
  -> *LA BOMBA BR YA ESTA DESACTIVADA*
  (PUMP BR IS ALREADY OFF)

## CONCLUSIONS

A model for a natural language interface of control centers has been proposed. A prototype of a NLI which accepts Spanish sentences corresponding to most possible dialogs between operators and control centers has been developed in PROLOG and has been applied to a pilot plant. The NLI could be greatly enhanced by introducing other techniques such as temporal reasoning in the explanation mechanism, and automatic planning and learning for treating goal oriented commands.

## ACKNOWLEDGEMENT

The authors would like to acknowledge CICYT for funding the work under grant ROB89-0614.

## REFERENCES

Clocksin, W.F., and C. Mellish (1984), *Programming in Prolog*, 2nd. Edition, Springer Verlag.

Gazdar, G., and C. Mellish (1989), *Natural Language Processing in Prolog*, Addison Wesley.

Kasturi, R., R. Fernández, M.L. Amlani, and W. Feng (1989). *Map Data Processing in Geographic Information Systems*. Computer, Dec. 1989, 10–21.

Kosy, D.W. (1989) *Application of Explanation in Financial Modeling*. In Widman, Loparo and Nielsen (Ed.), *Artificial Intelligence, Simulation and Modeling*, Wiley Interscience.

Martin, T., J. Kivinen, J.E. Kijnsdorp, M.G. Rodd, and W.B. Rouse (1990). *Appropiate Automation – Integrating Technical, Human, Organizational, Economic, and Cultural Factors*. Preprints 11th IFAC World Congress, vol. 1, 47–65.

Myers, B.A. (1989). *User–Interface Tools: Introduction and Survey*. IEEE Software, Jan. 1989, 15–23.

Niemann, H., G.F. Sagerer, S. Schröder, and F. Kummert (1990). *ERNEST: A Semantic Network System for Pattern Understanding*. IEEE Transactions on Pattern Analysis and Machine Intelligence, Sep. 1990, 883–905.

Rich, S.H., and V. Venkatasubramanian (1987). *Model–Based Reasoning in Diagnostic Expert Systems for Chemical Process Plants*. Comput. chem. Engng. vol. 11, No. 2, 111–122.

Rodríguez Hontoria, H. (1989). *GUAI: un generador automático de interfases en lengua natural*. Univ. Politécnica Cataluña.

Romano, J.M.G., Ternero, J.A., and Camacho, E.F. (1991). *Natural Language Interface for Process Control Centers*. Preprints 3rd IFAC International Workshop on Artificial Intelligent in Real Time Control.

Selfridge, M. and Vannoy, W. (1986). *A Natural Language Interface to a Robot Assembly System*. IEE Journal of Robotics and Automation, vol. RA-2, No. 3, 167–171.

# QUALITATIVE SUPERVISION OF NAVAL DIESEL ENGINE TURBOCHARGER SYSTEMS

**J.M. Marchal\* and E.F. Camacho\*\***

*\*Esc. Superior de la Marina Civil, Universidad de Cadiz, Spain*
*\*\*Escuela Superior de Ing. Industriales, Universidad de Sevilla, Spain*

Abstract This paper presents a qualitative model the diesel engine turbocharger system of a ship. The paper also shows how qualitative models can be use for an intelligent monitoring of the process concerned.

Keywords: Qualitative modelling, causal propagation, supervision, engine monitoring.

## INTRODUCTION

One of the most useful fields of application of expert systems is intelligent monitoring of complex processes. Intelligent monitors are supposed to assist control center operators by performing, among other things, the following functions:

1. Presenting relevant information about the present state of the process to the operator.

2. Diagnosing faults, if any, that lead the process to its present state.

3. Predicting possible future states or faults if certain actions are (are not) taken.

4. Giving advice about possible actions to be taken.

All these functions need analysis and interpretation of sensor data to determine their meaning in order to explain what is (or may be) taking place in the process. It is clear that this type of interpretation must be based on a profound knowledge of the process. Two things are needed for this:

- A representation of the knowledge of the process.

- The possibility of reasoning with this knowledge.

The knowledge of a process can be represented by heuristic rules and/or by models (Travé-Masuyés,1990). The interest of models for knowledge representation is their inherent possibilities for reasoning (Davis and Hamscher,1988). This must comprise not only knowledge of the separate parts, but also of how they are connected and about how they work together.

Object-oriented programming languages are good tools for representing this type of knowledge. Concepts of parts, components and their relationships are easily coded in these types of languages. On the other hand, Qualitative Simulation (Kuipers,1986) seems to be the appropiate technique to perform the causal reasoning needed in some of the functions mentioned above.

This paper presents a qualitative model of a ship diesel engine system. The paper also shows how Qualitative Modelling can be used for an intelligent monitoring of the process concerned. Artificial Intelligence Techniques have been applied to ship engine monitoring in the past (Katsoulakos et al., 1989), although the approach used was *rule based* and did not consider qualitative behaviour. The main objective of this work is to develop qualitative behaviour models of the main engine a ship that could be used for monitoring, failure detection, diagnosis, prediction and instruction.

The system developed has been implemented in *SMALLTALK* (Goldberg and Robson,1984), which is a general purpose object-oriented language and thus allows Qualitative Simulation and time causal reasoning to be integrated into a more general reasoning system.

The paper is organized as follows. Section 2 describes the process concerned and the model developed. Section 3 is dedicated to the qualitative propagation algorithm while section 4 explains how the object - oriented approach has been used. Some simulation runs showing malfunction detection are presented in section 5. Section 6 is dedicated to presenting some concluding remarks.

## MODEL DESCRIPTION

The engine room of a ship considered is composed of the following subsystems: The main engine, the main engine turbocharger, the auxiliary engine, the lubricating subsystem, the salt water system, the combustion feeding system, the compressed air system etc.. The engine room of a ship is a fundamental part of the same and has, therefore, to be continuously supervised.

The process considered, see figure 1, corresponds to

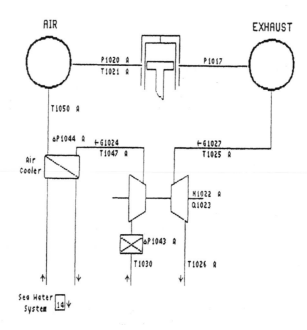

AIR                                                    EXHAUST

P1020 A        P1017
T1021 A

T1050 A

ΔP1044 A   ←G1024        ←G1027
           T1047 A        T1025 A

Air
Cooler

                                        H1022 A
                                        Q1023

                        ΔP1043 A

                T1030       T1026 A

Sea Water
System   14

Figure 1: Ship main engine turbocharger system

a ship main engine turbocharger system. Naturally aspirated engines draw air of the same density as the ambient atmosphere and this density determines the maximum weight of fuel that can be burned in the cyclinders and therefore the maximum power obtained. If the air density is increased, by a compressor, the amount of air is increased and more weight of fuel can be effectively burned and the power developed also increases. This procedure is implemented in most modern diesel engines by using exhaust gas turbocharging where exhaust gases are used to power the compressor. A substancial amount of the total heat energy is wasted to the exhaust gases, and although it is relatively inexpensive to drive the compressor directly from the engine by gear, an increase in power is obtained by using the exhaust gases to drive the compressor.

The inlet air is filtered and goes through the compressor. As the temperature of the inlet air after being compressed is too high to go into the cylinders, it has to be cooled down. This is accomplished by an air cooler using sea water as a coolant. Some of the surplus energy of the exhaust gases is used to power the turbine coupled with the compressor as indicated before.

Temperatures of gases in ship diesel engines are very valuable sources information for monitoring their conditions. A model of the behaviour of the gases was considered to be a good tool for the supervision of ship engine rooms. In this sense, a qualitative model of the turbocharger subsystem was developed. A modular and hierarchical decomposition of the system was established. This way of representing the system adapts to physical reality, topology, the operator's mental models and allows for easy generalization when representing the global complexity of a ship engine room.

The main parts of the model presented are:

- Filters
- Air Cooler
- Turbine-compressor
- Cylinders
- Receiver

The main variables taken into account by the model are: cylinder inlet air pressure and temperature, air-cooler inlet air temperature and pressure drop, seawater inlet temperature, turbine and compressor temperatures and exhaust pressure and temperatures.

All these components have been modelled according to Kuipers (1984) although the idea used for their aggregation in order to form the the *system* is nearer to the component ontology used by De Kleer (1984). The same applies to the concepts of connections, causality and heuristics used.

The models have been obtained using physical laws and heuristic rules, given by the experts. The heuristic rules are used to resolve the ambiguities originated in the qualitative simulation. Notice that the models given corresponds to quasi stationary conditions.

Fig 2 shows the model of the air cooler. The difference ($DtTaw$) of inlet air temperature ($TiE$) and sea water temperature ($Tswi$) multiplied by the heat transfer coefficient determines the heat flow ($QE$). The heat flow is also related to the air flow ($GairE$) and the difference ($DtTE$) between inlet air temperature ($TiE$) and outlet air temperature ($ToE$). The air flow is also related to the difference ($DtPE$) between the inlet air pressure ($PiE$) and outlet air pressure ($PoE$) as indicated.

The qualitative model of the cylinders can be seen in Fig. 3 The air flow ($Gair$) multiplied by the fuel to air ratio ($Rm$) will determine the oil consumption ($Gcb$),

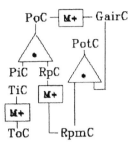

Figure 4: Compressor Model

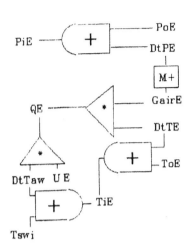

Figure 2: Cooler Model

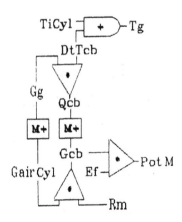

Figure 3: Cylinder Model

which in turn if multiplied by the cylinder efficiency ($Ef$) will generate the output power ($Pot$). The exhaust gas flow ($Gg$) is related to the air flow and the difference ($DtTcb$) of the gases ($TgCyl$) and air inlet temperature ($TiCyl$) as shown.

Figure 4 shows the qualitative model of the compressor. The power ($PotC$) depends of the product of the compressor speed ($RpmC$) and the air flow ($GairC$) through the compressor, which in turn is related monotonically to the compression ratio ($RpC$). Inlet ($PiC$) and outlet ($PoC$) air pressures are related by the compression ratio. Inlet ($TiC$) and outlet ($ToC$) air temperatures are also related monotonically.

## QUALITATIVE PROPAGATION BETWEEN COMPONENTS.

We consider a system to be a collection of interconnected components and the interaction between them to be a form of directional causality. Thus the behaviour of the system is obtained from the behaviour of each component and the connections between them (Williams,1990).
The qualitative simulation (Kuipers 1984;Kuipers 1986) begins with the propagation of the known information, or known disturbances, to the system through the constraints, in order to complete the description of the direction of change for each variable, at a given time-point.
The algorithm for the propagation of a disturbance in a component to the whole system is composed of the followings steps:

1 Set initial conditions.

2 Form a queue of components to be propagated (QCP) with the component which caused disturbances.

3 While queue QCP is not empty repeat:

    3.1 Get the first component of the queue QCP. Name this the active component. Remove it from queue QCP.

    3.2 For the active component propagate through the constraints of its qualitative models the information known about it.

    3.3 If there is any variable with ambiguity at the end of the propagation process use heuristic rules to solve it and propagate the results.

3.4 Get the possible states for active component and filter them by testing the consistency with the neighboring components in the system, using topology.

3.5 Using the connections apply causality to propagate the changes in the shared variables with other components and add these components to the QCP queue.

## OBJECT-ORIENTED APPROACH.

An object-oriented approach has been used to implement the model, to propagate disturbances and for qualitative simulation. This approach has the following advantages:

- modularity, with its possibility of description of complex systems.

- hierarchy of models and levels of description.

- good framework for the natural description of components.

- easy developement of prototypes.

A version of the QSIM algorithm for qualitative simulation, proposed by Kuipers (Kuipers 1986), has been implemented in SMALLTALK-V2 language.
T developed are the following:

A brief description of the more important classes developed is given in the following:

**System Class.** This establishes the set of component models and connections which define the system. It expresses the topology of the system, the aggregation of the components and the hierarchical organisation between them.

**Model Class.** This contains the set of variables and constraints which caracterize the structure of the model of a component.

**Conection Class.** This class expresses the causality and the link between components. It supports the continuity and compatibility conditions (conservation of matter for fluid systems, conservation of energy for thermal system,..) (De Kleer and Brown, 1984).

**Variable Class.** The objects of this class are variables, continuous real-valued functions of time, with a qualitative value and tendency, an ordered value space and a set of possible transitions from their current qualitative state.

**Constraint Class.** This contains a hierarchy of the different kind of constraints between variables.

**Transition Class.** This class describes the two types of qualitative state transitions for a variable:

P-transitions,moving from a point to an interval, and I-transitions, moving from an interval to a point.

**Tuple Class.** Each object of this class contains the set of pairs and triples generated for each constraint with the transitions of its arguments.

**State Class.** Each object of this class is the qualitative state of a model and its qualitative time description ( interval or time-point).

**Interpretation Class.** The object of this class are the possible new qualitative states resulting from applying the QSIM algoritm.

**Prediction Class.** This class contains the objects resulting from qualitative simulation. They are composed of an initial state and the set of states the model can reach until a distinguished time-point.

## APPLICATION EXAMPLE

As an application example the effect of a cooler soiling process on the diesel engine is analyzed. The engine speed is considered to be constant and the air filter is not taken into account for the sake of clarity.

The initial value of each variable is set by a landmark (written between brackets), which defines a qualitative value for the time points or by a pair of landmarks for the time intervals. Thus, the air flow in the cylinders $GairCyl$ get a landmark value ($Gairt$) corresponding to to the full ahead engine regime. This landmark is placed in the following ordered space:

$$[ zero....Gairst.....Gairt....+inf ]$$

where: $Gairst$ is the minimal air flow to ensure a complete combustion. So if the air flow is less than $Gairst$ the perfomance of the engine decreases and a different qualitative region is reached.
We consider a full ahead quasi-static engine regime. The engine speed, the efficiency of the combustion process, the efficiency of the turbine, the input sea water temperature and the input compressor temperature are supposed to be constant for this qualitative state. This is a common situation. The air cooler soiling process is represented by assigning a decreasing tendency to the coefficient of heat transfer.

The initial situation of the air cooler is given by:

| Variable | Name | Value | Tend. |
|---|---|---|---|
| Inlet air temp. | $TiE$ | $(ToCt)$ | std |
| Inlet sea water temp. | $Tswi$ | $(Tswi0)$ | std |
| Heat transfer coeff. | $UE$ | $(UE0)$ | dec |
| Inlet air pressure | $PiE$ | $(PoCt)$ | std |

From this situation it can be inferred by propagation that the heat flow in the cooler decreases. By applying the heuristic :

*if the heat flow QE changes, then the temperature difference DtTE in the cooler will change in the same direction*

Thus, we obtain:

| air temp. diff. | DtTE | (DtTE0) | dec |
|---|---|---|---|

that being propagated implies:

| cooler outlet air temp. | ToE | (ToEt) | inc |
|---|---|---|---|

Within this situation there are three possible states for the cooler that correspond to the ambiguity in air flow GairE.

a)

| GairE | (Gairt) | inc |
|---|---|---|
| PoE | (PoEt) | dec |
| ToE | (ToEt) | inc |

b)

| GairE | (Gairt) | std |
|---|---|---|
| PoE | (PoEt) | std |
| ToE | (ToEt) | inc |

c)

| GairE | (Gairt) | dec |
|---|---|---|
| PoE | (PoEt) | inc |
| ToE | (ToEt) | inc |

After checking the consistency of each possibility in the receiver model it is noted that c) is the only one that is consistent with it.
The output variables of the cooler (pressure *PoE*, temperature *ToE* and air flow *GairE*) are inputs for the receiver model, and this implies:

| air density | Dair | (Dairt) | dec |
|---|---|---|---|

The cylinder variables already defined, for being shared with the receiver are:

| cylinders inlet air temp. | TiCyl | (ToEt) | inc |
|---|---|---|---|
| air flow | GairCyl | (Gairt) | dec |

Taking into account that for this qualitative region the efficiency of the cylinder is kept constant as well as the power PotM, the cylinder model implies:

| exhaust gas flow | Gg | (Ggt) | dec |
|---|---|---|---|
| exhaust gas temp. | Tg | (Tgt) | inc |

In the turbine we can deduce that the power PotT decreases by applying, apart from the qualitative model, the heuristics:

a) *if the exhaust gas flow changes, then the power will change in the same direction.*
b) *if the exhaust gas temperature changes, then the temperature difference between input and output in the turbine will change in the same direction.*

The latter is of the KVL type of heuristic proposed by De Kleer (De Kleer 1984)
In the compressor the variables which have already been stated by their connections with the cooler and the turbine are:

| air flow | GairC | (Gairt) | dec |
|---|---|---|---|
| power | PotC | (PotTt) | dec |

The input temperature is considered constant and, for the supposed regime, the influence of a change in the compression ratio *RpC* upon the output temperature *ToC* has been excluded. In the compressor we have applied the heuristic:

*if the power PotC changes, then the air flow and the speed RpmC will change in the same direction.*

The compressor model implies:

| comp. outlet air pressure | PoC | (PoCt) | dec |
|---|---|---|---|

The compressor-cooler connection implies that the input pressure of the latter *PiE* changes its initial tendency of (*std*) to *dec*, so that it is necessary to propagate with the new tendency, and apply a new heuristic:

*if the input cooler pressure PiE changes, then the output cooler pressure will change in the same direction.*

With this heuristic a new tendency (*dec*) is obtained for the pressure in the receiver (it was *inc* before) that is consistent with its qualitative model and that does not produce new changes in the rest of the components. To summarize, the new situation of the main engine turbo-charger system can be characterized by the following variables:

| air receiver pressure | PCol | (PoEt) | dec |
|---|---|---|---|
| turbine speed | RpmT | (RpmTt) | dec |
| exhaust gas temp. inlet turbine | TiT | (Tgt) | inc |
| exhaust gas temp. outlet turbine | ToT | (ToTt) | inc |
| comp. inlet air temp. | TiC | (Tsm) | std |
| cooler inlet air temp. | TiE | (ToCt) | std |
| cooler outlet air temp. | ToE | (ToEt) | inc |
| air receiver temp. | TCol | (ToEt) | inc |
| air flow | Gair | (Gairt) | dec |
| cooler diff. pressure | DtPE | (DtPEt) | dec |

Thus, we have an initial state for the qualitative simulation of all the components. As we are at a time-point the next qualitative state will be a time-interval, where the variables will be reach their next landmarks (for example *Gairst* for the air flow *Gair*). The new qualitative states will appear and we must change some qualitative models (for example the cylinder model if

Table 1: Numerical Results Obtained with the Simulator

| | | clean cooler | dirty cooler |
|---|---|---|---|
| air receiver pressure | $PCol$ | 1.603 bar | 1.410bar |
| turbine speed | $RpmT$ | 6535 rpm | 6039 rpm |
| exhaust gas temp. inlet turbine | $TiT$ | 360.8 °C | 420.8 °C |
| exhaust gas temp. outlet turbine | $ToT$ | 269.8 °C | 329.8 °C |
| compressor inlet air temp. | $TiC$ | 36.7 °C | 36.9 °C |
| cooler inlet air temp. | $TiE$ | 129.7 °C | 130.8 °C |
| cooler outlet air temp. | $ToE$ | 35.9 °C | 62.9 °C |
| air receiver temp. | $TCol$ | 43.7 °C | 71.2 °C |
| air flow | $Gair$ | 14.02 ton/h | 11.81 ton/h |
| cooler diff. pressure | $DtPE$ | 128 mmW | 91 mmW |

the air flow *Gair* reaches the *Gairst* landmark). However, in monitoring tasks on ships these qualitative changes are not usually allowed to happen, and it is more important to identify the previous disturbance process.

Some numerical simulation tests were carried out in the DPS 100 of NORTHCONTROL trainning simulator in order to in order to verify the results obtained by the qualitative simulation. The results of the numerical simulation are indicated in table I. The information given in the table corresponds to the numerical values of the main variables, in a quasi stationary regime, when working with a clean and a dirty cooler. As can be seen, the numerical results agree with the qualitative results given before.

## CONCLUSIONS

Qualitative modelling is a good alternative for tasks as supervision, diagnosis, instruction, and, in general those that can be aided by techniques based upon knowledge.
The application of heuristic rules would has shown to be very useful in reducing the ambiguities of the qualitative models. Several examples have been presented. In the same way the relationships between compatibility and consistency, that can be established among the components, through the connections, reduces these ambiguities.
The object oriented approach has shown to be quite adequate to represent complex systems, and a structure of classes for qualitative simulation has been proposed.

## ACKNOWLEDGEMENT

The authors would like to acknowledge CICYT for funding this work.

## REFERENCES

Davis R. and W. Hamscher (1988) *Model-based reasoning: troubleshooting*, in H.E. Shrobe (Ed.). Exploring Artificial Intelligence, Chap.8.Morgan Kaufmann, San Mateo, California, pp 297-347.

De Kleer J.(1984). *How circuit work*, in D.G. Bobrow (Ed.) Qualitative Reasoning About Physical Systems. North-Holland, Amsterdam, pp 205-280.

De Kleer J. and J.S Brown (1984). *A qualitative physics based on confluences*, in D.G. Bobrow (Ed.) Qualitative Reasoning about Physical Systems. North-Holland, Amsterdam, pp 7-84.

Goldberg A.] and D. Robson (1984) *Smalltalk-80: The language and its Implementation*. Reading, MA. Addison-Wesley.

Katsoulakos P.S., J.Newland, J.T. Stansfield and T. Ruxton (1989) *Monitoraggio, Raccolta Dati e Sistemi Esperti per la Diagnostica delle Avarie di Machina*, Tecnologie per il Mare, L'Automazione Navale, pp 38-42, march.

Kuipers B. (1984) *Commonsense reasoning about causality: deriving behaviour from structure*, in D.G. Bobrow (Ed.) Qualitative Reasoning About Physical Systems. North-Holland, Amsterdam. pp 169-204.

Kuipers B. (1986) *Qualitative simulation*, Artificial Intelligence 29, 289-338.

Travé-Masuyés L.,A. Missier and N.Piera (1990) *Qualitative models for automatic control process supervision*, Proc. of the 11$^{th}$ IFAC World Congress, Vol. 7, pp 198-203, Tallin, Estonia.

Williams B.C. (1990) *Temporal qualitative analysis: explaining how physical systems work*, in D.S. Weld and J. de Kleer (Ed.). Readings in Qualitative Reasoning About Physical Systems, Chap.2. Morgan Kaufmann, San Mateo, California. pp 133-177.

# QUALITATIVE/QUANTITATIVE MODELS DESCRIPTION LANGUAGE

## V. Thomas-Baudin and J. Aguilar-Martin

*LAAS-CNRS, 7 Avenue du Colonel Roche, 31077 Toulouse Cedex, France*

**Abstract.** The aim of this research is to built a language for quantitative/qualitative representation in expert supervisory process control. The specification of such a language led to the definition of ALCMEN (Automaticians Language for Causal Modelisation of Expert kNowledge). Its purpose is to represent in a qualitative way imprecision and uncertainty, as well as classical numerical equations, in order to solve the communication problem between process experts and control engineers; it appears as a network of interconnected blocks and a list of structured variables.

Two principal choices characterize the structure of ALCMEN:

    1) A causal network is the underlying structure of any model represented using ALCMEN.

    2) Qualitative variables have as semantic reference subsets of quantitative spaces connected to the physical variables: i.e. absolute referenced magnitudes, not relative order of magnitude.

A first approach for the mixed (qualitative/quantitative) Knowledge representation has been programmed using the Real Time Expert system development tool G2. It has been applied to the modelization of a discontinuous process in the framework of the Esprit project IPCES (Intelligent Process Control by means of Expert Systems). The definition of qualitative operations as well as the organization of the different workspaces will be described on this example.

**Keywords.** Knowledge representation, mixed quantitative/qualitative models, expert supervisory process control, artificial intelligence.

## ALCMEN-Language    global description.

ALCMEN, i. e. Automaticians Language for Causal Modelisation for Expert kNowledge is an attempt to build a language able to handle some imprecision and uncertainty,.... but in a precise manner, and then to solve the communication problem between Process Experts and Control Engineers.

The purpose of ALCMEN is to give a sound foundament for the description of processes enhancing the modelization of causal relations, as well as to develop a practical computer tool for representing processes from qualitative and quantitative descriptions by experts. It takes advantage of the features of an object-oriented computer language.

ALCMEN appears as a network of interconnected blocks and a list of structured variables such that they are related through causal relations.

Each block exhibits causes, effects , and the variables assigned to those roles may be qualitative i.e. symbolic, or quantitative i.e. numerical, they can be conditionned by distiguished variables called conditions or "parameters".

Each causal relation represents the knowledge proposed by the expert, it can be given either by equations, qualitative tendencies or constraints, equilibriums, etc ......

## Basic elements description.

The basic concepts of this language are: variables, relations and blocks:

- A *BLOCK* is a causal sub-graph composed by interconnected qualitative and quantitative relations, static or dynamic, with 3 associated lists of variables, inputs, outputs and conditions.

- A *VARIABLE* is a structured object, its attributes are:
  - <u>type</u> : symbolic, numeric or mixed (both).
  - <u>rank</u> : the set of its possible or attainable values
  - <u>disinguished subsets</u> of values : target values, acceptable and unacceptable values, forbiden values.
  - numerical-symbolic and symbolic-numerical <u>conversion</u>, if its type is "mixed"
  - the list of Blocks where it is an <u>input</u>
  - the list of Blocks where it is an <u>output</u>
  - the list of Blocks where it is a <u>condition</u> parameter
  - A *RELATION* is an elementary block, that cannot be decomposed. The following types of relations are considered:

  - static qualitative
  - dynamic qualitative
  - dynamic quantitative
  - numeric-symbolic converter
  - symbolic-numeric converter

The relations can be chosen in a pre-established set of standard relaions or programmed on purpose.

With the informations given in the definitions of variables and blocks, a graph is built by the structural analyzer of ALCMEN and analyzes its coherence with respect to the connections, the type of the variables and the completude of the graph; e.g. if a variable appears only as input of blocks and is not one of the external causes of the graph, an error message must appear, if another variable is a symbolic output in one block and a numeric output in another, there must be a symbolic-numeric converter between the two blocks.....

## Causal concepts.

**ALCMEN** captures the causal influence of a CAUSE on an EFFECT and provides the possibility of parametrizing that relation by one or many CONDITIONS.

The exact representation and functon of CONDITION variables is left to be actualized for each causal relation, according to the most explicit available knowledge, its description may either use natural language ( French, English , ...) or mathematical language ( formulae, boolean , ...).

*RELATIONS* may be defined as mathematical formulae if possible, nevertheless as ALCMEN pretends to complete classical simulation languages by representing uncertainties or undeterministic knowledge by mixing Qualitative and Quantitative knowledge, it is proposed,of course, that the Qualitative features are to be used only if mathematical relations are not suitable or are unknown.

Numerical relations are equations in the static case and time recurrent equations in the numerival case.

Causal relations can be SISO (single input-single output) or MISO (multiple input-single output)

The four basic SISO relations are (fig 1):

    1) increasing function: when x increases y increases
    2) decreasing function: when x decreases y decreases
    3) maximum de-tuning function: when x moves off its nominal value, y decreases
    4) minimum de-tuning function: when x moves off its nominal value, y increases

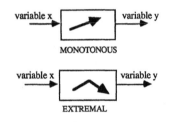

<u>figure 1</u>

For the de-tuning function the nominal or extremal value must be given, and CONDITION may act on this extremal value.

It may be noticed that this 4 relations could be mathematically reduced to 2: *monotonous* and *extremal*.

## Combination of qualitative variables.

The MISO relations represent the simultaneous effect of two or more causes. It is a situation clearly solved for analytical relations but not for qualitative environments. Similarly to the SISO Relations, it is attempted to give a typology. For that the MISO relations are decomposed into SISO relations and *COMBINATION*.

The present state of ALCMEN uses as a basis for combinations the connective functions of fuzzy logic called t-norms, applied to multi-valued symbolic variables. The connection relations so constructed are associative and enable the conection of any number of variable by iterating the function defined for only 2 inputs. In figure 3 are represented the 2 most important combination functions, related to fundamental non standard logics.

Qualitative variables are here assumed to take their values in ordered sets of linguistic labels, represented in figure 2 by 5 grey levels.

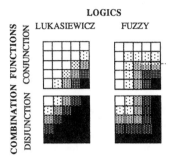

figure 2

Another class of combination functions is directly related to arithmetic operations. Let us assign to any qualitative (symbolic) variable its index i.e. the ordinal of the linguistic label representing the value of the variable in its ordered set. Then the *DIFFERENCE* between symbolic variables is the integer, positive or negative, obtained by the difference of their indexes. Consequently the *SUM* of a symbolic variable and an integer returns a symbolic variable such that its index is the minimum between the sum index+ integer and the maximum (or minimum) index in the set of labels..

### Dynamic qualitative relations:

Given a sampled time base, the most elementary dynamical relation is the unit delay. As in sample data numerical systems, the dynamics are therefore represented by unit delays with feedback. The delay operator D realizes the memorization of a qualitative value of a variable during the unit sampling interval, and DD is its inverse $\nabla = D^{-1}$: so $X(t+1) = D\ X(t)$ and $X(t) = \nabla\ X(t+1)$

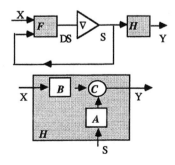

figure 3

As shown in figure 3, a qualitative dynamic system can therefore be writen with the two equations: $Y = H(S)$ and $DS = F(S, X)$, and it can be split into 3 SISO relations and one connective as shown in figure 4.

Tendencies are special causal relations that act on differences, they use therefore the arithmetic combination functions, as shown in figure 4

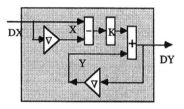

figure 4

### G2 prototype-0 for ALCMEN.

In the IPCES project (Intelligent Process Control by means of Expert Systems), we have a process specification, described by large number of physical constants, process parameters as well as very many qualitative and quantitative model specifications. In this paper, we take some of those sub-models showing the use of ALCMEN.

One of the IPCES applications is the Philips phosphor screening process. We just pick up a subset of quantitative model specification in the
flowcoating green. For each relation, the name of the block is given between brackets. We have a relation describing the evolution of the thickness of sediment layer:

$$hs(t) = \int_{ts1}^{t} \frac{dhs}{dt}(t)\ dt$$

where :
ts1 : start of settling
    dhs/dt : rate of change of thickness of sediment layer due to settling[rate_hs]

$$\frac{dhs}{dt} = \frac{as}{9} * \frac{\dfrac{\rho_p * r^2}{\rho_s(t)-1) * Ks(t)}}{\dfrac{F}{(Vp(t)-1) * g * \cos\alpha(t)}}$$

    as : grain shape factor in Strokes' formula (parameter)
    $\rho_p$ : specific density of green phosphor (parameter)
    $\rho_s$ : specific density of suspension [spec_den]

213

$\rho_s(t) = Vp(t) * \rho_p + [1-Vp(t)] * \rho_f(t)$

$Vp(t)$ : phosphor volume fraction in suspension at time t

F : packing density of phosphor grain (parameter)

$\rho_{f(t)}$ : specific density of PVA-solution at time t

r : average grain size of phosphor in suspension

$Ks(t)$ : kinematic viscosity of suspension at time t.

g : gravity (parameter)

$\alpha$: screen tilt angle (parameter)

So, for each relation we can describe a block. For example, the block describing the specific density of suspension, named **spec_den**:

We represent a part of this matrix:

| | | blocks | | | |
|------|---|--------|---------|----------|-----------|
| var. | I | st_set | rate_hs | spec_den | thickness |
| ------- | I | -------- | ------- | -------- | --------- |
| cs | I | param | - | - | |
| td | I | param | - | - | - |
| td1 | I | param | - | - | |
| ts1 | I | output | - | - | input |
| as | I | - | - | - | param |
| $\rho_p$ | I | - | param | param | param |
| $\rho_s$ | I | - | input | output | - |
| ... | I | | | | |

When a variable is used both in qualitative and quantitative manner, a last field appears, giving the used conversion method. Those informations are stored as, for example :

| DYNAMIC BLOCK | spec_den | Specific density of suspension |
|---|---|---|
| Vp | INPUT | Phosphor volume fraction in suspension |
| $\rho_p$ | PARAM | Specific density of green phosphor |
| $\rho_f$ | OUTPUT | Specific density of PVA-solution |
| $\rho_s$ | OUTPUT | Specific density of suspension |

$$\rho_s(t) = Vp(t) * \rho_p + (1 - Vp(t)) * \rho_f(t)$$

An example of a variable of this block is: variable Vp

| QUANTITATIVE VARIABLE | Vp | Phosphor volume fraction in suspension | |
|---|---|---|---|
| | UNITS | RANGE | |
| | % | [ 0.5 ... 1 ] | |
| ZONES | target | constraint | |
| | 0.05 | F < Vp | |
| BLOCKS | input | output | param |
| | rate_hs spec_den | phos_vol_frac | - |

For each block, and each variable, the ALCMEN user must give informations. The list of input or output or param blocks for the variables, are build using the initial informations given by the description block.

The system builds a matrix A, where raws are variables, columns are blocks, and A(raw,column) is INPUT, OUTPUT or PARAM.

| Numeric/symbolic interface | SYMBOLS C° |
|---|---|
| | freeze |
| | 18 |
| | cold |
| | 21 |
| METHOD: | well |
| | 23 |
| segmentation | hot |
| | 26 |
| | alarm |
| | 33 |
| | panic |

Knowing all those informations, we can build the interconnected net of blocks. (figure 5)

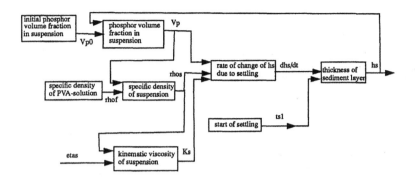

interconnected-net of some blocks of an IPCES example

Figure 5

For a first prototype of ALCMEN, we have choosen the real time expert system development tool G2. G2 allows us verifying some rules used to built blocks and variables, and also it offers a well interface .

Some workspaces of the first implementation of ALCMEN using G2 are presented on the following figure. (figure 6) We have the main menu, and some workspaces used for the creation of blocks and variables. This version of ALCMEN allows to see the complete block list, and also to save and restore a prevous description of models.

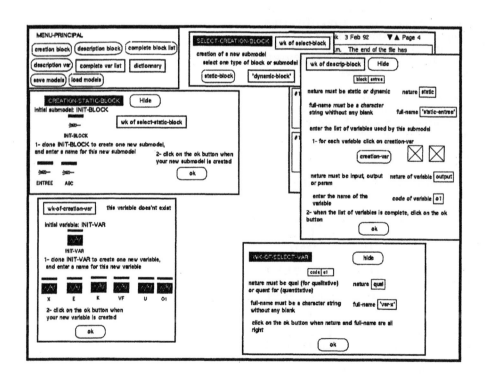

Figure 6

# CONCLUSION

A model description language for the qualitative/quantitative representation and simulation of an industrial process is proposed here. It has been specified in the framework of the ESPRIT project called IPCES (Intelligent Process Control by means of Expert Systems). It consists on the description and knowledge representation about a process in TV screens manufacturing. In a linguistic and logic description of the chain, first described by Petri nets, it appeared that most of the known relations could not be represented due to a lack of quantitative knowledge; therefore an object oriented tool for real time expert systems development has been used and gave rise to the ALCMEN philosophy.

## REFERENCES

[J. AGUILAR-MARTIN & Ph. DESROCHES & V. THOMAS-BAUDIN, ALCMEN: a Language for Qualitative/Quantitative Knowledge representation in Expert Supervisory Process Control. Proceedings of the 2nd annual conf on AI, Simulation and Planning in High Autonomy Systems, pp 80-87, IEEE Comp. Soc. press, Cocoa Beach, Florida, April 1-2, 1991,

[J. AGUILAR-MARTIN & N. PIERA i CARRETE, Les Connectifs Mixtes: de Nouveaux Opérateurs d'Association des Variables dans la Classification Automatique avec Apprentissage.In Data Analysis and Informatics, pp. 253-265, Edited by E. DIDAY et al., ELSEVIER SCIENCE PUBLISHERS, 1986]

[P. J. RAMADGE & W. M. WOHNAM, Supervisory Control of a Class of Discrete Processes, SIAM J. Control and Opt. vol 25 N° 1 pp. 206-230, 1987]

[M. SAMAAN, M. DUQUE, M. M'SAAD, A real time supervision system for adaptive control, IFAC- Glasgow 1989].

[L. TRAVE-MASSUYES & N. PIERA, The orders of magnitude as qualitative algebras, 11th Intnal. Joint Conf. on A. I., Detroit, pp1261-1266, Aug. 1989]

[V.THOMAS-BAUDIN, EXPERT SUPERVISORY CONTROL, European GENSYM users society meeting, Paris, Sept. 1991]

[R. VALETTE, V. THOMAS & S BACHMANN, SEDRIC: un simulateur à évènements discrets basé sur les réseaux de Petri, Revue APII, vol 19, pp. 423-436 - 1985]

# MODELING PERSPECTIVE FOR QUALITATIVE SIMULATION

### K. Bousson and L. Trave-Massuyes

*LAAS-CNRS, 7 Avenue du Colonel Roche, 31077 Toulouse Cedex, France*

**Abstract:** The purpose of qualitative simulation is to derive all the possible behaviors of
an ill-known dynamical system from its structural description. The ability to operate with
incomplete knowledge is one of its impressive trumps in comparison with classical quantitative
simulation of flabby numerical models. However qualitative simulation often yields spurious
predictions. This paper proposes a modeling methodology based on term rewriting and the use
of experiential knowledge to reduce the occurrences of spurious behaviors while using Qsim.
The proposal is then applied to the simulation of a fed-batch fermentation process and is proved
to eliminate considerably the spurious behaviors.

**Keywords:** Qualitative Simulation, Qsim, Spurious Behaviors, Modeling, Fed-batch fermentation process.

## 1 Introduction

One of the most difficult problems when controling or supervising an incompletely known dynamical system is to predict its behavior. The trouble is that some of the system's variables may be difficult to measure on-line or are not measured (e.g. simply because appropriate sensors are not available). The recurring escapisms control engineers use to solve the problem is to better the description of the system and to estimate non-mesurable variables and parameters. Although the use of estimators enables to tackle most situations efficiently, it requires having a precise and complete formalization of the system.

In biotechnology, devising a precise and complete model for a given process is rather impossible because 1) the process models usually contain variables and parameters strongly varying in time, 2) experiments are not reproducible, 3) the process dynamics is non-linear and non-stationary, 4) all the process variables and parameters are not known. Therefore biotechnological models are inaccurate and incomplete in essence.

More recently, research has been done in artificial intelligence to reason about incompletely known dynamical systems in the framework of qualitative physics (Bobrow, 1984; Weld and De Kleer, 1989). The goal of qualitative physics is to achieve predictive and explanatory reliability similar to quantitative physics while neither necessarily processing real-valued data, but qualitative ones, nor requiring complete and precise model of the domain of interest. For instance, during a fermentation process, one must be able to predict the behavior of the biomass growth rate in mean-term given a qualitative description of the process variables and parameters. One component of qualitative physics is qualitative simulation which is aimed

at simulating systems' dynamics from their qualitative descriptions. Hence qualitative simulation fits with reasoning about biotechnological systems. Meanwhile, the serious limitation of qualitative simulation is that it often yields spurious behaviors in addition to the actual behaviors of the system. Spurious behaviors are those behaviors satisfying the qualitative model (i.e. the qualitative counterpart of the numerical model) of a system but are not actual behaviors for any real-world physical system which agrees with the qualitative model. In non-causal approaches, modeling systems to be qualitatively simulated always uses first principles alone and neglects the use of expert knowledge about the problem domain and some other useful aspects of modeling. When first principles underlying a system are accurate and complete the use of expert knowledge is not necessary for qualitative simulation, but otherwise the simulation turns quickly on a jigsaw puzzle in that the simulation process burts in considering multiple behaviors. Hence additional strategies are needed.

This paper tackles the problem of modeling complex systems for which inaccurate and incomplete first principles as well as experiential knowledge are available for the purpose of qualitative simulation. For instance, a set of ordinary differential equations representing a fermentation process is the result of an extrem simplification of the fermentation dynamics since the phenomena actually involved in the process are not thoroughly known; instead, experiential knowledge, non formalizable in terms of numerical equations, is available to reason about it in most situations. The paper proposes a modeling methodology for reducing the occurrence of spurious behaviors while using the well-known Qsim simulator (Kuipers, 1986). Our discussion will be divided as follows. First, we present a synopsis of qualitative simulation in general and briefly the Qsim formalism. Next we present a modeling methodology for Qsim. Finally, we apply the proposed methodology to a fed-batch fermentation process and point out the ad-

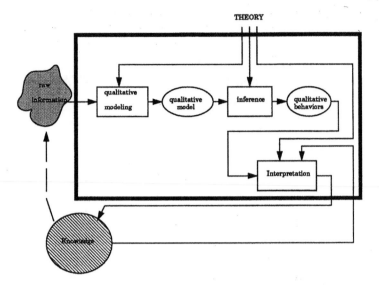

Figure 1: Tasks underlying qualitative simulation

vantages of that methodology with respect to the existing ones.

## 2 Qualitative Simulation

The purpose of qualitative simulation is to derive all the possible behaviors of an ill-known dynamical system from its structural description. The ability to operate with incomplete knowledge is one of its impressive trumps in comparison with classical quantitative simulation of flabby numerical models. The tasks underlying qualitative simulation can be described according to the following steps (fig. 1):

- Raw information - a set of equation-based or observed data describing the behavior of a system - is translated into a qualitative model using a set of constraints to express the relevant qualitative information it carries in it relatively to a theory.

- The qualitative model is then simulated by means of an inference system based on the theory. This task yields a description of the system's qualitatively interesting changes over time: the qualitative behaviors.

- Finally, these behaviors (the simulation results) are interpreted using the items of the theory and the knowledge about the system for the purpose of predictions and explanations.

The qualitative modeling, inference and interpretation tasks may be regarded as a single task block, say qualitative simulation block, as depicted in figure 1 by the thick border block. Actually, the ultimate aim of qualitative simulation is to enrich the knowledge people have about a physical system by feeding the qualitative simulation block with raw information. In other words the knowledge available after having interpreted simulation results must be less incomplete than that needed for the modeling and interpretation tasks.

For example, consider a set of differential equations modeling a fed-batch fermentation process. These equations are modeled qualitatively then they are simulated to obtain a set of possible behaviors of the process variables. The simulation results must enable to know more about the fermentation process than that needed to model it, in the way to reason about it more efficiently.

Significant headway has been made in qualitative simulation through many approaches (De Kleer and Brown, 1984; Forbus, 1984; Kuipers, 1986). All of them share the aforementioned synopsis (figure 1). We briefly present hereafter the Qsim approach.

Qsim is concerned with qualitative abstraction of ordinary differential equations modeling physical systems and predicts all the possible qualitative behaviors of those systems. In the Qsim ontology, a *parameter* is a continuous differentiable real-valued function of time having a finite number of critical points. A parameter value is described at any time with respect to its order relations with a totally ordered set of *landmark values*. The landmark values for a parameter always include zero, the values of the parameter at the beginning and the ending of the time interval over which it ranges, and the values of the parameter at each of its critical points. Landmark values are the parameter's values, given as symbols, which are supposed to be relevant to the human observer. A *distinguished timepoint* of a parameter is a time-point at which the parameter value equals a landmark value. A *qualitative value* is either a landmark value or the open interval between adjacent landmarks. The Qsim process may discover new landmark values and add them to the set of pre-specified ones. The *qualitative state* of a parameter over time is a pair of qualitative value and qualitative derivative (i.e increasing, steady or decreasing). The behavior of a parameter is a sequence of qualitative states alternating between states at distinguished time-points, and states on intervals between distinguished time-points.

Referring to the synopsis, a qualitative model in Qsim formalism corresponds to a set of parameters and associated landmark values, and a set of constraints describing the qualitative interrelationships between those parameters. The constraints used in Qsim are arithmetical such as $add(x,y,z), mult(x,y,z), minus(x,y)$ or differential: $deriv(x,y)$ or functional: $M^+(x,y)$ (respec. $M^-(x,y)$) which means that $x$ and $y$ are linked to each other by an increasing (respec. decreasing) monotonic function. The simulation results, that is, the possible behaviors of a system are represented as paths in a graph whose nodes are states on time-points and edges are states at time-intervals. The graph may have more than one starting nodes each of which corresponds to a possible initial state of the system's dynamics.

The concepts and methods used in Qsim formalism are most impressive. They are proved to achieve qualitative prediction tasks while avoiding precise and equation-based formulations of problems and being computationally less intensive than the quantitative simulation approaches. Meanwhile, the main limitations of Qsim (and all the qualitative simulators in existence so far) are the predictions of spurious behaviors. In the Qsim algorithm spurious behaviors originate from the following features:

- *Lack of feedback analysis*: The occurrence of a change in a parameter's behavior may return in time to influence the source of that change. Feedback appears as a ubitiquous phenomenon in engineering systems (e.g. analog circuits, harmonic oscillators, traffic systems, and so on ...) and most of physical properties depend on it. As pointed out by Williams (1986), the delay through a feedback loop may cause a quantity to remain steady. Since Qsim does not cope with feedback analysis it is forced to envision different values for a parameter even if it remains steady due to some feedback loop.

- *Local search strategy*: The Qsim algorithm is essentially local in the sense that successor states are generated taking into account the sole information about the current state. That strategy proceeds in the spirit of quantitative simulation for in quantitative simulation, states sum up the parameters' histories; that is not the case in qualitative simulation because of the incompleteness of the information about the system and/or the deep behavioral abstraction carried by the qualitative model. The fact that histories are not accounted for leads to predict behaviors which could have been avoided. Indeed, some of the predicted behaviors may be incompatible with respect to the past values. As Qsim is not concerned with histories, those past-incompatible behaviors cannot be detected, hence they are unfortunately considered as being actual behaviors.

- *Total ordering on events*: The temporal representation in Qsim suffers from both inaccuracy and incompleteness in that time is described qualitatively and durations of phenomena are not treated. The brittle representation of time in Qsim formalism and, following Hayes (1985), the fact that a system's behavior is described as a sequence of states lead to put a total ordering on events. Since events occur in time but also in space, the total ordering on them constrains to predict multiple behaviors some of which are not realistic.

# 3   A modeling methodology

A great number of spurious behaviors undoutedly induces a prohibitive amount of interpreting results. To enhance the reliability of Qsim output data subject to interpretation, various techniques have been proposed elsewhere for firing spurious behaviors. The resolution of feedback effects

has been studied by Rose and Kramer (1991) by structuring models causally. Williams (1986) has built a temporal constraint propagator based on the concept of concise history which is characterized by the fact that every point in it where two episodes meet symbolizes a change in the value of the parameter. Weld (1988) has tackled the temporal problem by comparative analysis, and Missier and Travé-Massuyès (1991) have studied the explicit computation of duration in Qsim. Although those proposals have been proved to reduce considerably the numbers of spurious behaviors, they are not implemented in Qsim. Taking Qsim as such, we believe that suitable modeling may avoid the envisionment of most of the spurious predictions for a wide spectrum of engineering systems. We propose a modeling methodology based on the use of *term rewriting* and *experiential knowledge*.

It happens that adding a few number of contraints to or removing them from the qualitative model gives rise to a considerable variation in the number of the predicted behaviors, hence in the number of spurious behaviors. Viewing a qualitative model as a mechanical system whose components are interconnected by means of constraints, the phenomenon just described means that the model is *structurally unstable*. Terms or equations, of a raw model can be rewritten in such a way to strengthen the qualitative model. Suitable term rewriting allows to make the qualitative model as structurally stable as possible.

Consider we have to simulate the equation $y = x^2 - x$ (i) with $x \geq 0$ and $dx/dt > 0$. By factorizing the equation, we get $y = x(x-1)$ (ii). Both forms (i) and (ii) are equivalent but modeling from (i) with Qsim gives more spurious behaviors than modeling from (ii).

When two parameters are linked by the relation $M^+$ it seems to be useful to substitute one of them to the other under certain conditions, but not all the time. For instance, in (i) we cannot replace $x^2$ by $x$; the problem is that doing so leads to the equation $y = 0$, an equation which is not equivalent to (i.e. does not have the same solutions as) the starting one. A substitution preserves equivalence if it does not interact with any other term of the same equation. We call *valid* substitution with respect to a given equation a substitution that preserves equivalence when it is performed in that equation.

We give the following rewriting procedure which is to be performed on raw (quantitative equation-based) models:

**1:** Factorize each equation.

**2:** Remove constant factors from each equation (i.e. replace each constant factor by the real number 1).

**3:** If $x$ is a parameter such that $M^+(x,t)$ where $t$ is a term in the raw model then replace $t$ by $x$ provided that this operation does not affect other terms of the same equation (i.e perform valid substitutions).

**4:** Model qualitatively the set of the obtained equations (using Qsim constraint relations).

When the quantitative model of the system is accurate, that procedure in cooperation with phase plane analysis, chatters resolution and other methods proposed in Qsim leads to significant reduction of spurious behaviors. The step 1 agrees with the method proposed by Charles, Fouché and Melin (1991) and that they used to draw useful conclusions about a non-trivial second-order non-linear system. In cases of inaccurate models, the procedure above has to be combined with experiential knowledge about the system if available.

Spurious behaviors occur when sufficient information to discard them is missing. This means that some relevant information may not be captured by the raw model or the qualitative model. Embedding expert knowledge to the qualitative model may make it more informative, and therefore rule out unwanted predictions.

Assume $C$ to be a constraint set from which a set $B$ containing all the actual behaviors of a system and eventually additional spurious behaviors may be obtained. We call such a set $B$ a *solution set* and write $C->B$ to denote that the behaviors in $B$ are obtainable from the set of constraints $C$. Since a constraint set is meant to give a set of behaviors we assume it to be *consistent* (i.e. any of the constraint it contains does not contradict any other one). When $B$ contains only the actual behaviors of the system we call it the *actual solution set*. The set $C$ is said to be *redundant* if there exists a constraint $c$ in $C$ such that $c$ can be derived from $C-\{c\}$. In such a case $C-\{c\}->B$.

**Definition:** A constraint set $C$ is termed *irreducible* if it is not redundant and yields a *solution set*.

**Proposition:** Let $\{C_n, n \geq 0\}$ be a non-decreasing sequence of irreducible constraint sets of a system converging to $C^*$. Then $C^*->B^*$ where $B^*$ is the actual solution set.

The proposition states that if an irreducible constraint set can be extended to a bigger one which is inextendable then that latter set yields the actual solution set. When the constraint set is obtained from first principles only, this proposition does not hold for it has been shown (Kuipers, 1986) that spurious behaviors may remain in the qualitative simulation results because of the abstraction of the models. Instead, if we appeal to experiential knowledge which are not described by the qualitative translation of the quantitative model, then we can at the limit fire all the spurious behaviors (theoretically). A trivial remark is that if $C_1$ and $C_2$ are two irreducible constraint sets of a physical system such that $C_1 \subset C_2$, $C_1->B_1$ and $C_2->B_2$ (where $B_1$ and $B_2$ are solution sets), then $B_2 \subset B_1$. Thus, the modeling based on experiential knowledge consists in two steps:

**1:** Build a qualitative model from raw information using the procedure earlier mentioned and modeling recipes proposed in Qsim.

**2:** Add to the qualitative model experiential knowledge that it does not capture.

# 4 Application

This section illustrates the modeling methodology we have just presented with a fed-batch fermentation process.

## 4.1 Process model

Cultures in biotechnological industries consist in the degradation of a substrate by a population of microorganisms (the biomass) into metabolites. The type of fermentation in biomass production depends on the way that fresh nutrients are fed into the fermentor. Fed-batch fermentation is characterized by the fact that it starts with an initial volume of nutrients and that additional nutrients are brought during the fermentation process when required. The behavior of the process is modeled from stoichiometric and mass-balance considerations. The model is a set of non-linear differential equations involving biomass, substrate, and dissolved oxygen:

$$dX/dt = (\mu - F)X \qquad (1)$$

$$dS/dt = -K_1\mu X + F(S_i - S) \qquad (2)$$

$$dD/dt = K_{La}(D_s - D) + F(D_i - D) - Q_{O_2}X \qquad (3)$$

where

| | |
|---|---|
| $X$ | biomass concentration |
| $S$ | substrate concentration |
| $S_i$ | influent substrate concentration |
| $D$ | dissolved oxygen rate |
| $D_s$ | dissolved oxygen saturation rate |
| $D_i$ | influent oxygen rate |
| $\mu$ | specific growth rate |
| $F$ | specific feed flow rate |
| $K_{La}$ | specific oxygen transfer rate |
| $Q_{O_2}$ | specific oxydation rate |
| $K_1$ | a non-negative constant |

The sole variables of the equations are $X$, $S$, $D$ and $\mu$, the other elements occurring in them are supposed to be positive constant parameters. We choose the specific growth rate from Monod's law:

$$\mu = \mu_{max}S/(K_s + S) \qquad (4)$$

where $K_s$ is a positive constant parameter.

## 4.2 Modeling and simulation results

For the purpose of modeling in Qsim formalism, we adopt the following notations:

$XP = dX/dt$, $MU = \mu$, $MUF = \mu - F$, $SP = dS/dt$, $DP = dD/dt$, $DP = dD/dt$, $PI = K_{La}(D_s - D) + F(D_i - D)$.

Applying the rewriting procedure described in section 3 to the equations (1) through (3) and considering equation (4), we obtain the following constraints:

$deriv(X, XP)$, $deriv(S, SP)$, $deriv(D, DP)$,
$M^+(S, MU)$, $M^+(S, MUF)$, $M^-(D, PI)$,
$mult(MUF, X, XP)$, $add(X, DP, PI)$.

That model yields a total of at least 25 behaviors (figure 2.a).

Now let us add experiential knowledge to the previous model. During a fermentation process the dissolved oxygen is controled so that it remains within an appropriate interval so that the biomass grows conveniently. But this is not always possible and it happens that the dissolved oxygen goes outside of the specified interval. Thus, to study the full behavior of the process, we simulate it when the dissolved oxygen is inside the appropriate interval as well as when it is outside of it. In the first case, assuming that we simulate the beginning of the process, Qsim finds one behavior (that corresponds to the first operating region in figure 2.b) when we add this experiential knowledge:
$decreasing(XP)$, $decreasing(DP)$, $increasing(SP)$.

In the second operating region we do the same using other experiential knowledge. The simulation yields three behaviors (second operating region in figure 2.b). The following conclusions from the simualtion are most interesting (figure 3):

- The process is under oxygen limited growth whereas it remains some substrate in the fermentor.

- The process is under substrate limited growth whereas it remains oxygen in the fermentor.

- The process is under both substrate and oxygen limited growth.

In fed-batch fermentation process, engineers usually choose the initial quantities of oxygen and substrate such that the limitation under substrate occurs after that in oxygen. Here Qsim predicts that the limitation under any one of these nutrients can occur earlier than the other according to the situations, that agrees with the reality.

Another non trivial conclusion is that the dissolved oxygen increases from the starting of the process, then decreases. This phenomenon is purely biological. At the starting of the process the activities of the microorganisms are slow whereas the oxygen flow rate is maintained steady, then the dissolved oxygen increases because it is less consumed by the microorganisms than it is fed into the fermentor. But as the process continues the microorganisms grow and consume much more oxygen. Hence the dissolved oxygen decreases.

The three behaviors predicted by Qsim are all actual.

# 5  Conclusion

In this paper, we have proposed a modeling methodology based on a term rewriting procedure and the use of experiential knowledge of the domain of interest while using the well-known qualitative simulator Qsim. The proposal allows to capture the maximum relevant information underlying a raw model of a given dynamical system. It has been shown to enable to fire spurious predictions considerably on non-trial systems. Modeling a fed-batch fermentation process using the proposed methodology has led Qsim to yield only the actual behaviors of that process.

### Acknowledgment

We are grateful to Jean-Philippe Steyer and the process experts at the Centre de Transfert en Biotechnologie et Microbiologie (CTBM) of UPS-INSA, Toulouse.

### References

Bobrow D. G. (1984), Qualitative Reasoning about Physical Systems, *Artificial Intelligence* **24**.

Charles A., Fouché P. and Mélin C. (1991), Recent Improvements in Qualitative Simulation, in *Proc. of the 13th IMACS World Congress*, pp. 1745-1746.

De Kleer J. and Brown J. (1984), A Qualitative Physics based on Confluences, *Artificial Intelligence* **24**, pp. 7-83.

Forbus K. D. (1984), Qualitative Process Theory, *Artificial Intelligence* **24**, pp. 85-168.

Hayes P. J. (1985), The Second Naive Physics Manifesto, *Formal Theories of The Commonsense World*, eds J. Hobbs and B. Moore, pp. 1-36.

Kuipers B. (1986), Qualitative Simulation, *Artificial Intelligence* **29**, pp. 289-338.

Missier A. and Travé-Massuyès L. (1991), Temporal information from order of magnitude reasoning, *Second annual conference on AI, Simulation and planning in High Autonomous Systems*, pp. 298-305, Cocoa Beach, Florida, USA.

Rose Ph. and Kramer M. A. (1991), Qualitative Analysis of Causal Feedback, *Proc. AAAI-91*, pp. 817-823, Anaheim.

Weld D. (1988), Comparative Analysis, *Artificial Intelligence* **36**, pp. 333-373.

Weld D. and De Kleer (1989), *Readings in Qualitative Reasoning about physical systems*, Morgan Kaufmann, San Mateo, CA.

Williams B. C. (1986), Doing time: Putting Qualitative Reasoning on former ground. *in Proc. AAAI-86*, 105-112.

Figure 2.a: Behavior tree without experiential knowledge.

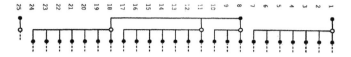

Figure 2.b: Behavior tree by means
of experiential knowledge.

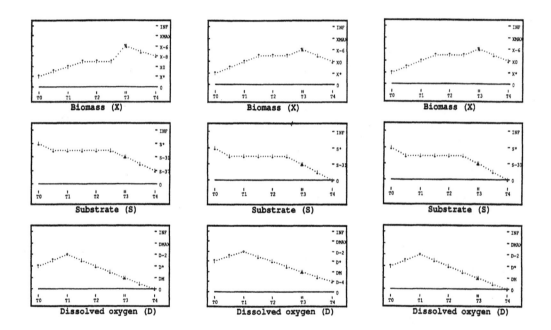

Figure 3: Behaviors of a fed-batch fermentation process.

# COMBINED QUALITATIVE/QUANTITATIVE SIMULATION MODELS OF CONTINUOUS-TIME PROCESSES USING FUZZY INDUCTIVE REASONING TECHNIQUES

**F.E. Cellier\*, A. Nebot\*\*, F. Mugica\*\*\* and A. de Albornoz\*\*\***

*\*Department of Electrical and Computer Engineering, The University of Arizona, Tucson, AZ 85721, USA*
*\*\*Institut de Cibernetica, Universitat Politecnica de Catalunya, Diagonal 647, 2NA. Planta, Barcelona 08028, Spain*
*\*\*\*Department d'Esaii, Universitat Politecnica de Catalunya, Diagonal 647, 2NA, Planta, Barcelona 08028, Spain*

**Abstract**. The feasibility of mixed quantitative and qualitative simulation is demonstrated by means of a simple hydraulic control system. The mechanical and electrical parts of the control system are modeled using differential equations, whereas the hydraulic part is modeled using fuzzy inductive reasoning. The mixed quantitative and qualitative model is simulated in ACSL, and the simulation results are compared with those obtained from a fully quantitative model. The example was chosen as a simple to describe, yet numerically demanding process whose sole purpose is to prove the concept. Several practical applications of this mixed modeling technique are mentioned in the paper, but their realization has not yet been completed.

**Keywords**. Modeling; simulation; mixed quantitative and qualitative models; inductive reasoning; forecasting theory; fuzzy systems; learning systems; artificial intelligence.

## INTRODUCTION

Qualitative simulation has recently become a fashionable branch of research in artificial intelligence. Human reasoning has been understood as a process of mental simulation, and qualitative simulation has been introduced as an attempt to replicate, in the computer, facets of human reasoning.

Qualitative simulation can be defined as evaluating the behavior of a system in qualitative terms (Cellier, 1991b). To this end, the states that the system can be in are lumped together to a finite (discrete) set. For example, instead of dealing with temperature as a real-valued quantity with values such as 22.0°C, or 71.6°F, or 295.15 K, qualitative temperature values may be characterized as 'cold,' 'warm,' or 'hot.'

Qualitative variables are variables that assume qualitative values. Variables of a dynamical system are functions of time. The behavior of a dynamical system is a description of the values of its variables over time. The behavior of quantitative variables is usually referred to as trajectory behavior, whereas the behavior of qualitative variables is commonly referred to as episodical behavior. Qualitative simulation can thus be defined as the process of inferring the episodical behavior of a qualitative dynamical system or model.

Qualitative variables are frequently interpreted as an ordered set without distance measure (Babbie, 1989). It is correct that 'warm' is "larger" (warmer) than 'cold,' and that 'hot' is "larger" (warmer) than 'warm.' Yet, it is not true that

$$\text{'warm'} - \text{'cold'} = \text{'hot'} - \text{'warm'} \qquad (1)$$

or, even more absurd, that

$$\text{'hot'} = 2 * \text{'warm'} - \text{'cold'} \qquad (2)$$

No '−' operator is defined for qualitative variables.

Time, in a qualitative simulation, is also frequently treated as a qualitative variable. It is then possible to determine whether one event happens before or after another event, but it is not possible to specify when precisely a particular event takes place.

The most widely advocated among the qualitative simulation techniques are the knowledge–based approaches that were originally derived from the *Naïve Physics Manifesto* (Hayes, 1979). Several dialects of these types of qualitative models exist (de Kleer and Brown, 1984; Forbus, 1984; Kuipers, 1986). They are best summarized in (Bobrow, 1985).

The purpose of most qualitative simulation attempts is to enumerate, in qualitative terms, all possible episodical behaviors of a given system under all feasible experimental conditions. This is in direct contrast to quantitative simulations that usually content themselves with generating one single trajectory behavior of a given system under one single set of experimental conditions.

## MIXED MODELS

In the light of what has been explained above, it seems questionable whether mixed quantitative and qualitative models are feasible at all. How should a mixed quantitative and qualitative simulation deal with the fact that the quantitative subsystems treat the independent variable, *time*, as a quantitative variable, whereas the qualitative subsystems treat the same variable qualitatively? When does a particular qualitative event occur in terms of quantitative time? How are the explicit experimental conditions that are needed by the quantitative subsystems accounted for in the qualitative subsystems?

Quite obviously, a number of incompatibility issues exist between quantitative and qualitative subsystems that must be settled before mixed simulations can be attempted. In a mixed simulation, also the qualitative subsystems must treat time as a quantitative variable. Furthermore, the purpose of qualitative models in the context of mixed simulations is revised. It is no longer their aim to enumerate episodical behaviors. Instead, also the qualitative models are now used to determine a single episodical behavior in response to a single set of qualitative experimental conditions.

Do so revised qualitative models make sense? It is certainly illegitimate to request that, because human pilots are unable to solve Riccati equations in their heads to determine an optimal flight path, autopilots shouldn't tackle this problem either. It is not sufficient to justify the existence of qualitative models by human inadequacies to deal with quantitative information.

Two good reasons for dealing with information in qualitative ways are the following:

1. Quantitative details about a (sub)system may not be available. For example, while the mechanical properties of a human heart are well understood and can easily be described by differential equation models, the effects of many chemical substances on the behavior of the heart are poorly understood and cannot easily be quantified. A mixed model could be used to describe those portions of the overall system that are well understood by quantitative differential equation models, while other aspects that are less well understood may still be representable in qualitative terms.

2. Quantitative details may limit the robustness of a (sub)system to react to previously unknown experimental conditions. For example, while a human pilot is unable to compute an optimal flight path, he or she can control the airplane in a much more robust fashion than any of today's autopilots. Optimality in behavior can be traded for robustness. A fuzzy controller is an example of a qualitative subsystem that is designed to deal with a larger class of experimental conditions in suboptimal ways.

Mixed quantitative and qualitative models may be used to address either or both of the above applications. However, in order to do so, it is necessary to devise qualitative modeling and simulation capabilities that are compatible with their quantitative counterparts and that can be used to represent qualitative subsystems as those mentioned above appropriately and in terms of knowledge available to the system designer at the time of modeling.

It is the purpose of this paper to describe one such mixed modeling and simulation methodology. In the advocated approach, the qualitative subsystems are represented (modeled) by a special class of finite state machines called fuzzy optimal masks, and their episodical behavior is inferred (simulated) by a technique called fuzzy forecasting. The overall process of qualitative modeling and simulation is referred to as fuzzy inductive reasoning.

## FUZZY RECODING

Recoding denotes the process of converting a quantitative variable to a qualitative variable. In general, some information is lost in the process of recoding. Obviously, a temperature value of 97°F contains more information than the value 'hot.' Fuzzy recoding avoids this problem. Figure 1 shows the fuzzy recoding of a variable called "systolic blood pressure."

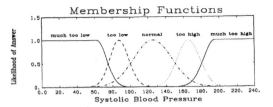

**Figure 1.** Fuzzy recoding.

For example, a quantitative systolic blood pressure of 135.0 is recoded into a qualitative value of 'normal' with a fuzzy membership function of 0.895 and a side function of 'right.' Thus, a single quantitative value is recoded into a triple. Any systolic blood pressure with a quantitative value between 100.0 and 150.0 will be recoded into the qualitative value 'normal.' The fuzzy membership function denotes the value of the bell–shaped curve shown on Fig.1, always a value between 0.5 and 1.0, and the side function indicates whether the quantitative value is to the left or to the right of the maximum of the fuzzy membership function. Obviously, the qualitative triple contains the same information as the original quantitative variable. The quantitative value can be regenerated accurately from the qualitative triple, i.e., without any loss of information.

Due to space limitations, details of how quantitative variables are optimally recoded into qualitative triples will not be given in this paper. These details are provided in (Li and Cellier, 1990; Cellier, 1991a).

## FUZZY OPTIMAL MASKS

A mask denotes relationship between different variables. For example, given the following raw data model consisting of five variables, namely the inputs $u_1$ and $u_2$ and the outputs $y_1$, $y_2$, and $y_3$ that are recorded at different values of time.

$$
\begin{array}{c|ccccc}
time & u_1 & u_2 & y_1 & y_2 & y_3 \\
0.0 & \cdots & \cdots & \cdots & \cdots & \cdots \\
\delta t & \cdots & \cdots & \cdots & \cdots & \cdots \\
2 \cdot \delta t & \cdots & \cdots & \cdots & \cdots & \cdots \\
3 \cdot \delta t & \cdots & \cdots & \cdots & \cdots & \cdots \\
\vdots & \vdots & \vdots & \vdots & \vdots & \vdots \\
(n_{rec} - 1) \cdot \delta t & \cdots & \cdots & \cdots & \cdots & \cdots \\
\end{array} \tag{3}
$$

Each column of the raw data model contains one qualitative variable recorded at different values of time, and each row contains the recordings of all qualitative variables at one point in time. The raw data matrix is accompanied by a fuzzy membership matrix and a side matrix of the same dimensions.

A mask denotes a relationship between these variables. For example, the mask

$$
\begin{array}{c|ccccc}
t \backslash^x & u_1 & u_2 & y_1 & y_2 & y_3 \\
t - 2\delta t & 0 & 0 & 0 & 0 & -1 \\
t - \delta t & 0 & -2 & -3 & 0 & 0 \\
t & -4 & 0 & +1 & 0 & 0 \\
\end{array} \tag{4}
$$

denotes the following relationship pertaining to the five variable system

$$ y_1(t) = \tilde{f}(y_3(t - 2\delta t), u_2(t - \delta t), y_1(t - \delta t), u_1(t)) \tag{5} $$

Negative elements in the mask matrix denote inputs of the qualitative functional relationship. The example mask has four inputs. The sequence in which they are enumerated is immaterial. They are usually enumerated from left to right and top to bottom. A positive element in the mask matrix denotes the output. Thus, Eq.(4) is simply a matrix representation of Eq.(5). The mask must have the same number of columns as the raw data matrix. The number of rows of the mask matrix is called the *depth* of the mask. The mask can be used to flatten a dynamic relationship out into a static relationship. The mask can be shifted over the episodical behavior. Selected inputs and outputs can be read out from the raw data matrix and can be written on one row next to each other. Figure 2 illustrates this process.

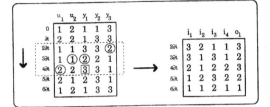

**Figure 2.** Flattening dynamic relationships through masking.

After the mask has been applied to the raw data, the formerly dynamic episodical behavior has become static, i.e., the relationship is now contained within a single row

$$o_1(t) = \tilde{f}(i_1(t), i_2(t), i_3(t), i_4(t)) \quad (6)$$

The resulting matrix is called *input/output matrix*.

How is the mask selected? A *mask candidate matrix* is constructed in which negative elements denote potential inputs, and the single positive element denotes the true output of the mask. A good mask candidate matrix for the previously mentioned five variable system might be

$$
\begin{array}{c|ccccc}
t \backslash x & u_1 & u_2 & y_1 & y_2 & y_3 \\
t - 2\delta t & -1 & -1 & -1 & -1 & -1 \\
t - \delta t & -1 & -1 & -1 & -1 & -1 \\
t & -1 & -1 & +1 & 0 & 0
\end{array} \quad (7)
$$

A mask candidate matrix is an ensemble of all acceptable masks. The optimal mask selection algorithm determines the best among all masks that are compatible with the mask candidate matrix. The mask of Eq.(4) is one such mask. The optimal mask is the one mask that maximizes the forecasting power of the inductive reasoning process, i.e., the mask that results in the most deterministic input/output matrix.

Due to space limitations, the details of how the optimal mask selection algorithm works are omitted from this paper. These details are also provided in (Li and Cellier, 1990; Cellier, 1991a).

## FUZZY FORECASTING

Once the optimal mask has been determined, it can be applied to the given raw data matrix resulting in a particular input/output matrix. Since the input/output matrix contains functional relationships within single rows, the rows of the input/output matrix can now be sorted in alphanumerical order. The result of this operation is called the *behavior matrix* of this system. The behavior matrix is a finite state machine. For each combination of input values, it shows which output is most likely to be observed.

Forecasting is now a straightforward procedure. The mask is simply shifted further down beyond the end of the raw data matrix, future inputs are read out from the mask, and the behavior matrix is used to determine the future output, which can then be copied back into the raw data matrix. In fuzzy forecasting, it is essential that, together with the qualitative output, also a fuzzy membership value

and a side value are forecast. Thus, fuzzy forecasting predicts an entire qualitative triple from which a quantitative variable can be regenerated whenever needed.

In fuzzy forecasting, the membership and side functions of the new input are compared with those of all previous recordings of the same qualitative input contained in the behavior matrix. The one input with the most similar membership and side functions is identified. For this purpose, a cheap approximation of the regenerated quantitative signal

$$d = 1 + side * (1 - Memb) \quad (8)$$

is computed for every input variable of the new input set, and the regenerated $d_i$ values are stored in a vector. This reconstruction is then repeated for all previous recordings of the same input set. Finally, the $\mathcal{L}_2$ norms of the difference between the $d$ vector of the new input and the $d$ vectors of all previous recordings of the same input are computed, and the previous recording with the smallest $\mathcal{L}_2$ norm is identified. Its *output* and *side* values are then used as forecasts for the *output* and *side* values of the current state.

Forecasting of the new membership function is done a little differently. Here, the five previous recordings with the smallest $\mathcal{L}_2$ norms are used (if at least five such recordings are found in the behavior matrix), and a distance–weighted average of their fuzzy membership functions is computed and used as the forecast for the fuzzy membership function of the current state.

More details of fuzzy forecasting are provided in (Cellier, 1991a).

## AN EXAMPLE

In the remainder of this paper, an example will be presented that demonstrates, for the first time, the process of mixed quantitative and qualitative simulation using fuzzy inductive reasoning. The example was chosen simple enough to be presented in full, yet complex enough to demonstrate the generality and validity of the approach. However, it is not suggested that the chosen example represents a meaningful application of mixed quantitative and qualitative simulation. The example was chosen to prove the concept and to clearly illustrate the procedure, not as a realistic application of the proposed technique.

Figure 3 shows a hydraulic motor with a four–way servo valve.

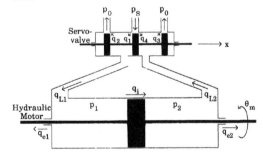

**Figure 3.** Hydraulic motor with a four–way servo valve.

The flows from the high–pressure line into the servo valve and from the servo valve back into the low–pressure line are turbulent. Consequently, the relation between flow and pressure is quadratic

$$q_1 = k(x_0 + x)\sqrt{P_S - p_1} \quad (9a)$$

$$q_2 = k(x_0 - x)\sqrt{p_1 - P_0} \quad (9b)$$

$$q_3 = k(x_0 + x)\sqrt{p_2 - P_0} \quad (9c)$$

$$q_4 = k(x_0 - x)\sqrt{P_S - p_2} \quad (9d)$$

The change in the chamber pressures is proportional to the effective flows in the two chambers

$$\dot{p}_1 = c_1(q_{L1} - q_i - q_{e1} - q_{ind}) \qquad (10a)$$
$$\dot{p}_2 = c_1(q_{ind} + q_i - q_{e2} - q_{L2}) \qquad (10b)$$

where the internal leakage flow, $q_i$, and the external leakage flows, $q_{e1}$ and $q_{e2}$, can be computed as

$$q_i = c_i \cdot p_L = c_i(p_1 - p_2) \qquad (11a)$$
$$q_{e1} = c_e \cdot p_1 \qquad (11b)$$
$$q_{e2} = c_e \cdot p_2 \qquad (11c)$$

The induced voltage, $q_{ind}$, is proportional to the angular velocity of the hydraulic motor, $\omega_m$

$$q_{ind} = \psi \cdot \omega_m \qquad (12)$$

and the torque produced by the hydraulic motor is proportional to the load pressure, $p_L$

$$T_m = \psi \cdot p_L = \psi(p_1 - p_2) \qquad (13)$$

The hydraulic motor is embedded in the control circuitry shown on Fig.4

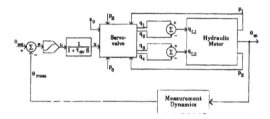

**Figure 4.** Hydraulic motor position control ciruit.

An ACSL program (MGA, 1985) was written that simulates the control system over 2.5 seconds. A binary random input signal was applied to the input of the system, $\theta_{set}$, and the values of the the control signal, $u$, the angular velocity, $\omega_m = \dot{\theta}_m$, and the torque, $T_m$, of the hydraulic motor were recorded for later reuse.

For demonstration purposes, it is now assumed that no knowledge exists that would permit a description of the hydraulic equations by means of a differential equation model. All that is known is that the mechanical torque, $T_m$, of the hydraulic motor somehow depends on the control signal, $u$, and the angular velocity, $\omega_m$.

In a mixed quantitative and qualitative simulation, the mechanical and electrical parts of the control system will be represented by differential equation models, whereas the hydraulic part will be represented by a fuzzy inductive reasoning model. The mixed simulation results will be compared with the previously obtained purely quantitative simulation results for validation purposes.

Optimal recoding would suggest that the three variables $u$, $\omega_m$, and $T_m$ be sampled once every 0.025 seconds if a mask depth of 3 is chosen. This value is deduced from the slowest time constant (eigenvalue of the Jacobian) to be covered by the mask. A more detailed explanation is provided in (Li and Cellier, 1990; Cellier, 1991a). Unfortunately, fuzzy inductive forecasting will predict only one value of $T_m$ per sampling interval. Thus, the overall control system will react like a sampled–data control system with a sampling rate of 0.025. Thereby, the stability of the control system is lost. From a control system perspective, it is necessary to sample the variables considerably faster, namely once every 0.0025 seconds.

Therefore, it was decided to choose the following mask candidate matrix

$$
\begin{array}{c c c c}
_t\backslash^x & u & \omega_m & T_m \\
t - 20\delta t & \begin{pmatrix} -1 & -1 & -1 \\ \end{pmatrix} \\
t - 19\delta t & \begin{pmatrix} 0 & 0 & 0 \\ \end{pmatrix} \\
\vdots & \vdots & \vdots & \vdots \\
t - 11\delta t & 0 & 0 & 0 \\
t - 10\delta t & -1 & -1 & -1 \\
t - 9\delta t & 0 & 0 & 0 \\
\vdots & \vdots & \vdots & \vdots \\
t - \delta t & 0 & 0 & 0 \\
t & -1 & -1 & +1 \\
\end{array}
\qquad (14)
$$

of depth 21. As mandated by control theory, the sampling interval $\delta t$ is chosen to be 0.0025 seconds. Yet, as dictated by the inductive reasoning technique, the mechanical torque, $T_m$, at time $t$ will depend on past values of $u$, $\omega_m$, and $T_m$ at times $t - 0.025$ and $t - 0.05$.

The optimal mask found with this mask candidate matrix is

$$
\begin{array}{c c c c}
_t\backslash^x & u & \omega_m & T_m \\
t - 20\delta t & 0 & -1 & -2 \\
t - 19\delta t & 0 & 0 & 0 \\
\vdots & \vdots & \vdots & \vdots \\
t - 11\delta t & 0 & 0 & 0 \\
t - 10\delta t & 0 & 0 & 0 \\
t - 9\delta t & 0 & 0 & 0 \\
\vdots & \vdots & \vdots & \vdots \\
t - \delta t & 0 & 0 & 0 \\
t & -3 & 0 & +1 \\
\end{array}
\qquad (15)
$$

In other words

$$T_m(t) = \tilde{f}(\omega_m(t - 0.05), T_m(t - 0.05), u(t)) \qquad (16)$$

The first 900 rows of the raw data matrix were used as past history data to compute the optimal mask. Fuzzy forecasting was used to predict new qualitative triples for $T_m$ for the last 100 rows of the raw data matrix. From the predicted qualitative triples, quantitative values were then regenerated. Figure 5 compares the true "measured" values of $T_m$ obtained from the purely quantitative simulation (solid line) with the forecast and regenerated values obtained from fuzzy inductive reasoning (dashed line).

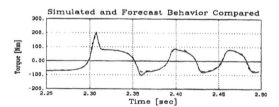

**Figure 5.** Simulated and forecast torque trajectories compared.

The results are encouraging. Quite obviously, the optimal mask contains sufficient information about the behavior of the hydraulic subsystem to be used as a valid replacement of the true quantitative differential equation model. Notice that the fuzzy inductive reasoning model was constructed solely on the basis of measurement data. No insight into the functioning of the hydraulic subsystem was required other than the knowledge that the torque, $T_m$, dynamically depends on the control signal, $u$, and the angular velocity, $\omega_m$.

In a mixed quantitative and qualitative simulation, the fuzzy inductive reasoning model was then used to replace the former differential equation model of the hydraulic subsystem while the electrical and mechanical subsystems were described using differential equations as before. The mixed model is shown on Fig.6.

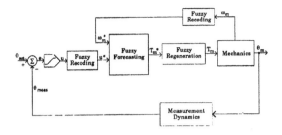

**Figure 6.** Mixed model of the hydraulic system.

The quantitative control signal, $u$, is converted to a qualitative triple, $u^*$, using fuzzy recoding. Also the quantitative angular velocity, $\omega_m$, of the hydraulic motor is converted to a qualitative triple, $\omega_m^*$. From these two qualitative signals, a qualitative triple of the torque of the hydraulic motor, $T_m^*$, is computed by means of fuzzy forecasting. This qualitative signal is then converted back to a quantitative signal, $T_m$ using fuzzy signal regeneration. The mechanical parts of the hydraulic motor are simulated by means of a differential equation model. The same holds true for the measurement dynamics.

Forecasting was restricted to the last 100 sampling intervals, i.e., to the time span from 2.25 seconds to 2.5 seconds. Figure 7 compares the angular position, $\theta_m$, of the hydraulic motor from the purely quantitative simulation (solid line) with that of the mixed quantitative and qualitative simulation (dashed line).

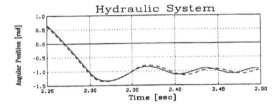

**Figure 7.** Comparison of quantitative and mixed simulations.

As was to be expected, the mixed model behaves like a sampled–data control system. The mixed simulation exhibits an oscillation amplitude that is slightly larger and an oscillation frequency that is slightly smaller than those shown by the purely quantitative simulation. Surprisingly, the damping of the mixed model is slightly larger than that of the purely quantitative model.

## CONCLUSIONS

The example demonstrates the validity of the chosen approach. Mixed simulations are similar in effect to sampled–data system simulations. *Fuzzy recoding* takes the place of analog–to–digital converters, and *fuzzy signal regeneration* takes the place of digital–to–analog converters. However, this is where the similarity ends. Sampled–data systems operate on a fairly accurate representation of the digital signals. Typical converters are 12–bit converters, corresponding to discretized signals with 4096 discrete levels. In contrast, the fuzzy inductive reasoning model employed in the above example recoded all three variables into qualitative variables with the three levels 'small,' 'medium,' and 'large.' The quantitative information is retained in the fuzzy membership functions that accompany the qualitative signals. Due to the small number of discrete levels, the resulting finite state machine is extremely simple. Fuzzy membership forecasting has been shown to be very effec-

tive in inferring quantitative information about the system under investigation in qualitative terms.

Due to the space limitations inherent in a publication in conference proceedings it was not possible to provide, in this paper, any details of the programs used for simulation. Fuzzy inductive reasoning is accomplished using SAPS–II (Cellier, 1987), a software that evolved from the General System Problem Solving (GSPS) framework (Klir, 1985, 1989; Uyttenhove, 1979). SAPS–II is implemented as a (FORTRAN–coded) function library of CTRL–C (SCT, 1985). A subset of the SAPS–II modules, namely the recoding, forecasting, and regeneration modules have also been made available as an application library of ACSL (MGA, 1986), which is the software used in the mixed quantitative and qualitative simulation runs. More details will be provided in an enhanced version of this paper that is currently being prepared for submission to a journal.

## ACKNOWLEDGMENTS

The authors are thankful to Dr. Rafael Huber of the Instituto de Cibernètica of the Universitat Politècnica de Catalunya for suggesting the problem discussed in this paper. His support of the project is gratefully acknowledged.

## REFERENCES

Babbie, E. (1989), *The Practice of Social Research*, 5th Edition, Wadsworth Publishing Company, Belmont, Calif.

Bobrow, D.G., ed. (1985), *Qualitative Reasoning about Physical Systems*, M.I.T. Press, Cambridge, Mass.

Cellier, F.E. (1991a), Continuous System Modeling, Springer-Verlag, New York, 755 p.

Cellier, F.E. (1991b), "Qualitative Modeling and Simulation — Promise or Illusion," *Proceedings 1991 Winter Simulation Conference*, Phoenix, AZ, pp. 1086–1090.

Cellier, F.E. and D.W. Yandell (1987), "SAPS–II: A New Implementation of the Systems Approach Problem Solver," *International J. of General Systems*, **13**(4), pp. 307–322.

de Kleer, J. and J.S. Brown (1984), "A Qualitative Physics Based on Confluences," *Artificial Intelligence*, **24**, pp. 7–83.

Forbus, K.D. (1984), "Qualitative Process Theory," *Artificial Intelligence*, **24**, pp. 85–168.

Hayes, P.J. (1979), "The Naïve Physics Manifesto," in: *Expert Systems in the Micro–Electronic Age* (D. Michie, ed.), Edinburgh University Press, Edinburgh, Scotland, pp. 242–270.

Klir, G.J. (1985), *Architecture of Systems Problem Solving*, Plenum Press, New York.

Klir, G.J. (1989), "Inductive Systems Modelling: An Overview," *Modelling and Simulation Methodology: Knowledge Systems' Paradigms* (M.S. Elzas, T.I. Ören, and B.P. Zeigler, eds.), Elsevier Science Publishers B.V. (North–Holland), Amsterdam, The Netherlands.

Kuipers, B. (1986), "Qualitative Simulation," *Artificial Intelligence*, **29**, pp. 289–338.

Li, D. and F.E. Cellier (1990), "Fuzzy Measures in Inductive Reasoning," *Proceedings 1990 Winter Simulation Conference*, New Orleans, La., pp. 527–538.

MGA (1986), *ACSL: Advanced Continuous Simulation Language — User Guide and Reference Manual*, Mitchell & Gauthier Assoc., Concord, Mass.

SCT (1985), *CTRL–C, A Language for the Computer–Aided Design of Multivariable Control Systems, User's Guide*, Systems Control Technology, Inc., Palo Alto, Calif.

Uyttenhove, H.J. (1979), *SAPS — System Approach Problem Solver*, Ph.D. Dissertation, SUNY Binghampton, N.Y.

# EXPERT REASONING WITH NUMERIC HISTORIES

## R. Milne, E. Bain and M. Drummond

*Intelligent Applications Ltd., Kirkton Business Centre, Kirk Lane, Livingston Village, West Lothian,*
*Scotland, EH54 7AY, UK*

Abstract. This paper describes an application developed for a major British Steel plant. It combines expert systems for process supervision with the on-line data acquisition system of the plant. The data from the plant is in a form of analogue and digital values. The diagnostics must reason about faults indicated by these values. A most important factor is that the faults are indicated by a time history, rather than immediate measurements.

As a result, extensive facilities have been developed to abstract the time histories of these numeric values into a qualitative form usable by the expert system. This abstraction process from numerical histories to qualitative states is one of the essential activities in combining numeric and qualitative reasoning.

The paper gives a description of the application and the numeric time history abstraction functionality.

Keywords. Expert Systems; qualitative reasoning; process monitoring; integration of numeric and qualitative reasoning.

## INTRODUCTION

Most plants need data interpretation and diagnosis. Significant benefits in terms of reliability and quality can be achieved if this is performed automatically (Milne, 1987a). Expert systems provide a very powerful means for achieving this (Milne, 1990a). They provide the appropriate mechanism for capturing the skill of an experienced engineer and turning that into an automatic computer program.

To be truly useful, these systems must be on-line (Milne, 1987b). These systems are able to acquire data continuously from the plant but need to pre-process the data to a form suitable for the expert system diagnosis. The expert system then checks for a large set of possible faults and can provide extensive diagnostic information (Milne, 1990b). Traditional expert system shells only provide a means of implementing the diagnostic rules; they do not provide the extra functionality needed to develop a real time system and interface to the plant (Jakob, 1990).

In order to support qualitative reasoning (Trave, 1989; Kuiper, 1986), it is necessary to provide a mechanism for data extraction. This mechanism must convert the raw numerical data into the appropriate qualitative states for the qualitative reasoning mechanism. This data extraction mechanism is also an essential part in facilitating the combination of numeric and qualitative reasoning (Charb, 1991).

In this paper, we discuss a strategy for abstracting the data to support expert systems and qualitative reasoning. A key aspect of this work is that the abstraction is based on numerical time histories of values, rather than just individual point values. The techniques described in this paper have been fully implemented on a large scale real-time expert system for British Steel at Ravenscraig in Scotland (Milne, 1990c). In this paper we briefly outline the application, the data abstraction and how they fit together.

## THE APPLICATION

We have developed a VAX based, large scale real-time expert system at British Steel Ravenscraig. It monitors the steel making process both to identify the faults they have now, and to predict faults that will occur in the future. The expert system is implemented as a rule-based system.

A ladle of molten iron (over 300 tons in this case) from the blast furnace is poured into a conversion vessel. The iron is then blasted with oxygen from a water cooled lance. Oxygen combines with the carbon impurities in the iron, producing carbon monoxide waste gas, leaving liquid steel. This process is called Basic Oxygen Steelmaking (BOS). It fits between the operations of the blast furnace, producing the molten iron and the mills where they roll it out into tubes, plates and a variety of beams.

The diagnostic system monitors the waste gas extraction and the primary process variables for the engineering (as distinct from production) areas. Any fault requires attention by an engineer and may also affect production quantity, quality and safety.

The system identifies those faults that need to be fixed now and those that will need attention in the near future. The engineering staff want to know "What do I fix now, what do I fix in the future and what should I do to prevent a predicted fault occurring?". The system provides various degrees of information. The senior managers are able to look and see high level information about the state of the plant and make strategic decisions about what has to be fixed and when, for example if there is a short shut-down. On the other hand, junior staff need able to look at a display, get all the relevant information about that fault and repair it without involving more senior staff.

This is not an experimental system, but is currently one of the largest real time expert systems developed in the UK. Such a large system is inevitably complex, comprising several cooperating sub-systems, including data acquisition and processing, expert system based fault diagnosis and prediction, fault management and display combined with extensive reporting facilities from the system databases.

The data acquisition system was designed to interface with the existing British Steel data sources. *Annie* (Milne, 1990a) provides the capability to couple the expert system with this data acquisition allowing real time diagnosis. In this paper we are only concerned with the data processing portions.

The primary output of the system is a list of plant faults. These faults are divided into two categories. Those which have been identified as already existing and those that are predicted to happen. The plant engineer is able to look at a summary of the number of faults in each sub-area of the plant. He is then able to examine for each fault, the trigger that caused the system to investigate it, and an explanation of the fault including the past values related to the particular control items. For example, the controller output time history would be included with a controller output fault. This provides for a more comprehensive explanation and background behind the fault.

Another important aspect is a measure of how this fault has affected the production. Some faults will occur continuously after their first occurrence, others may occur in an intermittent fashion, depending on the actual parameters and state of the steel making process. The expert system output includes; a summary of each fault, of how long since it first occurred and how many steel batches were affected by this fault. This is particularly useful in spotting intermittent faults. It is also possible for the engineer to look at the history of faults in the database to see for example, what other occurrences of a controller valve being stuck have happened over the past several months.

A major aspect of the system is it's ability to examine long term trends and predict faults. For example, consider the control cone. To control the air flow, a large cone is moved up and down. One parameter that can be measured over a period of time is the build-up of particles on the outside of the cone. It gradually acquires materials on the side that clog it up. The system can trend over time with regard to how high the cone is and look at its average position. As it gets built up it must be opened further and further each time. We can then trend how long until the cone is fully open, at which point it is not useful for control as it can only close in the one direction. At that point it has to be repaired and the system can now forecast how many days until they must be accomplished.

The outputs of the expert system analysis are directed to a number of places. In the first instance they are stored in a database on the primary diagnostic VAX, so that past histories and reports can be produced. There are a variety of reports available including; end of shift, end of week and end of campaign reports. The screen fault displays are available on a number of terminals throughout the engineering portion of the steel works.

## THE DATA ACQUISITION STRATEGY

The system being described is implemented on a VAX cluster under VMS. This is a multi process computer and as a result, is able to perform several tasks simultaneously. The data acquisition is primarily performed by an attached special processor. This processor acquires data from throughout the plant and makes it available over the VAX network to the expert system VAX. A high priority process runs continuously acquiring data over the network and placing it into the memory of the VAX. Currently, the data acquisition is cycling approximately every three seconds.

The heart of the data abstraction is the *Annie* software from Intelligent Applications. It takes care of the management and access to the incoming data.

The *Annie* data acquisition process places the incoming data in a large circular buffer. This buffer is configurable, but holds approximately two minutes worth of data. The buffer cycles continuously, inputting new data with the appropriate time stamps. The expert system is then at its leisure to remove data from this buffer.

Because the data required may be spread out through the two minute history, it requires special abstraction functions to properly access the input. In support of the time processing, there is a database description for each individual channel. It contains information with regard to how fast it is updated, how long values must be at certain levels to be considered consistent, as well as thresholds for normal, high and low values. One of the difficulties of any plant is filtering out short term unimportant changes or random variations in this signal level. As part of this database description, the time limit over which a channel is considered stable is identified.

This data acquisition strategy represents a realistic industrial situation. The data acquisition was specified as standard within the steel plant. It would not have been appropriate for the expert system software to attempt to take control of data acquisition and specify when values should be measured. It is important to recognise that the data acquisition goes to a number of supporting computers and not just the expert system. It would have been totally inappropriate to have a specific scheduler to acquire data when the expert system wanted it or on the expert system schedule. That would not have been consistent with the rest of the plant wide data acquisition systems.

The expert system itself is implemented in the Nexpert system shell. The current rulebase contains 1,500 rules covering 450 faults, cycling every 15 seconds providing fault output to a system of databases and fault displays. The *Annie* software is responsible for the interface between Nexpert and the data acquisition system.

## DATA ABSTRACTION

The simplest and a very common form of access for incoming data from the expert system is to ask for its current value. The *Annie* system is able to provide the current numeric value, if that is requested. However, for large industrial plants many values change quite rapidly and would appear to be slightly unstable. *Annie* uses the short term history to help stabilise these values. For example, a signal would only be considered to be qualitative high, if it went high and remained high for 30 seconds. In a similar way, a digital would only be considered to be ON if went to the ON position and remained there for 10 seconds.

The exact time periods for the latching mechanism are configurable by a database. The majority of the diagnostic rules rely on the qualitative state rather than the exact numeric values. The thresholds for defining these qualitative states are also contained in a separate database. This allows the user to adjust the definition of the qualitative states without requiring training or without affecting the expert system itself.

A key observation is that only some of the data is actually used by the rules. The pre-processing step filters out the data which would never be used, and abstracts it to a more useful form. The rulebase is also structured so that a trigger event will begin a more indepth analysis of a particular set of problems. Many of the more complex functions are only used when such a trigger event has occurred. This results in an efficient processing of the incoming data and reduces un-needed analysis.

Many of the rules are oriented around a time history of the values. Particular events such as a flow setpoint altering significantly, trigger the expert system to check a set of possible faults. In this case, *Annie* provides extensive functionality to provide access to the history of values. Because the system has planned ahead to acquire the data in a suitable form, these functions are able to look back and determine the pattern at the time the rulebase runs.

For qualitative reasoning systems, it is usually important to know whether a value is increasing or decreasing, increasing rapidly, decreasing rapidly etc (Trave, 1990). These qualitative values are determined by looking at the time history of the variable. For this application we have developed a library of useful abstraction functions.

As an example rule, consider the gas analyser on the flare stack. For safety reasons, it is important to detect whether the gas analyser is faulty. In a normal situation the mixture of a gas such as CO is constantly varying. The *Annie* time history function **percent variation less than for** is used to determine whether the CO level is constantly changing. If the gas analyser is changing by less than a certain percentage, then we can deduce that it has failed. This can also be confirmed by comparing the percent variation of other gas analyses in other parts of the system. The percent variation function examines the two minute time history of the data. It looks at the way it has been changing over that period and the amount of variation which has occurred. To implement this looking back is relatively straight forward, however, to have implemented this as a constant analysis as data was arriving, would be extremely difficult. Table 1 provides a list of further examples of the *Annie* time history functions. Multiple signal versions of many of these are also available.

### TABLE 1 Annie Data Abstraction

```
        equal_for
        constant_for
        less_than_for
        greater_than_for
        less_equal_for
        greater_equal_for
        channel_increased
        channel_decreased
        value_between_for
        number_outside_limit
        number_above_limit
        number_below_limit
        percent_variation_lt_for
        max
        min
        gradient
        mean
        integrate
        last_change_of
        at_status_for
        last_time_to
        response_time
        read_time
        read_system_time
        read_rules_loop_time
        start_timer
        read_timer
        timer_running
        timer_status
```

As will be observed by an examination of Table 1, there are large varieties of ways to evaluate the history of a parameter. Different fault diagnoses require different historical analysis of the data. We follow a specific approach in order to select which of these abstraction functions should be developed. During the knowledge engineering process, the plant engineers describe in their own words how portions of the equipment should change over time. Although there are many possible variations, the total of these is finite and is often fairly small. They essentially represent a qualitative language to describe the data (Milne, 1987a).

As a result of the knowledge engineering process, we identified this qualitative vocabulary. The items of the vocabulary are the abstraction functions needed, so that the expert system can be developed in the same mental image as that used by the plant engineer. Given the basic mechanisms provided by *Annie*, it is relatively straight forward to implement many of these functions or to extend the functionality of the system. It can be noted that in the current implementation, many of the qualitative abstract functions provide more complex analysis of the data than used by most current qualitative reasoning applications.

### EXAMPLE DIAGNOSTIC RULE

The following is an example of the type of rule which can be implemented; Imagine the fault to be detected is a control valve, being stuck or too slow to respond. If a flow rate increases significantly above the set point, the controller output should decrease and as a result the valve position decrease.

Figure 1 shows an example rule to investigate this as well as the time history of the data. In this case the dashed lines indicate correct operation of the central system while the solid branch illustrates this particular fault. Once the expert system has found that the flow rate is too high and the set point is OK then it checks to see if the controller output to altered to attempt to reduce the flow. If the flow rate is too high and the controller output has decreased to close the valve and the valve position is not closing, then the control valve is stuck.

In order to determine whether the control output has decreased, *Annie* looks back over the last two minutes of data, and through one of its abstraction functions, is able to determine that the control output is gradually decreasing. It is then able to look at the valve position over the last two minutes and determine by how much it has decreased. If the valve position has not decreased significantly after being instructed to close then we would say that it is stuck or has a slow response. Elaboration of the rules in this area would allow us to differentiate between these cases. It should be noted that by looking at the time history, this is easy to detect. To try and determine from the current values could be difficult and unreliable.

### SUMMARY

In this paper we have described a large scale real-time application. One of the key aspects of this system was the ability to abstract the numeric histories of data into qualitative state. These qualitative states include not only a qualitative current value of the parameter, but more importantly, qualitative states based on the time history of the data. The key element of the system that made this possible, was the use of time histories of the incoming data and the development of suitable abstraction functions. The data is pre-processed and stored in special history areas. The diagnostic rules are then able to examine this history with powerful data abstraction capabilities in order to determine how the system has behaved over time. This provides an effective strategy for interfacing qualitative reasoning systems with numeric input.

### REFERENCES

Charb, P. (1991). A Quasi Qualitative Formalism for Fault Diagnosis as a Solution to Numeric-Symbolic Interface Problems. DSS&QR 91, Toulouse.

Jakob, G. (1990). Situation Assessment for Process Control. IEEE Expert, pp. 49-59.

Kuiper, B.J. (1986). Qualitative Simulation. Artifical Intelligence.

Milne, R.W. (1987a). Artificial Intelligence for On-line Diagnosis, IEE Proceedings, Vol. 134, Pt. D, No. 4th.

Milne, R.W. (1987b). On-Line Artificial Intelligence.

The 7th International Workshop on Expert Systems & Their Applications, Avignon, France.

Milne, R.W. (1990a). Diagnostic Strategies. Systems & Control Encyclopedia Supplementary Volume 1, Madan G. Singh, Editor in Chief, pp. 155-160. Pergamon Press Publications.

Milne, R.W. (1990b). Expert Systems On-line. Systems & Control Encyclopedia Supplementary Volume 1, Madan G. Singh, Editor in Chief, pp. 243-251. Pergamon Press Publications.

Milne, R.W. (1990c). Case Studies in Condition Monitoring. Knowledge-Based Systems for Industrial Control, IEE Control Engineering Series 44. (J. McGhee, M.J. Grimble and P. Mowforth, Editors), pp. 255-266. Peter Peregrinus Ltd Publications.

Milne, R.W. (1990d). Monitoring Process Control Systems. Advances in Engineering Software, Computational Mechanics Publications, vol 12, No. 3, pp. 129-132.

Trave, L. (1989). The Order of Magnitude Models as Qualitative Algebras. IJCAI 1989, Vol. 2 pp. 1261-1266.

Trave, L. (1990b). Qualitative Reasoning: Methods, Tools and Applications. MQ&D Project Report, co-ordinator Rapport du pole AUTOMATISATION INTEGREE du CNRS.

## FIGURE 1.: EXAMPLE DIAGNOSTIC RULE

### INVESTIGATE CAUSE OF HIGH FLOW RATE

IF    Setpoint NOT too high

AND    Controller output has decreased to close the valve

AND    Valve is NOT already fully closed

AND    Valve has not moved in response to controller

THEN   Cause is    Control valve stuck, or slow response

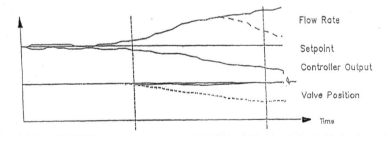

Copyright © IFAC Intelligent Components and Instruments
for Control Applications, Malaga, Spain, 1992

# A PERFORMANCE EVALUATION SYSTEM OF PROCESS CONTROL

J. Quevedo, T. Escobet, J.A. Gallardo, J.C. Hernandez and J. Armengol

*Universidad Politecnica de Cataluña, C/Colom, 11, 08222 Terrassa, Spain*

Abstract. In this work we present a system for evaluating, comparing and reporting the perfomance of different commercial controller in real time. The system has a test bench with four real processes that cover a wide range of industrial processes and allow analysis of single and multivariable controllers. On the basis of the acquisition and processing of the information in real time, the system assist the operator to supervise the process and in to take decisions. So far, a series of test and criteria for evaluating the perfomances of the controllers has been defined with regard to the self-tuning function. On the basis of the caracterization of the process in open loop, a suitable structure of the controller is recommended. Finally, the first evaluation results of an adaptive commercial controller are presented by means of the process of temperature control; these will soon be increased to cover other process and controllers for suitable comparison and analysis.

Keywords. Perfomance analysis; supervisory control; controllers; PID control; process parameter estimation; thermal variables control.

## INTRODUCTION

The correct selection and configuration of a process controller is crucial for obtaining certain functional performances of the process being controlled, such as consumption, stability, speed and accuracy. Generally, the responsable operator of the process faces with an installation where the sensors and associated actuators are already incorporated, having no more freedom than the choice of the controller structure (feedback, feedforward, cascade, ratio, etc.) and tuning (either manual or automatic) the parameters in order to attain the desired performances.

The importance of correctly tuning the parameters of a controller and the development of proven automatic tecniques of self-tuning and adaptive control, leading to an easy implementation in microprocessors has given rise to the rapid and massive development of commercial controllers with automatic tuning and/or adaptive control incorporated, this has become an inseparable feature of modern digital controllers.

At the same time, the users may believe that all self-tuning automatic controllers provide the same or similar perfomances for a given process and that

their field of action is extensible to all types of processes with same perfomances. In fact, this is not at all true, and therefore it becomes necessary to provide the users with information on the real capacities and possible limitations of these controllers.

There are a good number of publications and works on this subject (Hang, 1991; Kaya, 1986; Krauss, 1984; Nachtigal, 1986; Quevedo, 1988), perfomance of different controllers and different functional test. Many of these works are limited to analyzing the results obtained in simulated processes, using analog or digital simulators (Hang, 1991; Nachtigal, 1986) whereas others (Krauss, 1984; Quevedo, 1988) merely provide the analysis of experimental results of one or two controllers.

Our purpose has been to create a real time evaluation system on the perfomance of any commercial controller, by completing a set of benchmark test with real laboratory processes wich provide useful information to the operator supervising the controlled process so that he make take this own decision (change the controller, structure, or once again activating self-tuning of the parameters) whenever necessary.

In order to carry this out, a test bench was set up with laboratory processes (temperature, level, flow, pH), connected to a data processing system for the acquisition, treatment and storage of information, with visualization of the results of the evaluation. Also defined was a series of standard test which prove highly useful for obtaining comparative results between different advanced controllers.

## STRUCTURE OF THE EVALUATION SYSTEM

The implemented system (Fig. 1) for evaluating the perfomance of the controllers is composed of:

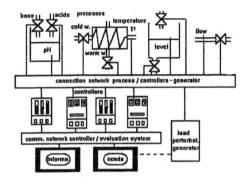

Fig. 1 The evaluation system set

● a set of four independent processes for level, flow, temperature and pH
● a network of connections between the sensors and actuators of the processes and controllers to be tested
● a personal computer for the adquisition, supervision and management of the results, with the different communication drivers with the controllers
● a personal computer for processing the information and presentation the results emerging from the evaluation
● a local communications network between the controllers and computers.

The system also includes a signal generator that causes disturbance of load in the different processes. Currently, the generator is activated manually although in future it will be synchronized and will thus become active automatically from the acquisition, supervision and storage computer.

## PROCESS BENCH

Our evaluation system comprises real and non-simulated processes which range from those which are relatively easy to control, such as the

temperature control of an interchanger and the level control of a tank, to more complicated processes such as the control of the flow trhough a pipe and the control of pH in a mixture of hydrochloric acid and sodium hidroxide.

The processes can run either independently or in combined form, creating different configurations of single variable or multivariable controls. For example, it is posible to control at the same time the temperature of cold water at the outlet of an interchanger, the flow of hot water circulating through the same one which could pour out into one of the tanks, and to control the level of this tank.

One of the feature of the processes selected is that they are all prepared to admit external load disturbance, for example in the temperature process (Fig. 2), one can act on the flow of hot water with a servovalve in order to control the temperature at the oulet. At the same time one can modify the flow of cold water with a noise signal generator and another sevovalve, creating load disturbances.

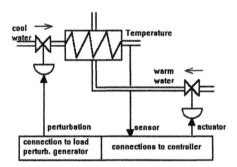

Fig. 2 Temperature process diagram

## ACQUISITION AND SUPERVISION SYSTEM (SCADA)

In order to carry out an evaluation system of controllers, it is essential to have a subsystem that performs functions of acquisition, processing and storage of the results that materialize from the evaluation test and the supervision and control functions in order to interact with the evaluation test. This is generically known as Supervisory Control and Data Acquisition (SCADA).

The choice of a SCADA system is not easy due to the considerable proliferation of commercial SCADA's. In general, the choice is based on criteria that evaluate hardware and software capacity.

The aspects to be born in mind with regard to the hardware are its versatility with regard to the computer, the possibility of working in single or multiple task at the same time, minimum size of the memory that the computer requires, and the communications and peripherals that may be connected to it.

The aspects to be taken into account concerning the software are that the system should accept parameters for a given application or allow new configurations for generating new applications; the programming language used by the software, ability of increasing the calculation and information processing functions, the list of communication drivers with different brands of controllers and PLC, as well as its capacity for connecting up with other commercial software packages: Data Base, Resident Modules, Lotus 1.2.3 ...

For our own evaluation system, the SCADA we adopted is the commercial package *PROCESSYN* by Logique Industrielle, which allows single or multiple tasks in personal computer with the MS-DOS or OS/2 operative systems. In our case, it was installed in a 486 computer with the MS-DOS V5.0 operative system. PROCESSYN is an application generator with which it is possible to programme the acquisition, supervision and control functions of our evaluation system; it is equipped with a resident module that permits exchange of information with other computers and peripherals. This SCADA is also equipped with multiple communications drivers with commercial controllers, amongst which are Protonic PS by Hartmann Braun, different models by Eurotherm, Microcor III by Coreci, TCS, Sauter, etc., thus making it particularly attractive for our application.

The application developed captures the following data from the controller with a frequency of 4 Hz: set point, control variable and process variable, wich are sent to the information and evaluation system for processing. It also permits acquisition or modification of the different control parameters (proportional gain, integration time, derivative time, manual or automatic mode, ...).

The SCADA also permits modification of the function of the controlled process, since it can change the set point and parameters of the controller as well as passing over from manual control (open loop control) to automatic control (closed loop control).

## EXPERIMENTAL TEST PROCEDURE

The benchmark test, as designed here, evaluates the self-tuning perfomance of commercial controllers with the processes mentioned above, covering a

wide range of real dynamic processes. The test conditions are based on certain values of set-point and load disturbances that are easily reproducible with all the controllers and processes to be tested, thus permitting a standard and systematic test and the possibility of comparing the results obtained with the different controllers.

The benchmark test consist of eight steps (Fig. 3):

1. Self-tuning of the controller on one of the processes, followed by the procedure as described by the constructor, with a set point value of 55%. The load will have a constant value of 50%. Acquiring the data on the control variable and on the process variable.

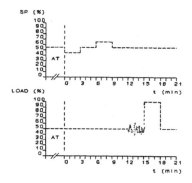

Fig. 3 Experimental test evolution

2. Changing the set point from 55% to 45% over period of time sufficient for reaching the new operating point (a time equal to the time constant plus the pure delay time of the process in open loop has been established).

3. Changing the set point from 45% to 55%, acquiring data on the control variable and on the process variable during an interval of time, and other conditions identical to those in the previous step.

4. Changing the set point from 55% to 65% under identical conditions.

5. Changing the set point from 65% to 55% and acquiring data on the process under identical conditions of load and time.

6. Applying a random disturbance load based on a white noise generator with gaussian disribution with an average value of 0% and standard deviation of 7% of the load, to which is to be added the constant disturbance of 50% of the load. Acquiring the same data under identical conditions.

7. Applying a load disturbance, increasing to 100%

the value of the load during the same interval of time as in the previous steps, which must be sufficient to reach the new operating point.

8. Applying load disturbance, reducing the value of the load to 50%. Acquiring the same data under identical conditions.

If the controller allows auto-adaptive adjustment of this parameters, steps 2 to 8 are repeated until the parameters remain fixed (at least 5 times), and the results are stored.

## PERFOMANCES EVLUATION AND INFORMATION SYSTEM

The evaluation system offers certain criteria for evaluating the function of the different processes controlled and subjected to the evaluation test, which will serve for comparing perfomances of the different controllers in closed loop. The evaluation criteria will quantify the accumulated control error, the consumption of energy of the control variable, the response speed of the process, and the overshoot and damping of the process variable.

Also, one important feature of the evaluation system is that it operates in real time and on the basis of the information that it processes, and assists the operator to supervise and modify the control of the process. Thus, for example, by determining the noise level of the process output and calculating the dynamic features (gain, time constant and pure delay time) of the process, the information system can recommend the operator to change the structure of the controller (P, PI, PD, PID) and also calculate the time period of application steps 2 to 8 of the evaluation test. This information is drawn up when the process operates in open loop as detailed below.

### Information in Open Loop

The vast majority of commercial controllers with self-tuning operate in open loop applying a step, a pulse or a square signal at the process input and capturing the process response by tuning the controller parameters. In this way, the information in open loop described below is obtained during the self-tuning process of the controller ( step 1 in the evaluation test, mentioned above); otherwise, a suplementary stage is made to work in open loop (manual operation of the controller).

The processes described are not accumulators of energy and therefore their impulse response is always positive, and the response to a step input is a monotonous incresed of exponential type, with the following approximate generic expression:

$$y_m(t) = y_0 + A(1 - e^{-(t-d)/T}) , \qquad \text{if } t > d \quad (1)$$

and,

$$y_m(t) = y_0 , \qquad \text{if } t < d \quad (2)$$

$$\text{for} \quad u(t) = B.u_{step} + u_0$$

Therefore, the approximate transfer function is:

$$G(s) = \frac{Y_m(s)}{U(s)} = \frac{A e^{-ds}}{B(1 + Ts)} \quad (3)$$

The parameters A, B, d, and T of the transfer function are calculated on the experimentals results of the output variable $y_r(t)$ during an application time of the step input u(t).

The algoritm calculating the model parameters uses the simplex search method (Nelder and Mead, 1964) that calculates the minimum of the norm of $(y_r - y_m)$, $y_r$ being the real output of the process and $y_m$ being the output of the model. This algorithm has been programmed in Matlab, and provides the gain of the process (A/B), the time constant (T), the dead time (d) and the normalized dead time ($\phi = d/T$). The noise of the process output can be characterized on the basis of the modelling error $(y_r - y_m)$ and it calculates the average value (m), the standard deviation (s) and the noise/signal ratio (r=s/A).

If the controller acting on the processes were to use self-tuning techniques based on the empiric Ziegler-Nichols formula, the operator would then receive information on the best structure of the controller (P,PI,PID or other) according to the apparent dead time ($\phi$), the noise level in the process variable (m,s). Thus, the Table 1 collects up the information (Astrom, Hang and Persson,1988) for defining the best structure of the controller, in accordance with these parameters and the required degree of control perfomance.

### Closed Loop Evaluation

In the steps 3,5,6 and 8 of test procedure, the following criteria are evaluated:

- IAE = $\int |e| \, dt$      :Integral Absolute of Error

- ISE = $\int |e^2| \, dt$      :Integral Square of Error

- ISU = $\int |u^2| \, dt$      :Integral Square of Process Input

236

TABLE 1 Structure Controller Recommendation

| $\phi$ | Tight control is not required | Tight control is required | |
|---|---|---|---|
| | | High measurement noise | Lou measurement noise |
| $\phi < 0.15$ | P | PI | P or PI |
| $0.15 < \phi < 0.6$ | PI | PI | PID |
| $0.6 < \phi < 1$ | I + FF$_r$ | I + FF$_r$ | PI or PID+FF$_r$+C$_r$ |
| $\phi > 1$ | I + FF$_e$ + C$_r$ | I + FF$_e$ + C$_r$ | PI + FF$_e$ + C$_e$ |

FF$_r$: FeedForward Compensation recommended
FF$_e$: Feedforward Compensation essential
C$_r$: Dead-Time Compensation recommended
C$_e$: Dead-Time Compensation essential

- ITAE $= \int t|e| \; dt$ :Integral Time of Absolute Error
- $\phi$ : Overshoot of the Step Response
- D : Damping of the Step Response
- T$_r$ : Rise Time (10% to 90%) Step Response
- T$_{at}$ : Self-tuning Time

The first criterion quantifies the accumulated regulation error since it is the integral of the difference between set point and the process output; the second criterion qualifies the large errors in control by raising their value to square; the ITAE criterion takes into account small errors that occur later on; the third criterion quantifies the energy consumption of the procces input; $\phi$, D and T$_r$ provide the caracteristic parameters of the time response of the process output. Finally, T$_{at}$ quantifies the time that the controller takes for self-tuning.

## EXPERIMENTAL RESULTS

This paper presents the first results of the implemented evaluation system, whose fundamental features have been described in the previous sections.

The experimental results concern the process of controlling the temperature of the heat exchanger by adjusting the hot water servovalve to regulate the temperature at outlet with the cold water flow acting as a load disturbance (Fig. 2).

The commercial controller tested is the Eurotherm, model 903, which has self-tuning and auto-adaptive adjustment of the parameters of a PID controller. When the operator setup the self-tuning, an on-off controll sequence is applied at the process input, and the natural response of the process is determined so that control parameters can be derived and the P, I and D actions are set automatically in a closed loop operation. The auto-adaptive algorithm is activated

when load disturbances cause deviations from set point in excess of a predefined trigger point. The algorithm then proceeds to analyze the closed loop response and applies expert rules to return the loop.

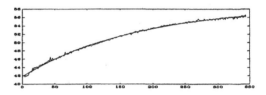

Fig. 4 Open loop response

In order to obtain the dynamic features of the process, an input step has been applied to the process in open loop (Fig. 4), acting from the SCADA, and the process output has been stored with a sampling period of 1 second. The routine considering the characteristic parameters of the process has given the results shown in Table 2.

TABLE 2 Estimation of the Parameters

| PARAMETERS | ESTIMATION |
|---|---|
| A | 16.8 °C |
| B | 9.5% |
| K | 1.77 °C/% |
| T | 178 s |
| d | 2 s |
| $\phi$ | 0.011 |
| m | -4.55 e-8 °C |
| s | 0.19 °C |
| r | 0.0113 |

TABLE 3 Evaluation Criteria in Closed Loop Operation

| STEP | IAE (s°C) | ISE (s°C²) | ITAE (s²°C) | ISU (s) | $T_r$ (s) | $\phi$ (%) | D (%) |
|---|---|---|---|---|---|---|---|
| 3 | 254.1 | 1497.0 | 8032.2 | 36.6 | 26 | 10.1 | - |
| 5 | 351.8 | 2538.4 | 10180.3 | 12.2 | 35 | - | - |
| 6 | 116.9 | 104.1 | 10197.4 | 17.8 | - | - | - |
| 8 | 115.5 | 142.1 | 6901.4 | 17.2 | - | - | - |

The low values obtained of the normalized dead time and the noise/signal ratio ($\phi$=0.011 and r=0.0113) allow us to conclude that a controller of P or PI type with empiric Ziegler-Nichols tuning is sufficient for controlling the process temperature in accordance with Table 1 proposed by Astrom.

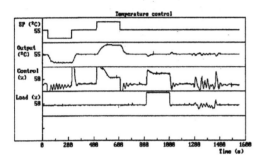

Fig. 5 Results of closed loop evaluation

On the other hand, the Fig. 5 synthesizes the experimental results of the 8 steps of the experimental test, with the Eurotherm model 903 and the temperature process in closed loop. The time required for carrying out self-tuning (step 1) was 166 seconds, and the Table 3 shows the numeric results of the evaluation criteria proposed.

## CONCLUSIONS

A useful system for evaluating the perfomances of advanced commercial controllers has been designed and shown in this paper. The evaluating system works on-line, acquiring the data of the controlled process and processing the information in real time, and consequently the system can assist the operator to supervise the process and take decisions.

The system is composed of four real processes which allow to analyze single and multivariable controllers.
A series of test and criteria for evaluating the self-tuning and auto-adaptive functions have been defined, the first evaluation results of an adaptive controller have been obtained and information about PID structure has been derived.

Now a useful system to do perfomance analysis and different advanced controllers of the market is at one's disposal, so interesting conclusions could be extracted to aid the users to take decisions.

Acknowledgement. This work has been supported by the Spanish Board for Research and Technology (CICYT) under contract ROB0736/91, and the paper has been partially supported by the Group for Study and Research in Automatic Control (CERCA).

## REFERENCES

Astrom, K.J., C.C. Hang and P.Peterson (1988). Heuristic for assessment of PID control with Ziegler-Nichols tuning. Technical Report LUFTD2/(TFRT-7404), Dep. of Automatic Control, Lund Institute of Technology, Lund.

Bristol, E.H. (1977). Pattern recognition: An alternative to parameter identification in adaptive control. *Automatica*, **13**,197-202.

Hang, C.C. and K.K. Sin (1991). A comparative perfomance study of PID auto-tuners. *IEEE Control Systems*,vol.**11** n.**5**,41-47.

Kaya, A. and S. Titus (1986). A critical perfomance evalaution of four single-loop self-tuning control products. *Proc. American Control Conference*, 1659-1664.

Krauss, T.W. and T.J. Myron (1984). Self-tuning PID controller uses pattern recognition approach. *Control Engineering*,vol.**31** n.**6**,106-111.

Nachtigal, C.L. (1986). Adaptive controller simulated process results: Foxboro EXACT and ASEA Novatune. *Proc. American Control Conference*, 1434-1439.

Nelder, J.A. and R. Mead (1964). A simplex method for function minimization. *Computer Journal*,**7**,308-313.

Quevedo,J., and other (1988). Prueba experimental de dos controladores adaptativos industriales. *Automatica e Instrumentación*,**178**,173-179.

# DECENTRALIZED CONTROL DESIGN FOR UNCERTAIN SYSTEMS USING MULTIMODELLING

**L. Bakule[1] and J. Rodellar**

*Departamento de Matematica Aplicada III, Universidad Politecnica de Catalunya, Gran Capitan S/N,
08034 Barcelona, Spain*

**Abstract.** Nonlinear composite control of a class nominally linear large scale systems is presented. Control strategies for decision makers using different models of the same system are derived. Interconnected systems with slow and fast dynamics and bounded uncertainties both in parameters and signals are considered for decentralized control design using practical stabilization. Conditions for the validity of this procedure are given and illustrated on a power system example.

**Keywords.** Decentralized control; decomposition; distributed control; feedback control; Lyapunov methods; nonlinear control systems; robustness; singular perturbations; stabilisers; load frequency control.

## INTRODUCTION

Situations in which strategies of various decision makers are designed using different models of the same system are a characteristic of large scale systems practice. Considering large scale system consisting of one slow and $N$ fast subsystems it is rational for a fast system controller to neglect all other fast subsystems and to concentrate on its own subsystem, plus the interaction with others through the slow core. It results in an particular model of the same system for each decision maker. This situation is called "multimodelling". These problems have been investigated for both deterministic and stochastic systems (Khalil and Kokotovic, 1978; Naidu, 1988). The inclusion of unknown but bounded deterministic uncertainties within the concept of multimodelling has not been considered yet. In this paper practical stabilization of singularly perturbed systems by Garofalo and Leitmann (1988) is employed to capture the multimodel nature of interconnected systems with slow and fast dynamics. The derived method is illustrated by a load–frequency control problem for a two–area power system.

## PROBLEM FORMULATION

A nominally linear uncertain system consisting of strongly coupled slow and weakly coupled fast subsystems is modelled by

$$\dot{x}_i = A_{11}(q)x + A_{12}(q)z + B_1(q)u + \omega_1(q),$$
$$\epsilon_i \dot{z}_i = A_{21}(q)x + A_{22}(q)z + B_2(q)u + \omega_2(q), \quad (1)$$

with

$$A_{12}(q) = (A_{o1}(q), ..., A_{oN}(q)),$$
$$B_1(q) = (B_{11}(q), ..., B_{1N}(q)),$$
$$A_{21}(q) = (A_{1o}(q), ..., A_{No}(q)),$$
$$A_{22}(q) = diag(A_{d1}(q), ..., A_{dN}(q)),$$
$$B_2(q) = diag(B_{21}(q), ..., B_{2N}(q)),$$
$$u = (u_1^T, ..., u_N^T)^T,$$
$$z = (z_1^T, ..., z_N^T)^T,$$
$$\epsilon = (\epsilon_1, ..., \epsilon_N),$$
$$\omega_2(q) = (\omega_{21}^T(q), ..., \omega_{2N}^T(q))^T,$$

where $(x^T, z^T)^T \in R^{n+m}, z_i \in R^{m_i}$ represents the vector of states, $u \in R^p$ is the control vector, $q \in R^q$ is the uncertainty, $\omega_1(q)$ and $\omega_2(q)$ are the external disturbances, and $\epsilon \in (0, \infty)$ is the vector of "small" singular perturbation parameters, where $\epsilon_i$ are supposed to be scalars of the same order. Matrices $A_{ij}(.), B_j(.), i, j = 1, 2$ and vectors $\omega_i(.), i = 1, 2$ are continuous in their argument. Suppose that the uncertainty $q(.) : R \to R^q$ is Lebesgue measurable and its value $q(t)$ lies in a prespecified bounding compact set $Q \subset R^q$ for all $t \in R$. Suppose that the following assumptions hold:

A1. (Matching assumption) There exist known constant matrices $\overline{A}_{ij}, i, j = 1, 2$ and full rank matrices $\overline{B}_j, j = 1, 2$ such that

$$A_{ij}(q) = \overline{A}_{ij} + \overline{B}_i D_{ij}(q), D_{ij}(0) = 0,$$
$$B_i(q) = \overline{B}_i + \overline{B}_i E_i(q), E_i(0) = 0, \quad (2)$$
$$\omega_i(q) = \overline{B}_i d_i(q), d_i(0) = 0, i, j = 1, 2,$$

where $D_{ij}(.), E_i(.)$ are matrices and $d_i(.)$ is a vector,

---

[1] On leave from the Institute of Information Theory and Automation, Czechoslovak Academy of Sciences, CS-182 08 Prague 8, CSSR

all of appropriate dimensions.

<u>A2.</u> The pairs $(\overline{A}_{di}, \overline{B}_i), i, j = 1, 2$ are controllable.

<u>A3.</u> The matrix $\overline{A}_{di}$ is nonsingular.

The assumption A1 reflects the qualitative knowledge of the system and it concerns the manner in which the uncertainty structurally enters into the system. The assumption A2 enables to assign arbitrary behaviour (at least stable one). The A3 assumption is the standard two–time scale control design requirement.

Suppose that there are generally $N$ decision makers for the system (1), where each of them knows only his simplified model. This model is constructed for the i–th decision maker by neglecting all fast dynamics $\epsilon_j, i \neq j$ and uncertainties $q(t)$ of all j–th fast subsystem makers, so that this model has the form

$$\dot{x}_i = A_{11}(q)x + A_{12}(q)z + B_1(q)u + \omega_1(q),$$
$$\epsilon_i \dot{z}_i = A_{io}(q)x + A_{di}(q)z_i + B_{2i}(q)u_i + \omega_{2i}(q), \quad (3)$$
$$0 = \overline{A}_{jo}x + \overline{A}_{dj}z_j + \overline{B}_{2j}u_j.$$

The design objective is to construct practically stable controller under such multimodel assumptions.

The concept of practical stability and a criterion to verify it is given for instance in Garofalo and Leitmann (1988).

## SOLUTION

Motivated by the single–parameter singular perturbation approach, we propose that each decision maker will use the two–time scale design method by Garofalo and Leitmann (1988). The principle of the multimodel solution can be fully covered by the case $N = 2$ (Khalil and Kokotovic, 1978). Following this approach, we have to solve two separate subproblems for the slow and fast subsystems.

The slow time scale subproblem of the i–th decision maker is obtained by setting $\epsilon_i = 0$ for the fast nominal system, i.e. for $q(t) = 0$ for the fast subsystem.

The fast time scale subproblem of the i–th decision maker is obtained by substituting the expression for the slow controller into the i–th fast subsystem component and solved for the slow state variable as a fixed parameter. The block diagonal structure of the fast subsystem components enables decentralized design of fast subsystems control (Bakule and Lunze, 1988). To employ the single parameter solution by Garofalo and Leitmann (1988), we use a joint time scale for fast subsystems. A condition for the composite controller is derived and an upper singular parameters bound is given. Therefore, the way of the solution summarizes only the necessary steps which are identical and emphasises new extensions when compared with the results by Garofalo and Leitmann (1988) and Khalil and Kokotovic (1978). First, slow controller is designed, then the design of fast controllers follows. The evaluation of conditions for the composite controller to be practically stable is the last design step.

The multimodel representation (3) results in the i–th decision maker simplified model as follows

$$\dot{x}_i = A_i(q)x_i + A_{oi}(q)z_i + B_{1i}(q)u_i +$$
$$+ \sum B_{1j}^i(q)u_j + \omega_1(q), \quad (4)$$
$$\epsilon_i \dot{z}_i = A_{io}(q)x + A_{di}(q)z_i + B_{2i}(q)u_i + \omega_{2i}(q),$$

where

$$A_i(q) = A_{11}(q) - \sum_{i \neq j} A_{oj}(q)\overline{A}_{dj}^{-1}\overline{A}_{jo},$$
$$B_{1j}^i(q) = B_{1j}(q) - A_{oj}(q)\overline{A}_{dj}^{-1}\overline{B}_{2j}.$$

## The slow controller design

This design step is the same for multimodeling with that one in Garofalo and Leitmann(1988). We summarize it. The design is performed for $q(t) = 0$ in the fast subsystem and all $\epsilon_i = 0$. The slow subsystem has the form

$$\dot{x}_s = A_o(q)x_s + B_o(q)u_s + \omega_1(q) \quad (5)$$

with

$$A_o(q) = A_{11}(q) - A_{12}(q)\overline{A}_{22}^{-1}\overline{A}_{21},$$
$$B_o(q) = B_1(q) - A_{12}(q)\overline{A}_{22}^{-1}\overline{B}_2.$$

It is decomposed employing the matching properties as follows

$$A_o(q) = \overline{A}_o + \overline{B}_1 D_o(q) + \overline{F}_o,$$
$$B_o(q) = \overline{B}_1 + \overline{B}_1 E_o(q) + \overline{G}_o, \quad (6)$$

with

$$\overline{A}_o = \overline{A}_{11} - \overline{B}_1(\overline{B}_1^T\overline{B}_1)^{-1}\overline{B}_1^T\overline{A}_{12}\overline{A}_{22}^{-1}\overline{A}_{21},$$
$$D_o(q) = D_{11}(q) - D_{12}(q)\overline{A}_{22}^{-1}\overline{A}_{21},$$
$$E_o(q) = E_1(q) - D_{12}(q)\overline{A}_{22}^{-1}\overline{B}_2 -$$
$$- (\overline{B}_1^T\overline{B}_1)^{-1}\overline{B}_1^T\overline{A}_{12}\overline{A}_{22}^{-1}\overline{B}_2,$$
$$\overline{F}_o = \overline{T}_1(\overline{T}_1^T\overline{T}_1)^{-1}\overline{T}_1^T\overline{A}_{12}\overline{A}_{22}^{-1}\overline{A}_{21}.$$

where $T_1 \in R^{nx(n-p)}$ is a matrix whose columns span the subspace $R(\overline{B}_1)^\perp$. Define further $k_{D_o} = max\|D_o(q)\|$, $k_{E_o} = max\|E_o(q)\|$, $k_{\overline{G}_o} = \|\overline{G}_o(q)\|$, $k_{\overline{F}_o} = \|\overline{F}_o(q)\|$, $k_{d_1} = max\|d_1(q)\|$. Assume further that the following assumption is satisfied.

<u>A4</u> There exists a positive constant $\alpha_s = \lambda_m(Q_s) - 2k_{\overline{F}_o}\|P_s\|$ with the positive definite solution $P_s$ of the Riccati equation

$$\overline{A}_o^T P_s + P_s\overline{A}_o + Q_s - 2\rho_s P_s\overline{B}_1\overline{B}_1^T P_s = 0 \quad (7)$$

with $\rho_s > 0$ and $Q_s$ symmetric positive definite. Moreover, $k_{E_o} < 1$.

Then the controller is designed using the following theorem.

**Theorem 1.** Suppose that the assumptions A1,A2,A4 are satisfied. A linear controller

$$u_s = -\frac{\gamma_s + \rho_s}{1 - k_{E_o}}\overline{B}_1^T P_s x_s = -\bar{\gamma}_s \overline{B}_1 \overline{B}_1^T P_s x_s \quad (8)$$

guarantees the practical stability of the closed loop system (5),(8) satisfying (6) for any $\gamma_s \in (\gamma_1, \gamma_2)$, where

$$\begin{aligned}
\gamma_1 &= \frac{1}{4}[\frac{2k_{D_o}^2}{c_1 \alpha_s} + \frac{k_{d_1}^2}{c_2}], \\
\gamma_2 &= \frac{(1 - k_{E_o})}{4}\frac{(1 - c_1)\alpha_s}{\|\overline{B}_1\|^2 \|\overline{G}_o\|\|P_s\|^2} - \rho_s
\end{aligned} \quad (9)$$

and the nonnegative constants $c_1, c_2$ such that $c_1 < 1, c_1 \neq 0$ if $k_{D_o} \neq 0, c_2 \neq 0$ if $k_{d1} \neq 0$, and $\gamma_1 < \gamma_2$. A ball of ultimate boundedness has the radius

$$\underline{d}_s = [\frac{\lambda_M(P_s)v_2}{\lambda_m(P_s)v_1}]^{\frac{1}{2}} \quad (10)$$

and with initial condition $x_s^o, \|x_s^o\| > \underline{d}_s$ every subset of radius $\overline{d}_s > \underline{d}_s$ is reached in a finite time not greater than

$$T_s = \frac{\lambda_M(P_s)\|x_s^o\|^2 - (\lambda_m^2(P_s)/\lambda_M(P_s))\overline{d}_s^2}{v_1(\lambda_m(P_s)/\lambda_M(P_s))\overline{d}_s^2 - v_2} \quad (11)$$

and thereafter remains within it. The constants $v_1, v_2$ are

$$\begin{aligned}
v_1 &= \frac{1}{2}\alpha_s(1 - \frac{\gamma_1}{\gamma_s}c_1 - \frac{\gamma_s + \rho_s}{\gamma_2 + \rho_s}(1 - c_1)) > 1, \\
v_2 &= \frac{\gamma_1}{\gamma_s}c_2 > 0.
\end{aligned} \quad (12)$$

**The fast controller design**

It is performed by substituting (8) into the fast part. Because of block diagonal structure of the fast subsystem, each fast controller can be designed independently. To solve the overall fast subsystem control design using the results by Garofalo and Leitmann (1988) we need a joint fast time scale. Therefore, by introducing this joint fast time scale as $\tau = t/\sqrt{\epsilon_1\epsilon_2} = t/\epsilon_T$ we obtain

$$\begin{aligned}
\frac{dz_f}{d\tau} &= T[A_{22}(q)z_f + B_2(q)u_f + F_o(q)x + \omega_2(q)] = \\
&= A_{22T}(q)z_f + B_{2T}(q)u_f + F_{oT}(q)x + \omega_{2T}(q),
\end{aligned} \quad (13)$$

where

$$\begin{aligned}
z_f &= z - h(x), h(x) = -\overline{A}_{22}^{-1}(\overline{A}_{21} - \bar{\gamma}_s \overline{B}_1 \overline{B}_1^T P_s)x, \\
T &= diag(\sqrt{\frac{\epsilon_2}{\epsilon_1}}I_{m_1}, \sqrt{\frac{\epsilon_1}{\epsilon_2}}I_{m_2}), \\
&= diag(t_1 I_{m_1}, t_2 I_{m_2}), \\
F_o(q) &= \overline{B}_2 H(q), \\
H(q) &= (A_{21}(q) - \bar{\gamma}_s \overline{B}_1 \overline{B}_1^T P_s) - \\
&\quad - A_{22}(q)\overline{A}_{22}^{-1}(\overline{A}_{21} - \bar{\gamma}_s \overline{B}_1 \overline{B}_1^T P_s) \\
&= \overline{B}_2[D_{21}(q) - \bar{\gamma}_s E_2(q)\overline{B}_1^T P_s) - \\
&\quad - D_{22}(q)\overline{A}_{22}^{-1}(\overline{A}_{21} - \bar{\gamma}_s \overline{B}_1 \overline{B}_1^T P_s)].
\end{aligned}$$

This global fast subsystem consists of two independent fast decision makers. Let us deal with the i-th decision maker now. Denote $k_{H_i} = max\|H_i(q)\|, k_{B_{2i}} = max\|E_{2i}\|, k_{D_{4i}} = max\|D_{di}\|, k_{d_{2i}} = max\|d_{2i}\|, i = 1, 2,$where $D_{22}(q) = diag(D_{d1}(q), D_{d2}(q))$. To construct practically stable controller for the boundary layer system we need the following assumption.

**A5.** $k_{B_{2i}} < 1, i = 1, 2.$

Making the boundary layer model practically stable we construct the controller using the following theorem. Note that $z_f = (z_{f1}^T, z_{f2}^T)^T, u_f = (u_{f1}^T, u_{f2}^T)^T.$

**Theorem 2.** Consider the system (13) with the controller

$$u_{fi} = -K_{fi}z_{fi} - \bar{\gamma}_{fi}(z_{fi}, x)\overline{B}_{2i}^T P_{fi}z_{fi}, i = 1, 2, \quad (14)$$

where $K_{fi}$ is such that $\overline{A}_{di} - \overline{B}_{2i}K_{fi}$ is asymptotically stable and

$$\bar{\gamma}_{fi}(z_{fi}, x) = \frac{\gamma_{f1i}\|z_{fi}\| + \gamma_{f2i}\|x\| + \gamma_{f3i}}{\|\overline{B}_{2i}^T P_{fi}z_{fi}\| + \delta_i} \quad (15)$$

with a parameter $\delta_i > 0$ and $P_{fi}$ the solution of the Lyapunov equation

$$(\overline{A}_{di} - \overline{B}_{2i}K_{fi})^T P_{fi} + P_{fi}(\overline{A}_{di} - \overline{B}_{2i}K_{fi}) = -Q_{fTi}, \quad (16)$$

where $Q_f = (Q_{f1}, Q_{f2}), Q_{fT} = TQ_f, Q_f > 0$ and

$$\begin{aligned}
\gamma_{f1i} &\geq \frac{k_{D_{di}} + k_{B_{2i}}\|K_{fi}\|}{1 - k_{B_{2i}}}, \\
\gamma_{f2i} &\geq \frac{k_{H_i}}{1 - k_{B_{2i}}}, \gamma_{f3i} \geq \frac{k_{d_{2i}}}{1 - k_{B_{2i}}}.
\end{aligned} \quad (17)$$

A ball of ultimate boundedness of the i-th subsystem has the radius

$$\underline{d}_{fTi} = [\frac{\lambda_M(P_{fi})}{\lambda_m(P_{fi})}]^{1/2}[w_{2i} + \sqrt{w_{2i}^2 + 4w_{4i}w_{3i}'}]/2w_{1i}, \quad (18)$$

where $w_{1i} = 0.5\lambda_m(Q_{fTi}), w_{2i} = \delta_i\gamma_{f1i}, w_{3i}' = w_{3i} + w_{4i}\|x\|, w_{3i} = \delta_i\gamma_{f3i}, w_{4i} = \delta_i\gamma_{f2i}$ and further with initial condition $z_{fi}^o$ such that $\|z_{fi}^o\| > \underline{d}_{fTi}$ a subset of every ball of radius $\overline{d}_{fTi} > \underline{d}_{fTi}$ is reached in a finite time not greater than

$$T_{FTi} = \frac{\lambda_M(P_{fi})\|x_{fi}^o\|^2 - (\lambda_m^2(P_{fi})/\lambda_M(P_{fi}))\overline{d}_{fi}^2}{w_{1i}(L_\lambda \overline{d}^2 - w_{2i}L_\lambda)^{1/2}\overline{d}_{fi} - w_{3i}'},$$

$$(19)$$

where $L_\lambda = \lambda_m(P_{fi})/\lambda_M(P_{fi})$ and thereafter remains within it. Note only here that this time is related to a joint fast time scale.

**Composite control and global system practical stability**

The way of evaluation of the global system practical stability using composite controller employes the results by Garofalo and Leitmann (1988) and extends them on the case of two fast controllers with different time scales. The global feedback system has the form

$$\dot{x} = (A_o(q) - \overline{\gamma}_s B_o(q)\overline{B}_1^T P_s)x_s +$$
$$+ F_{12}(x, z, q)(z - h(x)) + \omega_1(q),$$
$$\epsilon_T \dot{z} = T[A_{22}(q) - B_2(q)K_f - \qquad (20)$$
$$- \overline{\gamma}_f(z - h(x))B_2(q)\overline{B}_2 P_f)(z - h(x)) +$$
$$+ F_o(q)x + \omega_2(q)],$$

where $\overline{\gamma}_f = (\overline{\gamma}_{f1}(z_1 - h_1(x)), \overline{\gamma}_{f2}(z_2 - h_2(x)))$.

Conditions for practical stability of this controller presents the following theorem.

<u>Theorem 3.</u> Suppose that the assumptions A1-A5 are satisfied for the system (1). Suppose that the following inequality

$$\beta\gamma_{f2} = \beta(\gamma_{f21} + \gamma_{f22}) < \frac{v_1}{\|P_s\|}, \beta = max\|B_1(q)\|, \quad (21)$$

holds. Then it is possible to find such $\epsilon^* = (\epsilon_1^*, \epsilon_2^*)$ for the system $(1),(8),(15)$, $\epsilon \in (0, \epsilon^*)$ that the controller $(8),(15)$ is practically stable, where

$$\epsilon_i^* = t_i^{-1}\epsilon_T^*, \epsilon_T^* = \frac{(v_1 - b_1)w_1}{(v_1 - b_1)a_2 + a_2 b_2}, c^* = \frac{b_2}{a_1 + b_2} \quad (22)$$

with

$$b_1 = \|P_s\|\beta\gamma_{f2},$$
$$b_2 = \|P_s\|\sum(\beta\gamma_{f1i} + max\|A_{oi}(q) - B_{2i}(q)K_{fi}\|),$$
$$a_1 = (\beta\gamma_{f2} + max\|A_o(q) -$$
$$- \overline{\gamma}_s B_o(q)\overline{B}_1^T P_s\|)\sum\|P_{fi}\overline{A}_{di}^{-1}(\overline{A}_{io} -$$
$$- \overline{\gamma}_s \overline{B}_{2i}\overline{B}_{1i}^T P_s\|,$$
$$a_2 = \frac{b_2}{\|P_s\|}\sum\|P_{fi}\overline{A}_{di}^{-1}(\overline{A}_{io} - \overline{\gamma}_s \overline{B}_{2i}\overline{B}_{1i}^T P_s)\|,$$
$$w_1 = \frac{1}{2}\sum\lambda_m(Q_{fTi}).$$

The complete system is practically stable with a ball of ultimate boundedness whose radius is given

$$\underline{d} = [\frac{\lambda_M(P(c))}{\lambda_m(P(c))}]^{1/2}[(\|m(c)\| + (\|m(c)\|^2 +$$
$$+ 4\lambda_m(M(c))l(c))^{1/2}/2\lambda_m(M(c))] \quad (23)$$

and a subset of every ball of radius $\overline{d} > \underline{d}$ with initial conditions $x_o, z_o$ such that $n_{xzo} = \|x_o\|^2 + \|z_o - h(x_o)\|^2 > \underline{d}$ is reached in a finite time not greater than

$$T = max\{t_1^{-1}, t_2^{-1}\}T_\epsilon,$$
$$T_\epsilon = \frac{\lambda_M(P(c))n_{xzo} - [\lambda_m(P(c))/\lambda_M(P(c))]\overline{d}^2}{R_\lambda\lambda_m(M(c))\overline{d}^2 - \sqrt{R_\lambda}\|m(c)\|\overline{d} - l(c)},$$
$$(24)$$

where $R_\lambda = \lambda_M(P(c))/\lambda_m(P(c))$. The symbols in $(23),(24)$ mean for $c \in (0, 1)$

$$P(c) = diag[(1 - c)P_s, cP_f],$$
$$m(c) = [(1 - c)b_3 + \frac{w_4}{\epsilon_T}c + c(a_3 + \frac{w_2}{\epsilon_T})]^T,$$
$$l(c) = (1 - c)v_2 - c\frac{w_3}{\epsilon_T},$$
$$M(c) = \begin{pmatrix} (1 - c)(v_1 - b_1) & \frac{-ca_1 - (1-c)b_2}{2} \\ \frac{-ca_1 - (1-c)b_2}{2} & c(\frac{w_1}{\epsilon_T} - a_2) \end{pmatrix},$$
$$b_3 = \beta\sum\gamma_{f3i},$$
$$a_3 = \beta\sum\gamma_{f3i} + max\|\omega_1(q)\|,$$
$$w_2 = \sum w_{2i}, w_3 = \sum w_{3i}, w_4 = \sum w_{4i}.$$

**Proof.** The single parameter single fast subsystem case reduces this theorem on Theorem 3 in Garofalo and Leitmann (1988) which is completely proved. It is based on two weighted Lyapunov functions by the constant c and follows the results by Saberi and Khalil (1984) to determine scalar $\epsilon^*$. Because of those proved theorem, we can present only the differences between the proofs of these two theorems.

The nonsingular fast time scaling $T$ transfers the original multiparameter singular perturbation problem into a single parameter singular perturbation problem which can be solved using the results by Garofalo and Leitmann (1988). This change to a joint time scale in (13) means that changes concern matrices $A_{22}, B_2$ which can be considered as changes on $Q_f$ only under the same $A_{22}, B_2$. It results via $w_1$ in the corresponding changes in $\underline{d}_f, T_F, \epsilon_T^*, \underline{d}, T_\epsilon$. Therefore the nonsingular rescaling of these quantities is necessary to return to the original problem solution, when the problem has been solved in the joint fast time scale.

The block diagonal structure of the fast controller resulting in independent fast controllers design means that instead of scalar $\gamma_f$ used in Garofalo and Leitmann (1988) we have a vector. Therefore, considering the fast subsystem Lyapunov function

242

for (20) as $W(x,z) = \frac{1}{2}(z - h(x))^T P_f(z - h(x)) = \frac{1}{2}(z_1 - h_1(x))^T P_{f1}(z_1 - h_2(x)) + \frac{1}{2}(z_2 - h_2(x))^T P_{f2}(z_2 - h_2(x))$ and using the property (for $A, B$ matrices or vectors of appropriate dimensions) $\|A + B\| \leq \|A\| + \|B\|$ in evaluating $\dot{W}(x,z) < 0$ results in $a_i = a_{1i} + a_{2i}, i = 1, 2, 3$ and $w_j = w_{1j} + w_{2j}, j = 1, ..., 4$. Analogously, considering the slow subsystem Lyapunov function for (20) as $V(x) = \frac{1}{2}x^T P_s x$ and evaluating its derivative $\dot{V} < 0$ we obtain $\dot{V} \leq -v_1\|x\|^2 + v_2 + \|x\|\|P_s\| \sum(\beta\gamma_{f2i}\|x\| + (\beta\gamma_{f1i} + max\|A_{oi} - B_{2i}K_{fi}\|)\|z_i - h_i(x)\| + \beta\gamma_{f3i})$. Denoting $b_i = b_{1i} + b_{2i}, i = 1, 2, 3$ then all $a_i, w_j, b_i$ correspond with the same terms in the proof of Theorem 3 in Garofalo and Leitmann (1988). Moreover, the terms for $\epsilon_T, P(c), m(c), l(c)$ are the same.

Comment. The requirement on the joint fast time scale is directly inserted into the design requirement on the $Q_f$ matrix. Therefore, the radius of the ball of ultimate boundedness can be adjusted by appropriate choice of $Q_{fi}, \delta_i, c$ and $\gamma_{f3i}, i = 1, 2$.

## EXAMPLE

Problem formulation. Suppose the load frequency control model of two–area connected power system with stiff tie–line. We use the simplified version of the model with stiff tie–line by Khalil and Kokotovic (1978). The importance of the stiff tie–line modelling has been recognized by Basanez et al.(1984). The slow state is frequency deviation and the fast subsystems states are turbine output variation and turbine valve position variation. The model consists of the power balance equation, the non–reheat steam turbine equation and the governor equation, respectively, as follows

$$T\dot{\Delta}f = -\beta\Delta f + (1 - \alpha)\Delta P_{G_1} - \alpha\Delta P_{G_2} +$$
$$+ (1 - \alpha)\Delta P_{d_1} - \alpha\Delta P_{d_2},$$
$$T_{t_i}\dot{\Delta}P_{G_i} = -\Delta P_{G_i} + \Delta v_i, \qquad (25)$$
$$T_{G_i}\dot{\Delta}v_i = -\Delta v_i + \Delta P_{c_i} - \frac{1}{r_i}\Delta f_i,$$

where $T = 20 = (1 - \alpha)T_1 - \alpha T_2$ =system inertia constant; $\alpha = T_1 Pr_1/(T_1 Pr_1 + T_2 Pr_2) = 0.2, T_i$ are inertia constants, $Pr_i$ is rated power; $\beta = 0.5(1 - \alpha)D_1 - \alpha D_2, D_i$ are load frequency constants; $\Delta P_{G_i}$ is turbine output; $\Delta P_{d_i}$ is load disturbance which is considered as disturbance inputs – constants; $T_{t_i}$ are time turbine constants; $T_{G_1} = T_{G_2} = 0.1$ are governor time constants; $\Delta v_i$ is the turbine valve position; $r_1 = r_2 = 0.25$ are speed regulations. All quantities are per unit and time constants in seconds. To put this system into the required singular perturbation form we choose $\epsilon_i = max\{T_{G_i}, T_{t_i}\}/T$. The slow state corresponds with $\Delta f$, the fast states of two subsystems each with $\Delta P_{G_i}, \Delta v_i$. Suppose that we study the possibility of the change of the load demand consumption by changing $D_2$. The standard way of control is via $\Delta P_{c_i}$. Suppose that the uncertainty is given by measurements in the slow subsystem and the second fast subsystem. Defining the state and control variables as

$$x = (\Delta f, \Delta P_{G_1}, \Delta v_1, \Delta P_{G_2}, \Delta v_2)^T, u = (u_1, u_2)^T. \qquad (26)$$

the model data are as follows

$$A_{11}(q) = (-0.025),$$
$$A_{o1}(q) = (0.04 \quad 0), A_{o2}(q) = (0.02 \quad 0),$$
$$B_1(q) = \overline{B}_1 = (0.05 \quad 0.001),$$
$$A_{1o}(q) = A_{2o}(q) = (0 \quad -0.4)^T,$$
$$D_{2o}(q) = (\omega_{2o}(q)),$$
$$\overline{\omega}_{2o} = max(\omega_{2o}(q)) = 0.1008,$$
$$A_{d1}(q) = A_{d2}(q) = \begin{pmatrix} -0.05 & 0.05 \\ 0 & 0.1 \end{pmatrix}, \qquad (27)$$
$$D_{d2}(q) = (0 \quad \omega_{d2}),$$
$$\overline{\omega}_{d2} = max(\omega_{d2}(q)) = 0.25$$
$$B_{21}(q) = B_{22}(q) = (0 \quad 0.1)^T,$$
$$\overline{\omega}_1 = max(\omega_1(q)) = 0.001.$$

Consider two cases: 1) $T_{t_1} = T_{t_2} = 0.2$. Therefore $\epsilon_1 = \epsilon_2 = \epsilon_T = 0.01$; 2) $T_{t_1} = 0.18, T_{t_2} = 0.222$ Therefore $\epsilon_1 = 0.09, \epsilon_2 = 0.1111, \epsilon_T = 0.01$.

The objective is to design practically stable controller using the derived method illustrating both fast subsystems independence and different fast time scales.

Solution. Consider first the design for single singular parameter case with two independent fast subsystems. It means to design sequentially slow, both fast subsystems control and evaluate the composite controller. This solution will be further used as a joint fast time scale solution in the case of the model with different time scales.

1. The slow controller design is performed for $Q_s = 2$. The corresponding matrices and their norms are $\overline{A}_o = -0.1518, D_o(q) = 0, E_o(q) = (0.0634 \quad -0.0317), \overline{F}_o = 0.0468, \overline{G}_o = (-0.0234 \quad 0.0117), k_{D_o} = 0, k_{E_o} = 0.0709, k_{\overline{G}_o} = 0.0262, k_{\overline{F}_o} = 0.0468, k_{d_1} = 0.4472, P_s = 6.588, \alpha(\overline{Q}_s) = 1.3835$. Therefore, $\rho_s = 0$. The slow subsystem has the form

$$\dot{x}_s = -0.1518x_s - (0.05 \quad 0.001)u_s =$$
$$= -01518x_s - \overline{\gamma}_s 0.0165x_s = -0.1696x_s. \qquad (28)$$

It is evident that $\gamma \in (0, \infty), c_1$ is arbitrary. Computing $\gamma_1, \gamma_2$ for $c_1 = 0, c_2 = 0.005$ results in $\gamma_1 = 0.1, \gamma_2 = 113.1836$. Selecting $\gamma_s = 1$, then $v_1 = 0.6911, v_2 = 0.0005, \underline{d}_s = 0.00851$ and

$$T_s = \frac{6.588\|x_s^o\|^2 - 6.588\overline{d}_s^2}{0.6911\overline{d}_s^2 - 0.0005}. \qquad (29)$$

The first fast controller has included no uncertainty. Therefore, $F_o(q) = 0, k_{H_1} = 0, k_{B_{21}} = 0, k_{D_{2i}} = 0, k_{d_{21}} = 0$. We require the pole shift into the position $-0.3, -0.2$ by computing $K_{f1}$ and choose $Q_{f1} = diag(20, 30), \gamma_{f11} = \gamma_{f21} = 0, \gamma_{f31} = 0.01, \delta_1 = 10^{-4}$. It results in the following controller

$$u_{f1} = -\begin{pmatrix} 7.5 & 3.5 \end{pmatrix} \begin{pmatrix} x_{13} \\ x_{12} \end{pmatrix} -$$
$$- \frac{0.01}{10(123.21x_{12}^2 + 475.04x_{13}^2)^{1/2} + 0.0001}, \quad (30)$$
$$\underline{d}_{f1} = 0.16 * 10^{-5},$$
$$T_{f1} = \frac{282.59\|z_{f1}^o\|^2 - 2.194\overline{d}_{f1}}{0.9482\overline{d}_{f1}^2 - 10^{-5}}.$$

where $x_{12} = x_2 + 4.0355x_1$, $x_{13} = x_3 + 4.0355x_1$. Note only that $u_1 = u_{s1} + u_{f1}$.

The second fast controller design includes uncertainties $\overline{\omega}_{2o}, \overline{\omega}_{d2}$. Choose $K_{f2} = K_{f1}$. The norms are as follows $k_{H_2} = 0$, $k_{B_{22}} = 0$, $k_{D_{42}} = 2.5$, $k_{d_{22}} = 0$. Note that $\beta = 0.05$. We choose $Q_{f2} = Q_{f1}$, $\gamma_{f12} = 0.1008$, $\gamma_{f22} = 0$, $\gamma_{f32} = 0$, $\delta_2 = 10^{-4}$. It results in

$$u_{f2} = -\begin{pmatrix} 7.5 & 3.5 \end{pmatrix} \begin{pmatrix} x_{15} \\ x_{14} \end{pmatrix} -$$
$$- \frac{0.108(x_{14}^2 + x_{15}^2)^{1/2}}{10(123.21x_{14}^2 + 475.04x_{15}^2)^{1/2} + 0.0001}, \quad (31)$$
$$\underline{d}_{f2} = 0.16844 * 10^{-4},$$
$$T_{f2} = \frac{282.59\|z_{f2}^o\|^2 - 2.194\overline{d}_{f2}}{(8.81\overline{d}_{f2}^2 - 8.18 * 10^{-5})^{1/2}\overline{d}_{f2}}.$$

where $x_{14} = x_4 + 4.0355x_1$, $x_{15} = x_5 + 4.0355x_1$. Note only that $u_2 = u_{s2} + u_{f2}$ with $u_{s1} = u_{s2}$.

The composite controller satisfies the condition (21) and the parameter $\epsilon_T^* = 0.012$, $c^* = 0.0048$. Note only that $T_\epsilon = T$. Selecting $c = 0.2$ corresponds with the positive definite matrix $P(c)$. A ball of ultimate boundedness has the radius $\underline{d} = 0.4612$ and

$$T_\epsilon = \frac{56.4(\|x_o\|^2 + \|z_o - h(x_o)\|^2) - 0.0881\overline{d}}{2.0444\overline{d}^2 - 0.4487\overline{d} - 4.233 * 10^{-4}}. \quad (32)$$

2. Consider now the solution of the second case with different singular parameters. Because $\epsilon_T$ in this case is the same as the single parameter in the previous case, we use this solution as the solution with the joint time scale. In our case $t_1 = 1.111$, $t_2 = 0.9$. Therefore, choosing $Q_s$ the same as in the previous case and

$Q_{f1} = 1/t_1 diag(20, 30) = diag(18, 27)$, $Q_{f2} = 1/t_2 diag(20, 30) = diag(22.22, 33.33)$ means to apply on the solution only the nonsingular rescaling. We obtain $\epsilon^* = (0.0101, 0.01332)$. A ball $\underline{d}$ is transformed into an ellipsoid with the same slow variable component as the ball and the components $\underline{d}_i = 1/t_i\underline{d}$. An analogous result holds for $T$ using (24) and $T_\epsilon$ given by (32).

## CONCLUSION

New decentralized nonlinear control design method is presented for uncertain multiparameter singularly perturbed systems using multimodelling. The presented method extends the recent results on multimodelling using practical stabilization by the inclusion of both parameters and signals unknown but bounded uncertainties. An illustrative example is supplied.

## ACKNOWLEDGEMENT

This work has been partially supported by the Dirección General de Investigación Científica y Técnica of the Spanish Government.

## REFERENCES

Bakule, L., and J. Lunze (1988). Decentralized design of feedback control for large–scale systems. Kybernetika, 3-6, 1-100.

Basanez, L., Riera, J., and J.Ayza (1984). Modelling and simulation of multiarea power system load-frequency control. Mathematics and Simulation, North–Holland, 54-62.

Garofalo, F., and G. Leitmann (1988). Nonlinear composite control of a class of nominally linear singularly perturbed uncertain systems. Variable StructureControl (ed. A.S.I. Zinober), IEE Press, London.

Khalil, H.K., and P. Kokotovic (1978). Control strategies for decision makers using different models of the same system. Trans. Automat. Control, vol. AC-23, 289-298.

Naidu, D.S. (1988). Singular Perturbation Methodology in Control Systems, IEE Press, London.

Saberi, A., and H.K. Khalil (1984). Quadratic–type Lyapunov functions for singularly perturbed systems. Trans. Automat. Control, vol. AC-29, 542-550.

# LOW COST ROBUST NON LINEAR CONTROLLERS

**A. Balestrino and A. Landi**

*DSEA - Department of Electrical Systems and Automation. University of Pisa, Via Diotisalvi 2, Pisa 56126, Italy*

Abstract    Most plants can be efficiently controlled by using simple low cost
controllers. Many variations of the classical PID regulator have been developed in
order to obtain high speed of response, steady state precision, disturbance rejection
and robustness with respect to parametric variations and unmodeled dynamics. In this
paper two classes of non linear plants with input saturation are considered. Class I
deals with first order plus time delay processes, class II deals with plants modelled by
a process with a dominant pair of complex conjugate poles. A variable structure
standard regulator (VSC) for class I plants and a robust oscillating controller (ROC) is
for class II plants are proposed. To improve the overall performance a ROC-PID
controller is also considered. Such controllers can be easily implemented by using low
cost microprocessors and their industrial use seems to be very attractive. A complete
set of simulation results is reported.

Keywords: Advanced controllers, Non Linear Plants, Robustness, Variable Structure
Control, Process Control

## INTRODUCTION

Industrial process controllers are usually of PID
type, in force of their simple and robust nature.
Beside standard regulators a variety of
configuration are now implemented by including
special features such as adaptation, self-tuning
and prediction. Non linear standard regulators
based on variable structure systems theory have
been examined in Balestrino (1989, 1990a).
These regulators are low cost and show high
insensitivity to external and parametric
disturbances. In this paper two classes of non
linear plants to be controlled are considered.
Class I deals with plants modelled by a first order
plus time delay process with input saturation, see
fig.1.

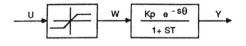

Fig.1. Class I plant

Class II deals with plants modelled by a process
with a dominant pair of complex conjugate poles
and with input saturation. In the second section
advantages and drawbacks of variable structure
non linear standard regulators are reviewed with
reference to class I plants.

In the third section a special nonlinear controller
is introduced for plants of class II. This
controller is obtained by a suitable switching
between two distinct controllers, one specially
designed for the transient and the other one for
the steady-state response. Performance and
robustness properties are analysed by means of
simulations.

## VS-PI CONTROL

A large class of non linear plants can be
controlled by using adaptive model following
control with non linear high gain subsystems. A
variable structure controller can be designed as in
Balestrino (1989, 1990b). The structure of the
controller is the classical one with proportional,
integral and derivative actions. The general form
of this controller is as follows:

$$C(s) = k_p(e)e + \frac{k_i(e)}{s} + k_d(e)s \qquad (1)$$

where e is the error variable, i.e. the difference between the desired output r and the actual response y:

$$e = r-y \qquad (2)$$

The coefficients in (1) are non linearly dependent on the error e; a particular simple choice is as follows:

$$k_p(e) = \frac{K_p}{c_p} \frac{c_p}{1+c_p|e|} \qquad (3a)$$

$$k_i(e) = \frac{K_i}{c_i} \frac{c_i}{1+(c_i e)^2} \qquad (3b)$$

$$k_d(e) = \frac{K_d}{c_d} \frac{c_d}{1+(c_d e)^2} \qquad (3c)$$

From (3a) we have

$$k_p(e)\, e = \begin{cases} k_p\, e & \text{if } c_p|e| \ll 1 \\ \dfrac{K_p}{c_p} & \text{if } c_p|e| \gg 1 \end{cases} \qquad (4)$$

i.e. the non linear proportional action reduces to the classical one if $c_p|e| \ll 1$, while in the limit $c_p|e| \gg 1$ we obtain a relay action. Of course $k_p$ can be chosen accordingly to the usual tuning rules (Ziegler, 1942); $c_p$ depends directly on the plant saturation level. Similar remarks hold for (3b) and (3c); the dependence on $(ce)^2$ ensures that the integral or derivative actions reduce or else vanish whenever $c|e| \gg 1$. In this way wind-up and instability problems are effectively removed.

The smoothed transition from relay to proportional action due to (3a) eliminates the drawbacks due to chattering. Of course chattering can be suppressed by other solutions as in Harashima (1985), Wang (1989), Balestrino (1988); however the solution given by (3a) is very simple and easily implementable.

With reference to the simplest control configuration

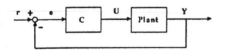

Fig.2. Feedback control configuration

shown in fig.2 note that for very large errors the I control action vanishes and the P action reduces to a bang-bang control; in this case the plant dynamics is open loop. Therefore we expect that this controller can successfully operate with stable plants. Indeed by simulations and experimental tests we have shown that non linear standard regulators work effectively with class I plants or more generally with plants having a non oscillatory indicial response.

Of course a variety of control schemes may be used. A typical configuration is feedback (FF) cascade control with prefiltering (PF) and feedforward (FF) control , see fig.3.

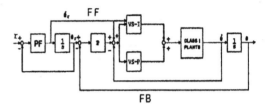

Fig.3. Typical control configuration

In Fig.4 ,5 are shown some speed responses (Balestrino,1989).

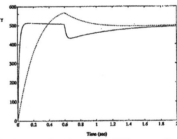

Fig.4. Conventional standard PI controller: speed response. Continuous line:minimum moment of inertia; dotted line: maximum moment of inertia

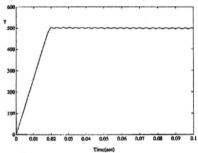

Fig.5.Speed response with a VS-PI controller: the minimum moment of inertia case and relay with a fixed amplitude

Responses to square and triangular waveforms corresponding to the configuration in Fig.3 are shown in figg. 6,7 (Balestrino,1990 b) where a constant disturbance is added to the output from t=5 sec on.

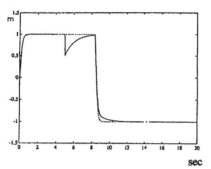

Fig.6.Response to square waveform

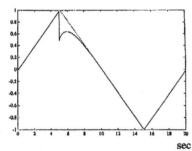

Fig.7.Response to triangular waveform

Chattering is completely suppressed; note that in fig.3 the innermost speed loop show a very high performance so that only a standard P-action is needed in the outer loop.

However other control configurations are possible, e.g. similar to adaptive model following control (Landau, 1974) . The structure shown in fig.8 has been used successfully in Balestrino (1991); in this case the process is a sinusoidal brushless motor, while the model is a DC motor and SAL represents the signal adaptation law.

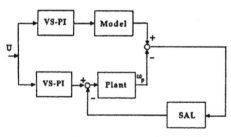

Fig.8. SAL AMFC

For plants of class I a Smith predictor control structure is shown in Fig.10; the tuning rules and the simulation results are reported in Balestrino (1990 a).

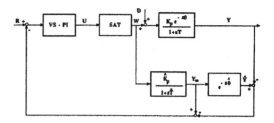

Fig.9. Smith predictor control structure

ROBUST OSCILLATING CONTROL

In many cases models of class I are not applicable, but we must use models of class II. Class II plants are shown in fig. 10.

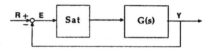

Fig.10

Such plants typically show a pair of complex conjugate poles and give rise to oscillatory responses; they are difficult to stabilise by using the nonlinear controllers presented in the previous section.

A natural question arises about the possibility of implementing simple robust controllers for such processes without derating the overall performance. A possible solution can be devised by taking into account that for a pure oscillatory plant with bounded inputs the maximum principle of Pontriaguine (Pontriaguine, 1974) suggests a bang-bang control. With initial zero conditions if the set point varies from zero to r with a plant gain equal to one then the optimal control reduces to feeding the plant input at r/2 until the output reaches the value r; then the plant input is set to zero (fig.11).

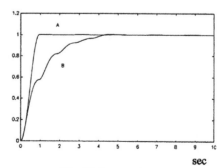

Fig.11 Step responses

247

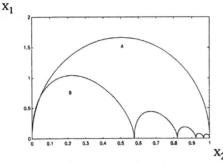

$X_1$

$X_2$

Fig.12 Phase plane plot

The output reaches its final value with zero steady state error with a dead-beat response; the transient duration is exactly one half of the oscillatory period of the plant. Of course if the gain is not known and the plant is not a pure oscillator the above control strategy is no more applicable. However we may fuzzify the procedure with some cautions in order to get a robust controller. Assuming a coarse knowledge of the plant, eventually by overestimating the plant gain and the frequency of natural oscillation, the plant input is chosen accordingly to the following rules:
1) if the desired output is $y_d = r$ then the input is

$$u = \min\left(\frac{|y_d|}{2K_e}, |l_s|\right) \text{sgn } y_d. \qquad (5)$$

until the error $e(t_1) = y_d - r$ attains its first minimum value;
$K_e$ is the overestimated plant gain and
$l_s$ is the maximum admissible input
2) having missed the target $y = y_d$ the plant input is set up accordingly to the general rule:

$$u = \min\left(\frac{|e(t_k)|}{2K_e}, |l_s|\right) \text{sgn } e(t_k) \qquad (6)$$

if $t_k < t < t_{k+1}$; $t_k$, $t_{k+1}$ are the instants where the error $e(t)$ attains its local minima. The procedure is illustrated in figg 11A, 12A for the ideal case of perfectly known gain and for the non ideal case (11B, 12B).
Remarks. Note that in every interval $(t_k, t_{k+1})$ the plant is open loop controlled. The durations $t_{k+1} - t_k$ depend on the plant and can be estimated about equal to $T_c/2$; where $f_c = 1/T_c$ is the frequency of the natural oscillation. The overall time response can be simply underestimated by $T_c/2$. Therefore if the plant output must be able to track a variable signal $y_d(t)$ we must take into account the bounds due to the plant structure. A typical bound is

$$|\dot{y}_d(t)| < \frac{2K_e l_s}{T_c} \qquad (7)$$

If the desired input $y_d(t)$ is known in advance then the control laws (6) can be modified as follows:

$$e(t_k) = y_d(t_{k+1}) - y(t_k) \cong$$
$$\cong y_d(t_k + T_c/2) - y(t_k) \qquad (8)$$

In the non ideal case, mainly for robustness problems arising from a not perfect knowledge of the plant, the above procedure is no more able to produce a dead-beat response. Moreover updating the control input at the instants $t_k$ is a form of natural sampling highly dependent on our ability to recognize local minima of the error $y_d$-y. This problem can be tackled by monitoring the error derivative, which sometimes is physically available, or a suitable estimate of $\dot{e}$. A possible test is made by monitoring when $|\dot{e}| < S$ with sign($e \dot{e}$) < 0, because it is not very reliable to detect zero crossings of the error derivative. This is true especially if noise is added to the output; in this case larger is the noise level, larger is the threshold to be used. Of course the implementation of this robust oscillating controller (ROC) is easily realized by using a microcontroller. Note also that the ideal case may be approached if the plant gain is exactly known. When the plant structure is as in fig.10 with at least a pole at the origin then the closed loop gain is exactly one independent of the saturation slope. In the other cases, also with variable gains, simple procedures can be used for estimating the static gain. Indeed by comparing the output sample at the sampling instants $t_k$ with the desired value the gain can be computed; if the signal variations are too small a suitable square wave signal may be added to the input for a short time. However any other procedure may be applied realizing a simple adaptive controller (AROC). In order that the controller algorithms can be kept as simple and robust as possible the adaptation or self-tuning properties must be introduced with care. Low cost requirements and simplicity of tuning and operation combined with previous issues make robust oscillating controllers (ROC) attractive. The transient response is fast and well shaped, almost always without overshoots; the major drawbacks are due to the properties of the steady-state response, e.g. the presence of a steady-state error or ripple. Being ROC suitable for the transient response it is apparent that, being the error and its derivative quite small, near steady-state the controller structure can be successfully commuted from ROC to a standard PID regulator, of course with bumpless transfer, see fig. 13.

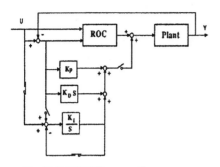

Fig.13 ROC-PID controller structure

When the control signal is produced by the standard regulator, the other controller is still operating albeit with its output disconnected; in this way the ROC is able to supervise the system performance and resume quickly its control activity.


SIMULATION RESULTS

In this section some simulations are reported in order to illustrate and clarify the behaviour of class II plants with robust oscillating controllers.

Example 1
The plant in fig.10 is described by

$$G(S) = \frac{k}{1+s^2} \tag{9}$$

Without saturation, the indicial response is shown in fig. 11 (case A) along with the phase plane trajectories in fig.12 (case A) and the ROC output is shown in fig.14.

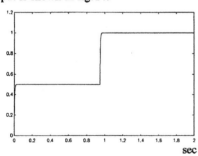

Fig.14 ROC output

Example 2
Let the plant be described by

$$G(S) = \frac{k\ (s-3)}{s^2+3s+2} \tag{10}$$

By using the robust oscillating controller we obtain the time indicial responses in fig. 15.

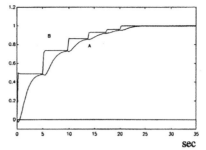

Fig.15 Case A: indicial response; case B: ROC response

Example 3a
The plant in fig.10 is described by

$$G(S) = \frac{k}{s^3+4s^2+6.25s} \tag{11}$$

with the nonlinearity equal to 0.5 sat( ).
The responses to a reference signal

$$y_d(t) = 1(t) + 1(t\text{-}30) \tag{12}$$

are shown in figg 16, 17.

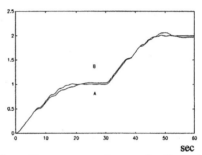

Fig.16 Time responses (case A: $k=5$, case B: $k=15$)

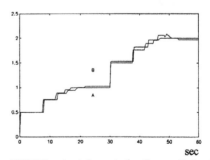

Fig.17 ROC output (case A: $k=5$, case B: $k=15$)


Example 3b
By using the more complex controller which combines ROC and PID in the case of the plant (11) the time responses to a reference signal

$$y_d(t) = 1(t)\text{-}2*1(t\text{-}30) \tag{13}$$

are shown in fig.18.
Case N is the nominal case, case U is the response to an uncertain plant with a 100% variation of its parameters.

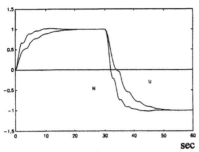

Fig.18 ROC -PID time responses

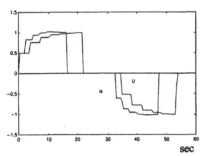

Fig.19 ROC output

In this case the saturation bound is set to 5. In fig.20 is shown the effect of the nonlinearity limiting the amplitude of the control input of the plant

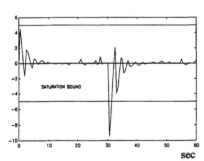

Fig.20 Saturation effect

## CONCLUSIONS

Many plants can be efficiently controlled by using simple low cost controllers. It is convenient to collect plants into classes so that for a particular class is possible to select a specified controller. In this paper we have shown that for two classes of non linear plants, class I and class

II, we can successfully utilize Variable Structure Standard Controllers (VSC) and Robust Oscillating Controllers (ROC) respectively. Such controllers can be easily implemented by using low cost microprocessors so that their industrial use seems to be very attractive.

## ACKNOWLEDGMENTS

The Authors acknowledge the contribution of dr.G.Battini and dr.G.Tonelli for their careful execution of the numerical simulations. This work has been developed with the financial support of MURST 40% and CNR-PFR.

## REFERENCES

Balestrino, A., G.Gallanti and D.Kalas (1988). On the implementation of variable structure control systems. IMACS Symp., 267-272.

Balestrino, A., M.Innocenti and A.Landi (1989). Variable structure conventional controllers. IFAC Proc. Low Cost Automation, W 187-191.

Balestrino, A., A. Brambilla, A. Landi, C. Scali (1990a). Non linear standard regulators. IFAC Proc.World Congress.

Balestrino, A. and A.Landi (1990b). Intelligent variable structure control in electrical drives. IEEE Proc. Int. Workshop on Intelligent Motion Control, 719-722.

Balestrino, A., A.Landi and G.Grandi (1991). Advanced control of high performance drives in robotics. COMCONEL 91, 199-202.

Harashima , F, H. Hashimoto and S. Kondo (1985). Mosfet converter-fed position servo system with sliding mode control. IEEE Trans. on IE, 32, 238-244.

Landau, I.D. and B.Courtiol (1974). Design of multivariable adaptive model following control systems, Automatica, 10, 483-494.

Pontriaguine, L., V.Boltianski, R.Gamkrelidze and E.Mitchtchenko (1974). Theorie matematique des processes optimaux. Ed.MIR, Moscow.

Wang, L. and D.H. Owens (1989). Self tuning/adaptive control using approximate models, Proc. ICCON, TP, 6-6.

Ziegler, J.G., and N.B. Nichols (1942). Optimum settings for automatic controller. Trans. ASME, 64, 759-768.

# A CRITERION FOR COMPARING THE IMPLEMENTATION QUALITY OF DIGITAL CONTROLLERS

## R. Caccia and A. de Carli

*Department of Computers and Systems Sciences, University of Rome "La Sapienza",
Via Eudossiana 18, I-00184 Rome, Italy*

Abstract: the implementation of digital filters and controllers implies some problems on the accuracy and the computing time connected to the word length, the frequency clock, the underflow and overflow handling. The ability of the designer is to select the structure of the algorithm which minimizes the computing time and maximizes the accuracy independently on the word length and the clock frequency.

In order to attain these goals an easy to use criterion is proposed in this papers. It can be applied without a specific background because it consists in interpreting the algorithm as a dynamic system and in evaluating its behaviour as the evolution of a dynamic system. By this approach the structure of the algorithm optimizing the implementation can be easily worked out and its validity is confirmed by an example.

Keywords: digital control, digital filters, finite word length effects, round-off noise, sensitivity.

## INTRODUCTION

Innovative controllers use microprocessors instead of analog or hybrid chips and implement advanced control strategies instead of conventional PID regulators. The innovative controllers fulfill the disturbance/output and the input/output requirements. This goal is attained by using a controller in the direct loop, a dynamic feedback and a predictor on the feed forward path.

The design of an innovative controller involves both the definition of the microcomputer hardware, and the design of the dedicated software implementing the desired control strategy. The main problem is to produce a control algorithm that fulfills the performance specifications without oversizing the microprocessor hardware. A skilled designer should therefore maximize the accuracy and minimize the computing effort. For the diffusion of digital controllers it is fundamental that the cost of the controller should match to the cost of the plant and to the benefits introduced by the advanced control strategies. It is therefore useful to get a general criterion for relating the accuracy to the structure of the algorithm.

Possible structures of the algorithms and different approaches for evaluating the accuracy are presented in many books [1,2,3] and papers [4,5]. The proposed criteria for evaluating the accuracy generally require a specific background and can be used by a limited set of specialists.

Intuitive approaches are therefore frequently applied and troubles due to an heuristic approach are in general overcame by oversizing the microprocessor hardware.

The aim of this paper is to present an easy to use criterion for minimizing the computing time and maximizing the accuracy independently on the word length and on the clock frequency.

To attain these results the behavior of the algorithm is assimilated to the evolution of a dynamic system and the truncation error to a white noise. Efficient suggestions about the accuracy are obtained by examining the structure of the dynamic system representing the algorithm and by referring to general criteria for evaluating its behavior. Before implementating of the algorithm, the proposed criterion can be validated by simulating the algorithm on a general purpose computers in which a suitable instruction takes into account the limited word-length too.

251

## DEFINITION OF THE PROBLEM

The mathematical model of a digital controller can be formulated by:

1) the z-transfer function, i.e.:

$$G(z) = \frac{b_{n-1}z^{n-1}+...+b_1 z + b_0}{z^n + a_{n-1}z^{n-1}+...+a_1 z + a_0} \quad (1)$$

2) the discrete convolution integral:

$$y(i) = T \cdot [g(0) \cdot u(i)+...+g(k) \cdot u(i-k)] \quad (2)$$

where $g(0), g(1),...,g(k)$ are the signifying samples of the impulse response;

3) the difference equation:

$$y(i) + a_{n-1}y(i-1)+...+a_0 y(i-n) = $$
$$= b_{n-1}u(i-1)+...+b_0 u(i-n) \quad (3)$$

4) the discrete time state space equations:

$$y(i) = c \cdot x(i)$$
$$x(i+1) = \Phi \cdot x(i) + \Psi \cdot u(i) \quad (4)$$

Equivalent representations of the input-output behavior of a digital controller are given by these mathematical models. Each one corresponds to a different structure of the algorithm with different computing time and accuracy. Since the formulation of the mathematical model determines the structure of the algorithm, a criterion for selecting the more suitable one should be applied to minimize the computing time and to maximize the accuracy, independently of the word-length and the clock frequency.

## STRUCTURE OF ALGORITHMS

Whichever the formulation of the mathematical model may be, the output variable is step-by-step computed by processing additions, products, and data shifts.

At $i$ step the output variable can be computed by:

1) processing the discrete convolution integral given by Eq. (2);

2) equating the difference between the linear combination of the previously sampled values of the input variable and the linear combination of the previously sampled values of the output variable. The number of the samples equals the maximum degree of the denominator polynomial or the order of the discrete time state space model;

$$y(i) = [b_{n-1}u(i-1)+...+b_0 u(i-n)]$$
$$-[a_{n-1}y(i-1)+...+a_0 y(i-n)] \quad (5)$$

3) computing the linear combination of the samples of inner variables x.

The value of each one is obtained as the difference between the present sample of the input variable and the linear combination of the previously sampled values of the inner variable. The number of the samples equals the maximum degree of the denominator polynomial or the order of the discrete time state space model;

$$y(i) = [b_{n-1}x(i-1)+...+b_0 x(i-n)]$$
$$x(i+1) = u(i) - [a_{n-1}x(i-1)+...+a_0 x(i-n)] \quad (6)$$

4) applying a canonical realization [6, 7], so as the characteristic matrices have coefficients different from zero and one equal in number to the coefficients of the transfer function. Therefore the number of products and additions does not change in the new formulation. Without any additional computation, the formalization of the transfer function given by Eq.(1) allows to define the following forms of the characteristic matrices:

a) observer form:

$$\Phi(T) = \begin{bmatrix} 0 & 0 & . & 0 & -a_0 \\ 1 & 0 & . & 0 & -a_1 \\ . & . & . & . & . \\ 0 & 0 & . & 0 & -a_{n-2} \\ 0 & 0 & . & 1 & -a_{n-1} \end{bmatrix} \quad \Psi(T) = \begin{bmatrix} b_0 \\ b_1 \\ . \\ b_{n-2} \\ b_{n-1} \end{bmatrix} \quad (7)$$

$$c = \begin{bmatrix} 0 & 0 & . & 0 & 1 \end{bmatrix}$$

b) controller form:

$$\Phi(T) = \begin{bmatrix} -a_{n-1} & -a_{n-2} & . & -a_1 & -a_0 \\ 1 & 0 & . & 0 & 0 \\ . & & . & . & . \\ 0 & 0 & . & 0 & 0 \\ 0 & 0 & . & 1 & 0 \end{bmatrix} \quad \Psi(T) = \begin{bmatrix} 1 \\ 0 \\ . \\ 0 \\ 0 \end{bmatrix} \quad (8)$$

$$c = \begin{bmatrix} b_{n-1} & b_{n-2} & . & b_1 & b_0 \end{bmatrix}$$

By some additional computations, two new sets of canonical forms can be obtained. By the following formulation of the transfer function

$$G(z) = \frac{b'_{n-1}(z^{n-1}+...+a_1)+...+b'_1(z+a_{n-1})+b'_0}{z^n + a_{n-1}z^{n-1}+...+a_1 z + a_0} \quad (9)$$

an observability and a controllability form are so obtained. The $b'_i$ coefficients are computed by applying the following relationship:

$$\begin{bmatrix} b'_0 \\ b'_1 \\ . \\ b'_{n-2} \\ b'_{n-1} \end{bmatrix} = \begin{bmatrix} 1 & a_{n-1} & . & a_2 & a_1 \\ 0 & 1 & . & a_3 & a_2 \\ . & . & . & . & . \\ 0 & 0 & . & 1 & a_{n-1} \\ 0 & 0 & . & 0 & 1 \end{bmatrix}^{-1} \begin{bmatrix} b_0 \\ b_1 \\ . \\ b_{n-2} \\ b_{n-1} \end{bmatrix} \quad (10)$$

c) observability form:

$$\Phi(T) = \begin{bmatrix} -a_{n-1} & -a_{n-2} & . & -a_1 & -a_0 \\ 1 & 0 & . & 0 & 0 \\ . & . & . & . & . \\ 0 & 0 & . & 0 & 0 \\ 0 & 0 & . & 1 & 0 \end{bmatrix} \quad \Psi(T) = \begin{bmatrix} b'_0 \\ b'_1 \\ . \\ b'_{n-2} \\ b'_{n-1} \end{bmatrix} \quad (11)$$

$$\mathbf{c} = \begin{bmatrix} 0 & 0 & . & 0 & 1 \end{bmatrix}$$

d) controllability form:

$$\Phi(T) = \begin{bmatrix} 0 & 0 & . & 0 & -a_0 \\ 1 & 0 & . & 0 & -a_1 \\ . & . & . & & . \\ 0 & 0 & . & 0 & -a_{n-2} \\ 0 & 0 & . & 1 & -a_{n-1} \end{bmatrix} \quad \Psi(T) = \begin{bmatrix} 1 \\ 0 \\ . \\ 0 \\ 0 \end{bmatrix} \quad (12)$$

$$\mathbf{c} = \begin{bmatrix} b'_{n-1} & b'_{n-2} & . & b'_1 & b'_0 \end{bmatrix}$$

Another set of canonical forms is obtained by decomposing the mathematical model in the diagonal block form. It is necessary at first to decompose the transfer function $G(z)$ in the summation of first or second order transfer functions, according to the poles of $G(z)$ being real or complex. This involves further computations of the transfer function, which can be formulated as follows:

$$G(z) = \sum_{i=1}^{k} G_i(z) + \sum_{j=1}^{n-2k} G_j(z) \quad (13)$$

in which:

$$G_i(z) = \frac{\beta_i(z - \alpha_i) + \gamma_i \omega_i}{(z - \alpha_i)^2 + \omega_i^2} \quad (14)$$

$$G_i(z) = \frac{r_j}{z - p_j}$$

e) diagonal controller form:

$$\Phi(T) = \begin{bmatrix} \alpha_1 & \omega_1 & . & 0 & . \\ -\omega_1 & \alpha_1 & . & 0 & . \\ . & . & . & . & . \\ 0 & 0 & . & p_1 & . \\ . & . & . & . & . \end{bmatrix} \quad \Psi(T) = \begin{bmatrix} 1 \\ 0 \\ . \\ 1 \\ . \end{bmatrix} \quad (15)$$

$$\mathbf{c} = \begin{bmatrix} \beta_1 & \gamma_1 & . & r_1 & . \end{bmatrix}$$

f) diagonal observer form:

$$\Phi(T) = \begin{bmatrix} \alpha_1 & -\omega_1 & . & 0 & . \\ \omega_1 & \alpha_1 & . & 0 & . \\ . & . & . & . & . \\ 0 & 0 & . & p_1 & . \\ . & . & . & . & . \end{bmatrix} \quad \Psi(T) = \begin{bmatrix} \gamma_1 \\ \beta_1 \\ . \\ r_1 \\ . \end{bmatrix} \quad (16)$$

$$\mathbf{c} = \begin{bmatrix} 0 & 1 & . & 1 & . \end{bmatrix}$$

Figg. 1 to 4 show the block diagrams connected to the previous mathematical models. Only the block diagrams representing the diagonal realizations are referred to the mathematical model of a second order digital controller.

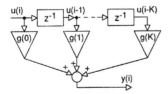

Fig.1 Block diagram for the implementation of a discrete convolution integral.

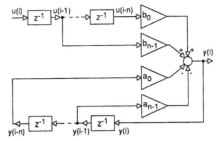

Fig.2 Block diagram for the implementation of an input/output difference equation.

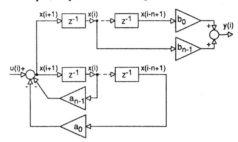

Fig.3 Block diagram for the implementation of an inner variable difference equation.

## EVALUATION CRITERIA

According to the previously indicated goals in the implementation of an algorithm, the elements influencing the computing time will be discussed before the elements determining the accuracy.

The computing time is evaluated by taking into account the number of additions and products requested from the computation of each sample of the output variable.

In the discrete convolution integral this number depends on the number of the samples of the impulse response which in turn depends on the width of the sampling time interval and on the duration of the impulse response. An algorithm implementing a discrete time convolution integral is known as a non-recursive algorithm because each sample of the output variable is computed in terms of the previously sampled values of the input variable only. The behavior of the algorithm is quite similar to the evolution of an open loop system.

An algorithm implementing the finite difference equation requires a number of additions and products proportional to the number of the input and output samples appearing in the formulation of the mathematical model. This number depends on the order of the characteristic matrices and on its realization and it is strongly reduced by applying a canonical realization.

The implementation of the algorithm connected to a finite difference equation or to a discrete-time state-

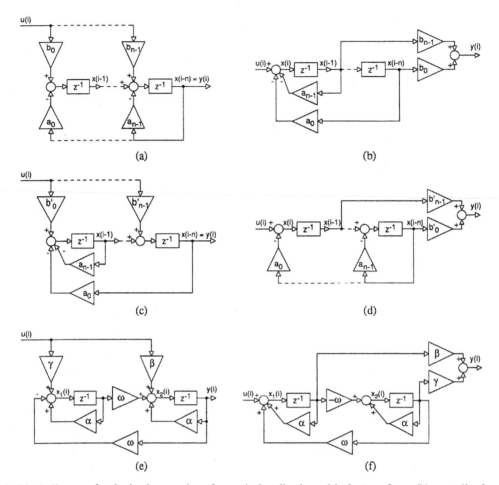

Fig.4 Block diagrams for the implementation of canonical realizations: (a) observer form, (b) controller form, (c) observability form, (d) controllability form, (e) diagonal observer form, (f) diagonal controller form.

space model is indicated as a recursive algorithm because the computation of each sample of the output variable is effected in terms of the previously sampled values of the input and of the output or the internal variable. The number of these samples is equal to the order of the discrete-time state-space model. The behavior of the algorithm is quite similar to the evolution of a dynamic system with an external feedback loop, some internal feedback loops, and some direct path.

For minimizing the computing time it is therefore necessary:
- to implement the algorithm by referring to a canonical realization or to a finite difference equation;
- to process only fixed point operations;
- to address the memory locations in the minimum number of clock cycles.

The improvement of the accuracy is attained by adopting different expedients. It is first necessary to avoid the attainment of the overflow for all the variables. In a fixed-point representation the

overflow is avoided by a suitable scaling of the input variable. This feature could however introduce offset errors of the inner variable due to the underflow. The effort is therefore concentrated in eliminating the origin of underflow and overflow errors.

Accuracy is very much improved by formulating the mathematical model so as all the variables and all the coefficients span in the same range. The most suitable formulation can be selected by referring to the above presented block diagrams and by taking into account that in a feedback system:
- the output variable from a comparator spans into a band narrower than of the two input variables ones;
- an internal feedback loop limits the range of the inner variable;
- each variable involved into the linear combination could range differently from the result.

Fig.3 shows that the formulation given by Eq.(3) is not indicated for improving the accuracy. The same situation occurs for the implementation according to the controllability form and the diagonal controller form, given by Eq.(8), Eq.(12) and Eq.(15)

respectively. Fig.2 and Fig.4 (a), (c), (e) show that the output variable of the feedback block diagram coincides with the output variable of the mathematical model. The formulation of the mathematical model connected to those figures is then to be the most indicated one for improving the accuracy.

Another element for evaluating the accuracy is given by examining the structure of the internal feedback loops. Fig.4(e) shows that an internal feedback loop involves each inner variable while Fig.2 and Fig.4(a) and Fig.4(c) show that the internal feedback loop involves only one inner variable. The mathematical model connected as shown in Fig.4(e) is therefore the most indicated one for limiting of the range of the inner variable.

Further elements for evaluating the accuracy consist in examining the effect of the finite word length on the output variable of the algorithm. Since:
- each product between a variable and a coefficient produces a new variable which should be truncated for its representation by a finite length word;
- the round-off error is functionally equivalent to add a white noise to the truncated variable;
- the effect of the white noise on the output of the algorithm is damped by the feedback;
the mathematical model connected to the block diagram of Fig.4(e) is really the most indicated one, since an inner feedback loop involves each product.

Accuracy can be finally improved by formulating the mathematical model so as coefficients span into the same range. To reach this goal it is suitable to decompose the transfer function into first or second order elementary transfer functions, according to real or complex poles. The decomposition can be done to obtain a parallel or cascade realization of the elementary transfer functions. i.e.:

$$G(z) = \frac{r}{z+p} + \sum_{i=1}^{int(n/2)} G'_i(z) = \frac{b^*_{1_i} z + b^*_{0_i}}{z^2 + a_{1_i} z + a_{0_i}} \quad (17)$$

or

$$G(z) = \frac{r^*}{z+p} \cdot \prod_{i=1}^{int(n/2)} \frac{b^*_{1_i} z + b^*_{0_i}}{z^2 + a_{1_i} z + a_{0_i}} \quad (18)$$

Since the absolute value of coefficient $a_{1_i}$ ranges between one and two, the term $a_{1_i} z$ can be expressed as twice $a'_{1_i} z$. It is then obtained:

$$G_i(z) = \frac{b_{1_i} z + b_{0_i}}{z^2 + a'_{1_i} z + a'_{1_i} z + a_{0_i}} \quad (19)$$

The above mentioned expedients allow to obtain that all coefficients range below a value lower than one and can be represented by a fixed point arithmetic without any further scaling.

EXAMPLE

The previous criteria have been applied to a digital controller having the following transfer function:

$$G(z) = \frac{.3848 z - .3753}{z^2 - 1.8953 z + .9048}$$

$$= \frac{.3848 (z - 1.8953) + .3540}{z^2 - 1.8953 z + .9048}$$

$$= \frac{.3848 (z - .9477) - .0822 * .1292}{(z - .9477)^2 + .0822^2}$$

It processes the derivative of an input signal within a limited bandwidth. Fig.5 shows the step and the harmonic response of this digital controller. The test of the accuracy has been effected by simulating a tracking operating condition.

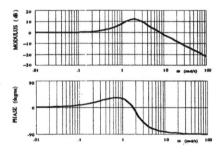

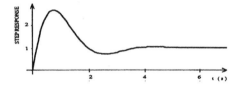

Fig.5 Step and harmonic response of a derivative digital controller.

The range of the inner variables has been computed first by implementing all the proposed formulations of the mathematical model. The offset error and the maximum range of the error have then been computed in correspondence of a word length of 8 and 16 bit microprocessor. Results are summarized in Table 1. These results indicate that some formulations of the mathematical model require the scaling of the input variable according to a high rate. In the implementation by an 8 bit microprocessor the offset error assumes so large value to discourage the application. Only the formulation given by the finite difference equation, the observer form, the diagonal forms can be used.

In the implementation by a 16 bit microprocessor the errors are always limited in correspondence of all the formulation of the mathematical model.

# TABLE 1

| | range of $x_1$ | range of $x_2$ | input variable scaling | offset error 8 bits | range of error 8 bits | offset error 16 bits | range of error 16 bits |
|---|---|---|---|---|---|---|---|
| Eq. (5) | -- | -- | 1/2 | 0.40 | ± 0.40 | 0.0012 | ± 0.0012 |
| Eq. (6) | [0,150] | [0,150] | 1/150 | -- | ± 1.70 | 0.0050 | ± 0.0050 |
| Eq. (7) | [-2.5,0] | [0,1.8] | 1/2.5 | 0.38 | ± 0.38 | 0.0020 | ± 0.0020 |
| Eq. (8) | [0,150] | [0,150] | 1/150 | -- | ± 1.70 | -0.0050 | ± 0.0050 |
| Eq.(11) | [0,1.4] | [0,1.8] | 1/3 | -- | -- | 0.0015 | ± 0.0015 |
| Eq.(12) | [-130,0] | [0,150] | 1/150 | -- | -- | 0.0015 | ± 0.0055 |
| Eq.(15) | [0,9] | [0,12] | 1/14 | -0.12 | ± 0.25 | -0.0005 | ± 0.0009 |
| Eq.(16) | [-6,2] | [-6,2] | 1/6 | -0.10 | ± 0.20 | -0.0005 | ± 0.0007 |

The minimum value of the errors occurs for the implementation according to the diagonal observer form in correspondence of both an 8 bit and 16 bit microprocessor. This result confirms the validity of the proposed evaluation criterion.

## CONCLUSIONS

The diffusion of digital controllers depends on the joint development of both the hardware and the software. These two aspects should be considered strictly connected if an optimized product should be realized. The optimization is highly justified when the digital controllers are applied in a mass-product.

In prototype applications it offers the most efficient way for avoiding troubles due to a heuristic design and by an oversize in the implementation.

Good results in the implementation of a digital controller can be attained by considering the problems of minimizing the computing time and the accuracy independently on the word length and on the frequency of the clock.

To guide a designer towards an efficient implementation of fixed parameter algorithm, a general criterion has been worked out by considering also the fundamentals of the System Theory and the Dynamics of Linear Systems. The generality of principles used does not require a specific background and should make easier the application.

A good implementation of a digital controller requires in fact a wide range set of knowledge in different topics. The implementations effected without these requirements could produce poor results and discourage the introduction of innovations.

## BIBLIOGRAPHY

[1] PHILLIPS C.L.; NAGLE Jr H.T.: *Digital Control System Analysis and Design*, Prentice-Hall, Inc, Englewood Cliffs, N.J., 1984.
[2] FAREGON C. Editor: *The Digital Control of Systems*,North Oxford Academic, 1989.
[3] ISERMANN R.: *Digital Control*, Spriger-Verlag, Berlin,1990.
[4] KOWALCZUK Z.: *Round-off noise in the digital implementation of DDC algorithms*, Preprints of 10th World Cogress on Automatica Control, Tallin 1990, vol. IV, pag. 108.
[5] LI G., GEVERS M.: *Minimization of Finite Wordlength Effects in Compensator Design*, Proceedings of the First European Control Conference, Grenoble 1991, vol. I, pag. 544.
[6] KAILATH T.: *Linear Systems*, Prentice-Hall, Inc, Englewood Cliffs, N.J., 1980.
[7] ACKERMANN J.: *Sampled data control systems*, Springer-Verlag, Berlin Heidelberg.

This paper has had the financial support of the Ministry of University and Technological And Scientific Researches in the research projects presented in 1991.

# FAULT DETECTION AND ISOLATION AND MODE MANAGEMENT IN SMART ACTUATORS

## M. Bayart, A.L. Gehin and M. Staroswiecki

*LAIL (URA CNRS 1440)/LAFA, U.F.R. D'IEEA, Université des Sciences et Techniques de Lille Flandres Artois, 59655 Villeneuve d'Ascq Cedex, France*

## ABSTRACT

This paper presents the generic functional architecture of a smart actuator. It mainly addresses the decomposition level relative to Fault Diagnosis and Isolation and details the mode management module.

**Keywords** : Smart Actuators, Fault Detection and Diagnosis, Mode management, Distributed Intelligence System.

## 1 Introduction

The evolution of sensors and actuators linked to recent development of microelectronics and communication gives rise to an increasing interest for Distributed Intelligence in Production Automation System.

Moreover, recent works lead to the concept of integrated automation : one doesn't only design the process control but a genuine Real Time Process Operating System (RTPOS), which provides helps to the different operators which intervene in the various stages of the plant life cycle.

The local processing abilities associated to sensors and actuators allow to entrust them with some RTPOS functions and confer a certain intelligence on them [1], [2].

In that sense, a smart actuator will simultaneously be producer and consumer of system's information. Its mission consists then on :

- acting on the process, according to the received orders,

- processing validation functions and providing operator's helps for control, maintenance and technical management,

- memorizing its state informations so as to participate to the real time distributed data base.

Therefore, the distribution of RTPOS functions at the instrumentation level requires communication possibilities in order to insure a correct synchronisation of the whole system. The field bus type link is possible if the instrumentation is equiped of communication capacities. This one allows thus to increase the parallel RTPOS execution possibilities and contributes therefore to the decrease of the interconnexion costs, thanks to the bus architecture [3].

The submitted results have been developed at the University of Lille (France) and validated through the working group "Actionneurs Intelligents" of the "Comité Interprofessionnel pour l'Automatisation et la MEsure" which inludes users, searchers and actuators constructors.

The paper, first develops the generic functional architecture of a smart actuator. This architecture is represented using the S.A.D.T. (Structured Analysis and Design Technique) formalism [4] in order to make appear for each functionality, its decomposition into elementary activities, for which we give the inputs, outputs, data processing and activation conditions.

The decentralization of the supervision system is also realized using local processing power for local Fault Detection and Isolation. So, in the second part, we first present the decomposition of a local F.D.I. algorithm which takes into account all the locally available informations. Then, we propose a generic mode management system of the smart actuator which requires the F.D.I. results

## 2 Architecture of a smart actuator

### 2.1. Hardware architecture

A smart actuator is composed of a mechanical part which allows to realize the action and some local intelligence which processes the available information.

In fact, the actuator has own sensors, that we call proprioceptive, which control its operation, for example, position sensor on a motor shaft or output torque sensor. This instrumentaton is already implemented and consequently can be used by Fault Detection and Isolation (F.D.I.) algorithms to supervise the actuator and to inform as early as possible, the control operator and other components of the automation system, of the actual state of the actuator.

On the other hand, maintenance and technical management operators can exploit those local informations in order to improve the security, availability and reliability of the whole system.

In that sense, smart actuators have communication possibilities with their environment.

*- Communication with the process*

Insofar as the local intelligence processes information provided by local sensors, it can also take into account data which are elaborated by sensors implemented in the near environment of the actuator. We call this instrumentation, exteroceptive sensors.

The information coming from such sensors may be transmitted to the smart actuator via :

- the field bus : they are in this case issued from smart sensors, whose information is periodically displayed to the adequate consumers.

- a direct link : those sensors are a part of the smart actuator, which is then considered by the consumers as the producer of their data.

*- Communication with the operators and the supervision system*

The smart actuator offers possibilities of local communication with the operators via a local interface, the communication with the control and other components is realized through filed bus type link. Fig. 1 illustrates the structure of a smart actuator, in the mechanical field.

### 2.2. Functional model

Intelligent devices are connected to the automation system by standardized field buses which permit to exchange informations. In that sense, the smart components really must be interoperable. This property allows the communication between several devices, issued from various constructors, on the same bus.

In that sense, we develop a generic functional architecture of a smart actuator, which can be used to normalize smart components.

The analysis of the operators'requirements, for various activities such as installation and configuration, control, maintenance and technical management [5], leads to define a set of functionalities that a smart actuator has to provide.

The architecture is represented using the SADT (Structured Analysis and Design Technique) formalism in order to define at different decomposition levels the organization of the modules independantly of hardware considerations.

In fact, according to the level of local intelligence a part or all the functions will be implemented at the actuator's level while others will be implemented in the numerical control system.

The functional decomposition makes appear for each functionality, its decomposition into elementary activities, for which we give the inputs, outputs, data processing and activation conditions.

The first decomposition level distinguishes the communication functions (input and output) and the processing ones, including data processing, power processing and data acquisition. Fig. 2 shows the functionnal architecture.

Fig. 3 describes the second decomposition level with the set of data processing functions while figure 4 corresponds to the application modules.

In this paper, we describe more precisely the Mode Management and Fault Detection and Isolation modules.

### 3 Mode Management module

### 3.2. Operating modes definition

Various works in France on running and stopping modes of a machine led to the elaboration of a standardised guide : the GEMMA (Guide d'Etudes des modes de Marche et Arrêt) [6] proposes arrangements of these modes based on a graphic representation.

This guide can be adapted to various machines requirements and is of use for specifications, in order to define all the operating cycles as well as the conditions of the transitions.

The procedures are classified according to four parts :
- verification and tests,
- normal operating,

- failures
- starting up and stopping operations.

These results easily allow to establish a precise description of a machine behaviours in state graph form [7].

The major inconvenient of these approaches is that the failures are considered only in certain operating modes, which are generally limited to normal operating and configuration.

In our approach, insofar as we want to integrate the production and technical management functions, it becomes necessary to consider the operating modes, in each step of the life cycle. We create a hierarchical organisation in which, for each step of life, given operating modes are defined.

These operating modes are function of the user mode and the actuator's state.

To each operating mode, corresponds a set of executable activities.

### a. Steps of the life cycle

The analysis of the life cycle of a smart actuator associated to the analysis of the users needs has allowed to distinguish the following steps [8]:
- constructors operations
- receipt and storage
- putting in service and configuration
- exploitation
- maintenance
- demolishing

The transitions from a step to another are realized according to defined rules. They can be realized automatically or in response to operators requests.

### b. User modes

In each step of the cyle life, we define a set of user modes of the actuator. They can be selected by the operator, and they correspond to the exploitation constraints. They are for example :
- disconnected
- connected
- in operation

The user mode "in operation" is decomposed into several modes according to the considered step of the life cycle. For example, in exploitation, we distinguish the "automatically", "step by step", "cycle by cycle" operatings.

The transition from one mode to another is realized according to the requested user mode which is asked by the operator and to the actual mode.

### c. Actuator's state

The F.D.I. algorithm can detect the components of the actuator which are in a faulty situation, for example a mechanical failure of the actuator, a bias or an abnormal noise on an analogic sensor of the instrumentation, a failure of one out of two redundant sensors, ... Several degradations can be found, in particular, some of them can have consequences on the safety of the actuator, process and humans. In this case, a fold back to security position has been defined and the faulty situation has for consequence to position the actuator in the security state (open or close for a valve, as example), by hardware configuration or software order.

According to the results of the F.D.I. algorithm, the actuator can be in one of the following states :
- normal : In this case, the actuator and its instrumentation are in nominal situation.
- degraded : In this case, the required activities in the considered step can nevertheless be realized according to the robustness of the smart actuator.

This situation justifies the transmission of a warning signal to the control, maintenance and supervision operators.
- out of order : In this case, the failure is such that the required activities in the considered step can't be realized.

An alarm signal is sent to control and maintenance operators and to the supervision system.

### 3.2. Operating modes management

We define an operating mode as the combination of the user mode and the actuutor's state.

At each time, the actuator is in a step of its life cycle, and in a given operating mode, to which corresponds a set of executable activities.

The authorization or the interdiction of an activity requested by the operator is managed by the mode management module.

For example, in the "in operation" mode and "exploitation" step, it is forbidden to write configuration information.

The other important role of this module is to change the operating modes according to the information delivered by the F.D.I. algorithm and the defined state graph.

From all the degraded states, the return to normal operating mode is realized after maintenance and eventually new configuration operations.

This state graph is specified for each given application, and the mode management module has to evaluate the firing of its transitions, via the processing of the operators's requests and of the FDI algorithm's results.

### 4 Local F.D.I. modules

According to the measures of the physical values from the exteroceptive and proprioceptive sensors and to the received control, this module uses the actuator's model in order to provide information about its actual state, its functional availability and elaborates the functional validation of the physical measures.

These data are used by the control operator and by the modules "to elaborate controls", "to elaborate informations" and " to manage modes".

A great number of works have been devoted to fault detection and isolation procedures based on global treatment of informations, using static or dynamic models of the whole process [9], [10]. These procedures can be applied in local intelligent devices [11], [12].

Indeed, the F.D.I. algorithm is decomposed into four functions which are detection, isolation, estimation and diagnosis. Fig. 5 shows this level.

The detection function receives images of physical values with a technological validation. This last concerns the conditions in which the information have been produced by the proprioceptive sensors and produced and transmitted for the data coming from exteroceptive sensors.

Then, according to the redundancy possibilities, this function provides functional validation concerning all the images of physical values if no-anomaly has been detected, or concerning only a part of them if any anomaly has been found.

The isolation function receives informations coming from the detection function and eventually identifies the failure by comparing the received information (the obtained signature) with known failures signatures.

The estimation function allows to estimate on one hand, the images of those physical values whose functional validation hasn't be obtained and on the other hand, according to theses values, other measures and wished position given by 'to elaborate controls" module, provides an image of the real position of the actuator.

The diagnosis function gives a diagnosis of the detected failure, when anomaly has been detected, according to available knowledge.

This information is provided to the mode management module for updating the actuator's state.

### Conclusion

The increasing interest for Production Automation System is, on one hand, the consequence of the concept of integrated automation, in which all the steps of life cycle from constructors'operation to demolishing via configuration, exploitation, maintenance, ... are taking into account.

On the other hand, it results from the progress of microelectronics and communication, which allow the development of smart actuators ans sensors.

Their local data processing abilities of smart instruments make easy realizable the distribution of RTPOS functions at the instrumentation level. In that sense, smart actuators not only act on the process but also perform state validation functions and provide different helps for the control, maintenance and technical management operators.

In the paper, we present the main decomposition levels of a generic functional architecture of a smart actuator, using the SADT formalism. Then, we detail the decomposition of the local Fault Detection and Isolation modules with take into account not only proprioceptive sensors but also information in the local environment of the smart actuator, via the exteroceptive sensors.

The role of the mode management module is very important insofar as it forbids activities according to the operating mode and the actual step of life cycle.

The implementation of such an architecture has been realized on a on-off valve at the "Laboratoire d'Automatique Fondamentale et Appliquée" in Lille University.

### REFERENCES

[1]    G.J. BLICKLEY, Valves join transmitters in getting smart, Control Engineering, January 1990.

[2]    H. GRIEB, E. LINZENKIRCHNER, "Intelligent actuators, High performance and reduced price". IFAC LCA, Milan, Italy, November 1989.

[3]. M. BAYART, M.STAROSWIECKI, "Smart actuators for distributed intelligent systems", IFAC DIS'91, Arlington, Virginia, 13-15 Août 1991.

[4] D.T. ROSS, "Structured Analysis : a language for communicating ideas", IEEE Transactions of software engineering, Vol. 3, N°1, 1977.

[5] Livre blanc "Les Actionneurs Intelligents" Edition CIAME, Paris, 1988

[6] ADEPA, Le GEMMA, "Guide des Modes de Marche et d'Arrêt", 1979.

[7] S. BOIS "Integration de la gestion des Modes de marche dans le pilotage d'un système automatisé de Production". Thèse de Doctorat, Lille, 28 Novembre 1991.

[8]. M. BAYART, M.STAROSWIECKI, "Smart actuators : generic functional architecture, service and cost analysis, SICICI'92, Singapore, 18 - 21 Février 1992.

[9]. R. PATTON, P. FRANCK, R. CLARK,"Fault diagnosis in dynamic systems. Theory and application". Prentice Hall International (U.K.), 1989.

[10]. J.J. GERTLER, "Survey on model-based failure detection and isolation in complex plants". IEEE Control System Magazine, December 1988, 3-11.

[11]. M. BAYART, P-H. DELMAIRE, M. STAROSWIECKI, "Fault Detection and Isolation in Smart Actuators", ACTUATOR 90, Bremen, FRG, 20-21 Juin 1990.

[12]. M. BAYART, M. STAROSWIECKI, "Functional analysis of smart actuators", COMADEM 91, Southampton,U.K., 2-4 Juillet 1991.

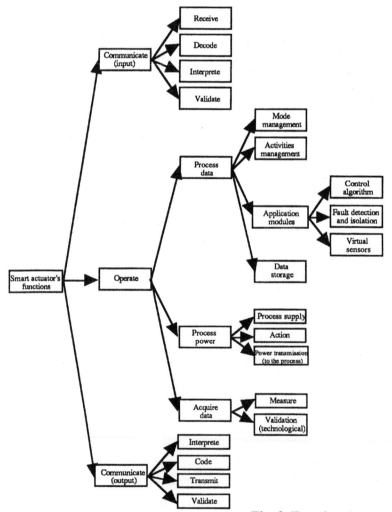

Fig. 2 Functional structure

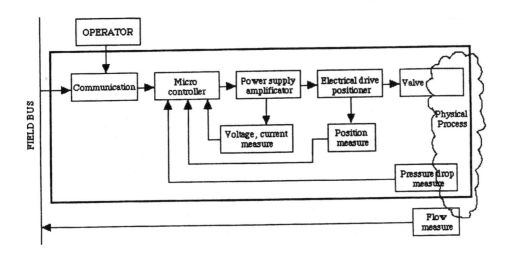

Fig. 1  Structure of a smart actuator:
Example of a valve

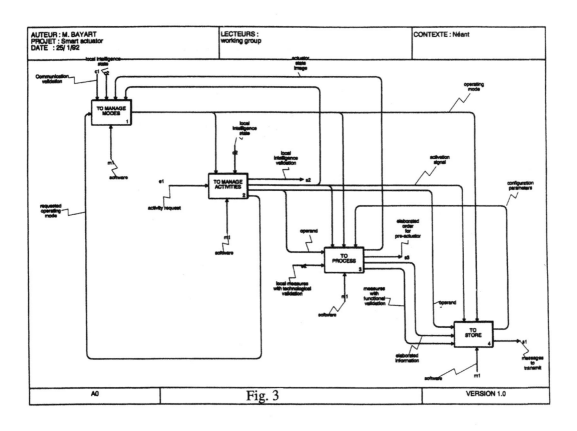

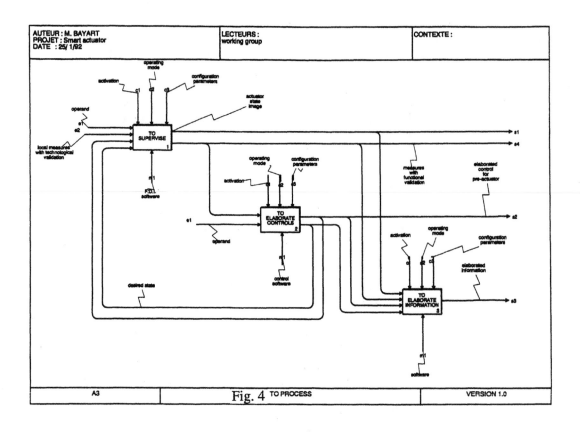

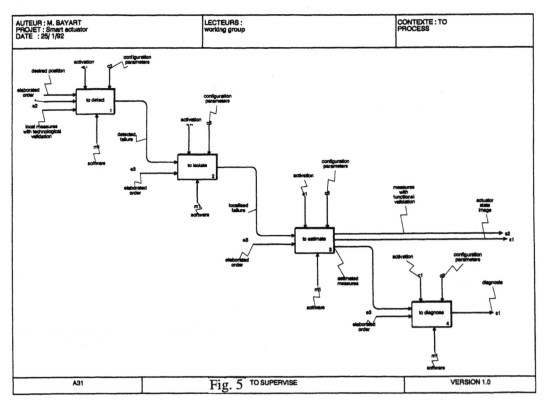

# HIERARCHICAL DATA VALIDATION IN CONTROL SYSTEMS USING SMART ACTUATORS AND SENSORS

J.P. Cassar, M. Bayart and M. Staroswiecki

*LAIL (URA CNRS 1440)/LAFA, UFR IEEA, Université des Sciences et Techniques de Lille,
59655 Villeneuve d'Ascq Cedex, France*

## ABSTRACT.

The distribution of the intelligence among all the components of the automation system allows to consider them as a connexion of a number of intelligent objects which produce or consume informations via a communication system associated with global management rules.
The exchanges between components require the validation of the transmitted or received data. The functional validation is based on the physical or analytical redundancy of available information which are transmitted via a direct link or via a field bus. It is achieved by a Fault Detection and Isolation algorithm.
This paper presents a hierarchical decomposition of this algorithm and shows how it can then be implemented into smart components.

**Keywords:** Fault Detection and Diagnosis, Distribued Intelligence System, Smart Actuators and Sensors.

## I. INTRODUCTION.

The recent developments in microelectronics and communication lead to a transformation of control systems. Sensors have first taken advantage of these evolutions to become smart sensors. Now, actuators benefit from these progresses, and happen to be associated with local data processing. This gives them new functionalities promoting them to the status ot "smart" actuators[1].

The local processing power allows the processed information to be more numerous, and increases the possibilities of the global system[2]. It leads to a distribution of the intelligence among all the components of the automation system transforming centralized supervision system into distributed intelligent ones[3].

On the other hand, the locally available information can be used by a local Fault Detection and Isolation (FDI) procedure so as to generate and validate the information concerning the actual state of the component ( actuator or sensor) and its operational ability.

The paper presents a decomposition of the global FDI algorithm which allows a hierarchical data validation . Several approaches for the implementation are discussed:
First we will consider that all the data used for the FDI purposes are also processed by the control algorithms; in the second case, some data aren't yet processed and their affectation can freely be achieved without any constraint issued from the existing implementation.
Under theses two hypothesis, the centralized and decentralized implementation schemes are compared.
The number of data to be transmitted for the FDI activity then constitutes a performance criterion of each implementation scheme.

## II. SMART COMPONENTS AND DATA VALIDATION.

The distribution of the intelligence among all the components of the automation system allows to consider them as a connexion of a number of intelligent objects which produce or consume informations via a communication system associated with global management rules [1],[4].
The exchanges between components require the validation of the transmitted or received data . We distinguish two levels of validation:
Technological validation concerns the conditions under which the information has been produced and/or transmitted. As an example: operation in an allowed temperature range.
Functional validation is interested in the content of information and uses Fault Detection and Isolation software [5],[6].

This last validation is based on the redundancy of locally available information which already exist in the component's instrumentation, and are provided by the proprioceptive sensors.

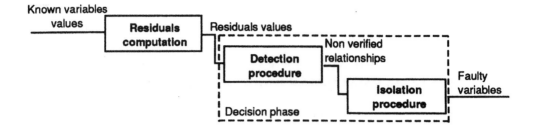

Figure 1.

On the other hand, available information in the components' environment can also be processed to improve the local FDI algorithm performances. They are transmitted via a direct link or via the field bus according to critical time and wiring cost constraints.

## III. DECOMPOSITION OF THE FDI ALGORITHM.

### III.1 - FDI system functional decomposition.

A great number of methods which have been developped for the FDI purpose in complex dynamical systems are model based.

The values of the known variables are processed, using model information, to obtain residuals whose value are close to zero in normal operation and significantly different from zero when a default occurs.

A decision procedure processes the values of the residuals to determine if the deviations from zero are the effect of failures or not. This is the detection phase. The isolation phase researches the sensors which could be failed.

Many methods can be used by the residual generation phase, one of them is based on the Analytical Redundancy Relationships: from the system model, some relationships between time sequences of available system variables (known variables) are found[7][8]. They can be obtained in the static case (parity space ) or in the the dynamic case (generalized parity space [9]).

The functional decomposition of the FDI algorithm can be then presented as in the figure 1[10].

### III.2 - Structure of a FDI system
#### III.2.1 - Definition

A model based FDI system is composed of a given set, $\mathcal{R}$, of analytical redundancy relations (ARR).

Each relation r out of $\mathcal{R}$ processes variables whose value is known, either via the measure performed by a sensor, or as the result of another computation.

The structure of an ARR r is named $S(r)$ and defined as follows :

$$S(R) = \{c \; / \; c \text{ is processed by the relation } r\}$$

The overall set of the known variables which are used by the FDI system is then :

$$\mathcal{C} = \cup \, S(r) \qquad r \in \mathcal{R}$$

#### III.2.2 - Hierarchical data validation

Let us consider a given set of ARR. Its structure may be represented by a binary matrix whose entries are defined as follows :

$$a_{ij} = 1 \qquad \text{iff} \quad c_j \in S(r_i)$$
$$a_{ij} = 0 \qquad \text{otherwise}$$

A permutation on the rows and columns of this matrix allows to obtain specific organizations of the variables $\mathcal{C}$ and the relations $\mathcal{R}$.

Let us consider two partitions, $\{\mathcal{R}_k\}$ and $\{\mathcal{C}_k\}$ of $\mathcal{R}$ and $\mathcal{C}$, with the same number of classes, ordered by the following rules :

$$(1) \qquad \mathcal{C}_i \subset S(r) \qquad \forall r \in \mathcal{R}_i$$

$$(2) \qquad S(\mathcal{R}_i) \backslash \mathcal{C}_i \subseteq \bigcup_{k-1}^{i-1} \mathcal{C}_k \quad \forall i$$

For such partitions, the structure matrix presents full diagonal blocs and is lower triangular. It allows a hierarchical approach to data validation through the following procedure :

- if the variables of $\bigcup\limits_{k-1}^{i-1} \mathcal{C}_k \cap S(\mathcal{R}_i)$ have been validated by the preceding steps, the step i, which computes the relations of $\mathcal{R}_i$, allows the validation of $\mathcal{C}_i$ (normal or faulty).

- if some variables of $\bigcup\limits_{k-1}^{i-1} \mathcal{C}_k \cap S(\mathcal{R}_i)$ have not been validated, the step i may allow the isolation of the faulty ones and the validation of the others, under the hypothesis that the variables of $\mathcal{C}_i$ are not faulty.

This procedure is summarized by the figure 2.

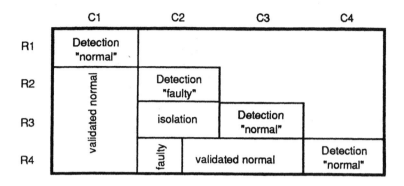

| | C1 | C2 | C3 | C4 |
|---|---|---|---|---|
| R1 | Detection "normal" | | | |
| R2 | validated normal | Detection "faulty" | | |
| R3 | | isolation | Detection "normal" | |
| R4 | | faulty | validated normal | Detection "normal" |

Figure 2.

A permutation which allows to obtain the desired structure is obtained through the use of a triangularisation algorithm [11], which creates diagonal blocs containing a maximum number of "1" entries. As several permutations may allow to obtain the same kind of triangular structure, the result is not uniquely defined.

The interest of such a structure is that it provides the order according to which the ARR have to be computed and allows the validation of a set of variables $C_i$ through the computation of a reduced number of relations $R_i$.

## III 3.- Implementation of the FDI sytem

### III.3.1 - Structure of the data processing system

Different application procedures process the available information : automatic control loops, smart actuators, fault detection and isolation, man-machine interface,....

We suppose that all the procedures which do not belong to the FDI system are already distributed among a given set of intelligent units, connected to a field bus communication system. The sensors are connected either to the processing units or directly to the communication system (smart sensors).

Let $T_i$ be a data processing unit and $V_i$ be the set of the variables provided by sensors which are connected to $T_i$. For those sensors, $T_i$ plays the role of a data concentrator; they are named the local variables of $T_i$.

Let V be the set of the variables produced by sensors which are directly connected to the field bus. Let q be the number of the processing units. Then,

$$V = \{V_1, V_2, ... V_q, V\}$$

constitutes a partition of the set $C_T$ of the variables which are processed by the control procedures. This implementation scheme is described on figure 3.

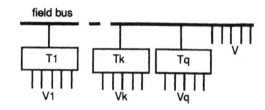

field bus

Figure 3.

### III.3.2 - Distribution of the FDI system

The residual computations derived from the set of the ARR may be realized using different schemes :
- a centralized computation, devoted to a specific FDI processing unit
- a centralized computation, devoted to an already existing unit $T_i$
- a distributed computation in which each unit $T_k$ processes a subset of ARR $R_k$

For the two centralized schemes, the communication cost can be evaluated through the set of the variables which have to be transmitted :

$S(R) \cap C_T$  in the first case

$[S(R) \cap C_T] \setminus V_i$  in the second one

### III.3.3 - Data validation in centralized schemes

Before it is used by a given unit, any data must be validated. In the centralized scheme, this leads to the following procedure :
- transmission of all the rough informations to the validation unit,
- centralized validation, through the FDI procedure,
- transmission of all the validated data to their respective consumers (the units $T_i$).

This procedure is illustrated on fig. 4 a.

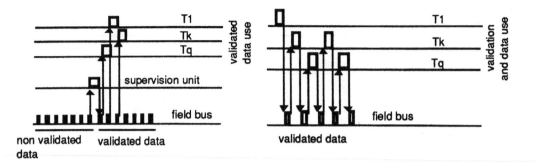

Figure 4.a.

Figure 4.b.

### III.3.4 - Data validation in decentralized schemes

Let us suppose that the partition V is compatible with a finer one which allows a hierarchical data validation. Then, the data can be localy validated in each unit. The validation procedure is described as follows :

- validation of the set of variables $C_i$ by the unit $T_i$, as soon as the set of variables $S(\mathcal{R}_i)$ becomes available from other units and local acquisition,
- transmission of the validated $C_i$ to those units which need them.

This procedure is illustrated on fig. 4 b.

### III.3.5 - Decentralization advantages

The distribution of the FDI procedure allows generally to use the existing data processing units, because of the low computational cost added to each unit. On the contrary, the centralized scheme leads to additional computational ressources, either to support the specific unit or to back up the existing unit in which the FDI procedure will be implemented.

The validated informations are immediately delivered to the other application procedures which are implemented in the same unit. No delay is induced for their use.

Only are transmitted on the bus those variables which are needed by other units : variables which are localy validated and used have not to be transmitted.

The variables are transmitted only once, under their validated form.

So, the distributed implementation scheme appears to be the most convenient one. However, it rests on the hypothesis that the partition V is compatible with a finer one which allows a hierarchical validation. Let us now study the possibility of such an implementation.

## IV - IMPLEMENTATION

Two cases will be considered, according to the fact that the data which are used for FDI purposes are, or not, also processed by the control algorithms.

### IV.1 - First case

We suppose first that $S(\mathcal{R}) \subseteq C_T$, i.e. that the data which are used by the FDI algorithms are already connected to the control units and to the field bus. So there exists a partition of the set of variables $S(\mathcal{R})$, which is a consequence of the partition V :

$$S = \{S_1, S_2, \dots S_q, S_{bus}\}$$
$$\text{with } S_k \subseteq V_k \qquad \forall k$$

One has then to find a permutation of the structure matrix, according to the criteria 1, 2 under the constraint :

$$C_i \subseteq S_l \qquad \forall i$$

which expresses that the decomposition $C$ is compatible with the partition S.

If such a permutation exists, we implement in the control unit $T_k$ the subset of those relations of $\mathcal{R}$, named $\mathcal{R}_1$, such that : $C_1 \subseteq V_k$

An example of such an implementation is given figure 5.

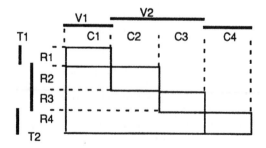

Figure 5.

266

If no permutation can be found, this means that for any decomposition, at least one class $C_k$ exists such that its variables are distributed over several classes of S :

$$\exists k, J(k) \qquad C_k \cap S_j \neq 0 \quad \forall j \in J(k)$$

The relations of $\mathcal{R}_k$ may then be implemented in any one of the control units which are associated to the $S_j$.

It is nevertheless possible to minimize the bus traffic, by the choice of the control unit for which $\mid C_k \cap S_m \mid$ is maximum. Let $T_m$ be such a unit. An example of such an implementation can be found below.

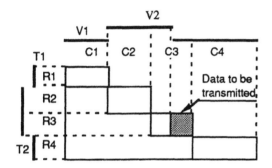

Figure 6.

The computation of the relations of $\mathcal{R}_k$ is achieved, in that case, similar to the centralized implementation :
- transmission to $T_m$ of those non validated data which are connected to other units (those data form the set $\cup [C_k \cap S_j]$ $\quad j \in J(k) \setminus \{m\}$)
- validation of the data in $C_k$
- diffusion of the validated data

### IV.2 - Second case

This case is a more general one, and supposes that some data are FDI specific, the associated sensors are not connected to any existing control unit. The set $S(\mathcal{R})$ is thus described by the following partition :

$$S = \{S_1, S_2, ... S_q, S_{bus}, S_N\}$$

with $S_k \subseteq V_k \qquad \forall k$

$S_{bus} \subseteq V_{bus}$

$S_N$ : non affected sensors

The same approach than above leads to research a decomposition which is compatible with the already fixed affectations :

$$C_i \setminus S_N \subseteq S_l \qquad \forall i$$

When associating the FDI relations to the control computations, the necessary variables of $S_N$ will be affected. To a given unit $T_k$ will be associated those $C_l$ such that :

$$C_l \setminus S_N \subseteq V_k$$

and the variables of $C_l \cap S_N$ will be connected to the unit $T_k$.

## V. CONCLUSION.

The decentralization of the control algorithms into smart components allows to envisage the decentralization of the FDI algorithm for data validation.

The hierarchical decomposition of the FDI process makes this kind of implementation possible and we have shown that this scheme is the more powerfull of two proposed ones according to the number of the data to be transmitted and the delay of availability of validated data.

The actual implementation can be realized in all the cases, even if the localization of the sensors doesn't match with any structural decomposition of the FDI system.

## REFERENCES.

[1] G.J BLICKLEY, "Valves join transmitters in getting smart", Control Engineeing, January 1990.

[2]. M. BAYART, M.STAROSWIECKI, "Smart actuators : generic functional architecture, service and cost analysis, SICICI'92, Singapore, 18 - 21 Februaryr 1992.

[3]. M. BAYART, M.STAROSWIECKI, "Smart actuators for distributed intelligent sytstems", IFAC DIS'91, Arlington, Virginia, 13-15 August 1991.

[4] H. GRIEB, E. LINZENKIRCHNER, " Intelligent actuators , High performance and reduced price", IFAC LCA, November 1989,Milan, Italy.

[5].M. BAYART, P-H. DELMAIRE, M. STAROSWIECKI, "Fault Detection and Isolation in Smart Actuators", ACTUATOR 90, Brême, FRG, 20-21 June 1990.

[6].M. BAYART, M. STAROSWIECKI, "Functional analysis of smart actuators", COMADEM 91, Southampton,U.K., 2-4 Jully 1991.

[7] COCQUEMPOT,V. , J. PH. CASSAR, M. STAROSWIECKI, "Generation of robust analytical redundancy relations," in Proceeding of the conference ECC91, Grenoble, France, July 2-5,1991, pp 309-314

[8] STAROSWIECKI, M., V. COCQUEMPOT, J. PH. CASSAR, "Generation of analytical redundancy relation in a linear interconnected system", in Proceedings of the IMACS-IFAC International Symposium MIN-S$^2$-90,,1990, Brussel, September 3-7,ppII-A-3-1 to 6.

[9] CHOW, E.Y. (1980). "Failure detection system design methodology," Sc. D. Thesis, Lab. Information and Decision system, M.I.T., Cambridge,M.A.

[10] GERTLER, J.J.,. "Survey on model-based failure detection and isolation in complex plants", IEEE Control System Magazine, pp 3-11, December 1988.

[11] RECH, C.,"Commandabilité et observabilité structurelle des systèmes interconnectés.", Thèse de doctorat, Université de Lille, France, June 1988.

# RELUCTANCE FORCE CHARACTERIZATION OF A THREE DEGREES-OF-FREEDOM VR SPHERICAL MOTOR

K.M. Lee* and U. Gilboa**

*The George W. Woodruff School of Mechanical Engineering, Georgia Institute of Technology, Atlanta, USA
**Armament Development Authority, Haifa, Israel

## Abstract

*The research presented here is to establish a basis for modeling and control of a variable-reluctance (VR) spherical motor which offers some attractive features by combining pitch, roll, and yaw motion in a single joint. The objective of this paper is to provide a good understanding of the magnetic fields and forces at play for realizing effective design and control of a VR spherical motor. The permeance-based model, which is commonly used in the stepper motor community to model the reluctance force of a stepper motor, was developed to model the reluctance of a VR spherical motor. Since the success of the permeance-based model depends on the assumed shape of the magnetic flux tubes, a finite element method was used in this study to provide physical insights of the magnetic flux patterns and to examine the reluctance force computed using the assumed flux shape and to illustrate the operational principle of a VR spherical motor.*

*__Key Words__: Robots, Spherical Motor, Variable-reluctance, Modeling, Computational Methods.*

## INTRODUCTION

The research efforts described here focus on the creation of an interesting three degrees-of-freedom (DOF) variable-reluctance (VR) spherical motor. Due to the ball-joint-like structure without the use of a speed reducer, the VR spherical motor results in a relatively simple joint kinematic and has no singularities in the middle of the workspace except at its boundaries. The VR spherical motor has potential applications such as laser and plasma cutting where high-speed, smooth, isotropic manipulation of the end-effector is required.

Recently, several design concepts of three DOF spherical motors have been presented. An induction spherical motor was conceptualized by Vachtsevanos *et al.* (1987) for robotic applications. However, realization of a prototype spherical induction motor remains to be demonstrated. The mechanical design of an induction spherical motor is complex. Laminations are required to prevent movement of unwanted eddy currents. Complicated three phase windings must be mounted in recessed grooves in addition to the rolling supports for the rotor in a static configuration. These and other considerations have led Lee *et al.* (1988) to investigate an alternative spherical actuator based on the concept of variable-reluctance stepper motors which are easier to manufacture. Hollis *et al.* (1987) developed a six DOF direct-current (DC) "magic wrist" as a fine-motion manipulator. An alternative DC spherical motor design with three DOF in rotation was demonstrated by Kanedo *et al.* (1989), which can spin continuously and has a maximum inclination of $\pm15°$ Although the DC spherical motor is characterized by its constructional simplicity, the range of inclination and the torque constant are rather limited. As compared with its DC counterpart, a VR spherical motor has a relatively large range of motion, possesses isotropic properties in motion, and is relatively simple and compact in design. The trade-off, however, is that sophisticated control scheme is required.

Variable-reluctance motors are actuated as a result of electromagnetic attraction between the rotor and the stator poles. The magnetic attraction or the reluctance force is created as the system tries to minimize the energy stored in the magentic field, or as the system tends to reduce the reluctance of the magnetic path. Lee and Kwan (1991) have presented the design concept the VR spherical stepper motor and developed the theory based on the local interaction between the adjacent stator and rotor poles to illustrate the concept feasibility. To allow for a relatively few but evenly spaced coils for smooth motion control of a VR spherical motor, Lee and Pei (1991) developed a method to examine the influences of the design configurations on motion feasibility.

A good understanding of the magnetic fields and forces at play is necessary to realize an effective design and control of an innovative VR spherical motor. For this reason, research efforts have been directed towards the characterization of the reluctance force of a VR spherical motor. The analysis presented here was performed by using both the permeance-based model (Chai, 1973) and the finite-element method (Sylvester and Ferrari, 1986). Unlike the previous study by Chai (1973) where the permeance-based model was developed for a single-axis VR stepper, a detailed study on the reluctance force characterization is presented here for a three DOF VR spherical motor. The permeance-based model has potential uses in design optimization, dynamic modeling, and motion control of a VR spherical motor. However, since the success of the permeance-based model depends significantly on the assumed shape of the magnetic flux tubes, a finite-element method has been used in this investigation to provide additional insights of the flux patterns on the torque generation and a means to examine the reluctance force computed by using an assumed flux path.

## THE VR SPHERICAL MOTOR

A schematic of a VR spherical motor is shown in Fig. 1, which consists of three sub-assemblies: the rotor, the stator, and the measuring mechanism. The rotor has m iron cores (rotor poles) imbedded in a smooth sphere. The rotor poles which may be designed with or without electromagnetic coils are radially oriented and meet at the rotor center. The stator is a hollow sphere which has n coils wound on magnetics iron cores (stator poles). The stator poles are threaded to the spherical shell such that the airgap may be easily adjusted. The stator poles are connected by the magnetic conductor layer in the stator shell to form a magnetic circuit. To evenly space the stator and rotor poles, the number of poles has been determined by Lee and Pei (1991) to be $2 \leq m < n < 20$. The detailed kinematic relationship which describes the orientation of the rotor as a function of encoder readings is given in (Lee and Pei, 1991). The stator coils can be energized individually using a control circuitry. As the stator coils are energized, a magnetic field is generated, which creates the magnetic energy in the airgap. The created energy is a function of the relative position between the rotor and the stator. Since the rotor can move freely inside the stator, the motion of the spherical VR motor is thus generated as the rotor tends to move to a position such that the energy in the airgap is minimized.

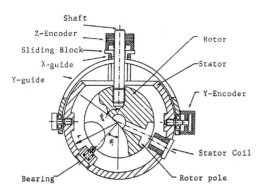

Fig. 1 Schematic of a VR spherical motor

## THE VARIABLE-RELUCTANCE MODEL

When a pair of poles of magnetic potential $V_1$ and $V_2$, are subjected to a differential displacement dx, an attraction is created as the system tries to minimize the stored energy $dW_m$ in the airgap between the two poles by reducing the reluctance of the magnetic path. From the principle of virtual work, the mechanical force f along the direction of x is given by

$$f = \frac{dW_m}{dx} = -\frac{1}{2}\left(\frac{1}{R}\right)^2 \frac{dR}{dx} \ (V_1 - V_2)^2 \tag{1}$$

where R is the reluctance at the airgap. Equation (1) illustrates that the reluctance force prediction of VR motors requires the reluctance model and its derivative with respect to the displacement x for a given potential difference to be characterized. Consider a differential flux tube of cross section ds and length $\ell$ between the two equipotential surfaces. The magnitude of the field intensity H is $(V_1 - V_2)/\ell$. The differential flux $d\Phi$ is $\mu_0 H ds$ where $\mu_0$ is the permeability of

air. The total flux $\Phi$ flows through the surfaces is the integral of $d\Phi$ over the entire equipotential surface S. Since $(V_1 - V_2)$ is a constant, we have

$$R = \frac{V_1 - V_2}{\Phi} = \frac{1}{\mu_0 \int_S \frac{ds}{\ell}}. \tag{2}$$

### Permeance-Based (PB) Model

Alternatively, the reluctance can be determined from its reciprocal of the permeance which is defined as

$$P = \mu_0 \int_S \frac{ds}{\ell}. \tag{3}$$

Equation (3) shows that the permeance is a function of geometry and that the computation requires the knowledge of the flux tube (i.e., S and $\ell$). When the airgap is much shorter than the dimensions of the adjacent pole faces, the magnetic flux $\Phi$ is constrained essentially to reside in the core and the airgap, and is continuous throughout the magnetic circuit. Thus, the permeance of the airgap can be approximated as

$$P = \frac{\mu_0 A_0}{g} \tag{4}$$

where $A_0$ is the overlapping area and g is the shortest distance between the two overlapping poles. The assumption implies that a zero overlapping area corresponds to a zero flux, and that the flux density distribution in the overlapping area is uniform. One of the most commonly used techniques to account for the fringing effects in modeling the permeance is to assume the shape of the flux tubes in Equation (3). A typical flux path for a given rotor position with respect to the stator coordinate frame is shown in Fig. 2 where the poles of a spherical motor are shown in conical shape for the simplicity of illustration.

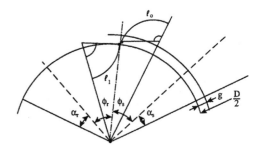

Fig. 2 An assumed flux path

The airgap that separates the rotor from the stator is divided into elements. The flux path from one element at a potential surface is connected to an element at the nearest surface by a straight line and/or a circular arc such that the flux enters into or emerges from the iron surface perpendicularly and does not cross other flux paths. For an elemental area on the spherical surface at the location defined by the spherical polar coordinate $(\theta, \psi, D/2)$ (Pual, 1981).

$$\delta A_{\ell k} = (D/2)^2 \; \delta \psi \; \delta \theta \; S_{\psi} \qquad (5)$$

where $\ell$ and k denote the indices along the $\theta$ and $\psi$ respectively, $S_{\psi}$ denotes the trigonometric sine function of the angle $\psi$, D is the diameter of the rotor, and $\delta \psi$ and $\delta \theta$ are the incremental angular displacements in the $\psi$ and $\theta$ direction respectively. The permeance of each airgap element derived from Equation (3) is given by

$$P_{\ell k} = \frac{\mu_o \; \delta A_{\ell k}}{\ell_o + g + \ell_1} \qquad (6)$$

where $\ell_o$ and $\ell_1$ are the arc lengths of the assumed path (Fig. 2) between the airgap element and the nearest stator and rotor poles, respectively. Let $C_{\ell k} (\theta, \psi)$ and $C_{rj}(\theta_{rj}, \psi_{rj})$ be the position vectors of the airgap element and the $j^{th}$ rotor pole respectively. The angle between these position vectors, $\phi_r$, can be determined from their dot (inner) product:

$$\cos \; \phi_r = \frac{C_{\ell k} \bullet C_{rj}}{(D/2)^2} \qquad (7)$$

Similarly, the angle between the position vectors of the $i^{th}$ stator pole and the airgap element is given by

$$\cos \; \phi_s = \frac{C_{\ell k} \bullet C_{si}}{(D/2)^2} \qquad (8)$$

where $C_{si}(\theta_{si}, \psi_{si})$ is the position vector of the $i^{th}$ stator pole. Since the stator poles and the airgap elements are fixed with respect to the stator frame of reference, only the arc length $\ell_1$ varies with the rotor position.

The arc length $\ell_o$ between the airgap element and the nearest stator pole is given by Equation (9):

$$\frac{\ell_o}{D} = \frac{1}{2} \; (1 + \frac{g}{D})(\frac{\pi}{2} + \phi_s - \alpha_s) \tan (\phi_s - \alpha_s) \frac{\text{sgn}(\phi_s - \alpha_s) + 1}{2} \quad (9)$$

where $\alpha_s$ is the half cone-angle of the stator. Similarly, the arc length $\ell_1$ between the airgap element and the nearest stator pole is given by Equation (10):

$$\frac{\ell_1}{D} = \frac{1}{2} \; (\frac{\pi}{2} + \phi_s - \alpha_r) \tan (\phi_r - \alpha_r) \; \frac{\text{sgn}(\phi_r - \alpha_r) + 1}{2} \qquad (10)$$

where $\alpha_r$ is the half cone-angle of the rotor pole.

## Flux Computation

With the assumed flux paths, the permeance of each airgap element may be calculated from Equation (6). The flux flowing through the airgap is determined by using magnetic circuit analogy with the following assumptions: 1) The iron reluctance is infinite as compared to the airgap reluctance. Thus, the predominant energy storage occurs in the airgap. The error introduced by this assumption depends on the geometrical dimensions of the structure and the permeability of the materials. 2) There is no saturation of iron elements in the system. This assumption is reasonable as

long as the coil excitations are limited so that the flux density in the iron is within the linear portion of the iron magnetization curve. 3) There are no magnetic flux leakages between the adjacent stator coils, between the adjacent rotor coils, or anywhere else in the system. The assumption implies that the spacing between any adjacent rotor (or stator) poles is much larger compared to the airgap. 4) The coil excitations are such that there are only attraction between the rotor and the stator poles. Repulsion between the stator and rotor poles generates significant leakage fluxes. This assumption is also a necessity in order for the assumption (3) to be reasonably stated. The permeance of the overlapping area is given by Equation (4) and the permeance of the fringing flux is accounted for by using Equation (6). A complete derivation of the overlapping area between any two overlapping circular poles in a spherical coordinate frame is given in (Lee and Pei, 1991).

An equivalent magnetic circuit of a VR spherical motor is given in Fig. 3. The magnetic flux $\Phi_{ij}$ flow through the airgap between the $i^{th}$ stator pole and the $j^{th}$ rotor pole can be determined from Equation (11):

$$\Phi_{ij} = P_{ij} [M_{si} + M_{rj} - V] \qquad (11)$$

where $P_{ij}$ is the permeance between the $i^{th}$ stator and $j^{th}$ rotor poles; $M_{si}$ and $M_{rj}$ are the magneto-motive-forces (mmf's) of the $i^{th}$ stator coil and $j^{th}$ rotor coil respectively, and V is the magnetic potential at the rotor core. Since

$$\sum_{i=1}^{m} \; \sum_{j=1}^{n} \; \Phi_{ij} = 0 \, , \qquad (12)$$

the magnetic potential V can be derived by substituting $\Phi_{ij}$ from Equation (11) into Equation (12), which leads to

$$V = \frac{\sum\limits_{i=1}^{m} \; \sum\limits_{j=1}^{n} P_{ij} (M_{si} + M_{rj})}{\sum\limits_{i=1}^{m} \; \sum\limits_{j=1}^{n} P_{ij}} \; . \qquad (13)$$

Thus, the magnetic core potential V can be determined once the permeance of each element is computed. The flux flowing through each of the airgap elements can be determined by using Equation (11) where $\Phi_{ij}$ and $P_{ij}$ are replaced by $\Phi_{\ell k}$ and $P_{\ell k}$, respectively.

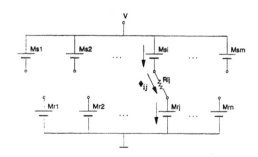

Fig. 3 Equivalent magnetic circuit

## Torque Prediction

The rotor is constrained by the stator and thus, the radial component of the reluctance force does not contribute to the motion. The reluctance forces along the directions of $\psi$ and $\theta$ are given as follows:

$$F_\psi = \frac{1}{2} \left( \frac{\Phi_{\ell k}}{P_{\ell k}} \right)^2 \frac{dP_{\ell k}}{d\psi} \qquad (14\,a)$$

and

$$F_\theta = \frac{1}{2} \left( \frac{\Phi_{\ell k}}{P_{\ell k}} \right)^2 \frac{dP_{\ell k}}{d\theta} \; . \qquad (14\,b)$$

The resulting torque contributed by all the airgap elements with respect to the stator coordinate frame is

$$T = \begin{bmatrix} T_1 \\ T_2 \\ T_3 \end{bmatrix} = \frac{D}{2} \sum_{\ell=1}^{M} \sum_{k=1}^{N} \begin{bmatrix} C_\theta C_\psi & -S_\theta & C_\theta S_\psi \\ S_\theta C_\psi & C_\theta & S_\theta S_\psi \\ S_\theta S_\psi & C_\theta & S_\theta S_\psi \end{bmatrix} \begin{bmatrix} -F_\theta \\ F_\psi \\ 0 \end{bmatrix} \qquad (15)$$

where M and N are the number of elements in $\theta$ and $\psi$ directions, respectively.

## MAGNETIC FLUX PREDICTION USING FINITE-ELEMENT (FE) METHOD

To provide additional insights and a means to validate the computed reluctance force, it is of interest to predict the flux patterns directly from the Maxwell's Equations which can be numerically solved by using a finite element method. The finite element solution yields the nodal potential and the average elemental flux density values. With the knowledge of the fields at hand, the flux through a surface can be calculated using the following summation:

$$\Phi = \sum_S (\underline{B}_i \bullet \underline{n}_i) \, \Delta S_i \qquad (16)$$

where $\underline{B}_i$, $\underline{n}_i$ and $\Delta S_i$ are the flux density at the centroid, the unit normal vector, and the surface area of the $i^{th}$ element. The reluctance can then be computed from Equation (2) and the reluctance force between the stator and the rotor poles from Equation (1) as a function of the displacement.

The finite-element models were computed using ANSYS finite-element package written by Swanson Analysis System. As it will be illustrated later, the computational results provide some physical insights to the design and control of VR motors without the 3D geometrical complications, that are inherent to the VR spherical motor. In addition, the 2D vector potential method explicitly computes the flux lines of the magnetic model and provides an effective means to validate the reluctance force computed using the assumed flux shapes.

## RESULTS AND DISCUSSIONS

For motion control of a VR spherical motor, both the torque prediction for a specified set of coil excitations and its inverse model which determines the input currents (or mmf's) required to obtain the desired torques are needed. The torque prediction is given by Equations (14)-(16) in the form of quadratic equations:

$$T_1 = \sum_{i=1}^{m+n} \sum_{j=1}^{m+n} \alpha_{ij} \, I_i \, I_j \; , \qquad (17a)$$

$$T_2 = \sum_{i=1}^{m+n} \sum_{j=1}^{m+n} \beta_{ij} \, I_i \, I_j \; , \qquad (17b)$$

$$T_3 = \sum_{i=1}^{m+n} \sum_{j=1}^{m+n} \gamma_{ij} \, I_i \, I_j \; , \qquad (17c)$$

where the coefficients $\alpha_{ij}$, $\beta_{ij}$ and $\gamma_{ij}$ are functions of the permeances and their derivatives. Also, they are functions of relative position between the rotor and stator. For (m+n) rotor coils, the solution to the inverse model may be determined from the three quadratic equations with (m+n) unknowns. This interesting feature is distinctively different from the popular three-consecutive-joint wrists based on traditional single-axis-motors or spherical motors of other types, which are typically characterized by their unique solutions to both their forward and inverse torque models. The nature of the multiple solution to the inverse model allows an optimal choice minimizing a specified cost function.

To illustrate the operational principle of a VR spherical motor, a 2D model was formulated by using the vector potential method as shown in Fig. 4 where the depth is infinite. The static force per unit depth between the stator and the rotor for a given coil excitation is computed using both the finite-element method and the permeance-based model. To make the two methods comparable, the product of the current density and the coil cross-sectional area is chosen such that it yields the desired magneto-motive-force (mmf). That is, the current density, $J = (mmf)/A$, where J in Ampere-turns/$m^2$; mmf is in Ampere-turns; and A is the cross-sectional area of the coil in $m^2$. In the simulation, the cross-section area of the coil on each side is 20mm x 3mm. Thus, 100 Ampere-turns are equivalent to $1.66E+6$ Amperes/$m^2$. Two different values of iron permeanability were used, namely, $\mu_i/\mu_o = 1E+3$ and $1E+7$.

TABLE 1 summarizes the computation result where the percentage errors of the reluctance forces listed are relative to that computed using the 2D finite element method. A few selected flux patterns computed using the finite-element method is given in Fig. 5. The excitation are indicated as positive if the mmf is directed toward the air gap. The magnitude of each excitation is 100 Ampere-turns. Except for Case (5), the permeance-based model yields relatively good approximation when the magnetic material has a very high permeability. There are three major assumptions which may accumulate significant errors in the permeance-based model: 1.) The model assumes no reluctance in the iron core. 2.) The model neglects leakage paths. 3.) The model is inaccurate in describing the flux paths. The contribution of the first source may be inferred by running a modified finite element model using an iron core of very high permeability. As shown in TABLE 1, the relative errors in Case (1) are 35.3% and 15.5% with the iron permeance of 1E+3 and 1E+7, respectively. The decrease in the relative errors can also be observed in all cases, which is consistent with the assumption made in permeance-based model that the reluctance of the iron is negligible or the permeability of the iron is infinite.

The relative contribution of the second and the third sources to the error is not as obvious. However, the assumption of no

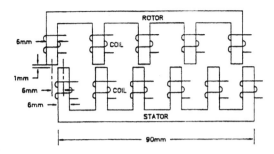

Fig. 4 Two-Dimensional Computation model

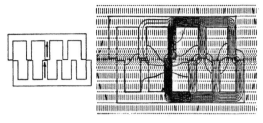

5(e) Case 5

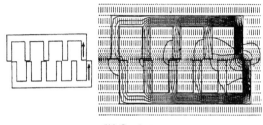

5(a) Case 1

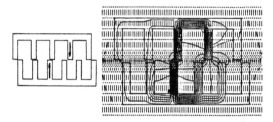

5(f) Case 6

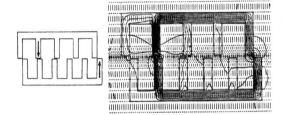

5(b) Case 2

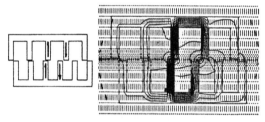

5(g) Case 7

Fig.5 Typical Flux Distribution

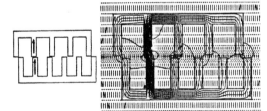

5(c) Case 3

TABLE I: COMPUTED RELUCTANCE FORCE

| Case | P-B. Model | F.-E. Model (ANSYS) | | Relative Error* | |
|---|---|---|---|---|---|
| | | $\mu_i / \mu_a = 1E+3$ | $\mu_i / \mu_a = 1E7$ | $E_{(1)}$ | $E_{(2)}$ |
| | (N/m) | (N/m) | (N/m) | (%) | (%) |
| 1 | 16.47 | 12.17 | 13.52 | 35.3 | 15.5 |
| 2 | 10.81 | 7.81 | 8.79 | 38.4 | 23.0 |
| 3 | 10.31 | 7.90 | 8.82 | 30.5 | 16.9 |
| 4 | 63.95 | 47.68 | 52.33 | 34.1 | 22.2 |
| 5 | -8.02 | -4.12 | -4.33 | 94.7 | 85.2 |
| 6 | -5.89 | -5.39 | -5.89 | 10.9 | 1.5 |
| 7 | -28.88 | -26.75 | -28.88 | 0.1 | -7.4 |

* $E_{(1)}$ and $E_{(2)}$ are the relative error of calculated reluctance forces using permeance-based model with respect to that computed using finite-element model with $\mu_i / \mu_a = $ 1E+3 and 1E+7 respectively.

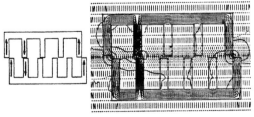

5(d) Case 4

magnetic leakages in permeance-based model implies that the flux would generally flow through the excited coils and returns through the remaining poles on both side of electromagnetic structure. As shown in the flux pattern computed for Case (5), the inaccurately assumed flux path may result in a relative error over 90%. Thus, the magnetic flux path should be selected to reduce unmodelled flux leakages. Repulsions between the stator and the rotor poles, which generate significant leakages, should be avoided in permeance-based model. To comply with the assumption of no repulsion between poles, there is a need for a procedure to check the relevance of the solutions in the permeance-based model of a VR spherical motor. Alternatively, the rotor poles may be designed to be constructed of magnetic material of high permeanability but to have no electromagnetic coils or permanent magnets. The elimination of coil excitations in the rotor also reduces the complexity of the design and simplifies the inverse model for real-time control of the VR spherical motor.

As illustrated in case (4) where the input excitation is tripled, the reluctance force increases by a factor of four as compared to that of case (1). The significant increase of the reluctance force is a direct result of a well-shaped magnetic flux path which not only utilizes all the rightward force generating airgaps, but also effectively eliminates the magnetic flux from flowing through the airgaps contributing to the generation of leftward forces. The VR spherical motor allows the input power to be distributed among several poles each of which contributes a fraction of the total mmf and thus requires a relatively low current per coil but a large surface area for heat dissipation.

It is interesting to note that there are infinite combinations of coil excitations for a given power input. Thus, the solution to the torque model may have multiple solutions if only a total input power is specified. This interesting feature is demonstrated in cases (1), (2) and (3) for generating the rightward force and cases (5) and (6) for generating the leftward force. As an example, case (1) results in a much higher reluctance force than that of cases (2) and (3) for the same power input. Thus, one would expect that there exists an optimal set of coil excitations which offers a specified torque with a minimum power. The optimal solution to the inverse problem is thus, a standard problem of constraint extrema. Introducing three Lagrange multipliers $\lambda_\ell$, $\ell=1,2,3$, the optimal input is the solution to the system of $(m+n+3)$ simultaneous equation of the form:

$$\frac{\delta E}{\delta I_k} + \sum_{\ell=1}^{3} \lambda_\ell \frac{\delta T_\ell}{\delta I_k} = 0, \quad \text{where } k=1,2,\ldots,m+n \tag{18}$$

and

$$T_\ell = \sum_{i=1}^{m+n} \sum_{j=1}^{m+n} (\alpha_{ij} \ I_i \ I_j)$$

where the energy cost function E is given by

$$E = \sum_{i=1}^{M} I_i^2 + \sum_{j=1}^{N} I_j^2 , \tag{19}$$

and $I_i$ and $I_j$ are the input currents to the $i^{th}$ stator coil and the $j^{th}$ rotor coil respectively.

## CONCLUSIONS

The modeling of the reluctance torque which is essential to the design and control of a VR spherical motor has been presented.

It has been shown that the finite-element simulation could be a useful tool not only to gain physical insights for the design but also a rational basis for the design and control of a VR spherical motor. The computation results show that the magnetic field patterns have significant influences on the maximum reluctance force obtainable. In addition, several interesting features uniquely characterize the VR spherical motor are highlighted. The VR spherical motor offers a potential advantage of distributing the input power among several coils, each of which contributes a small fraction of the total mmf's required to generate a specified torque, and thus, it allows a relatively low current per coil but a large surface area for heat dissipation. Since there are multiple solutions to the selection of the coil excitations, the torque generation of a VR spherical motor for a specified power input depends on the flux patterns which are functions of both the distribution of the coil excitation and the displacement between the stator and the rotor poles. The flexibility to control the reluctance force of the spherical motor using multiple coil excitation allows an optimal electrical input vector to be chosen to minimize a pre-selected cost function. It is expected that the analysis offered in this paper would serve as a basis for design optimization, dynamic modeling, and control of a wide range of VR motors.

ACKNOWLEDGEMENT

This research is supported under the National Science Foundation grants DMC-8810146 and DDM-8958383.

REFERENCES

Chai, H. D. (1973). Permeance Model and Reluctance Force Between Toothed Structures. Theory and and Applications of Step Motor, Kuo, B.C. (Ed.), West Publishing.

Hollis, R. L., Allan, A.P., and Salcudan, S. (1987). A Six Degree-of-freedom Magnetically Levitated Variable Compliance Fine Motion Wrist. Proc. of the Int. Symp. on Robotics Research.

Lee, K.-M. and Kwan C.-K. (1991). Design Concept Development of a Spherical Stepper Wrist Motor IEEE J. of Robotics and Automation, Vol. 7, No. 1, 175-181.

Lee K.-M. and Pei, J. (1991). Kinematic Analysis of a Three Degrees-of-freedom Spherical Wrist Actuator. Proc. of Fifth Int. Conf. on Advanced Robotics.

Lee, K-M., Vachtsevanos, G. and Kwan C-K. (1988). Development of a Spherical Stepper Wrist Motor. J. of Intelligent and Robotic Systems, Vol. 1, 225-242.

Paul, R. (1981) Robotic Manipulators, MIT Press.

Swanson Analysis System, Intro. to ANSYS.

Sylvester, P. and Ferrari, R. L. (1986) Finite Elements for Electrical Engineers, Cambridge Univ. Press, New York.

Vachtsevanos, G., Davey K., and Lee, K.-M. (1987). Development of a Novel Intelligent Robotic Manipulator. Control Systems Magazine, 9-15.

# ADVANCED SENSORS FOR THE PROCESS INDUSTRY

## M. Heitor

*Department of Mechanical Engineering, Instituto Superior Tecnico, Technical University of Lisbon,
A. Rovisco Pais, 1096 Lisboa Codex, Portugal*

*Abstract:* New sensor methodologies for on-line measurements of process streams with emphasis on hot, dirty, hostile
and multi-phase environments are revised and discussed in this paper taking into account the current requirements for the
integration of advanced sensors in CIME environments. Emphasis is on optical sensors making use of optical fibers and
solid-state devices, as well as on image processing methodologies which have been developed and implemented to
improve process monitoring and control. In addition, the development of data analysis strategies on the top of the new
measurement capabilities are briefly outlined because they give a considerable level of *"intelligence"* to the sensors.

Keywords:                 applied metrology; process control; optical diagnostics; smart sensors.

## INTRODUCTION

Reduced maintenance, improved reliability and self-checking
has been shown during the last decade (e.g. Shaw, 1980;
Higham, 1989) to be the most important features to improve
confidence in process measurements. In addition, the need of
improved industrial competitiveness together with the current
energy/environment related problems has led to the
development of flexible, small size and affordable sensors for
applications in dirty and harsh environments. This has
generalized the use of non-intrusive optical techniques, such
as those widely found to monitor physical, chemical and
biological phenomena (e.g., Nielsen, 1990), and their
implementation in the process industry has been quickly
extended from simple inspection tasks (e.g. Cielo, 1990) to
the detection of controlling parameters, such as flow,
temperature and pressure (e.g. Ebbeni, 1988). This paper is
aimed to review and discuss the implementation of these
novel sensor methodologies which may allow the
implementation of optimized control strategies based on non-
centralized schemes, as analysed in detail by Heitor (1991).

Almost irrespective to which approach is selected to improve
industrial competitiveness, success or otherwise is based on
some process measurement, involving predominantly either
temperature, pressure, level, flow or mass. Shaw (1980) and
Higham & Medlock (1983) have compared exhaustively the
relative importance of the various types of available sensors
in the process industry and figure 1 summarizes their analysis
in terms of relative number and initial cost of equipment. The
figure shows that the greatest perceived need for process
measurements is in flow technology followed by pressure
measurements, while temperature has been an important tool
for process monitoring. It should also be pointed out that the
requirements for advanced sensor systems refer nowadays
usually to on-line measurements of the composition of
process streams, particularly in hot, dirty, hostile and multi-
phase environments. This is because these environments, and
in particular those involving two phases, encompass a very
wide range of industrial processes in which the dispersed
phase (i.e., contaminant, particle or bubble) plays a critical
role. To better understand or control such processes, whether
in or out of the industrial environment, knowledge of the
characteristics of the discrete phase is essential. For these
purposes, the use of non-invasive instrumentation is highly

desirable, but must couple with the requirements for
flexibility, compatibility, accuracy, reliability and low price.
Current solutions to these problems with increased
possibilities for implementation in the process industries are
reviewed in this paper.

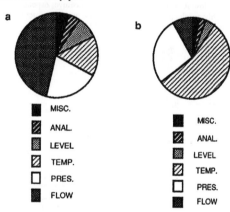

Figure 1. Relative importance of the process measurements,
Higham & Medlock (1983)
a) Process Control;
b) Monitoring

The following section discusses novel sensor methodologies
based on optical diagnostics and employing fiber optics and
semiconductor materials. Section 3 describes the use of image
processing to help improve process engineering and section 4
discusses the implementation of "intelligent" methodologies
in the sensor systems, which are associated with the novel
development of distributed and dispersed control architectures.
The last section summarizes the main conclusions of this
review.

OPTICAL SENSORS FOR ADVANCED PROCESS
MONITORING AND CONTROL

The recent review of Heitor (1991) shows that significant advances have occurred in flow, pressure and temperature measurements in the recent past and a wide range of meters is now commercially available, as described for example by Miller (1983), Higham (1987), Furness and Heritage (1989) and Furness (1990). The most important aspects of the current evolutionary trends include the use of optical devices to detect and monitor process variables (e.g. Reay & Pilavachi, 1989). Their development has relied on the increased use of lasers, which have become reliable and cheap. Also, the need to improve reliability in detecting the continuous motion and/or random fluctuations of most industrial processes or products in the production line has led to the development of optical sensors not only because of their non-intrusive/non-contact nature, but also because of their high response speed, intrinsic spatial and temporal resolution and increasing ruggedness. This has been possible thanks to the development of compact and affordable components such as diode lasers, optical waveguides or fibers and infrared detectors (e.g., Soares et al, 1988).

To obtain local measurements it is necessary the presence of a transducer at the particular measuring point and laser-Doppler velocimetry accomplishes this. It requires that the location of the focal point of a light beam and, more commonly, the intersection of two light beams at their focal point be accurately known, as shown in figure 2. Although details are not presented here, the sensor can be regarded as consisting of the combined effects of scattering particles and the incident light in the intersection region of the two laser beams. The light scattered from the particles in the stream should be conveniently collected and focused onto a photodetector, which provides a Doppler signal with a frequency proportional to the instantaneous velocity of the particles. Directional ambiguity can be easily solved making use of acousto-optic modulation. Interface with the flow is avoided, but particles must be added to the flow and be small enough to follow it. This may be difficult to achieve, especially in highly accelerated air flows, althouh a number of devices is available, as described by Durst et al (1981).

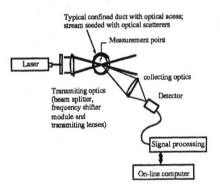

Figure 2. Diagram of typical arrangement of laser Doppler velocimeter.

In the context of flow measurement, a laser velocimeter apart from providing detailed knowledge of the time-averaged and turbulent velocity characteristics upstream and downstream a measuring device (e.g. Durão and Heitor, 1990), can also improve understanding of the fluid dynamics processes typical of current meters and to guide the design of new devices. Examples of this latter use can be found in the measurements reported by Durão et al (1988) around a square obstacle similar to those used in vortex flowmeters, or in the study of Emilia and Tomasini (1986) in the vicinity of a orifice plate meter.

The use of forms of laser velocimetry and other light scattering techniques to characterize two phase flows have experienced significant improvements over the past decade. Advances in the understanding of the light scattering phenomena has led to the development of new measurement concepts and improvements in those already in use, as reviewed for example by Bachalo (1988), Tayali & Bates (1990) and Taylor (1991).

The above paragraphs have shown that realiable optical techniques do exist to provide accurate local information of fluid flow characteristics for single phase flows and some dispersed two-phase flows. At present, available instrumentation is particularly suitable for measurements in well-controlled laboratory environments, although many novel arrangements has been proposed since the initial descriptions of the seventies. The extension of these techniques to industrial processes requires the development and practical implementation of flexible and small instruments making use of components such as semiconductor lasers and photodiodes, as described by Dopheide et al (1988, 1989) and Bopp et al (1989). Only recently, components suitable for LDV (see for example Kauffman, 1988) have become commercially available and allowed to constructed optical systems which are substantially reduced in cost and size compared with those represented in figures 2. Unlike conventional units, the size of the present systems (see figure 3) is not determined by the space occupied by the light source and the detector but rather by the sizes of other optical components, such as the lenses and the beam splitter. As an example, Bopp et al (1989) described a semiconductor based LDV system which houses both the focusing and the receiving optics in an assembly of dimensions 40mmx40mmx230mm. The construction is modular and allows for an easy replacement of parts.

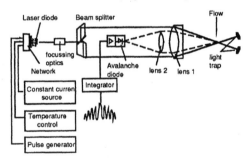

Figure 3. Schematic diagram of miniature, semiconductor-based laser-Doppler velocimeter

Another important achievement to improve the flexibility of laser velocimeters and their consequent application to industrial environments has consisted in the use of optical fibers. The fibers are used to link the light source to the transmission optical components and/or to link the collection optics to the photodetector. A recent practical application has been described by Bopp et al (1990) to measure low volumetric flow rates in pipes. It consists of a conventional LDV system built in a measuring head with overall dimensions of 200x70x50 mm, which is connected to a laser source and photodetector through appropriated glass fibers.

Optical fibers have had also an increased application on on-line optical inspection procedures, which have increased the reliability of automatic inspection of continuously moving products, as reviewed by Cielo (1990). Recent applications in process measurements include infrared temperature measurements, either in furnaces (Krapez et al, 1990) or during spray depositions (Moreau et al, 1990), in-situ film thickness monitoring in deposition processes (e.g. Severin & Severijons, 1990) and spatially-resolved flow measurements for single- and two-phase flows (e.g., Paul & Kychakoff, 1986; Nakatani et al, 1988; Bopp et al, 1990).

Fiber-optic pyrometry will find an increased range of applications in the measurement of high industrial temperatures, such as those typical of melting processes in which the temperature of the melt range from 1100 °C to 1400 °C and should be measured reliably and accurately. For example, the hostile conditions often found in metallurgical operations make such measurements difficult to obtain, once the submerged thermocouples used in conventional techniques have a tendency to break or dissolve. Alternatively, temperatures can be measured from above the melting surface, but flames, dust and slag emissions tend to interfere. It should be noted again that commercial pyrometers are affected by important errors related to variations of the surface emissivity, which may change unpredictably depending on the metal composition, surface finish and degree of surface clearness or oxidation. To overcome these problems, Krapez et al (1990) have reported recently the use of an integrating cavity infrared temperature fiber-optic sensor, whose design is shown in figure 4. It consists in a double-wavelength sensor coupled to a fiber optic probe to transmit the infrared radiation from the metal surface within the furnace to the detector unit, which is installed in a protected environment outside the furnace. The sensor head comprises and integrating cavity made of two gold-plated mirrors at a small angle to the metal plate in such a way that the wedge-covered sheet has an emissivity close to unit.

Optical fibers can also be used as pressure sensors, and because they are essentially passive devices, they are useful in hazardous environments where sparks cannot be tolerated or electromagnetic interference might hamper the sensing operation. An example of a pressure sensitive fiber is described by Ebbeni & Sendrowicz (1988) based on incoherent diffraction effects.

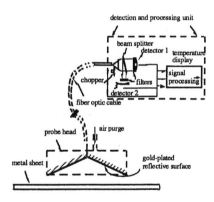

Figure 4. Schematic diagram of the fiber optic in-furnace temperature sensor developed by Krapez et al (1990)

IMAGE PROCESSING

With the advent of digital image analysis during the last several years, the utility of visualization methods in the analysis of industrial processes has been enhanced considerably. Over the past few years the technology available to facilitate real-time digital imaging has developed from large, highly specialized mainframe computing facilities to microcomputer-based boards and dedicated software. At the same time, solid-state video devices (such as CCD cameras) have become common place, increasing in both resolution and reliability, namely in comparison with well established laser scanning techniques (e.g. Schmalfub, 1990). As a result, these techniques can now be used as effective and efficient means of obtaining quantitative information of typical industrial processes.

Example of current application areas for image processing range from machine vision for robotics and inspection activities in manufacturing (e.g. Dickmans & Zapp, 1987; Costeira et al, 1989; Almeida et al, 1990), through oceanography and earth resources planning to medical imaging (e.g. Blattenbauer & Kim, 1989). Early attempts to use vision system techniques in industrial processes include the works of Iso et al (1987) in steelmaking and those of Lilja et al (1986) and Jadoul (1990) in industrial process furnaces, although the applications are restricted to the provision of information describing a furnace state. On the other hand, Farmer et al (1992) developed a new methodology to control glass melting processes making use of a furnace viewing system, as schematically shown in figure 5 and described in detail by Victor et al. (1991). The vision system processes acquired digitized images in the combustion chamber and uses the resulting information for control and classification procedures. Geometric features such as flame area, centre, orientation, symmetry axis and length are conveniently quantified by the present system. The values of these features can be used directly in the low-level control loops (e.g. to control the flame length or to control separately the fuel flow to the burners) or as inputs to a classification stage. In the classification stage, flames are classified according to the burners operating conditions. At a first stage (off-line) the classifier learns the classes of flames to be separated in the future, and in a second stage (on-line) it will identify the different operating conditions. The vision system is also able to quantify the amount of unmelted materials and to describe their motion in the melting tank as well as monitoring the activity of the bubblers. This information can be supplied to a KBS that accordingly will produce messages for the human operator or commands to the low-level control system.

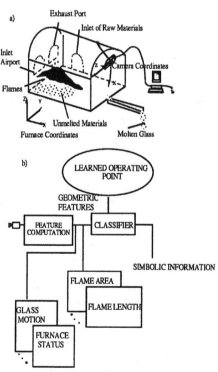

Figure 5. Schematic diagram of image analysis to monitor and control industrial furnaces
a) The installation of the system in the furnace;
b) Block diagram of the vision system

In the context of flow measurement the use of image analysis has become particularly important in the study of multi-phase flows, as discussed for example by Hesselink (1988) and Tayali & Bates (1990). The wide availability and decreasing cost of video recording machines and charge-coupled device (CCD) cameras, has led to increasing uses of techniques such has Particle Image Velocimetry (e.g. Adrian, 1988). These techniques can also be used to extract particle size, and a typical arrangement is shown in figure 6. The images of the particles in the laser sheet are recorded by the camera and subsequent analysis of consecutive frames can be used to compute the velocity and size as well.

A technique with better depth of field than those described in the previous paragraphs is holography. It involves the storage of the size and relatve position data of a dynamic three-dimensional distribution of particles. Reproduction of stationary images by ordinary photography can then be carried out. Lee & Kim (1986) review both in-line (Fraunhofer) and off-axis holographic techniques and possible applications for the analysis of multi-phase flows with industrial relevance.

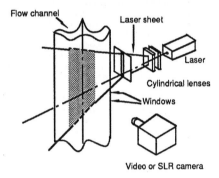

Figure 6. Typical arrangement of particle image velocimetry, PIV, for quantitative flow analysis

### DATA PROCESSING, INTELLIGENT SENSORS AND THE IMPLEMENTATION OF ADVANCED CONTROL STRATEGIES

Although the process industries have been very reluctant during the last decades to make changes to their tradicional or established technologies, the integrity and reliability of current digital and signal processors has led in the last years to the development and application of new ideas and concepts either in connection with instrumentation or for process control, as well as the co-ordination and integration of process operations (e.g. Astrom et al, 1986). Process control has evoluted from single loop local control, through centralized single loop control, to centralized distributed control, as discussed by Higham (1987). Current trends appear to be devoted to the development of dispersed control architectures in which either individual loops or a group of loops are controlled locally, but managed from a central location.The control requirements are, therefore, broken down into functionally independent tasks, which are connected by a communications bus and all actions are coordinated, but not controlled, by a system director module.

Progress in the development of standards and new technologies is encouraging the proliferation of advanced sensors, especially applied to dispersed control schemes. In this context, a fieldbus standard will allow the easy interconnection of the various subsystems of a dispersed control system in such a way that avoids the need for special interface hardware and software (e.g. Wood, 1988; Pleinvaux & Decotignie, 1988).

The implementation of these novel control methodologies implies necessarily the use of a *"new generation"* of sensors able to provide the local controllers with the required level of *"intelligence"* (e.g. Kerlin, 1986). This has been analysed among others by Higham (1989) and Laforie & Hubin (1990) and has relied on current advances in the electronic industries. The initial benefits have been derived from the improved understanding of sensor performance and actual operating conditions, but future developments will focus on the extension of this work to condition monitoring and process plants through the analysis of the signal which arrives at the interface between the process and the sensor. Following the definition of Kerlin (1986) a *"smart sensor"* is that with built-in capability to facilitate adjustment to operating range, capability to compensate for effects of environmental conditions, capability to compute required functions of the measurand (e.g. square root) and/or capability to perform self-diagnosis.

At present some commercially available sensors provide only some of these capabilities, such as remote adjustment of zero and span, on-board temperature corrections and/or linearization logic. A different approach is the use of resonant sensors, or sensors which operate in the time/period domain, as discussed in detail by Higham (1989). Resonant sensors have been particularly applied to sense pressure and are characterized by the high signal-to-noise ratio which can be achieved in the transmission of their signal and by the relatively low power level required for their operation. The transfer function of the measurement system can be determined under essentially ideal conditions on completion of its manufacture and the information, which can be regarded as its signature, stored in memory. When it is installed and the associated process plant is set in operation, a new signature can be determined and stored in the software. Subsequent changes in the signature provide a warning of variations in the conditions or status of the instrument.

In the field of flow measurement the strategies under analysis here are particularly suitable to be applied in turbine and vortex flowmeters through detailed inspection of the spectral distribution of the output of the sensors. Higham et al (1986) and Fell (1988) have introduced the concept of Sparse Fourier Transform as a mean of calculating the power spectra of pulse signals and examine the information content in the response of turbine and vortex flowmeters with special emphasis on the physical causes of the interpulse interval variations, their effects on flowmeter performance and their detection using the pulse signal spectrum. Extension of this technique to electromagnetic flowmeters working with A.C. excitation was considered by Fell et al (1988) and proved to be able to provide reliable warnings of flowmeter faults, to diagnose unusual flow conditions, or to correct the flowmeter reading when operating under non-ideal conditions.

### CONCLUSIONS

Advanced sensors for process measurements involving predominantly flow, pressure and temperature are briefly described and is shown that results can be established with equipment commercially available, although the need for improved reliability, the ability to withstand harsh environments, and improved repeatibility have resulted in the development of novel technologies. The most important aspects of the current evolutionary trends include: i) dedicated optical devices to detect and monitor process variables making use of optical fibers and semiconductor materials; ii) real-time digital image analysis to provide quantitative information of processes; and iii) improved data analysis to create *"intelligent"* sensors capable of adjusting their operating range, compensate for effects of environmental conditions and to perform self-diagnosis. This allows the integration of data acquisitoon techniques, process control and information management into a single plant wide network.

# REFERENCES

Adrian, R.J. (1988). *"Statistical Properties of Particle Image Velocimetry Measurements in Turbulent Flows"*. In: *"Laser Anemometry in Fluid Mechanics - III"*, ed. Adrian et al. LADOAN, Lisboa.

Almeida, A.I., Nunes, U.C., Dias, J.M., Araújo, H.J. and Batista, J. (1990). *"A Distributed System for Robotic Multi-Sensor Integration"*. Industrial Metrology, 1, pp. 217-229.

Astrom, K.J., Anton, J.J. and Arzen, K.E. (1986). *"Expert Control"*. Automatica, 22, (3), pp. 277-286.

Bachalo, W.D. (1988). *"Complete Particle Field Characterization with Optical Diagnostics"*. Proc. ICALEO'88, Laser Institute of America, vol. 65, pp. 229-262.

Balttenbaeur, J.A. and Kim, Y. (1989). *"Bringing Image Processing into Focus"*. Mechanical Engineering, July 1989, pp. 94-56.

Bopp, S., Durst, F. Muller, R., Naqwi, A., Tropea, C. and Weber, H. (1989). *"Small Laser-Doppler Anemometers Using Semiconductor Lasers and Avalanche Photodiodes"*. In: *"Laser Anemometry in Fluid Mecahnics - IV"*, ed. by Adrian et al, pp. 315-337, Springer Verlag.

Bopp, S., Durst, F., Holweg, J. and Weber, H. (1990). *"A Laser-Doppler Sensor for Flowrate Measurements"*. Flow Meas. Instrum. 1, pp. 31-38.

Cielo, P.G. (1990). *"On Line Optical Sensors for Industrial Material Inspection"*. In: *"In-Process Optical Measurements and Industrial Methods"*, pp. 22-38. Edited by H.A. MacLeod and P. Langenbeck, SPIE Proceedings Series, Vol. 1266. SPIE, Washington.

Costeira, J.P., Victor, J.A.S. Fernandes, J.P. and Sentieiro, J. (1989). Ver - *"A Stereo Vision System for Mobile Robots"*. IMA - Conference on Robotics, Longhborough, U.K.

De Witt, D.P. and Albright, L.F. (1986). *"Measurement of High Temperatures in Furnaces and Processes"*. A.I. Ch. E. Symp. Series, vol. 82, Nº 249. American Institute of Chemical Engineers, New York.

Dickamns, E.D. and Zapp, A. (1987). *"Autonomous High Speed Rad Vehicle Guidance by Computer Vision"*. Proc. NATO - Advanced Research Workshop on Mobile Robotics Implementation. Oporto, Portugal, May 1987.

Dopheide, D, Faber, M., Reim, G. and Taux, G. (1988). *"Laser and Avalanche Diodes for Velocity Measurement by Laser Doppler Anemometry"*. Exp. in Fluids, 6, pp. 289-297.

Dopheide, D., Taux, G., Reim, G. and Faber, M. (1989). *"Laser-Doppler Anemometry Using Laser Diodes and Solid-State Detectors"*. In: *"Laser Anemometry in Fluid Mechanics III"*, ed. by Adrian et al, pp. 19-36. LDOAN, Lisboa.

Durão, D.F.G. and Heitor, M.V. (1990). *"Intercomparison of Flow Measurements in Pipe Components"*. Proc. Intl. Conf. on Flow Measurements of Commercially improtant Fluids. London, Feb. 28 - March, 1. IBC Tech. Services Ltd.

Durão, D.F.G., Goulas, A., Heitor, M.V., and Teijema, J. (1988). *"Guidelines For Flow Measurements"*. to be conducted under the Community Bureau of Reference of the Commission of the European Communities. C.E.C. - BCR Report.

Durst, F., Melling, A. and Whitelaw, J.H. (1981). *"Principles and Practice of Laser-Doppler Anemometry"*, 2nd ed., Academic Press, London.

Dybbs, A. and Edwards, M.V. (1985). *"An Index Matched Flow System for Measurements of Flow in Complex Geometries"*. In: *"Laser Anemometry in Fluid Mechanics - III"*, ed. Adrian et al, pp. (?). LADOAN, Lisbon.

Ebbeni, J. (1988). *"Overview of Optical Methods in Metrology"*. In: *"Laser Technologies in Industry"*, ed. Soares et al, pp. 66-82, SPIE Proc. Series, Vol. 952. SPIE, Washington.

Ebbeni, J. and Sendrowig, H. (1988). *"Pressure Sensors Based on Incoherent Diffraction Effect"*. In: *"Laser Technologies in Industry"*, ed. Soares et al, SPIE Proc. Series, vol 952, pp. 393-399. SPIE, Washington.

Emilia, G. and Tomasini, E.P. (1986). *"On the Evaluation of Flow Meter Accuracy in Pulsating Flow Using a LDA System"*. Proc. 3rd Intl. Symp. on Appl. of Laser Anemometry to Fluid Mechanics. Lisbon, July 7-9.

Farmer, D., Heitor, M.V., Sentiero, J. and Vasconcelos, A.T. (1992). *"AIMBURN - A Multi-Sensor Methodology for the Analysis of Combustion Chambers and the Intelligent Control of Large Industrial Glass Furnaces*. Proc. European Seminar on Umproved Tech. for the Rational Use of Energy in the Glass Industry". 4-6 Feb., Wiesbaden, Germany.

Fell, R. (1988). *"Pulse Signal Analysis - Opportunities for Flowmeter Enhancement"*. School of Control Engineering, University of Bradford, England.

Fell, R., Ajayi, A.A. and Tsiknakis, E. (1988). *"Signal Processing for Flow Measurement"*. School of Control Engineering, University of Bradford, England.

Furness, R.A. (1990). *"Modern Developments in Flow Measurements"*. Proc. Intl. Seminar on Sensors, Inst. Sup. Ind. de Liége, Liége 1-2 February, 1990.

Hesselink, L. (1988). *"Digital Image Processing in Flow Visualization"*. Ann. Rev. Fluid Mech., 20, pp. 421-485.

Heitor, M.V. (1991). "Advanced Sensor Systems for the Application of CIME Technologies in the Process Industry: A Review. Industrial Metrology, 2 (1), pp. 1-31.

Higham, E.H. (1987). *"An Assessment of the Priorities for Research and Development in Process Instrumentation and Process Control"*. The Institute of Chemical Engineers - Industrial Fellowship Report. Rugby, England.

Higham, E.H. (1989). *"A Route to Better Process Measurements"*. I. Chem. E. Symposium Series, Nº 115, pp. 321-333.

Higham, E.H. and Medlock, R.S. (1983). *"The Relevance of Optical Transducers to the Process Industries"*. Proc. Intl. Conf. on Optical Techniques in Process Control, The Hague. BHRA/SIRA.

Higham, E.H., Fell, R. and Ajaya, A. (1986). *"Signal Analysis and Intelligent Flowmeters"*. Measurement and Control, 19, (5), pp. 47-50.

Hottel, H.C. (1986). *"Background and Prespectives on Temperature Measurements in Furnaces"*. A.I. Ch.E. Symp. Series, vol. 82, pp. 1-22.

Iso, H. Yoshida, T. Yoneda, M. Sakane, K., Atsumi, T. and Sakamashi, N. (1987). *"Instrumentation and Control for a Refining Process in Steelmaking"*. IEEE Control Systems Magazine, October 1987, pp. 3-8.

Jadoul, P. (1990). *"Application d'une Camera CCD à la Thermographie d'um Four de Verrier"*. Proc. Capteus, Liege, Feb. 1-2, 1990.

Kaufman, S.L. (1988). *"Alternative Lasers for Laser-Doppler Velocimetry"*. Proc. ICALEO'88, pp. 36-46. Instrum. Lasers of America, Vol. 67.

Kerlin, T.W. (1986). *"Process Temperature, Pressure and Flow Measurements: Opportunities and Challenges"*. ASME, F.E.D. Vol. 44, pp. 1-4.

Krapez, J.C., Bélanger, C. and Cielo, P. (1990). *"A Double-Wedge Cavity Pyrometer for Aluminium Sheet Temperature Measurement"*. To appear in Phys. E: Sci. Instrum.

Laforie, P. and Hubin, M. (1990). *"Les Capteurs Intelligents"*. Proc. Intl. Seminar on Sensors, Inst. Sup. Ind. de Liége, Liége 1-2 February, 1990.

Lee, Y.J. and Kim, J.H. (1986). *"A Review of Holography Applications in Multi-Phase Flow Visualization Study"*. J. Fluids Eng., 108, pp. 279-288.

Lilja, K. Gleus, A. and Sutinens, A. (1986). *"Image Processing for the Control of Burning Processes"*. Processing in Industrial Applications Proceedings, pp. 81-89. 1st. IFAC Workshop, ESPOO, Finland.

Miller, R.W. (1983). *"Flow Measurement Engineering Handbook"*. McGrow-Hill Book Co.

Moreau, C., Cielo, P., Lamontagne, M., Dallaire, S. and Vardelle, M. (1990). *"Impacting Particle Temperature Monitoring During Plasma Spray Deposition"*. To appear in J. Phys. E: Sci. Instrum.

Nakatani, N. Maegowa, A., Izumi, T., Yamada, T. and Sokale, T. (1988). *"Advancing Multi-Point Optical Fiber LDV's Vorticity Measurement and some New Optical Systems"*. In: *"Laser Anemometry in Fluid Mechanics - III"*, ed. Adrian et al, pp. 3-18. LADOAN, Lisboa.

Nielsen, H.O. (1990) *"Environment and Pollutioon Measurement Sensors and Systems"*. SPIE Proceeding Series, Vol. 1269. SPIE, Washington.

Nouri, J.M., Whitelaw, J.H. and Yianneskis, M. (1988). *"A Refractive-Index Matching Technique for Solid/Liquid Flows"*. In: *"Laser Anemometry in Fluid Mechanics - IV"*, ed. Adrian et al, pp. 335-346. LADOAN, Lisboa.

Paul, P.H. and Kychakoff, G. (1986). *"A Miniature Fiberoptic Probe for Optical Particle Sizing"*. Proc. ICALEO'88, Laser Institute of America, vol. 58, pp. 122-128.

Pleinvaux, P. and Decotignie, J.D. (1988). *"Time Critical Communication Networks: Fieldbus"*. IEEE Network, Vol. 2/3, May 1988, pp. 55-63.

Reay, D. and Pilavachi, P.A. (1989). *"Needs for Strategic R & D in Support of Improved Efficiency in the Process Industries"*. Report EUR 1192OEN. Commission of the European Communities, Directorate-General for Science, Research and Development.

SchmalfuB, H.J. (1990). *"Laser Scanner Versus CCD Camera: A Comparison"*. Industrial Metrology, 1, pp. 155-164.

Severin, P.J. and Severijns, A.P. (1990). *"In-Situ Film Thickness Monitoring in CVD Processes"*. To appear in J. Electrochem. Soc.

Shaw, R. (1980). *"The Direction of Instrumentation Technology in the Eighties"*. Inst. M.C., November 1980.

Soares, O.D.D., Almeida, S.P. and Bernardo, C.M. (1988). *"Laser Technologies in Industry"*. SPIE Proceedings Series, Vol. 952. SPIE, Washington.

Tayali, N.E. and Bates, C.J. (1990). *"Particle Sizing Techniques in Multiphase Flows: A Review"*. Flow Meas. Instrum., 1, pp. 77-105.

Taylor, A.M.K.P. (1991). *"Optical Diagnostic for Two-Phase Flows"*. In: *"Combusting Flow Diagnostics"*, ed. Durão et al, Kluwer Academic Publ.

Victor, J.A.S., Costeira, J.P., Tomé, J.A. and Sentieiro, J. (1991). *"A Computer Vision System for Analysis and Calssification of Physical Phenomena Inside Glass Furnaces"*. CG Intl.'91,. Visualization of Physical Phenomena. MIT, Cambridge, Massachussets, June 22-28, 1991.

Wood, G.G. (1988). *"International Sandards Emerging for Fieldbus"*. Control Engineering, Oct. 1988, pp. 22-25.

# DEVELOPMENT OF INTELLIGENT FLOWMETERS
# THROUGH SIGNAL PROCESSING

### J.E. Amadi-Echendu*, H. Zhu** and D.P. Atherton**

*Thames Polytechnic, London SE18 6PF, UK
**University of Sussex, Brighton BN1 9QT, UK

Abstract. In the current evolution of instrumentation for process
control, there is considerable emphasis on the development of 'smart'
or intelligent instruments. This can be justified from an hierarchical
viewpoint because the performance of any control system is very
dependent on the quality and integrity of the measurement signals. It
can also be justified from a 'cost of ownership' viewpoint because it
facilitates optimisation of the signal levels for process management
and reduces the inventory of spare equipment for maintenance.

The present trend in the development of transmitters is to include a
microprocessor which optimises the accuracy of measurement, applies
self-diagnostic routines and provides means for remote setting of zero
and span. The overall design of the transmitters is such that the
output signal represents the mean value of the measurand but this does
not fully reflect the dynamic response of the sensor to the process.
On the other hand, our own research has shown that useful information
regarding the status of the process plant can be retrieved by applying
well-established signal processing techniques to the basic or
unconditioned signals from the sensors. This paper describes how these
techniques can be applied to improve the quality of signals available
from turbine and vortex flowmeters.

Keywords: Sensors, signal processing, failure detection, flowmeters

## INTRODUCTION

Most conventional process instruments are
designed for use in a process control
loop. Higham (1987) points out that for
this purpose, it is widely accepted that
there is no need to transmit measurement
signals having frequency components above
1 Hz. Consequently, there has been no
incentive to design the interfaces
between the process and the sensors, the
sensors themselves and the associated
signal conditioning circuits so that the
measurement signals reflect the full
dynamic response of the sensor to the
process. As a result, much useful
information regarding the operational
status of the process plant is ignored.

Several recent studies (Amadi-Echendu and
Hurren, 1990; Amadi-Echendu and Zhu,
1991; Higham et al, 1986; Hurren et al,
1988) have shown that useful information
regarding the condition of the process
plant and associated equipment can be
derived by analyzing both the

conventional and unconventional signals
from flowmeters. Although these papers
describe the application of various
signal processing techniques, there has
not been a previous attempt to
qualitatively compare and analyze
simultaneous signals from two different
types of flowmeters installed in series
in a flow loop. This paper compares the
performance of a turbine flowmeter with
that of a vortex flowmeter when both
instruments are installed in series in a
flow loop and the unconditioned sensor
signals are subjected to the same signal
analysis technique. It should be
emphasised that these instruments are
ideally suited for studying the response
of the measurement system to the flow
regime in which they are operating,
principally because the sensing mechanism
is immersed in the flowing fluid.
Therefore, errors due to an interface
between the process and the sensor are
avoided. The results demonstrate that,
although the unconditioned output signals

from both types of instrument are approximately sinusoidal, spectral analysis of the output from the turbine flowmeter provides a better discrimination of the plant conditions than the corresponding information from the vortex flowmeter.

## FLOWMETERS AND EXPERIMENTS

### Turbine and Vortex Flowmeters.

Turbine flowmeters are principally used for batch or fiscal measurements, and vortex flowmeters for flow measurements. In both types of flowmeter, the signal from the sensor is approximately sinusoidal and for batch or fiscal measurements it is conditioned to a train of pulses, each pulse representing a discrete volume of fluid. For flow measurements, the sensor signal is converted into a direct current in the range 4 to 20 mA and the time constant of the conditioning circuitry is limited to about one second.

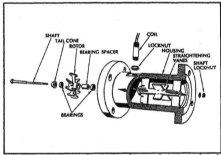

**Fig. 1. An exploded view of a typical turbine flowmeter. (Courtesy of Foxboro Company)**

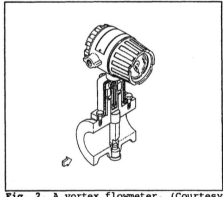

**Fig. 2. A vortex flowmeter. (Courtesy of Yokogawa Corporation of America)**

The turbine flowmeter used in the experiments was a KDG Mobrey Type M2/1000, which comprises six blades. An exploded view of a typical turbine flowmeter is shown in Fig. 1. As the fluid flows through the body of the flowmeter, it impinges on the blades, causing the rotor to rotate. The angular velocity of the rotor $\omega_r$ is detected by the movement of the blades past an electromagnetic sensor mounted on the outside of the flowmeter body.

Vortex flowmeters depend for their operation on the shedding of a train of vortices from a suitably shaped bar which is located diametrically across the conduit carrying the flowing fluid, as shown in Fig. 2. The flowmeter used in the experiments was a Foxboro E83L-01S. In this instrument, the vortices are created by a tee-bar bluff body and sensed by a piezoelectric differential pressure sensor. The sensor is located on the outside of the main body of the flowmeter, but connected, via ports, to the downstream section where the pressure fluctuations due to the vortices is maximum.

In operation, the unconditioned output, $y(t)$, from the sensor in both types of flowmeter is approximately a sinusoidal signal whose frequency $f$ is proportional to fluid velocity $v$. In the turbine flowmeter, $f$ is equal to the inverse of the mean blade crossing time and the level of the signal depends on the electromagnetic coupling as well as the rate of flow. In the vortex flowmeter, $f$ is the inverse of the mean time during which a vortex is sensed by the piezoelectric differential pressure sensor and the level of the signal depends on the sensitivity of the sensor as well as the rate of flow. However, as the formation of the *von Karman* vortex street is a random process, the basic signal from a vortex sensor exhibits random variations in both amplitude and frequency. Hence, even at constant rate of flow, the signals from both types of flowmeter are modulated by the prevailing process conditions.

### Experiments

For this paper, flow measurements have been made with the commercially available flowmeters described above. However, the particular instruments used have been slightly modified to gain access to the unconditioned sensor signals. For both types of flowmeter, the signals were taken directly from the sensors and connected to the data recording system.

A series of experiments were carried out to compare the simultaneous measurement signals from the flowmeters connected in series in a nominal 25.4 mm size flow loop as shown in Fig. 3.1. Each flowmeter was fitted with the recommended lengths of straight pipe, both upstream and downstream. One centrifugal pump and two positive displacement pumps were used. One positive displacement pump was fitted with a pair of tri-lobe rotors and the other was fitted with a pair of bi-wing rotors. Figures 3.2 and 3.3 are line diagrams depicting the structure of the pairs of rotors. A pulsator assembly was also included in the flow loop so that the flow regime could be perturbed at known frequencies and amplitudes which were independent of the rate of flow.

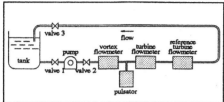

**Fig. 3.1.** A schematic of the flow loop.

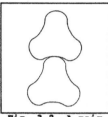 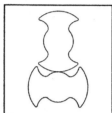

**Fig. 3.2.** A pair of tri-lobe rotors.

**Fig. 3.3.** A pair of bi-wing rotors.

The process fluid was water and tests were made at various flow conditions. For each condition, the signals from both flowmeters were simultaneously recorded on two channels of a TEAC model RD-110T PCM data recorder. The signals were subsequently played back to an Analog-to-Digital converter unit installed in a PC, prior to signal analysis.

## SIGNAL PROCESSING

The signal $y(t)$, from each channel of the recorder was converted to a digital data set $y(N)$. The length of each set was 2 seconds, comprising a sequence of $N$ ($N = 4096$) data points sampled at the rate of 2048 Hz. For each experiment condition and for each flowmeter, at least three data sets were collected. The power spectrum for each data set was computed from

$$P_{yy}(k) = |\frac{1}{N} \sum_{n=1}^{N} W_n \, y_n \, e^{-j\frac{2kn\pi}{N}}|$$

$$W_n = 0.54 - 0.46 \, cos(2\pi\frac{n}{N-1}); \quad n = 1, 2, \ldots, N$$

using the *hamming* window sequence $w_n$ and a radix-2 FFT algorithm. Thus, the frequency resolution of the spectrum plots is 0.5 Hz but, a square root function has been used to scale the magnitude in normalised arbitrary units.

## RESULTS

For brevity, only the results of four specific experiments are summarized. In the first experiment, the centrifugal pump was used to establish reference plant conditions for comparison with the later experiments. The second experiment was carried out with the pulsator assembly installed in the flow loop and with the centrifugal pump as the flow generating device. In the third experiment, the pulsator assembly was removed and the centrifugal pump was replaced by the positive displacement pump with the tri-lobe rotor. Similarly, the fourth experiment was carried out using only the positive displacement pump with bi-wing rotor as the flow generating device. For each of these experiments, the rate of flow was fixed at 93 litres/minute using a 40 mm reference turbine flowmeter, and the positions of the 25.4 mm flowmeters under test were interchanged so that both instruments were subjected to the same process conditions.

In the graphs that follow, the first half of each figure shows a plot of the time series of the signal whereas the second half shows a plot of the power spectrum computed using equation 1. For each instrument, four figures are shown representing sample results obtained from the corresponding four experiments.

## Analysis of Signal From Turbine Flowmeter.

The signal from a turbine flowmeter depends on the rate of change of flux caused by the blades rotating past the magnetic pick-up. As each of the $N_B$ ($N_B = 6$ in this case) blades of the rotor crosses the magnetic field setup by the sensor, a pulse similar to a sinusoid is generated. Thus, for each rotor cycle of the instrument used here, six of these pulses are generated. Amadi-Echendu and Zhu (1991) have found out that for an instrument with perfect geometry, a perfectly periodic but non-ideal sinusoidal waveform would be generated. Perfect geometry implies that the rotor comprises perfect bearings with the blades equally spaced around the rotor and of the same length and cross-sectional area; moreover, all manufacturing tolerances must be equal to zero.

**Centrifugal pump only.** Figure 4.1 shows the sample result obtained with the turbine flowmeter operating under reference conditions with the centrifugal pump as the flow generating device. The first half of the figure shows a time series graph plotted using 205 data points and depicts 40 cycles representing 6.67 rotor cycles. The graph of the power spectrum is plotted as shown in the lower half of the figure and reveals a dominant spectral peak (and its second harmonic) at six times ($6x$) the rotor frequency $x$. Subharmonic peaks are also visible at $1x$, $2x$, $3x$, $4x$, and $5x$. The pattern of

283

subharmonic peaks also repeats for the second harmonic of the dominant frequency.

For a turbine flowmeter with perfect geometry, the sensor signal is periodic but not exactly sinusoidal. This feature causes the harmonics of the dominant peak to be present in the spectrum but, with negligible power. Imperfections in meter geometry cause the subharmonic peaks to be present and the pattern of these peaks give an indication of the type of fault in the instrument.

Centrifugal pump and pulsator in flow loop. Fluctuations in the flow caused either by the pump or a pulsating (vibrating) device do not appear to distort the signal as shown in the time series plot in Fig. 4.2. However, such effects are detected as sidebands of the peaks in the reference spectrum. This is evident by comparing the spectral plots in Figs. 4.1 and 4.2. The graph in the lower half of Fig. 4.2 shows that a 20 Hz pulsation has been introduced in the flow using the pulsator assembly. This pulsation frequency is not apparent as sidebands of the harmonic and subharmonic peaks because of relatively little power in these peaks. There is also a loss in the amount of power in the dominant peak which has apparently been gained by the sideband frequency components.

It is important to appreciate that the pulsator assembly was used to introduce pulsations which were independent of the rate of flow. Thus, the frequency of the pulsations injected with the pulsator assembly can be used as the reference to determine or calibrate the frequencies of the pulsations introduced by other equipment installed in the flow loop.

Positive displacement pump in flow loop. In contrast to the effect of the pulsator assembly, the positive displacement pump fitted with either a tri-lobe or a bi-wing rotor introduces pulsations which vary in accordance with the rate of flow. At a constant rate of flow and with reference to Figs. 3.2 and 3.3, one would expect the positive displacement pump fitted with any of the rotors to introduce pulsations at frequencies which are harmonics of the pump rotation frequency. Furthermore, the dominant frequency introduced by the pump should correspond to the rate at which 'packets' of fluid are discharged for every cycle of pump shaft rotation. It can be inferred from examining the structure of the rotors that for every pump rotation cycle, six packets of fluid should be discharged by the tri-lobe rotor and four packets by the bi-wing rotor.

Pump fitted with tri-lobe rotor. As shown in Fig. 4.3, the time series plot is apparently similar to those shown earlier in Figs. 4.1 and 4.2. However, the graph of the spectrum in Fig. 4.3 shows that the tri-lobe pump introduces significant pulsations which are evident as the sidebands of the significant frequency peaks when compared to the reference spectrum shown earlier in Fig. 4.1. By comparing the spectra in Figs. 4.2 and 4.3, the indication is that the pump introduced pulsations in the flow at two dominant frequencies, one near 20 Hz and the other near 40 Hz, thus implying that the pump rotor frequency is about 6.6 Hz. (The actual speed of the pump was measured as 391 revolutions per minute.) Also in Fig. 4.3, other harmonic peaks of the pump frequency are visible, causing a cluster of frequency components to surround each of the peaks seen in the reference spectrum shown earlier in Fig. 4.1. In Fig. 4.3, this effect is particularly significant around the dominant peak at 400 Hz.

Pump fitted with bi-wing rotor. The time series plot in Fig. 4.4 shows no apparent difference when compared to time series plots in previous figures. In a similar manner to the tri-lobe, the bi-wing rotor causes the pump to introduce pulsations at a dominant frequency of about 30 Hz implying that the pump rotation frequency is 7.5 Hz. ( The actual speed of the pump was measured as 446 revolutions per minute.) The cluster of other frequencies around 400 Hz are barely visible because of the relatively low power associated with the flow pulsations caused by the bi-wing rotor pump.

By comparing the profiles of the two types of rotor lobes, these results complement the expectation that a pump fitted with a bi-wing rotor would generate a more steady flow situation than the same pump fitted with a tri-lobe rotor.

Analysis of Signal From Vortex Flowmeter.

The unconditioned signal from a vortex flowmeter is derived from the time taken to shed the vortices created by a bluff body which obstructs the fluid flow process. Although the formation of vortices is a complex fluidic phenomenon, the rate at which vortices are shed is remarkably periodic. Thus, the basic signal from the vortex sensor appears resonably periodic with some random variations in amplitude and frequency. If fluctuations are introduced in the flow by other equipment installed in the flow loop, one would expect such fluctuations to affect the vortex formation/shedding process, and hence to be able to detect the effects by analysing both the time series and the spectrum of sensor the signal.

The graphs in the first half of Figs. 5.1 through 5.4 show the time series plots of the signal from the vortex sensor with the flowmeter operating under the various experiment conditions described earlier. All four pictures appear very similar and depict a periodic sinusoidal signal with low frequency random variations in both amplitude and frequency. The graphs in the lower half of these figures show the spectrum of the signal from the vortex sensor for the same four experimental conditions.

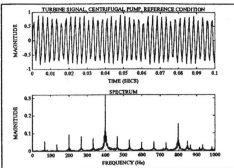

**Fig. 4.1.** Results of turbine flowmeter, centrifugal pump only.

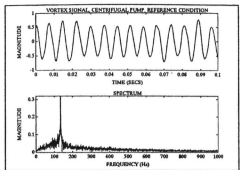

**Fig. 5.1.** Results of vortex flowmeter, centrifugal pump only.

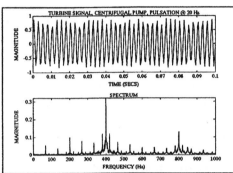

**Fig. 4.2.** Results of turbine flowmeter, pulsator operating.

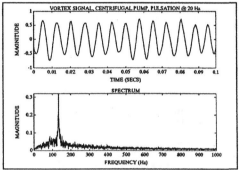

**Fig. 5.2.** Results of vortex flowmeter, pulsator operating.

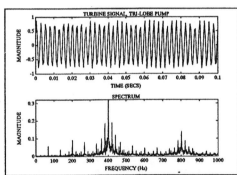

**Fig. 4.3.** Results of turbine flowmeter, tri-lobe pump.

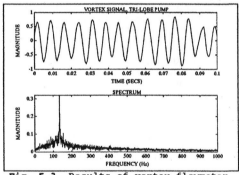

**Fig. 5.3.** Results of vortex flowmeter, tri-lobe pump.

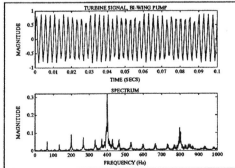

**Fig. 4.4.** Results of turbine flowmeter, bi-wing pump.

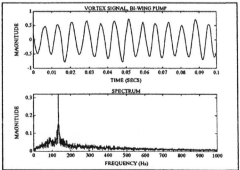

**Fig. 5.4.** Results of vortex flowmeter, bi-wing pump.

There is no apparent difference between the spectrum pictures shown, that is, the appearance of the spectrum pictures looks the same regardless of whether or not the flow is steady or pulsating. Although intuition and experience indicate that the formation and shedding of vortices should be sensitive to flow conditions, this method of analyzing the signal does not provide sufficient evidence to support that such dynamic behaviour exists.

## DISCUSSION

From the results presented here, it appears that with the turbine flowmeter, the spectrum of the signal can be analyzed and used to detect the difference between steady and fluctuating flow conditions, particularly when the flow is pulsating. This is because fluctuations in the flow tend to *modulate* the operation of the sensing mechanism in the turbine flowmeter. Consequently, modulation of the operation of the turbine rotor also manifests as modulation in the signal produced by the electromagnetic sensor. Hence, the signal produced is sufficiently sensitive to spectral analysis for one to distinguish between the spectral peaks caused by the sensing mechanism and those caused by the difference between steady and pulsating flow.

On the other hand, the results presented indicate that the signal produced by the vortex sensor appears to be rather insensitive to spectral analysis for one to discern the difference between the effects of steady and pulsating flow conditions. It is possible that pulsations in the flow generally tend to lock on to the process of vortex formation and shedding. Hence, the signal produced may not necessarily comprise discernible modulating frequency components.

Current work suggests that by applying a different method of signal processing to the signal from the vortex sensor, it may be feasible to distinguish between the effects of steady and perturbed flow. Preliminary results have shown that a statistical analysis method can be used to differentiate between the effects of steady and perturbed flow.

## CONCLUSIONS

The work presented in this paper demonstrates that information reflecting the dynamic response of a flowmeter to the process conditions can be obtained by analyzing the unconditioned signal from the sensor by applying suitable signal processing techniques. For the turbine flowmeter, analysis of the spectrum of the signal reveals both a 'signature' of the instrument and the difference between the effects of steady and pulsating flow

conditions. For the vortex flowmeter, the means for discriminating such effects may not be readily obtained from analyzing the spectrum of the signal but probably, by statistical analysis methods.

It is important to appreciate that signal processing techniques can be applied to extract information which can be used to enhance the performance of turbine and vortex flowmeters. Such performance enhancement should improve the integrity of the measurement information presented to a supervisory control system.

## ACKNOWLEDGEMENT

The authors wish to acknowledge the collaboration with ICI Engineering in this work, and their financial support since February 1989 as part of the ESPRIT II--KBMUSICA project.

## REFERENCES

Amadi-Echendu, J. E., and P. J. Hurren (1990). Identification of process plant signatures using the flow measurement signal. *IEEE Transactions on Instrumentation and Measurement*, Vol. 39, pp. 416-420.

Amadi-Echendu, J. E., and H. Zhu (1991). On monitoring the condition of turbine flowmeters. Proceedings of *IEEE* Instrumentation and Measurement Technology Conference, Atlanta. pp. 112-116.

Higham, E. H., R. Fell, and A. A. Ajayi. (1986). Signal analysis and Intelligent Flowmeters. InstMC J. of Meas. and Control, Vol. 19, No. 5.

Higham, E. H. (1987). An Assessment of the priorities for research and development in process instrumentation and process control. An IChemE Industrial Fellowship Report, The Institution of Chemical Engineers, UK.

Hurren, P. J., J. E. Amadi-Echendu, and E. H. Higham (1988). Condition Monitoring using a resonant-wire differential pressure sensor. BHRA J. of Condition Monitoring, Vol. 2, No. 1. pp. 17-25.

# ACCURACY ENHANCED MACHINING WITH A MAGNETIC SPINDLE

## D.K. Anand* and M. Anjanappa**

*Mechanical Engineering and Systems Research Center, University of Maryland-UMCP,
College Park, MD 20742, USA
**Mechanical Engineering, University of Maryland-UMBC, Baltimore, MD 21228, USA

Abstract. A magnetic bearing spindle, with its unique features, retrofitted to existing machine tools has been shown to minimize tool path error while maintaining high metal removal rate. An error minimization controller based on a general and flexible accuracy enhancement methodology has been formulated, implemented, and demonstrated. Using feedforward compensation, active magnetic bearing spindle error compensation of several sample error sources was experimentally evaluated. The work reported in this paper shows that by using error maps and magnetic spindle control the static and dynamic tool path error minimization is both practical and achievable.

Keywords. Magnetic Suspension, Error Compensation, Machine Tools, Actuators, Adaptive Control

## INTRODUCTION

The use of magnetic bearing spindle in machining can not only provide the benefits of high speed machining but, can also minimize tool path error. Tool path error is defined as the position vector difference between the programmed and the actual tool path. Accuracy can be enhanced by utilizing the unique features of the magnetic bearing spindle to compensate for tool path errors in real time. The current research project at the University of Maryland, investigates this approach by the use of error maps and spindle control of an S2M magnetic spindle retrofitted to a Matsuura CNC machining center. The following tasks defines the complete scope of the project;

- Generate static and dynamic tool path error map for end milling operations,
- Develop and implement a methodology for controlling a magnetically suspended spindle to minimize deterministic tool path error, and
- Experimentally test and validate the control system and algorithm using a CNC vertical machining center fitted with a magnetically suspended spindle.

This paper presents a detailed discussion of the project.

## TOOL PATH ERROR

The accuracy and surface finish of a machined part is a function of tool path error. Previous research suggests separating errors as either 'cutting force independent' (CFI) or 'cutting force dependent' (CFD). CFI errors are those errors that occur in the absence of metal cutting (i.e., dry run) while CFD errors can be directly linked to the metal cutting. Each of these errors can further be classified as deterministic and non-deterministic in nature. The deterministic errors, both static and dynamic, are those errors that reoccur when an identical set of input parameters exist on a given machine tool structure. The non-deterministic errors (such as errors due to chatter), are defined as those errors which occur when a random input is presented to the machine tool and are not considered here. All these errors are discussed in detail in [Anand et al 1986]. For purposes of this research the deterministic errors are classified as shown in Fig. 1.

Static loading produces CFI error resulting in static/geometric errors which show up as positional inaccuracies due to errors in production and assembly of the elements which are used in the machine construction. Dynamic loading results in CFI errors dependent on the acceleration and deceleration of machine components. It produces transient and steady state trajectory errors as well as

feedrate dependent over-shoot and under-shoot errors. By including the effect of feedrate on positioning errors, it is possible to combine static geometric errors and over-shoot/ under-shoot errors into one error class called 'geometric position error'. Thermal loadings, due to internal and external heat sources, introduce error in the relative position between the tool and workpiece.

Separation of the influence of each loading condition on the CFD errors imparted to the workpiece is difficult at best primarily due to the complex nature of cutting process. Traditionally CFD errors are minimized by error avoidance. One deterministic CFD error, considered here, is the ramp error. Ramp error is the result of the deflection of a compliant work piece by cutting forces and is defined as the difference between the thickness at the top and the thickness at the bottom of a thin rib [Anand et al 1986], which is often encountered in the manufacture of microwave guides. Since the geometry of the work piece is known, it is possible to relate the ramp error to the forces imparted by the cutter.

## MAGNETIC BEARING SPINDLE

A magnetic spindle currently available for use on machine tools was developed and built by Societe de Mecanique Magnetique (S2M) of France. At present there are three models of magnetic spindles available for milling purposes whose speeds range between 30,000, and 60,000 rpm with a rated horsepower between 20 and 34 [Field et al 1982].

A magnetic spindle consist of a hollow shaft supported by contactless, active radial and thrust magnetic bearings, as shown in Fig. 2 [SKF 1981]. In operation, the spindle shaft is magnetically suspended with no mechanical contact with the spindle housing. Position sensors placed around the shaft continuously monitor the displacement of shaft in three orthogonal directions. This information is processed by a control unit and any variation in the position of the shaft are corrected by varying the current level in electro-magnetic coils, thereby forcing the spindle shaft to its original position. Conventional ball bearing are provided to serve as touchdown bearings in case of a power failure.

There are numerous advantages in using magnetic spindles over conventional spindles [Anand et al 1986]. Two unique features that are used to advantage in this research work are,

- Built-in 3-dimensional force and position sensors, and
- Ability to translate and tilt the spindle shaft within air gap restriction (±0.005 inches of translation and 0.5 degrees of tilt

possible), with no effect on the performance.

Several investigators have used magnetic spindles [for example Nimphius 1984, Aggarwal 1984, and Schultz 1984]. Their primary focus was to use the magnetic bearing spindle to improve metal removal rate. In the present work it is suggested that the above two advantages of using a magnetically controlled spindle [to improve tool path errors] can take precedence over the advantage of high metal removal rate.

Derivation of an analytical model of the S2M magnetic levitation system was required to analyze the magnetic spindle capabilities and to support control system synthesis. This process combined engineering analysis with experimental data to generate a model which closely replicates the actual system.

The magnetic levitation system is comprised of four major components, viz, PID controller including sensor signal conditioning, power amplifier and bearing stator windings, magnetic levitation forces, and spindle rotor rigid body dynamics. The model was implemented using the ACSL simulation language. This simulation model has a one-to-one correspondence with the actual structure of the S2M B25/500 magnetic spindle system. The simulation includes control signal and coil current limits, coil current rate limits, magnetic saturation, nonlinearities, and losses, air gap reluctance variations and constant flux compensation. The modelling of amplifier dynamics was based on time and frequency domain experimental data and builds on the steady state parameter identification. Equations for spindle rotor was derived from a simple inertial model, subjected to small angle gyroscopic effects. Detailed description of the building of analytical model is available in [Zivi 1990].

Model validation was performed using experimental large signal step response and small signal frequency response data. As shown in Fig. 3, the simulation closely tracks the experimentally observed large signal step response. The small signal simulation validation results of Figure 4 compare experimental and simulation closed loop Bode frequency response. To obtain the experimental frequency response characterization, an HP 3582A spectrum analyzer was used to inject small ($25\,mV\,rms$) sinusoidal excitation, into the S2M PID controller feedback loop through the customary user spindle position command interface. Spindle displacement response was measured using the S2M sensed spindle position instrumentation interface. Simulation based frequency response was obtained from a linear state space model extracted from the nonlinear simulation using the ACSL numerical Jacobian solver. Prior to computation of the Jacobian, the ACSL equilibrium finder was used to ensure that the simulation was properly trimmed to steady state conditions with a null input command. The resulting state space linear model was imported into the MATLAB linear analysis environment to compute the Bode response. The gain increase in the 125 $Hz$ range results from the derivative feedback and the spindle rotor first bending mode.

### RESEARCH FACILITY

To accomplish the objectives of this project, first of all, a CNC machine fitted with a magnetic spindle was needed. In addition, such a system must provide an user interface whereby the user can tap into the current status and be able to command the translation and tilt of the spindle rotor on-line in real time. As there is no such machine available, an S2M-B25/500 magnetic spindle was retrofitted to an existing Matsuura MC500 CNC machining center at the University of Maryland.

Considerable effort was invested in the establishment of the magnetic spindle research facility. Figure 5 shows the primary elements of the facility along with their interconnections. Hardware interfacing details has been reported in [Woytowitz et al 1989]. The functional requirements can be summarized as, providing the operational control necessary to operate the CNC mill with the magnetic bearing spindle, implementation of safety interlocks, communication interfacing necessary for real-time process monitoring, and coordination necessary to implement the error correction scheme.

Central to this research is the implementation of the error minimization controller which will be discussed in detail in a later section of this paper. Various on-line process parameters are provided to the error minimization controller as inputs to the control algorithm. Available input parameters include, x, y, and z axis

position and velocity commands, spindle displacement and bearing forces, NC part program commands, and thermal conditions. In this paper, the x,y, and z axis of a machine tool refer to the motion axis of table, cross-slide and the spindle head respectively. Extraction of real-time x, y, and z axis position and velocity command data from the Yasnac 3000G CNC controller of Matsuura CNC machining center was achieved by various hardware and software debugging methods.

### ERROR MINIMIZATION CONTROLLER

It is possible to minimize the deterministic tool path errors, either by pre-compensating or by on-line correction, if they can be identified and quantified. Pre-compensation of tool path errors (primarily used to correct CFI errors) consists of determining the errors committed by the tool and implementing compensation schemes which rely on data recorded off-line. For example, Tlusty [1971] uses a semi-automatic master part (trace test) to measure errors. Dufour et al [1980] use an error matrix of coordinate corrections to improve the accuracy of large NC machine tools. Donmez [1985] refined this approach by implementing statistic principles to determine the characteristics of geometric positioning errors. This same methodology has been applied by Zhang et al [1985] to improve the accuracy of coordinate measuring machines at the National Institute of Standards and Technology. In [Anjanappa et al 1988] it was shown that cutting force independent errors can be measured and pre-compensated on a vertical machining center reliably.

The on-line correction (primarily used to correct transient portion of CFD errors) schemes, typically, monitor the machining status and adaptively control the appropriate machining parameters. For example, DeVor et al [1983] present a model describing the compliance due to cutting forces of a thin web work piece. Watanabe and Iwai [1983] reports application of adaptive control to increase the accuracy of the finished surface in end milling. The deflection of the spindle nose is used to compute the necessary change required in feedrate to maintain the error within limits.

Several researchers have investigated the application of incremental displacement actuators for error compensation. Kouno [1984] implemented a piezoelectric incremental error compensation using LVDT measured position feedback. Anjanappa et al [1987] modeled the cutting process as a discrete stochastic system and an on-line optimal controller was developed for maintaining surface roughness within specified values.

The on-line correction techniques discussed so far are found effective in reducing machine tool errors. However, the requirement of attaching highly sensitive instruments to moving machine elements in a manufacturing environment makes the approaches difficult, expensive, and hence, impractical for many applications.

To overcome these limitations, the research facility setup at the University of Maryland uses the two unique features of the magnetically suspended spindle which provides an ideal situation for error minimization.

The error minimization methodology developed for this work is shown in Figure 6. The error correction is achieved by superimposing magnetic bearing spindle movements (high resolution and wide band width incremental movements) upon nominal machine table movements. The post processed M&G codes are downloaded to the machine controller, which sends out motion commands for x, y, and z movements. The real time position information from the machine controller is fed to the error minimization controller to generate incremental motion signals proportional to the error (taken from a priori obtained error map) at that position. The displacement-bias thus induced is used to translate and tilt the spindle-tool system within its air gap to achieve enhanced accuracy. This methodology requires resolution of several issues:

- On-line determination of (apparent) ool path errors,
- Calculation of corrective action,
- Actuation of corrective action, and
- Coordination of machining and compensation.

Error determination is based on pre-calibrated representations codified in an 'error map' formulation. On-line monitoring of the machining process, using interfaces described in earlier section, provides the independent parameters, used to extract real-time error estimates, from the error map representation. These perturbational corrective actions are implemented by displacing the spindle within the magnetic bearing air gap. Since the machining process is reviewed as a sequence of tool path trajectories, coordination is provided by downloading a representation of the part program to the error minimization controller. Inclusion of handshaking functions, allows the CNC and error minimization controllers to operate in a coordinated fashion.

Figure 7 shows the physical structure of the system. The portion of the figure, above the dotted line, represents the conventional CNC milling machine. The accuracy enhancement, shown below the dotted line, includes the error minimization controller the magnetic spindle system, and the required integration and coordination logic.

The error minimization controller has been implemented using an Intel 80386-based real-time computer system. Real-time input/output is provided by analog-to-digital and digital-to-analog boards. The control program is partitioned into two primary components. Non real-time processing is coded in Intel FORTRAN and supports the primary operator interaction and processing of files. Real-time processing is accomplished using Intel PL/M language. The FORTRAN main program calls PL/M procedures to transfer data structures and initiate real-time processing. Once the real-time processing has completed, the FORTRAN main program resumes.

The currently defined error representations are:

- Terminal point error = f(position, nominal feedrate)
- Trajectory error = f(position, feedrate)
- Ramp error = f(position, feedrate)
- Dynamic error = f(position relative to activation)

In the current implementation, the first three error representations are mutually exclusive while the fourth provides an additive y-axis correction. The first three error representations are implemented as two dimensional error maps defining a axial or ramp error in the terms of the position and velocity along the axis. The control task has been developed as a distinct, self contained module to facilitate future enhancements and extensions.

## EXPERIMENTAL WORK

The net tool position displacement relative to the machine table for each axis is determined from seven error terms, which are measured under static conditions using the Hewlett Packard 5528A-based laser measurement system. Although there are 18 error terms for a true 3-axis simultaneous motions, only 7 terms are needed for motions in two axis (x and y axis) which covers most of the prismatic part machining operations. Four types of measurements were conducted, viz, axial position, straightness, angular, and Squareness (see reference Anjanappa et al 1988 for details) to obtain all the seven error terms. Each error term was measured at 0.0254m increments of table motion at five commonly used feed rates (0.254, 0.762, 1.27, 2.54 $m/min$ and Rapid (5.08$m/min$)) in positive and negative directions. Six sets of data are recorded at each feed rate and the standard deviation are computed to assure that the averages of the errors represent a repeatable error.

The resulting geometric position errors were represented in the form of an error matrix. This matrix consists of scale and straightness errors for each axis for each node for five feedrates and for positive and negative approaches. Since error data at several feedrates is present, errors due to the servo drive and electronic control system are included in the map. In summary, the geometric position errors possess a position, direction, and feedrate dependence. More detailed information on experimental data can be obtained from reference [Woytowitz et al 1989].

Thermal deformation induced errors, with the magnetic spindle, is complicated since it has air cooling and extensive cooling water circuits around the magnetic coils. Tests were performed to determine the thermal deformation errors along the x and y axis. Only the errors due to changes in ambient conditions and servo motor heat sources were found to be of importance. Further, since the spindle bore/tool is centered about the x-axis of spindle housing, only the y-axis errors are discussed. Several 'recovery from cool down' and servos on 'warm up' tests were conducted to obtain the y-axis error plot.

Using the data from these tests a linear model of the thermal deformation error along the y-axis based on the change in the bulk spindle temperature and the change in the ambient machine temperature was constructed. By monitoring the above two temperatures, the quasi-static thermal error can be compensated by moving the spindle shaft inside the air gap of the research facility.

## VALIDATION

Several error compensation experiments were performed using the research facility to demonstrate and evaluate the error minimization methodology. Non-cutting terminal point error (CFI errors) compensation was performed first using the geometric position error map generated earlier. The HP 5528A laser metrology system was used to measure milling machine table position during this test. The laser optics were affixed to the spindle tool holder, providing a direct measurement of relative axial table position. From machine home, the table position command was stepped, in 25.4 $mm$ increments, across the positioning range. Figure 8 presents the measured x-axis errors before and after compensation at a nominal feedrate of 2.54 $m/min$. The result indicates terminal point accuracy improvement factor of about five. Although Woytowitz et al [1989] reported the standard deviation of the terminal point errors to be approximately one tenth the magnitude of the error, the positional behavior, shown in Fig. 8, retains larger systematic errors. Since the calibrated 3 sigma positional uncertainty of the spindle position is approximately 2.54 $micron$, these systematic compensation errors were assumed to be caused by long term drift in the error phenomena. To evaluate the short term terminal point error compensation capability, the laser metrology procedure of error map generation was repeated at a single, 2.54 $m/min$ feedrate. With a one day turn around, between metrology and compensation, the results of Figure 9 were obtained. This shows more than an order of magnitude improvement. Continuing metrological investigations, by other researchers, are currently in progress to refine the error characterization.

One CFD error compensation demonstrated is the correction of a linear distortion of a prismatic part. As shown at the top of Figure 10, the nominal test part geometry is rectangular, 25.4 $mm$ by 114.30 $mm$. However, the the programmed shape was distorted into a trapezoid which specifies a linear variation in width from 25.60 $mm$ (+100 $microns$ per side) to 25.20 $mm$ (-100 $microns$ per side). The error map, required to remove the distortion, specifies a simple linear y-axis error as a function of x-axis position and direction of feed. Bottom of Fig. 10 shows the error compensation results.

This work has reported initial error compensation experiments and demonstrated a general accuracy enhancement methodology. Future qualitative improvements resulting from metrological refinements and enhanced magnetic spindle performance are anticipated.

Accuracy enhancement performance is limited by the quality of the error representation and by the ability to effect the proper compensation. Using a priori obtained error map, the quality of the error representation is dependent on two primary factors, viz; identification of the error phenomenon independent parameters and statistical properties of the error phenomenon. Recent metrological experience indicates that the CFI error repeatability may approach values of an order of magnitude better than the CFI accuracy. However, in order to exploit this fully under realistic conditions, improved error characterization is required.

Compensation implementation limitations may be characterized in terms of bandwidth and accuracy. Relative to tool path trajectory dynamics (<10 $Hz$), the 125 $Hz$ bandwidth of the existing S2M B25/500 is sufficient for effective error compensation. In order to improve the error compensation capability, of magnetic spindles, the effective stiffness must be improved.

## CONCLUSIONS

This research has, for the first time, established the ability of a magnetic bearing spindle to provide accuracy enhancement through incremental error compensation. It is shown that they can provide simultaneous high speed, error compensation, and process monitoring capabilities. A general and flexible accuracy enhancement methodology has been formulated implemented and demonstrated. The hierarchical control structure, using an AMB spindle, is an effective method for both static and transient dynamic error compensation.

## ACKNOWLEDGEMENTS

This project was supported by the National Science Foundation through grant NSF 8516218 and the Engineering Research Center at the University of Maryland. In addition, this project was supported, in part, by the David Taylor Research Center Independent Research Program, sponsored by the Office of the Chief of Naval Research under task area ZR-000-01-01. Contributions from Dr. J.A. Kirk, Dr. E.L. Zivi, M. Woytowitz and S. Shyam towards this project is acknowledged.

## REFERENCES

Aggarwal, T. (1984). Research in Practical Aspects of High Speed Milling of Aluminum, Cincinnati Milacron Technical Report.

Anand, D.K., Kirk, J.A., and Anjanappa, M. (1986). Magnetic Bearing Spindles for Enhancing Tool Path Accuracy, Advanced Manufacturing Processes, 1:121-134.

Anjanappa, M., Kirk, J.A., Anand, D.K. (1987). Tool Path Error Control in Thin Rib Machining, Proc. of 15th NAMRC, Bethlehem, PA, 485-492.

Anjanappa, M., Anand, D.K., Kirk, J.A., Shyam, S. (1988). Error Correction Methodologies and Control Strategies for Numerical Controlled Machining, Control Methods for Manufacturing Processes, DSC-Vol. 7, ASME publications, 41-49.

DeVor, R.E., Sutherland, J.W., Kline, W.A. (1983). Control of Error in End Milling, Proc. of 11th NAMRC, 356-362.

Donmez, M. A. (1985). A General Methodology for Machine Tool Accuracy Enhancement Theory-Application and Implementation, Ph.D. Dissertation, Purdue University.

Dufour, P., Groppetti, R. (1980). Computer Aided Accuracy Improvements in Large NC Machine Tools, Proc. of 21st Int. MTDR, 611-618.

Field, M., S.M. Harvey, J.R. Kahles (1982). High Speed Machining Update, 1982: Production Experiences in the U.S.A., Metcut Research Associates Inc., Cincinnati, Ohio, USA.

Kouno, E. (1984). A Fast Response Piezoelectric Actuator for Servo Correction of Systematic Errors in Precision Machining, Annals of CIRP, 33:369-372.

Nimphius, J.J. (1984). A New Machine Tool Specially Designed for Ultra High Speed Machining of Aluminum Alloys, High Speed Machining, ASME Publications, 321- 328.

Schultz, H. (1984). High-Speed Milling of Aluminum Alloys, High Speed Machining, ASME publications, 241-244.

SKF (1981). Active Magnetic Bearing Spindle Systems for Machine Tools, SKF Technology Services Report.

Tlusty, J. (1971). Techniques for Testing Accuracy of NC Machine Tools, Proc. 12th Int. MTDR Conf., 333-345.

Watanabe, T., Iwai, S. (1983). A Control System to Improve the Accuracy of Finished Surfaces in Milling, Journal of Engineering for Industry, Transactions of ASME, 105:192-199.

Woytowitz, M., Anand, D.K., Kirk, J.A., and Anjanappa, M. (1989). Tool Path Error Analysis for High Precision Milling with a Magnetic Bearing Spindle, Advances in Manufacturing Systems Engineering, PED-Vol. 37, ASME Publications, 129-142.

Zhang, G., Veale, R., Charlton, T., Borchardt, B., Hocken, R. (1985). Error Compensation of Coordinate Measuring Machines, Annals of CIRP, 34:

Zivi, E. (1990). Robust Control of Magnetic Bearing Spindle for Milling Tool Path Error Minimization, Ph.D. Dissertation, Univ. of Maryland, USA.

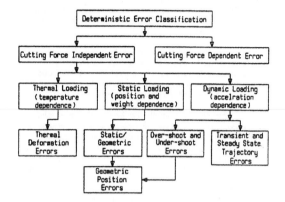

Fig. 1. Tool Path Error Classification

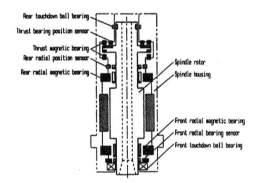

Fig. 2. Magnetic Bearing Spindle Configuration

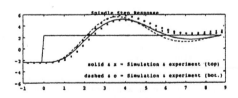

Fig. 3. Spindle Step Response
Horizontal: Time (*ms*)
Vertical: Displacement (×25.4 *microns*)

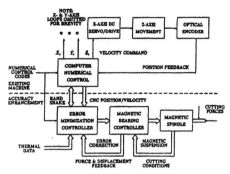

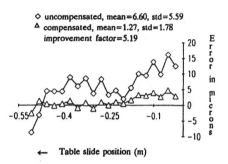

Fig. 7. Physical Structure of the System

Closed Loop Spindle Magnitude Response

Closed Loop Spindle Phase Response

Data markers denote
experimental results

Fig. 4. Spindle Frequency Response
Horizontal: Frequency (*Hz*)
Vertical: Top-Magnitude (*dB*)
Bottom-Phase (*degrees*)

◇ uncompensated, mean=6.60, std=5.59
△ compensated, mean=1.27, std=1.78
improvement factor=5.19

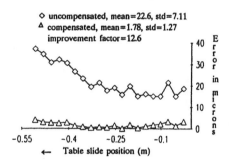

← Table slide position (m)

Fig. 8. X-axis, 2.54 *m/min*, Terminal Point Error Compensation

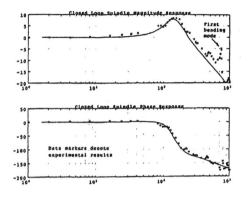

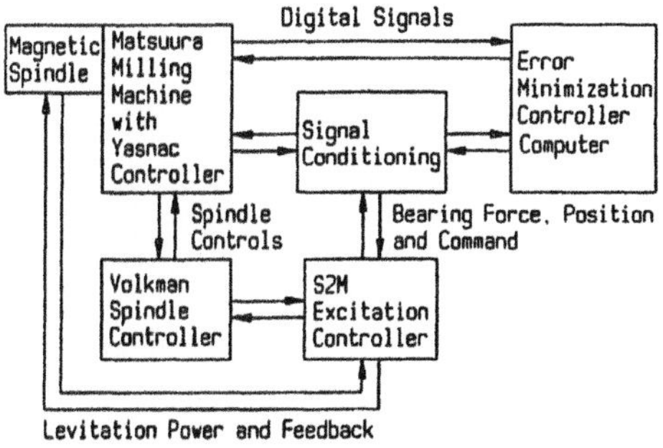

Fig. 5. Primary Elements of Research Facility

◇ uncompensated, mean=22.6, std=7.11
△ compensated, mean=1.78, std=1.27
improvement factor=12.6

← Table slide position (m)

Fig. 9. Short Term X-axis, 2.54 *m/min* Terminal Point Error
Compensation

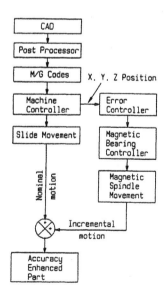

Fig. 6. Error Minimization Methodology

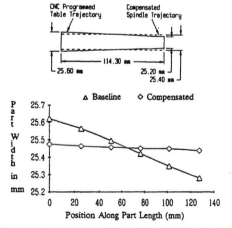

△ Baseline   ◇ Compensated

Fig. 10. Trapezoidal Test Part and Error Compensation Results

# A HIGHLY-INTEGRATED INTELLIGENT SENSOR FOR GAS-LIQUID-CHROMATOGRAPHY

**W.A. Halang and M.G.J.M. Lenferink**

*Department of Computing Science, University of Groningen, P.O. Box 800, 9700 Av. Groningen, The Netherlands*

**Abstract.** The hardware design and ASIC realisation of a multichannel interfacing unit is described. This unit is used in a PC-based chromatography system for the acquisition of data from a number of analytic instruments. The attached analogue signals are partially integrated, digitised, buffered in a FIFO memory for speed compensation purposes and, finally, provided to the PC via a serial asynchronous RS-232-C connection. Based on the acquired integrals, the evaluation software employs a novel numerical method for extracting the parameters of the chromatographic peaks. A standard graphical user interface is employed for the visualisation of the chromatograms being processed.

**Keywords.** Gas-liquid-chromatography; integrating sampling; sensor data fusion; intelligent sensor; application specific integrated circuits; analogue-digital conversion; computer interfaces; data acquisition; data handling; data reduction; numerical analysis; sampled data systems; signal processing; splines.

## INTRODUCTION

Gas-liquid-chromatography (GLC) is an important and frequently used method for the separation of gaseous or liquid substance mixtures into their components. It allows the detection of very small concentrations. A chromatograph yields as output information an extinction-time-function, by the interpretation of which the desired quantitative results are obtained (cp. Fig. 1). Increasingly, GLC is used as a process-integrated analytical technique, i.e., samples are not processed by a dedicated laboratory in a batch mode any more, but immediately when they become available, because the results are needed for real time process control. Examples for application areas are all kinds of chemical processes and environmental pollution control.

Initially, mainframe computers were employed for the evaluation of GLC runs. Presently available chromatographic data systems are based on either mini-computers, integrators, or on dedicated microcomputers directly integrated into the analytical instruments. Since all of them are not being produced in large volumes, their prices are relatively high. Furthermore, due to the workload caused by the applied sampling method, the dedicated microcomputer systems can handle only a few data channels. Characteristic for all these data systems is that the inputs are connected via analogue transmission lines which are susceptible to noise, that point measurements are used to sample the analogue signals, and that the sampling frequencies have to be about 7 times as high as required by the signals' information content in order to be able to smooth out the noise contained in the sampled data with the help of digital filters. Based on this data acquisition method, after having performed digital filtering, the evaluation software has to investigate each point for peak evaluation purposes, viz., whether it is an extremum, an inflection point or part of the baseline. For the latter's determination, all known evaluation algorithms use a certain assumption on the behaviour of the baseline or request the user to define it with a graphic cursor.

In contrast to this, the here presented GLC evaluation method only employs low-cost standard personal computers equipped with standard software, in particular a graphic user interface. This is achieved by utilising novel, spline-function based processing algorithms and by the implementation of the signals' analogue-to-digital conversion, their preprocessing, and temporary buffering in a highly integrated application specific integrated circuit (VLSI ASIC).

The ASIC was designed with CAD software of Mentor Graphics in connection with standard cell libraries, test, simulation, and layout programs of Philips for an 1.5 $\mu$m CMOS process. It contains some 30,000 gate functions. A dual channel version with reduced data buffering capacity was derived from this design, which was well suited for prototype realisation within the PICO project ("Project Integrated Circuits Onderwerpers") of the Dutch Ministry for Education and Science (Kok and Marinus, 1990).

## NUMERICAL METHODS

Our new procedure for the acquisition and computer internal representation of chromatograms is not based on point measurements, but on local integrals. Since noise is inherently suppressed by integration (Halang, 1984), considerably less raw data are required. Experience shows that the data can be further processed without any numerical filtering. This holds especially true if no analogue transmission lines are used. Therefore, low-cost single-chip voltage-to-frequency converters with high dynamic range, which are mounted adjacent to the analytic instruments, are employed to carry through analogue-to-digital conversion. As outputs they generate impulse trains, which can be transmitted via simple twisted wire pairs and which, after a reconstruction being free of information loss, are counted at a central location to produce digital values. This counting is equivalent to the integration of the original signal-time-functions. Thus, we obtain a very low-priced data acquisition characterised by insensitivity to noise.

In order to reduce the complexity of the evaluation software, the empirical signal-time-functions are approximated by quadratic spline functions with respect to equidistantly spaced nodes. The approximation is performed by the linear and positive operator introduced by Halang (1980), which employs the sampled local integrals and requires no further calculations. The operator has a number of advantageous properties (Halang, 1984). Important for sampled data processing is that it is "variation diminishing", i.e., inert in relation to changes in the data. Hence, the operator is intrinsically immune against noise and spikes, and erroneous data have only a locally bounded effect. Furthermore, the interpretation of curves necessary for the determination of the chromatographic parameters is considerably facilitated by the spline representation selected: since the splines reduce to quadratic polynomials between the nodes, only these polynomial pieces need to be inspected for extrema and inflection points. Thus, the computational complexity for the detection and evaluation of chromatographic peaks is by about the factor 30 less than for conventional methods. To perform the baseline correction, a one-sided $L^1$-approximation from below is carried out. This method does not involve any assumption on the form of the baseline and is based on a fast revised simplex algorithm.

For acquiring and processing a chromatographic signal-time-function $c(t)$ on the time interval $[0, T]$ we partition the interval with a suitably selected $n > 0$ into partial intervals of length $h = T/n$. The local integrals of $c(t)$ measured relative to this partition are used as coefficients

$$a_i = \frac{1}{h} \cdot \int_{(i-1)\cdot h}^{i\cdot h} c(t)dt, \ i = 1, ..., n,$$

and $a_0 = a_1$, $a_n = a_{n+1}$

in the approximation function

$$s(t) = \sum_{i=0}^{n+1} a_i \cdot b\left(\frac{t}{h} - i + \frac{1}{2}\right)$$

of $c(t)$ defined with the quadratic B-spline

$$b(x) = \begin{cases} \frac{1}{2}(x + \frac{3}{2})^2, & -\frac{3}{2} \le x \le -\frac{1}{2} \\ \frac{3}{4} - x^2, & -\frac{1}{2} \le x \le \frac{1}{2} \\ \frac{1}{2}(x - \frac{3}{2})^2, & \frac{1}{2} \le x \le \frac{3}{2} \\ 0, & \text{otherwise} \end{cases}$$

This linear and positive spline-approximation operator has been investigated in great detail by Halang (1980, 1984). It possesses a number of advantageous properties. Important for sampled data processing is its "variation diminishing" property and the local boundedness of the effect of erroneous data. Furthermore, the chosen computer internal representation of $c(t)$ considerably facilitates its interpretation which is necessary for the determination of the chromatographic parameters. For, as a spline function, $s(t)$ reduces in each of the subintervals $[(i-1)h, ih]$, $i = 1, ..., n$, of $[0, T]$ to a quadratic polynomial and its continuous first derivative $s'(t)$ consists of straight line segments. The latter's zeros and, hence, the extrema of $s(t)$ or $c(t)$, respectively, can be determined easily with little effort. This holds even more for the inflection points of $s(t)$, which must coincide with some of its nodal points $ih$, $i = 0, ..., n$. In contrast to this, the classical evaluation method requires a much higher effort: the considerably more numerous data points need to be filtered numerically and the curve must be smoothed in order to prepare for the numerically unstable calculation of the curve's first two derivatives in each data point, which is necessary for the determination of the extrema and of the inflection and baseline points.

All baseline correction procedures described in the literature make certain assumptions on the shape of the baseline. For a general purpose software package this approach is not feasible. Therefore, we are using here the same criterion as the human eye in order to distinguish between a baseline and the analytic signal superimposed to it: the baseline is a function which varies more slowly with time than the analytic signal. This property is reflected by prescribing the following general form for the baseline to be determined

$$u(t) = \sum_{i=0}^{m+1} g_i \cdot b\left(\frac{t}{H} - i + \frac{1}{2}\right)$$

where $l$ and $m$ are integers such that $n = l \cdot m$ and $H = l \cdot h$. Since at no point in time it may hold $u(t) > s(t)$, the baseline must be closely fitted to $s(t)$ from underneath. The most suitable measure for the kind of fit, which we are seeking, appears to be the $L^1$-norm, which reduces in the considered special case to

$$\int_0^T [s(t) - u(t)]dt.$$

Hence, the baseline is determined by minimising the

294

above expression with respect to the free parameters. An algorithm solving this problem was first described by Marsaglia (1969). It works iteratively and can be shown to converge in not more than $2m + 4$ steps. In each of these iteration steps a new parameter vector is calculated, which defines a certain function $\hat{u}(t)$. Then, the maximum difference $s(t) - \hat{u}(t)$ for $0 \leq t \leq T$ needs to determined. The latter is carried out approximately, in order to reduce the computing expense. To this end, before starting the iteration, a sufficiently fine mesh of values of the function $s(t)$ is calculated and stored. Then, in each iteration step, the function $\hat{u}(t)$ is evaluated for the same abscissae and the maximum ordinate difference is searched. The expense of evaluating $\hat{u}(t)$ is relatively low, because for any abscissa not more than three of the horizontally shifted basis functions are non-vanishing, and since their values are only needed at certain points. Therefore, the latter can be calculated and stored in advance. After convergence of the iteration procedure, the actual baseline correction is carried out making use of the fact that $u(t)$ is also a quadratic spline with respect to the partition $ih$, $i = 0, ..., n$. Hence, first the following second representation of $u(t)$ is calculated

$$u(t) = \sum_{i=0}^{m+1} d_i \cdot b(\frac{t}{h} - i + \frac{1}{2})$$

before constructing the baseline-corrected chromatographic extinction-time-function by subtraction of the respective coefficients $a_i$ and $d_i$, $i = 0, ..., n + 1$.

## SENSOR IMPLEMENTATION BY AN ASIC

The data acquisition module to be described in the sequel independently and continuously performs the integrations of chromatographic functions occurring in the above representation. The peripheral was designed with the objectives of minimising the implied processor load and of relaxing the timing constraints in relation to fetching the generated data values. The only interfacing requirement of the unit at the attached computer is a RS-232-C serial line. The module is designed to handle a maximum of 16 channels, i.e., chromatographic detectors, which is sufficient for most laboratories. However, if more channels are needed, the hardware design can be extended easily. Since a dynamic range of $10^6$ is desirable as is bipolar operation, there is a need to use at least 23 bits representing signed integers. Therefore, 24 bits are chosen as length of the data words, which is a multiple of the industry standard 8 bits byte. The module was realised as an application specific integrated circuit (ASIC). The configuration in which it works is depicted in Fig. 2.

In order to avoid analogue transmission, which generally increases the noise level, the actual analogue-to-digital conversion of the time-dependent signal is continuously carried out by voltage-to-frequency converters (VFC), which are mounted directly adjacent to the signal sources at the analytic instruments. The VFCs must be operated in a bipolar way to detect negative voltages. This is implemented here by coding negative voltages with twice the pulse width of positive voltages. A VFC that meets all these requirements was described by Halang (1987). The digital impulse trains generated by up to 16 chromatographic detectors, which are easy to transmit and whose repetition frequencies are proportional to the analytical signals, are then fed into the data acquisition and preprocessing module.

As shown by Fig. 3, the ASIC comprises 7 functional blocks, which will be outlined in the sequel. For a fully detailed description, we refer to Kok and Marinus (1990) and Lenferink (1990). The frequency generator provides a system clock of 4.9152 MHz, from which all further clocks and control signals are derived.

Associated with each channel is a counter-storage module, where the impulses transmitted from the VFCs are received. Since these impulses may lose their steep edges on the transmission lines, the pulses are re-shaped on entering the unit using Schmitt trigger gates. Unless these gates are closed by a control signal, the pulse trains are counted in 22 bits long up/down-counters during every integration period, whose length is pin-selectable in the range from 0.25 bis 4 sec, in order to determine the local integrals. The counter length of 22 bits was selected to accommodate for the 1 MHz maximum output frequency of contemporary VFCs during a 4 sec period. Two further bits each serve for the representation of sign and channel state (on or off). Correspondingly, each channel generates 3 bytes of data in every integration period. The counting direction is determined for every single pulse by interpreting the pulse length (long for negative and short for positive voltage). A channel is off, if no pulse arrives during an integration cycle. At the end of any integration period, the counters' contents are latched in buffer registers, and the counters are reset and re-started. The contents of the latches are then written into a first-in-first-out buffer memory, which is to compensate for variations in the attached computer's service rate, i.e., to soften the computer's timing requirements.

The word width of the FIFO is 1 byte. For the determination of its capacity the transmission rate of the UART and the maximum non-availability period of the attached computer need to be considered. The latches of the counters are organised as 3 independent bytes each and are connected to the FIFO input port via an 8 bits wide bus. A control unit handles the transmission of the 48 bytes of information gathered in every integration cycle into the FIFO shortly after the cycle's completion. Its main element is a 48 bits long shift register, through which a single bit of value 1 is shifted. In each position it enables the tri-state transceiver output to the bus of exactly one byte of the buffer registers.

The module's final stage is a universal asynchronous receiver transmitter (UART), which operates at 9600 Bd (or 2400 Bd in connection with a modem) and which communicates via a RS-232-C interface to a computer's serial, asynchronous input. The transmission of a data byte is requested by the computer through sending a certain character. If the FIFO is not empty, its first byte is read out and loaded into the UART, from where it is sent out under control of the transmission frequency generator and with addition of a start and 2 stop bits.

The above description reveals that there is only one quite soft temporal constraint imposed on the processor service by the interfacing unit, i.e., the FIFO must be emptied before it overflows. In the case of our implementation, this takes less than 2.5 sec for a 2K FIFO and a transmission rate of 9600 Bd. For a 1 sec integration period it lasts more than 40 sec to produce 2K of data. This results in a processor load of less than 6.5 %.

## EVALUATION SOFTWARE

The software was implemented on a standard PC without the need for a real time operating system or any assembly language programming, just using C and FORTRAN, and the graphical and multi-tasking features of MS-Windows. One objective in designing the multichannel integration unit was to keep the necessary program support as small as possible, which will become evident in the following.

Data arriving after initialisation of the ASIC are collected in a communication buffer. When this is full, a task is activated, which compares the new channel states with the previous ones in order to perform the activities necessary when a run on a channel is started or stopped, respectively. The integrals measured for those channels that are turned on are converted to floating-point representation and multiplied by a scaling factor finally yielding the coefficients

$$\frac{1}{h} \cdot \int_{i \cdot h}^{(i+1) \cdot h} c(t)dt$$

These are then brought to buffers associated with the individual channels. When required by the application, full buffers are written to mass storage.

The evaluation program can be started when a chromatographic run on one of the serviced channels is completed. As first processing step the baseline correction subroutine is called. It calculates a sufficiently dense sequence of points describing the chromatogram, which is required for the approximate solution of the one-sided $L^1$-approximation problem. Then the baseline parameters are determined by invoking the subroutine which performs the dual simplex algorithm as adjusted to the considered application. Finally, the actual correction is carried out under utilisation of the above mentioned second representation of the baseline. These two subprograms comprise 35 and 53 FORTRAN statements, respectively.

The subsequently called subroutine analyses the chromatographic signal-time-functions in order to recognise peaks and to calculate their parameters. Owing to the application of spline based methods, it is considerably shorter than functionally equivalent programs known from the literature: it has just 125 statements. The processing depends on ten parameters of peak geometry, which need to be a priori set by the user for each channel and/or type of analysis. After appropriate variable initialisations, the routine calculates in a loop the function and derivative values in all mesh points of the spline function representing the considered chromatogram. Between these points the spline reduces to a quadratic polynomial each. A marker determines whether the logic searches for a peak start, a peak top, or a peak end. Accordingly, certain criteria are evaluated and checked. If they are fulfilled, peak parameters determined by the present state of the marker are calculated and the marker changed to direct the search to the criterion expected to be fulfilled next. Backtracking in this search procedure can occur if a local maximum is not high enough for a peak or if the width of a peak at its base is so small that it is most likely a spike. When a peak has been recognised, its area is determined and stored together with its retention time in appropriate arrays. In the case that a group of fused peaks is encountered, the following methods are applied to estimate individual peak areas. For very small peaks found on the slopes of bigger ones the area is calculated, which is enclosed by the envelope of the two peaks and a tangent onto this envelope. The tangent passes through the local minimum between the two peaks, which is located on one side of the small peak's top, and touches the envelope on the other. For fused peaks of comparable size the areas bounded by the envelope and straight lines connecting the starting point of the peak group, local minima, and the peak group's end point, respectively, are determined. The area below this polygonal function is finally distributed to the single peaks of the group proportionally to the areas they have above the polygon ("democratic distribution"). As outputs the evaluation program provides the number of peaks in the chromatogram, the sum of peaks areas, and two arrays containing the retention times and the areas of the different peaks. These data are stored together with the coefficients describing the chromatogram's spline function on disk to be used later by graphic inspection and postprocessing programs. The routine could easily provide further peak parameters like height and width at half height, but these are generally not needed any more.

All other programs of the implemented package make use of the graphical facilities of MS-Windows. Upon request a display task is invoked. Its purpose is to display the chromatogram running on a specified channel as far as it has already been acquired at the time of request. After chromatograms have been fully acquired, they are further processed with a program supporting visual inspection and input of additional information. For example, the graphic output of a general overview of a chromatogram and of enlarged sections thereof can be requested. The further processing of a chromatogram can be restricted to the peaks within a time interval or a certain minimum peak height. Spikes not recognised by the evaluation program can be removed. As required for retention index calculation purposes, peaks of some known compounds can be marked as internal standards. Finally, the input of text parameters for every chromatogram is supported.

## IMPROVEMENTS ACHIEVED

From the conceptual point of view, the here described interface unit provides high inherent noise suppression, because the system is based on directly measured local integrals as empirical data set, and since A/D-conversion can be located very closely to the voltage sources of the analytic instruments. To achieve the same noise reduction as with digital filtering of point measurements, 7 times less data values need to be gathered (Lenferink, 1990). One or more of the interface units can be connected via standard RS-232-C-adapters to a PC, which serves for the data evaluation. The approach of using integrating interface units allows to handle many channels due to the relaxed time constraints for fetching the data and less computation requirements of the evaluation algorithms.

This PC-based laboratory instrumentation system provides the opportunity to develop in a comfortable environment user specific software for the interpretation of the results. Such a system has approximately the same power as a minicomputer and more capabilities than an integrator for a much lower price. Furthermore, the general purpose PC provides additional possibilities to apply it in a laboratory environment, viz., for calculations, sample administration, word processing, and so on.

Customarily, a separate strip-chart recorder is connected to each chromatographic channel for plotting the measured signal in dependence on time. These can be saved, since the PC is able to display any channel's actual accumulated signal-time-function on a screen at any time upon request. Hence, for the price of just a few strip-chart recorders otherwise needed, the user obtains a complete powerful data system.

The software could be implemented without the need for a real time operating system or any assembly language programming, just using the features of FORTRAN and MS-Windows. The numerical method applied for the detection and evaluation of peaks reduces the calculation effort by about the factor 30 and has other favourable properties like intrinsic immunity against noise and spikes. The baseline correction scheme is applicable for any kind of gas or liquid chromatography. It does not require from the user to specify a certain shape for the baseline and adjusts itself with aging chromatographic columns.

In summary, the described ASIC and PC based GLC evaluation system achieves the following improvements:

- low cost data acquisition and processing
- the PC needs neither analogue input nor real time operating system
- analogue to digital conversion directly at the analytic instrument
- insensitivity to and inherent suppression of noise
- low sampling rates allowing multichannel operation
- evaluation method with low calculation requirements
- general purpose baseline correction scheme

## ACKNOWLEDGEMENT

We are deeply indebted to Messrs. E. Kok and H. Marinus for transforming a conventional hardware design into a VLSI design. The ASIC development was carried through under the direction of Ir. F. A. van Dijk of Rijkshogeschool Groningen.

## REFERENCES

Halang, W. A. (1980). Approximation durch positive lineare Spline-Operatoren konstruiert mit Hilfe lokaler Integrale. *Journal of Approximation Theory* 28, 2, 161 – 183.

Halang, W. A. (1984). Acquisition and representation of empirical time functions on the basis of directly measured local integrals. *Interfaces in Computing*, 2, 345 – 364.

Halang, W. A. (1987) A voltage-to-frequency converter design without inherent linearity error suitable for bipolar operation. *Computer Standards & Interfaces* 6, 221 – 224.

Kok, E., and H. Marinus (1990). *VLSI-DESIGN — een multichannel chromatograaf interface.* 2 volumes. B.Eng. Thesis E-9075. Rijkshogeschool Groningen, Electrical Engineering.

Lenferink, M. G. J. M. (1990). *A PC-Based Chromatography System with Buffered Interface.* M.Sc. Thesis. University of Groningen, Department of Computing Science.

Marsaglia, G. (1969). *One-sided approximation by linear combinations of functions.* Boeing Scientific Research Laboratories, Seattle.

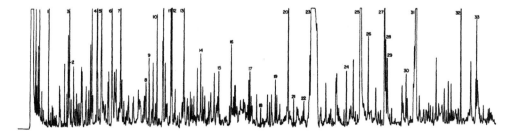

Fig. 1. A typical gas chromatogram

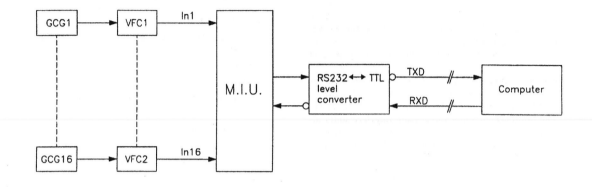

**Fig. 2. Structure of the data acquisition system**

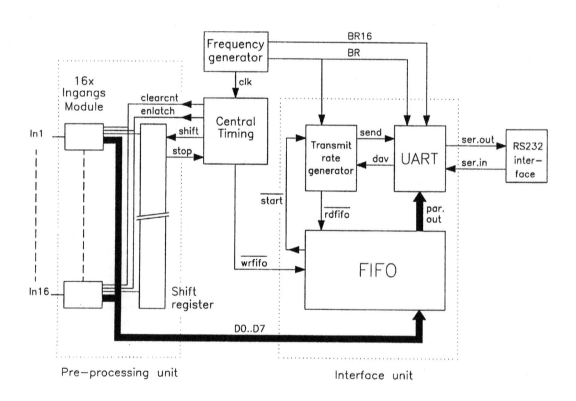

**Fig. 3. Functional modules of the ASIC**

# AN INTELLIGENT ULTRASONIC SENSOR FOR RANGING
# IN AN INDUSTRIAL DISTRIBUTED CONTROL SYSTEM

### G. Benet, J. Albaladejo, A. Rodas and P.J. Gil

*Departamento de Ingenieria de Sistemas, Computadores y Automatica, Universidad Politecnica de Valencia, Spain*

**Abstract.** In this paper, an intelligent instrument for ranging purposes is described. It is based on the precise determination of the time of flight (TOF) of an ultrasonic pulse to evaluate the distance from an object. Ultrasonic sensors offer high reliability, wide measurement ranges, reduced sensitivity to EM noise, and the ability to operate in dusty and dirty environments. This instrument uses digital signal processing techniques in order to improve the accuracy of the measurements and to increase its capabilities. The digital signal processing is managed by an 8-bit microcontroller. The communication tasks through a fault-tolerant real-time LAN, are carried out by an specific processor in order to discharge the main processor.

**Keywords.** Ultrasonic ranging, Time-of-flight evaluation, Signal conditioning, Analog-digital conversion, Digital signal processing, Intelligent instruments, Distributed Control.

## INTRODUCTION

The use of ultrasonics in distance measurement and proximity sensing in air is well-known and has been studied for many years. The noncontacting techniques of measurement are preferred in industrial automation processes because its inherent simplicity and accuracy. In these techniques, ultrasonic sensors offer high reliability, wide measurement ranges, reduced sensitivity to EM noise, and the ability to operate in dusty and dirty environments (Canali, 1982; Hickling, 1986; Marioli, 1988).

There are, however, some potential drawbacks in ultrasonic air applications that must be taken into account to achieve sufficient accuracy in the measurement. First, absorption in air is strongly dependent on frequency; the measurement range and the beam angle decrease as the frequency increases. Thus, for distance measurement in air, low-frequency low-bandwidth transducers are indicated. Moreover, propagation of ultrasonic waves in air depends also on environmental parameters as air pressure, air temperature or atmospheric turbulence (Lynnworth,1989), being the air temperature the most important in

an industrial ambient. Nevertheless, the influence of temperature can be easily compensated (Canali, 1982) or by measuring the temperature and computing the new value of the sound speed, or by comparison with a reference echo.

Time of flight (TOF) measurement techniques are often used to compute the distance between the transducers and an object. A short train of ultrasonic waves is generated by the transmitter; it travels through the air and is reflected by the target surface back to the receiver. This TOF elapsed between the transmision and the reception of the wave train is a measurement of the path covered by the ultrasonic wave.

In practice, the total error featured by these noncompensated ultrasonic systems is about tens of millimeters using low-cost piezoelectric transducers at frequencies around 40-50 kHz. Although this error may be tolerable for large distances, it seems unacceptable for distances under 1m.

Depending on the applications, the same transducer may be used for both transmit and receive purposes.

Alternatively, we can use separate transducers for each job. In this work, we have preferred this last choice, because its inherent flexibility.

The use of microprocessors in computing the distance makes possible to improve the accuracy of measurements and to increase the capabilities of the instrument, allowing us to obtain more elaborated data from the ultrasonic echo. Thus, we can measure the distances corresponding to multiple objects that produce multiple echoes. Also, it is possible to obtain other related data from these distance measurements, as the mean relative speed of an object detected or the violation of any proximity limit.

Another aspect to be considered is the output produced from the instrument to the control system. Frequently, the distance measured by the sensor is converted into a voltage or intensity and thus transmitted as an analog signal, with the disadvantages that it involves. The actual trend towards an intelligence distribution between the elements of the whole control system implies intelligent sensors with a certain level of data processing, capable of comunicate the results of this processing through a Standard Bus or Local Area Network (LAN).

## TRANSDUCER SIGNAL PROCESSING

Good accuracy in the determination of the time at which the reflected train of pulses reaches the receiver (TOF), is essential to achieve enough precision in the distance measurement. Unfortunately, this is not an easy task, due to the shape of the signal supplied by the transducer.

We can model this waveform as a damped sinusoid, as suggested by Parrilla (1991), and its envelope can be expressed as

$$v(t) \approx V_0 t^m e^{-t/h}, \qquad (1)$$

where $m$ and $h$ are constants, dependent on the transducers and on the number of cycles transmitted per pulse train. Typical values for $m$ range between 1 and 3, providing good approximations. In Fig. 1 is depicted a typical envelope of an ultrasonic echo.

Usually, most of ultrasonic systems employ specific integrated circuits, designed to convert the pulsed input signal into an edge that is used to find the TOF (NATIONAL, 1982; TEXAS, 1991a, 1991b). These IC's do not detect the true TOF. Instead, they generate the edge at the signal transition through a fixed threshold, thus loosing some initial cycles that do not reach the threshold amplitude. However, if the peak amplitude, or the ratio between the threshold and the peak amplitude were constant, this added delay would be also constant and compensable, thus yielding accurate measurements.

Unfortunately, the pulse amplitude is strongly dependent on the distance and on the reflection surface characteristics, and cannot be easily modeled. Thus, these TOF measurements made with a fixed threshold will be affected by a nonconstant offset and will give us large errors. In practice, the total error featured by these noncompensated ultrasonic systems is about tens of millimeters using low-cost piezoelectric transducers at frequencies around 40-50 kHz. Although this error may be tolerable for large distances, it seems unacceptable for distances under 1m.

In a recent study (Parrilla, 1991), several digital processing methods for increase the accuracy of the range measurements were compared using a synthetic approach. All these techniques featured reduction of the errors by several orders of magnitude in relation to simple threshold crossing techniques. In that study was also established that digitizing resolutions higher than 6 bits do not improve significantly the precision attained in most of these algorithms.

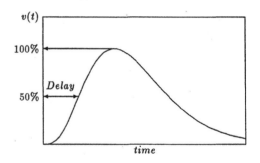

Fig. 1. Aspect of the echo envelope. Triggering with a threshold voltage produces a time delay in relation to the origin.

With the above considerations in mind, we have adopted a mixed technique to be used in the processing of the ultrasonic signal received, in order to determine the TOF with a sufficient degree of accuracy. In our instrument, the input signal is first amplified with a digitally programmable gain amplifier, then it is full-wave rectified and its envelope extracted by means of a low-pass filter. All these previous opperations are carried out by analog circuits and they give us the envelope of the ultrasonic echo. Inmediatelly after, this signal is digitized with a high-speed, 8-bit A/D converter and stored in the main processor memory.

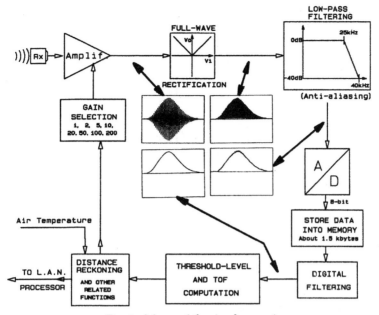

*Fig. 2. Scheme of the signal processing.*

With this digitized representation of the envelope, a digital filter is applied in order to to suppress the noise and the residual ripple. Then, the peak amplitude of the envelope is obtained and also the threshold voltage is computed as the 50 % of the peak value, obtaining the intersection point with the envelope by means of linear interpolation. This value of the TOF obtained has an added delay $t_d$ that is theoretically constant, because the ratio between the threshold and the peak amplitude is constant (0.5). Thus, compensation of this constant added delay is an easy task for the main processor, provided that we can obtain this $t_d$ in a previous calibration stage of the sensor. A simplified scheme of the signal processing is depicted in Fig. 2

This approach has some advantages:

- The analog processing is minimum and introduces delays that are both reduced and amplitude independents.

- The threshold value of 50 % of peak value has been chosen because it is easy to compute, and also because in this zone of the envelope, the shape is very linear, and the linear interpolation gives us a good approximation, as we can see in the Fig. 1.

- The load of the main processor is low, enabling real-time response of the sensor and additional features as multiple echo management or target speed calculus.

Another common source of error in the distance mea-
surement is the temperature dependence of the propagation speed of ultrasonic waves through air. Fortunately, this dependence is well known (Lynnworth, 1989), and can be theoretically computed for ideal gases as:

$$c = (\gamma RT/M)^{1/2} \qquad (2)$$

where $\gamma$ = specific heat ratio $C_p/C_v$, $R$ = gas constant, $T$ = absolute temperature and $M$ = average molecular weight. That is, $c$ exhibits a square root dependence with absolute temperature $T$. However, in practice, we can approximate the Eq. (2) by an straight line at temperatures between -10°C and 40°C. So, we can use the next equation for obtain the value of $c$:

$$c \approx 167.2 + 0.6T \qquad (3)$$

being $c$ expressed in m/s and $T$ in Kelvin. Thus, we can assume that an increment of one degree in $T$ produces an increment of 0.6 m/s in $c$. This approach is enough accurate for our purposes and simplifies the calculations. In the operation of the instrument, the ambient temperature is periodically sensed and the corresponding value of $c$ is obtained using Eq. (3).

## IMPLEMENTATION

In Fig. 3 is depicted a simplified scheme of the ultrasonic sensor that we have developped. This sensor has been designed as a part of an industrial distributed control system, whose architecture is described in (Serrano, 1992).

301

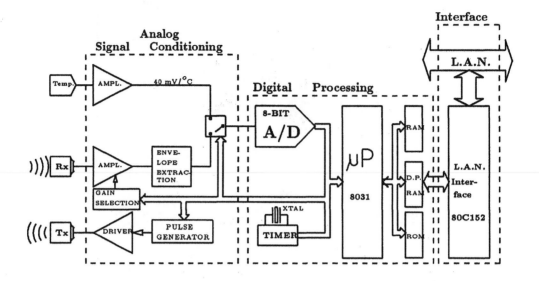

*Fig. 3. Simplified scheme of the intelligent ultrasonic sensor*

The circuit uses two ultrasonic transducers, for receive and transmit, respectively. They are low-cost piezoelectric devices adjusted for 40 kHz. Also, there is a transducer for sensing the air temperature. This sensor is the LM335, a low-cost integrated circuit which supplies 10 mV/$^\circ K$ with an accuracy of $1^\circ C$. The ultrasonic transmitter is driven by a sixteen-cycles burst signal, whose start is triggered by the main processor.

The ultrasonic echo received is amplified with a controlled-gain amplifier. The gain of this amplifier is selected by the main processor from among 8 possible values : 1, 2, 5, 10, 20, 50, 100 and 200. The amplified signal is full-wave rectified by an active rectifier, and its envelope extracted with an peak-tracking circuit. This envelope is digitized with the circuit MAX 150, a high-speed 8-bit A/D converter. This converter is also used to digitize the temperature signal supplied by the temperature sensor.

The digital processing block has been designed around the Intel 8031, a 8-bit microcomputer for embedded industrial applications, running at 12 MHz. This main processor carries out the following steps:

- Sets the initial value of the amplifier gain.

- Resets the time counter and sends the ultrasonic pulse.

- Obtains the peak value of each envelope received and modifies the gain if necessary.

- Interpolates the envelope and obtains the threshold-crossing point.

- Computes the TOF and its compensation for each echo detected.

- Computes the distance of each echo, using the air temperature measured to correct the sound speed $c$.

- Stores the results and relevant data for each echo.

- Computes the mean speed of each echo if exists possible correlation with the previous measurement.

- Interprets the commands and messages received from the system controller (SC) through the communications processor.

- Sends the requested data to the SC through the communications processor.

- Manages the self-calibration process, by measuring the distance from a known object.

The L.A.N. interface block is built around the processor 80C152, which is a derivative of the 80C51 microcontroller that includes one CSMA/CD data link, enabling user custom protocols, and supports three different modes of collision resolution, having also DMA capability. The architecture of the fault-tolerant LAN is described with more detail in (Serrano, 1992). The communication between the two processors is implemented with a dual-port RAM.

The circuits of this intelligent instrument have been designed taking into account future ampliations of transducers for transmit and receive, with tridimensional ranging purposes, although the present work deals only with the one-dimensional ranging.

## CONCLUSION

In this paper, we have presented an intelligent ultrasonic sensor that offers some main contributions to the actual state of the art:

- First, the accuracy in the determination of the TOF has been improved, obtaining the envelope of the input signal and digitizing it with enough time and voltage resolution. With this digitized envelope, and its subsequent digital process one can obtain more accurate values of the TOF, in order to eliminate an important source of error in the classic measurement process. In fact, this technique reduces in one the order of magnitude of the measurement error, featuring total errors about 1 mm for the 40 kHz transducers.

- Also, a software temperature compensation has been introduced in the instrument, measuring the air temperature, digitizing it and making the opportune correction in the value of the sound speed $c$. Other parameters as air pressure or turbulence, have minor influence and we have assumed they are almost constant in an industrial environment.

- Both previous functions are carried out by a microprocessor that, in addition, is also capable of undertaking some basic functions such as: target speed calculus, readings storage, out of limits detection and alarm generation, and so on. These additional features performed by this instrument ease the tasks of the process controller and gives to it a more elaborated and more reliable information. Also, all these capabilities can accomplish real time requirements, provided the relatively low load of the main processor caused by the low frequency of the signals involved, and the simplified algorithms of signal processing.

- Finally, this instrument has a communications specific processor for interfacing with a Local Area Network (LAN). It processes and gives the adequate format to all messages transmitted or received by the sensor. We preferred the addition of another specific processor instead of overload the main processor with the communications task, in order to achieve an optimum performance of the instrument.

## REFERENCES

Canali, C. and others (1982). A temperature compensated ultrasonic sensor operating in air for distance and proximity measurements. *IEEE Trans. Ind. Electron.*, IE-29, 336–341.

Hickling, R., and S. P. Marin (1986). The use of ultrasonics for gauging and proximity sensing in air. *J. Acoust. Soc. Amer.*, 79,(4).

Lynnworth, L.C. (1989). *Ultrasonic Measurements for Process Control: Theory, Techniques and Applications.* Academic Press, San Diego.

Marioli, D., E. Sardini and A. Taroni (1988). Ultrasonic distance measurement for linear and angular position control. *IEEE Trans. Instrum. Meas.*, 37(4), 578–581.

NATIONAL Semicond. Corp. (1982). LM1812, Ultrasonic Transceiver. In *Linear Databook*, Part 9: Industrial Blocks, pp.77–84.

Parrilla, M., J. J. Anaya and C. Fritsch (1991). Digital signal processing techniques for high accuracy ultrasonic range measurements. *IEEE Trans. Instrum. Meas.*, 40(4), 759–763.

Serrano, J.J., M. Sánchez and J.A. Gil (1992). A fault tolerant real time LAN for process control applications. Presented at *IFAC Symposium on Intelligent Components and Instruments for Control Applications SICICA '92*. Málaga, Spain. May 20-22.

TEXAS Instr. (1991a). SN28828, Sonar Ranging Module. In *Linear Circuits Databook. Special Functions*, Vol.1, Part.4, pp.67–71.

TEXAS Instr. (1991b). TL852, Sonar Ranging Receiver. In *Linear Circuits Databook. Special Functions*, Vol.1, Part.4, pp.155–159.

# ON THE FUSION OF DISPARATE SENSORY DATA

## H. Xu

*Department of Flexible Production Systems, CP 106, University of Brussels, Avenue F. Roosevelt 50,
B-1050 Brussels, Belgium*

Abstract. Multisensor system has received increasing attention in recent years. Multisensor data fusion can overcome the limitation of the individual sensory measurements. Not all sensors can provide complete estimates. The fused data are not always independent. This paper proposes a general methodology for multisensor data fusion. The sensory data may be partial or indirect. The proposed method deals with disparate sensory data and provides general, sensor independent and practical solutions. The correlations between fused data are taken into account. For a multisensor system, this method can implement the fusion of derived information in an efficient manner and give more accurate estimates. A case study as well as Monte Carlo simulations illustrate application of the presented method to greatly improve the location estimation for a mobile robot system by using various sensors.

Keywords. Robots; sensors; estimation; data fusion; simulation; bar code scanners; cameras; ultrasonic transducers.

## INTRODUCTION

In order for a robot to execute any sort of intelligent tasks, it is essential to maintain a world model and to keep the accurate locational estimates of the robot as well as other objects around it. The intelligent robot should not rely on the a priori world description since this model is usually incomplete and inaccurate, and cannot reflect the frequent change of the environment. To perceive the environment and to keep an accurate locational estimate, the sensors are usually required. Since single measurement is usually imprecise and partial, using redundant measurements have received increasing interest in the robotic research. The redundancy can be achieved either by one sensor repeating its measurement for more times at a fixed position, or one sensor measuring at different positions, or measurements from multiple sources. Redundant measurements by one sensor may reduce the uncertainty of individual measurement and the fusion of multiple sensory data can overcome the limitation of individual sensors.

Various techniques have been developed to represent the uncertainty and fuse the estimates of disparate sensors. A scalar error estimation was used by Chatila and Laumond (1985) as the uncertainty measure to combine individual estimates; Brooks (1985) employed a min/max

error bounds approach, while other approaches based on probability and geometric reasoning have been also widely studied (Durrant–Whyte, 1986; McKendall and Mintz, 1988; Preciado and others, 1991; Sabaster and Thomas, 1991; Smith and Cheeseman, 1987; Watanabe and Yuta, 1989). For a survey, see Hackett and Shah (1990). But there has been less attention paid to the incomplete information handling and no general method to integrate the partial and indirect sensory measurements. Actually, there are a number of sensors whose measurements either are partial estimate or provide implicit information. A relocalization system using bar code scanner and position reference beacons, for example, cannot produce a locational estimate but only constraints to the current location if there is not enough beacons perceived on the spot. In addition, when the probability model is used for data fusion, the independence between the fused data is always assumed to reach the solution. This assumption simplifies the problem as well as the solution. But we believe that this assumption is not always respected and the correlation of the fused data is not always negligible. On the other hand, handling dependency wouldn't substantially complicate the formulation and the solution of the data fusion problem. The statistical tests on the experimental data of our used sensor indicate the correlation and it is believed that this is not the only case in the real applications. This paper proposes a general methodology for disparate sensor data fusion. The solutions for data fusion, based on the three primitive sensory data, are presented in next section. Then the correlations between the fused data are taken into account. A mobile robot is taken as example in section 5 to illustrate application of presented method to improve location estimation by handling incomplete sensory information and the results of Monte Carlo simulation are finally shown.

## THE PRIMITIVE DATA TYPE

Disparate sensors provide different types of sensory data. Generally, the data of most sensory measurements can be classified into three basic data types, that is, complete, partial and indirect data.

Some of sensing techniques can produce full dimensional estimates about location or relationship. A widely used one is deadreckoning by optical encoders. For a wheeled mobile robot, for example, encoders are usually equipped to sense wheels' movement and on the basis of this measurements to estimate the position and orientation of the mobile robot. Uncertainty of this estimate increases with the travelled distance due to various errors. Generally, if any two individual sensor systems give their complete locational estimates, fusion of these estimates may yield a new estimate with reduced uncertainty. A variety of approaches have been developed to solve this type of problem. A short review is given in the following section.

Certain of sensory measurements cannot yield full dimensional estimate but only some elements. For instance, if a robot is moving on the plan, its location can be described by a three dimensional vector $(x, y, \theta)$ where $(x, y)$ are coordinates in some reference frame and $\theta$ indicates the orientation of the robot. A radio guided system(RGS) equipped on this robot may give position estimate $(x, y)$. While a gyroscope based sensor may give another element of the location vector, that is the orientation $\theta$. These estimates can also be inaccurate.

The third type is concerned with those sensory measurements which cannot produce explicit location estimate. Actually this type of data usually can be formulated as constraints to the location in a form of:

$$G(X, S) = 0 \qquad (1)$$

where X denotes the estimated location vector; S, scalar or vector, is the data from sensory measurement and G is a function or function vector.

## THE FUSION TECHNIQUES

This section presents and discusses available solutions for fusion of different types of sensory data. We suppose that the systematic error can be identified and corrected by calibration and the uncertain data can be described by following model:

$$X_i = X + \in_i \qquad (2)$$

where X denotes n dimensional vector to be estimated, $X_i$ is the estimate obtained by sensor i while $\in_i$ is estimation error. In this section the fused sensory data are considered independent of each other. Moreover, we suppose $\in_i$ has Gaussian distribution, zero mean and covariance matrix $C_i$, or $\in_i \sim N(0, C_i)$. This implies that $X_i \sim N(X, C_i)$.

### Fusion of Complete Data

Various approaches have been introduced for fusion of complete data. Watanabe and Yuta (1989) employed Maximum Likelihood Estimation (MLE) to fuse the estimates obtained by deadreckoning and by external sensor. Smith and Cheesman (1987) defined a merging estimation process. By means of KALMAN filter theory, the merging process fuses two sensory data and generates new estimate. Bayesian theory has been widely used for decision making on uncertain circumstance. It has been introduced to fuse multiple sensor data (Durrant–Whyte, 1986). One of the sensory data can be regarded as prior estimate and others are used as evidences to give posterior estimate.

If $X_1$ and $X_2$ are two complete location estimate, $C_1$ and $C_2$ are their covariance matrices, under the probability assumption stated at the beginning of this section, these fusion techniques reach same result which is

$$X_3 = (C_1^{-1}+C_2^{-1})^{-1} (C_1^{-1}X_1+C_2^{-1}X_2) \qquad (3)$$

and the covariance matrix of the new estimate

$$C_3 = (C_1^{-1}+C_2^{-1})^{-1} \qquad (4)$$

### Extended Maximum Likelihood Fusion of Partial Data

Data Type II is the special of Type I. In this case, sensors produce only some elements of estimate vector, say k dimensional estimate $X_{2k}$ as well as k*k covariance matrix $C_{2k}$, here k<n. Now suppose that $X_1$ is Type I data from some sensor, that is, an n dimensional location estimate. $X_1$ and $X_{2k}$ can be fused by the following formulas of Extended Maximum Likelihood Fusion (EMLF):

$$X_3 = (C_1^{-1}+C_2^{-1})^{-1} (C_1^{-1}X_1+C_2^{-1}X_2) \qquad (5)$$

$$C_3 = (C_1^{-1}+C_2^{-1})^{-1} \qquad (6)$$

where

$X_2$ is n dimensional vector consisting of $X_{2k}$ and zeros
$C_2^{-1}$ is n*n matrix which is composed of $C_{2k}^{-1}$ and zeros

Recall the example of the mobile robot with Radio Guided System, n=3 and k=2 for this example. Suppose that there are an existing location estimate $X_1$ and another estimate $X_{2k} = [x_2 \ y_2]^T$ from RGS, as well as 3*3 covariance matrix $C_1$ and 2*2 covariance matrix $C_{2k}$. The EMLF gives new estimate by (5) and (6), for this example $X_2 = [x_2 \ y_2 \ 0]^T$ and

$$C_2^{-1} = \begin{bmatrix} C_{2k}^{-1} & & 0 \\ & & 0 \\ 0 & 0 & 0 \end{bmatrix}$$

In general, this method can be applied to fuse any data with different dimensions.

## Constrained Maximum Likelihood Fusion of Data and Constraints

In the second section, we represented the data Type III as constraints like $G(X, S) = 0$. We present now the Constrained Maximum Likelihood Fusion (CMLF) to fuse this type of data with others. If $L_1(X)$ and $L_2(X)$ are the likelihood functions resulted from the existing estimate $X_1$ and the sensor measurement vector $S_1$ respectively, then the following mathematical programme corresponds to this data fusion problem:

maximize $L_1(X, S)$ and $L_2(X, S)$

subject to the constraint $G(X, S) = 0$      (7)

The discussion in this paper is limited to linear programming problem, that is, $G(X, S)$ is linear function of X and S. In case of nonlinear constraints, we suppose that the errors with $X_1$ and $S_1$ are small enough so that $G(X, S)$ can be linearized at $(X_1, S_1)$. We first consider only one constraint equation which is in a form of $G(X, S) = H^T X + R^T S + a = 0$.

With Gaussian distribution, the solution of the problem (7) can be given by following formula of Constrained Maximum Likelihood Fusion:

$$X_2 = X_1 - \frac{C_1 H G(X_1, S_1)}{H^T C_1 H + R^T C_s R} \tag{8}$$

together with the covariance matrix

$$C_2 = C_1 - \frac{C_1 H H^T C_1}{H^T C_1 H + R^T C_s R} \tag{9}$$

When sensory measurements result in more than one constraint equation, CMLF can be used repeatedly to handle multiple constraints one by one. The alternative solutions are reducing the number of constraints by elimination or applying following vector form of formula of CMLF for multiple constraint equations:

$$X_2 = X_1 - C_1 H (H^T C_1 H)^{-1} H^T X_1 - C_1 H (H^T C_1 H)^{-1} A \tag{10}$$

$$C_2 = C_1 - C_1 H (H^T C_1 H)^{-1} H^T C_1 \tag{11}$$

where

$$H = [H_1 \ H_2 \ ... \ H_k], \quad A = [a_1 \ a_2 \ ... \ a_k]^T$$

It is noted that with formula (8) and (9) for one constraint or sequential processing of simultaneous constraints, the matrix inversion is avoided which may reduce the computational overhead.

As different types of sensory data are available, by

means of the methods presented in preceding subsections, they can be fused in any order, that is, the result is independent of fusion order.

Moreover, Type I and Type II data can be regarded as specials of Type III data since they can be represented as constraints in some ways. This leads to a general representation and solution.

## TREATMENT OF CORRELATION

In this section, we consider the correlation between the fused data and propose the solution based on the Maximum Likelihood Estimation. We always suppose the Gaussian distribution and zero mean of the fused data. When $X_i$ and $X_j$ are correlated, we suppose that they have joint Gaussian distribution and can be described completely by their mean vector and the $n_i + n_j$ dimensional covariance matrix where $n_i$ and $n_j$ are dimensions of $X_i$ and $X_j$, respectively. This covariance matrix consists of two diagonal matrix blocks which are the covariance matrices of $X_i$ and $X_j$, and the other two matrix blocks describing the correlation between $X_i$ and $X_j$.

## Type I and Type II

In last section, it is known that the fusion of two independent explicit estimates can be done by means of MLE which gives (3) and (4).

When the correlation between $X_1$ and $X_2$ is taken into account, the following formula should be used

$$X_3 = \frac{1}{2} ((C^{-1})_{11} + (C^{-1})_{12} + (C^{-1})_{21} + (C^{-1})_{22})^{-1} *$$

$$((C^{-1})_{11}(2X_1) + (C^{-1})_{12}(X_1 + X_2) + (C^{-1})_{21}(X_1 + X_2) + (C^{-1})_{22}(2X_2)) \tag{12}$$

where C denotes the $2n*2n$ covariance matrix of $[X_1^T \ X_2^T]^T$ which includes four $n*n$ matrix blocks. $(C^{-1})_{ij}$ represents the (i, j) block of the inversion matrix of C.

Generally, for m correlated fused data which correlation is represented by their covariance matrix $C_m$, and the update is $X_N$, the general fusion formula can be written as below:

$$X_N = \frac{1}{2} \left[ \sum_{i=1}^{m} \sum_{j=1}^{m} (C_m^{-1})_{ij} \right]^{-1} \left[ \sum_{i=1}^{m} \sum_{j=1}^{m} (C_m^{-1})_{ij} (X_i + X_j) \right] \tag{13}$$

Furthermore, introduce

$$\mathbf{X_m} = [ \ X_1^T \ ... \ X_m^T \ ]^T \tag{14}$$

$$A_i = \sum_{j=1}^{m} (C_m^{-1})_{ij} \tag{15}$$

and

$$Q = \frac{1}{2} \left( \sum_{i=1}^{m} A_i \right)^{-1} [A_1 + A_1^T \dots A_m + A_m^T] \qquad (16)$$

then we have the compact form,

$$X_N = Q\mathbf{Xm} \qquad (17)$$

and the update of the covariance matrix

$$C_N = QC_mQ^T \qquad (18)$$

## Fusion of Constraints

The existing estimate $X_1$ can be correlated with the sensory measurement S which imposes constraint $g(X, S) = 0$ due to the correlation between sensory measurements. For example, let $X_1$ be the update by using $S_i$, $X_1$ and $S_i$ are correlated. When $X_2$ becomes the current estimate coming from $X_1$ and should be fused with $S_j$ from the same sensor, $X_2$ and $S_j$ should be correlated if $S_i$ and $S_j$ are correlated.

To deal with the correlation, introduce

$$Z = [X^T \ S^T]^T \qquad (19)$$

If $X_1$ and $S_1$ are existing estimate and a sensory measurement, respectively, then

$$Z_1 = [X_1^T \ S_1^T]^T \qquad (20)$$

Accordingly we define $C_{z1}$ to be the covariance matrix of $Z_1$. If

$$Q = [H^T \ R^T]^T \qquad (21)$$

then the constraint equation becomes

$$G(Z) = Q^T Z + a = 0 \qquad (22)$$

With these definitions, the problem can be formulated as

maximize $\qquad L(Z)$

subject to the constraint $G(Z) = 0 \qquad (23)$

Solving this mathematical programme gives following general CMLF formula

$$Z_2 = Z_1 - \frac{C_{z1}Q \ G(Z_1)}{Q^T C_{z1} Q} \qquad (24)$$

together with the covariance matrix

$$C_{z2} = C_{z1} - \frac{C_{z1}QQ^T C_{z1}}{Q^T C_{z1} Q} \qquad (25)$$

It can be observed that the formulas (24) (25) coincide with (8) (9) if X and S are uncorrelated. In this case, the non–diagonal blocks of $C_z$ are zeros. Actually, the formulation and solution in last section can be considered as the special cases of this section.

## APPLICATION: A MOBILE ROBOT EXAMPLE

Determining the location of a mobile robot is important in autonomous navigation. For a wheeled mobile robot, the basic technique is deadreckoning which estimates the current position and orientation relative to a initial state by sensing wheels' rotation. As robot's moving, the estimation error by deadreckoning is accumulated and the estimation accuracy decreases. This leads to the requirements for positioning technique with external sensors. Various sensors can be used for improvement of location estimation. In this section, we consider a mobile robot equipped with encoders attached to its wheels and external sensors on board, and illustrate multisensor fusion to improve the location estimation.

## Estimation by Deadreckoning

The location of a mobile robot can be represented by a three dimensional vector $(x, y, \theta)^T$. The distance which robot has travelled and the orientation angle changed can be calculated from the incremental changes of the odometric measurements of both wheels' motion for every sampling interval. Let

$P(n)$ and $M(n)$ the location vector and the vector to represent the real travelled distance and the orientation change, respectively;

$P_e(n)$ and $M_e(n)$ the location estimate and the measurement of travelled distance and the orientation change by internal sensor. Then

$$P(n+1) = f[P(n), M(n)] \qquad (26)$$

$$P_e(n+1) = f[P_e(n), M_e(n)] \qquad (27)$$

If errors of $P_e(n)$ and $M_e(n)$ are small enough, $P(n+1)$ can be linearized at $P_e(n)$ and $M_e(n)$. Similar to the position estimation formula in Watanabe and Yuta (1989), we have following relation

$$P(n+1) = P_e(n+1) + \delta P(n+1) \qquad (28)$$

and the estimation error is

$$\delta P(n+1) = F_p(n)\delta P(n) + F_m(n)\delta M(n) + \in(n) \qquad (29)$$

where $F_p$ and $F_m$ are Jacobian matrices of $P(n+1)$ with respect to $P(n)$ and $M(n)$ at the point $(P_e(n), M_e(n))$ while $\in(n)$ represents the approximation and truncation errors.

The uncertainty of location estimation can be described by covariance matrix $C_p$ which is calculated in a recursive way:

$$C_p(n+1) = F_p(n)C_p(n)F_p(n)^T + F_m(n)C_m(n)F_m(n)^T + C_\in(n) \qquad (30)$$

where $C_m$ is the covariance matrix of the measurement error, $C_\in$ denotes the covariance matrix of the computational noise.

## Beacon Reference System for Relocalization

The beacon reference system usually consists of perception sensors on the robot and a number of beacons which are fixed in the surroundings beforehand . More information about robot's location is acquired by perception of beacons and enables a more accurate location estimation. The number of beacons perceived at certain moment depends on the beacon placement and the effective range of the perception sensor. Three or more beacons can result in a three dimensional location estimate $(x, y, \theta)$ (Type I data) by triangulation. This estimate can be fused with deadreckoning estimate by means of the methods for complete data fusion. A complete explicit location estimate may be unavailable for some positions due to degeneration of triangulation algorithm, robot's high speed motion as well as lack of beacons. In this case, perception can only produce constraints.

Measurements of the sight angle to the beacons. The sight angle to the beacon can be measured by a bar code scanner or a camera. Suppose $(x_b, y_b, \alpha)$ be the sensory information obtained from relocalization system by perception of certain beacon where $(x_b, y_b)$ is the beacon's position and $\alpha$ denotes the sight angle to perceive the beacon relative to robot's heading. This information is usually known with uncertainty. The following constraint equation is produced by perception of one beacon

$$G(X, S) = y - y_b - (x - x_b)\tan(\theta + \alpha) = 0 \qquad (31)$$

$S = [x_b \ y_b \ \alpha]^T$ for this example.

Measurements of the distance to the beacons. The distance to a beacon may be estimated by using a camera or specially designed beacon system. In this case, the constraint is given by perception of one beacon which is

$$G(X, S) = (y - y_b)^2 + (x - x_b)^2 - d^2 = 0 \qquad (32)$$

where d denotes the distance and $S = [x_b \ y_b \ d]^T$.

## Measurements of the Distance to a Wall

The ultrasonic sensor is widely used for obstacle detection in the robotic system. When the robot is following a wall, the sonar ring installed on the robot can detect the distance to the wall and keep the robot at certain distance away from the wall. This distance measurement can help the location estimation as well. Suppose that the wall can be identified and approximately represented by a line equation $y = kx + b$ and that the distance to the wall is d, then fusion of the constraint:

$$G(X, S) = y - kx - b - d*SQRT(1 + k^2) = 0 \qquad (33)$$

can update the location estimate. Note $S = [k \ b \ d]$ and that these parameter and measurement may be noisy.

## Monte Carlo Simulation

To test the effect of the methodology for data fusion, Monte Carlo simulations have been run for fusion of Type I and Type III data. Position estimation and computation of covariance matrix are done for a sequence of fictitious measurements of distance travelled

by right and left wheels. Perceptions are made at selectable moments. The constraints which result from sensory readings are in the form of (31) (32) (33). The fusions of estimates and constraints are implemented by means of formulas (8) (9). All the sensory measurements as well as the beacon and wall parameters are blurred by the random noises.

The typical simulation results have been shown in Fig. 1 to Fig. 3 for fusion of the deadreckoning estimate and the measurements from different external sensors. Ellipses are the regions of 95% confidence. The robot intends to follow the straight line trajectory. As passage of time, the estimation error by deadreckoning grows and the difference between the estimate and the actual location increases. If there were no other sensor, the robot would drift away from its expected trajectory. Once the perception is done, the estimate is updated and the difference between estimate and the actual location is reduced. Continual perceptions can keep the robot within a small deviation from its intended trajectory. The different external sensory data are fused separately in Fig. 1 to Fig. 3. Their different effects can be observed from the figures. Combined use of different sensors can compensate with each other and enables better performance.

Although Gaussian distribution is assumed in the error model, these techniques work quite well even when the non–Gaussian noises are introduced in the simulation. For example, for the measurements with random uniform noises, little performance degradation has been observed.

The measurements from a real laser scanning sensor have been applied to the simulation and good performance can be reached although the statistical tests on these data show them non–Gaussian and slightly correlated. When the correlation is treated by using the formula (24) and (25), with the well adjusted correlation factors, considerable improvement (up to 50%) in the maximum error can be observed. It should be mentioned that the variance as well as correlation are application and sensor specific, and are usually determined by the knowledge and the experiments of a real system.

The accuracy of the estimate update depends on the accuracy of the sensory measurements and that of the parameters of the perceived objects like beacon's position and wall equation. The beacon placement and the perception strategy can also have something to do with the resulted trajectory. A position accuracy of a few millimetre can be reached with a good sensor system and perception design.

## SUMMARY

In this paper, we discuss a general technique for fusion of different type of sensory information. Based on the definition of three primitive data type, the solution for fusion of disparate data is presented. On the basis of Maximum Likelihood Fusion, we developed Extended Maximum Likelihood Fusion (EMLF) for partial data and Constrained Maximum Likelihood Fusion (CMLF) for constraints. We deal with the correlation of fused data and propose the solution. By the presented methodology, sensor data fusion can be implemented accurately and efficiently. A case study and Monte Carlo simulations have shown the application and the performance of CMLF to greatly improve the location estimation for a mobile robot system with various sensors.

## REFERENCES

Brooks, R.A. (1985). Visual map making for a mobile robot. *Proc. IEEE Int. Conf. on Robotics and Automation*, pp. 824–829, St. Louis, Missouri.

Chatila, R. and J–P.Laumond (1985). Positioning referencing and consistent world modeling for mobile robots. *Proc.*

*IEEE Int. Conf. on Robotics and Automation*, pp. 138–145, St. Louis, Missouri.

Durrant–Whyte, H. F. (1986). Consistent integration and propagation of disparate sensor observations. *Proc. IEEE Int. Conf. on Robotics and Automation*, pp. 1464–1468, San. Francisco, California.

Hackett, J.K. and M.Shah (1990). Multi–sensor fusion: a perspective. *IEEE Int. Conf. on Robotics and Automation*, pp. 1324–1330, Cincinnati, Ohio.

McKendall, R. and M.Mintz (1988). Robust fusion of location information. *Proc. IEEE Int. Conf. on Robotics and Automation*, pp. 1239–1244, Philadelphia, Pennsylvania.

Preciado, A., D.Meizei, A.Segovia and M.Rombaut (1991). Fusion of multi–sensor data: a geometric approach. *Proc. IEEE Int. Conf. on Robotics and Automation*, pp. 2806–2811, Sacramento, California.

Sabaster, A. and F.Thomas (1991). Set membership approach to the propagation of uncertain geometric information. *Proc. IEEE Int. Conf. on Robotics and Automation*, pp. 2806–2811, Sacramento, California.

Smith, R.C. and P.Cheeseman (1987). On the representation and estimation of spatial uncertainty. *Int. Journal of Robotics Research 5*, 56–68.

Watanabe, Y. and S.Yuta (1989). Estimation of position and its uncertainty in the dead reckoning system for the wheeled mobile Robot. *Proceedings of 20th ISIR*, pp. 205–212.

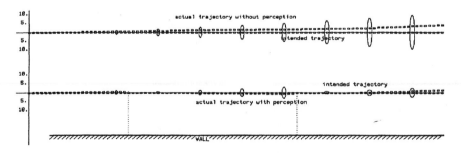

Fig. 1. Fusion of the sight angle to beacon

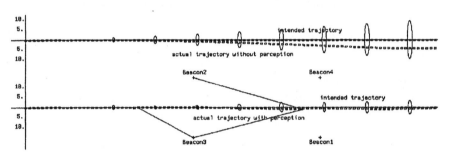

Fig. 2. Fusion of the distance to beacon

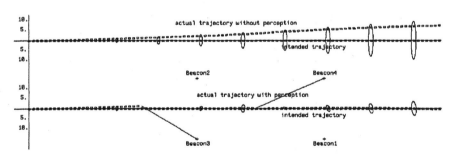

Fig. 3. Fusion of the distance to a wall

# VIRTUAL SENSORS IN FLEXIBLE ASSEMBLY

**J.J. Rowland and H.R. Nicholls**

*Department of Computer Science, University College of Wales, Aberystwyth, Dyfed SY23 3BZ, UK*

**Abstract** The paper describes our work in EUREKA/FAMOS project 'InFACT' (EU 321), a collaborative project concerned with the design and implementation of a fully integrated robotic assembly workcell that allows quick change of product. The computing architecture for the workcell is based on transputers. Our own contribution, described here, is the sensor integration system, which coordinates all the sensory information in the machine and provides it at an appropriate level of abstraction to the supervisor software, which controls and schedules the assembly tasks. The sensor integration system is based on the concept of 'virtual sensors', a modular design approach with considerable versatility of application.

**Keywords.** Robot; Assembly; Virtual Sensors; Logical Sensors.

## BACKGROUND

The work described here builds on our previous work on virtual sensing, which was inspired by the logical sensor concept described by Henderson (Henderson & Shilcrat, 1984; Henderson, Hansen & Bhanu, 1985; Luo & Henderson, 1986). Earlier work at Aberystwyth has also contributed (Hardy, Barnes & Lee, 1987).

Previously we have explored the extension of the logical sensor concept to combine parameters from different sensors to derive 'new' measured parameters, calculated from individual 'raw' parameters (Nicholls, Rowland & Sharp, 1989; Rowland & Nicholls, 1989). Henderson appears only to have considered this aspect in relation to stereo vision and we use the term 'virtual sensors' to distinguish our approach.

In the work described here we have further extended our virtual sensor concept to embody the temporal aspects of sensor information, providing virtual sensors that can detect and respond to temporal sequences of sensor events, as well as combining the outputs of different physical or other virtual sensors.

### The InFACT Machine

InFACT was a collaborative project involving eight partners. The aim was to design and implement a fully integrated robotic assembly machine (Seeley & Williams, 1990), based on earlier work by Williams (1985). The machine is designed for automated assembly of light electro-mechanical goods such as, for example, automotive light clusters, electrical wall sockets, and electrical meters. It will provide users with versatile, economic, operation and, to this end, many elements of the machine are generic across assembly tasks and product ranges so as to minimise product-specific tooling and to maximise versatility.

The InFACT machine is based around two gantry-based Cartesian manipulators known as the *main* and *power operations* manipulators. The main manipulator is equipped with a turret-type end-effector with exchangeable fingers; it can also accommodate an enhanced end-effector, currently under development, in which the grippers on the turret may themselves be exchanged. The power operations manipulator has available a range of tools for operations such as press fits and screw driving. Nuts, screws, and other small parts are supplied by linear vibratory feeders; larger parts or subassemblies are pre-loaded into pallets and brought from the supply zone on a track-based shuttle which also returns pallets of completed assemblies.

The software architecture (Figure 1) is based on a hierarchy of authority. At the highest level is the Human Computer Interface (HCI). Beneath this, and taking instructions from it, is the supervisor, which has sole authority to command all actuators, by issuing commands to *action modules*. The supervisor accesses the sensors in the machine via the sensor integration system which, in turn, has sole authority to access the sensing functions provided by the action modules.

The action modules are hardware/software modules that each control a specific machine function (such as a manipulator, the shuttle, a group of feeders); they do not inter-communicate and cannot initiate action themselves. Action modules each have an actuation or sensing function, or both; those action modules that are concerned with actuation also provide sensory information about the actuators they control.

The computing power for control of the machine is provided by a specially designed transputer architecture. Each of the action modules contains one or more transputers; the supervisor runs on a T800 transputer, as does the sensor integration system, which also has the provision for a second transputer to host the more time-critical operations. The language occam 2 was used for implementation of all software in the machine, which was developed with the Inmos PC-based Transputer Development System.

## SENSORS AND SENSING STRATEGY

Part of the philosophy of InFACT involves minimising uncertainty prior to assembly and accommodating uncertainty during assembly. The use of pallets and track feeders for parts delivery minimises feeding errors, positional uncertainties, and orientation errors, and the use of end-effector compliance seeks to accommodate any remaining uncertainties. This philosophy is reflected in the sensing strategy, so that only relatively simple physical sensors are required; there is, for example, no vision sensing.

However, the flexibility and resilience required in the more advanced assembly operations for which the machine is intended necessitate the ability to detect and, where possible, resolve such errors as these. This is achieved by measurement tasks that utilise a combination of binary touch-probe sensors, some involving up to four probes, and end-effector position sensing. The sensing on the compliant mount of the main end effector permits the measurement of insertion force, and sensors on the vibratory feeders indicate presence/absence of a part in the pick position. Other sensors are concerned with confirming the current state of the machine itself returning, for example, information on the brakes, the axis end-stops, the type of fingers currently mounted, and the status of the safety systems.

Although the philosophy involves making good use of relatively simple sensing, there are a great many diverse sensors in the machine. In spite of their diversity they are made available to the supervisor in a uniform way by the sensor integration system, which was our own contribution to this project. The sensor integration system permits abstraction of sensor data, presenting sensors as *virtual* entities whose physical sensing details are hidden from the supervisor. The Sensor Integration System is specifically designed to provide the sensing facilities needed by the machine supervisor during the assembly operations specified by the user via the Human Computer Interface (HCI).

### The Virtual Sensors.

Virtual sensor functions can be derived from a combination of the current values of a number of sensors and from sequences of sensor events, or other temporal relationships. The present design accommodates five classes of virtual sensor:

**State** is the most primitive virtual sensor class. State sensors return the current value of a virtual sensor and are used for determining the current state of the machine. For example, state sensors are available that indicate which of the feeders contain parts and are available for use. There are also state sensors

that provide the coordinates of the manipulator end-effectors; these state sensors always return a six-vector $(x, y, z, rx, ry, rz)$ even though not all of the degrees of freedom are implemented in the present version of the machine.

**Event** sensors are defined in terms of state sensors and provide notification to the supervisor that a pre-specified change of value of a state sensor has occurred (or, where the state sensor returns a vector, a single component of the vector must be specified). Event sensors therefore detect transients and are used either to confirm that an expected event has taken place, or may indicate that an undesired, or error, condition has arisen. An event sensor defined on a touch probe can be used to confirm correct orientation of certain assembly components – where failure of the event to occur during a particular trajectory indicates either incorrect orientation or absence of the component. An event sensor defined upon the $z$-axis compliant deflection can be used to terminate a guarded move. Other event sensors may warn of undesired conditions, such as that a feeder has no more parts.

**Trigger** sensors form a higher virtual class. They return the current value of one sensor when another sensor undergoes a pre-specified change. This is a temporal combination of state and event sensing, and trigger sensors are therefore defined in terms of a pre-existing state sensor and a pre-existing event sensor (which is itself defined in terms of a state sensor). The principal use of the trigger class is in measurement; for example by using a trigger sensor that returns the end effector coordinates when a specified touch probe is activated the precise position, in the gripper, of the gripped object can be determined. By using several such triggers simultaneously more detailed measurements can be made, effectively 'imaging' the shape of the object at a low, but often sufficient, resolution (Nicholls & Hardy, 1992).

**Trend** sensors store, either upon supervisor request or upon occurrence of a specified event, successive values of specified state sensors and thereby provide a log of sensory data for subsequent statistical processing to give information on machine performance or product quality. Upon interrogation, the trend sensor returns the mean and standard deviation of the sampled data. It is also possible effectively to define an event sensor upon the output of a trend sensor so as to provide warning that a systematic drift in the sampled value such that the value is likely to become unacceptable. This is referred to as 'drift' sensing. Although our design contains provision for these virtual sensor classes they have not yet been implemented.

Virtual sensors can therefore be defined in terms of other virtual sensors to form a hierarchy of increasing functionality and provide the supervisor with powerful functions for direct use within assembly generics. The application of virtual sensing techniques with the ability to respond to temporal sequences of sensor events ensures that sensor data is passed to the supervisor in a relevant form, and only *when* it is relevant. Presenting the sensing capabilities of the machine to the supervisor as virtual sensors, at a high level of abstraction compared with the physical sensors, allows the supervisor to control the wide variety of sensors in a uniform way. It also allows the sensor integration system to provide sensing facilities that do not relate directly to physical sensing devices; for example, manipulator position vectors can be ascribed fixed values to represent degrees of freedom that are absent; when machine enhancements provide additional degrees of freedom, sensor integration does not need to change.

Two **health** sensors are available to the supervisor; these check the integrity of the computational aspects of the sensor systems. The health sensors effectively belong to the **state** class but are restricted in that no other virtual sensors may be defined in terms of the health sensors. One of the health sensors tests a single state sensor per action module as a means of checking that the action module is present and operational; the other checks each potential state sensor in turn to ascertain that the value returned is within its acceptable range and that no error messages are generated.

### Sensor Command/Response Languages

Supervisor – Sensor Integration. We have developed a command/response language that permits creation and control of virtual sensors by the supervisor within the five virtual sensor classes. A related language, almost a subset, is used for communication between the sensor integration system and the action modules.

The supervisor must 'define' virtual sensors before they can be used and may 'destroy' them when they are no longer required. The supervisor can thus bring virtual sensors into and out of scope according to the needs of the current assembly task and this adds to the integrity of the system through detection of erroneous attempts at sensor access. The parameters supplied with the 'define' command specify the properties of the virtual sensor that is to be defined. Different classes of virtual sensor require different parameter sets for their definition, and are controlled via further commands that are appropriate to the class, as indicated below.

**State** sensors may be defined with a 'define state' command that takes as parameter one of the 'fundamental system sensors', which correspond to a low-level abstraction of the physical sensors themselves. On receipt of this command the sensor integration system checks that the command and the fundamental sensor are valid and, if so, issues a 'sensor-id'. This is a unique identifier for that virtual sensor and is used by the supervisor to reference the virtual sensor in subsequent commands and to identify the source of responses. Once defined, the virtual state sensor may be sent an 'interrogate' command, whose parameter is the sensor-id, to which the response contains the current value of that sensor. When the sensor is no longer required the supervisor sends a 'destroy' command, whose parameter is the sensor-id. Following the 'destroy' the sensor-id becomes invalid and further attempts to access the virtual sensor result in error messages.

**Event** sensors are constructed upon, and specified in terms of, state sensors that the supervisor has defined (but not yet destroyed). The 'define' command for an event sensor therefore takes as parameters the sensor-id of a state sensor, along with the threshold value and a comparison operator. Sensors that return a numeric data type may take the operators 'less than' or 'greater than';

where the data is of integer type the operators 'equal' and 'not equal' are also available. Boolean sensors may take the operators 'equal' or 'not equal'. Following validation, the sensor integration system responds with a sensor-id for the new virtual event sensor. Once defined, the event sensor may be sent a 'monitor' command, with the (event) sensor-id as parameter, which causes the sensor to issue a 'monitor.fired' response in the event of the specified condition being satisfied; following this, monitoring ceases until a new 'monitor' command is issued by the supervisor. An 'abort' command, issued by the supervisor, causes the sensor to cease monitoring. This is used to cancel event sensors that have not 'fired' but are no longer relevant; the sensor may be subsequently re-armed with a new 'monitor' command. When the event sensor is no longer required, a 'destroy' command is issued.

**Trigger** sensors are defined in terms of already defined state and event sensors. The commands are similar to those of event sensors, except that 'define' takes as parameters simply the sensor-id of a state sensor and the sensor-id of an event sensor. In place of the 'monitor.fired' response, the trigger sensor returns the value of the specified state sensor.

Other classes of virtual sensor can be controlled by similar commands and parameter sets.

Because of the hierarchical dependency amongst virtual sensors, checks are made to ensure that no sensor can be destroyed while there is a higher level virtual sensor (such as an event sensor or trigger sensor) currently defined upon it. The rigorous checks on the supervisor commands are supported by a comprehensive set of error responses provided by the sensor integration system to permit the supervisor to detect and recover from syntax errors and parameter errors.

Sensor Integration – Action Modules. The language that supports commands and responses between the sensor integration system and the action modules is rather simpler because only state and event sensors are handled by the action modules themselves. Higher level virtual sensor classes are implemented entirely within the sensor integration system. The reasoning behind this is straightforward: the 'fundamental sensors' (more or less corresponding to the physical sensors), upon which the state sensors are based, are connected directly to the action modules. Event sensors operate by continually accessing state sensors and performing comparisons between the sensory data and the threshold value specified for the event sensor; to avoid excessive communications overhead this *monitoring* is done within the action modules. Higher classes operate either as a result of a stimulus provided by an event sensor or as a direct result of a command from the supervisor. They are also likely to use sensors provided by more than one action module and so they are more appropriately handled at the higher level.

Furthermore, at the level of the action modules, all possible state sensors (corresponding to all the available 'fundamental' sensors) are considered to be permanently defined. Consequently only three sensory commands need be accepted by the action modules namely 'interrogate' (for state sensors), 'monitor' and 'abort' (for event sensors). Because there is no 'define' command at this level the 'monitor' command takes as parameters the threshold and operator. All messages passing between sensor integration and the action modules carry the appropriate virtual sensor-id, which permits the sensor integration system and the action module to relate messages to specific virtual sensors.

Rigorous checking is provided by sensor integration for the detection of syntax and parameter errors in the responses received from the action modules; if an error is found, a message indicating 'action module failure' is passed to the supervisor.

## Sensor Integration Architecture

Whilst the action modules themselves are functional modules in the conventional sense, each corresponding to a functional element of the machine, the sensor integration system does not use this type of functional modularity. In contrast to previous approaches which have organised the software as functional modules that each correspond to a virtual sensor function, we have, within the sensor integration system, used modules that each correspond to a sensor management function (Figure 2).

There are concurrent modules for management of each of the virtual sensor classes: state, event, trigger, etc. An 'interpreter' module handles the dialogue with the supervisor, checking the validity of commands and distributing messages to other sensor integration modules, as well as returning response and error messages to the supervisor. Other modules each handle a class of virtual sensors: there is a state module, an event module, a trigger module, etc. The modules communicate with each other via a communications module, and associated asynchronous buffering, which together form a communication spine that also routes the relevant messages to and from the action modules. Although the modules are therefore connected to the communications spine in the form of a linear network their hierarchy is imposed through restrictions on intermodule communication. The interpreter is the only module that can initiate a dialogue with all other sensor integration modules; in contrast, the state module can initiate a dialogue only with an action module. A comparison of figures 2 and 3 illustrates this hierarchical relationship.

All messages entering the sensor integration system, from the supervisor or from the action modules, are checked for validity of the sensor-id, sensory data, or command parameters that they carry. A comprehensive set of error messages informs the supervisor of any errors found.

The *variant protocol* feature of occam 2 is used for all messages so that errors in message structure and type are detected at compile time. The use of variant protocols for message type checking as well as for checking message structure does lead to large numbers of variants and consequently to large CASE statements each time the protocol is input to a module. The absence in occam 2 of variant data structures is a serious omission from the language and leads to extremely cumbersome sections of program code each time a variant protocol is handled. For example, asynchronous buffers, relatively simple in concept, become major software items where variant protocols with a large number of variants are used, purely because of the non-availability in the language of variant data structures. The restricted data structure facilities of occam 2 have also resulted in more general difficulties

because of the highly data-oriented nature of this application.

All messages handled by the communications module are copied to a 'diagnostics port' to which may be connected a PC containing a transputer. This allows all messages to be displayed or to be logged to disc. This is intended primarily as a diagnostic aid but it can also provide useful data on sensor usage.

The design of the sensor integration system is such that it can be distributed across two transputers to accommodate more stringent timing constraints.

## CONCLUSIONS

Use of the virtual sensor concept in the InFACT machine has led to some simple, yet important and powerful, enhancements to the concept; they involve the specification of classes of virtual sensor based on temporal relationships. This has permitted design of a versatile command/response language for the control of sensors. It also permits the sensor integration system to be designed on a modular basis that reflects sensory management functions rather than the sensory functions themselves, experience having shown that designs on the latter basis can quickly become unmanageable.

The work described in this paper has allowed us to move further towards establishing general design principles for (virtual) sensor systems. The eventual aim is a design that allows:–

- a general ability to specify any desired virtual sensor function in terms of other appropriate virtual sensors that have been already defined, .

- ease of configuration for different applications.

At the time of writing, the prototype InFACT machine is undergoing extended testing to improve the reliability of the basic control systems and mechanical components. Its effectiveness in the assembly of specific products is also being evaluated, at this stage via custom-designed operation sequences rather than sequences based on generic operations. Market research indicates that the first commercial models should operate in this way, with the more advanced software features added later to form an enhanced machine range. The sensor integration system has been extensively tested, using harnesses that represent accurately the eventual machine environment; experiences indicate that the facilities offered by our virtual approach to sensing are appropriate to an assembly machine of this type.

## ACKNOWLEDGEMENTS

The U.K. contributions to the InFACT project, including the work described in this paper, received financial support from the U.K. Department of Trade and Industry, under EUREKA/FAMOS project number EU 321.

## REFERENCES

Hardy, N.W., D.P. Barnes, and M.H. Lee. Declarative sensor knowledge in a robot monitoring system. In U. Rembold & K. Hormann, editor, *NATO ASI Series, 29, Languages for Sensor-Based Control in Robotics*, pages 169–187. Springer Verlag, Berlin, 1987.

Henderson, T., C. Hansen, and B. Bhanu. The synthesis of logical sensor specifications. In *Proc. SPIE Vol. 579 Intelligent Robots and Computer Vision*, pages 442–445, 1985.

Henderson, T.C., and E. Shilcrat. Logical sensor systems. *J. Robotic Systems*, 1(2):169–193, 1984.

Luo, R.C., and T.C. Henderson. A servo-controlled robot gripper with multiple sensors and its logical specification. *J. Robotic Systems*, 3(4):409–420, 1986.

Nicholls, H.R., and N.W. Hardy. Distributed touch sensing. In H.R. Nicholls, editor, *Advanced Tactile Sensing for Robotics*, chapter 6. World Scientific Co Pte Ltd, Singapore, 1992. In press.

Nicholls, H.R., J.J. Rowland, and K.A.I. Sharp. Virtual devices and intelligent gripper control in robotics. *Robotica*, 7(3):199–204, June 1989.

Rowland, J.J., and H.R. Nicholls. A modular approach to sensor integration in robotic assembly. In E.A. Puente and L. Nemes, editors, *Information Control Problems in Manufacturing Technology*, pages 371–376. IFAC, Pergamon Press, Oxford, England, 1989.

Seeley, C., and A.M. Williams. Generic assembly - the InFACT way. In *Proc. 11th Int. Conf. on Assembly Automation*, Dearborn, Michigan, 1990. SME/IFS(Pubns.), Kempston, Bedford, UK.

Williams, A.M., et al. A flexible assembly cell. In *Proc. 3rd Int. Conf. on Automated Manufacturing*, pages 57–66. IFS(Pubns.), Kempston, Bedford, UK, 1985.

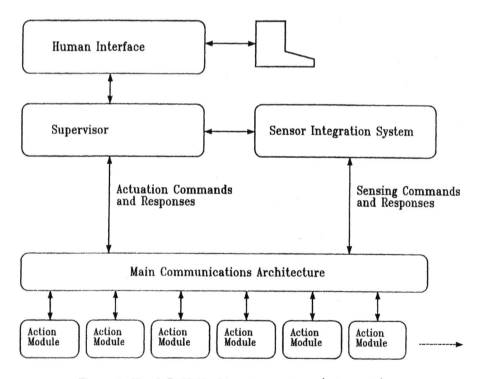

Figure 1. The InFACT Machine Architecture (schematic).

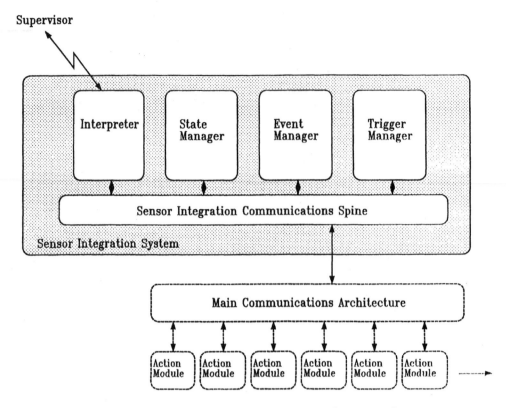

Figure 2.   The Sensor Integration System (schematic).

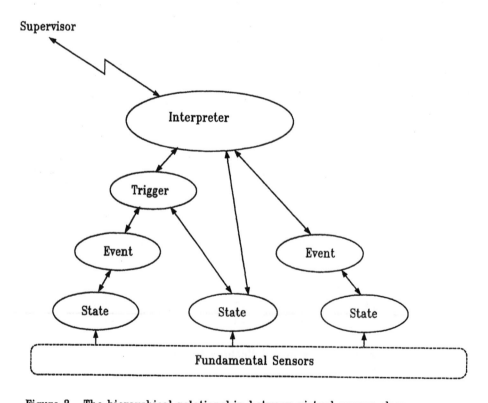

Figure 3.   The hierarchical relationship between virtual sensor classes.

# SENSOR FUSION INTO A MANUFACTURING CELL CONTROLLER USING CONSTRAINT MODELLING TECHNIQUES

**R.T. Rakowski, G. Mullineux and A.W. Ansari**

*Manufacturing and Engineering Systems Department, Brunel University, Uxbridge, UB8 3PH, UK*

Abstract. The U.K. Manufacturing Industry has an increasing need for greater automation for improved design, control and production. These needs have spurred on developments of existing I.T. systems, research in artificial intelligence, and new information structures and search procedures to enable efficient real-time operation. The solution of real engineering problems will require the integration of sensor signal processing techniques with knowledge based systems. The paper highlights the importance of collecting sensor information as part of a robot cell-control strategy and proposes a new model for a cell-controller using constraint modelling techniques. The advantages of such a system, based on the RASOR programming language, and its relevance to practical applications are presented. A practical demonstrator for the cell controller is being created around the problem of handling of food pieces by a robot gripper integrated with tactile sensors. This example will represent the various inaccuracies which can occur when a robot is required to work in an uncertain environment.

Keywords. Artificial intelligence; constraint theory; data handling; data reduction; knowledge engineering; robots; sensor fusion.

## INTRODUCTION

Modern robots possess capabilities of vision, speech, and can sense objects by touching. Grippers with integrated sensors still have many problems in handling objects especially if the robotic system is to be designed to operate in an unstructured environment. Much research effort is being directed into gripper technology to provide generic gripper devices. These new devices should not only be versatile but be capable of fulfilling the semi-autonomous features of advanced robotic systems. To this end it is essential that the enabling technologies for sensor fusion and real time decision making are also developed. Over the years various handling techniques have been adopted which reflect the different types of sensors used within the gripper arrangement, for example optical sensors, proximity and range sensors, tactile sensors. Recently magnetoresistive (MR) tactile sensors have been fabricated and tested in the department of Manufacturing and Engineering Systems at Brunel University (Adl and colleagues, 1990). The sensor configuration of two force and two slip sensing elements in an area of 1cm², show a very good quality of performance. These sensors are incorporated into a parallel jawed servo-driven gripper specially designed for the handling of fragile or compliant objects, where the need for normal as well as shear force information is of utmost importance. The handling of, say, a vegetable piece or meat piece with the inevitable variation in both dimensions and mechanical properties, will require sophisticated handling routines and error recovery strategies.

This variation in an objects properties will be termed the uncertain environment. In this context the term has a number of components.

(a) The characteristics of an object to be handled are not clearly defined and contain a number of variations.
(b) Misplaced or misaligned presentation of parts for assembly/handling, and

(c) The intrusion into the workspace of foreign bodies, or dropped parts.

## THE GRIPPER CONTROL PROBLEM

Obviously any complex handling routine will require information from a wide range of sensors including a vision system. The main problem with real time control will be to act upon these sensor signals in a prioritised way enabling fast response times to be achieved. Because this research project will investigate the use of constraint theory as the decision making engine, the controller demonstrator will focus only on the integration of tactile sensor information. The MR sensors in the gripper fingers will be used to sense the forces exerted by the object to be handled. For example if slip or shear forces, exerted by the object on the fingers of a gripper, are detected corrective action might need to be taken to prevent handling failure. The phenomenon of conversion of force into an electrical signal is achieved by the action of the MR transducers. The transduction principle is based on changes in an external magnetic field producing changes in the resistivity of the MR element. The change in resistivity produces a corresponding change in output voltage and it is this signal that can then be processed and interpreted. After completing a processing cycle, a signal is transmitted to the gripper servo-motor to move the gripper actuating mechanism or to incrementally increase the motor torque. This is done in order to make the applied grip on the object sufficiently tight to avoid dropping the object. These sequential events look so simple but in the real world, particularly in an uncertain environment, they become very complex.

## DESIGN OF CELL CONTROLLER

An analysis of a cell controller to operate in such an uncertain environment, incorporating a servo-driven gripper of a robot, is

317

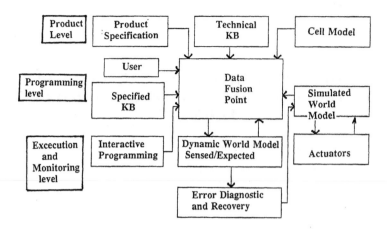

Fig. 1. Model for Data Fusion System.

the main objective of this research. By constructing such a cell controller using constraint modelling techniques a new approach to general sensor fusion will be formulated.

To date, cell controllers have been designed using different techniques such as batch processing (Appleton, 1990), hierarchical structures (Elkins and Maimon, 1987), and using information system concepts (Camarinha-Matos and Steiger-Garcoa, 1988). Camarinha-Matos and Steiger-Garcoa (1988) describe a robot based assembly programming system where problems at the cell-controller design and planning phases are discussed. An information system has been utilized as the organising kernel for the global system. To incorporate systems flexibility, methods for the introduction of error recovery have also been considered. In addition systems design must take into account the technological as well as the economic constraints at the planning phase. Process planning is also preceded by careful product design which is interconnected with feedback from the system simulation and cell design process. In designing the systems cell specification, information about the robot, feeders, sensors, cell layout, geometric and kinematic models should be included. The process specification should contain information about task level operations and their operandi, information about tools, fixtures and technical constraints such as tolerances, maximum allowed torque forces, grasp forces etc.

Considering all this information, a system model for a data fusion system is presented in Fig. 1. Here three levels of hierarchy of product level, programming level and execution level have been specified. To obtain an implicit solution, information about the product specification, technical knowledge base, and the cell model are supplied. For an explicit solution, information from sensors about the dynamic world model are compared with actual sensed and expected signals. If any error is detected it will be sent to obtain an error recovery procedure from the simulated world model. In the case of a system crash in programming or time scheduling, user interrupt and interactive facilities are also provided at the programming level.

In terms of hardware design the conceptual knowledge required for a generalized flexible system tool to evaluate alternative configurations in a hardware cell has been proposed by Dooner and De Silva (1990). At present various computer architectures are in use within a manufacturing cell to implement communications at various levels. These architectures are serial, parallel or a mixture of the two. The serial method is in broad terms cheap but slow, whereas the parallel method is fast but costly. There is also a trend to provide a mixed mode combination of serial and parallel architectures (Hwang and Briggs, 1985). For the purpose of a gripper controller, it is proposed to use a serial or mixed mode of communication. The controller will need to use the minimum number of sensors with a minimum software implementation time, improving the real time operation for the picking and placing of objects. The hardware configuration used at the gripper control level is shown in Fig. 2. Here the four sensors on each gripper finger can be identified and accessed.

## DATA FUSION

Data fusion is a process performed on multi-sensor data at several levels, each of which represents a different level of abstraction of data. The process of data fusion includes detection, association, correlation, estimation and combination of data. The results of data fusion include state and identity at the lower levels, and assessment of the overall tactical situation at the higher levels. This approach will be adopted for the purpose of the servo-motor gripper controller. At the lower level the strategy for detection of data, association of data and state (position of gripper, and grip) is made. At the higher level, an assessment of object state (object movements, due to weight or friction, and object type) is calculated and necessary procedures are adopted to avoid object mishandling.

The application of quantitative sensors is normally found in the area of distributed sensing and control, although they can also be useful in expert systems, behavioural modelling, and structural modelling. A decentralised architecture for multi-sensor data fusion has been discussed in detail by Edward and Llinas (1990). This architecture has no central processing, no centralised communication medium and does not use hierarchy. Each sensor node has its own processing element and its own communication method and extended Kalman filter equations have been used. Data fusion is achieved using a recursive object architecture which could lead to a real time expert system being built for command and control purposes.

In various universities in the U.K., research in the field of sensor fusion is in progress. Durrant-Whyte (1991) looks at sensor data fusion from two perspectives, one quantitative and the other qualitative. Quantitative data fusion methods are well defined and understood. For the purpose of developing design architectures, Kalman filters, geometric modelling, probabilistic estimation are all used. The areas of application of these methods are in distributed sensing and control. In qualitative sensor data fusion the methods, although well defined, still have some inherent problems. These methods are beneficial in the fields of expert systems, behavioral modelling and structural modelling.

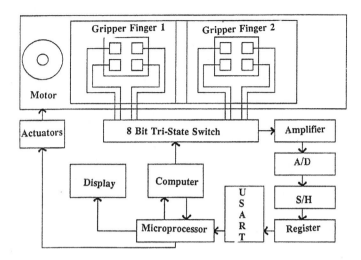

Fig. 2. Hardware Configuration at the Gripper Control Level.

The use of constraint modelling techniques for a servo-driven gripper controller is new, in the sense that it uses both geometric modelling and behavioral modelling methods, with the constraint modeller selecting the easiest strategy at its disposal. It is thus possible that the benefits and disadvantages of the two methods could be averaged, as suggested by Durrant-Whyte (1991).

The RASOR language is employed to achieve the objective of developing intelligent control, particulary when considering an uncertain environment. The intelligent controller will need to consider all the physical constraints used as well as taking into account all sensor signal information. The signals obtained from sensor bridges are identified individually or collectively for the purpose of detecting any error. The signals so obtained, once identified, are then associated and correlated. The advantages of other control systems can be utilized to achieve optimal control. For this purpose any subsidiary information from other resources should also be available from knowledge base or expert systems. Knowledge engineering and broad information about the general nature of the object to be handled will also play a part in selecting the most suitable handling strategy.

At present many other data fusion techniques are in use for example Rule Bases Systems, Fuzzy Sets, Kalman Filtering, Baysian Methods and Neuron Networks. We shall propose a constraint modelling solution by using the Simplex Method, as widely used within the RASOR programming solution of constraints (Medland, 1988; Leigh and colleagues, 1989; Mullineux, 1988).

In addition to developing an intelligent cell controller, we are also proposing a controller for a servo-motor driven gripper using RASOR software through constraint modelling. Sripada, Fisher and Morris (1987) have combined AI techniques and Fuzzy Logic for real-time servo-control. The artificial intelligence controllers take typically 2 to 4 seconds to complete a control cycle. Here a Knowledge Based (KB) system is used to control rules for input switching and for setting the rules for learning and/or adjusting switching parameters. The AI servo controllers use a smooth control action and give a controlled variable response that is as good as or better than that given by the widely used PID algorithm.

## KNOWLEDGE BASED CONTROL

The industrial automation achieved through the use of KB systems is usually presented to increase flexibility and to cope with the complexity of interactions in automatic operator expertise, automatic complex plans and the extended range of automatic tasks. But limitations which exist in conventional methods, invalid modelling assumptions, sub-optimal performance, lack of knowledge, restricts the expansion of such system. Although these techniques provide a unique and practical solution to manufacturing tasks, they have difficulties in implementing heuristic methods to obtain their goals. This is because a full and complete knowledge of the system is essential. Further problems occur when the operating environment is uncertain or ill defined, requiring complex modelling and control algorithms. Hence to provide satisfactory solutions to this problem, symbolic computation is adopted.

### RASOR

RASOR is an implicit or task oriented language and it should, therefore, be easy to implement the functions of a robot in an application oriented manner. By assigning constraint modelling methods to the gripper controller task it is hoped to nullify the inaccuracies associated with the uncertain environment through the use of sensor information. RASOR allows the user to build functions that involve algebraic expressions in named variables. A particular command within RASOR is "rule" and this is used to specify the constraint. RASOR actually searches for feasible solutions to a number of constraints presented to it. This differs from a normal expert system. Here the computer and user both learn together.

In the case of an expert system the rules are previously known. In design tasks, where previous rules are generally not known, constraints on the range of allowable design solutions are compared with initial and emerging rules. The redesign of a guillotine mechanism (Leigh and colleagues, 1989) has shown the power of constraint modelling techniques for solving problems in real industrial applications. The major benefit of the constraint modelling method are that rules can be built up gradually by the user as the understanding of the problem increases. In a robotic assembly cell it is easier to establish cell boundary conditions and operating constraints rather than establish rules for every possible eventuality. It was decided to

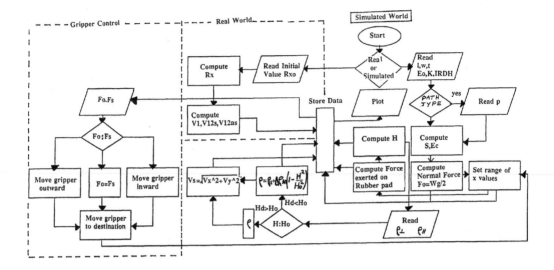

Fig. 3. Flow Diagram for Gripper Control.

use RASOR as the methodology for creating a handling/cell controller because of the following advantages.

1. Easy language to learn and modify.
2. Structural type .
3. Decision ability to control graphics.
4. Make decisions on the basis of solutions of equations.

### INTELLIGENT CONTROL

The development of RASOR for the dynamic control of geometric entities has, through the menu facility, also made it a suitable selection for data fusion or blackboard concepts. The solution of constraints using the Simplex Method is a key component within the RASOR programming language.

In a servo-gripper handling sequence the constraint of Force versus Voltage developed across the bridge of sensors is computed. Since the MR sensor produce a change in resistance when the object is gripped, the force interpretation of the signal is applied to the adaptive controller. The torque developed by the servo-motor must be optimised to prevent object damage or object slippage. Next, constraints of object position and orientation and gripper finger movements are introduced. Then volume movement of the gripper in three dimensions are utilized to satisfy the requirements of the pick-and-place operation. In this way more and more constraints are developed and controlled at the higher levels of the system control hierarchy as recommended by the National Bureau of Standards (Washington) (Haynes and colleagues, 1984).

The flow diagram for the gripper controller using the RASOR language is shown in Fig. 3. We have seen that the property of the MR element plays an important role in controlling the movements of the gripper. The change in force, magnetic field, resistance of the MR element, and voltage can either be simulated or measured from the real world model. The results obtained from the two models are compared to provide information for an error recovery routine needed to move the servo-motor in the correct direction.

Figure 3 shows that the gripper controller consists of three parts; Simulated Model (SM), Real World Model (RWM) and Gripper Control (GC). In SM the characteristics of the rubber pad ($E_o$, K, IRHD, l, w, t) are read interactively or from stored information files. By setting the range of possible

displacements x, the normal force ($F_o$) exerted by an object on the rubber pad is computed. The next step is to compute the change in magnetic field (H). To calculate the electrical resisisivity (ρ) of the MR element, it is required to read perpendicular resistivity($\rho_{por}$), and parallel resistivity ($\rho_{par}$) interactively or from stored information. The difference in these two values will give the maximum change in resistivity ($\Delta\rho_{max}$) expected. If the demagnetising field $H_d$ and anisotropy field $H_k$ are known, the internal field $H_o$ can be calculated. By comparing Ho and H, the resistivity can be calculated by using the following formula:-

$$\rho = \rho_{par} + \Delta\rho_{max} (1-H^2/H_o^2) \qquad (1)$$

Hence the change in resistance and voltage could be calculated and sent to a stored information file for further processing and control.

The error obtained from RTM and SM is balanced through constraint modelling using RASOR programming. Finally, the error detected is sent to the GC to start a recovery routine. It should be noted that the force balancing conditions are obtained from the force exerted by an object $F_o$ and the force sensed by the sensor $F_s$. A backward-error-recovery technique may be applied which is based on a formal description of the recovery process within a total system of interactive (communicating) processes (Zielinski, 1987). When these conditions are satisfied the gripper is then able to move to the destination required to complete the pick-and-place cycle.

### RESEARCH PROPOSAL CONCLUSIONS

The project is concerned with aspects of a robot cell controlling program depending on available computer hardware and operating systems. Expert systems where solutions are task oriented are a favoured approach to this problem. For example a research group at the University of Karlsruhe (Rembold, 1987) have developed a KB expert system. The essential feature of this system is the use of implicit programming which is textual and can be developed off-line, without directly interfering with the robot. The KB of such a system constitutes rules and data. The control of the KB is carried out by an inference engine, on rules (conditions and consequences) similar to the cause and effect basis. The disadvantage of such a system is that it is limited in dealing with "uncertainties" in a real situation. To cope with this sensors can be used to handle

320

uncertainties in assembly tasks. This suggests that programming in terms of task achieving behavioral modules would be beneficial.

Research work at Brunel University has so far shown that constraint modelling techniques using the RASOR programming method meets both the demands of task orientation and NBS standards.

## REFERENCES

Adl, P., Memon, Z.A., Mapps, D.J., Rakowski, R.T. (1990). Serpentine Magnetoresistive Elements for Tactile Sensors Applications. *IEEE Transactions on Magnetics*, Vol:26 Iss 5, pp 2047-2049.

Appleton, E. (1990). Robotic Assembly of Aerospace Fabrications- A case study. *ARA 3rd National Conference on Robotics*, Melbourne, Australia.

Camarinha-Matos, L.M., Steiger-Garcao, A. (1988). An Integrated Robot Cell Programming System. *Int. Symp. and Exposition on Robots*, Nov. Sydney, Australia.

Dooner, M., and De Silva. (1990). Conceptual Modelling to evaluate the Flexibility Characteristics of Manufacturing cell designs", *MATADOR 90*,Manchester, England.

Durrant-Whyte, D. (1991). Intelligent Control. *Computing and Control Division Colloquium*, IEE, 4 February, England, U.K.

Edward, L.W., and Llinas, J. (1990). *Multisensor Data Fusion*. Artech House Boston-London.

Elkins, K.L., Maimon, O.Z. (1987). *Information Flow Controller and its Robotic Sub-System*, Elsevier Science Publishers B.V.(Norh Holland).

Haynes, L.S, Barbera, A.J., Albus, J.S., Fitzgerald M.L,Cain,H.G. (1984). An Application Example of NBS Robot Control System. *Robotics & Computer-Integrated Manufacturing*, Washington, USA.

Hwang, K., Briggs, F.A. (1985). *Computer Architecture and Parallel Processing*, McGraw-Hill International Edition.

Leigh, R.D., Medland, A.J., Mullineux, G., Potts, I.R.B. (1989). Model Spaces and their Use in Mechanism Simulation. *Proc. I.Mech.E. part B: Journal of Engineering Manufacture*, 203, pp 167-174.

Medland, A.J. (1988). Approaching CAD Through Constraint Modelling. *CAM-I Design Automation Workshop on Current Trends in Product Modelling and Design and Automation*, Dorset Institute, 7 March.

Mullineux, G. (1988). The Introduction of Constraints into a Graphics System. *Engineering with Computers 3*, 201-205.

Rembold, U. (1987). Progamming of Industrial Robots: To-day and in the Future. *Languages for Sensor-Based Control in Robotics*, Springer-Veslag, Berlin.

Sripada, N.R., Fisher, D.G., Morris, A.J. (1987). AI Applications for Process Regulation and Servo Control. *IEE Proceedings Part D*. vol. 134, No. 4.

Zielinski, K. (1987). *Error Recovery Point Management for Dynamically Identified Recoverable Atomic Actions*. Proc. IEE. Vol. 136, Pt. D, No. 1.

# A TESTBED FOR RESEARCH ON MULTISENSOR
# OBJECT RECOGNITION IN ROBOTICS

### J. Neira, L. Montano and J.D. Tardos

*Departamento de Ingenieria Electrica e Informatica, Centro Politecnico Superior de Ingenieros,*
*Universidad de Zaragoza, Maria de Luna 3, E-50015 Zaragoza, Spain*

**Abstract.** In this paper we propose an object recognition system to be used as a research
and development tool to implement and evaluate multisensor perception and recognition
strategies. The system is constituted by a blackboard in which solution state information
is maintained, a database of models of the objects that the system can recognize, a set
of perception and recognition agents that update the blackboard, and a coordinator that
controls the execution of agents to achieve the goals of the system. Our approach to recognition, the way the system can be used to implement and evaluate this approach, and some
fundamental issues to investigate are discussed.

**Keywords.** Object recognition; blackboard architecture; knowledge engineering; robots;
sensors.

## 1 INTRODUCTION

Identifying and locating objects in the workspace of
an industrial robot is a very complex problem in two
aspects: first, unstructured environments confronts us
with a very large solution space, even if only a limited
set of known objects is considered; second, information extracted from the environment is generally low
level, subject to uncertainty in measurement and ambiguity in interpretation.

For these reasons, researchers are directing their efforts towards the application of Artificial Intelligence
problem solving methods that combine explicit knowledge representation schemes with efficient knowledge
application strategies and appropriate uncertainty
management.

One of these problem solving methods is the *blackboard model* [Nii 1986], which proposes an organization of data by the decomposition of the solution space
in a hierarchy which gives a description of the space
at different levels of representation, aggregation or abstraction. The data structure that supports this hierarchy is called *the blackboard*. Specific domain and
generic strategic knowledge are codified as independent, self-contained and anonymous *knowledge sources*
or *agents*. The blackboard acts as a working database
used to maintain the solution state. Each knowledge
source analyzes the blackboard to detect situations in
which it can contribute information to improve the
solution state. A *control mechanism* decides which
area of the blackboard is the most promising to focus
on, and which knowledge source should be allowed to
execute next.

The blackboard model has been used to develop a
wide variety of systems in many areas of expertise
(see [Engelmore 1988] for a review). Its fundamental
advantages are, in the first place, its *modularity* and
*extensibility*. The addition or removal of knowledge
sources does not affect other elements of the system.
In the second place, the model prompts the development of *uniform* and *knowledge source independent*
mechanisms to represent and manipulate information.
A third advantage is the *flexibility of control*, specifically the capacity to reason in an *opportunistic* way
(applying the most promising methods to the best
data at each moment). Finally, since the knowledge
sources are limited to data-directed interactions, there
is a great opportunity for *parallelism*.

All these advantages make the blackboard model very
adequate for the analysis of different combinations
of sensors and the evaluation of different perception
and recognition strategies, in the development of an
efficient object recognition system. Some examples
of robotic perception and recognition systems based
on this model are [Paul 1986], [Harmon 1986] and
[Berge-Cherfaoui 1991].

This paper is organized as follows: given its importance in recognition, the next section is devoted to
object modeling. Section 3 contains a description of
the system structure: the blackboard, the perception
and recognition agents, and the coordinator. The fundamental approach to object recognition and the way
in which the system implements it are discussed in
this section. Finally, in section 4 we outline some fundamental issues to be investigated using the system.

## 2 OBJECT MODELING FOR RECOGNITION

Sensors can give information at parametric and geometric levels. In simple situations, where objects are not overlapping and in one of their known stable positions, parametric information can lead to successful recognition. In more complex situations, the information provided by parametric sensors can be used to determine regions of interest in the scene. Geometric recognition is based on the search for consistency between features observed by sensors and geometric models of objects. To support the different stages of the recognition process, rich, robust, and sensor independent object models are required.

We use a hybrid model, that describes the parametric and geometric properties of an object. In the parametric model, an object is characterized by a set of descriptors specified for each of its stable positions. We use 2D global shape descriptors: area, perimeter, maximum and minimum radii, and moments of image. In general, these descriptors are sensor and view dependent, and very sensitive to occlusion. However, since parametric recognition can be very fast, this model is used in the early stages of recognition.

The geometric model of an object is a three level description that allows the system to reason at different levels of analysis:

**Object level.** It includes a general description of shape, object symmetries, and a list of other objects that have similar shape. Symmetries are used to rule out equivalent hypotheses, and similarities can direct the system to possible alternative interpretations of a set of observations. Each model includes a list of the object's most relevant features, as well as of its surface-edge-vertex representation.

**Compound feature level.** Relevant features of an object are specified at this level. At present, compound features include coplanar edges, corners, dihedrals, and circular and polyhedral pegs and holes. Recognition focuses on compound feature to speed up the process.

**Simple feature level.** It is a surface-edge-vertex representation of an object extracted from a CAD system. We consider straight and circular edges, and planar, conical, cylindrical and spherical surfaces. Geometric properties of features, such as length and area, geometric relations between features, such as angles and distances, as well as topological relations (adjacency) and directions of visibility of each feature are specified at this level.

Each feature is scored with a *relevance coefficient*, related to its relative scarceness in the models, its size, and the amount of location information that it contributes (related to the degrees of freedom of object location that the feature determines). Relevance is used for feature selection to guide the recognition process.

The model database is encapsulated with a *kernel* of access functions where the system can obtain information such as:

- Which objects have edges of length $l$?

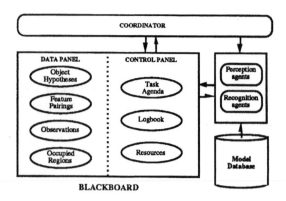

Fig. 1: System structure

- Which objects have two edges at distance $d$ and angle $\alpha$?

- Which edges of object $m$ are adjacent to edge $e$?

- To recognize object $m$, which features should be searched for first?

- If object $m$ is at location $t$, where should feature $f$ be located?

This kernel is accessible to all agents and constitutes the interface between the model database and the system.

## 3 SYSTEM STRUCTURE

The system has four fundamental components (Fig. 1): *the blackboard*, which stores sensorial data and solution state information, *the model database*, a set of *perception and recognition agents* that perform their tasks consulting the model database and updating the blackboard, and a *coordinator*, that controls the activity of the system.

In the sequel, each of the fundamental components of the system is explained with greater detail.

### 3.1 The blackboard

The blackboard is divided in two panels: the *data panel* and the *control panel* (section 3.3). The data panel contains scene information collected by perception agents, and hypotheses about the objects that may have originated this information, generated and updated by recognition agents. This information is classified at four levels:

**Occupied regions level.** This level gives a gross description of regions of the scene where the presence of objects has been detected. These regions are used to concentrate perception activities on them.

**Observations level.** Observations are data collected by perception agents, and can be of one of the following types:

- *Simple or compound features*, characterized by their location and geometrical properties, such as length, area, radius, etc.

| Feature Pairing | |
|---|---|
| *slot* | *valueclass* |
| id: | (identifier) |
| created: | (timestamp) |
| modified: | (timestamp) |
| model.feature: | (feature) |
| observation: | (observation) |
| estimated.location: | (uncertain.location) |
| credibility: | (float) |
| state: | (one.of predicted verified failed) |

Fig. 2: Representation of a feature pairing

| Object-Location Hypothesis | |
|---|---|
| *slot* | *valueclass* |
| id: | (identifier) |
| created: | (timestamp) |
| modified: | (timestamp) |
| model.object: | (object) |
| estimated.location: | (uncertain.location) |
| support: | (list.of pairings) |
| potential.support: | (list.of features) |
| failed.support: | (list.of pairings) |
| credibility: | (float) |
| reachable.credibility: | (float) |
| state: | (one.of accepted rejected pending) |

Fig. 3: Representation of an object-location hypothesis

- *Geometric subfeatures*, that give partial information about the localization of a geometric feature, such as a point belonging to a plane, the normal of a plane, or a point lying on an edge.

- *Parametric descriptors*, that are measurements of global properties of objects, such as those mentioned in section 2.

Observations are tagged with a timestamp, viewpoint of acquisition, and a reliability coefficient that states the confidence that the agent has in the observation not being spurious. A general method for the representation of the location of any geometric observation and its uncertainty is used: the *Symmetries and Perturbation model* (SPmodel) (a formal presentation of this model can be found in [Tardós 1992]). The SPmodel is a probabilistic model whose main advantage is its generality: it is valid for any object, geometric feature or sensorial observation.

**Feature Pairing level.** This level represents hypothesized *pairings* that relate observations to model features. Each pairing (Fig. 2) is given a *credibility coefficient* that scores its quality, and is related to the relevance of the model feature, the reliability of the observation, and its location uncertainty.

**Object Hypothesis level.** This level contains *object-location hypotheses*, which state the identity and localization of an object present in the scene. Each hypothesis (Fig. 3) is related to a set of feature pairings that constitute its *support*. The total credibility of this support becomes the credibility of the hypothesis. Model features that have not been paired, but should be visible for the hypothesized location of the object, are *potential support* of a hypothesis. In the same way, model features that have been predicted but could not be observed in the scene constitute its *failed support*. Their potential contribution to the credibility of the hypothesis allows to estimate the *reachable credibility* that the hypothesis could attain. This information is used to decide whether further verification for a hypothesis should be done.

As the model database, the blackboard can be accessed using a *kernel* of functions that allow to create and update blackboard elements, and pose queries such as:

- Which observations are located near observation $o$?

- Which pairs of observed circles are nearly *concentric* and *coplanar*?

- Which are the alternative feature pairings of observation $o$?

- Do hypotheses $h_1$ and $h_2$ state alternative locations for the same object?

- Are hypotheses $h_1$ and $h_2$ compatible?

The information contained in this panel of the blackboard is used and updated by the agents when performing perception and recognition tasks, as explained in the next section.

### 3.2 The agents

An agent is a specialist in carrying out some task. It contains all knowledge related to the resources, algorithms, and strategies that are involved in performing that task in a specified way, and it competes with other agents in performing that task. Agents have two main functions:

- *Task evaluation.* Given the characteristics of a proposed task, the agent evaluates the feasibility of undertaking it, and responds with an affirmative or negative answer, along with the list of resources that it would require, and an estimation of its expected response time, accuracy and reliability. The coordinator uses this information when considering competing agents to select the most adequate to perform a task.

- *Task execution.* When an agent is selected and allowed to execute, it performs the necessary operations, which can include interacting with the model database and blackboard kernels, giving instructions to sensing devices for sensor positioning and acquisition, processing sensorial information, and launching other tasks.

Agents are divided into two groups: *perception agents* and *recognition agents*. In the next two sections the characteristics, purpose and some examples of these groups of agents are given.

### 3.2.1 Perception agents

A perception agent implements a data acquisition strategy. When evaluating a proposed task, it decides

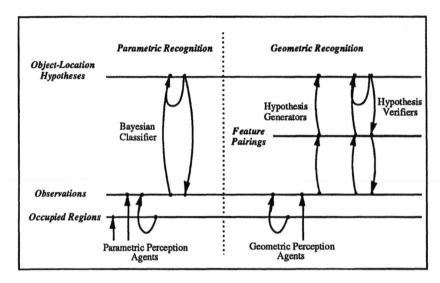

Fig. 4: Blackboard data panel and agents

its feasibility depending on its scope of observation relative to the scene region to be scanned, and estimates its response time, accuracy and reliability, based on its knowledge about its performance and on whether there is any a priori information that allows the use of goal-directed perception strategies. Perception agents are particular in the sense that they use and can monopolize *system resources*. These are:

- *Sensing devices*. At present, sensing devices include a fixed camera, a force/torque sensor mounted on the robot wrist, an ultrasound proximity sensor mounted on the robot hand, and several infrared and laser proximity sensors mounted on the fingers of the gripper. A camera mounted on the robot hand will be available soon.

- *The robot*. Since there are several sensors mounted on the robot, it is considered a resource that one of the sensors monopolizes when active.

- *The workspace*. The workspace of the robot is a resource in the sense that some sensors (like global 2D vision) need it to be clear when operating, while the ones that use the robot invade it when working. Thus, several agents that need the workspace clear but do not invade it can be working in parallel, but only one workspace-invading agent can work at a time.

Perception agents perform one of the following tasks (Fig. 4): *identification of occupied regions*, in which they must obtain parametric descriptors of objects to be be used to attempt parametric recognition, or at least to establish regions of interest; *feature extraction*, in which, without any a priori information, the agents must scan a region in search for some type of geometric features; *feature location uncertainty reduction*, in which the location of a feature is refined to reduce the uncertainty in the location of the hypothesized object; and *feature verification*, in which an agent must locate a specific feature at a hypothesized location.

In our system, several agents can compete to perform a task. Consider for example an edge extraction task. The set of perception agents competing to extract edges from the scene can be:

- STEREO.VISION.EDGE.EXTRACTOR: Stereo vision needs the workspace free but does not monopolize it. It has global scope, but cannot reach occluded areas. Acquisition can be fast, but the correspondence problem gives low reliability to results.

- MOBILE.VISION.EDGE.EXTRACTOR: Since the mobile camera is mounted on the robot hand, it monopolizes the workspace. Since it is mobile, it can have access to occluded areas, and it can be positioned to obtain several images which result in highly reliable observations, at high cost.

- RANGE.FINDER.EDGE.EXTRACTOR: As stereo vision, it does not need the robot arm. It is less accurate than stereo vision, but observations are more reliable.

- INFRARED.PROXIMITY.EDGE.EXTRACTOR: Infrared proximity active sensing can be highly reliable in local extraction, but without any a priori information it has very limited use.

The combination of several perception agents allows to implement multisensor perception strategies. An example is the use of vision and proximity for edge detection. Object edges can be located first using the edge detector agent of a 2D vision system. A 2D vision edge extraction agent generates a first observation containing information that can be used by a proximity sensor edge verifier to control a non-contact compliant motion using proximity sensors [Sagüés 1991] to obtain two 3D points lying on the edge. These three observations can be integrated into one consensus location estimation of the edge.

326

### 3.2.2 Recognition agents

A *recognition agent* performs a task related to parametric or geometric recognition (Fig. 4). Parametric information is used by a *Bayesian classifier*, in which each stable position of an object is modeled as a different class. As it has been said before, given the higher complexity of geometric recognition techniques, parametric recognizers can be more efficient in simple situations. If the recognition credibility obtained is not enough, the generated hypotheses are used to guide the acquisition of geometric information, allowing the use of goal-oriented perception strategies.

Geometric recognition is based on a generation-verification scheme in which *hypothesis generators* heuristically select a set of observations to generate initial object hypotheses. To verify each generated hypothesis, *hypothesis verifiers* implement different strategies, as for example:

- MODEL.DRIVEN.VERIFIER: If there is no good sensor information, and the hypothesized object model has easily observable and highly relevant features, a *model driven* procedure is used, in which a new feature of the object is selected and its location in the scene predicted. The feature is then searched for in the data or tried to be obtained by a sensor. Such a scheme is used in [Faugeras 1986] and [Lozano-Pérez 1987].

- DATA.DRIVEN.VERIFIER: When the set of available observations is small, and observations are close to the hypothesized object location, pairing the available observations with visible model features avoids the prediction of features that may not be visible or are costly to observe.

In both cases, when a matching observation is found, it is used to improve the estimation of object location and to increase the credibility of the hypothesis.

The generation of a large number of false hypotheses is one of the most complex problems that the recognition process faces. In order to reduce it, some strategies that are subject of actual research are discussed in section 4.

### 3.3 The control panel and the coordinator

The fundamental entity that the blackboard control panel contains is the *task agenda*. A task is a goal proposed to the system, that an agent or set of agents is capable of attaining. In that sense, a task is a statement of *what* to do, while agents know *how* to do it (there is a *one-to-many* relation between a tasks and the agents that compete to perform it). The four fundamental tasks that can be proposed to the system are:

- *Describe scene:* Identify and locate all the objects present in the workspace of the robot.

- *Find object:* Supposing that object $m$ is present in the scene, obtain its accurate location.

- *Verify object:* Supposing that object $m$ is at location $l$, decide whether the hypothesis is correct.

- *Verify feature:* Supposing that object $m$ is at location $l$, verify that feature $f$ is at its corresponding location.

The execution of an agent can in turn generate new tasks in the agenda. For example, the model driven hypohesis verifier can generate a feature extraction task if the model feature it is seeking is not found among the available observations. When a task is generated, *task parameters* and *task requirements* must be specified. Task parameters are elements in the blackboard related to the task, such as the hypothesis to verify, the observation to refine, etc. Task requirements are maximum response time, desired accuracy, and desired reliability of the results.

Processing the agenda is the main responsibility of *the coordinator*. It acts as an agent arbiter to opportunistically select the most appropriate agent for each task. The coordinator also records the activity of the system in the *logbook* and maintains *resource state* information in the blackboard control panel. The general operation of the system follows these steps:

- At startup, each available agent declares the type of task that it can potentially carry out.

- The user defines the goal of the system by creating a task in the agenda. The existence of a task in the agenda initiates a cycle in which:

  1. The coordinator consults with each potential agent the feasibility of performing the specified task. Agents answer with an affirmative or negative response, along with an estimation of response time, accuracy and reliability.

  2. The most suitable agent that satisfies task requirements is selected and allowed to execute. If no agent can accomplish the task, the coordinator can try to relax task requirements.

  3. The execution of the agent may create new tasks in the agenda. The coordinator chooses one, and goes back to the first step.

  4. If an agent fails to accomplish a task, the coordinator must try to execute the next best agent for that task. If there are no more available agents, the task is considered failed.

The process continues until there are no pending tasks in the agenda, and there are no alternatives for failed tasks.

### 3.4 System implementation

The system is being implemented as a distributed system composed of a PUMA 560 robot, a sensor controller subsystem where the perception agents are executed, and a Sun workstation, where the model database, the blackboard and the perception agents reside (KEE, from Intellicorp, has been chosen as the development environment). Sensing devices are those enumerated in section 3.2.1.

### 4 DISCUSSION

Most object recognition systems obtain scene information using one or two types of sensors and limit the matching process to one type of feature. The system proposed in this paper allows the implementation and evaluation of different recognition methods based on parametric or geometric information given by any combination of sensors using different types of perception strategies.

Since parametric recognition has a well established theory, we concentrate our research on geometric recognition. Our goal is to reduce the complexity of the process with the use of two types of strategies: the application of geometric constraints and the use of feature selection heuristics.

**Geometric constraints.** The number of potential matches of the initial set of features can be greatly reduced using simple geometrical constraints based on properties that are invariant with object location such as dimensions of features and angles and distances between them [Grimson 1990]. However, this type of restrictions are not tight: they may be satisfied by certain incoherent hypotheses. To reduce the influence of this problem, an alternative is to generate object-location hypotheses from the smallest set of observations that allows to estimate the object location. This strategy allows the application of location-dependent geometric constraints, such as the relative location of each feature with respect to the object, its spatial extension, or its visibility for the hypothesized object location, that are stronger than location-independent constraints.

The early calculation of object location also allows to predict the location of new features that can be searched using goal-directed perception strategies, that can reduce the cost of acquisition. For example, the extraction of an edge with stereo vision may be time-consuming due to the correspondence-search problem, while the verification of a predicted edge can be very fast. We intend to establish the advantages of this approach.

**Feature selection heuristics.** The selection of the initial set of features for hypothesis generation may be based on their relevance, that is related to the scarceness of the type of feature in the object models, their size and the amount of location information that they contribute. To obtain good accuracy in object location, features with low location uncertainty are preferable, and singular configurations as near-parallel edges should be avoided. Also, it is interesting to choose a set of proximal observations, in order to increase the probability that they come from the same object. The number of candidate hypotheses generated and then the complexity of the recognition process can be greatly reduced by focusing on compound features because possible matchings are reduced in orders of magnitude, and the probability that the set of features belong to a single object is higher [Tardós 1991].

During hypothesis verification, various aspects, in addition to relevance, should be taken into account for the selection of features, such as feature visibility, estimated cost and reliability of observation, and discriminating power between competing hypotheses.

Representation methods for these types of information, and the evaluation of these strategies constitute future work in the development of our recognition system.

## ACKNOWLEDGMENTS

This work was partially supported by CONAI-DGA, project IT-5/90 and CICYT, project ROB91-0949.

## REFERENCES

[Berge-Cherfaoui 1991] V. Berge-Cherfaoui and B. Vachon. A multi-agent approach of the multi-sensor fusion. *Fifth Int. Conf. on Advanced Robotics (ICAR'91)*, pages 1264–1269, Pisa, Italy, June 1991.

[Engelmore 1988] R. Engelmore and T. Morgan, editors. *Blackboard Systems*. Addison-Wesley Publishers, 1988.

[Faugeras 1986] O. D. Faugeras and M. Hebert. The representation recognition, and locating of 3D objects. *Int. J. Robotics Research*, 5(3):27–52, 1986.

[Grimson 1990] W.E.L. Grimson. *Object Recognition by Computer: The Role of Geometric Constraints*. The MIT Press, Massachusetts, 1990.

[Harmon 1986] S.Y. Harmon, G.L. Bianchini, and B.E. Pinz. Sensor data fusion through a distributed blackboard. *IEEE Int. Conf. on Robotics and Automation*, pages 1449–1454, San Francisco, California, 1986.

[Lozano-Pérez 1987] T. Lozano-Pérez, J.L. Jones, E. Mazer, P.A. O'Donnel, and W.E.L. Grimson. Handey: A robot system that recognizes, plans and manipulates. *IEEE Int. Conf. on Robotics and Automation*, pages 843–849, Raleigh, North Carolina, 1987.

[Nii 1986] H. P. Nii. Blackboard systems: The blackboard model of problem solving and the evolution of blackboard architectures. *A.I. Magazine*, 7(2):38–53, 1986.

[Paul 1986] R.P. Paul, H.F. Durrant-Whyte, and M. Mintz. A robust, distributed sensor and actuation robot control system. *Robotics Research: The Third International Symposium*. MIT Press, 1986.

[Sagüés 1991] C. Sagüés and L. Montano. Active sensing strategies with non-contact compliant motions for constraint-based recognition. *IFAC/IFIP/IMACS-Symposium on Robot Control SYROCO'91*, pages 295–300, Vienna, Austria, September 1991.

[Tardós 1991] J.D. Tardós. *Integración Multisensorial para Reconocimiento y Localización de Objetos en Robótica*. PhD thesis, Dpto. de Ingeniería Eléctrica e Informática, University of Zaragoza, Spain, Febrero 1991.

[Tardós 1992] J.D. Tardós. Representing partial and uncertain sensorial information using the theory of symmetries. *IEEE Int. Conf. on Robotics and Automation*, Nice, France, May 1992. (To appear).

# DISTRIBUTED REAL-TIME SYSTEMS FOR
# MANUFACTURING AUTOMATION

## H. Rzehak

*Universitat der Bundeswehr Munchen, Werner-Heisenberg-Weg 39, D-8014 Neubiberg, Germany*

Abstract: In manufacturing automation systems of distributed application processes interacting across a local area communication network are coming up with increasing importance. Standards (e.g. MAP and MMS) have been established for this kind of systems. For applications with real time constraints these standards are not fully satisfying. The paper gives some answers to key questions concerning achievable message delays and precautions to be met for getting a predictable behaviour of the whole system. Concerning aspects of the Manufacturing Message Specification (MMS) are studied in more detail.

Keywords: Distributed Real-Time Systems; Manufacturing Automation; Concurrency Control; Communication Architecture; Performance of Communication Systems

## 1. INTRODUCTION

In manufacturing automation we are more and more faced with systems of distributed application processes interacting across a local area communication network. Standards have been established for communications and interfaces (e.g. MAP [8]). With the Manufacturing Message Specification (MMS, ISO 9506) we have a hopefully suitable specification of communication services for application programmers.

Using these communication standards for applications with real-time constraints one may be faced with new problems originating from unknown facts or a pure understanding what's going on in a distributed system. Some key questions are

- What is the message delay from application process to applications process, or at least can we give some rough estimates?

- What can we do to improving the real-time behaviour, e.g. using a reduced communication architecture (Mini-MAP, PROWAY, PROFIBUS)?

- Due to varying message delays we introduce a kind of nondeterminism into the system which should be predictable. How can we overcome this?

## 2. REAL-TIME PROPERTIES OF COMMUNICATION SYSTEMS

Except some papers of the author and his group [2], 5], [13], [14], [16] only a few work based on measurements has been published on real-time properties of ISO communication networks. The work of the author includes an assessment of the reduced (three layer) communication architecture as proposed e.g. in PROWAY [12]. This gives realistic measures on the communication overhead and achievable message delays.

To show the order of magnitude, we present some results of measurements on a MAP-network (Carrier Band Version with 5 MHz transmission speed). Table 1 give results for the message delay ($T_D$) and the response time ($T_A$) in a server-client-model for a user on top of the transport layer (TP), the network layer (NET), the logical link layer (LLC1), the medium access layer (MAC) for layer-to-layer communication in a network without further load. In all these measurements the service time on the server side is set to be zero.

TABLE 1 Message delay and response time in a network without further load

| Layer | $T_{D, Layer}$ | $T_{A, Layer}$ |
|---|---|---|
| Medium Access (MAC) | 3 ms | 6 ms |
| Logical Link Control (LLC1) | 4,5 ms | 9 ms |
| Network (NET) | 6 ms | 12 ms |
| Transport (TP) | 9 ms | 22 ms |
| Application Entity (estimated; e.g. read simple variable) | 20 ms | 48 ms |

With load on the network, the measured time increases drastically. This is shown in Fig.1. In addition the response time of the transport system in a reduced architecture (LLC3) is shown. The difference between TP and LLC3 demonstrates the benefits of the reduced architecture.

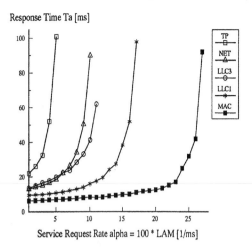

Response Time Ta [ms]

Service Request Rate alpha = 100 * LAM [1/ms]

Fig. 1    Response Time for increasing Load

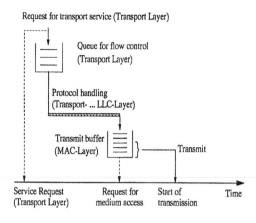

Fig. 2    Sources of Time Delays within a Transmitting
          Controller

## 3. IMPROVING REAL-TIME BEHAVIOUR

The time spent for communication may appear unexpected
long to some people and in some cases it will be unacceptable too. Looking for improving real time performance we
should keep in mind that the transmission time on the physical layer is only 1.7 ms for the results shown in Fig. 1. Therefore we cannot achieve better results by improving the
transmission time only. We have to improve the communication controller. For a state-of-the-art controller (commonly
used VLSI-Chips for the medium access protocol; MC68020
with 12,5 MHz for protocol handling of the other layers) as
used in our reference installation it is not possible to speed up
the network by some degrees of magnitude using today's
technology.

Figure 2 shows the major sources of time delays within a
transmitting communication controller. The processing time
depends on the processor and the management of the protocol tasks mostly performed by some kind of real-time operating system, whereas the queuing delays depend additionally
on the complexity and some special attributes of the protocol
itself.

For reducing the time spent on communication first of all we
have to improve the controller (see [15]). This can be
achieved by using

- faster processors;
- more than one processor;
- special VLSI-chips.

To reduce the complexity of the communication protocol
(and the associated overhead) for real-time applications, we
propose a reduced architecture hooking the application layer
on top of the LLC layer. This is in accordance with the
PROWAY-LAN [12] and the Mini-MAP concept (see [5],
[6], [8]). For reliable communication, we need acknowledged
LLC services (LLC-Type 3) instead of the not acknowledged
connection-less services (LLC-Type 1), and the missing network layer restricts communication to a single segment,
because routing is not possible and the receiving node must
be addressed by its LLC service access point. The transport
protocol of layer 4 is dispensable because with the LLC-Type
3 services we have a realiable communication. With respect
to the missing layers 5 and 6, a predefined context (e.g. transfer syntax) has to be used (see [5], [6]).

## 4. NONDETERMINISM IN DISTRIBUTED APPLICATIONS

### 4.1 The Role of Message Delays

Time constraints in a distributed system include the demand
on limited message delays for communication. Bounds for
message delays have to be guaranteed and shall be sufficiently small for a certain application. In other words, we
need a deterministic behaviour of the communication system.
For the token bus the maximum delay caused by the MAC-
layer is known [13] and depend on some protocol parameters.
In addition the queuing delays on the queue for transport
service requests (see Fig. 2) have to be restricted. Generally
message delays are arbitrary (within possibly known upper
bounds) and must not be neglected. The state of a process
sensed in a remote node may have changed before we receive
the state information. In addition the time need may vary
considerably, and therefore the time order of events rising in
different nodes does not always appear to be the same from
node to node. These facts introduce some kind of nondeterminism in a distributed application and it is the job of the
programmer to eliminate this influence.

We have seen that it is not possible to get a timely view over
the global state of a distributed application. The local clock
of an arbitrary node is a special case of a local state variable.
With respect to varying message delays, we cannot determine
the current local clock on a remote node. Furthermore, we
cannot suppose that all local clocks coincide. Synchronizing
local clocks is a difficult and expensive task to perform if the
tolerance is in the range of 1 ms or even less. Solutions with
a global clock (available by each node) are possible on the
restricted area of a job shop floor, but need additional measures and precautions with respect to a breakdown of this central clock. A special case of a global clock is the use of a
radio timing signal. To receiving such a signal on a job shop
floor may be a hard task or even impossible. Generally, for
clock resolutions of one second or more it may work quite
well. But for resolutions of one millisecond or less the uncertainties in signal delays of the radio wave and within the
receiver are of the same magnitude or more and therefore
clocks will run asynchronously. Generally, we are forced to
solve problems with local clocks, and in special cases the
application programmer is faced with the above-mentioned
relativistic event ordering.

### 4.2 Conflicts and Concurrency Control

Application entities may be mutual dependent and therefore need a coordination. We have to distinguish two reasons:

- The entities are causal dependent and need an execution in a well defined sequence.

- The entities should be executed as concurrent activities, but this results in conflicts.

For causal dependent entities we need services for starting, suspending and continuing of remote entities, which may be called explicitly or event driven. Appropriate services provided by the Manufacturing Message Specification (MMS, see [8], [16]) are reviewed in the next section. Using these services we have to keep in mind the relativistic time order of events mentioned above. E.g. if we start two activities in different nodes in a given order the execution may start really in the reverse order.

Conflicts may arise from the dependencies among the manufacturing processes running concurrently under computer control or from the common use of resources (global data, files, devices, etc.) within the computer network. Unresolved conflicts may lead to incorrect results or disturbances and faults during the manufacturing process. An example concerns the track-keeping vehicles as shown in Fig. 3.

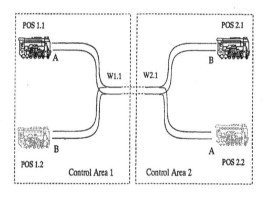

Fig. 3    An example for the need of concurrency control

An application process that moves engine A from position POS 1.1 to POS 2.2 has to perform the following sequence of actions:

Scan if target position is empty;
If not, wait until it is empty;
Make reservations for target position;
Scan if track to target position is empty;
If not, wait until it is empty;
Make reservations for the track to the target position;
Set switches;
Start drive;
Control velocity;
Stop drive if engine is in final position.

If simultaneously another application process on a different node has to move engine B from position POS 2.1 to POS 1.2 using the same sequence of actions (changing the variables "engine" and "position") we have a good chance that both asking processes will receive the state "free", because none of the processes has finished its reservation and both processes are trying to make reservations for the same resources. This conflict can be resolved if status sensing and reservation making are both performed in an indivisible (atomic) action.

These conflicts arise if resources inside or outside the computer must be used exclusively by a process during work on it. Such a situation may be unavoidable for physical reasons or because the status of the resource may be changed during further usage. Furthermore, it may be necessary to reset the resource to an earlier status if concurrent processes cannot deliver a required partial result. An example is to put back a workpiece that cannot be tooled further.

## 5. MMS AND DISTRIBUTED APPLICATIONS

The Manufacturing Message Specification (MMS, ISO 9506) is a specification of standardized communication services for implementing distributed applications. This section gives a survey of MMS describing basic principles and kinds of services. As an example, we describe the means for concurrency control offered by MMS to solve problems such as those shown in Fig. 3.

MMS is based on a virtual manufacturing device (VMD), which is thought to be some kind of abstraction of a real manufacturing device. A VMD may represent a manufacturing cell or a production line. It is part of the application, as well as part of the application layer (ISO Reference Model for Open Systems Interconnection). The operations on elements of the VMD have to be performed by operations on the real hardware of the target node (executive function). MMS can distinguish between several application entities, each covering a part of the VMD.

A VMD consists of the following elements (see Fig. 4):

- an executive function mapping the services of the VMD onto the local resources;

- at least one domain, which is a logically consistent part of the functionality of the VMD; e.g., a VMD "manufacturing_cell" may have a domain "robot" and a domain "milling_machine";

- if necessary, a virtual file store (using FTAM) for local files of the VMD;

- within the domains, several objects like programs, variables, events, semaphores, and journals accessible by MMS-services.

Objects like variables and events may be defined globally for the VMD or locally for a domain.

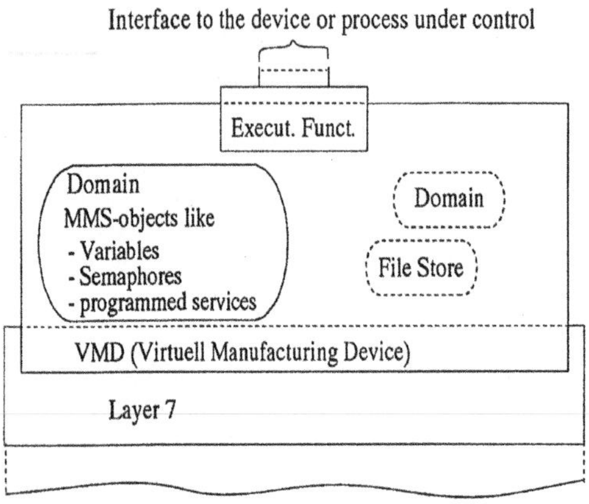

Interface to the device or process under control

Fig. 4    Structure of a VMD

MMS-services are composed of the following groups:

1. Context Management
   - Initiate, Conclude, Abort, Cancel, Reject
2. VMD-Support
   - Status, GetNameList, Identify, Rename, Unsolicited Status
4. Program Invocation Management
   - Create/Delete-ProgramInvoca-
     tion, Start/Stop/Resume/
     Reset/Kill GetProgramInvocation Attribute
5. Variable Access
   - Read, Write, ScatteredAccess
6. Semaphore Management
   - Define/Delete-Semaphore, TakeControl,
     RelinquishControl,
     ReportSemaphoreStatus, AttachToSemaphore
7. Operator Communication
   - Input, Output
8. Event Management
   - EventCondition, EventAction, EventEnrollment,
     EventNotification, ...
9. Journal Management
   - InitializeJournal, WriteJournal, ReadJournal,
     ReportJournalStatus
10. File Management
   - Obtain File, File Get
   Using FTAM:
   - File Open/Read/Close
     File Rename, File Delete, File Directory

Nearly all MMS-services are confirmed services, i.e., an acknowledgement is returned to the invoking application entity. Thus a sequentialization of events in different nodes is possible. Methods for resolving conflicts are described in the following section.

Some well known basic concepts exist for resolving conflicts like those in the examples of the section 4.2. We consider these concepts to be logically equivalent because one can simulate each of them by another. MMS provides the semaphore concept, and therefore we will show how this works, recognizing that some problems remain with the use of semaphores. The semaphore concept basically solves the problem of mutual exclusion. In the most simple form, a (binary) semaphore is a synchronizing variable with two states: free and blocked. It is accessible by two procedures:

- With P(sema) or REQUEST(sema), a process enters the critical region, and

- with V(sema) or RELEASE(sema), it leaves the critical region.

Associated with each semaphore is a queue for waiting processes requesting to enter a "critical region", if the semaphore is blocked.

The structure of a program looks like this:

```
.
P(sema)
.  (statements of the critical region)
V(sema)
.
```

A "critical region" of a process is the dynamic section of an application process during which it uses a resource (protected by the semaphore) exclusively. All processes using the same resource must use the same semaphore-variable for entering (or leaving) their critical region.

For a binary semaphore the access procedures P(sema) and V(sema) are defined as follows:

```
procedure P(sema);
begin

  if sema = free            | atomic
  then sema := blocked;     | action
  else "put calling process on a waiting-queue";

end; /* procedure P */

procedure V(sema);
begin

  if "waiting-queue is empty"
  then sema := free;
  else "continue a waiting process and discard it
        from the waiting-queue";

end; /* procedure V */
```

It is important that checking the condition (if sema = free) and changing the value from free to blocked must be done in an indivisible atomic action in order to prevent checking by more than one process prior to changing the value to blocked.

Usually the implementation of a semaphore uses a special instruction (the test-and-set instruction) for testing the old value and setting the blocked status. This instruction is not interruptable and therefore performs the necessary atomic action. It works only if we have a common memory for all processes. Solutions by algorithms exist for implementing semaphores in a real distributed system, but they are time consuming and cause considerable communication overhead. In MMS the semaphore management is based on a semaphore manager established for each semaphore on a single node. Processes trying to use a semaphore have to direct their requests to the corresponding semaphore manager. If it is free, the requesting process gets back an is_free message as a special pattern (token). To release, the token must be returned to the semaphore manager at the end of the critical region.

MMS provides semaphores in a more general form of counting semaphores. The semaphore manager can issue at most n tokens having in mind that the corresponding resource may have at most n users simultaneously. To granting further requests, tokens have to be returned by some using processes. If n=1, we have the above-mentioned binary semaphore. The actions of the semaphore manager are shown in Fig. 5. It acts as a logically not interruptable process to perform the necessary atomic action. For convenience, named resources may be attached to a semaphore (pool semaphore) with automatic requesting and releasing if the resource is used.

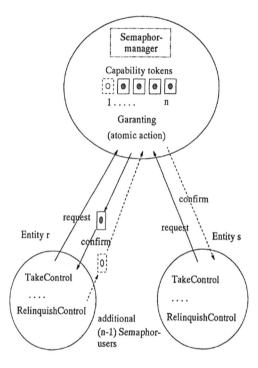

Fig. 5    Actions of the Semaphore Manager

The semaphore manager normally belongs to the application layer; however, delegation to an application entity is possible. In the first case, the semaphore manager can answer requests without user interaction, but an application entity may be invoked for performing some supplying tasks. In our example for concurrency control (see section 4.2), the belonging semaphore manager can give the privilege of handling a switch to that application entity to which the token has been sent.

Now we can give a sketch for the activity of moving the engine in our example (see Fig. 3):

Engine A:

    :
/* Make reservations for track section;
    S_track_1, S_track_2; semaphore, non-automatic */

TakeControl (S_track_1);
    /* Make reservations for track in control area 1;
      calling process gets handling privilege
      for switch W1.1 */
TakeControl (S_track_2);
    /* Make reservations for track in control area 2;
      calling process gets handling privilege
      for switch W2.1 */
Set switches;
Start drive;
Control velocity;
If engine has left control area 1:

    RelinquishControl (S_track_1);
    /* Handling privilege for switch W1.1 released */
If engine is in final position:

    Stop drive;

    RelinquishControl (S_track_2);
    /* Handling privilege for switch W2.1 released */
    :

The reservations take place similarly for moving engine B. To avoid deadlocks, the requests for semaphores must be made in the same sequence, that is S_track_1 prior to S_track_2. The example should show how the means for the concurrency control provided by MMS work.

Generally the semaphore concept has some drawbacks. First, all conflicting activities have to know about each other because they have to use the same semaphore. This is hardly to realize in an open environment, in which application entities may be added and discarded (capability for incremental expansion). Ideally an application entity to be added only should know about services to be imported from other entities and nothing else. A step in the right direction is the pool semaphore supported by MMS. A resource may be attached to a pool semaphore controlling the access to this resource. The operations for requesting and releasing the semaphore are performed implicitly on behalf of the accessing entity. Other conflicting situations have to be resolved in accordance with some agreements of the concerning entities.

Secondly problems arise if an entity does not leave the critical region, and the used semaphore may remain blocked forever. Mostly this happens as a result of program bugs in fairly unusual situations, but there are other reasons too. MMS provides two means for resolving. We can specify a timeout for releasing a blocked semaphore, or a monitoring task can sense the status of the semaphore without changing it and perform appropriate actions. Using the first means we are forced to specify a reasonable timeout condition, using the second means we have to decide, whether the blocked status of the monitored semaphore is a defective one or not. In both cases there are no general rules to follow, and it is not obvious how a normal or defective situation looks like.

Thirdly the semaphore concept is generally not fault tolerant. As we have seen in the above paragraph an abnormal termination of an activity within its critical region could be handled individually, the breakdown of the node hosting the semaphore manager is not tolerable by any means.

# 6. CONCLUSIONS

We have discussed some problems concerning distributed real-time systems for manufacturing automation. There are solutions, e.g. the MAP-specifications, which basicly meet the needs of real-time applications, but message delays and response times are not quite satisfactory for a couple of applications. Our detailed studies show that improvements of some order of magnitude are not possible with today's technology. The transfer rate on the medium is not the bottleneck. For real-time applications we strongly recommend the reduced architecture (like Mini-MAP).

For application oriented services in a distributed manufacturing environment MMS is a hopefully accepted standard. After a survey on the services we have shown that MMS has the necessary features for the coordinating dependent and concurrent activities. In the opinion of the author the means for concurrency control provided by MMS are not the only and best one but may have the advantage being a commonly understood concept.

# REFERENCES

[1]    Ciminiera, L.; A. Valenzano:
       Acknowledgement and Priority Mechanisms in the 802.4
       Token Bus; IEEE Trans. on Industrial Electronics Vol. 35
       (1988), No.2

[2]    Elnakhal, A.E.; I. Kulka; H. Rzehak:
       MAP und MMS: Realzeitumgebung für die Ferti-
       gungsautomatisierung; Fachzeitschrift industrie-elektro-
       nik+elektrik, 35. Jahrgang 1990, Nr. 9, Hüthig

[3]    A.E. Elnakhal; H. Rzehak:
       Untersuchungen zur Implementierung einer verkürzten
       Kommunikationsarchitektur (Mini-MAP-Konzept);
       GI/GMA-Fachtagung "Prozessrechensysteme '91", Berlin,
       26. - 27. Feb. 1991; Informatik-Fachberichte Bd. 269, S.
       238 - 250, Springer-Verlag

[4]    A.E. Elnakhal:
       Effiziente Kommunikationsarchitekturen für zeitkritische
       Anwendungen in lokalen Rechnernetzen; Dissertation,
       Universität der Bundeswehr München 1991;

[5]    European MAP Users Group;
       User Requirements for Communications in Time Critical
       Applications; EMUG Report Nr. TC 89/5, Sept. 14, 1989

[6]    Factory Automation Interconnection System (FAIS) -
       FAIS Cell-Net Implementation Specification (Version
       1.0);
       International Robotics and Factory Automation Center
       (IROFA), Tokyo, 1990

[7]    R. Jäger; H. Rzehak:
       Performance Analysis and Real Time Assessment of the
       Prioritized Tokenbus Access Method According to the
       Manufacturing Automation Protocol (MAP); 2nd Int. Con-
       ference on Software Engineering for Real Time Systems;
       Cirencester, UK, 18.-20. Sep. 1989; IEE Conference
       Publication Number 309, S. 189-193

[8]    Manufacturing Automation Protocol - A Communication
       Network Protocol Specification for Open Systems Inter-
       connection - Version 3.0; (c) 1988 General Motors Corpo-
       ration, Warren, published as MAP/TOP 3.0 Specification
       by EMUG (European MAP Users Group)

[9]    Marathe, Madhar V.; Robin A. Smith:
       Performance of a MAP Network Adapter; IEEE Network,
       May 1988, Vol. 2, No. 3, pp. 82-89

[10]   Mortazavi, B.:
       Performance of MAP in the Remote Operation of a CNC;
       IEEE Transaction on Software Engineering, Vol. 16, No.
       2, Feb. 1990, pp. 231-237

[11]   Murata, M.; H. Takagi:
       Two-Layer Modelling for Local Area Networks; IEEE
       Transaction on communications, Vol. 36, No. 9, Sept.
       1988, pp. 1022-1034

[12]   Process Communications Architecture;
       Draft Standard ISA-DS 72.03 - 1988; Instrument Society
       of America

[13]   Rzehak, H.; A.E. Elnakhal; R. Jäger:
       Analysis of Real-Time Properties and Rules for Setting
       Protocol Parameters of MAP Network; The Journal of
       Real-Time Systems, Vol. 1, No. 3, pp. 221-241, 1989

[14]   Rzehak, H.:
       Three-Layer Communication Architecture for Real Time
       Applications; Proceedings of the EUROMICRO'90
       Workshop on Real Time, June 6-8, 1990; IEEE Computer
       Society Press, pp. 224-228

[15]   H. Rzehak:
       Echtzeitkommunikationssysteme; Einführungsvortrag für
       ein Fachgespräch über Echtzeitkommunikationssysteme
       auf der GI-Jahrestagung 1991, Darmstadt, 14. - 18. Okt.
       1991, Informatik-Fachberichte, Bd. 293, J.Encarnacao
       (Hrsg.), S. 631 - 642, Springer-Verlag

[16]   Rzehak, H.:
       Distributed Systems for Real Time Applications Using
       Manufacturing Automation as an Example; in "Real-Time
       Systems, Engineering and Applications", M. Schiebe, S.
       Pferrer (Eds.), Kluwer-Verlag 1991

[17]   Schutz, H.A.:
       The Role of MAP in Factory Integration; IEEE Trans-
       actions on Industrial Electronics, Vol. 35, No. 1, Feb. 1988

[18]   Valenzano, A.; L. Ciminiera:
       Performance Evaluation of Mini-MAP Netowrks; IEEE
       Transactions on Industrial Electronics, Vol. 37, No. 3, Juni
       1990, pp. 253-258

[19]   Watson, K.S.; D. Janetzky:
       Token Bus Performance in MAP and PROWAY; IFAC
       Workshop on Distributed Computer Control Systems,
       Mayschoss (FRG); Sept., 30 - Oct., 2, 1986

# UNIVERSAL WORK CELL CONTROLLER FOR FLEXIBLE AUTOMATION WITH ROBOTS

**E. Freund and H.J. Buxbaum**

*Institut für Roboterforschung (IRF), Universität Dortmund, Otto-Hahn-Strasse 8, W-4600 Dortmund 50, Germany*

Abstract. Robots in the industrial factory plant build the backbone for CIM, which is outlined to be
the factory automation strategy of the future. In this paper a universal work cell controller for robotic
work cells in flexible manufacturing applications is described. The concept is based on a strategic
architectural model for hierarchical manufacturing control, where it focuses on the process sequence
coordination in the bottom hierarchical layers. The cell controller coordinates the components in the
work cell and offers an open interface to control and planning systems beyond cell level. Full
production flexibility by individual product identification as well as universal cell configurability are
characteristic for the cell controller system.

Keywords. CAM; automation; work cell controller; flexible manufacturing; integrated plant control;
robots

## INTRODUCTION

The major applications of robots are production plants for
large batches with very small product variations. The actual
trend to small batches, large product variaty and far-
reaching considerations to individual customer requests
shows the necessity to develop new factory automation
strategies with the aim to enlarge the production flexibility.
A migration from normal machine control technology to a
comprehensive factory wide planning, coordination and
control structure is indicated. The robot itself is a suitable
device for the purpose of flexible automation, if only
because of its kinematic structure.

This paper presents an universal work cell control concept
for flexible automation systems and the use of the work cell
controller in different applications. Today a wide range of
different work cell controllers for special applications are
available. They are strictly dedicated to the connected
manufacturing devices (lack of universability) and closely
programmed for the associated production process (lack of
flexibility). More intelligent tasks like re-equipment or error
recovery must be done manually. The significant differences
between those conventional concepts and the new approach
described here are the universal applicability and the support for flexible automation.

Basing on a generative model of a function unit hierarchy
first a strategic architectural model for manufacturing control structures is proposed. The essential meaning of the
model coincides with comparable approaches by Albus
(1981), Hidde (1990) and Pritschow (1990). Subsequently the
bottom layers of the manufacturing contol model are closer
examined and the general structure of a flexible work cell
is shown. A coordination strategy which allows the demand

for flexibility up to batch-one manufacturing and an appropriate cell controller system are introduced. The individual
functionality of the factory floor devices is hidden for the
cell controller. To preserve this anonymity all communication with the devices is done by the use of a formal standard
interface. Specific device drivers are necessary to convert
the standard communication on cell controller level to
individual machine code. Universal applicability is given as
long as the factory floor devices have communication facilities.

The advantage of this strategy is the possibility of
standardization already on cell controller level. The concept
and the structure of the cell controller's application
software as well as the interfaces to hierarchically higher
production planning and scheduling systems must not vary
depending on the application. This is a basic requirement
for standardization of manufacturing control structures.

## HIERARCHICAL MANUFACTURING CONTROL

For factory automation a manufacturing control system is
necessary. Such a system has the function of planning manufacturing activities and of controlling the production
equipment in the factory floor to ensure that the products
ordered by the customer can be delivered on schedule.
Fig. 1 shows the desired transparency of the information
flow from the customer order through the manufacturing
control system to the production equipment on the factory
floor. The feedback from each point is included. Besides
controlling the information flow, the manufacturing control
system also provides coordination mechanisms for the
internal flow of material and goods.

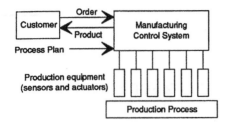

Fig. 1: Organization of manufacturing control

All activities in a factory are services and can be structured as function units. The architecture of this structure should be clearly laid out to ensure the possibilities to introduce effective strategies and concepts for manufacturing automation. Fig. 2 shows a generative model of a function unit hierarchy. All services (function units) are categorized to different hierarchical layers. Each function unit is assigned tasks from a function unit in the layer above and in turn delegates sub-tasks to function units layered below. A manufacturing control strategy should be applied according to this concept.

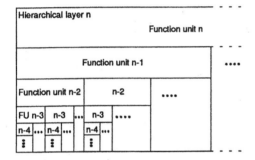

Fig. 2: Generative model of a function unit hierarchy

Basing on the generative model fig. 3 shows a 6 layer strategic architectural model for manufaturing control. The 6 layers cover the full range from production and factory systems (top) down to the sensors and actuators in the factory floor (bottom). The layers also extend over various time horizons which encompass both long-term planning and real-time processing.

Fig. 3: Architectural model for manufacturing control

To allow the information flow the devices in the several layers of the manufacturing control model need communication opportunities to send and receive data. For the material flow at least some of the devices need material dispatch and receive opportunities. Therefore the manufacturing control hierarchy must be combined with data communication facilities, which are needed for the infor-

mation flow, and with product transport facilities, which take over the part of carrying out and coordinating the material flow. Both are also hierarchically structured facilities but mutually independent from the manufacturing control model.

The data communication hierarchy should accord with the ISO 7 layer OSI (open systems interconnection) model to make adaptions and extensions easier. Basing on the ISO model all communications can be standardised.

The hierarchical model of the internal transport system has three layers. The *sensor and actuator layer* enables the material displacement from one node to another through a transport medium (standard transport). The *material transport layer* controls standard transports. Each part is run through a sequence of one or several standard transports along a transport route. The *routing control layer* manages the different routes from dispatcher to receiver. Routing control is not necessary if the transport system does not offer alternative routes, e.g. in a simple flow line production with station-to-station transport tasks.

## STRUCTURE OF THE FLEXIBLE WORK CELL

A work cell is a cooperating machinery unit in the factory floor which carries out entire manufacturing operations autonomously. The objective of those operations is to bring a product from one defined state of manufacturing to the next. A work cell usually consists of a few automation devices such as robots, machines and transport systems (production equipment).

In flexible systems this equipment is used for families of products. The variants of the product families can be manufactured by means of programming or data transfer. It is necessary to have the opportunity to re-equip the devices, so that product changes do not require a lot of time. The overall operational live time of the production equipment should exceed the life cycle of the actual product.

The process activities within the flexible work cell can be devided to three fundamental functions

- product identification
- internal logistics
- production process.

The *product identification* must be done everytime a new product arrives in the work cell. The tasks of *internal logistics* are the transport and the storage of products and parts within the work cell. The *production process* itself is delegated to the production facilities, which are often robots for reasons of flexibility. Fig. 4 shows the characteristic control hierarchy in a flexible manufacturing work cell.

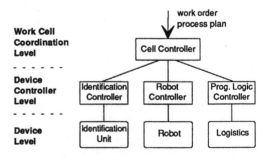

Fig. 4: Hierarchical layers in the flexible work cell

An exemplary layout of a flexible work cell is given in fig. 5. At the import point all products, which are reaching the work cell, are identified. Via a node in the transport system the product can be transported either to workplace A or to workplace B. To reduce idle times caused by transportation the robot can perform its operations on both workplaces similarly. After being processed by the robot the products leave the work cell.

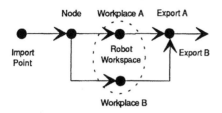

Fig. 5: Layout of a flexible work cell

## TASK OF THE CELL CONTROLLER

The device controllers in the flexible work cell are coordinated by a cell controller, which is an order driven system based on process plans. The workorders are emitted by a hierarchical higher authority, which may be a control computer system in an upper layer of the factory control hierarchy or a human operator as well. A workorder specifies the identification range of products which belong to this order and includes a reference to a process plan. The process plan points out all process steps to be done on a product which fits in the accompanying workorder's identification range. Related to the process plan in each step a program is started on the associated device controller. After the ready message is received, the execution process continues with the next step. Decisions depending on the execution result of the previous step allow the handling of error states as well as intelligent processing.

The number of different products which can be handled by the work cell simultaneously is not limited by the cell controller. Nethertheless there are physical limitations like the number of work places and buffers in a real work cell.

The complete production process is described by process plan or process scheme. Both content the same information - the sequence of the concerned automation devices and corresponding parameters for the device controllers - in different forms of representation. The process scheme shows the process sequence in a schematic graphical form, the process plan is the state table representation of this sequence. For a batch of unique products - as a single variation of the well known product family - a particular process plan (or scheme) is needed.

Fig. 6 shows two different process schemes for the examplary work cell. After the product identification on the automation device *Transport* the machine program *Node* is performed. In this example this results in transporting the product to the node. The product is transported to *Workplace A* (left scheme) in the next step. *Robot* performs its program *Prog.XYZ* on the product before it leaves the work cell in the last step. In case of a robot error during execution an *Operator* call is generated before the product leaves the work cell.

The second scheme shows the use of the other path via workplace B as a bypass in transport system. This is useful for products which are in the product flow but do not need to be handled by this work cell.

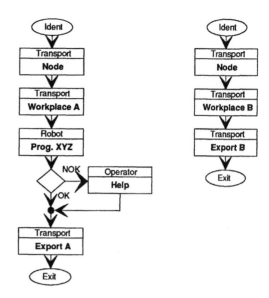

Fig. 6: Process schemes

In table 1 the process plan according to the first (left) process scheme is shown. The process plan is the internal data representation of a process scheme for use by the cell controller. Each row contains the information to perform a process step in the work cell. The first two columns contain the process step number and the result status of the previous step, so that conditional branches through the process plan are possible. The next two colums contain the device driver name and the machine program to be executed there for this step. The last column references the counter number of the next process step.

TABLE 1  Process Plan

| Proc. Step No. | Condi-tion | Device Driver Name | Machine Program | Next Proc. Step |
|---|---|---|---|---|
| 1 | | Transport | Node | 2 |
| 2 | | Transport | Workplace A | 3 |
| 3 | | Robot | Prog. XYZ | 4 |
| 4 | NOK | Operator | Help | 5 |
| 4 | OK | NOP | NOP | 5 |
| 5 | | Transport | Export A | End |

The demand for flexibility makes an assignment of an arriving product to the corresponding process plan inevitable. This assignment is done by product identification at the very beginning of the manufacturing process for each unique product. The workorder links the appropriate process plan to a range of product identification numbers. The internal sequence of identification and assignment is illustrated in fig 7.

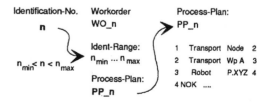

Fig. 7: Identification and assignment sequence

After identification of the product number $n$ all existing workorders are scanned for a matching identification range. Assume that workorder $WO\_n$ has an identification range from $n_{min}$ to $n_{max}$ and $n$ fits in this range ($n_{min} < n < n_{max}$). The selected workorder $WO\_n$ then links the product $n$ to the process plan $PP\_n$. The processing sequence in the work cell starts immediately according to process plan $PP\_n$.

## MODE OF OPERATION

The cell controller works in two different modes (fig. 8)

- data management mode
- run mode.

The internal mode handling is coordinated by the user interface. Users can either be the work cell operator in the factory floor or an information system on a higher coordination level, e.g. a shop or line controller.

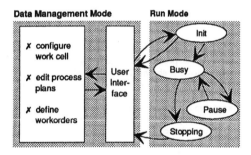

Fig. 8: Modes of the cell controller

The data management mode attends the information definition phase so that all necessary information is available on run time. The arranging of work cell configuration, the preparing of process plans and the definition of workorders are supported.

The entry point for the run mode is the *Init* state. All initializations have to be done here before the production process will be enabled. It is the aim to ensure that all components are working and the needed machine programs are available on run-time. The initialization has 3 different phases

- establishing of the communication paths: All ports are opened to realize the device specific external interfacing defined in the work cell configuration data.
- initialization of the device controllers: Via the communication paths the device controllers are set to an automatic mode, so that a remote start of different machine programs will be possible on run-time. The initialization of the device controllers normally implicits the initialization of the accompanying device. E.g. calibration or reference movements will be done here.
- checking the machine program existance: A lot of machine programs are usually referenced in the available process plans. To avoid run-time problems due to non-existing machine programs it is checked here, that the device drivers have access to the needed programs.

If an error occurs during initialization the run mode stops at once. If everything works okay, the cell controller state changes to *Busy*. The product identification device will be connected. If products reach the work cell they will be identified and handled according to the appropriate process plan. The production process is running.

To enable short breaks in the production process a state change to *Pause* can be requested by the operator. Temporary the state remains *Busy* but if the actual machine programs are finished on the device controllers no further actions will be taken by the cell controller. If everything has stopped in the work cell the state changes to *Pause* finally. On operators request the state changes back to *Busy* and the production process continues.

The operator can stop the run mode at any time. The state changes to *Stopping* then. The product identification device will be disconnected to avoid further entrance of products to the work cell. For the products which are already in the work cell the process sequence will be properly finished. The run mode stops as soon as the last product leaves the work cell.

The significant variance of the diverse automation devices and the demand to universal applicability leads to a control concept based on a library of specific device drivers. For each device in the work cell configuration a specific driver has to be installed, so that the cell controller consists of two independent components

- the common cell controller kernel
- an amount of different device drivers.

The device driver offers a formal standard interface to the common cell controller kernel. The abstraction of the individual functionality of the device controller preserves the anonymity of the factory floor components on the level of the cell controller kernel. The communication between kernel and driver can be limited to a few standardized elements that way.

The common cell controller kernel coordinates the process in the work cell on device driver level according to the current valid process plan. The formal standard intefaces of the device drivers allow standardization already on the level of the cell controller kernel. The cell controller's application software structure and the interfaces to higher production planning and scheduling systems do not vary depending on application.

For the definition of this formal standard interface the well known client-server model is used. The client in this model takes the active part in the communication relationship and sends requests and commands to the server. The server reacts on this by executing and reporting results. Client and server often have more than one communication relationship at the same time. A client can control several servers. A server usually acts as client too - on a hierarchically lower level.

The device driver is the server in the kernel-driver relationship, but also has a client functionality for communication to the external device controller. A request message from the kernel activates the driver to perform the requested action, which usually is also a communication task where now the driver is client and the external device controller is server. When the requested action is done, the driver sends a response message back to the kernel.

Becker (1991) defines three general cell communication services (GCS) as the formal standard interface between kernel and driver in a client-server model. Fig. 9 shows this internal communication model of the cell controller.

CELL CONTROLLER                    DEVICE

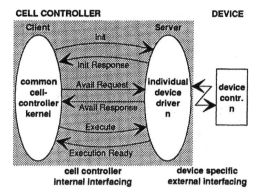

Fig. 9: GCS client-server model

The GCS services are

- Device initialization services *Init* and *Init Response*
- Machine program management services *Avail Request* and *Avail Response*
- Machine program invocation services *Execute* and *Execution Ready*.

Table 2 shows the GCS services classification.

### TABLE 2 GCS Services Classification

| GCS Service | Message Type | Origin | Run Mode Type |
|---|---|---|---|
| Init | Request | Client | Init |
| Init Response | Response | Server | Init |
| Avail Request | Request | Client | Init/Busy |
| Avail Response | Response | Server | Init/Busy |
| Execute | Request | Client | Busy/Stopping |
| Execution Ready | Response | Server | Busy/Stopping |

All device specific adaptions must be implemented in the individual device drivers. The complexity of the drivers can range from simple transmission of the GCS services up to extensive communication algorithms for e.g. the use of standardized communication protocols like MMS. Due to the free programmability of the driver user defined coordination strategies for e.g. multi-robot systems can be implemented here as well.

To connect a new device controller always an appropriate driver has to be installed, that abstracts the device's functionality to GCS. For this reason a virtual device driver can be defined as a state model where the state changes accord with the GCS protocol. Fig. 10 shows the state changes in the virtual device driver. All state changes in the driver are initiated by requests of the kernel or responses by the driver. A real driver must adapt the individual device controller functionality to this common state model.

After power on or reset the driver is in the state *Not Initialized*. An *Init* request message from kernel changes the state to *Initializing*, the driver now initializes the communication path, the device controller and the device. Afterwards an *Init Response* message is sent to the kernel and the state changes to *Idle*. In this state the device driver is ready and waiting for execution. An *Avail Request* message from the kernel sets the driver's state to *Program Management*. In this state all file handling functions are done by the driver before the *Avail Response* message is sent to the kernel and the state changes back to *Idle*. An *Execute* message from the

kernel changes the state to *Busy*. Now the driver will run the appropriate machine program on the device controller. Afterwards an *Execution Ready* response message is sent to the kernel and the status is changed from *Busy* to *Idle*.

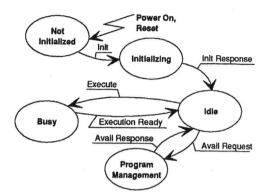

Fig. 10: State changes in the virtual device driver

### APPLICATIONS

The universal work cell controller concept has been verified in various simulations on a 3-D work cell simulation and animation tool. A pilot assembly work cell has been build up as a laboratory testbed for implementation and tuning of the cell controller software. An industrial application for a welding work cell in a flexible production of steel furniture is in development.

Fig. 11 shows a simulation hardcopy of a flexible assembly work cell in a printed circuit board production line. The assembly cell consists of an identification system, a line-structured continuous product transport system and a SCARA type assembly robot with a components magazin. The robot performs flexible pick and place jobs on different components and boards.

This layout takes into consideration, that some boards must not be handled by the robot. For those boards a bypass structure is added to the transport system, so that the assembly process can be continued without interruption. The work cell has defined import and export points for the boards. The product flow direction is displayed from the left side to the right. Incoming boards will be sensed and identified automatically by a camera system. This enables the individual processing for each product (local flexibility). The bypass structure allows to remove certain products from the sequence which do not need to be handled by this work cell (global flexibility).

Essentially the pilot laboratory assembly work cell is been build up according to this simulation study. The mechanical flexibilty is increased by the use of two robots instead of one and a common gripper changing system. The degree of flexibility is even higher as in the simulation study, so that a wider range of products can be covered. Next to simple pick and place jobs also difficult sensor supported assembly tasks can be carried out. The handling of the product carriers is done by a line-structured continuous pallet carrier system, which was layouted as in the simulation study. Parts handling is supported by a third robot. The product identification is done by automatical barcode reading. All machine programs are managed by the cell controller exclusively, so that, in conjunction with the gripper changing system, the re-equipment of the robots can be done seperately and full automatically.

339

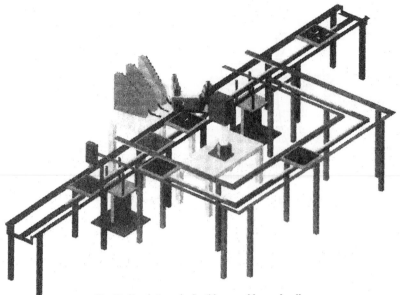

Fig. 11: Simulation of a flexible assembly work cell

Fig. 12 shows a welding robot in an industrial flexible work cell application. This work cell manufactures component parts and groups for steel furniture. The process is controlled by a robot (and welding) controller and a PLC for gadget control, which will be connected to the cell controller. The tasks of the cell controller are

- operator support on process and re-equipment
- flexibility support by identification of components, orders and tools
- machine program management and download
- tools management
- gathering, sampling and preprocessing of machine and process data for diagnosis and offline analysis tasks.

Fig. 12: Industrial welding work cell

## CONCLUSION

Full production flexibility via free programmability of the process sequence in conjunction with individual product identification on the one hand and universal cell configu-

rability via integration of device specific driver software on the other hand are characteristic for the cell controller system. The universal configurability is shown by different simulations and applications. For the pilot laboratory assembly work cell an increase of production flexibility could be achieved. The reasons are, next to the individual product handling via barcode identification, mainly a notable reduction of re-equip time on product changes. So a reasonable batch-one production is possible.

The cell controller concept is independent of the process or the product and therefore convertable to other fields of application. So the standardization of the control information flow is possible already on the level of the cell controller. Nethertheless the open interfaces permit easy link up to the logistic and production control systems which are already employed in a factory.

## REFERENCES

Albus, J. B., Barbera, A. J., Nagel, R. N. (1981). Theory and practice of hierarchical control. *National Bureau of Standards*, Washington DC.

Becker, M. (1991). Entwurf und Implementierung eines Leitstandes für Arbeitszellen flexibler Fertigungssysteme. *Institut für Roboterforschung*, Dortmund.

Buxbaum, H.-J., Hidde, A. R. (1990). Flexible Zellensteuerung - Bestandteil eines produktunabhängigen Fabrikautomatisierungskonzepts. *Werkstattstechnik 80*, 133-136, 262-264.

Freund, E., Buxbaum, H.-J. (1991). Robotergestützte flexible Montage graphisch simulieren. *Arbeitsvorbereitung 28*, 29-32.

Freund, E., Buxbaum, H.-J. (1991). Flexible assembly cell simulation. *6th Int. Conf. on CAD/CAM, Robotics and Factories of the Future*, London.

Hidde, A. R. (1990). Continuous flow of information in the area of factory automation by standardized interfaces and open networks. *22nd Int. Symp. on Automotive Technology and Automation*, Florence, 1265-1272.

Pritschow, G. (1990). Automation technology - On the way to an open system architecture. *Robotics and Computer Integrated Manufacturing*, Vol. 7, No. 1/2, 103-111.

# PETRI NETS IN A KNOWLEDGE REPRESENTATION
# SCHEMA FOR THE COORDINATION OF PLANT ELEMENTS

**J.L. Villarroel, M. Silva and P.R. Muro-Medrano**

*Dpto. de Ingenieria Electrica e Informatica, Centro Politecnico Superior, Universidad de Zaragoza,
Maria de Luna 3, E-50015 Zaragoza, Spain*

**Abstract.** This paper is focused on the real-time coordination of manufacturing systems. A software architecture for real-time control of complex manufacturing systems is presented. The architecture follows a hierarchical approach for control ranging from real-time coordination to long term scheduling, where the different layers are sharing a common knowledge base. With this purpose, a knowledge representation schema for manufacturing systems is also proposed. This schema integrates knowledge representation techniques based on frames, high level Petri nets to describe plant behavior and it follows an object oriented programming methodology. The coordination function is materialized by making a coordination model evolve (a Petri net based representation is underlying in that model).
**Keywords.** Petri nets, Knowledge engineering, Manufacturing processes, Control systems, Software development.

## INTRODUCTION

A manufacturing system pursues the objective of elaborating a serie of products to the satisfaction of the client (which is measured basically by the product's quality and delivery date) while making a profit for the manufacturer. To achieve these objectives, the manufacturing system must possess a considerable degree of automation and flexibility. This requires an integrated computer control system where issues like hierarchy, complexity, concurrency, optimization, capacity, uncertainty and feedback are central. Researchers have agreed in the hierarchical decomposition of manufacturing systems control. In the hierarchical decomposition adopted in our work the following levels are considered: *production planning, operation scheduling, coordination of plant elements* and *local control.*

In this paper, our attention is focused on the real-time coordination of manufacturing systems. To carry out the different coordination functions, a precise model of the manufacturig system behavior is needed. The high degree of concurrency between manufacturing system elements and their number, make the coordination model very complex. In the approach presented here, the specifications of the factory behavior are modelled by Petri nets, the interpretation of the Petri net model provides the guidance to carry out the coordination task. The use of Petri nets and high level Petri nets for the coordination functions has received a large attention in the technical literature in recent years. The most relevant approaches, related with the work presented here, have been conducted by two european research groups: 1) [Valette, 1988] and [Ben, 1991] propose the utilization of tools from the Petri Net family to implement and analyze the coordination and monitoring functions. 2) CASPAIM, [Gentina, 1987] and [Castelain, 1988], represents another important approach where Petri nets are used for the design of control systems in manufacturing ap-

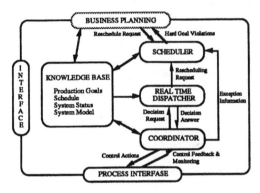

Figure 1: Control system global architecture.

plications. Coordination decision problems are solved by a higher control level implemented using rule based techniques. The Petri net model uses an special interface to communicate with this decision system.

## CONTROL SYSTEM ARCHITECTURE

The functionality of the global control system (planning, scheduling and coordination) can be implemented in two different ways: 1) a single software module which integrates all functions; or 2) a set of communicating software modules. The first approach produces a more compact software, however the second approach has been selected here for flexibility and modularity, with this option each function can be designed and implemented separately with specific techniques and can therefore be more efficient and each function can be altered without modifying the rest.

Figure 1 outlines the main components of the global control system architecture adopted in our approach. The architecture is structured in four subsystems

341

sharing a common knowledge base. The basic functions of *coordination system* or *coordinator* are the coordination and monitoring plant elements. In this framework, the responsability of the coordinator is to manage and supervise the execution of local controllers tasks. Plant monitoring (data collection and exception handling) is carried out by an internal monitoring module. The coordinator evaluates which operations can start at any given time and which of them can do it simultaneously.

The *dispatching subsystem* is responsible for providing solutions to these decision problems in real time. The coordinator sends messages posing decision problems and the dispatching responds with a specific solution. The dispatcher makes final real-time decisions and solve discrepancies between the system status and the schedule by means of a flexible interpretation of the schedule (a similar approach is considered in [Valette, 1988]).

The *scheduling subsystem* is responsible for generating the schedule of operations for each period of production time using the available information about orders, inventory, plant characteristics and resources, production objetives, etc. We also assume an ability to incrementally revise the current schedule in situations where significant discrepancies between the schedule and the current factory state are detected.

The main function of *Bussiness planning subsystem* is the production planning. From the client's orders, the state of warehouses and the manufacturing capacity of the plant, a list of production orders is elaborated for each period of time.

Finally, *global knowledge base*, is a common store of the system declarative knowledge (factory characteristics and status, productions plans, schedules, constraints, heuristics,...). This information is shared by the four subsystems of the control architecture but each subsystem has its particular vision.

## KNOWLEDGE REPRESENTATION SCHEMA

An important step in the development of the control system presented here, was the definition of a knowledge representation schema for CIM to be used for a variety of applications, ranging from plant coordination to production planning. Production goals, current production schedule, system status and system model are the main informations to be stored in the common knowledge base. In this way, it must represent production orders, manufacturing operations, produts, materials, manufacturing resources, ... and their relationships. There are a lot of non trivial concepts and a complicated relationship structure. To deal with this representation problem a knowledge representation schema, called KRON (Knowledge Representation Oriented Nets) has been adopted [Muro, 1990, Villarroel, 1990]. KRON is based in the following main concepts: frames, high level Petri nets (HLPN) and object oriented metodology.

In KRON, a model is composed by objects which can contain data, procedures, relations and metaknowledge. In addition to these features generally sup-

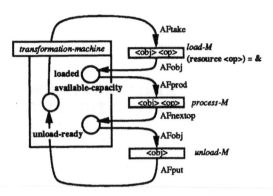

Figure 2: Underlying Petri net modelling the **transformation-machine** prototype behavior.

ported by object oriented languages, a set of semantic primitives implementing the high level Petri net formalism are included. The HLPNs provide the mechanism to describe the internal behaviour of dynamic objects and the interactions between them. This paper will not present the High Level Petri Nets (HLPN) formalism (the reader is referred to [Jensen, 1991]).

There are some knowledge representation schemas similar to our approach. The Object Petri Nets [Sibertin, 1985] are the most referenced schema in related technical literature: a system model is composed by a set of related objects (which define the HLPN marking) and a HLPN which especifies the system global evolution. In this approach, the HLPN are no fully integrated in the object oriented environment. Other interesting approach is the one proposed in [Baldassari, 1988] where the internal behaviour of objects and their interaction are specified by HLPN. However, this approach has not the power of frame based schemas to represent relations (different from dynamic interactions) between objects. Finally, [Micovsky, 1990] proposes another interesting frame and HLPN based representation schema, but the object oriented advantages are missing.

We will illustrate the proposed representation schema by creating the model to represent a typical manufacturing                                        machine, which will be called **transformation-machine**. A **transformation-machine** model can be represented as an object (frame) containing information about its attributes, constitutive parts and structure. Characteristics such as abstraction hierarchy relations, predicted information and data collection, physical characteristics, interface features and graphic representation characteristics are also represented in the object definition.

In addition to the previous features, the dynamic behaviour must be represented. Figure 2 shows the HLPN graphical specifications for the dynamic characteristics of an isolated prototype for **transformation-machine**. The behaviour of this physical entity is defined such as can load parts to be processed and it has a limited available capacity.

Three places (**available-capacity**, **loaded** and **unload-ready**) and three transitions (**load-M**, **process-M** and **unload-M**) are shown in figure 2. Tran-

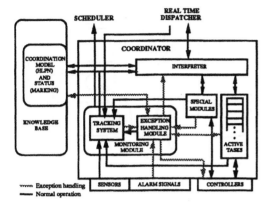

SCHEDULER  REAL TIME DISPATCHER

COORDINATOR

INTERPRETER

COORDINATION MODEL (HLPN) AND STATUS (MARKING)

KNOWLEDGE BASE

SPECIAL MODULES

TRACKING SYSTEM

EXCEPTION HANDLING MODULE

MONITORING MODULE

ACTIVE TASKS

Exception handling
Normal operation

SENSORS   ALARM SIGNALS   CONTROLLERS

Figure 4: Coordinator Subsystem

sitions in the model are associated with system actions used to change the state, which is itself represented by the net marking. Tokens or colors can be identified in this case with products which evolve in the production plant. The presence of a token in a place is indicated by the presence of the frame name as a state slot value.

As an example for this and further sections, let us create now the model of a manufacturing work cell. To simplify, we consider a workcell C1, with three machines M1, M2 and M3, a random access store ST and a robot R for palleting. Firstly, machines, stores and the transport system of the cell are separately defined. Machines M1, M2 and M3 can be defined as instances of the transformation-machine prototype. Following a similar method, ST and R must be defined as instances of isolated prototypes modelling a random access store and a robot. The structural description of the workcell is completed by establishing interfaces between machines, store and the transport system, defining in this way the flow of parts between them. This is done by synchronizing the appropiate actions. Figure 3 shows the HLPN modelling the dynamic behavior of the complete workcell C1.

## COORDINATOR

*Coordination* and *monitoring* of plant elements are the basic functions of the *Coordinator Subsystem*. In order to achieve these goals, the coordinator communicates with local controllers through a set of signals:

- *Controller signals.* The coordinator sends *orders* to the controllers for the execution of activities. When the activities are successfully finished *activity-end* messages are sent in return. *Exception* messages are also sent to the coordinator if any error or exception arise during a order execution.

- *Sensor signals.* Set of incoming signals from sensors complementing coordinator information about plant evolution.

- *Alarm signals.* Alarm signals (such as machine and part breakdowns, abnormal stoppages) can

be sent to the coordinator by operators or special supervisor systems.

The coordination method is based on the *interpretation* of the dynamic behaviour specifications represented in the system model. The coordination function is materialized by making a coordination model evolve, which in turn makes the production system evolve. This is a closed-loop control where the actions are the orders sent to the controllers and the feedback is the composition of the activity end, exception, sensor and alarm signals. Decision problems, which are not solved by the coordinator, can appear during the evolution of the coordination model. That decision problems are posted to the dispatcher which is the responsible to provide its solution.

Plant *monitoring function*, basically consists in two activities: data collection and exception handling Figure 4 illustrates the internal architecture of the coordination subsystem, its main modules are explained in the following paragraphs.

## INTERPRETER

The *interpreter* subsystem is the main responsible of the coordination function. Basically, the interpreter is a centralized, concurrent and interpreted implementation of a HLPN [Colom, 1986], that is, a module which materialize the coordination model evolution by interpreting a data structure.

Transition firing modes represent production system activities. The firing a transition with respect to a firing mode implies the execution of a piece of code which implements the communication protocol between coordinator and controllers. These pieces of code are configured as tasks to allow their concurrent execution. Figure 5 shows a typical example of Ada task used to implement the code associated to a firing mode.

The communication between the coordinator and the local controllers is implemented by the following handshake protocol:

1. A coordinator task communicates with a controller and sends to it an order to run a specific control program. If the control program is a Petri Net implementation, the order can be seen as an event which enables the firing of a start transition.

2. The task waits for the controller answer. *Activity End* message means that the order has been successfully completed, the monitorig module is reported by the task about the end of the operation, it synchronizes with the interpreter (to establish the end of transition firing) and finishes its execution. If the order has been abnormally finished, an error code is received by the task. In this case, the task does not establish the synchronization with the interpreter but it transfers control to the monitoring module in order to manage the exception.

There are another communication protocol, between interpreter and dispatcher for dispatching problem solving purposes. These problems, from the coordination point of view, are located in: 1) *Conflicts.* They

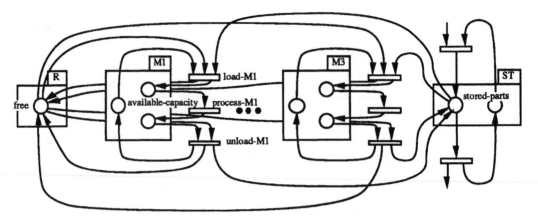

Figure 3: Structure of Petri net underlying in the model of workcell C1 at the machine level of abstraction.

represent situations where there exist several excluding firing modes. These problems arises for example when: a product can be loaded in several resources, a resource is required for several operation executions, ... 2) *Transitions* modelling operation starts. If all preconditions for an operation to start are satisfied, when it will be executed (inmediately or some time in the future) must be decided. These decisions are generally based on previsions provided by the current production schedule.

Specific coordination realized with techniques others than those based on HLPN may be required by certain elements of the production system. Thus for example, it may be important to perform the internal management of an automatic warehouse with a specific control system. All what is required in this case is the construction of the interface between this system and the HLPN interpreter, which will support the dialog regarding request, input and output of parts from the warehouse.

## MONITORING MODULE

In the proposed architecture, monitoring activities (data collection and exception handling) are realized by the *Monitoring Module*. This module is composed by two main components: tracking system and exceptions handling module.

The *tracking system* receives information from the transition firing tasks, special modules, the dispatcher and, directely, from the plant floor throught sensors and other automatic information gathering devices. The tracking system generates statistics and historical data reports from the physical components (e.g. utilization of tools by machines, down times, set up times, etc.), operations (e.g. processing times) and orders (e.g. satisfaction of due dates). It also manages incoming information from sensors and the dispatcher. This information is used by the tracking system to compute the values of events which constitute additional preconditions for transition firings. The information comming from the dispatcher provides the solutions to the decision problems (generally scheduling problems).

The *exceptions handling module* is in charge of han-

dling the exceptional situations arisen in the production plant and it is itself decomposed in three parts: detection, diagnosis and action. *Detection* function are distributed by the coordination system modules which have relation with the production plant (active tasks, special modules, tracking system and exception handling module). For example, a task associated with a firing mode can realize the supervision (e.g. time out as the example of figure 5) of order execution by a local controller. *Diagnosis* and *action* functions are centralized in exception module. This module receive messages from coorditator's modules and directly from the plant. Messages describe: machine stoppages (breakdown or manteinance), parts breakdown, comunication failure, accidents, etc. Exception module will try to recover the arisen error or, if it is not posible, to lead production system to a secure state. In this sense, exception module can send emergency stoppage orders to controllers and abort the execution of tasks associated with transition firing. In any case, this module must guarantee the coherence of model and production system states in exceptional situations, and inform to scheduler for a possible replaning of operation schedule. [Sahraoui, 1987] proposes the use of observers and watch dogs over the net evolution for detection, artificial intelligence and Petri nets techniques for diagnosis, and net reconfiguring (change of net structure or marking) for action. This approach for monitoring can be easily integrated in the framework proposed here.

## COORDINATION EXAMPLE

In order to illustrate the coordination ideas presented above, let us return to the example presented in section . The coordination function is based on the transition firing of the model. For example, transition load-M1 is modelling the loading of parts in machine M1. In order to load a part in the machine, several constrains must be satisfied: the robot must be available (a token in place free of the robot state object), the part must be in the store (part markig object must be present in place stored-parts of the store state object) and the machine must be free (any token in place available-capacity of the machine state object). Moreover, the specified operation on the part

```
task load-M1 ;
task body load-M1 is
  --
begin
loop
  -- Interpreter orders the execution of assoc.
  -- activity to (part, op) firing mode
  interpreter.start(load_M1)(part, op) ;
  declare
  time_out_M1: exception ;
  time_out_ROBOT: exception ;
begin
  -- Task sends order to M1 and ROBOT cont.
  M1_controller.begin_op (part, op) ;
  ROBOT_controller.begin_op (part, op) ;
  T := clock ;
  -- The task waits for end  op messages
  -- from M1 and ROBOT controllers
  select
  ROBOT_controller.end_op (message_ROBOT) ;
  T_ROBOT := T - clock ;
  select
  M1_controller.end_op (message_M1) ;
  or
   delay max_proc_time ( op,M1) - T_ROBOT;
   raise time_out_M1 ;
   --  op on M1 exceded the max proc time
  end select ;
  or
  delay max_proc_time ( op,ROBOT) ;
  raise time_out_ROBOT ;
  -- op on ROBOT exceded the max proc time
  end select ;
  case message(message_M1, message_ROBOT) is
  when end_op_OK =>
   tracking_system.end_task (load_M1,part, op);
  when error_M1 => raise exception_M1 ;
  when error_ROBOT => raise exception_ROBOT ;
  when general_failure =>
   raise general_exception;
  end case ;
  -- The  op has been completed successfully
  interpreter.end_task (load_M1, part,  op);
exception
  -- An exception has occurred in op execution
  when time_out_M1 =>
   exception_handling.time_out(M1);
  when time_out_ROBOT =$>$
   exception_handling.time_out(ROBOT);
  when exception_M1 =>
   exception_handling.cont_exc(M1);
  when exception_ROBOT =>
   exception_handling.cont_exc(ROBOT);
  when general_exception =>
   exception_handling.gen_exc(load_M1);
 end ;
 end loop ;
end ;
```

Figure 5: Example of Ada task associated to a firing mode.

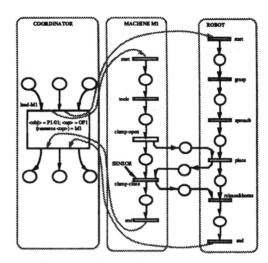

Figure 6: Firing of transition load-M1.

must be able to be carried out on the machine (predicate (ressource <op>) = M1 must be satisfied).

A transition firing, with regard to a firing mode, supposes the execution of a task which communicates with the local controllers. The activity modelled by transition load-M1 involves the cooperation between two resources, the robot and machine M1. For the machine to be loaded, the robot must pick up a part from the store and start the approach to the machine M1. Concurrently with that process, machine M1 tools must be prepared for operation and its subjection clamps be opened. Then, the robot is allowed to locate the part into the clamps. This action is detected by a sensor which is activating the subjection. When the part is graspped, the robot must release it and go to a home state. Figure 6 shows a Petri net representing the described local control function.

Loading part P1-01 on machine M1 involves the firing of transition load-M1 and, consecuently, the execution of the task (see figure 5) associated to that firing mode. Orders to the robot and M1 controllers are sent by the task. These orders enable the firing of start transitions. Its own net models are executed by each controler until transition end is reached. At this moment, an end activity message is sent to the coordinator. On finishing transition load-M1 firing, part P1-01 will be loaded on machine M1 and the robot will become free.

The exception handling operation will be illustrated using the same example. Let us suppose the following exceptional situation: the robot grasping operation fails and the approach to M1 is made with no part. Both controllers, from machine and robot, reach a deadlock because no part presence is detected by the robot hand sensor. This fact implies that transition clamp-close from M1 can not be fired (see figure 6). Following a typical monitoring technique, a watchdog can be included in the corresponding active task. When operation maximum time has been reached (detection), the arisen error will be communicated by the task and the task will be finish. At that moment the following actions will be carried otu by the exception handling module. Firstly, orders will be sent to the local controllers to conduct the resources to a secure state. After that, an alarm signal will be sent to the production plant to notify the operator about the arisen error. This notification is a request for the operator to provide the adequate correcting actions to physically correct the error, in this case the operator must place the part correctly. Additionally, the exception handling module must be notified about the exception origin (*diagnosis*). The next action consists on recovering the previous marking to the exception which represents the cell actual state (robot free, machine M1 free and part P1-01 in the store). At this point, the interpreter will try to fire the transition again to continue in normal operation.

## CONCLUSIONS

This paper deals with real time software for the coordination of complex manufacturing systems. A framework for the software architecture for the planning, scheduling, and coordination functions is proposed.

A knowledge representation schema has been developed to be used as a global knowledge base. This schema integrates frames, high level Petri nets and follows the object oriented programming paradigm. Manufacturing system's models are rapidly prototyped within this schema. Entity behaviour specification, concurrency, entity synchronization, model analysis, ... are facilitated by the underlying Petri net.

The paper is focused on the coordination function. System coordination is mainly made following the behavioural specifications provided by the Petri net. Ada is used to create the final control software where real time and concurrency constraints must be guided by the software execution. Actually, real time coordination software is created separately. A special Petri net implementation in Ada is used but the translation from the knowledge representation schema to Ada is made manually. At this moment, only dynamic behaviour features are translated to the Ada implementation but most of the rest is not taken into account. Research is conducted to use Ada for the complete development. Other coordination functions, such as monitoring, are also considered.

A prototype version of the knowledge representation schema has been already implemented in a 1186 Xerox machine running LOOPS. The definitive version is being implemented in a Sun workstation running KEE. The real time functions are being implemented in Vax 750 using Ada.

## ACKNOWLEDGEMENTS

This work has been supported in part by project ROB91-0949 from the Comisión Interministerial de Ciencia y Tecnología of Spain and project IT-10/91 from the Diputación General de Aragón.

## REFERENCES

[Ben, 1991] S. Ben Ahmed, M. Moalla, M. Courvoisier, and R. Valette. Flexible manufacturing production system modelling using object petri nets and their analysis. In *Proc. of IMACS Symposium MCTS. Lille, France*, pages 553–560.

[Baldassari, 1988] G. Baldassari and G. Bruno. *Advances in Petri Nets 1988*, chapter An Environment for Object-Oriented Conceptual Programming Based on PROT Nets, pages 1–19. Number 340 in Lecture Notes in Computer Science. Springer Verlag.

[Castelain, 1988] E. Castelain and J.C. Gentina. Petri-nets and artificial intelligence in the context of simulation and modelling of manufacturing systems. In *Proc. of IMACS International Symposium on System Modelling and Simulation, SMS'88, Cetraro, Italy*, pages 28–33, September 1988.

[Colom, 1986] J. M. Colom, M. Silva, and J. L. Villarroel. On software implementation of petri nets and coloured petri nets using high-level concurrent languages. In *Proc. 7th European Workshop on Application and Theory of Petri Nets, Oxford*, pages 207–241, July 1986.

[Gentina, 1987] J. C. Gentina and D. Corbeel. Hierarchical control of flexible manufacturing systems (fms). In *Proc. of IEEE International Conference on Robotics and Automation. Raleigh, North Carolina*, pages 1166–1173, March 1987.

[Jensen, 1991] K. Jensen and G. Rozenberg, editors. *High-level Petri Nets*. Springer-Verlag, Berlin, 1991.

[Muro, 1990] P.R. Muro-Medrano. *Aplicación de Técnicas de Inteligencia Artificial al Diseño de Sistemas Informáticos de Control de Sistemas de Producción*. PhD thesis, Dpto. de Ingeniería Eléctrica e Informática, University of Zaragoza, June 1990. A PhD thesis.

[Micovsky, 1990] A. Micovsky, L. Sesera, M. Veishab, and M. Albert. Tora: A petri net based tool for rapid prototyping of fms control systems and its application to assembly. *Computers in Industry*, 15(4):279–292, 1990.

[Sahraoui, 1987] A. Sahraoui, H. Atabakhche, M. Courvoisier, and R. Valette. Joining petri nets and knowledge based systems for monitoring purposes. In *Proc. of IEEE International Conference on Robotics and Automation. Raleigh, North Carolina*, pages 1160–1165, March 1987.

[Sibertin, 1985] C. Sibertin-Blanc. High-level petri nets with data structures. In *Proc. of Workshop on Applications and Theory of Petri Nets. Finland*, June 1985.

[Valette, 1988] R. Valette, J. Cardoso, H. Atabakhche, M. Courvoisier and T. Lemaire. Petri nets and production rules for decision level in fms control. In *Proc. of IMACS 12th World Congress on Scientific Computation. Paris*, volume 3, pages 522–524, 1988.

[Villarroel, 1990] J.L. Villarroel. *Integración Informática del Control de Sistemas Flexibles de Fabricación*. PhD thesis, Dpto. de Ingeniería Eléctrica e Informática, University of Zaragoza, September 1990. A PhD thesis.

# IMAGE PROCESSING IN COMBUSTION MANAGEMENT

## J. Hirvonen* and K. Ikonen**

*Technical Research Centre of Finland, Laboratory of Electrical and Automation Engineering,
Otakaari 7B, SF-02150 Espoo, Finland
**Imatran Voima oy Research and Development Division, P.O.X. 112, SF-01601 Vantaa, Finland

Abstract.    This paper describes an image processing system that calculates real time information
from the combustion process itself. The applications of the system on burner and supplemental fuel
adjustment, slag formation, and ignition trend monitoring are also discussed. Finally, two new types
of combustion control based on the information given by the developed system are discussed.

Keywords.    Image processing, management systems, power station control, combustion, digital
computer applications

## INTRODUCTION

Modern power plants are among the most highly automated and centrally controlled production facilities.
Increased energy costs have placed demands on power
producers for increased burning efficiency, high equipment availability, low maintenance costs, and safe operation. Environmental constraints have imposed lower limits on emissions which require the producers to pay
further attention to the combustion characteristics in their
plants. Although central control, automation, and monitoring systems have helped to satisfy many of these concerns, the energy efficiency goals continue to rise and
emission standards get continually stricter.

In solid fuel combustion systems the common factor in
tackling these problems is improving the quality of the
burning. Combustion management is therefore one of the
most critical components in power production. The management and control of the burners are, however, only as
good as the information provided by the existing instruments. Flame data is one of the most important information sources, but conventional flame monitoring systems
are rudimentary in operation and they serve primarily as
safety devices. These systems are essentially no more
than flame detectors which indicate the presence or the
absence of the flame.

Some of the more advanced systems use a combination of
detectors to provide information on flame "flicker" which
is a basic indicator of flame stability. In almost all of
these cases, however, the information cannot be used for
fine tuning burners or optimizing combustion. Boilers designed to burn solid fuels such as pulverized coal often
have stability problems due to variations in the fuel quality. Flame detectors can do little to help to regain stability, so additional oil support is often used which becomes

expensive.

## SYSTEM CONFIGURATION

To achieve real time information from the combustion
process itself a new type of system was defined and developed. The developed system is depicted in Fig. 1. and it
consists of the following parts:

1)    The camera subsystem (not shown in Fig. 1) contains separate cameras for each monitored burner. The air
cooled semiconductor cameras are set perpendicularly to
the burners.

2)    The flame analyzing subsystem contains one specially designed flame image preprocessing card (analyzing
unit, AU, which is depicted in Fig. 2) for each camera.
These cards are essentially independent of each other but
can exchange information.

The cards analyze images from the cameras, each card
processes several images in one second so that true on-
line information is obtained.

For reliability purposes, the cards contain self diagnostic
features and there is one reserve part (CU) for each four
analyzing cards all the time on-line. The CU collects the
flame data from the AU's and sends it to the PC. The
cards can also be replaced during normal operation without loss of the flame detection (the tasks of the missing
card are automatically given to the reserve part).

3)    The burner management subsystem consists of a PC,
monitors, and a track ball. It collects and stores the information supplied by the analyzing units. It also displays the
collected information to the operators, maintains user in-

347

terface and configuration tools. Furthermore it stores the trend information into a fixed disk for later inspection whenever the boiler trips for any reason. It can also save the same information by user command e.g. for later comparison of fuel quality.

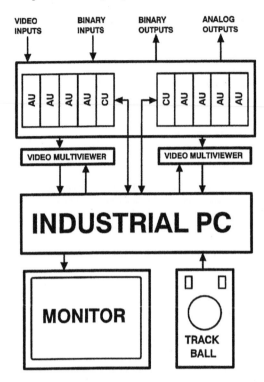

Fig. 1 The DIMAC system

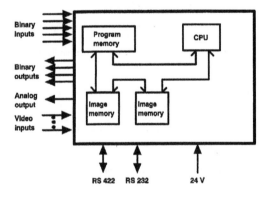

Fig. 2 Preprocessor unit

## APPLICATIONS

For the time being there are four installations of the DIMAC system in different types of power plants.

Next some applications of the developed system on one pulverized peat and coal fired plant with an opposed wall burning furnace will be discussed.

### Rauhalahti

The first system was installed in Rauhalahti in December 1988. Rauhalahti is an opposed wall fired plant with 265 MW fuel effect and it burns peat and coal. Coal is primarily used to replace oil as support fuel and to ensure a supply of economical fuel in the event of any failure in peat delivery and handling.

The installed system consists of eight cameras and AU's, two CU's, and a burner management subsystem.

The analyzing units calculate from the flame images the following parameters: ignition front, ignition distance from burner, flame stability, average and total irradiance of the flame and background as well as ignition threshold irradiance.

### Burner Adjustment

As the flames are analyzed individually, adjustments to the burners can be made faster, when either the quality of the fuel or load of the boiler changes.

### Supplemental Fuel Adjustment

Information on individual burner stability allows supplemental firing to be adjusted on an individual basis. The result is often greatly reduced consumption of more costly secondary fuel.

### Ignition Trend

The DIMAC system provides information to the operator on changes in the location of the ignition point and flame stability. Operators can observe trends that would lead to the loss of ignition in the future. Burners and fuel system can be adjusted to prevent loss of ignition.

### Slag Formation

Slag forming in the burner zone and its influence to flame form is discovered at an earlier stage.

## CONTROL SCHEMES

The following chapters discuss two possible control applications of the DIMAC-system. These ideas will be tested in the future.

### Modified Excess Air Control

New firing technology (firing with low excess air) together with the variations in firing rate or in the quality of the fuel may cause stability problems in pulverized fuel firing. These stability problems may cause the loss of the flame and a boiler trip or even furnace explosion.

In Fig. 3 the flame stability values of one particular

burner are shown when the firing parameters are changed. In the beginning of the period, support oil stabilizes the flame. When the support oil firing is terminated the flame stability decreases but is still rather high. It can be clearly seen that the flame stability decreases drastically as the fuel feed rate of the burner is low (in this case lower than 40 %).

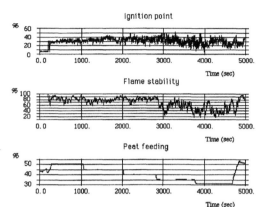

Fig. 3 Ignition point and flame stability when peat feed rate has been changed

Flame stability values can be used to control the support fuel firing and excess air of the furnace. If the flame stability of one particular burner is low the support firing can be started in that burner. This can be done automatically or manually.

In Fig. 4 there is an example of the excess air control of the furnace. The set point of the $O_2$ controller is calculated as a function of the flame stability values. When the flame stability values of the burners are high the set point of the $O_2$ controller is low. This ensures small flue gas loss and low $NO_x$ formation. On the other hand, when the flame stability is low due to the firing rate or the fuel quality, the set point of the $O_2$ controller has to be increased. This ensures that low excess air cannot cause the loss of the flame in the furnace.

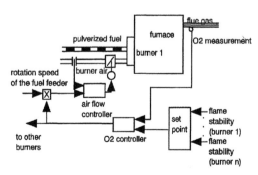

Fig. 4 Modified excess air control scheme

## Burner Air Correction Control

The pulverized fuel feeding is usually volumetric feeding and the density of the fuel varies greatly. Also due to the piping geometry, pulverized fuel is usually distributed

unequally between the burners in the same mill (Lawn, 1987). The fuel distribution is among other things a function of the firing rate. Due to the fact that the pulverized fuel mas flow cannot usually be measured on-line, the set point of the air flow controller is usually calculated according to the rotation speed of the fuel feeder and the burner air to the burners of the same mill is distributed equally. This causes an error in the set point of the burner air flow. These facts causes that the flames in some burners burn with too high excess air and some burners operate with too low excess air. Flames which burn with too high excess air contribute unnecessary $NO_x$ formation, and flame burning with too low excess air contribute combustibles in ash. Too low excess air may also cause stability problems in the flame of the burner.

Fig. 5 shows the flame irradiance values when the peat feeding has been changed. The burner flame irradiance is a function of the burner excess air and the burner fuel rate. Thus the flame irradiance values can be used to correct the burner air flow.

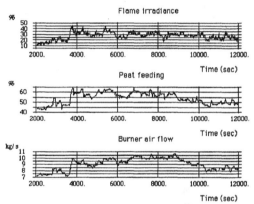

Fig. 5 Flame irradiance when peat feed rate has been changed

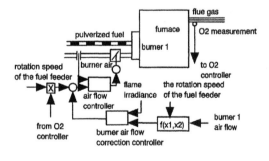

Fig. 6 Burner air correction control scheme

Fig. 6 shows an example of the burner air correction control.

The air flow to one particular burner is adjusted if the burner flame irradiance is not equal to the set point. The set point of the flame irradiance is a function of the burner air flow and the burner fuel mass flow. Due to the fact that the burner fuel mass flow cannot be usually measured on-line, an estimate must be used. The rotation speed of the fuel feeder can be used as an estimate of the burner fuel mass flow. The dependence of the set point of the flame irradiance on rotation speed of the fuel feeder and the burner air flow is evaluated by the measurements of these variables and the flame irradiance. These measurements have to be done in several operational points. During the evaluation of the dependence of the flame irradiance, burner fuel mass flow has to be measured off-line.

## CONCLUSIONS

In this paper a new image processing system that calculates real time information from the combustion process itself was presented. The applications of the system on burner and supplemental fuel adjustment as well as on slag formation and ignition trend monitoring were also discussed. Finally, new types of combustion control based on the information given by the developed system where proposed. The proposed control schemes have yet not been tested in real power plants. This work will be carried on in the future and it will eventually lead to a better combustion control.

## REFERENCES

Lawn, C. J. (1987). *Principles of Combustion Engineering for Boilers*, Academic Press, pp 50-58.

# QUALITY CONTROL OF FERRITE CORES THROUGH ARTIFICIAL VISION TECHNIQUES

**P. Campoy, J.C. Fernandez, J.M. Sebastian and R. Aracil**

*Department of Automatica, DISAM, Polytechnical University of Madrid, Spain*

Abstract. The present paper describes the system developed for detecting visual flaws in ferrite cores through artificial vision techniques. The physical dimension of the inspected parts, their high variety of types and their low cost have strongly determined the developed system. The whole system is described throughout the paper, analyzing specially the lighting system, the image processing algorithms and the image interpretation criteria. The obtained results are also described at the end of the paper.

Keywords. Artificial vision. Image Processing. Quality Control. Pattern Recognition.

## INTRODUCTION

The process to be automatized is the fault detection in the visual inspection of ferrite cores. During the manufacturing and manipulation of the parts, they can be totally broken or have some flaws, which make them not acceptable in a quality control inspection, and which have therefore to be detected.

The present quality control policy implies a visual inspection of 80% of the production which makes an average amount of 300 millions parts per year. This large amount of inspected parts gives to the quality control inspection an important role in the final cost. Additionally the more competitive market gives to the quality of the product an more relevant importance, where its automation avoids the human faults due to the monotony of the visual inspecting job. Additional advantages of the automation of the visual inspection are (Artley; Fu; Batchelor):

-On-line management of the defects.

-Automation of statistical and durable data.

-Integration to the manufacturing system, in order to obtain an on-line control of the process.

The automation of the visual inspection as the last part of the manufacturing automation has an increasing market in the last years, due to the new techniques in artificial vision and particularly to the more powerful hardware for image processing at reasonable prices (Batchelor; Bolles).

## SYSTEM SPECIFICATION

The specification of the system are determined by the features of the ferrite cores, by the defects to be detected and by the data of their production. The specifications of the system concerning the two first aspects are (figure 1):

-Cylindrical shape of the ferrite cores (with or without internal hole)

-Variable size of the parts between some millimeters and a few centimeters

-Uniform black tonality all around the ferrite core

-Defects may appear in the base or laterally all round the ferrite core close to the base

-Defects (flaws of the ferrite cores) don't have a different visual gray level

The specifications concerning the production

data are:

> -High production, which implies a short inspection time per part (about 750 ms)
> -Low cost of the product and therefore of its quality inspection
> -High variety of ferrite cores' sizes. This fact implies a high flexibility of the system

Fig. 1. Aspect of the ferrite cores to be inspected.

## DESCRIPTION OF THE SYSTEM

An overview of the system is shown in figure 2. The main subsystems are: the positioning and lighting system, the image acquisition and digitalizing system and the software structure (including image processing).

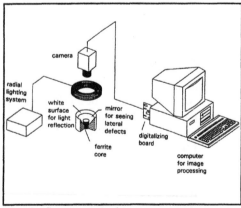

Fig. 2. System overview.

### Positioning and Lighting System

The first problem for the visual inspection

arises from the fact that defects may appear in the base as well as all round the lateral of the ferrite core. The simultaneous inspection of all the defects in one half of the part is solved by positioning the ferrite core within a conic mirror, as shown in figure 3. In this way the image taken from above contains one base of the ferrite core as well as its lateral reflected on the mirror (see figure 4).

The lighting system is chosen to enhance the flaws from the background, allowing a visual detection of the defects through image processing techniques. This aim is achieved through a ring of light transmitted by optical fiber like in figure 3.

The ring of light is reflected on the upper part of the cone where the ferrite core is positioned, in such a way that the lateral of the core is uniformly lighted but in the flaws which lay in shadow. This lighting system allows a visual detection of defects as shown in figure 4.

The cone (containing the mirror and the white surface for light reflection) where the part is inserted is changed accordingly to the different types and sizes of ferrite cores. The distance **d** to the light source is also changed, in order to adjust the reflecting angle of figure 3 in concordance to the new angle ß. This change of cone and adjustment of the distance to the light source gives the system the desired flexibility for the different types and of ferrite cores.

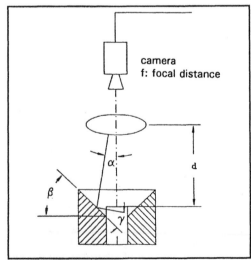

Fig. 3. Conic mirror and ferrite core positioned into it.

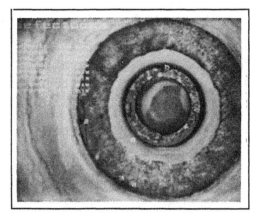

Fig. 4. Defects detected on a ferrite core.

## Acquisition and Digitalizing System

The acquisition system is one CCD matrix camera of 756x581 pixels which gives out in the standard black and white video format the image taken to the ferrite core. For the new industrial prototype two cameras are required for taking one image from each side of the ferrite core, once it is positioned in a double cone as shown in figure 5.

The digitalizing and hardware processing system is a commercial board (MATROX-MVP) placed on a AT compatible computer. This board digitalize the video signal into 512x512 pixels matrix and allows a hardware processing of most of the standard image processing functions.

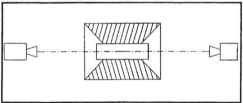

Fig. 5. Double cone-two cameras system for inspecting both sides.

## SOFTWARE STRUCTURE

An overview of the software structure is shown in figure 6.

The first module is a user friendly interface for defining the different types and sizes of ferrite cores. This module also helps the user to define and adjust the variable parameters of the positioning and lighting system according to each ferrite core, as well as the rejection criteria according to the detected defects. This

data is stored in the Ferrite Core Data Base for its ulterior on-line use by the module for Defects Detection.

The module of Defect Detection is an image processing module which extract the flaws from the image of each inspected ferrite core. The processing of the image is carried out in two regions of interest (ROI) which are defined in the Ferrite Core Data Base and which match respectively to the base and reflected lateral of the part. In these ROIs an image segmentation algorithm based in thresholding is carried out for flaw detection. Next module correspond to the measurement of size and position of the detected defects. In this module close flaws are associated as one defect and the resulting defects are quantify in size and position relative to the base. The size of the area in which all defects are considered as only one is an parameter adjustable by the user. This parameter is set according to the defect criteria of each type of ferrite core.

Last software module is the decision module of the quality of the inspected ferrite core. This module take into account the criteria given by the user and stored in the Ferrite Core Data Base,which determines the quality of each type of ferrite core in function of the size and number of defects detected by the prior module. The output of this module is the action over the actuators in order to reject the fault parts. This module also stores statistical data of the production and quality control of ferrite cores.

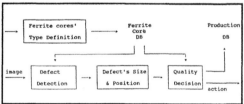

Fig. 6. Software structure.

## CONCLUSIONS

In this paper a system for an automatic fault detection in the visual inspection of ferrite cores has been presented. The main features of this system are related to the inspected parts, which have radial symmetry, are small, have a low price and there are a high variability of them. The presented system is an on-line, flexible quality control system, which also can be used for other products of the manufacturing industry with similar features.

The obtained results for this application are:

* The developed software allows a high flexibility in inspecting different ferrite types by easily changing of the conic mirror set. Presently up to 50 different types can be inspected.

* The above mentioned flexibility is achieved by an user friendly interface which easily allows the definition for new ferrite types in terms of physical features and inspection parameters.

* The system accuracy is about 0.25 $mm^2$, standard, $1mm^2$ worst case, depending on the ferrite-core size.

* The inspection time, per part, is mainly restricted by the mechannical device which takes 1 second for positioning each ferrite under the image acquisition system.

## REFERENCES

Artley, J.W. (1982). Automated visual inspection systems can boost quaility control affordably. Ind. Eng. 14, n° 12, 28-32.

Batchelor, B.G., Cotter, S.M., Heywood, P.W., and Moot, D.H. (1982). Recent advances in automated visual inspection. Proc. SPIE, Robot vision and Sensory Controls, 392, 307-326.

Batchelor, B.G., Bowman, C.C., Chow, K.W., Goodman, S., McCollum, A.J., and Rowland, S. (1986). Developments in image processing for industrial inspection. Proc. SPIE, Automated Inspection and Measurement, 730.

Bolles, R. (1981). An overview of image understanding applications to industrial problems. Proc. SPIE, Techniques and Applications of Image Understanding 281, 134-140.

Fu, K.S. (1983). Computer vision for automatic inspection. Proceedings, Robotic Intelligence and Productivity Conf., Detroit, Michigan, Nov. 1983, pp.7-15.

Sebastián, J.M., Campoy, P., Aracil, R. (1989). Artificial vision for automated quality control in the canned food industry. An European Workshop on Automation in the Food Industry, Dublin, Nov. 1989.

# DEVELOPMENT OF A COMPUTER VISION SYSTEM TO TRIDIMENSIONAL MEASURE OF LAMINATION STEEL PROFILES

**A. Diez\*, J. Cancelas\*, J. Diaz\*, C. Nerriec\* and F. Obeso\*\***

*\*Department of Electrical, Electronic of Computers and Systems, University of Oviedo, Spain*
*\*\*Process Informatic (Ensidesa), Spain*

Abstract. We measure all the dimensions of one H profile of a rolling mill, reposed and cold, by means of using two matricial video cameras and a software package developed for this application. The software package develped identifies the intersection of the profile edges with a thin line drawn by a laser beam. We made the integral calibration of the system, and we get, in the worst case, an accuracy of millimetre tenths. The designed prototype has been validated for real applications in Ensidesa company. On the basis of this project we can attempt the measurement of these profiles, when hot and on the rolling mill.

Keywords. Image processing, Steel industry, Computer applications, Data acquisition, Heuristic programming

## INTRODUCTION

When the National Steel Company ENSIDESA asked the Oviedo University System Engineering Area to develop a prototype for the measurement of the tridimensional dimensions of the lamination steel profiles, that measurement was done by hand using gauges.

Due to the low efficiency of this manual method and the need of automating the process, several automated measuring solutions were considered.

Nowadays there have been developed many measuring methods based on the use of laser beams. These methods a precise and reliable, but they are not suggested for tridimensional measures, being the main application fields surface analysis and one dimensions or two dimensions measures.

Another candidate method was the use of ultrasonic waves, but they present the same disadvantages as laser beams.

Lastly, the third method, that is showing a great development, are artificial vision systems.

Because the goal is to do 3-D measures, it is necessary to use two cameras and apply stereoscopic techniques to obtain every interesting dimension of the profile. It is possible with these techniques, knowing the relation between the points of the left image and the point of the right image, to fix the X, Y, Z, coordinates in an absolute reference system. Once the coordinates of a few points of interest are known the obtaining of the profile dimension is obvious.

## CAMERAS CALIBRATION

### Overview

A video camera makes a transformation from the ordinary three-dimensional space in which the objects are, to a bidimensional space, that of the photographed or digitalized image. If the lens keeps as straight lines the projections of the three-dimensional straight lines, the transformation appears as an ideal conic perspective.

When we measure the distance between two pixels of the image, we are using a flat coordinates system whose axis are arbitrarily placed on the image (usually it is used a horizontal X axis, growing left to right, and a vertical Y axis, downwards growing, and the coordinates origin is placed at the top-left corner of the image). Though we can suppose a third axis (Z) orthogonal to the two previous ones, this coordinates system would not necessarily coincide with the three-dimensional coordinates system in relation to which we want to take the measurements. However it is possible to establish a relation between both of them, taking into account:

- The perspective transformation due to the lens.
- A change of scale (due to the difference between the three-dimensional world units, normally millimetres, and the image units, normally pixels).
- A rotation of the camera coordinates system, in relation to the world reference framework.
- A translation of the camera coordinates system, in relation to the world reference framework.

In Fig. 1 the camera coordinates system (X1, Y1, Z1), the image coordinates system (u,v) and the world reference framework (**X**, **Y**, **Z**) can all be seen.

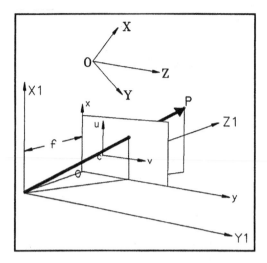

Fig. 1.     Camera model: Shows the relation between the image coordinates system and the exterior coordinates system.

The relationship between the image coordinates system (u,v) and the world coordinates system (**X, Y, Z**) can be stated in the following matrix equation:

$$\begin{pmatrix} u \\ v \\ 1 \end{pmatrix} = \begin{pmatrix} \alpha_u \vec{r}_1 + u_0 \vec{r}_3 & \alpha_u t_x + u_0 t_z \\ \alpha_v \vec{r}_2 + v_0 \vec{r}_3 & \alpha_v t_y + v_0 t_z \\ \vec{r}_3 & t_z \end{pmatrix} \begin{pmatrix} X \\ Y \\ Z \\ 1 \end{pmatrix} \tag{1}$$

Where $\vec{r}_1$, $\vec{r}_2$, $\vec{r}_3$ are the rows of the rotation matrix (each of them made of three components); $\alpha_u$, $\alpha_v$ are scale factors which include the effect of the perspective and the change of units; and $t_x$, $t_y$, $t_z$ are the components of the vector that shows the translation of the image coordinates system in relation to the world coordinates system. We will call the 4x3 transformation matrix that can be seen in Eq. (1) **M**.

Given the coordinates (u,v) of one point on the image, it is not possible to determine with just one camera the spatial coordinates of the point, as in the equation system originating from Eq. (1), there are just two equations linearly independent, but three unknown quantities (**X, Y** and **Z**).

If we have a pair of cameras and the transformation matrix $M_1$ and $M_2$ are know for each of them, it is now possible to determine the three-dimensional coordinates of one point, starting from the coordinates of that point in the image that each camera takes, as now there are four equations and three unknown quantities.

The main problem lies in finding the most accurate computing of the $M_1$ and $M_2$ matrices for a fixed setting of the cameras. This problem is named

calibration, and consists of giving to the computer a set of 3-D and 2-D coordinates (that means, X, Y, Z, u and v data from several points) from which the required parameters ($\vec{r}_1$, $\vec{r}_2$, $\vec{r}_3$, $t_x$, $t_y$, $t_z$, $\alpha_u$ and $\alpha_v$) are deduced. (All of them make a whole of twelve unknown quantities).

Solution Method.

We shall write the **M** matrix as follows:

$$M = \begin{pmatrix} l_{11} & l_{12} & l_{13} & l_{14} \\ l_{21} & l_{22} & l_{23} & l_{24} \\ l_{31} & l_{32} & l_{33} & l_{34} \end{pmatrix} \tag{2}$$

Letting $\vec{l}_1 = (l_{11}\ l_{12}\ l_{13})$, $\vec{l}_2 = (l_{21}\ l_{22}\ l_{23})$, $\vec{l}_3 = (l_{31}\ l_{32}\ l_{33})$ (that means, each of the **M** rows without its last component, the fourth one), and letting $\vec{X}_i = (X_i\ Y_i\ Z_i)$ be the 3-D coordinates of a point, and $(u_i, v_i)$ be the 2-D coordinates of its projection on the image, which are both already known, then from Eq. (1), the following equation system is posed:

$$\left. \begin{aligned} \vec{l}_1 \vec{X}_i - u_i \vec{l}_3 \vec{X}_i + l_{14} - u_i l_{34} = 0 \\ \vec{l}_2 \vec{X}_i - v_i \vec{l}_3 \vec{X}_i + l_{24} - v_i l_{34} = 0 \end{aligned} \right\} \tag{3}$$

For each point added the system increases by two equations, that is way six points at least are required to solve the twelve unknown quantities $l_{ij}$. In practice, in order to minimize the measurement errors in the 3-D and specially 2-D coordinates, many more than six points are used (we have used 48). That gives a system with more equations than unknown quantities, which should be solved by numerical methods. These methods should minimize the solution error, which means, if we write the system in the following way $A\vec{l} = 0$ (letting $\vec{l} = (\vec{l}_1^T, l_{14}, \vec{l}_2^T, l_{24}, \vec{l}_3^T, l_{34})^T$), the optimum solution would be the one that minimizes $[A\vec{l}]$.

As Faugueras and Toscani (1986) say, imposing the restriction $[\vec{l}_3] = 1$ (that it is kept though we might transform the coordinates system, which is very useful), and rewriting de system in the following way

$$B\vec{X}_9 + C\vec{X}_3 = 0 \tag{4}$$

letting $\vec{X}_9 = (\vec{l}_1^T, l_{14}, \vec{l}_2^T, l_{24}, l_{34})^T$ and $\vec{X}_3 = (\vec{l}_3)^T$ the optimum solution becomes:

$$\left. \begin{aligned} \vec{X}_9 &= -(B^t B)^{-1} B^t C \vec{X}_3 \\ \vec{X}_3 &= \tfrac{1}{\lambda}(C^t C - C^t B (B^t B)^{-1} B^t C) \vec{X}_3 \end{aligned} \right\} \tag{5}$$

If we pose the second equation in the form $\lambda \vec{X}_3 = D\vec{X}_3$

it becomes obvious that $\vec{X}_3$ should be an eigenvector of the **D** matrix. More precisely, it is proved that it is the eigenvector corresponding to the minor eigenvalue (in absolute value) of **D**. Once we have obtained this eigenvector, we will have three components of the matrix **M**, and by means of the first equation of Eq. (5) it is possible to find $\vec{X}_9$, which will give us the remaining nine components.

From the matrix **M** it is possible to determine the rotation matrix of the camera, as well as the translation vector and the internal parameters $\alpha_u$, $\alpha_v$. The same calculation process must be followed for the second camera.

## Implementation

For the implementation of this method several programs have been developed. The first of them makes the acquisition of data, for which the operator must place a "calibration plane", which consist of a plate holed in known coordinates points, under the cameras. In addition, the operator must type in the 3-D coordinates of these points. The computer digitalizes the scene for both cameras and searches the projections of these points in each camera image. That gives the computer the coordinates (u,v).

Afterwards, with these data the **B** and **C** matrix of Eq. (4) are built up according to the following expression:

$$B = \begin{pmatrix} X_i & Y_i & Z_i & 1 & 0 & 0 & 0 & 0 & -u_i \\ 0 & 0 & 0 & 0 & X_i & Y_i & Z_i & 1 & -v_i \\ \vdots & & & & \vdots & & & & \vdots \end{pmatrix} \quad (7)$$

$$C = \begin{pmatrix} -u_iX_i & -u_iY_i & -u_iZ_i \\ -v_iX_i & -v_iY_i & -v_iZ_i \\ \vdots & \vdots & \vdots \end{pmatrix} \quad (8)$$

Operating with these matrices, the matrix **D** of Eq. (6) is calculated. Next the **D** eigenvalues are calculated by means of the Mires method. The software implemented searches for the minor eigenvalue and its associated eigenvector (which is $\vec{X}_3$, as we already know). With that we calculate $\vec{X}_9$ by means of Eq. (5). With these vectors (two for each of the cameras) the calibration matrix $M_1$ and $M_2$ are built and saved on disk in order to be used by the measure system, which will be described later.

## PROFILE MEASURING

### Overview

It is necessary to find out the profile dimensions shown in Fig. 2, from the profile images taken by each camera. This creates two problems: On one hand it is necessary

to search for the minimum number of three-dimensional points which give us the required information. On the other hand we must create a method to look for the projection of these points on the images of each camera.

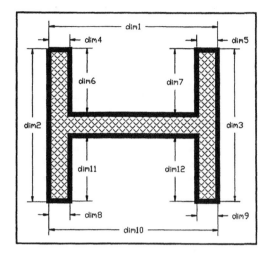

Fig. 2.    Profile dimensions to obtain.

Once these problems have been solved the calculation of the profile dimensions seems trivial, As it is possible to deduce its three-dimensional coordinates from the flat coordinates of these points in the images (by means of the $M_1$ and $M_2$ matrix previously calculated), and once we have the coordinates, the dimensions are calculated as distances between points.

We have decided to use as "key points" for the dimensions calculation, the ones labelled as 4, 5, 6, 7, 8 and 9 in Fig. 4.

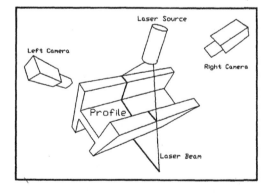

Fig. 3    Disposition of the cameras and the laser source for the profile measurement.

To identify these points in the image, in the most unmistakable form, we have used a laser beam for the profile illumination (see Fig. 3). This technique draws a very thin line on the profile core, therefore the method to be implemented becomes the search for certain

357

straight lines and their intersections.

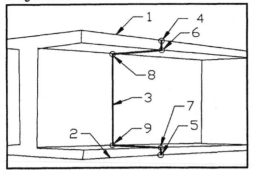

Fig. 4.    Schematic image from one camera, where the points and lines to locate are labelled. The line drawn by laser is represented in double width.

### Search of the Points in the Images

The method implemented to locate straight lines, was created for this specific application. This is not a general method, like the Hough transform or other algorithms for straight lines detection, it presupposes a very specific image model , which means: the line drawn by the laser beam on the profile core must be, approximately, perpendicular to the profile wings, and must appear in the image with a slope near to the vertical, and approximately in the middle of the image. Moreover, in the image the edges corresponding to the upper and lower wings of the profile must appear. These assumptions reduce the computational load of the process remarkably, allowing the time required for obtaining the straight lines to be decreased.

To give priority to the calculation speed over the algorithm generality, we have used different methods to obtain each straight line and each point of the image, as we will explain in following paragraphs.

Acquisition. The images are digitalized in a format of 512 x 512 pixels and 256 grey levels (in linear correspondence with the brightness of each point), for each camera.

Wings edges. As preprocess we apply a Sobel filter for horizontal edges, followed by a decrease in the brightness level (for noise reduction) and a slimming in the horizontal lines, through a displacement and subtraction of the images. After this, we have a binary image (with just two grey levels, 0 and 255) in which we must search for the lines 1 and 2 of Fig. 4. To do so, the image is dredged in columns, starting from the middle. For each of the columns, we search for the brightest points in the upper and lower sides of the image, their coordinates are stored in two separate arrays (one for the upper line and the other for the lower line). With these arrays we calculate the regression straight line of the points. In order to minimize the error, we repeat the calculation, but this time only with the points nearest to the straight line previously calculated as first approximation. The equations of the

straight lines 1 and 2 (for each camera) are obtained and saved.

Core line detection. This is the number 3 line in Fig. 4. We apply a Sobel filtering for vertical edges on the image as a preprocess. Next, starting at one side of the image, we dredge the image by rows, searching for pixels whose value exceeds a threshold value. When one is found, the following pixel in which the value decreases is searched for, and its coordinates are stored in an array. This technique finds the edge of the laser beam line, since the Sobel filtering produces a gradient image. With the values stored in the array, we calculate the regression straight line of these points, and we make, as in the preceding case, a more precise approximation after the elimination of the furthest away points. The equation of the straight line is stored (one for each camera).

External points of the wings. Now the question is to determine the points labelled as 4 and 5 in Fig. 4. To do so we calculate the intersection of the two lines previously found; we search for the region with maximum average in the proximity of these intersection points, following the wings lines direction. In this region there is a great concentration of bright points, corresponding to the laser beam mark. The coordinates of these points, which are found this way, are saved (they are two in each image).

Internal points of the wings. To locate these points (6 and 7 in Fig. 4) we start from the coordinates of the external points of the wings, previously found, and from a profile image on which we have applied the vertical Sobel filtering. Over this image we move a small window, starting from the external points of the wings and going towards the centre of the image. In this window the average bright is computed, as well as the relation between the central point brightness and the brightness of its adjacent pixels, in order to determinate the point in which the laser beam mark changes its direction. This is the required point, once it is found its coordinates are saved (they are two in each image).

Core points. When the core straight line is searched (seen before), and some furthest away points are discarded after the first approximation, we search among the "survivors" those two points whose vertical coordinates are respectively maximum and minimum. These coordinates are saved on disk.

### Reconstruction of the 3-D Coordinates

From 2-D coordinates of the points previously found for each camera, together with the matrix $\mathbf{M}$ and $\mathbf{M'}$ previously obtained (see calibration paragraph), we must calculate now the three-dimensional coordinates of these points, in order to get the real dimensions of the profile.

It is necessary to remark that the matrix $\mathbf{M}$, obtained by the Faugueras-Toscani method, has the constraint $\|\vec{t}_3\|=1$. This means that there are a scale factor k, coming true:

$$k \begin{pmatrix} u \\ v \\ 1 \end{pmatrix} = M \begin{pmatrix} X \\ Y \\ Z \\ 1 \end{pmatrix} \qquad (9)$$

Then, the unknown k is added for each camera to the X, Y and Z unknowns. If we pose the following equations system:

$$\left. \begin{aligned} k \begin{pmatrix} u \\ v \\ 1 \end{pmatrix} &= M \begin{pmatrix} X \\ Y \\ Z \\ 1 \end{pmatrix} \\ k' \begin{pmatrix} u' \\ v' \\ 1 \end{pmatrix} &= M' \begin{pmatrix} X \\ Y \\ Z \\ 1 \end{pmatrix} \end{aligned} \right\} \qquad (10)$$

we have the data: u, v, u', v' (2-D coordinates of the point "seen" for each camera), M, and M' (calibration matrix for each camera), and we have the unknowns: X, Y, Z (3-D coordinates of the point), k and k' (scale factors for each camera).

As we see, there are six equations but five unknowns quantities. If all the involved values were exacts, and the system had solution, it would be clear that the solution of five of the equations should satisfy the sixth. Then we could scorn any of the equations and solve the system made up of the five remaining ones. However, in our case there are measurement errors in the u, v, u', v' data, inherent in the digitalization process, that is why we will not be able to apply this method.

From the Eq. (10) the pose of the following equation system is immediate.

$$\begin{pmatrix} -u & 0 & l_{11} & l_{12} & l_{13} \\ -v & 0 & l_{21} & l_{22} & l_{23} \\ -1 & 0 & l_{31} & l_{32} & l_{33} \\ 0 & -u' & l'_{11} & l'_{12} & l'_{13} \\ 0 & -v' & l'_{21} & l'_{22} & l'_{33} \\ 0 & -1 & l'_{31} & l'_{32} & l'_{33} \end{pmatrix} \begin{pmatrix} k \\ k' \\ X \\ Y \\ Z \end{pmatrix} = \begin{pmatrix} -l_{14} \\ -l_{24} \\ -l_{34} \\ -l'_{14} \\ -l'_{24} \\ -l'_{34} \end{pmatrix} \qquad (11)$$

We deduce from the last equation of this system the following expression:

$$|l'_{31}X + l'_{32}Y + l'_{33}Z + l'_{34} - k'| \qquad (12)$$

which must equals zero if X, Y, Z and k' satisfy the five previous equations. Then, our policy will consist on solving the system constituted by the first five equations of Eq. (11), and later substitute the solution in Eq. (12), to obtain an estimation of the error within the result. The values of the data u, v, u' and v' are this way adjusted in little increments until Eq. (12) value reaches an acceptable small value which we have fixed to 0.001). The adjusts in u, v, u' and v' are made under the care that new values must satisfy the equation of the straight line to which the involved point belongs.

### Dimensions Calculation.

Once we have the 3-D coordinates of the points labelled 4, 5, 6, 7, 8 and 9 in fig. 3, the calculation of the profile dimensions showed in fig. 2 is immediate. It is enough to calculate the distances between the points, remembering that in the adopted world coordinates system the Z coordinate measures the distance to the plane in which the profile lies. Taking this into account we deduct immediately the following relationship.

$$\begin{aligned} dim1 &= \sqrt{(X_4 - X_5)^2 + (Y_4 - Y_5)^2} \\ dim2 &= Z_4 \\ dim3 &= Z_5 \\ dim4 &= \sqrt{(X_4 - X_6)^2 + (Y_4 - Y_6)^2} \qquad (13) \\ dim5 &= \sqrt{(X_5 - X_7)^2 + (Y_5 - Y_7)^2} \\ dim6 &= dim2 - Z_8 \\ dim7 &= dim3 - Z_9 \end{aligned}$$

To obtain the remaining dimensions of the profile it is necessary to turn it around and repeat the execution. There is another possibility not implemented yet which consist on another couple of cameras placed below the profile (in this case the surface on which the profile lies must obviously be transparent).

The developed software package will finally plot on the screen a skeleton representation of the profile, with the found dimensions.

## CONCLUSIONS

The hardware we have used is made of two solid state cameras (which have been chosen because of its toughness as contrasted with the vidicon type, also they do not have parabolic distortion). They are connected, by means of a multiplexer board, to a digitizer board of 512 x 512 pixels and 256 grey levels. We also own another board which had been specially designed for image process, which it was capable of making a certain type of convolutions and operations in a very short time. This was as well connected, by means of an own bus, to the digitizer board. This equipment was connected to a 80386 computer with a floating point co-processor.

The developed software runs under the MS-DOS and uses a set of special libraries for the interface with the digitizer and processor boards.

As we have said in the introduction, the prototype has been developed for the National Steel Company ENSIDESA, and it has been tested on H profiles. In these profiles the least tolerance is the one corresponding to the wing and the core, that in the worst of the cases cannot be more than ± 1.0 mm. (UNE 36527, 36528 and 36529 rules). With the developed algorithms the measurement error is always inferior to 0.6 mm, that is why the quality control is guarantied. The accuracy could be increased by the use of a major resolution.

The time required for the calculation of the profile dimensions (once we already have the calibration matrices of the cameras) is of about 50 seconds, counting from the moment in which the program has been started until the moment in which the measured profile drawing appears on sight. This is a little time, as it is a static measurement.

## REFERENCES

Basseville, M. (1982). Contribution à la détection séquentielle de ruptures de modèles statistiques. Thesis presented to L'Université de Rennes 8 June 1982.

Faugueras, O. D., and G. Toscani (1986). The Calibration Problem for Stereo. Proc. Computer Vision and Pattern Recognition, June, 15-20.

Fu, K. S., R. C. González, and C. S. G. Lee (1987). In McGraw-Hill Ed., Robótica, Detección, Visión e Inteligencia.

Gennery, D. B. (1979) Stereo-Camere Calibration. Proc. Image understanding Workshop, November, 101-108.

Isaguirre, A., P. Pu, and J. Summers (1985). A New Development in Camera Calibration: Calibrating a pair of mobile cameras. Proc. Int. Conf. Robotics and Automation, 74-79.

Lyvers, E. P., and O. R. Mitchell (1989). Subpixel Measurements Using a Moment-Based Edge Operator. IEEE Transactions Pattern Analysis and Machine Intelligence, December, 1293-1309.

Machuca, R., and A. L. Gilbert (1981). Finding Edges in Noisy Scenes. IEEE Transactions Pattern Analisys and Machine Intelligence, January, 103-111.

Reimar, K. L., and Y. T. Roger (1988). Techniques for Calibration of the Scale Factor and Image Center for High Accuracy 3-D Machine Vision Metrology. IEEE Transactions Pattern Analysis and Machine Intelligence, September, 713-720.

Treiber, F. (1989). On-Line Automatic Defect Detection and Surface Roughness Measurement of Steel Strip. Iron and Stell Engineer, September, 26-33.

Yakimovsky, Y., and R. Cunnigham (1978). A system for extracting three dimensional measurements from a stero pair of TV cameras. Comput. Graphics Image Processing, vol 7, 195-210.

# PATTERN RECOGNITION APPLIED TO FORMATTED
# INPUT OF HANDWRITTEN DIGITS

**F.A. Vazquez and R. Marin**

*Department of Systems Engineering and Information Languages and Systems, University of Vigo, Spain*

**Abstract :** The system presented in this paper uses Optical Character Recognition (OCR) techniques to allow massive capturing of handwritten data from product demand information contained in forms filled out by vendors in their visits to clients. OCR is confronted through feature extraction followed by classification using neighrest neighbor clustering and/or exaustive search (Q-analisis [7] ). Details are given on features and both classification methods are compared as far as error rate, speed and memory requirements are concerned.

**Keywords :** Optical Character Recognition, Pattern Recognition, Image Processing, Learning Systems, Artificial Intelligence, Cybernetics, Textile Industry, Computer Interfaces, Handwritten Digits, Data Adquisition.

## INTRODUCTION

Formatted input consists in reading information that is spatially distributed in a predictable manner. Usually, a form will consist of a set of delimiters (possibly boxes) that contain characters belonging to a certain set.

In order to develop a system capable of transfering the data contained in a form to a database, the image data contained in each box must be extracted from it and latter recognized as to correspond to a certain character.

Recognition is specially difficult when characters are handwritten and even the best OCR systems, humans, present an error rate of 4% in absense of context [6] .

Many digitizing-table based systems have been developed for formatted data introduction. The main problem with these systems is that digitizing tables are uncomfortable and expensive compared to paper. Paper based systems have been developed by AT&T [2,3], Hewlett Packard and a number of Japanese Companies but a need exists for Off The Shelf products that easily adjust to any company's needs.

Neural Network based systems like AT&T's ZIP-CODE reader [2,3] and the Necognitron [4] seam to work very well but are not oriented toward formatted input.

## NEW CONTRIBUTION

This paper is a summary of a project developed by the Department of Systems Engineering and

Information Languages and Systems of the University of Vigo as part of a CIM project for a textile company called Pili Carrera S.A.. Its mission is to provide massive capturing of data concerning production needs and trends.

Before developing this system, data capturing was the great bottle-neck of the company´s production cycle.
Without this data, production needs and trends must be guessed. Acurrate guessing is extremely difficult in textile markets and usually produces situations of excess stock (possibly non-reusable).

The project consists of a system that reads formatted forms which contain handwritten digits in a collection of small boxes. The system includes a user interface for correction of possible errors and queries a data base to guarantee data consistency.

This system has the following advantages over existing systems :

• It is paper based, that is, no expensive and uncomfortable digitizing tables are needed for input.
• Its user interface is totally integrated in the company's needs and extremely intuitive.
• It uses database query capabilities to assure data consistency and to correct errors produced both during recognition and data introduction.
• It can doubt on a decision and, instead of making an immediate hard decision for a digit keeps a record of possible alternative candidates. This information can be used during various stages of the system. This state of doubt is reflected by the user interface.
• A tool has been developed that allows the design

of any particular form using a graphical builder. That makes the system sufficiently Off the Shelf and available for any user without need for special installation or maintenance.

## SYSTEM OVERVIEW

The system can be separated into the following subsystems .

- Form reader
- OCR (Optical Character Recognition)
- Consistency check and contextual error correction
- Error Correction

Since the company´s original forms only contained numerical information, character recognition has been reduced to digit recognition. This limitation is not inherent to the system and any set of characters can be used. It is true though, that features have been chosen to enhance digit recognition and that it is much easier to obtain a low error rate if the set of characters from which to decide is small.

## FORM READER

The mission of the form reader is to extract the image data inside the boxes that constitute the form. These boxes should contain either a digit or a blank space.

The main difficulty involved in developing this subsystem is providing immunity to rotations, translations and spatial deformations produced by the scanning system ( a scanner or a FAX ) and dealing with large amounts of image data ( 300 d.p.i for an A4.).

The image data extracted by this subsystem is fed to the OCR subsystem. Once all of the boxes have been processed, it provides a table containing the data in the form.

This table will be used by the subsystems that follow it. The forms used where for a company in the textile sector which receives data from their vendors and consist of a header containing the client and form numbers and the date, also, a series of lines that contain a product number, color number, and number of units per size to be ordered (Fig. 1 ).

## OCR

The OCR subsystem processes the image data provided by the form reader to determine the character to which the data most resembles.

The OCR used in this case is feature based, that is, compresses the image data that describes the character into a feature vector which is later used to classify the character. The feature vector consists of a 54 dimensional binary vector which includes closed contours, straight segments, curves, number of horizontal and vertical crossings, number of strokes, etc..

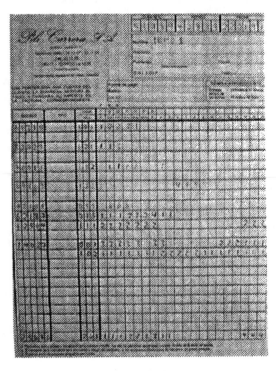

Fig. 1 Typical Form

Two types of clasification methods have been developed : the first one is a classical theoretical decision method which makes decisions based on the statistics of feature occurences for each character [1] ; the other method reviews all the history of a learning base searching for exact occurences of the feature vector.

Both methods construct a merit vector in which each component corresponds to a character of the set and its value is greatest for the component to whom the sample character most resembles.

Syntactical information [1] is also used in detecting some geometrical and topological features and the decision space is generated by learning from experience.

## CONSISTENCY CHECK AND CONTEXTUAL ERROR CORRECTION

The subsystem destined to consistency check and contextual error correction makes sure that non existent client or product numbers introduced or interpreted are detected and corrected and that the sizes and colors interpreted for a certain model exist. This can avoid errors due to wrong introduction or interpretation.

## ERROR CORRECTION

The error correction subsystem provides a friendly mouse driven user interface which allows correction of introduction or interpretation errors not detected previously.

This subsystem is intended to be used during an initial stage until the system is considered very reliable and should tend to be used only for visualizing results or forms.

## FEATURE VECTOR

The Feature Vector is formed by 54 binary components indicating the absense or existence of a feature. The following types of features are considered (see Fig. 2 for examples) :

• Closed Contours. Can be small, large or huge and the first two can be centered in the upper, middle or lower parts.
• Concavities . Can be oriented towards the right, left, top or bottom or can be straight lines, and can be located in the upper, middle or lower parts.
• Horizontal Crossings. Mean number of times a horizontal line crosses the character. The number of crossings can be less than once or more than once. For example : A four or a nine have more than one mean horizontal crossings in the top and one in the bottom.
• Vertical Crossings.
• Loose extreme locations. Can be left or right and in the upper, middle or lower parts.
• Number of Strokes. Can be one or more than one.
• Stroke crossing locations. Can be left or right and in the upper, middle or lower parts.
• Horizontal lines. Can be in the upper, middle or lower parts.
• Zig-Zags.

Some other properties have been used that are inherant to the way segmentation was done. More details can be found in [5] or by contacting the authors.

## THEORETICAL DECISION

The learning base is a 10 by 54 array where each component represents the probability that a certain character presents a certain feature. This is equivalent to 10 vectors or points in a 54 dimensional space.

Each sample vector is compared to the 10 vectors or points in the decision space to find the neigherest neighbor. The merit vector components are a measure of how close the sample vector is to each candidate character. This is done by calculating the normalized inner product of the sample vector with each of the 10 learning base vectors.

## EXAUSTIVE SEARCH

The learning base is formed by 54 bit vectors where each bit represents the existence or absense of a feature and each vector corresponds to a sample vector created during the learning process.

This method creates a merit vector where the value

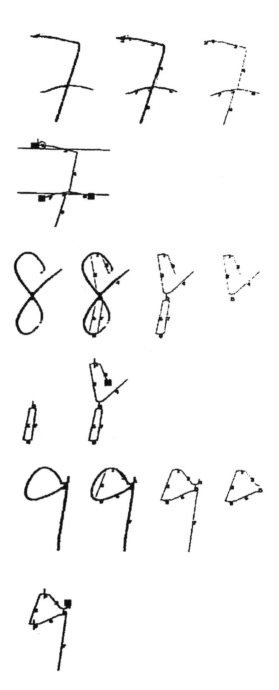

Fig 2. Some Features

of each component is :

$$M[i] = \frac{1}{ns_i} \sum_{j=0}^{di} w_j n_{ij}$$

where :
• nsi : number of samples in the learning base for

363

character i.
- di : maximum Hamming distance from the sample vector to any vector in the learning base that corresponds to character i.
- wj : weight for a Hamming distance of j.
- nij : Number of vectors in the learning base that are at distance j and correspond to character i.

The weights should decrease with distance so that small distances are favored over large ones, making them increase would generate a cost vector instead of a merit vector.

If $wj > nsi * wj-1$ for all j then, the method compares number of 0 distances and, in case of tie, compares number of 1 distances, and so on.

## THEORETICAL DECISION
### vs
### EXAUSTIVE SEARCH

Theoretical decision is faster and its data base is of a fixed sized and usually smaller than the other method´s data base. These favorable conditions are achieved by compressing all the experience in ten vectors.

The price to pay for this accumulative procedure is information loss and thus, an increase in error rate. A 12% error rate was achieved with theoretical decision while it dropped to 0.25 % with exaustive search.
These where achieved with a 50 Kb data base for exaustive search and a 4 Kb data base for theoretical decision. This project was developed in a UNIX environment where disk spaces under 1 Mb are far from being considered large.

Thus, the only compensation for error rate that could only be considered is speed. This brings us to a range of solutions :

- Classification could be a batch process and the user interacts with the error correction subsystem after classification, consistency check and contextual error correction. This is totally feasable in the case we have worked on and is the solution that has been implemented. Thus classification is the primary classification method and theoretical decision and some syntactical properties are taken into account in case of doubt.

- For interactive recognition, exaustive search seems slow and theoretical decision with or without exaustive search in case of doubt may be considered.

### References

- [1] King Sun Fu, " Syntactic Pattern Recognition and Applications " , Prentice-Hall.
- [2] Y. LeCun, B. Boser, J.S. Denker,..., " Backpropagation applied to handwritten zip code recognition ", Neural Computation, vol. 1, no. 4, pp. 541-551, 1989.
- [3] Y. LeCun, B. Boser, J.S. Denker,..., " Handwritten digit recognition with a backpropagation network ", Advances in Neural Information Processing Systems, vol. 2, pp. 396-404, 1990.
- [4] K. Fukushima, " Necognitron: A hierarchical neural network capable of visual pattern recognition", Neural Networks, vol. 1, pp. 119-130, 1988.

- [5] F. Vazquez, " P.F.C.: Sistema de Reconocimiento Automatico de Pedidos Manuscritos ", E.T.S.I. Telecomunicación Vigo, 1991.

- [6] Suen, Berthod & Mori, " Automatic Recognition of Handprinted Characters: The State of The Art ", Proc. IEEE, VOL.68, NO.4, April 1980, pp. 469-487.

[7] Ollero A. "Contribuciones al Análisis y Optimización Multicriterio de Sistemas Complejos". Tesis Doctoral. Universidad de Sevilla, 1980 .

# A 3-DIMENSIONAL ROBOTIC VISION SENSOR

T.B. Karyot

*Faculty of Aeronautics, Istanbul Technical University, Maslak, 80626 Istanbul, Turkey*

**Abstract**. A video camera based vision system is very helpful for
some robotic vision problems. But for real time object recognition
and localisation requirements, video data cannot be treated fast
enough to be feeded back into the control loop. In this work a new
approach and a new related sensor is developed in order to acquire
three dimensional position information about the external world of the
robot. The developed sensor uses active triangulation technique and
makes use of coherent monochromatic light. The sensor is very compact
and may be integrated within the end effector of the robot. The device
may deliver punctual three dimensional information of profile data,
within a field of wiew of 40 degrees and a range of up to three meters.

**Keywords**. Sensors; Position sensing; Robot vision; 3-D perception;
Cameras; Localisation; Opto-electronics

## INTRODUCTION

The utilisation of classical robot
manipulators is restricted to a well
structured environment. Even very
small differentiations in the structure
of the surroundings, may result to
failures of the tasks done by the robots.
To avoid the inadequate action of the
robots, intelligent motion control
structures have to be used within
robotic systems. These structures are
developed in hardware and software
level and uses appropriate artificial
intelligence techniques. The raw data
required by these structures are
collected through different sensory
inputs. These inputs are provided by
a wide range of sensors that are
shown in Fig. 1.

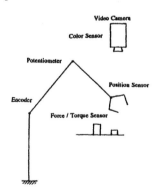

Fig. 1. Multiple Sensors Used
In A Robot

Today's advanced and future's
sophisticated robotic applications
require a steady flow of reliable
informations, generated by different
sensors. Correctly fused sensor data
will accelerate the control structure,
and the robot may gain an adaptive
behaviour. Hence adequately equipped
robots may move freely, even in an
unstructured and unknown environment.
For the new generation robot systems,
non contact multi sensing of the
external world, is going to become a
must.

The sensors used in robotic
applications may be divided into two
classes:
 -internal state sensors,
 -external state sensors.

Internal state sensors gives information
of the state of the robot itself, such as
position, velocity, acceleration of joint
angles or displacements. External state
sensors, on the other hand, allow a
robot to interact with its environment
in a flexible manner. They deal with the
detection of variables such as proximity,
range, location and orientation of
external objects in the environment of
the robot in two or three dimensional
space. Fig.2. illustrates the classical
internal and external state sensory data
integration in the control loop.
The second class of sensors may be
classified as contact and noncontact.
The contact sensors are of force-torque
sensor type and the most developed one
is the artificial skin. The noncontact
sensors utilises the response of a
special detector to the variations of
different reflected radiations,
especially in optical, electromagnetic
and ultrasonic spectrum.

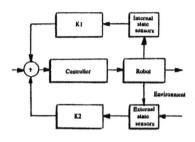

Fig. 2. Sensory Control

External sensing is used for object identification, recognition, measurement, handling and ultimately for robot guidance.

The noncontact sensing in the optical domain may be realised in passive or active manners.

The passive techniques include:

- Texture gradient analysis
- Occlusion analysis
- Photometric analysis
- Focusing
- Triangulation

The active techniques include:

- Structured lighting
- Time of flight measurement
- Inductive, Capacitive or
                   Hall effect analysis

Structured lighting is itself an active triangulation method and uses the projection of an illuminated ray, plane, grid or any combination of well ordered masks. The sensing element is generally a camera and the necessary information is extracted by time consuming heavy software processing.

The aim of this study is to design a three dimensional vision device that can provide only the necessary information of the position of external objects . Special attention was paid to get an end-effector holded, low weight, compact and easily installable product.

THE SENSING SYSTEM

The Principle
-------------

The technique used in this work is based on the principle of triangulation and is illustrated in Fig. 3. The coordinates of a point P in three dimensional space, observed by two optical observers, may be given in the reference frame r0 as:

$$P(r0) = M1 * t1 \qquad (1)$$
$$P(r0) = M2 * t2 \qquad (2)$$

where

M1 and M2 are the transformation matrices
t1 and t2 are the readings of the respective optical observers.

The problem is to match the image pixels of the point P in both image frames of the observers.

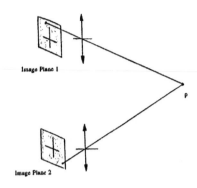

Fig. 3. Triangulation

In active triangulation, one of the observers is in fact a marker of known orientation. Thus the data t1 in Eq.(1) becomes a known, controllable quantity. Physically such a system can be realised by using an orientable light projector and a matrice of light detectors.

For robotic applications, the end-effector guiding vision system requires mainly a clear and distinct picture of the near field. This field begins from just the tips of the effector and spans up to one or two meters. The optical depth of field for such small distances is very limited. Therefore a big problem arises : how to focalise the optical sensing surface, to the correct depth, in real time ? This depth is itself a quantity to be measured by the sensor, as shown in Fig. 4.

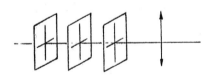

Fig. 4. Different Image Planes

Proposed Methodology
--------------------

In order to resolve this question, a new approach is defined. Instead of using the images of different scene crosses, perpandicular to the optical axis of the observers, we will benefit from only one planar scene. This plane will begin from the tips of the end effector and will lie longitudinally to it as shown in Fig. 5. In this way, one of the values to be measured, the distance in Y axis, will be kept constant. Therefore the unique plane for a given:

$$Y = Yo \qquad (3)$$

may be expressed as:

$$(A*X) + (B*Yo) + (C*Z) + D = 0 \qquad (4)$$

The interesting near field part of this plane, may now be focused onto a sensing surface by using a correctly tilted lens system.

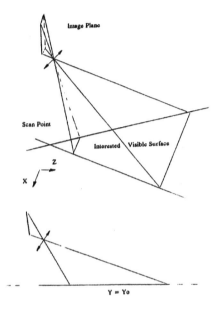

Fig. 5. The Modified Triangulation System

## The Physical System

A narrow beam of low energy light radiation is projected towards the scene to be observed, trough an orientable mirror mounted on a controllable galvanometer. Any diffuse reflection of this ray may be collected by an optical system. In order to avoid the falsifying interaction of uncontrolled brilliant sources on the collecting system, the light source is selected to be a laser illuminator and an interference filter of emitted wavelength is used in front of the sensing area.
The orientation of the laser beam is controlled trough a galvanometric scanner connected to the analog output port of a computer. The instantaneous analog output dictates the angular position of the galvanometer, thus the orientation angle φ of the laser beam. The transformed observer, i.e. the marker, may then be modelled with the equation of a line:

$$Z = \text{cotan } \phi * X \qquad (5)$$

For the sensing element a Position Sensitive Detector PSD, of square surface is utilized. Fig. 6. shows the equivalent electronic circuit of the PSD.

A spot of light on the PSD's surface produces a source of current, Is, by photovoltaic effect. Assume that the external circuit of the PSD is balanced with load resistances Rx and Rz. The generated currents Ix1, Ix2 and Iz1, Iz2, are collected by four electrodes. These currents are respectively proportional to the spot's physical position x and z on the PSD's surface and are given in Eq. (6).

$$X = \frac{Ix1 - Ix2}{Ix1 + Ix2} \quad ; \quad z = \frac{Iz1 - Iz2}{Iz1 + Iz2} \qquad (6)$$

with

$$Ix1 + Ix2 = Is \qquad (7)$$

$$Iz1 + Iz2 = Is \qquad (8)$$

Now, referring to Eq. (2) the new observer may be modelled as :

$$P(r0) = M * t \qquad (9)$$

where

t is a column vector with parameters x and z.

By solving together the Eqs. (5) and (9), one can find the physical values of the coordinates X and Z of the reflection causing point P, Y being equal to Yo, a known parameter.

The observed planar scene projection on the PSD surface is given in Fig. 7. As one may easily observe, the sensitivity of the sensor is better for nearer reflections than for distant ones.

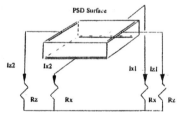

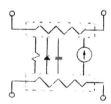

Fig. 6. The Position Sensitive Detector

367

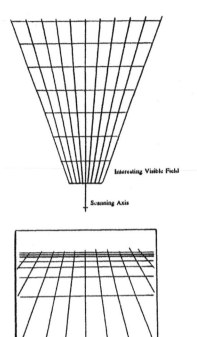

Fig. 7. The Projection of the Visible
Field onto the PSD Surface

## Application Results

The robot and computer connections of
the sensor is shown in Fig. 8. The
light beam orientation is computed and
dictated to the galvanometer through the
analog output port while the analog
position data from the PSD is captured
by the analog input port of the robot
contol computer. The stabilisation time
of the galvanometer mirror assembly
depends on the angle of deflection. An
angular motion of 2 degrees require 600
microseconds, while a deflection of 40
degrees necessitates 800 microseconds
for the assembly to be settled. The
signal stabilisation time for the PSD
is 100 microseconds. The analog to
digital conversion time of the computer
is 40 microseconds. In total, less than
one millisecond is sufficiant to get a
reading from the sensor.
The resolution of the angular motion of
the galvanometric scanner and the
resolution of the PSD being much less
than that of the marker light beam, the
global resolution of the sensor is equal
to the thickness of the light beam. For
well focused laser spots, the resolution
obtained is 0.1 millimeter.
In order to get the complete profile of
an object, a sweep in Y axis will be
necessary. The articulation of the end
effector holding the sensor may be
activited in this case.
The sensor is inefficient when used for
measuring the coordinates of specularly
reflecting surfaces. It may be used
effectively with diffusing textures.

## Extensible Ideas

Another possible variation of this
sensor may be realised by utilising a
CCD sensing element in place of the
PSD. The time required for image
acquisition may be reduced in this case,
by using a light plane instead of a
light beam. Thus a cylindrical lens may
produce the necessary planar
illumination. By thresholding the
received signal to only two levels,
light or dark and by reading the CCD
data trough a DMA, direct memory access
mechanism, one will even accelerate the
video data acquisition time.
We are actually developing the
prototype of this second system.

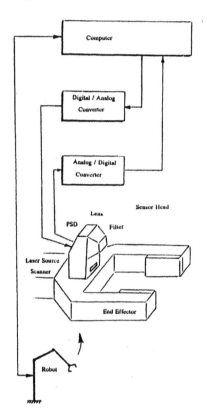

Fig. 8. The System Configuration

### CONCLUSION

The original principle and the design
technique of a three dimensional
coordinate extracting and profile
detecting device is presented. The main
advantage of this sensor is that it may
deliver the required data in one
millisecond. The user who needs only
the coordinates of a limited number of
some specific points in the environment
for robot guidance, may benefit from
this sensor and will not be bored by
the immense quantity of unnecessary
video image data.
The total weight of the sensor is in the
order of hundred grams and may be easily
integrated into the end effector of a
robot.

REFERENCES

Albus, J.S., Haar, R., Nashman, M.
(1981). Real time three dimensional
vision for parts acquisition.
Proceedings of SPIE., Vol. 283,
p 56-60.
General Scanning. Technical Data Sheets.

Hamamatsu. PSDs Technical Data Sheets.

Häusler, G., Hermann, J., Weissmann, H.
(1989). Neue 3D - Sensoren mit
nützlichen eigenschahten. In Gruen,
A., Kahmen, H. (Ed.), Optical 3-D
Measurement Techniques, Wichmann
p 57-66.
Harashima, F., Hashimoto, H., Kubota, T.
(1990). Sensor based robot control
systems.In Kaynak, O. (Ed.),
Proceedings of IEEE Int. Workshop
on Intelligent Motion Control., Vol.
1, p PL1-PL10.
Kanade, T., Asada, H. (1981). Noncontact
visual three dimensional ranging
devices. Proceedings of SPIE., Vol.
283, p 48-53.
Karyot, T. B. (1985). Conception et
realisation d'un proximetre telemetre
pour utilisation en robotique. Thesis
in Ecole Nationale Superieure de
l'Aeronautique et de l'Espace., Vol.
114.
Kent, W., Wheatley, T., Nashman, M.
(1985). Real time cooperative
interaction between structured light
and reflectance ranging for robot
guidance. Robotica, Vol. 3.
Morander, E. (1980). The optocator. A
high precision, non contacting
system for dimension and surface
measurement and control. Proc. Fifth
Intern. Conf. on Automated Inspection
and Product Control., p 393-396.
Strand, T. C. (1985). Optical three
dimensional sensing of machine
vision. Optical Engineering, Vol. 24.
United Detector Technology. PSDs
Technical Data Sheets.

# A REAL TIME VISION SYSTEM AS AN AID IN LEARNING TASKS IN ROBOTICS

**A.B. Martinez, J.M. Asensio and J. Aranda**

*Universidad Politecnica de Catalunya, Pau Gargallo, 5. 08028 Barcelona, Spain*

Abstract. This paper presents how vision can aid a learning system during the training phase. The learning system has no initial knowledge about the process, the task is to manipulate a planar object by tilting a tray and place it in a desired position and orientation.
The vision system will be used as an accurate instrument for measuring parameters of the process as well as it will have real time capabilities for recording data that occur at much more frequency that can be stored with a commercial board.
This work was carried out at the Manipulation Laboratory in Carnegie Mellon University during the summer of 1991.

Keywords. Real time, vision, learning, tracking, planning.

## INTRODUCTION

This paper describes how vision is integrated in a learning system to improve its performance. Providing robots with sensors allows them to interact more directly with the environment and makes them more flexible in new situations. In this application, the vision system will be used as an accurate instrument able to deliver data about a phenomenon or process. It makes sense that the robot learns by itself the description of its available actions. Other way, the approach to sensorless in robotic systems is to have a detailed description of both the working environment and the laws that are relevant to the problem, which is more rigid.

So, providing to the robotics system with the learning and vision system capabilities will allow it to learn its own actions autonomously. The robot starts with no model of its available actions, and acquires a model through its own experience.

A first attempt was a vision system that provided a snapshot of the initial and final conditions of the phenomenon or process, so the learning system could extract consequences about the effect of its actions. Despite of the fact that the consequences would be right, multiple sources of information about the actions are lost when observing only the initial and final conditions.

Next section presents a vision system that is able to keep track in real time of the phenomenon and provides the learning system with data about the phenomenon every 16.6 ms (60 times per second). Learning will occur in a more knowledge-rich environment.

## EXPERIMENT DESCRIPTION

The task is to place in a desired position and orientation objects by tilting a tray where they lie, see figure 1.

The tray is attached to a PUMA 560 industrial robot and it may be tilted using the wrist motions. Every tilt causes the object to slide and rotate while making contact with the tray walls and corners (Erdman, 1988)

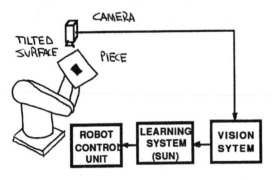

Figure 1 System configuration

The system has no initial knowledge of physical laws or physical properties of the objects, and acquires a model through its own experience. The system works in two modes.

In *learning mode,* the tray is tilted with a random sequence of movements and the effects of each tilt are observed by the vision system. By associating actions and their effects on the object, the learning system acquires a model of the process. Other self-training learning strategies have been implemented as shown in (Christiansen, 1991)

In *performance mode,* a planner use the model given by the learning system to generate a program in the form of a sequence of actions (tilt angles) that achieve a given goal. The planner and learning system run over a SUN Work Station and send the program to the robot control unit to execute it.

The vision system provides the robot with sensory information about the effects of its actions on the objects. The initial attempt used a very discretized model of world state. The tray was divided into nine sectors describing the location of the object centroid. Object orientation was horizontal or vertical. This information was read before and after an action to feed the learning system.
As noticed in (Christiansen, 1991) a nearly continuous description of the world state is available from the camera image, and should be used.

The implemented vision system can observe the accurate x,y coordinates and orientation of the object, as well as its linear and angular velocity and acceleration. The vision system provides the learning system with all the above information every 16.6 ms.

## VISION SYSTEM

### Vision System configuration

Vision is a natural mean for measuring position, orientation and velocity of moving objects.

As we mentioned in the last section, the phenomenon studied consists in the discovery of the laws that govern the slicing and rotation of a piece in a tilted surface. In order to provide the learning system with simple data, as well as to achieve real time performances, the vision system used a binary image.This approach will simplify information processing enormously allowing the vision system to give, at each point in time (every 16.6 ms), the centre of gravity and the orientation of the object. Velocity and acceleration of the centre of gravity as well as its angular velocity and acceleration can be extracted from this information.

The vision system has the following architecture ,see figure 2.

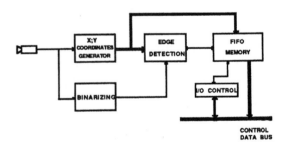

Figure 2 Vision System Architecture

The main goal of the vision system is the aquisition of the coordinates of the object contour. These coordinates have to be inside of the window that is tracking the object.

The x;y coordinates are aquired by a computer through a FIFO memory which makes the function of buffer. This was needed because of different speed rates between video and memory access time.

## Hardware description for real time operation.

In order to achieve real time performance, we have to reduce the flow of data without losing the more relevant information. Two important steps have to be considered to achieve this reduction: binarizing and edge detection. There is plenty of bibliography about those two issues.(Ballard, 1982)(Horn, 1986)... In this section we describe our approach and how we solve it.

The basic architecture for edge detection is shown in figure 3

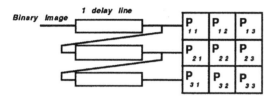

Figure 3 Edge detection Window

The 3x3 window scans all the image in real time. Edge detection can be achieved as edge = P23 $\oplus$ P12 + P23 $\oplus$ P21, but it has the inconvenient of being too much sensible, for instance, a single point in P22 will be seen as an edge.

In order to avoid the sensibility problem, we use the 3x3 pixel window to address a 512 x 1 bit ROM, which discriminates whether the window is placed at the boundary or not.

We have to say that, working with binary images and because of the advances in Programable Logic Devices, the above hardware was implemented in just one integrated circuit of less than 6000 gates.

## Camera set up.

In the first attempt, we only wanted to know the initial and final position of the object. We could have maintained the camera fixed when tracking the moving object but for giving the absolute spatial coordinates of the piece a lot of calculation has to be made.

In the experiment under investigation, we didnot need absolute positioning in space, we wanted 2D position of the object in the surface that is tilted by the robot arm. So, we decided to place the camera fixed to the tray. This way, the field of view of the camera always contains the whole surface (figure 4).

The main advantage is that camera calibration is easier.

## CAMERA CALIBRATION

To be able to sense the object position, it is necessary to go through a calibration phase which relates pixel locations in the image to physical locations on the tray plane. The calibration scheme (Tsai], 1987), computes the following parameters:

Figure 4 Camera attached to the robot arm

f:      focal length of the camera.
s:      a fudge factor to compensate for disparate digitalization timings between the camera and the frame grabber.
k1,k2:  lens distortion parameters.
R:      A 3x3 rotation matrix that describes the transformation from the camera frame to the world frame.
T:      A 1x3 translation matrix that denotes the translation vector between the above two frames.

It is necessary to provide input to this algorithm in the form of a set of training points for which both spatial coordinates in world frame and image coordinates in the image space are known. For objects that lie in a plane, it is only necessary to provide training points that all lie in this plane. This is accomplished by using an image calibration grid located on the tray as shown in figure 5

The location of the centroid of each of the circles on the grid is known a priori through careful measurement, while the corresponding centroids of the circles in image space must be computed. At present, the correspondence problem of deciding which circle in 3D space corresponds to which circle in the image space is solved by hand. The result of this process is a set of coordinates for every circle on the calibration grid:

row,col:       the centroid in image space.

x,y,z:       the centroid in world space.

The calibration scheme uses these training points as input to non linear minimization scheme to estimate the parameters $(f,s,k1,k2,R,T)$. Once these are known, given an $(i,j)$ pair in image space, the corresponding point in world space can be solved by eqns. 8(a) and 8(b) in Tsai's report. Essentially, two relationships $f(x,z)$, $f(y,z)$ are obtained.

In our case the object being tracked moves in a plane so the z parameter is known and it suffices to simultaneously solve for x and y in the two equations above.

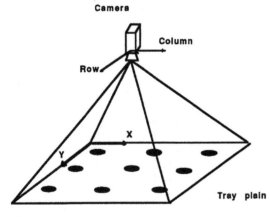

Figure 5 Training points

## TRACKING

The vision system works on a PC.Bus.The PC reads from the vision board the list of image coordinates of the object contour.

Based in this information the PC calculates the center of gravity and the orientation of the piece and sends this data to the learning system via Ethernet. All this proces takes no more than 16.6 ms and it runs as follows.

Initial conditions are not requiered by the tracking algorithm. It can detect any object in the tray and follows it without initial knowledge of its position, orientation or shape.

At every field, position and orientation can be calculated from the list of edge coordinates in a simple way as shown in (Horn, 1986). Center of shape $(X,Y)$ is just the average of x and y coordinates:

$$X = \sum_{i=1}^{n}(x_i)/n$$

$$Y = \sum_{i=1}^{n}(y_i)/n$$

where n is the number of edge pixels.

Orientation is given by how the shape is distributed arround its center. That is to say, to find the axis of least second moment

$$x_i'= x_i - X$$
$$y_i'= y_i - Y$$

$$a = \sum( (x_i')^2)$$
$$b = 2*\sum( (x_i'*y_i'))$$
$$c = \sum( (y_i')^2)$$

$$\tan 2\emptyset = b / (a-c)$$

Aditional calculus of linial and angular velocity and acceleration are made by measuring position and orientation increments between video samples.

After that, the PC transforms the above data from image coordinates to world coordinates and it sends them to the SUN workstation which runs the learning system.

The PC has a short time to display this data in its own monitor before a new contour image is ready on the video board.

## ACKNOWLEDGE

Our most heartfelt thanks to Professor Mason for giving us the opportunity to collaborate in the tray tilting problem.

We also would like to thank the research assistant Pere Castellsagués for his help in the implementation of the board and its debugging.

## REFERENCES

Christiansen, A.D.,Mason M.T.,Mitchell T.M. (1991). Learning reliables manipulations strategies without initial physical model. Proceedings of theInternacional Conference on Robotics and Automamation, Cincinnati, OH,1990.

Erdman, M.A., Mason, M.T. (1988). An exploration of sensorless manipulation. IEEE Journal of Robotics and Automation

Tsai, R.Y. (1987). A Versatile Camera Calibration Techniques for High Accuracy 3D Machine Vision Metrology Using Off-the-self TV Cameras and Lenses. IEEE Journal on Robotics and Automation. Vol RA-3. nº 4.

Ballard, D.H , Brown C.M.(1982). Computer Vision. Prentice Hall.

Horn, B.K.P.(1986). Robot Vision. MIT Press.Mcgraw-Hill.

# A VISION PERIPHERAL UNIT BASED ON A
# 65x76 SMART RETINA

**P. Nguyen\*, T. Bernard\*, F. Devos\*\* and B. Zavidovique\*,\*\***

*\*Lab. Systeme de Perception, ETCA, 16bis, Av. Prieur de la Cote d'Or, 94114 Arcueil Cedex, France*
*\*\*Institut d'Electronique Fondamentale, Bat 220, Université de Paris Sud, 91405 Orsay Cedex, France*

**Abstract**.    Smart retinas are devices including on a single chip optical sensors
and computing facilities. This proximity provides advantages for compactness and
communication speed between transducer matrix and on-chip processing units. This
paper presents the retina concept, control and programmation aspects of the device
built in our laboratory, and a few algorithm considerations. The hardware is based
on a VLSI retina chip containing a 65x76 mesh array of boolean processors.

**Keywords**. Parallel processing; Sensor; Image processing; Computer architecture;
Computer peripheral equipment.

## INTRODUCTION

Today's performances of computer vision systems remain very limited with respect to application's requirements, specially in real time
domain. This is why a lot of work has been
successfully devoted, for about ten years, to
design specific image processing architectures,
most of the time, by exploiting the peculiarities of algorithms. The development of novel
concepts is required to fill the gap faster than
would allow the regular increase in speed and
scale integration of circuits.

Classical visual perception performed by
computers can be decomposed as a chain of
processes, (Fig.1) starting with transducing
of the light phenomena and finally leading to
decision. In this scheme, process named 'low
level' extracts from images context meaningful objects. Once the information has been
filtered and concentrated into structural or semantic knowledge, the 2-D topology of the image disappears. Then the high-level processing can start. At this point objects become
arbitrary graphs, whose processing on multi-

processor architectures poses various connec-
tivity and programmability problems.

## RETINA CONCEPT

Low-level processes are characterized (or even
sometime defined) by their shift invariance.
Another correlative aspect of this level is the
huge amount of raw data being moved inside
the computing system. Thus low level image
processing strongly benefits from massively
spatial parallelism The second link in the
above computer vision chain (Fig.1), where
pixel data are sequentially sent out of the acquisition circuit after or before A/D conversion, appears as a bottleneck or even more
fundamentally as a harmful topological break
between both parallel acquisition and processing. For us this simple observation is only the
emergent part of the iceberg. It justifies the
attempt to implement on a single monolithic
VLSI chip (cf Fig.2), a 'rough but complete'
vision system, intimately associating opto-
electronic and analog devices with analog-to-
digital converters and minimal digital proces-

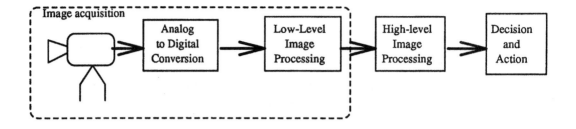

Fig.1: Classical Computer Vision.

sors interconnected with their closest neighbors. We call it a 'retina' or 'smart retina' to insist on the fact that our device is a computing machine (driven by a program) and not only a light to electron transducer.

Of course our retinas are not a panacea for the computer vision problem, but they turn out to be particularly well balanced vision systems:

- The continuity of the 2-D topology suppresses the communication bottleneck.

- Image sequences from the photosensors are known to be locally correlated both in the space and time domain. This can advantageously be exploited at the analog level to provide on chip encoding of the image flow into compact digital representations. For the sake of topology and compacity, we chose a binary image representation, based on a one-to-one mapping between analog and binary pixels. If the sole spatial correlation is taken advantage of, the analog-to-binary conversion is called 'halftoning'.

- Unlike the specific analog processing layer, the digital layer allows arbitrary interactions between pixels because the binary support of the information now enables cheap data movements. This is achieved by iterative image shifting, such that any shift-invariant boolean function can be computed. Those processings are called NCP standing for Neighborhood Combinatorial Processings.

- NCP's are well-adapted to low-level image processing. More generally, NCP's allow the implementation of a 'rough but complete' type of vision, for which NCP

algorithms results can be output from the retina in a concentrated fashion (such as the image integral, higher order moments, or sparse pixel coordinates) thus avoiding a potential communication bottleneck with the external world.

Another important feature of retinas is their high vision speed. Because of the massive spatial parallelism and the easy communication between layers, several thousand of images can be processed per second! This means that most usual events become quasistatic. Algorithmic consequences are impor-

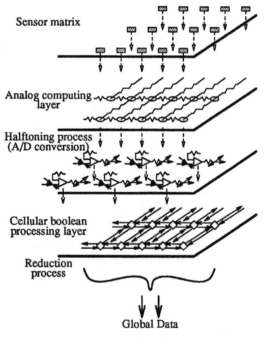

Fig.2: The Retina Concept.

tant. From a control theory point of view, the feedback loop of the system is about 2 orders faster than in today's computer vision systems. For example, a complex problem like

378

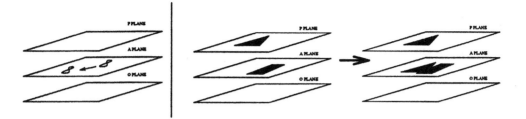

Fig.3: Shifting Image, Logical OR.

matching successive images of a moving scene, is reduced to its simpler expression when the sampling frequency is high enough. We actually rely on the retina high vision and processing speeds, and the subsequent simplification of vision problems to compensate for the overwhelming compacity constraint. Finally, the retina can be considered as a structural vision coprocessor. All semantics is subcontracted to a controller, which can be nothing more than a boolean pattern base manager. Such a scheme can be applied to many well delimited cases (from alarm or detection applications up to target tracking and more ...), where vision within the retina amounts to some tolerant dot pattern matching.

## VLSI IMPLEMENTATION

Bringing retina concept into play with commercial C-MOS technology requires drastic choice, and deep thinking about usefulness of each transistor on each layer. Since versatility of the digital data path is our first center of interest, analog computing layer has not been integrated into our first attempt. The A/D conversion layer has been reduced to a system which enables thresholding and Bayer's ordered dither halftoning of analog optic images.

The chip is composed of 4940 (65x76) cells including a photodiode, an minimal A/D converter and a Processing Element (P.E.) with 3 registers. Control signals generation for the set of P.E. is provided by a unique external device since the machine is designed for a Single Instruction Multi Data (SIMD) working mode. Further description of the VLSI implementation can be found in Patent(91) and Bernard(91).

## PROGRAMMING MODEL

The association of the VLSI retina chip with its control signal generator can be understood as a microprocessor working on matrices and including a 5000 ways unidirectional analog input port (the matrix of sensors). Programming model offers 3 register memory planes containing 3 images. Each plane can be shifted, logically complemented, set to 1 or reset to 0, loaded from the optical input port or from an electrical serial port. Permutations and two variables boolean operations (AND, OR) can also be performed between any two planes.

Those computing resources are sufficient to program the whole family of NCP operator, since normal disjonctive form of boolean fonction needs only two intermediate variables to store partial product and partial sum during valuation time.

Capability of the programming model relies on a few but well chosen command set. The strange appearance of their choice results from trade-offs between universality, VLSI implementation constraint and execution speed of NCPs.

Each cell contains three boolean registers named P, A and O. Shift invariance of operator leads to usages of relative indexation of registers. The P register of the first northern eastern neighbor of the local cell will be designated by $P_{1,1}$. $P_{0,0}$ will be short to P. Thus shifting the P plane to the south (plane A and O being unchanged) will be written $(P \leftarrow P_{0,1}, A \leftarrow A, O \leftarrow O)$ or abbreviated into $(P_{0,1}, A, O)$. Shift sensitive commands not described here and used for serial input-ouput purpose cannot be described under that form.

| | |
|---|---|
| **North** $\equiv (P_{0,-1}, A, O)$ | shifting P plane to the north |
| **South_west** $\equiv (P, A, O_{1,1})$ | shifting O plane to the south_west |
| **East_and_perm** $\equiv (O_{-1,0}, P, A)$ | circular permutation and shift of O plane to the east |
| **Perm** $\equiv (O_{-1,0}, A, P)$ | permutation of P and O planes |
| **CEast_and_perm** $\equiv (O_{-1,0}, \overline{P}, A)$ | circular permutation, complementation of P plane and shift of O plane to the east |
| **CP** $\equiv (P, P, O)$ | copy of P plane to A plane |
| **CPN** $\equiv (P_{0,-1}, \overline{P}, O)$ | copy of $\overline{P}$ to A plane and shifting P plane to the north |
| **And** $\equiv (P, A.P, O)$ | logical AND between P and A plane |
| **Cand_swO** $\equiv (P, A.\overline{P}, O_{1,1})$ | logical AND between $\overline{P}$ and A plane, shift of O plane to south west |

Fig.4: Basic Command set

The list of basic command set of Fig.4 is sufficient to construct the above mentioned programming model. The existence of the circular permutation named $\pi$ and provided by a South_west, East_and_perm, North sequence implies equivalence of the three planes. $\pi \equiv$ (O , P , A). Therefore being able to shift the P plane to the north means being able to shift any plane to the north. For example, shifting O can be done thanks to a $\pi$-North-$\pi^2$ sequence. Shifting the O plane to the East is provided by East_and_perm followed by $\pi^2$. These remarks enable shifting of any plane in any directions. Logical complementation of P, A and O being unshifted, can be produced by the sequence South_west-CEast-North followed by $\pi^2$. Two-variables boolean operations can be deduced from the basic command set following the same thought process.

**FIRST EXPERIMENTAL STEP**

The aims of this first step were to test and also to run representative applications on the retina VLSI chip. The control signal generator has been emulated by way of a 'homemade' analyzer and pattern generator along with a package of C programs. Memories, 32 bits wide and 256k deep, send control signals to and record activities of the VLSI with a time slice of 50 ns. Contents of memory is also read or written from the workstation. Once compiled, a retina program is loaded into the pattern generator. Results are analyzed by the host computer for test or display purpose. Examples of applications of this set of devices

Fig.5: Images extracted from the retina: Thresholding, Edge detection, Bayer's halftonig

are given in Fig.5. Edge detection is computed through mathematical morphology in 40 cycles of 50 ns i.e. 2 $\mu$s. This is of course due to the massive spatial parallelism of the computing layer. For readers not familiar with mathematical morphology, roughly speaking, a pixel is thought of belonging to an edge if it is set to 1 and if one of it neighbors is equal to 0. In 4-connexity this leads to the boolean fonction $f = a\overline{b} + a\overline{c} + a\overline{d} + a\overline{e}$ where b, c, d and e are the four neighbors of a.

Above mentioned Bayer's halftoning is based on periodic thresholding matrixes (see Bayer 73). On Fig. 5, a 2x2 matrix providing 5 grey levels was programmed, but it is of course possible to use a larger one.

**VISION PERIPHERAL UNIT**

In spite of the interest of the first control unit, it has seemed necessary to progress to a higher level of control of the retina chip. Indeed, many classes of vision algorithms need quick feedback between raw or processed global informations and the low level process running

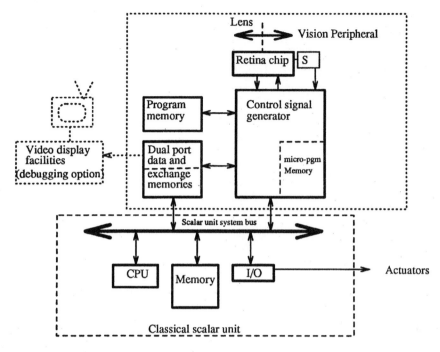

Fig.6: Rought vision system, using retina unit as vision peripheral unit

on the smart sensor.

In a dedicated machine, processing global informations can be done through dedicated logic 'glue', losing that way versatility. But to build a 'rough but complete' vision system for research purpose, it seems preferable to help the retina chip (which is specialized in NCP calculation) with another computing unit. Frontiers of activities of this non NCP unit cannot be precisely defined. This is the reason why we have chosen to use a classical scalar subsystem based on a commercial microprocessor.

Several ways are available to join the two processing units. A first one is to create, with the retina, a numeric coprocessor-like device. The smart sensor will therefore become an extending part of the numeric data-path of another CPU. This solution is quite attractive, mainly because the provided mono-task programming model will be easier to dominate than a distributed machine. Despite of this advantage, drawbacks may be quite penalizing.

- Incompatibility between bus instruction bandwidths of, in one hand, a universal purpose microprocessor, and in the other hand, a data processing chip consuming

a new instruction each 50 or 100 ns.

- Under utilization of data-paths computing power. Task parallelism will indeed be poorly used. Low and high level processes cannot run simultaneously on a mono-task model. Therefore, a dramatic decrease of the retina efficiency is to be expected in most applications.

Former remarks lead us to think about a distributed computer. Typical, well spread, architectures are organized around a multi-master bus; the two computing subsystems talking to each other by way of a global memory. Since the TCL unit has to provide the scalar unit with global information on the scene being observed, the amount of data being transfered should be small, as well as the amount of global memory required. Thus the universality of a multimaster bus (which is in general big "logic glue" consumers) may not be essential. This is the reason why designing the retina based unit as a peripheral system (or at least a slave card on a VME bus) seems to be an attractive trade-off between complexity, efficiency, and versatility.

At program loading time, the vision peripheral subsystem is configured (or programmed) by the scalar unit. Reconfiguration remains,

Fig.7: Halftoned image and anwers of the 'automaticaly constructed' region tracking NCP

of course, also possible at run time. The aims of this machine is the study of real time NCP based algorithms and of data exchange modes between vision peripheral and high level processing units. Its architecture remins of (may be presumptuously) the organization of biological systems where animal retina provides precomputation of raw data before sending it to visual cortex.

## NCP CONSTRUCTION

Advanced applications require automatic construction of NCP. Algorithms described by Catoni (91), Weill (91) are executable on the above mentioned architecture. Figure 8 shows a graphic representation of a monomial NCP (i.e. only a product of term). Points of the neighborhood which are present as direct term in the product are in white, while complementary terms are in black. Points non used in the monome are in grey. This NCP was automatically constructed to localize two regions (women eyes) of an halftoned image. Boolean functions constructed this way caracterize indeed the textural statistics of the tracked regions. Results are presented on Fig.7

## CONCLUSION

This is, to our knowledge, the first system based on a monolitic, digital smart vision sen-

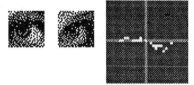

Fig.8: Regions appointed to be tracked (eyes) and graphic representation of NCP

sor of this size. Plenty of work remains to be done, in the study of a versatile analog computing layer, and also in the algorithmic domain. We think that the presented machine, with software tools being under development will provide an efficient support for further research on these fields.

## REFERENCES

Bayer, B.E. (1973). An optimum method for two-level rendition of continuous tone pictures. *IEEE Int. Conf. Commun. 1973.*
Bernard, T. (1991). A 65x76 VLSI retina for rough vision. *Workshop on computer architecture for machine perception, CAMP91*
Catoni O. (1991). Learning algorithms for patern recognition on halftoned images. *submitted to "Networks"*
Patent (1991) Réseau bidimentionnel de mémorisation et de traitement booléen d'images. *french licence n° 91-13703, 6 nov. 1991*
Weill J.C. (1991) ETCA Internal report

# THE USE OF IMAGE PROCESSING IN SATELLITE ATTITUDE CONTROL

## D. Croft

*EMBL, Meyerhofstrasse 1, 6900 Heidelberg, Germany*

**Abstract.**   Accurate orientation control is essential for any communication satellite. One way of achieving this is to point a video camera down at the Earth, and compare the observed position of the Earth with its ideal position. If there are any errors, the satellite's thrusters are activated to correct them. Although this idea is simple in theory, there are many problems in practice. This paper looks at one of those problems - finding the position of the Earth in the image obtained from the camera. This has been broken down into two subproblems: i) locating the Earth/space boundary and ii) using data gathered from i) in estimating an Earth centre position.
Finding the Earth/space boundary is non-trivial, because the Earth is in general only partially illuminated, and there are various forms of noise present in the image. Also, the image processing must be carried out in real time, since it is being used as part of a control system.
Given a (generally incomplete) set of points on the Earth/space boundary, finding a centre position is fairly straightforward. Two approaches are described here, and accuracy comparisons made.

**Keywords.**   Boundary detection, circle, domain constraints, image processing, nonlinear least-squares, satellite attitude control, subpixel accuracy, thresholding, tracking.

## INTRODUCTION

Global communications systems are coming to rely more and more on satellite links for conveying data. Generally, these satellites are put into geostationary orbits [1]. This makes it easy to set up ground stations, since an antenna can be pointed skywards in a fixed direction. However, in order to maintain a good signal-to-noise ratio, the satellite's own antenna must maintain its pointing direction accurately. This means that the orientation (or *attitude*) of the satellite, relative to the Earth, must be kept as constant as possible.

Attitude control is generally done by observing the Earth and comparing its position with the expected position when the satellite is ideally oriented. If there is a discrepancy, one or more of the satellite thrusters can be activated to reduce it to zero.

Traditionally, this has been done by scanning the Earth in the infra-red, and comparing the East and West extremes of its perimeter with the edges of the image. The problem with this technique is that it uses equipment with moving parts.

An optical method for obtaining Earth position discrepancy data has been proposed (Beard et al, 1990), using a camera on a chip with a wide-angle lens to observe the Earth. Typical Earth images are shown in Figs. 1 and 2. An edge detector would find discontinuities in the image; the discontinuity of most interest is the Earth/space boundary. Data from the edge detector would be fed into a circular matched filter. The strongest point in the filter's output corresponds to the Earth's centre position.

Edge detection and matched filtering over entire images are computationally expensive. It has recently been sug-

gested [2] that if there were constraints inherent in the domain, it might be possible to simplify the algorithms used. Two significant constraints are explored in this paper: using an annulus to limit the range over which image processing is performed, and using localised thresholding.

The specifications for the attitude control system give maximum pointing errors. Thus, if the system is working properly, the actual region of the image within which the Earth/space boundary will be found is quite small (only a few pixels). Any image processing can, therefore, be confined to an annulus, defined by the maximum pointing errors.

Boundary detection could be done by applying an edge detector within the annulus. However, a much faster method is to use a thresholding tracker to follow around the boundary. Threshold values can be calculated locally, within segments of the annulus, so that gradations in illumination over the Earth's surface can be taken into account.

Centre estimates can be made from points collected from the boundary. Two methods have been tried: a geometrical method, and a nonlinear least squares method. It is assumed in both cases that the visible portion of the Earth's boundary is a circular arc.

## EARTH/SPACE  BOUNDARY  DETECTION

Introduction

The first step in locating the position of the Earth as seen by the satellite's on-board camera is to find the boundary between the edge of the Earth and deep space. To do this, advantage is taken of the fact that the Earth is lit by the sun, whereas space is nearly black.

---

[1] A geostationary orbit is in the equatorial plane, and has a period of exactly one sidereal day, so that the satellite remains above the same point on the Earth's surface at all times.

[2] by Dr C.N. Duncan, of the Meteorology Department at the University of Edinburgh

Figure 1: METEOSAT full Earth image

Figure 2: METEOSAT partially illuminated Earth image

However, there are a number of problems. Firstly, a perfect circular image of the Earth is almost never available; there is usually some degree of self-shadowing, where only part of the Earth is illuminated. In the worst case, the sun is behind the Earth, and all that can be seen is a thin band where sunlight has diffused through the atmosphere. Secondly, images will be subject to noise. In part, this will be thermal and digitisation noise. However, the camera will also be exposed to damaging radiation, which can result in some pixels being permanently on or permanently off.

## Constraining the Problem

As already mentioned, the Earth's centre will never, under normal conditions, drift by more than a few pixels from the centre of the image. Hence, we can construct an annulus within which we can expect to find the Earth/space boundary (see Fig. 3). Any image processing can be confined within this boundary, rather than being applied to the entire image.

This also means that when the visible Earth boundary deviates from the circular (as it will in any partially illuminated images of the Earth), very little of the non-circular portion of the boundary will be included within the annulus. Excluding this portion of the boundary will greatly improve the accuracy with which the centre position can be estimated.

## Tracking

Tracking algorithms use local information to follow around discontinuities in an image - eg. an intensity discontinuity, such as the Earth/space boundary discussed in this paper (see Rosenfeld & Kak, 1982, for a more detailed discussion). They can be a faster way of finding edges that conventional mask convolution methods, since they use information already gathered to extend an edge.

Tracking has two phases: scanning and following. During the scanning phase, the system will look along a straight line through the image until it finds a discontinuity that will make a suitable start point. From this start point, the system will attempt to follow along an edge.

As a tracker follows an edge, it marks already tracked points, so that they will not be considered again. In order to grow the track, the pixels around the last point on the track are examined pairwise. If the intensity of one of the pair is above a given threshold (ie. it is part of the Earth) and the intensity of the other is below the threshold (ie. it is in space), then the above threshold point is marked as being on the Earth/space boundary, and becomes the new end of track[3]. The above procedure is then repeated. The track will thus grow until either i) there are no pairs that meet the "one above threshold and one below threshold" criterion, or ii) a previously tracked point is encountered.

## Spokes and Thresholds

Two questions arise from the previous subsection: how are the straight lines for the first phase of tracking laid down on the image, and how are thresholds for distinguishing Earth from space determined?

The answer to both of these questions is: spokes. A set of evenly spaced spokes, radiating from the centre of the image, is notionally constructed. The portion of each spoke actually used by the system only extends a few pixels each side of the annulus (see Fig. 3).

---

[3] Another approach, which has also been examined, is to look at the *difference* in intensity between the pixels in a pair, and compare this with a "difference threshold". Only the absolute intensity criterion is discussed in this paper, for the sake of brevity.

Thresholds are selected "locally" for each spoke. A histogram is constructed from the intensities of all the pixels lying under a spoke. Then, the *modal* method is used for finding a threshold (see Rosenfeld & Kak, 1982). That is, the threshold intensity is selected as being the histogram minimum point lying between the two largest peaks on the histogram.

As currently implemented, the system has 32 spokes. Starting with spoke 0, scanning is done by looking at the pixels along the spoke sequentially, until an over-threshold pixel is encountered. This pixel becomes the start point for the edge following phase of tracking; it will track as far as possible in both clockwise and anticlockwise directions from each spoke around the Earth/space boundary. The boundary points gathered in this manner are passed onto the centre locating algorithm.

## Some Embellishments

One of the big advantages of selecting thresholds based on local conditions is that there are significant intensity variations over a normal Earth image, so that different threshold values apply along different segments of the boundary. However, because the number of pixels on which the histograms are based is small, noise can cause significant peaks, and erroneous threshold selection can sometimes occur. Another problem is that there can be substantial jumps in threshold value from one spoke to the next. This means that there will be discontinuities in the tracked boundary, possibly giving rise to systematic errors in the centre estimate.

To reduce the effects of local errors, two techniques were investigated: smoothing and interpolation.

Threshold smoothing looks at the threshold found for a given spoke, and compares it with the thresholds on the immediately adjacent spokes. If it is intermediate between them, it is left unchanged. If it is greater than both or less than both, then it is adjusted to be the mean of the three threshold values. This tends to reduce the influence of "outlier" threshold values.

Threshold interpolation takes the thresholds on two adjacent spokes, and interpolates the thresholds linearly between them. Thus, instead of a sudden step in threshold between two spokes, there is a gradual change. This has the advantage of giving only a single track around the Earth/space boundary.

Smoothing and thresholding can be combined to enhance performance.

## Results

A number of images, both synthetic and "real" (derived from METEOSAT pictures of the Earth) were used to test the boundary detecting algorithms (see Croft, 1991). Here, results from just one of these tests, on a METEOSAT image will be presented. This image is one of those "worst case" ones, where the sun is behind the Earth, and all that is visible is a thin strip of atmosphere (see Fig. 4).

The results shown in table 1 give statistical comparisons of the distance of the points *actually* found on the boundary from the "ideal" circular boundary, centred on the centre of the image (distance values are in pixels). Results from a Sobel edge detector are also included for reference.

In this experiment, threshold interpolation produced a significant improvement in accuracy when compared to the unaided thresholding tracker. Over a larger number of experiments, interpolation plus smoothing tends to give the most accurate results (see Croft, 1991, for more discussion of this).

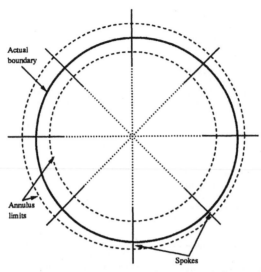

Figure 3: Boundary detection using spokes and annulus

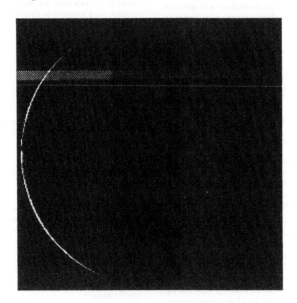

Figure 4: METEOSAT test image

Table 1: Effects of Tracking Algorithm on Edge Point Location Accuracy

| Boundary pt extraction method | point count | mean pt err | SD pt err |
|---|---|---|---|
| Sobel + modal thresholding | 328 | 0.962 | 0.626 |
| Thresholding tracker | 135 | 0.526 | 0.315 |
| Thresholding tracker + s | 140 | 0.546 | 0.331 |
| Thresholding tracker + i | 48 | 0.394 | 0.262 |
| Thresholding tracker + s + i | 111 | 0.466 | 0.314 |

(where + s indicates smoothing and + i indicates interpolation).

# CENTRE LOCATION

## Introduction

To a very good approximation, the Earth is a sphere. Hence, the image of the Earth as seen from a satellite will be circular. Knowing the height of the satellite above the Earth's surface, and the angular field of view of the camera, it is easy to calculate the radius (in pixels) of the Earth as it should appear in the image. Having tracked the Earth/space boundary, we also have a list of points around the circumference of the Earth. In general, these will not span the entire circumference; there will be gaps, small ones due to noise, and large ones due to the Earth's partial illumination.

How can these points be used to locate the Earth's centre? Two techniques are considered in this paper, least squares and geometric.

## Least Squares

The Levenberg-Marquardt nonlinear least squares method, as described in Press et al, 1988, can be used in centre location with a few minor modifications. The user must supply a function that, given a value of the independent variable, x, computes:

1. y, and
2. $\frac{\partial y}{\partial a_i}$, where $a_1, a_2, \ldots, a_N$ are the N parameters to be found for the nonlinear equation $y = f(x : a_1, a_2, \ldots, a_N)$.

We can express the equation for a circle as:

$$(x_c - x)^2 + (y_c - y)^2 = R^2$$

where R is the radius of the circle, $(x_c, y_c)$ is it's centre coordinate, and $(x, y)$ is a point on its circumference. We can rearrange this equation to get a solution for y, with x the independent variable:

$$y = y_c - \sqrt{R^2 - (x_c - x)^2}$$

The parameters we are interested in finding are the centre coordinates; we can write $a_1 = x_c$ and $a_2 = y_c$. By differentiation, we get:

$$\frac{\partial y}{\partial a_1} = 1$$
$$\frac{\partial y}{\partial a_2} = \frac{(x_c - x)}{\sqrt{R^2 - (x_c - x)^2}}$$

## Geometric

Given two points, $(x_0, y_0)$ and $(x_1, y_1)$, on the circumference of a circle, plus the radius of the circle, R, it is possible to calculate the centre position of the circle.

If we write:

$$u = x_1 - x_0$$
$$v = y_1 - y_0$$
$$w = (x_0^2 - x_1^2) + (y_0^2 - y_1^2)$$

then we can show that $y_c$ can be obtained by solving the quadratic:

$$(v^2/u^2 + 1)y_c^2 + (2x_0 v/u + vw/u^2 - 2y_0)y_c +$$
$$(x_0^2 + y_0^2 - R^2 + x_0 w/u + w^2/(4u^2)) = 0$$

and $x_c$ can be calculated from:

$$x_c = \frac{-[w + 2y_c v]}{2u}$$

We simply pick pairs of points from the list of Earth/space boundary pixels, and slot them into these equations. By using a large number of pairs, and averaging the $x_c$ and $y_c$ values obtained, we can get an accurate centre estimate. For maximum accuracy, the angle subtended by the two boundary points and the centre of the image is selected to be as near to $90^\circ$ as possible.

## Results

The centre location algorithms were tested with a number of synthetic images, to observe the effect of pixel position and intensity noise. Both geometric and least-squares proved robust in the presence of noise - see Croft, 1991, for full details. The effects of using different tracking algorithms were also investigated. For the METEOSAT image mentioned in the boundary detection sections (see Fig. 4), the errors in centre estimate are shown in table 2.

In this example (and in several other experiments), the least squares method obtained the centre location exactly. The geometric method works best if thresholds have been both smoothed and interpolated at the tracking stage.

## CONCLUSIONS

A system for identifying Earth/space boundary pixels in satellite based images of the Earth has been implemented, using computationally cheap image processing techniques. Two methods for using this data to locate the Earth's centre coordinates have been described. These coordinates can usually be determined to an accuracy of better than 0.1 pixels, even in the presence of significant noise. Threshold smoothing and interpolation at the tracking stage help to boost the accuracy with which the boundary is detected. The least squares centre finding algorithm usually gives a more accurate estimate than the geometrical algorithm.

Improvements could be made on the system as it currently exists, eg. it could be speeded up significantly by using integer instead of floating point arithmetic. Other constraints on the tracked points might be implemented, eg. rejecting portions of arc that do not fit within certain curvature limits (see Pavlidis & Horowitz, 1974). It would be instructive to estimate the comparative computational complexity of the geometric and the least-squares centre finding algorithms. Experimentation with other centre locating techniques, eg. Tamas, 1991, would be interesting.

It may also be possible to generalise the method to cover a broader range of problems. Eg., a "pseudo annulus" could define a band around an arbitrary shape, limiting the amount of computation needed to find its boundary. If the object were moving, so that it was in a different position in each frame, then knowledge of its position in the previous frame and some approximate velocity vector would give the position in the current frame. Hence, the object could be tracked. Such ideas have potential applications in robotics and in cloud tracking.

## ACKNOWLEDGEMENTS

*I would like to thank Mitch Harris for many useful suggestions, and my tutor Charles Duncan for keeping me on the right track. I would also like to thank the Meteorology Department at the University of Edinburgh for their support in this work.*

Table 2: Effects of Tracking Algorithm on Centre Location Accuracy

| Boundary pt extraction method | Least squares posn err | geometric posn err |
|---|---|---|
| Thresholding tracker | 0.000 | 0.879 |
| Thresholding tracker + i | 0.000 | 1.072 |
| Thresholding tracker + s | 0.000 | 0.971 |
| Thresholding tracker + i + s | 0.000 | 0.911 |

(where + s indicates smoothing, + i indicates interpolation and the "error" is the distance, in pixels, of the estimated centre from the centre of the image).

## REFERENCES

C.I. Beard, S.D. Hayward, M. Caola, C.N. Duncan and I. MacLaren (Sep. 1990). *Interim report on mosaic earth sensor feasibility study*, British Aerospace (Space Systems) internal document TP 8760.

D. Croft (1991). *An automated technique for finding earth centre in satellite images*, MSc report, Dept. of Meteorology, Univ. of Edinburgh, Scotland.

T. Pavlidis and S.L. Horowitz (Aug. 1974). *Segmentation of plane curves*, IEEE Trans. on Computers, vol C-23, no. 8, pp 860-870.

W.H. Press, B.P. Flannery, S.A. Teukolsky and W.T. Vetterling (1988). *Numerical recipes in C*, pp. 540-547, Cambridge University Press.

A. Rosenfeld and A.C. Kak (1982). *Digital picture processing*, Academic Press.

A. Tamas (1991). *On Circles Recognition*, Vision Interface '91.

# IMPACT SOUND CONTROL OF ROBOTIC MANIPULATORS

**H. Wada\*, T. Fukuda\*\*, K. Kosuge\*\*, F. Arai\*\*, H. Matsuura\*\* and K. Watanabe\*\*\***

*\*Toyoda Automatic Loom Works Ltd., 2-1 Toyota-cho, Kariya 448, Japan*
*\*\*Department of Mechanical Engineering, Faculty of Engineering, Nagoya University, Furocho-1, Chikusa-ku, Nagoya 464, Japan*
*\*\*\*Department of Mechanical Engineering, Faculty of Science and Engineering, Saga University, Honjomachi-1, Saga 840, Japan*

Abstract.   This paper presents a robotic control method to reproduce the impact sound emitted by the collision between the endpoint of the robotic manipulator and the object. Because the impact phenomenon occurs in a very short period of time, the feedback control of impact sound in very difficult. Instead of a feedback control method, a learning control algorithm is proposed to control the impact phenomena. The learning control algorithm modifies the reference signal to the servo controller of the robot manipulator so that the desired impact sound is reproduced. The algorithm is based on the optimization of the least-squares criterion of learning error and does not require the dynamic model of the impact phenomena. The algorithm is applied to the planar manipulator with one d.o.f. driven by a pneumatic actuator. Experimental results illustrate the effectiveness of the proposed algorithm.

Keywords.   Robots; Impact force; Impact sound control; Learning systems; Iterative methods.

## INTRODUCTION

Much research has been done for the control of robotic manipulators, which is position control or force control et.al. Little attention has been paid to the transient phenomena from position control to force control and vice versa. The transient phenomena involve the collisions between the manipulator endpoint and its working environment, and the control of the collisions is one of the key issues for the practical applications of force control.

Recently, some research works have been performed with respect to the control problem of collision phenomena. In modeling of impact, Wang and Mason(1987) studied the planer impact of two objects, and Kahng and Aminouche(1988) derived the maximum impact force equation for two-body collision problems. Zheng and Hemami(1985) derived the mathematical model of a robot collision, and proposed the assembly method of two arms using impulsive information(Zheng,1987). Yousef-Toumi et.al.(1989) valideted an analytical model of impact experimentally. In stable control of robotic manipulator with impact, Khatib and Burdick(1986) adopted velocity damping to dissipate excessive impact energy arising between

end-effector and objects. Mills(1990) developed a dynamic model including the effect of impact, and proposed a discontinuous control for the transition from noncontact to contact of robotic manipulator. Shoji and Fukuda et.al.(1990,1991) adopted a Heltz-type model with energy loss to collision dynamics, and proved the stability with respect to control system of robotic manipulator with collision phenomena. In reducing impact force, Walker(1990), Gertz. et.al.(1991) proposed the strategy using the redundancy. In this paper, we propose a control algorithm to reproduce the impact phenomenon so that we can utilize the impact phenomena in the applications of robotic manipulators, such as hammering and nailing. Impact control is very important for the tasks that require impulsive forces. In general, an impact phenomenon is not easy to observe. In this paper, the impact phenomenon is controlled based on the impact sound.

The feedback control of impact phenomena is very difficult, since the collision phenomena occur in a very short period of time. Moreover, it is very difficult to identify the dynamics of the impact phenomena. The control algorithm ,which we propose in this paper, is based on the learning control. The in this paper, is based on the learning control. The learning control algorithm modifies the reference signal to the servo controller so that the desired impact phenomenon is reproduced.The algorithm controls the collision, without using the real time feedback of the impact phenomena. The algorithm is based on the optimization of the least-squares criterion of learning error and is designed based on the impulse response of the system. The design of the control law does not require the dynamic model of the impact phenomena(Watanabe et.al.,1991). The proposed control algorithm is experimentally applied to the impact sound control between a manipulator with one d.o.f. and its environment. Experimental results illustrate the effectiveness of the proposed control algorithm.

## IMPACT SOUND CONTROL

An impact phenomenon is not generally easy to observe, because it occurs in a very short period of time. The impact phenomena are often accompanied with sounds. In this paper, the impact phenomenon is controlled based on the sound emitted by the impact. It is assumed that only the magnitude of impact sound is dealt with, but that the frequency and tone are not dealt with. The real-time feedback control is one way to control the collision, but the extremely high sampling rate is required to do so. We propose a learning control algorithm to control the phenomena in this paper. A learning control algorithm controls the output of the system by repetitively modifying the reference signal to the system so that the output error vanishes. It controls the output of the system in a feedforward manner and does not require the real-time feedback of variables related to the impact phenomenon. Fig.1 shows the structure of the proposed control system of the impact phenomena based on impact sounds.

## CONTROL ALGORITHM

In this section, a learning control algorithm for the impact sound control is proposed. The algorithm is based on the method proposed in [7]. Let the input-output relation of the impact phenomena be described as follows:

$$\mathbf{y} = \mathbf{L}v \quad (1)$$

where

$$\mathbf{L}^T = [J(0), J(1), ..., J(N)] \quad (2)$$

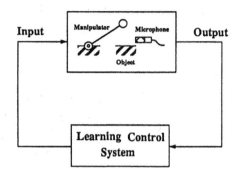

**Fig.1** Impact Sound Control

$$\mathbf{y}^T = [y(r), y(r+1), ..., y(N+1)] \quad (3)$$

$\mathbf{L} \in \Re^{(N+1) \times 1}$ is a vector, whose element is a Markov parameter, and $\mathbf{y} \in \Re^{(N+1) \times 1}$ is the output signal of impact sound. Note that $r$ is a positive integer which denotes the delay of responses. $v$ is the magnitude of the impulse input to the system. Given the desired setpoint or reference values such that

$$\mathbf{y}_d{}^T = [y_d{}^T(r), y_d{}^T(r+1), ..., y_d{}^T(r+N)] \quad (4)$$

and the accurate value of $\mathbf{L}$, we can determine a unique input to make $v$ as follows:

$$v = \mathbf{L}^{-L}\mathbf{y}_d \quad (5)$$

$$\mathbf{L}^{-L} \triangleq [\mathbf{L}^T\mathbf{L}]^{-1}\mathbf{L}^T \quad (6)$$

However we cannot determine it because $\mathbf{L}$ generally contains uncertainties. Therefore we apply to solve $v$ in a recursive form.

In order to use the weighted least-squares method, we assume that the unknown input is constant at any learning stage, so that we can introduce an identification model given by:

$$v_{k+1} = v_k \quad (7)$$

$$\mathbf{y}_d = \mathbf{L}v_k \quad (8)$$

where $k$ is the number of trials. Then, the optimal input that minimizes the weighted least-squares criterion

$$J = \|\mathbf{y}_d - \mathbf{L}v_k\|_V^2 \quad (9)$$

is given by

$$\hat{v}_k = \hat{v}_{k-1} + \mathbf{K}_k e_{k-1} \quad (10)$$

$$\hat{v}_{-1} = 0 \quad (11)$$

$$e_{k-1} = y_d - y_{k-1} \qquad (12)$$

where $y_{k-1}$ is an actual output, i.e. $L\hat{v}_{k-1}$. The matrix $K_k \in \Re^{1\times(N+1)}$ is the gain, which is obtained iteratively from the following relations:

$$K_k = P_{k-1}L^T[LP_{k-1}L^T + V^{-1}]^{-1} \qquad (13)$$
$$P_{-1} = \alpha I \qquad (14)$$

$$P_k = [I - K_k L]P_{k-1} \qquad (15)$$

The system expressed by eqs.(7), (8) is time-invariant, and is not stabilizable, because an uncontrollable mode exists on the unit circle. Therefore, we add a zero mean noise processes $w_k$ with positive definite covariance $W$ to eq.(7):

$$v_{k+1} = v_k + w_k \qquad (16)$$

Defining

$$W = GG^T \qquad (17)$$
$$L^T V L = F^T F \qquad (18)$$

and assuming that $(I, G)$ is reachable and $(F, I)$ is detectable, a unique positive definite stabilizable solution $P_s(> 0)$ is obtained from the following in algebraic Riccati equation:

$$P_s = P_s - P_s F^T \{F^T P_s F + I^{-1}\}^{-1} F P_s \\ + GG^T \qquad (19)$$

Therefore, the steady-state gain is given by

$$K = P_s L^T \{LP_s L^T + V^{-1}\}^{-1} \qquad (20)$$

Consequently, the learning rule in steady-state is obtained by:

$$\hat{v}_k = \hat{v}_{k-1} + Ke_{k-1} \qquad (21)$$
$$\hat{v}_{-1} = 0 \qquad (22)$$

The structure of the present learning controller system is shown in Fig.2, and the block diagram of learning controller is also shown in Fig.3.

New we explain the method of a learning control for impact sound control in the case of impact of M times. In this case, the matrix L shown in eq.(1) is transformed into eq.(23).

$$y = Lv \qquad (23)$$

where

$$L = \begin{bmatrix} J(0) & \cdot & 0 & \cdot & 0 \\ \cdot & \cdot & \cdot & \cdot & \cdot \\ \cdot & \cdot & 0 & \cdot & \cdot \\ J(k) & \cdot & J(0) & \cdot & \cdot \\ \cdot & \cdot & \cdot & \cdot & 0 \\ J(l) & \cdot & \cdot & \cdot & J(0) \\ \cdot & \cdot & \cdot & \cdot & \cdot \\ J(N) & \cdot & J(N-k) & \cdot & J(N-l) \end{bmatrix}$$
$$(24)$$

$$y = [y(r), y(r+1), ..., y(r+N)] \qquad (25)$$
$$v = [v(1), v(2), ..., v(M)] \qquad (26)$$

where $L \in \Re^{(N+1)\times M}$ is a matrix, and M denotes the number of times of impact. $v \in \Re^{1\times M}$ show's velocity just before impact in each time. Thus, following the same manner as shown in the case of impact of once, the learning control low is derived as follows:

$$\hat{v}_k = \hat{v}_{k-1} + Ke_{k-1} \qquad (27)$$
$$\hat{v}_{-1} = 0 \qquad (28)$$

where

$$K = P_s L^T \{LP_s L^T + V^{-1}\}^{-1} \qquad (29)$$
$$P_s = P_s - P_s F^T \{L^T P_s L + V^{-1}\}^{-1} F P_s \\ + GG^T \qquad (30)$$

Note here that the dimensions of matrices $K$ and $P_s$ is $\Re^{M\times(N+1)}$ and $\Re^{M\times M}$, respectively.

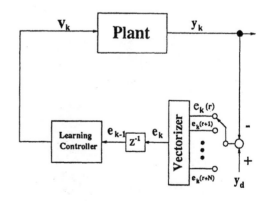

**Fig.2** Structure of learning control system

EXPERIMENT

Experimental Setup

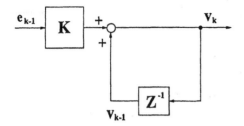

Fig.3 Block diagram of learning controller

The experimental setup is shown in Fig.4. The two pneumatic actuators (Rubbertuator, Bridgestone) are used to drive the arm, that is the difference of the output torques generated by these actuators through the pulley is used to drive the arm. We assume that the torque generated by an actuator is proportional to the input voltage to the controller of the actuator. The difference of the input voltages,

$$dv = (V_1 - V_2)/2 \qquad (31)$$

to these actuators is the input to system. The actual input to each actuator is calculated as follows:

$$V_1 = V_0 + dv \qquad (32)$$
$$V_2 = V_0 - dv \qquad (33)$$

where $V_0$ is the offset voltage to the controller.

Fig.4 Experimental setup

Filtering of impact sound

The sound, which is emitted by the collision phenomenon and recorded by a microphone, contains noises. We analyzed the signal from the microphone. Fig.5 shows the power spectrum of the impact sound of our experimental system. In our system, the sound was characterized by the frequency around 244Hz. Throughout the experiments, the sound signal was filtered using band pass filter. The control algorithm was calculated based on filtered signal.

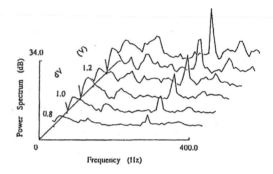

**Fig.5** Power spectrum

Experimental Method

The proposed control algorithm uses the impulse response of the system. We first got the impulse response of the system. Based on the impulse response, we constructed the matrix L of eq.(2), solved the Riccati equation, and gain matrix K.

Experimental were carried out as follows: 1) we put the manipulator arm at the its initial position, 2). the manipulator was controlled by the input signal calculated by the learning control algorithm until the collision was detected, 3) the input to the system at the next trial was calculated using eq.(21), 4) we put the manipulator at its initial position again and go to the step 2 to repeat the learning process.

Experimental Results

The desired impact sound is shown in Fig.6. The experimental results in v'=1.0 are shown in Figs.7, 8 and 9, the results with v'=0.01 are shown in Fig. 10 and 11, and the results with v'=0.0001 are shown in Fig.12 and 13. Fig.14 shows the mean-squares error of the output for each experiment. As shown in Fig.14, the output error decreases as the number of trials increase. The results illustrate the effectiveness of the proposed control algorithm.

## CONCLUSIONS

We proposed a learning control algorithm for the control of collisions between the manipulator and its environment. The proposed control algorithm reproduces the desired impact phenomena using sound information. The control algorithm is based on the learning control using weighted least-squares method. The algorithm is designed based on the impulse response of the system and does not require the details description of the model. The algorithm was experimentally applied to the planar manipulator with one degree of freedom driven by a pneumatic actuator. Experimental results illustrated the effectiveness of the proposed algorithm. The proposed control method can be applied to a series of impact phenomena without modification.

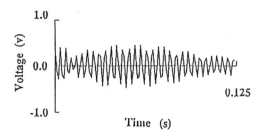

**Fig.8** Experimental results in v'=1.0(k=2)

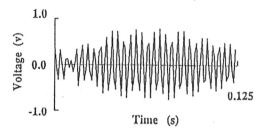

**Fig.9** Experimental results in v'=1.0(k=6)

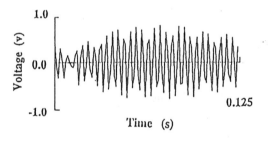

**Fig.6** Desired impact sound

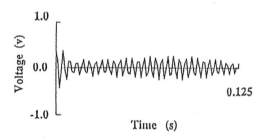

**Fig.10** Experimental results in v'=0.01(k=2)

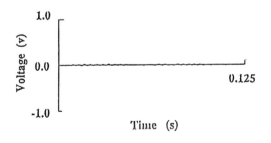

**Fig.7** Experimental results in v'=1.0(k=0)

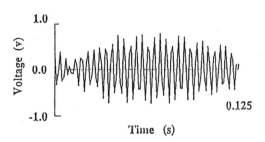

**Fig.11** Experimental results in v'=0.01(k=6)

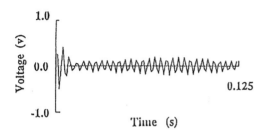

**Fig.12** Experimental results in v'=0.0001(k=2)

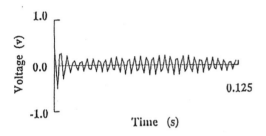

**Fig.13** Experimental results in v'=0.0001(k=6)

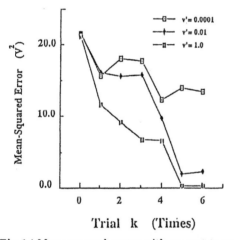

**Fig.14** Mean-squared errors with respect to various weights

REFERENCES

Fukuda.T., Shoji.Y. and Inaba.M. (1990). Stable Position and Force Control of Robotic Manipulator with Collision Phenomena, Proc. of the 11th IFAC World Congress, Automatic Control in the Service of Mankind, 9, pp.262-267.

Gertz.M.W., kim.J.O. and Khosla.P.K. (1991). Exploiting Redundancy to Reducing Impact Force, Proc. of the IEEE International workshop on Intelligent Robots and Systems (IROS'91), pp.179-184.

Kahng.J. and Anirouche.F.M.L. (1988). Impact Force Analysis in Mechanical Hand Design - Part I, International Journal of Robotics and Automation, 3-3, pp.158-164.

Khatib.O. and Burdick.J. (1986). Motion and Force Control of Robotic Manipulators, Proc. of the IEEE International Conference on Robotics and Automation, pp.1381-1386.

Mills.J.K. (1990). Manipulator Transition To and From Contact Tasks : A Discontinuous Control Approach, Proc. of the IEEE International Conference on Robotics and Automation, pp.440-446.

Shoji.Y., Inaba.M. and Fukuda.T. (1991). Impact Control of Grasping, IEEE transaction on industrial electronics, 38-3, pp.187-193.

Walker.I.D. (1990). The Use of Kinematic Redundancy in Reducing Impact and Contact Effects in Manipulation, Proc. of the IEEE International Conference on Robotics and Automation, pp.434-439.

Wang.Y. and Mason.M.T. (1987). Modeling Impact Dynamics for Robotic Operations, Proc. of the IEEE International Conference on Robotics and Automation, pp.678-685.

Watanabe.K., Fukuda.T. and Tzafestas.S.G. (1991). An Iterative Learning Control Scheme Using the Weighted Least-Squares Methods, Journal of Intelligent and Robotics Systems, 4-3, pp.267-284.

Yousef-Toumi.K. and Gutz.D.A. (1989). Impact and Force Control, Proc. of the IEEE International Conference on Robotics and Automation, pp.410-416.

Zheng.Y.F. and Hemami.H. (1985). Mathematical Modeling of a Robot Collision with its Environment, Journal of Robotic Systems, 2-3, pp.289-307.

Zheng.Y.F. (1987). Two Robot Arms in Assembly by Impulsive Information, Journal of Robotic Systems, 4-5, pp.585-603.

# COMPARATIVE STUDY OF THREE ROBUST CONTROL SCHEMES FOR SINGLE-LINK FLEXIBLE ARMS WITH FRICTION IN THE JOINTS

### V. Feliu*, K.S. Rattan** and H.B. Brown***

*Dpto. Electricidad, Electronica y Control, ETSI Industriales Uned, Madrid, Spain
**Department of Electrical Engineering, Wright State University, Dayton, OH 45435, USA
***Robotics Institute, Carnegie Mellon University, Pittsburgh, PA 15213, USA

**Abstract.** Tip position control of flexible arms with friction in the joints is studied in this paper. The control scheme is based on two nested loops: an inner loop to control the position of the motor and an outer loop that controls the tip position. The design of the inner loop is quite straightforward. However, the the outer loop may be designed several different ways. Three methods are proposed here to control the tip position: a feedforward based control method, a method that eliminates the principal natural frequency of the beam, and a method based on the sensing at several intermediate points of the mechanical structure of the arm. In the ideal case (when there are no perturbations), these three methods produce the same dynamic behavior of the arm. This paper compares their performances when perturbations and unmodelled dynamics are present. In particular, three cases are considered, i.e., external disturbances in the tip, unmodelled dynamics in the inner loop, and changes in the carried load. Analytic and simulation studies have been carried out to compare these methods and the results show that the third method which uses sensing at the intermediate points of the arm, had the best performance in all three cases.

**Keywords.** Flexible manipulators; robotics; robust control; feedforward control; decoupling control; identification.

## INTRODUCTION

The control of the tip position of flexible arms with friction in the joints is studied in this paper. This problem may be important for lightweight flexible arms or for heavy flexible arms moving at low speeds. A robust control scheme is proposed here. The scheme is different from the other existing methods (Cannon, 1985; Harahima, 1986; Kotnik, 1988; Matsuno, 1987; Siciliano, 1986) in that the position of the motor is the control signal instead of the current of the motor as used in other methods. The proposed scheme is composed of two nested loops: an inner loop that controls the motor position and an outer loop that controls the tip position.

The design of the inner loop, consisting of a simple *P.D.* controller, is quite straightforward (Rattan. 1988). However, the design of the outer loop may be carried out several different ways. Three methods are proposed here: a feedforward based control scheme (Feliu, 1989), a method that removes the first natural frequency of the beam (Rattan, 1990), and a method that uses position measurements at intermediate points of the beam (Feliu, 1992) apart from the tip and motor positions. All three methods have been experimentally tested exhibiting a similar behavior when tracking a specified trajectory. In order to decide which one is best, we studied the sensitivity of these methods to several disturbances. In the ideal case, when there are no perturbations, these three control schemes produce the same dynamic behavior of the arm. This paper compares their performances when perturbations and unmodelled dynamics are present. In particular, three cases are considered: external disturbances in the tip, unmodelled dynamics in the inner loop, and changes in the carried load.

The motor position control loop is described next followed by a brief description of the three methods for the control of tip position. Then, a comparative study is carried out and some conclusions are drawn.

## MOTOR POSITION CONTROL LOOP

The motor position control loop corresponds to the inner loop of Fig. 1. We want to achieve two objectives when designing a controller for this loop:

1. Remove the modelling error and the nonlinearities introduced by the Coulomb friction and changes in the coefficient of the dynamic friction,

2. Make the response of the motor position much faster than the response of the tip position control loop (outer loop in Fig. 1).

The fulfilment of the second objective allows us to substitute the inner loop by an equivalent block whose transfer function is approximately equal to one, i.e., the error in motor position is small and is quickly removed. This simplifies the design of the outer loop as illustrated in the next section. The differential equation relating the angle of the motor to the applied current can be written as

$$Ki = J\frac{d^2\theta_m(t)}{dt^2} + V\frac{d\theta_m(t)}{dt} + C_t(t) + \text{CF}, \qquad (1)$$

where K is the electromechanical constant of the motor, $i$ is the current of the motor, $\theta_m(t)$ is the angle of the motor, J is the polar moment of inertia of the motor and hub, V is the dynamic friction coefficient, $C_t(t)$ is the coupling torque between the motor and the link (the bending moment at the base of the link). CF is the Coulomb friction and $t$ is time.

To simplify the design of the inner loop, the system described in equation (1) can be linearized by compensating for the Coulomb friction and decoupled from the dynamics of the beam by compensating for the coupling torque. This is done by adding, to the control current, the current equivalent to these torques (Fig. 2) and is given by

$$i_c(t) = (C_t(t) + CF \text{ (sign of motor velocity)})/K . \quad (2)$$

The magnitude of the Coulomb friction, CF, is identified from the spectral analysis of the motor position and the current signals. The details of the identification method can be found in Feliu (1988). The coupling torque $C_t(t)$ can be calculated two ways: from the strain gauge measurements at the base of the link or by the difference of the measurements of the angles of the motor and tip. The second approach is used in this paper. Since the single-mass beam is nearly massless, we can assume that the coupling torque $C_t(t) = C(\theta_m(t) - \theta_t(t))$, where C = (3.E.I)/L, is a constant that depends on the stiffness (E.I) and length L of the arm. For a two-mass beam, the coupling torque is more complex as compared to the single-mass case. But because of the limited computational capabilities of the microprocessor used for this study, a simpler coupling term of the form $C_t(t) = \dot{C}(\theta_m(t) - \theta_t(t))$ was implemented. After compensating for the friction and the coupling torque, the transfer function between the angle of the motor and the current can be written as

$$G_m(s) = \theta_m(s)/i(s) = (K/J)/s(s + V/J). \quad (3)$$

The block diagram of the inner loop control system is shown in Fig. 2 (discrete control version). The series and feedback controllers ($G_{c1}$ and $G_{c2}$, respectively) are designed so that the response of the inner loop (position control of the motor) is significantly faster than the response of the outer loop (position control of the tip) and without any overshoot. This is done by making the gain of the series controller large and is limited only by the saturation current of the servo amplifier. It was shown (Rattan, 1988) that, in theory, this gain could be made arbitrarily large even in the case of the arm being a nonminimum phase system. It was also shown that a large gain in this loop reduces the effects of nonlinearities caused by friction.

When the closed-loop gain of the inner loop is sufficiently high, the motor position will track the reference position with very little error. The dynamics of the inner loop may then be approximated by '1' when designing the outer loop controller.

## TIP POSITION CONTROL SCHEMES

The three schemes proposed here are composed of two basic loops: an inner loop that controls the motor position and an outer loop that controls the tip position. All three schemes use the same inner loop, but they differ in the way the tip position is controlled. The schemes are represented in Figs. 3-5. The first scheme is the feedforward based control scheme (Scheme 1), the second is based on the method that removes the first natural frequency of the beam (Scheme 2), and the third scheme uses sensing at intermediate points of the beam (Scheme 3).

These three methods are based on the assumption that the response of the motor $\theta_m$ (after having closed the motor position inner loop), to changes in its reference $\theta_{mr}$, is faster than the beam dynamics. They differ in the importance given to the "a priori" knowledge of the plant (represented by the model) relative to the importance given to the on-line measured signals. The use of the flexible arm model in the

control reduces the number of variables to be measured. A way of using this model is to implement a controller with a feedforward component. This term can be computed off-line and allows us to simplify the feedback component of the controller that uses on-line measurements. Some advantages from the stability point of view have been reported (Khosla, 1986) for these schemes when applied to rigid arms. This seems to be also true for the flexible arms. However, these schemes are more sensitive to changes in the parameters of the plant and is especially true in flexible arms because of their undamped nature.

All three schemes have a feedforward component, but the first scheme is the one that relies most on the model of the arm. All the driving action is carried out by the feedforward term in this scheme and is generated from the desired tip position trajectory and the model of the arm. The measurements of the tip position are used only to correct small perturbations when following the desired trajectory.

The second scheme uses measurements of the tip position to compensate for the tracking errors (as in the first method), but these measurements are also used to cancel the first natural frequency of the beam which is the dominant frequency in most of flexible beams. The positive unity gain feedback loop shown in Fig. 4 always removes the first vibrating mode independent of the transfer function of the beam. Higher modes are canceled by implementing the inverse of the minimum-phase factors of the transfer function resulting after having closed the positive feedback loop. The feedforward term used here is much simpler than the one used in the first scheme. Now, part of the driving action is generated from the actual measurements of the tip position through the positive unity feedback signal.

The third scheme uses sensing at several points of the structure allowing us to simplify the feedback controller. It includes the feedforward term of the first scheme. But now the feedback loop does not depend completely on the model because it uses multiple sensing. This scheme assumes that the dynamics of the flexible beam may be expressed by

$$\mu \frac{d^2\Theta}{dt^2} = A\Theta + B\theta_m, \quad (4)$$

where $\Theta \in \Re^{n \times 1}$ represents the positions (angles) of the $n$ measured points of the beam (these points are labeled from 1 to $n$ in the sense of growing distances from the motor (tip position variable is $\theta_n$)); $A \in \Re^{n \times n}, B \in \Re^{n \times 1}$ are parametric matrices independent of the tip payload; and $\mu \in \Re^{n \times n}, \mu = diag(m_1, m_2, \ldots, m_n)$ is a parametric diagonal matrix. $m_i$'s are parameters independent of the tip payload for $i < n$. with $m_n$ being the tip payload. This model is exact in the case of lumped-mass flexible arms, meaning $m_i$'s are the lumped masses and is a reasonable approximation in the case of distributed-mass flexible arms. In Fig. 5, the block named "Beam" represents the dynamics of the beam and is given by

$$\Theta(s) = (\mu s^2 - A)^{-1} B\theta_m(s). \quad (5)$$

Expression (5) can be easily obtained from (4). The feedback control law for this scheme is $\gamma(s) = (\Lambda_1 + \Lambda_2 s)(\Theta_r(s) - \Theta(s))$, where $\Lambda_1, \Lambda_2 \in \Re^{1 \times n}$ are the coefficients of a P.D. like controller, and $\Theta_r(s)$ is the desired trajectory for the vector $\Theta$ of beam intermediate positions. This desired trajectory vector is generated by passing the tip reference $\theta_n$ through the column vector $\Upsilon(s)$ that relates the tip angle with the intermediate measurements as $\Theta(s) = \Upsilon(s)\theta_n(s)$. $\Upsilon(s)$ can be obtained from model (4) as

$$\Upsilon(s) = \frac{(\mu s^2 - A)^{-1} B}{C_n (\mu s^2 - A)^{-1} B}, \qquad (6)$$

where $C_n = \begin{pmatrix} 0 & 0 & \dots & 0 & 1 \end{pmatrix}$. As shown in Fig. 5, $P_p$ is the reference for the tip position. If we implement a model-based feedforward term, the motor angle reference given by this term would be $\theta_{mr} = G_b^{-1}(s) P_p(s)$. This motor reference may be used if $G_b(s)$ is minimum phase. But very often, flexible arms have non-minimum phase zeros and then $G_b^{-1}(s)$ becomes unstable producing an unbounded $\theta_{mr}$ signal. In order to overcome this, $P_p$ has to be passed through a filter that includes the non-minimum phase zeros of $G_b(s)$ in its numerator. This filter may be designed by minimizing the differences between $P_p$ and the filtered signal (Feliu, 1989, 1990; Rattan, 1990). But here, we use a filter which is just a normalized term whose factors are the right half-plane zeros of $G_b(s)$ and can be written as

$$F(s) = \prod_{i=1}^{n_1} (1 - a_i^{-1} s), \qquad (7)$$

where $a_i$'s are the $n_1$ right half-plane zeros. This is not an optimal filter but is designed to keep $\theta_{mr}$ bounded, which is what we need to compare the three methods.

## COMPARATIVE STUDY

In the ideal case (when there are no perturbations), all three schemes give the same response. Differences among the three control schemes appear when there are perturbations that have to be compensated by a feedback controller $R(s)$. Three disturbances (perturbations in the tip position, unmodelled dynamics in the inner loop, and changes in the payload) were used to compare these methods. Analytic and simulation studies were carried out and the results show that the method that uses sensing at several points of the flexible link had the best performance in all three cases.

### Perturbation In The Tip Position

The perturbation in the tip position $\varepsilon(s)$ is added to the reference motor position, $\theta_{mr}(s)$, to include disturbances in the tip and motor positions as shown in Figs. 3-5. The transfer functions between the tip position and the disturbance for schemes 1-3 can therefore be written as

**First scheme:**

$$\frac{\theta_n(s)}{\varepsilon(s)} = G_1(s) = \frac{G_b(s)}{1 + G_{c3}(s) G_b(s)}, \qquad (8)$$

**Second scheme:**

$$\frac{\theta_n(s)}{\varepsilon(s)} = G_2(s) = \frac{G_b(s)}{(1 - G_b(s))(1 + \frac{F(s) G_{c3}(s)}{s^2})}, \qquad (9)$$

**Third scheme:**

$$\frac{\theta_n(s)}{\varepsilon(s)} = G_3(s) = \frac{G_b(s)}{1 + G_b(s)(\Lambda_1 + \Lambda_2 s) \Upsilon(s)}. \qquad (10)$$

where $R_1$, $R_2$, $\Lambda_1 + \Lambda_2 s$ are the feedback controllers of the schemes 1, 2 and 3 respectively. Assuming that the order of the denominator of $G_b(s)$ is $n_g$, expression (9) shows that $n_g - 2$ poles of the closed-loop system are fixed in scheme 2, and are given by the zeros of $(1 - G_b(s))/s^2$. These fixed poles are normally far away from the origin and are ,therefore, less important as compared to the dominant poles that can be assigned using the controllers. If the system has less than 2 positive zeros, two dominant poles may be arbitrarily assigned by a $P.D.$ controller: $R_2(s) = r_{2,0} + r_{2,1} s$. Let us de-

fine the index $\rho_{i,j}(s) = G_i(s)/G_j(s); i, j \leq 3$ to compare the three schemes. Assuming that the controllers do not increase the order of the system, we have

$$\rho_{2,1}(s) = \frac{G_2(s)}{G_1(s)} = \frac{s^2 (1 + R_1(s) G_b(s))}{(s^2 + F(s) R_2(s))(1 - G_b(s))}$$

$$= \frac{\prod_{i=1}^{n_g}(s - \eta_i^1)}{(1 - r_{2,1} a_1) \prod_{i=1}^{n_g}(s - \eta_i^2)} \qquad (11)$$

$$\rho_{3,1}(s) = \frac{G_3(s)}{G_1(s)} = \frac{1 + R_1(s) G_b(s)}{1 + G_b(s)(\Lambda_1 + \Lambda_2 s) \Upsilon(s)} = \frac{\prod_{i=1}^{n_g}(s - \eta_i^1)}{\prod_{i=1}^{n_g}(s - \eta_i^3)} \qquad (12)$$

where $\eta_i^j$ is the $ith$ closed loop pole of scheme $j$ and $\rho_{3,2}$ may be obtained from

$$\rho_{3,2}(s) = \rho_{3,1}(s)/\rho_{2,1}(s). \qquad (13)$$

Expression (11) is used to compare the frequency characteristics of schemes 1 and 2. If $|\rho_{2,1}(j\omega)| < 1$, then scheme 2 attenuates the perturbation $\varepsilon$ more than scheme 1 at frequency $\omega$. If $|\rho_{2,1}(j \cdot \omega)| > 1$, then scheme 1 attenuates the perturbation more than scheme 1 at this frequency. Expressions (12) and (13) are used the same way.

### Unmodelled Dynamics

We consider the dynamics of the inner loop $M(s)$ in this section. In order to compare the effects of $M(s)$ on the overall performance of the three schemes, we assume small variations in the parameters $\alpha_i, i \leq n_\alpha$ around the parameter values vector $\alpha^0$ that gives the ideal $M(s) = 1$. Thus, we can use the sensitivity function (Rattan, 1988):

$$S_{T,\alpha_i} = \left| \frac{\partial T(s)}{\partial \alpha_i} \cdot \frac{\alpha_i}{T(s)} \right|_{\alpha = \alpha^0}, \qquad (14)$$

where $T(s)$ is the transfer function that relates the output $\theta_n(s)$ with the input $P_p(s)$, and $\alpha_i$ is the parameter of $M(s)$ that varies. The scheme most appropriate for a design assuming $M(s) = 1$, will be the one that exhibits the smallest sensitivity to changes in the parameters $\alpha$. First, we analyze the effects of $M(s)$ on the open-loop control (just using the feedforward terms of schemes 1-3). Then, we study the complete schemes that include the feedforward as well as the feedback terms.

#### Open-loop control

Schemes 1 and 3 use the same feedforward term and intermediate sensing is only used in scheme 3 for the feedback control. Therefore, we only need to compare schemes 1 and 2. From Figs. 3 and 4 we get

**First scheme:**

$$\frac{\theta_n(s)}{P_p(s)} = T_1(s) = F(s) M(s), \qquad (15)$$

**Second scheme:**

$$\frac{\theta_n(s)}{P_p(s)} = T_2(s) = F(s) \frac{M(s)(1 - G_b(s))}{1 - M(s) G_b(s)}. \qquad (16)$$

And the sensitivity functions with respect to a parameter $\alpha_i$ of $M(s)$ are

397

$$S_{T_1,\alpha_i} = \frac{\partial M(s)}{\partial \alpha_i}\alpha_i^0 \qquad (17)$$

$$S_{T_2,\alpha_i} = \frac{1}{1 - G_b(s)}\frac{\partial M(s)}{\partial \alpha_i}\alpha_i^0 \qquad (18)$$

for the first and second scheme respectively.

The beam does not exhibit any deflection when the arm is stopped. Applying the *Final Value Theorem* (Kuo, 13), we have that $G_b(0) = 1$ and $lim_{\omega \to 0} \mid G_b(s)/(1 - G_b(s)) \mid = \infty$ in (18). Consequently, comparing expressions (17) and (18), we get that scheme 1 is much better than scheme 2 at low frequencies (up to the first natural frequency), while at high frequencies both schemes are similar ($lim_{\omega \to \infty} G_b(s) = 0 \Rightarrow lim_{\omega \to \infty} 1/(1 - G_b(s)) = 1$). At medium frequencies, both sensitivity functions are of the same order. Low frequencies dominate in the spectrum of reference signal $P_p$. Therefore, open loop control of scheme 1 is more insensitive to the dynamics of the inner loop $M(s)$ than that of scheme 2.

### Closed-loop control

From Figs. 3-5, using expression (14) and assuming $\alpha = \alpha^0$ (or $M(s) = 1$), we get

**Scheme 1:**

$$\frac{\theta_n(s)}{P_p(s)} = T_1(s) = F(s)M(s)\frac{1 + G_b(s)R_1(s)}{1 + G_b(s)M(s)R_1(s)} \Rightarrow$$

$$S_{T_1,\alpha_i} = \frac{1}{1 + G_b(s)R_1(s)}\frac{\partial M(s)}{\partial \alpha_i}\alpha_i^0. \qquad (19)$$

**Scheme 2:**

$$\frac{\theta_n(s)}{P_p(s)} = T_2(s) = F(s)(1 - G_b(s))(1 + \frac{R_2(s)F(s)}{s^2})$$

$$\frac{M(s)}{1 + M(s)(F(s)R_2(s)\frac{1-G_b(s)}{s^2} - G_b(s))} \Rightarrow$$

$$S_{T_2,\alpha_i} = \frac{1}{(1 - G_b(s))(1 + \frac{F(s)R_2(s)}{s^2})}\frac{\partial M(s)}{\partial \alpha_i}\alpha_i^0. \qquad (20)$$

**Scheme 3:**

$$\frac{\theta_n(s)}{P_p(s)} = T_3(s) = F(s)\frac{M(s)(1 + G_b(s)(\Lambda_1 + \lambda_2 s)\Upsilon(s))}{1 + G_b(s)(\Lambda_1 + \Lambda_2 s)\Upsilon(s)} \Rightarrow$$

$$S_{T_3,\alpha_i} = \frac{1}{1 + G_b(s)(\Lambda_1 + \Lambda_2 s)\Upsilon(s)}\frac{\partial M(s)}{\partial \alpha_i}\alpha_i^0. \qquad (21)$$

We define indexes $\tau_{i,j}(s) = T_i(s)/T_j(s); i,j \leq 3$. Therefore, if $\mid \tau_{i,j}(j\omega) \mid < 1$, then scheme $i$ is less sensitive to unmodelled dynamics as compared to scheme $j$; and scheme $j$ is less sensitive as compared to scheme $i$ if $\mid \tau_{i,j}(j\omega) \mid > 1$. Using (19)-(21), we get that

$$\tau_{i,j}(s) = \rho_{i,j}(s); \quad \forall i,j. \qquad (22)$$

### Changes In The Payload

The sensitivity function used to study the effects of changes in the payload is given by

$$S_{T,m_n} = \mid \frac{\partial T(s)}{\partial m_n} \cdot \frac{m_n}{T(s)} \mid_{m_n = m_n^0}, \qquad (23)$$

where $m_n$ is the tip payload, and $m_n^0$ is the nominal tip payload for which the controllers have been designed. Therefore, for

**Scheme 1:**

$$S_{T_1,m_n} = -s^2 C_n(\mu s^2 - A + R_1(s)BC_n)^{-1}C_n^T m_n^0$$

$$= \frac{1}{G_b(s)(1 + R_1(s)G_b(s))}\frac{\partial G_b(s)}{\partial m_n}m_n^0 \qquad (24)$$

**Scheme 2:**

$$S_{T_2,m_n} = -s^2 C_n(\mu s^2 - A - BC_n + R_2(s)\frac{F(s)(1 - G_b(s))}{G_b(s)s^2}BC_n)^{-1}C_n^T m_n^0$$

$$= \frac{1}{G_b(s)(1 - G_b(s))(1 + \frac{F(s)R_2(s)}{s^2})}\frac{\partial G_b(s)}{\partial m_n}m_n^0 \qquad (25)$$

**Scheme 3:**

$$S_{T_3,m_n} = -s^2 C_n(\mu s^2 - A + B(\Lambda_1 + \Lambda_2 s))^{-1}C_n^T m_n^0$$

$$= \frac{1 - G_b^2(s)(\Lambda_1 + \Lambda_2 s)\frac{\partial \Upsilon(s)}{\partial m_n}/\frac{\partial G_b(s)}{\partial m_n}}{G_b(s)(1 + G_b(s)(\Lambda_1 + \Lambda_2 s)\Upsilon(s))}\frac{\partial G_b(s)}{\partial m_n}m_n^0. \qquad (26)$$

If we define indexes $\lambda_{i,j}(s) = T_i(s)/T_j(s); i,j \leq 3$, we have

$$\lambda_{2,1}(s) = \rho_{2,1}(s) \qquad (27)$$

$$\lambda_{3,1}(s) = \rho_{3,1}(s)(1 - G_b^2(s)(\Lambda_1 + \Lambda_2 s)\frac{\frac{\partial \Upsilon(s)}{\partial m_n}}{\frac{\partial G_b(s)}{\partial m_n}}) \qquad (28)$$

Simulation studies were carried out for one and two vibration modes flexible arms and the results show that the scheme 3 performs better than scheme 1 at low frequencies (Feliu, 1989). It was also shown that scheme 3 also performs better than scheme 2 for both low and high frequencies (Feliu, 1989).

## Conclusions

Three schemes to control the tip position of flexible arms with friction in the joints have been presented and compared. These schemes are based on a two nested loop structure. The inner loop controls the motor position and is common to the three schemes.

The three schemes give the same response in ideal conditions, but they differ in their behavior when disturbances are present. Three types of disturbances were used to compare these schemes: perturbations in the tip position, unmodelled dynamics in the inner loop, and changes in the payload. The second type of perturbation occurs because of our particular control structure which assumes that the response of the motor position loop is instantaneous compared to the response of the outer loop while in reality there is always a delay or a time constant.

It was found that scheme 3 is less sensitive to all of the disturbances, followed by scheme 2. Scheme 1 is the most sensitive. Minimum and non-minimum phase examples have been developed to illustrate the above statements. Analytical and simulation results show that:

- When the system has only one vibrational mode, all three schemes are equivalent.

- Having fixed the closed-loop dominant poles, the sensitivity characteristic of scheme 3 may be considerably improved by placing the secondary closed-loop poles as far away as possible from the origin.

- Secondary poles of scheme 2 are fixed which is a disadvantage compared to scheme 3.

- Scheme 2 exhibits the worst characteristic of the three schemes at high frequencies.

- Agreement between the results of the theoretical analysis and the simulations shows that this sensitivity analysis is a valuable tool to compare control schemes for robots.

## REFERENCES

Cannon, R.H., and E. Schmitz (1985). Precise Control of Flexible Manipulators. Robotics Research, 841-861.

Feliu, V., K. S. Rattan, and H. B. Brown (1988). Model Identification of a Single-Link Flexible Manipulator in the Presence of Friction, Proceedings of 19th Annual ISA Modelling and Simulation Conference, Pittsburgh, Pa.

Feliu, V., K.S. Rattan, and H.B. Brown (1989). A New Approach to Control Single-Link Flexible Arms. Part II, Technical Report CMU-RI-TR-89-14, Carnegie Mellon University, Pittsburgh, Pa.

Feliu, V., K.S. Rattan, and H.B. Brown (1992). Modelling and Control of Single-Link Flexible Arms with Lumped Masses, ASME Journal of Dynamic Systems, Measurement, and Control, In press.

Harahima, F, and T. Ueshiba (1986). Adaptive Control of Flexible Arm using the End-Point Position Sensing. Proceedings Japan-USA Symposium of Flexible Automation, Osaka, Japan.

Khosla, P.K. (1986). Real-Time Control and Identification of Direct-Drive Manipulators, Ph.D. Thesis, Robotics Institute, Carnegie Mellon University.

Kotnik, P., S. Yurkovich, and U. Ozguner (1988). Acceleration Feedback for Control of a Flexible Manipulator Arm, Journal of Robotic Systems, 5, 181-196.

Kuo, B.C. (1982). Automatic Control Systems, Prentice-Hall.

Matsuno, F., S. Fukushima, and coworkers (1987). Feedback Control of a Flexible Manipulator with a Parallel Drive Mechanism. International Journal of Robotics Research, 6, 76-84.

Rattan, K.S., V. Feliu, and H.B. Brown(1988). Identification and Control of a Single-Link Flexible Manipulator, Proceedings 27th IEEE Conference on Decision and Control, Austin, Texas.

Rattan, K.S., V. Feliu, and H.B. Brown,(1988). A Robust Control Scheme for a Single-Link Flexible Manipulator with Friction in the Joints, Proceedings 2nd Annual USAF/NASA Workshop on Automation and Robotics, Dayton, OH.

Rattan, K.S., V. Feliu, and H.B. Brown(1990). Tip Position Control of Flexible Arms Using a Control Law Partitioning Scheme, IEEE International Conference On Robotics and Automation, Cincinnati, OH.

Siciliano, B., B.S. Yuan, and W.J. Book(1986). Model Reference Adaptive Control of a One Link Flexible Arm. Proceedings 25th IEEE Conference on Decision and Control, Athens, Greece.

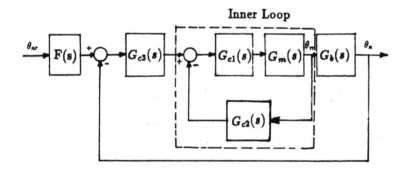

Fig. 1. General robust control scheme for flexible manipulators.

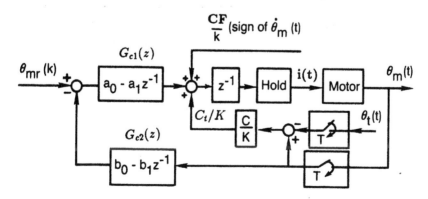

Fig. 2. Inner loop (motor position) control scheme.

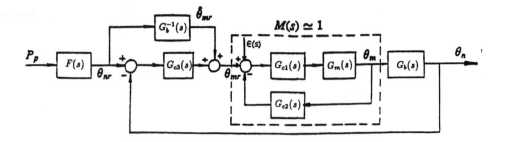

Fig. 3. Feedforward Based Control Scheme.

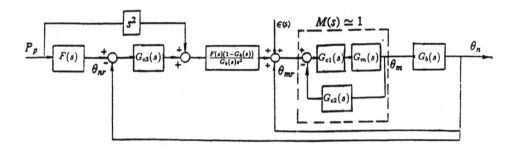

Fig. 4. Scheme that Cancels the First Natural Frequency.

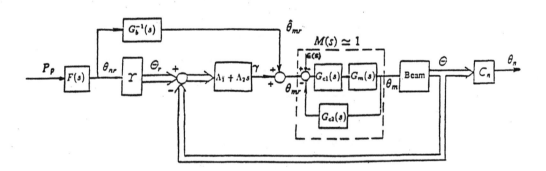

Fig. 5. Multiple Point Sensing Control Scheme.

# A TARGET TRACKING CONTROLLER FOR A REAL TIME POSITION AND ORIENTATION MEASUREMENT SYSTEM FOR INDUSTRIAL ROBOTS

**H. Gander, J.P. Prenninger and M. Vincze**

*Institute for Flexible Automation, Vienna University of Technology, Austria*

*Abstract.* This paper presents a target tracking controller for a dynamic 6 degree of freedom measurement system for industrial robots. The measurement system uses the beam of a laser interferometer and a mirror mounted on a cardan joint to follow a retroreflector, which is mounted to the robot´s endeffector. The position of the endeffector is measured with an accuracy in the range of micrometers and is computed in real time. The target tracking controller uses the computed position of the robot to extract the current velocity and acceleration of the robot´s endeffector and to estimate the positions of the next time intervals. Thus control algorithms similar to predictive control can be used to drive the motors of the cardan joint in a way that enables the measurement system to track the position of highly dynamic robots.

*Keywords.* Position and orientation measurement, state estimation, prediction, target tracking, maneuver detection, predictive control, digital control;

## INTRODUCTION

The application of robots in assembly grew rapidly in recent years and is still growing. Accompanying came the demand and first installments of off-line progamming systems. One of the critical items in off-line programming is the accuracy of the robot. Contrary to repeatability this characteristic measure is not necessary in teaching the movements of a robot. Therefore developers basically worked on improving repeatability. Though this also increases accuracy it can be said that the accuracy of an industrial robot is generally at least a magnitude lower than its repeatability.

Another factor for this bias in research and development was the absence of appropriate measurement systems. Repeatability can be easily metered with local position and orientation measurement equipment. Accuracy needs a system that reaches the entire workspace of the robot and works in a well established reference coordinate system. Since recent applications such as lasercutting or

metrology robotics also need high path accuracy the measurement system must be able to track the robot´s movement.

Due to this challenging conditions dynamic accuracy measurement systems are rare and limited to either straight lines, slow motion or low accuracy (Jiang, Black, Duraisamy, 1988).

The subsequently introduced Laser Tracking System (LTS) overcomes most of the shortcomings of other systems. One critical point in obtaining the required tracking capability is the control of the cardan joint mounted mirror. Since the motion of the robot cannot change drastically within a few sampling steps the application of estimation and prediction algorithms for tracking the robot´s endeffector is feasible (Berg, 1983). The prerequisite, the availability of the robot position in real-time, could be attained with the already in real-time calculated measurement data.

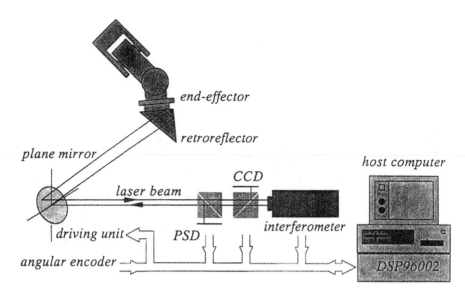

Fig. 1. Functional principle of the LTS.

## FUNCTIONAL PRINCIPLE OF THE LASER TRACKING SYSTEM (LTS)

Figure 1 shows the functional principle of the LTS. The beam of a high speed HeNe-laser interferometer is deflected by a plane mirror mounted on a cardan joint and hits the retroreflector, which is mounted to the robot's endeffector. The beam is reflected parallelly, is again deflected by the plane mirror and reaches the interferometer. The parallel displacement of the incoming laser beam, that is twice the distance between the laser beam and the centerpoint of the retroreflector, is measured with a position sensitive diode. Ideally, the laser beam hits the centerpoint of the retroreflector and there is no parallel displacement. When the robot starts moving the laser beam does not hit the centerpoint, thus there is a displacement of the laser beam. This displacement can be regarded as the tracking error and is minimized by the tracking controller through turning the axes of the cardan joint. This tracking error must not exceed 2 mm, otherwise the incoming laser beam does not hit the PSD and track is lost (larger PSD's can not be used due to insufficient dynamics and accuracy). This fact directly restricts the maximum allowable dynamic potential of the robot during measurement, especially the maximum acceleration. Thus a controller has to be developed, that keeps the tracking error as small as possible.

The data from the interferometer, the angular encoders and the PSD are used to determine the position of the robot's endeffector in real time with sampling rates up to 10 kHz (Gander and colleagues, 1991; Vincze, Gander, Prenninger, 1991). The specifications of the LTS regarding to position measurement shows Table 1.

*TABLE 1 Specifications of Position Measurement*

| measurement accuracy | 50 μm |
|---|---|
| repeatability | 10 μm |
| sampling rate | max. 10 kHz |
| measurement volume | > 1 m³ |

The orientation of the retroreflector can be extracted by analyzing the reflected laser beam with a vision system (Prenninger, Gander, Vincze, 1991). The specifications of orientation measurement are presented in Table 2.

*TABLE 2 Specifications of Orientation Measurement*

| measurement range | ± 30 degree |
|---|---|
| measurement accuracy | better than 10 arcsec |
| resolution | 1 arcsec |

## TARGET TRACKING CONTROLLER

The target tracking controller comprises both the hardware and the software that ensures that the laser beam follows the movements of the retroreflector, i.e. the robot. The specifications of the tracking controller are shown in Table 3.

*TABLE 3 Specifications of the Target Tracking Controller*

| maximum velocity of the robot | 6 m/s |
|---|---|
| maximum acceleration of the robot | 100 m/s² |

402

$$\begin{bmatrix} x \\ y \\ z \end{bmatrix} = \begin{bmatrix} -1/2 + \cos^2\beta + \sin^2\alpha \sin^2\beta & \cos\alpha \sin\beta \cos\beta & -\cos\alpha \sin\alpha \sin^2\beta \\ \cos\alpha \sin\beta \cos\beta & 1/2 - \cos^2\beta & \sin\alpha \sin\beta \cos\beta \\ -\cos\alpha \sin\alpha \sin^2\beta & \sin\alpha \sin\beta \cos\beta & 1/2 - \sin^2\alpha \sin^2\beta \end{bmatrix} \begin{bmatrix} 1 \\ \Delta y \\ \Delta z \end{bmatrix} \tag{1}$$

x, y, z    Cartesian coordinates of the robot´s endeffector,
$\alpha, \beta$    angles of the axes of the plane mirror,
l    distance between the retroreflector and the plane mirror,
$\Delta y, \Delta z$    2 dimensional displacement of the laser beam on the PSD;

A block diagram of the hardware is shown in Fig. 2.

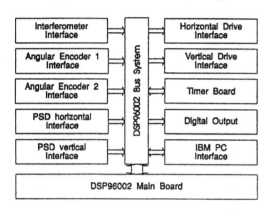

Fig. 2. Block diagram of the hardware of the tracking controller.

The hardware is based on a Digital Signal Processor (DSP) DSP96002 from Motorola. This DSP offers an arithmetic speed of 50 MFLOPS at 33 MHz (60 MFLOPS at 40 MHz). Via interfaces the DSP has access to the data from the interferometer, the PSD and the angular encoders. Additionally there are interfaces to the host computer (IBM - PC), to the power electronics of the motor drives and to the digital output of the position. The timer board generates programmable sampling times for the real time position computation and the target tracking controller. The sampling time currently used is 100 μs.

The power electronics of the motors control the angular velocity of the two axis. Stepping motors are used to turn the mirror. Since the moment of inertia of the mechanics and the friction of the gears must be taken into account, a torque observer is realized in hardware. During regular operation the torque control has no influence on the operation of the power electronics. Immediately before the maximum torque of the motors is reached (during acceleration) the angular velocity is reduced in order to prevent slipping out of step (Kerschbaum, 1991).

Several programs are running "simultaneously" (interrupt - controlled) on this hardware (see Fig. 3). First the position of the robot´s endeffector is computed in real time using the data from the interferometer, the angular encoders and the PSD. On one hand the actual position can be read out for purposes of measurement. On the

other hand the computed position can be used by the software modules of the target tracking controller (state estimator, predictor, maneuver detector, predictive controller) to ensure optimal tracking.

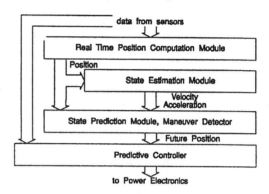

Fig. 3. Software modules of the target tracking controller.

### Real Time Position Computation

The real time position computation module may be considered as a part of the measurement system rather than a part of the tracking controller. With a sampling rate of 10 kHz the data from the interferometer, the angular encoders and the PSD are read in and the position of the centerpoint of the retroreflector is computed in real time according to Eq. (1).

Systematic errors of the position sensitive diode due to nonlinearities are corrected in real time using a table with precalibrated values. Additionally the wave length of the laser interferometer is computed using the actual temperature, air pressure and vapour pressure of water.

The position in Cartesian coordinates is stored in registers that are accessible from outside (digital output). The format of the x-, y- and z-coordinate is IEEE floating point single precision. The real time position computation needs about 25% of the computational power of the DSP96002 (programmed in assembly language) at a sampling rate of 10 kHz.

The computed position (and the data from the sensors) is also an important information for the target tracking controller and is used to improve the tracking performance.

## State Estimator

The state estimator uses the actual and the past positions of the robot´s endeffector to estimate the velocity and the acceleration of the robot (see Eq. (2), (3)).

$$\dot{x}(k) = \frac{x(k) - x(k-1)}{T} \qquad (2)$$

$$\ddot{x}(k) = \frac{\dot{x}(k) - \dot{x}(k-1)}{T} = \frac{x(k) - 2\,x(k-1) + x(k-2)}{T^2} \qquad (3)$$

where

$x(k)$   is the actual position at time k (without measurement noise),

$\dot{x}(k)$   is the mean velocity in interval k,

$\ddot{x}(k)$   is the mean acceleration in intervals k and k-1,

T   is the sampling time.

Due to the small values of the velocity and the acceleration (for example: an acceleration of 10 g = 1 μm/sampling interval$^2$) the effects of measurement noise (partly due to quantization) on the state have to be eliminated by filtering (see Eq. 4).

$$\hat{X}(k) = \hat{X}(k-1) + K\,[\ X(k) - \hat{X}(k-1)\ ] \qquad (4)$$

where

$\hat{X}(k)$   is the velocity or acceleration estimate at time k,

$X(k)$   is the computed velocity or acceleration based on the measurements at time k,

K   is the filter gain.

The filter gain K is a constant or is computed in real time (Chang, Tabaczynski, 1984). The tradeoff is performance versus real-time computational requirement. For the state estimator of the LTS a constant filter gain is used. Consequently the filter can be represented by a transfer function with low pass characteristic.

## Predictor and Maneuver Detector

The predictor estimates the future position of the robot´s endeffector using the current state (measured position, estimated velocity and acceleration) of each coordinate and a model for the robot (see Fig. 4, Eq. (5)).

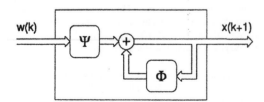

Fig. 4 Model for the robot.

$$x(k+1) = \Phi\,x(k) + \Psi\,w(k) \qquad (5)$$

with

$$x(k) = [x, \dot{x}, \ddot{x}, y, \dot{y}, \ddot{y}, z, \dot{z}, \ddot{z}]^T \qquad (6)$$

$$w(k) = [w_x, w_y, w_z]^T \qquad (7)$$

where

$x(k)$   is the state vector (see Eq. (6)) at time k,

$\Phi$   is the state transition matrix,

$w(k)$   accounts for unknown target maneuvers (see Eq. (7)),

$\Psi$   is the maneuver/state transition matrix and

T   denotes transposition.

The state transition matrix and the maneuver/state transition matrix for the third order state vector are shown in Eq. (8) and (9).

$$\Phi = \begin{bmatrix} 1 & T & T^2/2 & & & & & & \\ 0 & 1 & T & & & & & & \\ 0 & 0 & 1 & & & & & & \\ & & & 1 & T & T^2/2 & & & \\ & & & 0 & 1 & T & & & \\ & & & 0 & 0 & 1 & & & \\ & & & & & & 1 & T & T^2/2 \\ & & & & & & 0 & 1 & T \\ & & & & & & 0 & 0 & 1 \end{bmatrix} \qquad (8)$$

$$\Psi = \begin{bmatrix} T^2/2 & & \\ T & & \\ 1 & & \\ & T^2/2 & \\ & T & \\ & 1 & \\ & & T^2/2 \\ & & T \\ & & 1 \end{bmatrix} \qquad (9)$$

where T is the sampling period.

Assuming constant acceleration (and no maneuvers) T can also be regarded as the look ahead time for the prediction of the position and velocity of the robot´s endeffector.

Since the robot does not move with constant acceleration, changes in the acceleration are detected as maneuvers, $w(k)$ accounts for these changes. Maneuver detection improves the tracking performance during a maneuver considerably (Bar-Shalom, 1982).

## Predictive Controller

In contrast to conventional control algorithms predictive controllers minimize the actual and the future tracking error using Eq. (6).

$$J = \sum_{i=1}^{m} [w(k+i) - y(k+i)]^2 + \sum_{i=1}^{n} \Delta u^2(k+j-1) \overset{!}{=} \text{Min} \quad (6)$$

where

| | |
|---|---|
| J | is the cost function that is minimized, |
| w | is the set value, |
| y | is the real value, |
| $\Delta u$ | is the change of the controller output, |
| m, n | are parameters (Pritschow, 1991). |

Since the future tracking errors can not be measured, a model of the controlled system is used for the prediction of the future output signals.

In the case of the LTS there are no exact future set values, but the predicted future position of the robot´s endeffector. This estimated future position is used to determine the orientation of the plane mirror for a tracking error equal to zero. Next, the angles of the axes that are necessary for the required orientation of the mirror are computed. This predicted future angles are regarded as the set point angles for the predictive controller of each axis. Thus the difference between the predicted angles and the real angles of the axes are minimized. The predicted values are updated every sampling period using the measured actual position of the robot´s endeffector.

Consequently the tracking performance that can be reached by the predictive controller mainly depends on the state predictor of the target tracking controller.

## CONCLUSION

The LTS described is used for the determination of robot position and path accuracy characteristics. The measurement procedure requires the precise tracking of the moving robot. Due to high acceleration and speed capabilities of modern robots this task needs an advanced control algorithm. We propose a target tracking controller that works similar to a predictive controller. With a predictor and maneuver detector the future position of the robot is estimated. The predictive controller then minimizes the actual and future tracking error. First results with a slightly simplified controller show the feasibility of the approach.

Further research will be made towards more complex algorithms to improve maneuver detection and state (especially position) prediction. The dynamic potential of the LTS will be increased by the use of fully digital motor control. The accuracy of position measurement will be increased using additional sensors to account for mechanical errors.

## ACKNOWLEDGEMENTS

The authors would like to thank Prof. Dr. G. Zeichen, chief of the Insitut für Flexible Automation for his support of the research work and graduate students R. Kerschbaum, H. Guglia, T. Frühbeck, R. Mayer, K. Filz for their help in various detail problems of the research work.

The research was supported by the Austrian national research fund "Fond zur Förderung der Wissenschaftlichen Forschung", projects P6796PHY and P8190TEC.

## REFERENCES

Bar-Shalom, Y., K. Birmiwal (1982). Variable dimension filter for maneuvering target tracking. *IEEE Transactions on Aerospace and Electronic Systems*, Vol. AES-18, pp. 621-629.

Berg, R. F. (1983). Estimation and prediction for maneuvering target trajectories. *IEEE Transactions on Automatic Control*, Vol. AC-28, pp. 294-304.

Chang, C. B., J. A. Tabaczynski (1984). Application of state estimation to target tracking. *IEEE Transactions on Automatic Control*, Vol. AC-25, pp. 98-109.

Gander, H., J. P. Prenninger, M. Vincze, G. Zeichen (1991). A new measurement system for advanced modelling and identification for robot control. *Symposium on Robot Control*, 16.-18.9.1991, Vienna, Austria.

Jiang, B. C., J. T. Black, R. Duraisamy (1988). A review of recent developments in robot metrology. *Journal of Manufacturing Systems*, Vol. 7, Nr. 4, pp. 339-357.

Kerschbaum, R. (1991). Entwicklung eines digitalen Reglers für das Lasertrackingsystem. *Diploma Thesis*, Institut für Flexible Automation, Technische Universität Wien.

Prenninger, J. P., H. Gander, M. Vincze (1991). Real time 6DOF measurement of robot endeffectors. *International Robots & Vision Conference*, 22.-24.10.1991, Detroit, Michigan USA.

Pritschow, G., M. Jantzer (1990). Hochgenaue Bahnbewegungen an Fertigungseinrichtungen, Prädiktive Regelungen, *Werkstattechnik 80*, Springerverlag.

Vincze, M., H. Gander, J. P. Prenninger (1991). Laser-Tracking System zur Verfolgung beliebiger Roboterbewegungen. *VDI Berichte Nr. 921*, pp. 93-102.

# AN INTELLIGENT PROCEDURE FOR
# COLLISION-DETECTION IN ROBOTIC SYSTEMS

### J. Tornero and E.J. Bernabeu

*Departamento de Ingenieria de Sistemas, Computadores y Automatica (DISCA),*
*Universidad Politecnica de Valencia, P.O.B. 22012, E-46071 Valencia, Spain*

**Abstract.** This paper presents a great improvement in the collision-detection problem by considering: 1.- A simple description of the objects in the world, 2.- A fast distance computational technique and 3.- A collision-detection procedure which reduces the number of distance computations required. A multi-level hierarchical structure and a novel three-dimensional object representation technique based on spherical objects are adopted for modeling the world. Fast algorithms for distance computation, based on geometrical considerations, have been developed and used in the hierarchical collision detection procedure. This technique has been applied to a complex robotic system consisting of two PUMA robots, each mounted on a three-degree of freedom platform used in order to study robotic assembly of structures in space.

**Keywords.** Collision-detection, Collision-Avoidance, Object-representation, Robotic-applications, Path planning

## INTRODUCTION

Collision-detection is an important task in many robotic applications. Distances between the elements of a robot manipulator, between elements of cooperating robot arms and between the robotic elements and objects in the environment may all need to be computed to guarantee collision free payload transportation and manipulation.

Since many elements of robotic systems are bounded by plane surfaces, polyhedra have been used to develop different kind of procedures for computing distances (Gilbert,1988) . The algorithms developed for these geometries need to deal with vertices, edges and planes. The large number of comparisons required for such elements is highly time consuming. Thus, some authors have recast the problem into a standard linear programming form. Sometimes, norm 1 and norm ∞ rather than Euclidean distances are used to save computational effort (Whitehead,1989).

Some authors have used collision-detection techniques for obtaining spatial occupancy enumeration and hence the model of the freespace. From this description collision-free paths using spatial planning can easily be obtained. Octrees, a special form of spatial occupancy enumeration employs a successive subdivision of space into octants. Faverjon (1984) describes an algorithm for transforming cartesian obstacles into obstacles in the space of the first three joints of a manipulator based on a hierarchical structure.

This paper presents a great improvement in the collision-detection problem by considering: 1.- A simpler description of the objects in the world, 2.- A fast distance computational technique and 3.- A collision-detection procedure which reduces the number of distance computations required.

In terms of world description, the problem is solved in two stages. First a multi-level hierarchical structure is considered for describing the over-all robotic system. As a second stage, a novel three-dimensional object representation technique is described for modeling the objects in the world. This technique (Tornero,1990,1991) is conceptually between the bubble model of O'Rourke and Badler (1979) and the generalized cylinders of Agin (1972). The model is derived from the basic idea that an object can be approximated by an infinite number of spheres. Hence it retains the advantages of the bubble model and avoids the complexity of the generalized cylinder.

A very fast distance computation technique for the new object representation technique based on geometrical considerations is also commented in the paper.

Finally, a hierarchical procedure for the collision-detection problem is presented. This procedure has been applied on a very complex robotic system giving exceptional numerical results.

## WORLD DESCRIPTION

A hierarchical structure is considered for describing the manufacturing and assembly area of the factory where static and mobile robots are working. This area is decomposed in cells, systems, subsystems and elements. Cells are the toppest level in the structure. The following levels in the structure correspond to systems for describing mobile robots, assembly lines, etc.; subsystems for robots, machine-tools, conveyor, etc.; and finally elements for the robot-links, components of machine-tools, etc. as can be seen in figure 1

For the purpose of collision-detection, the different objects in the world need to be modeled. Each object can be represented by one o several models depending on the accuracy required. Generally, the complexity of the models used is directly related to the exactness of the representation obtained. Any attempt to make the model less complex leads to some kind of approximation. There is a trade off depending on how much simplification is acceptable for the accuracy required versus the speed up in the collision detection.

In addition, several objects in one level can be modeled separately or globally at the toper level. For example, the links of a robot can be modeled at the "Element" level or globally at the "Subsystem" level.

From the infinite number of possible volumes, we have considered spherical-object representation to constitute a set of volumes to solve the collision-detection problem.

The modeling technique is based on the idea that any object can be approximated by an infinite number of spheres. For a particular object many different sets of spheres can be generated, depending of the degree of accuracy required. In general accuracy is directly related to the complexity of the process for obtaining the set of spheres. Later on, we will introduce the

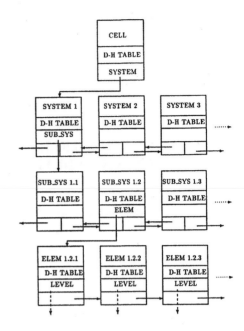

Figure 1: Multi-level hierarchical structure

idea of degree-of-freedom as a first measurement of this complexity.

The way of generating a set of spheres for a given object is provided by introducing the concept of dynamic-spheres. A dynamic-sphere is defined as a sphere whose center P(x,y,z) can move in a three-dimensional space and whose radius is a function of the position at each moment R(x,y,z).

The object model corresponds to the volume swept by the dynamic-sphere when its center moves in a bounded subspace. This subspace can be defined by

1. A set of inequality constraints,e.g., $x_0 \leq x \leq x_1$.

2. A set of functional constraints in the form $f_i(x, y, z) = 0$.

The number of functional constraints will determine the degrees of freedom for the center of the dynamic-sphere. For example, a two-degree of freedom geometry is obtained when one functional constraint is introduced (e.g. $f_1(x, y, z) = 0$ ). In parametric representation, that is

$$P = P(\lambda), \quad R = R(\lambda) \qquad (1)$$

The complexity of the volume swept by the dynamic-sphere depends on the functions $P(\lambda_i)$ and $R(\lambda_i)$, as well as the range of the $\lambda's$.

The simplest class of objects for the one-degree of freedom geometry is obtained by considering linear functions. We define a spherical-cone as the volume swept by a dynamic-sphere whose center and radius

Figure 2: Spherical-cones

Figure 3: Spherical-planes

obey linear functions in the parameter $\lambda$, that is

$$P = P_0 + \lambda P_1; \quad R = R_0 + \lambda R_1 \qquad (2)$$

Particular cases of this geometry for $\lambda$ varying within a given range, such as spherical-cylinders, etc., can be seen in Figure 2.

For a spherical-cone i, functions $\mathbf{P_i}(\lambda_i)$ and $\mathbf{R_i}(\lambda_i)$, can be written in terms of the radius and center of the end-spheres at the two extremes of the volume, $(\mathbf{P_{i0}}, \mathbf{R_{i0}})$ and $(\mathbf{P_{i1}}, \mathbf{R_{i1}})$, in the following way,

$$\left\{ \begin{array}{l} P_i = P_{i0} + \lambda_i \ (P_{i1} - P_{i0}) \\ R_i = R_{i0} + \lambda_i \ (R_{i1} - R_{i0}) \end{array} \right. \qquad (3)$$

$$\lambda_i \in [0,1]$$

The degree and the angle of convergence of a spherical-cone can be defined respectively by

$$\eta_i = \frac{(R_{i1} - R_{i0})}{\mid P_{i1} - P_{i0} \mid}, \quad \alpha_i = \arcsin \eta_i \qquad (4)$$

For values of degree of convergence equal or greater than 1, $(\eta_i \geq 1)$, the volumes degenerate into a simple sphere equal to the largest end-sphere.

For the two-degree of freedom case, the simplest geometry which can be represented is a dynamic-sphere moving tangentially between two planes, which is called as a spherical-plane. In a similar way as a spherical-cone is bounded by two end-spheres, a spherical-plane is bounded by four end-spherical-cones which constitute the sides of the plane. A volume i of this kind can be described by two spherical-cones with a common end-sphere, $[(\mathbf{P_{i0}}, \mathbf{R_{i0}}), (\mathbf{P_{i1}}, \mathbf{R_{i1}})]$ and $[(\mathbf{P_{i0}}, \mathbf{R_{i0}}), (\mathbf{P_{i2}}, \mathbf{R_{i2}})]$, which constitute two continuous sides. The volume is then generated from a dynamic-sphere which follows the following equations

$$\left\{ \begin{array}{l} P_i = P_{i0} + \lambda_{i1}(P_{i1} - P_{i0}) + \lambda_{i2}(P_{i2} - P_{i0}) \\ R_i = R_{i0} + \lambda_{i1}(R_{i1} - R_{i0}) + \lambda_{i2}(R_{i2} - R_{i0}) \end{array} \right.$$

$$(5)$$

$$\lambda_{i1} \& \lambda_{i2} \in [0,1]$$

According to Equations 5, the center of the dynamic-sphere moves inside a parallelogram, which can be easily cut-off by lines, defined as inequality constraints, in the form of linear combination of parameters $\lambda's$,

$$a_1 \cdot \lambda_1 + a_2 \cdot \lambda_2 \leq 1 \qquad (6)$$

Several kinds of spherical-planes can be obtained with this two-degree of freedom geometry as depicted in figure 3. Spherical planes are bounded at least by three spherical-edges which are in the class of spherical-cones or one-degree of freedom geometries. These spherical-edges intersect in spherical-vertices which corresponds to the zero-degree of freedom geometries.

Dynamic-spheres with three degrees of freedom are not considered in this paper, given that the most outstanding advantage of this object representation technique comes from the possibility of generating three-dimensional object models with less than three degrees of freedom. However, the basic idea can be easily extended to dynamic-spheres moving in three or even more degrees of freedom. Extra degrees of freedom can be introduced in order to increase the accuracy of the representation or to consider aspects such as the time-factor for describing moving objects.

All the entities in the hierarchical structure are modeled by spherical-volumes. A given object or set of objects can be modeled by different spherical-volumes with different degree of accuracy. For entities at the "Element" level, volumes used will depend on the shapes and dimensions of the objects. Toper entities, such as subsystems, systems and cells, will be modeled taking into account, in addition to shapes and dimensions of the objects included, the configuration at each instant.

The use of spherical-volumes for modeling at one level of the structure makes easier the computation of spherical-volumes at higher levels. In particular, a robot-arm described as an entity at the "Subsystem" level can be modeled as a spherical-volume including all the control-spheres used for modeling the elements of the robot at the "Element" level.

## DISTANCE COMPUTATION BETWEEN SPHERICAL VOLUMES

The use of spherical-volumes makes easier the computation of the shortest distance between objects.

The distance function between two spheres, $S_i$ and $S_j$, is defined as

$$d(S_i, S_j) = f\{\mid P_i - P_j \mid -R_i - R_j\} \qquad (7)$$

where $f\{x\} = x$ if $x \geq 0$
$\qquad\quad = 0$ otherwise

The problem of finding the shortest distance between two bodies whose volumes have been generated by the movement of two dynamic-spheres along their trajectories, $S_i$ and $S_j$, can be stated as an optimization problem as follows,

$$\min_{\alpha'_i s, \alpha'_j s} d(S_i, S_j) \qquad (8)$$

The problem can be expressed in terms of finding two spheres, each belonging to a distinct spherical-volumes, with the shortest distance between them.

In spite of the fact that many numerical optimal techniques can be applied to the new formulation of the problem, in the paper a very fast geometrical solution, presented in Tornero (1991), has been adopted.

The algorithms for the geometrical solution have been tested on a standard SUN SPARCstation 1, a RISC-based workstation, using sets of one hundred randomly generated and positioned volumes. The minimum, maximum, and average times in milliseconds for computing distance between all the combinations of spherical-objects introduced in the theory have been recorded in Table I.

| Objects Compared | Min. | Max. | Ave. |
|---|---|---|---|
| sphere/sphere | 0.05 | 0.05 | 0.05 |
| sphere/sph-cylinder | 0.14 | 0.14 | 0.14 |
| sphere/sph-cone | 0.25 | 0.26 | 0.26 |
| sphere/sph-plane | 0.65 | 0.91 | 0.87 |
| sph-cylinder/ sph-cylinder | 0.20 | 0.23 | 0.23 |
| sph-cylinder/sph-cone sph-cone/sph-cone | 0.65 | 1.57 | 1.05 |
| sph-cylinder/sph-plane* sph-cone/sph-plane* | 0.72 | 3.86 | 2.82 |
| sph-plane*/sph-plane* | 2.81 | 9.09 | 7.01 |

(*) For these computations, planes with 4 vertices have been considered
TABLE I: Distance Computation Times for randomly generated sets of spherical-objects.

In Tornero (1990,1991), it was studied the self-collision detection problem for a stand-alone PUMA robot-arm, giving a cost of less than 2.30 milliseconds according to values from Table I.

## COLLISION-DETECTION PROCEDURE

For the purpose of collision-detection, the hierarchical structure mentioned above is expanded by considering different models in accordance with the degree of accuracy required. Depending on the relative position between entities,

1. we select between local or global models. For example, the links of a robot can be modeled separately at the "Element" level or globally at the "Subsystem" level just as one object.

2. for each level, a model with appropriate accuracy is also selected

For the particular case commented above, the collision-detection procedure starts by checking global models for the robot-arm at lowest accuracy (e.g. sphere). If collision is detected better global models with higher accuracy are considered (e.g. spherical-cones, spherical-planes, etc. ). When the highest level of accuracy has been reached, local models in the robot-arm, describing the elements, are considered, starting with their lowest accurate representation. The procedure ends when no collision occures or when local models at highest accuracy have been checked.

The collision-detection procedure manages the multi-level hierarchical structure obtaining reductions in computational time around 90% on average as will be seen in the following section.

## APPLICATION AND SIMULATION RESULTS

A robotic testbed platform, used at CIRSSE[1] to study robotic assembly of structures in space, has been modeled in order to thoroughly test the entire theory. The robotic system described by spherical-volumes for an accuracy of 3.5 cm at the "element level" is shown in figure 4

The setup consists of a PUMA 560 and a PUMA 600 each mounted on a three-degree of freedom platform. Each platform can be independently translated along a common track. Each platform has two rotational degrees of freedom, rotation about a vertical axis and tilt about a horizontal axis.

The links of the robot-arm are modeled by spherical objects described with respect to their local coordinate systems. A kinematical representation based on homogeneous transformation matrices using a modified form of the Denavit-Hartenberg parameters (Craig,1986) is used in order to determine the position of the robot links. Characteristic spheres (control-spheres) in each link are described with respect to its local coordinate system.

Checking the robotic system for collisions consists of two steps. First, the characteristic points of the volumes in local coordinates are translated into world coordinates based on the joint angles and D-H parameters forward kinematics. Second, distances between

---

[1]Professor Tornero was Visiting Researcher, during 1990, at CIRSSE (Center for Intelligent Robotics Systems for Space Exploration), Rensselaer Polytechnic Institute, Troy, New York, USA.

Figure 4: Robotic system: Cooperative robot arms created from zero, one, and two degrees of freedom volumes

spherical-objects are computed using the appropriate algorithms.

For the overall robotic system, using the volumes shown in Figure 4, the number of comparisons required are 1 sphere to sphere, 14 sphere to sphere-cylinder, 11 sphere to spherical-plane, 20 spherical-cylinder to spherical-cylinder, 25 spherical-cylinder to spherical-plane and 12 spherical-plane to spherical-plane. This gives minimum, maximum and average times in milliseconds of 66, 220 and 170 respectively.

This computing times are exceptionally low as a consequence of the distance times obtained for spherical-objects and shown in Table II. However, withe the use of the multi-level hierarchical structure, the overall computational time can drastically be reduced. Just working at the "Element" level (local models) and with the set of different accurate models shown in Table I, we have obtained minimum, maximum and average times in milliseconds of 1.98, 6.60 and 5.10 respectively. This implies a reduction of 97% in computational cost. The reduction in time will be higher if "Subsystem" and "System" levels are considered.

| Code | Elements | Accuracy | | |
|------|----------|------|------|------|
| | | L0 | L1 | L2 |
| $E_0$ | Platform | Sph. | | |
| $E_1$ | Link 0 | Sph. | Sph-cyl. | |
| $E_2$ | Link 1 | Sph. | Sph-cyl. | |
| $E_3$ | Link 2 | Sph. | Sph-cone | Sph-plane |
| $E_4$ | Link 3 + Wrist | Sph. | Sph-cone | Sph-plane |
| $E_5$ | Gripper | Sph. | | |

TABLE II: Spherical-objects used in the robotic system application.

In figure 5, we can see the reduction in time for each set of elements $E_i$. It is interesting to underline that the reduction is more important when dealing with more complex spherical-models. The reductions are 88% for $E_0$, 96% for $E_1$, 96% for $E_2$, 99% for $E_3$, 99% for $E_4$ and 88% for $E_5$, The values have been obtained from 16,384 different configurations uniformity distributed in the range of joint variables.

## CONCLUSIONS

The paper has presented an efficient collision detection procedure based on a multi-level hierarchical structure for describing a complex robotic system.

All the entities in the hierarchical structure (objects or set of objects) have been represented by spherical models. This models are very simple in terms of mathematical formulation and they make easier the development of distance computation algorithms.

Spherical objects have been presented in previous papers, however, in this paper a new, still simple and more general mathematical formulation have been considered. This new formulations makes easier the generation of different models for different level in the structure. The implementation of distance computation algorithms base on numerical methods is also easier with the new representation.

The collision detection procedure based on the multi-level hierarchical structure is highly efficient when dealing with complex objects, for which different models with different accuracy can be considered. The reduced time consuming required makes this procedure useful for real-time applications.

411

Figure 5: Reduction in computational time for the robotic application

Further work will focus on using other kinds of functions for the dynamic-spheres, with the intent of increasing the variety of volumes generated. In addition, the movements of the elements will be also studied and modeled as more complex spherical volumes.

## REFERENCES

E. Gilbert, D. Johnson, and S. Keerthi, (1988)
"A Fast Procedure for Computing the Distance Between Complex Objects in Three-Dimensional Space,"
IEEE Journal of Robotics and Automation, vol. 4, No. 2, pp. 193-203.

J. Whitehead and K. Kyriakopoulos, (1989)
"Efficient Implementation of Linear and Quadratic Programming Algorithms for Minimum Distance Estimation Between Solids," Tech. Rep. CIRSSE-TR-89-22, Center for Intelligent Robotics Systems for Space Exploration, Rensselaer Polytechnic Institute, Troy, NY.

B. Faverjon, (1984)
"Obstacle Avoidance using an Octree in the Configuration Space of a Manipulator,"
IEEE Int. Conf. on Robotics, (Atlanta, Georgia), pp. 504-512.

J. Tornero, G. Hamlin and R.B. Kelley, (1990)
"Efficient Distance Functions Using Spherical-Objects and Their Application to the Two-Puma Platform System,"
Tech. Rep. CIRSSE-TR-90-64, Center for Intelligent Robotics Systems for Space Exploration, Rensselaer Polytechnic Institute, Troy, NY.

J. Tornero, G. Hamlin and R.B. Kelley, (1991)
"Spherical-Object Representation and Fast Distance Computation for Robotics Applications", IEEE
Int. Conf. on Robotics and Automation, (Sacramento, California), Vol. 2, pp.1602-1608.

J. O'Rouke and N.I. Badler, (1979)
" Decomposition of Three-Dimensional Objects into Spheres,"
IEEE Trans. PAMI, vol. 1.

B. Again, (1972)
Representation and Description of Curved Objects. PhD thesis, AIM-173, Standford AI Laboratory.

J. Craig, (1986)
Introduction to Robotics. Mechanics and Control. Addison-Wesley Publishing Company.

# REACTIVE LEVELING ON THE AMBLER
# WALKING MACHINE

P.V. Nagy, P. Gonzalez de Santos[1] and W.L. Whittaker

*Field Robotics Center, The Robotics Institute, Carnegie Mellon University, Pittsburgh, PA 15213-3890, USA*

Abstract. Walking robots depend on the terrain that they are stepping on for support. When walking on unstructured natural terrain these supports may fail -- endangering the safety of the robot. A reactive control method, reactive leveling, has been developed for such an event. It is shown that reactive leveling works in bringing the machine to level from unstructured support failures. However, the sensors used to measure body inclination should be chosen carefully in order to obtain appropriate mechanism response.

Keywords. Walking machines; attitude control; reactive control; terrain interaction; sensors.

## INTRODUCTION

Robot technology that utilizes active suspension is fairly immature, with not much experience other than a handful of machines that mainly traverse laboratory floors, e.g. the TITAN III (Hirose, 1985). Some machines, such as the OSU Adaptive Suspension Vehicle (ASV) have ventured outdoors (Waldron, 1988). There is a need to be able to traverse rugged and natural terrain autonomously with a high degree of reliability. Use of an active suspension will make this possible by adapting to terrain irregularities and reacting to terrain drop-out. Application domains for this technology include planetary exploration and agriculture.

The Planetary Rover research group at Carnegie Mellon University has developed and built a walking machine, AMBLER, that is appropriate to the needs of planetary exploration (Bares, 1989). The AMBLER has a mass of 2100 kg machine and stands 4-6 m tall. The AMBLER is configured as an orthogonal walker; horizontal actuations and link motions are decoupled from vertical actuations and motions. With this configuration, a subset of actuators support the body without propelling it forward, while another subset propels the body, with no gravity loading on them, for a level machine. This arrangement lowers power consumption, and simplifies the autonomous planning functions.

To simplify control, and to lower peak power requirements, control is separated into the vertical direction and the horizontal plane. Thus the machine is not propelled when the body is leveling, and vice-versa; the vertical actuators are locked-out during a body propulsion. This control scheme has been implemented on the AMBLER, allowing it to traverse benign terrain reliably and predictably. By benign terrain, we refer to terrain in which solidly supporting footholds exist. However, for exploring general, unstructured planetary terrain, such as that found on Mars, the terrain conditions may be less favorable. When climbing up a slope, it is possible that the slope might fail, especially if it is comprised of loosely consolidated sand. If a large number of feet have toe-holds on rocks, it is possible to slip off of these rocks. These drastic examples of foot support failures require appropriate reactive control, which is the subject of this work.

## THE REACTIVE LEVELING ALGORITHM

The reactive leveling algorithm was developed to respond to support failures. Without knowing which legs failed, nor by how much, this algorithm succeeds in bringing the machine to level, thereby increasing its energy stability margin (the potential energy required to tip the machine over). With this algorithm, inclinometers are incorporated in the control loop. After they are read, the vertical leg extensions required to level the body are calculated.

---

[1]Visiting Scientist at the Field Robotics Center. He is with the Instituto de Automatica Industrial-CSIC. 28500 Arganda del Rey. Madrid. Spain.

These extensions are used to determine the ratio of the leg velocity commands. The ratio of the commanded velocities equals the ratio of the calculated extensions. The speed that the machine levels is determined by the velocity that the leg with the maximum speed is commanded to travel. The value of this maximum velocity is a dynamically-set variable that is a function of the body tilt. The further the machine is tilted the faster this speed is. The leg velocities are recalculated each time the control loop is executed, so the machine slows down as it nears a level position. A deadband is used to prevent the machine from hunting about this position.

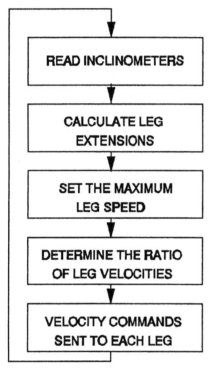

Fig. 1 The Reactive Leveling Algorithm.

The vertical leg extensions required to level the body are calculated using the same kinematic simplifications used by Klein (Klein, 1980) and others (Gorinevsky, 1990), (Ishino, 1983). There are other leveling methods that utilize more kinematic information (Gonzalez, 1991); however, these equations suffice for this application. Referring to Fig. 2a, the change in vertical leg extension for leg $i$ to level the body from an angle PITCH is:

$$\Delta z_{i_{PITCH}} = -DP_i \sin(PITCH) \qquad (1)$$

Similarly, to level the body from an angle ROLL (see Fig. 2b), the vertical leg extension of leg $i$ have to change in:

$$\Delta z_{i_{ROLL}} = DR_i \sin(ROLL) \qquad (2)$$

For PITCH and ROLL rotation these length changes can be superimposed, i.e.:

$$\Delta z_i = DR_i \sin(ROLL) - DP_i \sin(PITCH) \qquad (3)$$

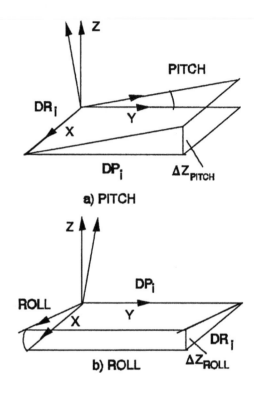

a) PITCH

b) ROLL

Fig. 2 Body Frame System showing PITCH and ROLL rotations.

This algorithm was successfully employed on a leveling testbed, a platform with six vertical actuators used in developing this algorithm (Nagy, 1991).

## REACTIVE LEVELING ON THE AMBLER

The reactive leveling algorithm was implemented on the AMBLER. Support failure stands were built to cause multiple support failures that caused the machine to tilt up to 12 degrees. The AMBLER has two stacks which have three legs each. If the support stands are placed under the legs of one stack, as shown in Fig. 3, then support failures that cause the body to roll up to 3.5 degrees occur. When the stands are placed under the front two legs (one from each stack) pitching failures of up to 12 degrees are accommodated.

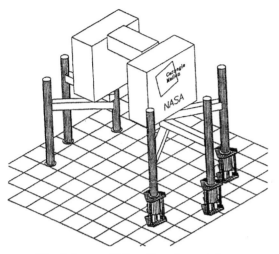

Fig. 3 Support Failure Experimental Set-up for the AMBLER.

Before performing support failure experiments, the reactive leveling algorithm was first tried with the body already in a tilted position to bring it to level. A typical attitude response before using a control deadband is shown in Fig. 4. The roll and pitch oscillate about the level position. A 0.5 degree deadband is added so that when the tilt is small, the velocities are all set to zero. With the deadband, the oscillations disappear, as shown in Fig. 5. Note that in both instances there is a false spike in the inclinometer reading at the onset of machine motion. The inclinometers that we are using are sensitive to horizontal linear acceleration, and attitude control of the AMBLER incurs these accelerations. Furthermore,

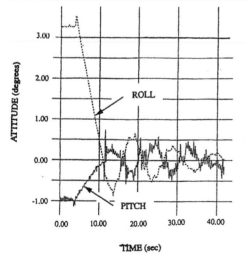

Fig. 4 Reactive Leveling to Level the Body with no DeadBand

the inclinometers only have a bandwidth of 1 Hz, which will also cause inaccuracies in registering tilt when the body is moving.

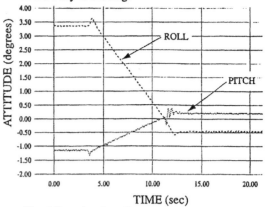

Fig. 5 Reactive Leveling to Level the Body with a Deadband.

The reactive leveling algorithm was then employed to react to a support failure of 3.5 degrees as shown in Fig. 3. This algorithm is invoked by a sensed body tilt of greater than 1 degree. The typical response of the inclinometers for this experiment is shown in Fig. 6. The AMBLER free-falls during the 3.5 degree drop, but the inclinometers first register a tilt of about five degrees in the opposite direction. As a consequence, the reactive leveling algorithm further tilts the body before the inclinometers finally register a tilt in the correct direction. The combination of these inclinometer noise spikes, low bandwidth, and slow control loop update times cause the machine to first drop further, then start to level and overshoot the level position, then finally stop leveling when the inclinometer readings (and tilt) are within the deadband.

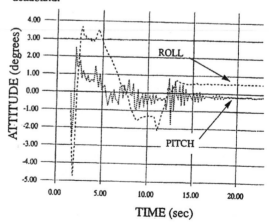

Fig. 6 Response of Inclinometers During a Support Failure.

In contrast, reactive leveling was usually successful

on the leveling testbed in responding to support failures, an example of which is shown in Fig. 7. The same inclinometers were used on this testbed, but they were mounted at about 1/7th the height they were on the AMBLER, and thus experienced 1/7th the linear acceleration. The inclinometer spikes occasionally affected mechanism response, but not nearly as bad as with the AMBLER. Another factor was that the control loop update with the leveling testbed was 1/50th of that used with the AMBLER. The current AMBLER real-time computing configuration was not designed for reactive control, and it will need to be upgraded for appropriate reactive control to be realized.

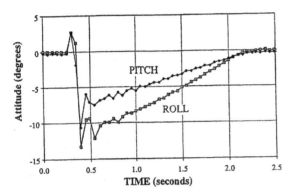

Fig. 7 Reactive Control on the Leveling Testbed.

To test the machine's capability to level the body we utilized our knowledge about how the supports would fail, thereby eliminating the need to read the inclinometers to ascertain body tilt. In this experiment the inclinometers were used to determine when support failure occurs, as the acceleration-induced spike was easily sensed. In doing this, the AMBLER was quickly re-leveled, bringing it to a more stable position. For the 3.5 degree rolling failure, leveling was achieved in 1.5 seconds. For the 12 degree pitching failure the response took 6 seconds. This demonstrated that the machine could respond in a timely manner if it had adequate tilt sensing.

## MODIFICATIONS TO THE REACTIVE LEVELING ALGORITHM

The reactive leveling algorithm succeeds in bringing the body to level without any knowledge of which supports failed, and by how much. However, this ignorance of the walker/terrain geometry also may allow some feet to remain in the air during this reactive control. Feet that are in the air do not contribute to the stability of the machine. Therefore foot contact sensing may also be employed in order to modify the commands sent to feet that are in the air. The reactive leveling algorithm works as before,

but legs in the air are not allowed to retract. This prevents legs that are already in the air from retracting further away from the ground surface. These legs should be commanded to servo downwards at a small velocity until ground contact is detected. After making contact again, these feet are then servoed in the same manner as the other ground-contacting feet by using the reactive leveling algorithm.

## CONCLUSIONS

The sensing and real-time computing currently used on the AMBLER preclude accurate sensing of tilt under dynamic conditions and incorporation of tilt feedback in the control loop. With knowledge of how the supports would fail, we were able to work around these problems and demonstrate in contrived examples that the mechanism can indeed respond quickly enough to a support failure, thereby increasing its survivability in such an event. The roll and pitch axes should be fitted with gyros to avoid acceleration-induced tilt signal peaks, and also to obtain higher sensor bandwidth. Inclinometers should be retained to obtain accurate tilt readings under static conditions. We further postulate the modifications to the reactive leveling algorithm in order to account for sensed foot contact. With this modification, the machine will also level, but will be increase its stability by bringing legs that are in the air back into contact with the ground.

## ACKNOWLEDGEMENTS

The authors acknowledge the support for this work provided by NASA under contract NAGW-1175. P. Gonzalez de Santos was funded by D.G.I.C.Y.T. (Spain) .

## REFERENCES

Bares J., et. al. (1989), "AMBLER: An Autonomous Rover for Planetary Exploration," IEEE Computer Magazine, June 1989, 18-26.

Gonzalez de Santos, Nagy P.V., and Whittaker W.L.(1991), "Leveling of the AMBLER Walking Machine: A Comparison of Methods," CMU Robotics Institute Technical Report CMU-RI-TR-91-13, July 1991.

Gorinevsky D.M. and Shneider A.Y.(1990), "Force Control of Legged Vehicles over Rigid and Soft Surfaces," International Journal of Robotics Research, Vol. 9, No. 2, April 1990, 4-23.

Hirose S., et. al. (1985), "TITAN III: A Quadruped Walking Vehicle," Proceedings of Robotics Research,

2nd International Symposium, MIT Press series in
artificial intelligence, 325-331.

Ishino Y., et. al. (1983), "Walking Robot for
Underwater Construction," Proceedings of the '83
ICAR, Tokyo, Japan, Sept. 12-13, 1983, 107-114.

Klein C.A. and Briggs R.L. (1980), "Use of Active
Compliance in the Control of Legged Vehicles,"
IEEE Transactions on Systems, Man, and
Cybernetics, Vol. SMC-10, No. 7, July 1980,
393-400.

Nagy P.V., Wu B.X. and Dowling K. (1991), "A
Testbed for Attitude Control of Walking Robots",
Proc. of the ISMM Int. Symposium on Computer
Applications in Design, Simulation and Analysis, Las
Vegas, NV, March 19-21, 1991, 120-123.

Waldron K.J., Vohnout V.J., and Murthy N.N.
(1988), "Terrain Interactions of Legged Vehicles,"
Proc. of the 19th ISIR, Sydney, Australia, Nov. 1988,
601-609.

# MOBILE ROBOT GUIDANCE BASED ON PREDICTIVE TRACKING

**J. Codina and J. Frau**

*Departamento ESAII, Facultat d'Informatica de Barcelona, Universitat Politecnica de Catalunya,
C/Pau Gargallo, 5. 08028 Barcelona, Spain*

Abstract. A system composed of two recognition modules which allow the tracking of a
predefined target on a multi-target environment has been installed on a mobile robot to
perform an image-based visual servoing. The tracking system performs an oriented search
based on the prediction of the motion of the target. Thus, the system is able to restrict the
searching window to a defined zone of the image where foreseeable the target will be
detected. The 3–D estimation of the target position is done from the location on the image
plane made by the individual tracking subsystems. In several trials the algorithm has shown
good results, computing the position in less than 100 milliseconds. The vision system
works only when the target is inside a restricted area, and the estimated position has always
some additive noise. The design of the control system had to decide between a low-pass
filter, and that means a real danger to loose the target, or to have fast response making the
system very sensitive to noise. A nonlinear system has been designed to solve this trade-
off. Basically, the time-constant increases with the error inverse. This configures a control
system with a special dead zone.

Keywords. Target tracking; Image processing; Motion estimation; Nonlinear control
systems; Mobile robots.

## INTRODUCTION

Automatic navigation is a technological challenge
with vital consequences for civilian areas such as
manufacturing, construction, traffic safety or security
as well as for military environments.

Eventhough a completely autonomous mobile robot
or, more specifically, a road following with decision
capabilities similar to human beings is beyond the
state of the current art, some semi-autonomous
human-directed mobile robots or autonomous with a
set of constraints have been approached.

Many positioning methods have been studied in the
past. Some of them are based on the "dead reckoning"
technique, that is, the estimation of the current
position (x,y,angle) of a wheeled mobile robot,
evolving on a plane, relative to an initial position by
accumulating the movement of the robot using the
measurement of wheels rotation. This approach
presents serious problems since the position
estimation error is accumulated as robot moving and
the accuracy of estimation decreases. To take care of
this problem, positioning methods using external
sensors are required.

Most of the positioning systems that use external
sensors need that the robot moves straight between
positioning process, restricting the robot action
(Murata,89). More general proposals have been done
by modifying the dead reckoning estimation
algorithm in the sense of incorporating asynchronous
information of the external sensor as moving. This
sensor can be a low dimensional and simple one
which allows one to keep a certain accuracy of the
dead reckoning estimation at alls positions.

The work carried out by Watanabe and Yuta (90) is
concerned with last approach, but dealing with the
position by sensing lighthouses that are set in the
environment as beacons and which location modify
estimation accuracy and efficiency. The strategy to
choose the effective points of beacons should be
studied in that case.

Other simple techniques have been developed which
take advantage of visual information for robot road
following. Morgan (90) faced the problem by
creating and updating a representation of the road
based on four parameters that describe the width,
direction and simple curvature of the road in a vehicle
centred (x,y,z) world coordinate system. The model

419

was created by tracking a set of measured edge points from frame to frame -assuming that robot motion is known- and using a weighted least squares process to find the four parameters of the road model. This method presents serious limitations when road junctions appear or other road vehicles occlude the scene.

Some years ago Wallace (87) presented a nice color vision based robot road following program which used an adaptive color model to classify pixels as road or shoulder features. The program tracked the shape of the road through the image sequence and allowed predicting the expected road position even in the presence of surface color variation, fluctuation of illumination conditions and deviation of sensor response. More recently Wallace included a texture energy measure in addition to the color features. Nevertheless, the method presents some problems caused by the attempt to fit straight road models to intersections.

Finally, the research realized by Dickmanns (90) is concerned with the derivation of a direct interpretation including spatial velocity components by smoothing integrations of prediction errors. The algorithms involved in the system provide obstacle detection and monocular relative spatial state estimation. Different architectures based on a MIMD parallel processing system have been developed allowing speeds up to 100 Km/h. In the mean time, control cicle times are near 0,1 seconds with a set of well-known today's microprocessors.

In this paper we present some aspects relating a predictive stereo-based tracking module as well as the control subsystem that have been designed and coupled in order to provide a visual control which creates a dynamic link between perception and action. The subjects presented can be considered as the continuation of previous works relating target tracking and mobile robot control which development can be found in (Amat,89) and (Frau,90a,90b, 90c,91a,91b,91c).

The paper is organized as follows. The approach we have considered suitable is briefly reviewed in the following section. After that, the predictive stereo-vision subsystem is evaluated taking into account the errors involved in the calculation of feature points that allow depth calculation. Finally, the control feedback through 3D motion estimation is given.

## APPROACH

In the context of mobile robots control it is common to distinguish between a high-level goal that is concerned with obtaining an accurate model of the environment and, on the other hand, a lower-level

objective focused on determining the current position within it. The former include the combination of multiple views to provide robust data,the identification and inclusion of novel features and the determination of free space within which the robot can move.

The low-level goal usually consists of a visual tracking that supplies the input control signals required to navigate an autonomous vehicle through an unstructured or partially structured environment. In the last case a set of scene beacons are strategically located.

Moreover, the temporal response required by each task is quite different. When using virtual servoing the sample rate or control cicle time is required to be much higuer than that at which the model of the environment needs to be updated. In addition, most of works that have been undertaken on autonomous navigation consider the dynamic tracking problem associated with following a path. This way, the detection/recognition module is hardly constrained because of the number of representation scenes considered in the analysis. Last approach permits modelling roads by means of simple parameters usually matched from edge points or straight lines.

Our proposal is not based on looking for roads or their features, but detecting and recognising a more general target -determined by the user- which model is a combination of geometric and kinematic features. Two cameras are located on a mobile robot in order to provide the signals to a couple of 2D-tracking subsystems that supply horizontal and depth coordinates of the target -with regard to the mobile robot reference- to the controller (Frau,90a,91a).

This way, the robot is required to track the target with a set of capabilities such as prudential distance - not necessarily constant-, follow the target with predictive performance and filter the data coming from vision module.

## QUANTIZATION ERROR INFLUENCE IN 2D-TRACKING FEATURES

The combination of geometric plus kinematic features which lead to a dynamic model can be divided into five levels at the end of which the vision subsystem provides the relative position -linear and angular- to the controller. The levels are: digitization, contour extraction and thinning of objects in the scene, detection/recognition of the target from geometric filtering and correlation matching and 2D motion estimation from the SMOP dedicated architecture (Frau,91c).

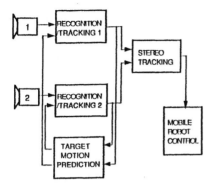

Fig.1. Block diagram of the visual servoing system.

The relative location of the target with respect to the robot is obtained from the couple of motion vector sequences:

$$z_1 = \left[x_c, y_c, \varnothing\right]^t_1 \qquad z_2 = \left[x_c, y_c, \varnothing\right]^t_2$$

where $(x_c, y_c)$ are the coordinates of the center of gravity (c.g) of the target projection on the image planes of cameras 1 and 2, and $\varnothing$ is the angle corresponding to the main axis of inertia of the target with respect to the baseline direction.

Consequently the c.g. is a virtual point that becomes the reference point from which it is attempted to solve the correspondence problem. Calculation of the c.g. coordinates is mainly disturbed by two main sources of noise: acquisition and preprocessing. Due to the important role that digitization error could play in calculating 3D coordinates of the target we present a short analysis of its impact in the computation of the c.g. of the target for each camera.

A realistic measure of error is the average or expected error. The formulation considered in the analysis is based on the theory developed by Kamgar-Parsi(89).

Assuming that $\partial$ is the quantization unit, the maximum quantization error in quantities $X_c$ and $Y_c$ is half the size of $\partial$.

Taking into account the expressions of $X_c$ and $Y_c$ obtained from the segmented image

$$X_c = \frac{\sum_{i=1}^{A} x_i}{A} \qquad Y_c = \frac{\sum_{i=1}^{A} y_i}{A} \qquad (1)$$

and the variance associated with these quantities (Frau, 91c)

$$\sigma^2 = \frac{1}{12}\left(\alpha_1^2 \cdot \partial_1^2 + ... + \alpha_A^2 \cdot \partial_A^2\right) \qquad (2)$$

where

$$\alpha_1 = ... = \alpha_A = \frac{\partial X_c}{\partial x_i} = \frac{1}{A} \qquad (3)$$

and $\partial_1 = \partial_2 = ... = \partial_A = 1$ .Consequently,

$$\sigma^2 = \frac{1}{12A} \qquad (4)$$

Assuming that the area of the target is large enough so that the Central Limit Theorem holds, we can apply the relationship between the average error and the standard deviation (Beers,57)

$$\frac{\sigma}{E\{\Delta X_c\}} = \frac{\sigma}{E\{\Delta Y_c\}} = \sqrt{\frac{\pi}{2}} \qquad (5)$$

Therefore, the average error estimate in calculating $X_c$ and $Y_c$ is given by

$$E\{\Delta X_c\} = E\{\Delta Y_c\} = \frac{\sigma}{\sqrt{p/2}} \cong \frac{0.23}{\sqrt{A}} \text{ pixels} \qquad (6)$$

The average error as a function of the area of the target is shown in figure 2. The larger the area, the smaller is the average error. Nevertheless, taking into account that the expression has been obtained assuming a number of pixels larger enough (i.e. ten) we can infer that the error decreases very slowly with the area.

As far as stereo triangulation is concerned, two identical pinhole cameras have been considered. We assume that the pixels are squares of size $l_p$ and focal lengths -f- of the two cameras are equal.

The geometry of the stereo setup is shown in figure 3. The notations are the following: $\Delta y$ is the baseline, $Z_E$ is the projection of Z -depth- on the epipolar plane (Llario,86).

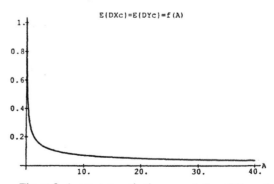

Figure 2. Average error in the computation of $X_c$ and $Y_c$ coordinates (c.g) as a function of the area of the target due to digitization.

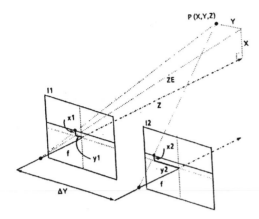

Figure 3. Stereo geometry assuming that motion of the cameras is usually in the same horizontal plane.

Depth information can be obtained by analysing the projection of the stereo geometry in the epipolar plane (figure 4).

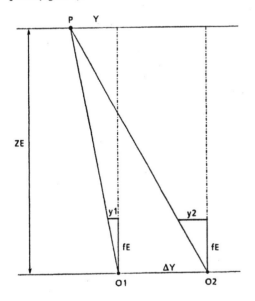

Figure 4. Projection in the epipolar plane.

From figure 4, one can easily show that

$$\frac{f_E}{y_1} = \frac{z_E}{y} \qquad \frac{f_E}{y_2} = \frac{z_E}{y+\Delta y} \qquad (10)$$

where $f_E = \sqrt{f^2 + x_1^2}$ is the projection of the focal length in the epipolar plane. Moreover, it can be shown that depth is obtained from the expression

$$Z = \frac{z_E}{f_E} f \qquad (11)$$

Hence, we can write last formula as a function of the

observed disparity $N_{p2}-N_{p1}$ where $N_p$ is the number of pixels and, therefore, $y_i = N_{pi}.l_p$. Consequently

$$N_{p2}-N_{p1} = \frac{f_E}{l_p \, z_E} \Delta y \qquad (12)$$

Finally, we obtain

$$Z = \frac{\Delta y}{(N_{p2} - N_{p1}) \, l_p} f \qquad (13)$$

The error in depth can be considered according to the study developed by Blostein and Huang (87). The probability of the quantity $\left| e_z \right| = \left| \Delta_z \right| / z$ , where $\Delta z$ is the error in z due to quantization and is the scaled error, can be writen as

$$\left| e_z \right| = \left| (y_1 - y_2) / (N_{p2} - N_{p1}) \right|$$

## CONTROL FEEDBACK THROUGH 3D–MOTION ESTIMATION

The 3-D tracking module has been used as a motion feedback in a mobile robot. The Y and Z coordinates of the target are supplied to a controller that tries to maintain the target's c.g inside an area around the geometric center of both image planes and to keep a prudential distance between the target and the mobile platform. As a result the mobile robot has to follow the target (Fig. 5).

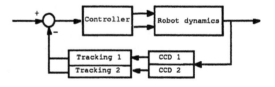

Fig. 5. System structrure

### The Control Problem

The mobile robot with the in-built tracking subsystem must follow the target with some conditions:

-Keep a prudential distance between the vision system and the target. If the target is too far it will be difficult to distinguish it, but if the distance is shorter than a lower limit, then the risk of collision could be high.
-Try to follow the target, not to catch it. The tracking system must be predictive, and the robot has to be always behind the target and "looking" at it.
-Filtering the results coming from the 3-D tracking module

One of the main problems when controlling the robot is that the direction it takes is the direction it looks at, and the camera's aperture angle (a) puts an upper limit on the maximum error allowed between the direction of the robot's speed vector and the angle it makes with the target relative position.

The input from the vision system is very noisy, mainly in the estimation of the distance between the target and the robot. This means that the control system must filter this noise. This filter could have a long time constant, but always avoiding that the error could grow over the limits of the vision system operability. So, when the error grows dangerously, this time constant must decrease to make the system more sensitive to changes on it.

If we are able to achieve this trade-off, then while the error will be inside the security area our system will have a smooth response, filtering the errors coming from the vision system. But if the target goes out from this security area, then it will react, even if the input was just noise.

That is why we have implemented a function with a gain that ranges from zero to one while the error is increasing. The function is smooth to avoid oscillations around some points. As soon as the nonlinearity gain is as maximum one, the stability of many systems will not be affected

The function is:

$$F_{\mu,n}(x) = x^{2n+1}/(x^{2n}+\mu^{2n})$$

where $\mu$ is the radius of the dead zone (Security angle). This function has values:

$$F_{\mu,n}(\mu) = 0.5\mu$$

$$F_{\mu,n}(-\mu) = -0.5\mu$$

$$F_{\mu,n}(0) = 0$$

$$F_{\mu,n}(x) = x \text{ if } x \gg \mu$$

If n is large enough, we can consider $F_{\mu,n}(x)=0$ x if in $(-\mu,\mu)$, and $F_{\mu,n}(x)=x$ if x in $(\mu,\infty)$ or x in $(-\infty,-\mu)$

The derivative of $F_{\mu,n}(x)$ is:

$$F_{\mu,n}'(x) = 1 + ((2n-1)x^{2n}\mu^{2n}-\mu^{4n})/(x^{2n}+\mu^{2n})^2$$

where for n=1

$$F_{\mu,1}'(x) < 1 \text{ if } x<\mu$$

$$F_{\mu,1}'(x) > 1 \text{ if } x>\mu$$

and if n>1

$$F_{\mu,1}'(x) < 1 \text{ if } x<\mu/(2n-1)^{1/2n}$$

The proposed control system block diagram is as follows:

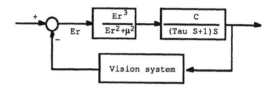

Fig 6. Control system diagram

This is a nonlinear system that can be linearized as shown:

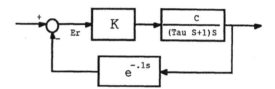

Fig 7. Linearized control system diagram.

and having a transfer function

$$H(s) = KC /((Tau\ s+1)s+KC\ e^{-.25\ s}).$$

This is a second-order, low pass filter with cut-off frequency KC/Tau vhere K is the gain of $F_{\mu,n}(e)$, and it can vary from 0 to 1. So that when the error is 0 the cut-off frequency is zero, and as the error increases, K tends to 1 and the system will be a low-pass filter with cut-off frequency KC/Tau.

Stability Criterion.

If the lineal equivalent system is stable then the non linear system, derived will also be stable. Despite the concept relating this sentence is simple the mathematical demonstration is difficult because we are in front of a non linear system.

We can use a modified version of the Nyquist criterion to state this property. We can decompose our system in two parts, the nonlinearity, and the system dynamics G(s) and calculate the describing function of the nonlinearity, N(Er). The Nyquist criterion states that there will be oscillations if D(jw) = -1 or as in this case N(Er)G(jw) = -1, instead. This means that there will be oscillations if G(jw) = -

1/N(Er) for some Er

And the system will be unstable if G(jw) encircles the -1/N(Er), or stable if -1/N(Er) is outside G(jw)

N(Er) is the first harmonic of the output of the nonlinearity, when the input is Er.cos(t).

In this case -1/N(Er) ranges from -1, when e is bigger than $\mu$ to minus infinite when Er tends to 0. So if G(jw) is on the right of -1 the system will be stable. And if G(s) is a minimum-phase stable system, then the nonlinearity will not affect the stability.

It is possible to compare this nonlinearity with a classical dead zone, for example:

a) $X = 0$      if $|Y| < K$
    $X = Y - K$    if $|Y| > K$
or
b) $X = 0$      if $|Y| < K$
    $X = Y$      if $|Y| > K$

Those systems have some disadvantages. Case a: the output is always less than it should be, so the effect of the dead zone is present for any value of the output. Case b: This function has problems when the error becomes bigger than K because there is a discontinuity. The function proposed avoids these problems. Varying the value of n, different shapes can be obtained, that can be useful for different problems. If we increase n to the infinite we obtain a system like b.

The controller equations for the left wheel speed (SpL) and for the right wheel speed (SpR) are:

SpL = LinSpeed + AngSpeed

SpR = LinSpeed - AngSpeed

LinSpeed is the linear speed of the mobile robot and its value depends on the target distance from the robot:

LinSpeed = $C_1 ( z^3 / ( z^2 + \mu_1^2))$
z is the distance error

AngSpeed = $C_2 (ang^3 /( ang^2 + \mu_2^2)$

AngSpeed is the difference of speeds between wheels. Ang is defined by the position vector of the target with respect to the robot and the robot longitudinal axis.

To have an appropriate response of the system it is necessary to choose the right values for the constants, and those depend on the system inertia, Tau and other characteristics. $C_2$ must be small, and depends on the distance between wheels. IF $Z_0$ is the desired distance between robot and target, then $\mu_1$ should be less than $Z_0/4$, because if $\mu_1$ is similar to $Z_0$ then we lose the security margin. The limit of $\mu_2$ depends on the camera aperture angle, and the stereovision system operability. The choice of n depends on how strict we want the dead zone, but as n increases the function loses smoothness, and can appear the oscillations around $\mu$.

Features of the Control System

Using $F_{\mu,n}$ we get a nonlinear system that has a special kind of dead zone with the following main features:

- Filters the noise
- The error tends to zero
- Dead zone
- A delayed response

All this is done in a specific way due to the fact that the system will not respond till the error is $\pm\mu$. So if the error is zero, and there is noise, then it will be filtered. If the error is near $\mu$ then the controller will only filter the positive part of the noise but not the negative one. This effect is normal, being at the edge of the permissible error. If there is an increase in the error the system can not wait two or three samples to be sure that it is not noise.

Due to the shape of $F_{\mu,n}$, in the dead zone (the function has low values but not zero), the error tends to zero.

The system has three areas of convergence: $-\mu$, 0, $\mu$. When the derivative of the system input is zero, the error decreases to zero. If the derivative is positive, then the error tends to be $\mu$ and the time the error needs to reach this value is the system delay. If the derivative is bigger than $\mu$, then the system can be considered just a linear integrator. For a negative derivative the quiescent point is $-\mu$.

CONCLUSIONS

In this paper we have studied some aspects relating a stereo-tracking vision system which allows to feedback a mobile robot and a nonlinear control which creates a dynamic link between perception and action. We have demonstrated the robustness of the motion vector computation with regard to digitization errors when the target that is going to track the robot has an area -on the image plane- large enough. Moreover, the relationship between the depth and the observed disparity corresponds to a equilateral hyperbola. This way, the larger the disparity, the shorter is the scaled quantization error

To use the results coming from the tracking system, a special control algorithm has to be designed to filter the noise but avoid losing the target from the scene.

A nonlinear system with a kind of dead zone, has been shown to have competence and good performance in these kind of systems.

ACKNOWLEDGEMENT

This work has been partialy supported by CERCA.

REFERENCES

Amat J., A. Casals (1989). "Real-time tracking of targets from a mobile robot", Proc. Int. Conf. on Intelligent Autonomous Systems-2, vol. 1, pp. 361-367.

Beers, Y. (1957) Introduction to the theory of error. Addison-Wesley.

Blostein, S.D., T.S. Huang, (1987) Error analysis in stereo determination of 3-D point positions. IEEE Trans. PAMI pp. 753-765

Dickmanns, E.D. (1990) An integrated spatio-temporal approach to automatic visual guidance of autonomous vehicles. IEEE Trans. on Systems Man and Cybernetics, vol. 20, n 6.

Frau, J., J. Codina, V. LLario (1990a). Motion estimation from target tracking. NATO ASI on expert systems and Robotics, Springer-Verlag vol F71, pp. 445-458

Frau, J., V. LLario (1990b). 3D-tracking and adaptive motion prediction of a target from a mobile robot. IEEE Int Conf. on intelligent motion control pp. 433-437

Frau, J., et al (1990c) Polynomial regression analysis for estimating motion from image sequences . SPIE symposium on andvances in intelligent systems -Mobile robots V- pp. 329-340

Frau, J., et al . (1991a). A robust tracking system for mobile robot guidance. Euriscon'91

Frau, J., V. LLario (1991b). Predictive tracking of targets using image sequences. IEEE IROS'91

Frau, J., S. Casas (1991c). SMOP system for target tracking purposes. ESPRIT-BRA Workshop on Specialized Processors for Real-time image analysis, Springer-Verlag (in press).

Kamgar-Parsi, B., B. Kamgar-Parsi (1989) Evaluation and quantization error in computer vision. IEEE trans. on PAMI, vol 11, n 9.

Llario. V. (1986). Aportacio a la resolució, en temps real, de la correspondència de punts característics en imatges seqüencials mitjançant l´anàlisi espai-temporal amb restriccio epipolar. PhD Thesis.

Morgan A.D. (1990). Road edge tracking for robot road following: a real-time implementation. Image and vision computing, vol 8, n. 3.

Murata, S., T. Hirose (1989). Onboard locating system of autonomous vehicle Proc. IROS´89, pp. 228-234.

Wallace, R.S. (1987) Robot road following by adaptive color classification and shape tracking. Proc. IEEE Int. Conf. on robotics and automation

Watamabe, Y., S. Yuta (1990). Positgion estimation of mobile robots vith internal and external sensors using uncertainty evolution technique. Proc IEEE int conf on robotics and automation.

# IMPROVEMENT IN ROBOT POSITION ESTIMATION
# THROUGH SENSOR CALIBRATION

**S.W. Sedas\* and J. Gonzalez\*\***

*\*Carnegie Mellon University, Field Robotics Center, Pittsburgh, PA 15213, USA*
*\*\*Universidad de Malaga, Facultad de Informatica, Plaza El Ejido, 29013 Malaga, Spain*

## ABSTRACT

*Spot laser range scanners are needed to build depth maps which are required for mobile robot navigation. These maps are used to locate potential obstacles in the path of the vehicle and to position the robot in a known environment. Errors introduced during the formation of the depth image degrade the accuracy with which a robot can estimate its position. This paper describes the beam orientation and range measurement errors that occur in a spot laser range scanner. It illustrates the use of high resolution calibration to extend the operating range of a sensor and to improve the accuracy of automatic position estimation. Experiments show substantial improvement in the positioning of a mobile robot platform.*

Keywords: *Sensors, Robots, Intelligent-Machines, Instrumentation, Position Control*

## INTRODUCTION

A mobile robot system operating in a known environment will need a range sensor to build a range model of its surroundings. Given a map of the environment, the robot can locate itself by estimating the correspondence between common features that are visible on both the range model and the pre-recorded map. The accuracy with which the robot can position itself does not only depend on the accuracy of the range sensor and on how well the system can identify pairs of corresponding features. It also depends on the resolution at which the sensor is calibrated and on the software's ability to correct for sensor calibration and operation errors.

The accuracy of the position estimation can be improved by calibrating the range sensor (beyond the limits described in the manufacturer's specifications) and by modifying the software to correct for calibrated errors. This paper gives an overview of beam orientation and range measurement errors that are present in spot laser range scanners and describes an experiment that was performed to illustrate how high resolution sensor calibration can substantially improve the positioning of a mobile robot.

## MOBILE ROBOT POSITION ESTIMATION

Consider the following scenario. A mobile robot configured with a radial laser range scanner is required to navigate within a known environment. The sensor scans a horizontal plane to generate a two dimensional cross-sectional view of the environment. Using this information, the robot must navigate safely and register itself against a pre-recorded electronic map of the environment. (See Figure 1)

The robot is equipped with an iconic based position estimation software module (Gonzalez,1991). This module estimates the position and orientation of the robot by computing the local correspondence between the range image and the pre-recorded map of the environment. Every point in the depth image (labeled $x_i$) is matched against a corresponding point (or feature) in the pre-recorded map (labeled $y_i$) reducing the estimation problem to that of finding a geometric transformation that will minimize the distance between each pair of matched points $(x_i, y_i)$.

The position and orientation of the robot given by the matrix T, can be obtained by solving the following least squares minimization problem (Gonzalez,1991):

$$\text{Minimize } \Sigma \, |x_i - T \, y_i|^2 \qquad \text{(EQ. 1)}$$

Note that the accuracy of the transformation T depends on:

- accurately measuring $x_i$, and
- searching for a correct match of $(x_i, y_i)$

Therefore the accuracy of T can be improved by improving the measurement of $x_i$ and the search for matching points $x_i$ and $y_i$.

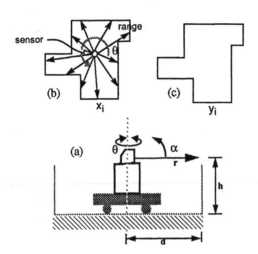

Figure 1 (a) A mobile robot configured with a single axis radial range sensor. (b) A two dimensional raster image of the environment characterized by the set of points $x_i$. (c) a pre-recorded map of the environment characterized by the set of points $y_i$

## SPOT LASER RANGE SENSORS

The configuration of a laser range scanner consists of a laser beam transmitter, a receiver and a beam orientation device. The transmitter emits a narrow beam of laser light. When the beam strikes an object in the environment, part of the energy is reflected back and detected by the receiver. By measuring the time of flight of the transmitted signal, the sensor can estimate the distance to the object. Pulse modulation, AM modulation and Frequency Modulation are three of the techniques that are used to time the return of the signal (Besel).

Through mirrors and other mechano-optical devices, a laser range scanner can orient the beam of light to point to different locations of the environment. By recording the range measured at different beam orientations, a range scanner can construct two and three dimensional raster representations of its surroundings. Figure 2 is an illustration of the CYCLONE, a single axis radial range sensor configured at the Field Robotics Center at Carnegie Mellon University, which uses a single rotating mirror to project the beam therefore generating a two dimensional raster image of a cross-sectional view of the environment (Figure 1-b).

## BEAM ORIENTATION AND RANGE ERRORS

The range and beam orientation angles of a scanned image are expressed in a local sensor reference frame. However most applications require that the data be expressed in a common reference frame. A simple trigonometric operation can be used to transform a point in the raster image into a global rectangular coordinate representation. However, nonlinearities in this transformation propagate small errors in the range and beam orientation to large errors in the cartesian representation. It is important to understand the cause and effects of these errors so that these can be compensated for in the application.

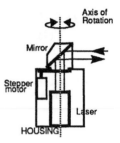

Figure 2 A Schematic Representation of the CYCLONE single axis radial range scanner (Singh, 1991).

### Beam Orientation Errors.

A laser beam range sensor has many components such as the laser transmitter, the mirror and the axis of rotation of the mirror which project and orient the laser beam. Errors in the manufacturing, positioning and configuration of each of these components will result in errors in the orientation of the beam. Although small beam orientation errors do not significantly affect the range measurement (assuming the beam strikes the correct target and that the nominal angle of incidence of the beam is zero), they do substantially affect the region of the environment that is scanned by the sensor.

The following experiment (Sedas,1991) will illustrate this effect. Position the Cyclone so that it scans a horizontal plane. Place a number of different cylinders each of a different radius centered around the sensor such that their central axis is perpendicular to the scanning plane (See Figure 3). For continuous mirror orientations($\theta$), compute and plot the height (h) at which the laser intersects each of the cylinders. Repeat the experiment for different tilts in the sensor housing, mirror and laser source. [1]

When no errors are present, the Cyclone will scan a flat surface perpendicular to the axis of rotation of the mirror. A tilt in the mirror causes the sensor to scan a conic surface. This is observed in the set of multiple lines that are displayed in the plots of Figure 9-a. As illustrated in Figure 9-b, the plot generated by a tilt in the sensor housing is a symmetrical sinusoid whose amplitude and phase change with tilt angle. The plot that is generated from a tilt in the laser source resembles a non-sinusoidal symmetric periodic function that increases with tilt (Figure 9-c). Finally, the plot corresponding to a tilt in both the laser source and the mirror approximates an asymmetric sinusoidal (Figure 9-d).

---

[1] Details of the model and simulation experiments have been reported in (Sedas,1992).

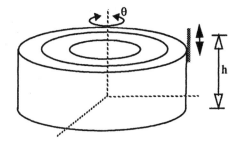

Figure 3 A set of cylinders used to measure beam orientation errors due to orientation errors in the housing, mirror and laser source.

It is important to calibrate the beam orientation errors as a function of mirror orientation and target distance as this would allow the system to minimize the apparent distortion caused by these errors which in actuality are a consequence of the interpretation of the raster image. The beam orientation can be calibrated by recording the height at which the beam strikes a set of cylinders of different radiuses. Although useful in tabular form, a user may want to compress the information by fitting first and second order models of the optical configuration of the sensor[2].

Range Errors.

The receiving end of a laser range sensor consists, among other things, of a photo detector and a lens. The lens collects as much of the reflected energy as it can and focus it onto the detector. The detector then converts this energy into electrical energy to be used by the timing circuitry. The amount of electrical energy that is generated is proportional to the amount of incident energy that is received by the sensor.

Laser range sensors are tuned to detect targets which are located within a specified minimum and maximum range distance. The amount of energy that reaches the sensor from a target located beyond the maximum range of operation may not be enough to be detected by the sensor. Conversely, the energy that reaches the sensor from target that is located below the minimum distance may saturate the sensor or the circuitry of the detector. Environmental conditions such as fog and rain may attenuate the signal and reduce the maximum effective range of the sensor.

Recent experiments have shown that the accuracy of a range scanner is affected by temperature (Kweon, 1991), target material properties and target orientation (Kweon,1991)(Sedas,1992). Infrared detectors are quantum detectors which operate on the following principle. The interaction of an incident radiation with electrons in a solid cause the electrons to be excited to a higher energy state; this excitation generates electric energy(Sowan,1971). Because of the material properties of the detector, only a small amount of infrared energy is required to excite the

2. see (Sedas,1992) for details

electrons. However thermal agitation of the solid will also excite the electrons generating electrical noise. Thus to lower the levels of noise, infrared detectors need to be cooled. Nevertheless, extreme changes in operating temperature can alter the measurement and operating characteristics of the sensor.

The accuracy and dynamic range of a laser range sensor is affected by the intensity of the return signal. This in turn is affected by the angle of incidence of the beam to the target, the roughness of the target surface and the target material properties. As seen by the unified reflection model (Nayar,1991) illustrated in Figure 4 the amount of energy that is reflected off a smooth target in the direction of the sensor is strongest when the target is perpendicular to the beam and decreases with increasing angle of incidence. Furthermore, at large incident angles, the relative intensity of the signal will increase with surface roughness, but for a small incidence angle, the relative intensity will decrease (because of light dispersion) with surface roughness. The experimental observations reported by Sedas and Gonzalez (1991) show that the accuracy of the range measurement decreases with an increasing incidence angle.

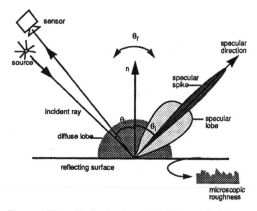

Figure 4 The unified reflection model. (Nayar, 1991) An incident beam of light will generate three types of reflection: a lambertian or diffuse reflection which has a uniform intensity in all directions, a specular lobe which has a narrow dispersion centered about the specular direction, and the specular spike which is a very sharp reflection in the specular direction.

In an effort to verify the range accuracy of the Cyclone, Sedas and Gonzalez (1991) performed a series of measurements during which they recorded the difference between the distance measured by the sensor and a true target distance for a wide range of target distances. Figure 5 is a plot of the range offset versus target distance).

Sedas and Gonzalez observed that the difference between the distance measured by the sensor and the true target distance (denoted as the range offset) was repeatable (well within the resolution of the sensor)

429

and could be used to correct measurement errors (even below the range of operation given by the manufacturer's specifications). Within the range of operation specified by the manufacturer, the range offset was a slowly varying linear function of target distance. However, at ranges below the minimum recommended distance of operation, the range offset decreased exponentially with target distance. For the reader's information, the manufacturer specifications indicate the range of operation of the sensor are between three and fifty meters and the accuracy within this range is twenty centimeters at a resolution of ten centimeters (Radartechnik).

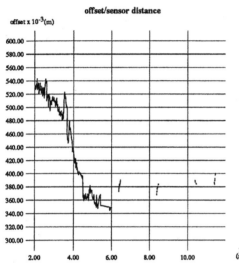

Figure 5 Range error vs. range characterization (Sedas,1991)

## IMPROVING ROBOT POSITION ESTIMATION

As was mentioned in a previous section one way to improve the accuracy of a robot position and orientation estimation is to reduce errors in range measurement. This can be done by calibrating the systematic errors of the sensor. The following experiment will illustrate how an accurate sensor characterization such as that which was performed by Sedas and Gonzalez (1992) can be used to extend the range of operation of a laser range finder and improve the accuracy of a robot position estimation. Only systematic range errors have been corrected, the correction for beam orientation errors has been left for a future exercise.

The results that were obtained in this experiment prove two things. First, that image warping caused by uncompensated systematic errors affect the accuracy that can be obtained by a position estimation software. Second, that modifying (or calibrating) the software to compensate systematic calibration and operation errors can substantially improve the accuracy of the application.

The testbed for these experiments consisted of the CYCLONE, a single axis radial laser beam range scanner mounted on top of a mobile robot platform (Fitzpatrick,1989), a mock-up environment made up of large wooden panels, an accurate map of the environment, a position estimation software provided by Gonzalez and collaborators(1992), and a numeric table which characterizes the systematic range errors of the CYCLONE for different measured target distances(Sedas,1992).

The robot was programmed to navigate along a predetermined path and to scan the environment at twenty different locations along the path. At each marked location, the position of the robot was surveyed and the magnitude of the error between the surveyed and estimated position was recorded. Figure 7 is an illustration of the task. The robot was programmed to make the scans at twenty equidistant locations along the path that is indicated by the arrow.

Figure 6 The locomotion emulator and CYCLONE range scanner

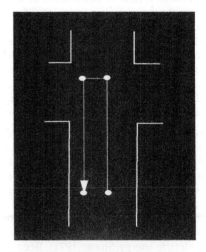

Figure 7 An illustration of the test site. The arrow indicates the planned trajectory for the mobile robot.

The position estimation software adjusted each range measurement by subtracting from it the a range error given by a *range error characterization function*. The experiment was repeated for three different characterization functions. The first function, labeled O1, is a piecewise approximation of the range offset characterization obtained by Sedas and Gonzalez (1992).(Figure 5) The second function, labeled O2, was defined by the constant error measured for targets located at a long distance from the sensor. The third one, labeled O3, was set at the error measured against targets that are located in close proximity to the sensor.

Figure 8 is a plot of the magnitude of the position estimation error (vertical axis) for each location at which the robot's position was estimated (horizontal axis) evaluated for the range error characterization functions ($O_1$, $O_2$ and $O_3$) that were defined above. From these results one can observe that the calibration constant $O_2$ is an accurate approximation of the range error for targets that are located at a large distance from the sensor. Similarly, $O_3$ is an accurate approximation of range error for targets that are near the sensor. However neither $O_2$ or $O_3$ are a good enough general approximation to be used for all targets. By contrast, one can observe from the position estimation errors in that $O_1$ is a good approximation of the range error at all target distances.

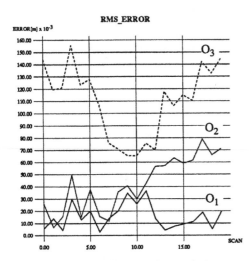

**RMS_ERROR**

Figure 8 Magnitude of position estimation errors for three range characterization functions

## CONCLUSIONS

The Cyclone range scanner is a sensor that is used in a number of autonomous vehicles at the Field Robotics Center. These vehicles rely on the accuracy of the sensor to avoid obstacles and position themselves in a known environment. In this front, the calibration of the Cyclone range sensor (Sedas,1992) had two major impacts. First, it affected the usable range of operation of the sensor therefore allowing the robots to navigate closer to walls and to other obstacles in the environment. Second the correction of systematic range errors substantially improved the position estimation of an autonomous vehicle. Experiments reported in this paper showed a factor of three improvement over worst case. Even more surprisingly is the fact that although the resolution of the Cyclone is ten centimeters and its accuracy twenty, the position estimation software was able to position the robot within 3 cm (worst case error) and 1.9 cm (average error) without using more than a single scan at each location[3].

A overview of beam orientation errors and their contribution to errors in range images was presented in the paper. However beam error compensation was not included in the experiment. Future work in automatic position estimation should take both beam and range errors into consideration to correct the depth map, to assign different weights in the error evaluation of (EQ. 1) and to direct the search for a set of matching features.

## ACKNOWLEDGEMENTS

Our appreciation to W. Whittaker and R. Krishnamurti for their continuous guidance, encouragement and support, T. Stentz who first suggested the need for this work, G. Shaffer who provided the interface to the Cyclone and E. Krotkov, M. Hebert, S. Shafer, R. Hoffman and T. Shoart who guided us during the initial phases of the Cyclone calibration.

## REFERENCES

Besl, J. (1988) "Range Imaging Sensors" Tech. Rept. Computer Science Department, General Motors Research Laboratories, Warren, Mi.

Fitzpatrick, K.W., Ladd,J.L (1989) "Locomotion Emulator: A testbed for navigation research." in *1989 World Conference on Robotics Research: The Next Five Years and Beyond*, May 1989.

Gonzalez, J., A. Stentz, A. Ollero, (1992) "Iconic Position Estimator for a 2D Laser Range Finder". *IEEE Int. Conference on Robotics and Automation*

---

[3.] This means that multiple scans were <u>not</u> used to improve the statistical value of each measurement.

Radartechnik & Electrooptik, "Laser-Rangefinder. Reference Manual", Reference Manual, Radartechnik & Elektrooptik, A-3754 Trabenreith 35, Austria.

Kweon, I., Hoffman, R., Krotkov, E.,(1991) "Experimental Calibration of the Perceptron Laser RangeFinder", Tech. Rept. Robotics Insitute, Carnegie Mellon University, CMU-RI-TR-91.

Nayar, S., Ikeuchi, K.,Kanade, T. (1991) "Surface Reflection: Physical and Geometrical Perspectives", IEEE Transactions on Pattern Analysis and Machine Intelligence, Vol. 13, No. 7, July

Sedas, S., Gonzalez, J.(1992) "Analytic and High Resolution Characterization of a Radial Range Sensor", CMU-RI-92 Robotics Institute, Carnegie Mellon University, Pittsburgh, PA

Shaffer, G., Stentz, A. "Position Estimator for Robotic Mine Equipment", Field Robotics Center, Carnegie Mellon University

Singh, S., West, J. (1991) "Cyclone: A Laser Scanner for Autonomous Vehicle Navigation," Tech. Rept. Robotics Insitute, Carnegie Mellon University, CMU-RI-TR-91-18, August

Sowan, F. A. (ed), (1971) *Applications of Infrared Detectors*, Mullard Limited, Mullard House, Torrington Place, London WC1E7HD

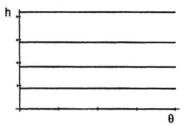

(a) height at which the beam strikes a set of cylinders observed for a single error in the orientation of the mirror.

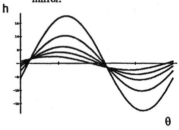

(b) height at which the beam strikes a cylinder observed for multiple errors in the orientation of the sensor housing.

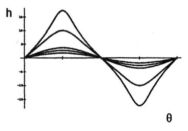

(c) height at which the beam strikes a cylinder observed for multiple errors in the orientation of the laser source.

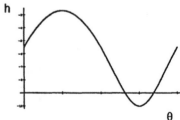

(d) height at which the beam strikes a cylinder observed for a simultaneous error in the orientation of the mirror and the laser source.

Figure 9 An observation of the effect of beam orientation errors caused by errors in the orientation of the sensor, mirror and laser source simulated for the CYCLONE radial laser range scanner (Sedas,1991).

# WORLD MODELLING AND SENSOR DATA FUSION IN A NON STATIC ENVIRONMENT. APPLICATION TO MOBILE ROBOTS

## L. Moreno*, E.A. Puente* and M.A. Salichs**

*Dpto. Ingenieria de Sistemas y Automatica, Universidad Politecnica de Madrid (UPM), Spain
**Dpto. Informatica y Automatica, Universidad de Murcia (UM), Spain

Abstract. We describe a world modelling method able to integrate static and moving objects existent in dynamic environments. The static world is modelled by using an occupancy grid. The method is capable of modelling several moving objects. Whereas measurements belonging to actual targets are processed using a Kalman filter to yield optimum estimates, all other measurements are used to create or maintain multiple hypothesis corresponding to possible mobile objects. The viability of the method has been tested in a real mobile robot. Portions of this research has been performed under the EEC ESPRIT 2483 Panorama Project.

Keywords. Autonomous mobile robots; environment modelling; static data fusion; dynamic data fusion; mobile objects detection.

## Introduction

Mobile robots must be able to operate in static and dynamic environments. Whereas research on static environments is extensive, research on dynamic environments is just beginning. A contributing factor to this situation is that solving problems in dynamic environments is much more difficult when compared to the static environment. In particular, the problem of modelling dynamic environments is important and challenging. For example, it may be desirable for a mobile robot to track several targets in order to avoid them, to plan a motion to a goal while avoiding moving objects, to follow the target, or to just perform any task while being aware of the number and state of each target. In addition, world modelling in the context of dynamic environments is more difficult when compared to static environments. Modelling and characterization of moving obstacles is crucial for further developments of mobile robots in dynamic environments.

Accordingly, one of the main functions performed by a mobile robot is to keep a model of all the moving objects of interest within the coverage region of its sensors.

The foremost difficulty in the application of non static world modelling involves the problem of associating measurements with appropriate mobile objects, especially when there are oclusions, unknown targets (requiring mobile object initiation), and false measurements (from noise and background). To model non static environments, it is firstly required the modelization of the static objects, otherwise it is not possible to discriminate between both.

## Static Environment

Two main approaches are widely used for the static world modelling problem: *geometric* modelling and *occupancy grid* modelling. Both methods requires a first comment:

### Geometric Models.

This group of techniques use observations of geometry features to compare and integrate observations, updating object locations to a world model (Leonard 1990, and Ayache 1989).

In this approach the integration mechanism try to combine any number of observation from sensors that provide measurements of different geometric features. Common to these approaches is the use of the Kalman filter to model and propagate uncertainty in both the position of the robot (Cheeseman 1986, and Crowley 1989) and the geometric features using covariance matrix. This is a powerful tool for dealing with noise. The experimental results presented by the research groups that uses this tecnique are all for well structured environments in most cases.

### Occupancy Grids.

This thecnique uses a probabilistic tesselated representation of spatial information. This method developed by Moravec and Elfes (Elfes 1987) based on the use of an occupancy grid of the environment, combines a probability of occupancy with a spatial uncertainty. Each cell hold an estimate of the confidence that is occupied. The uncertainty surrounding an object position is represented by a spatial distribution of these probabilities within the occupancy grid.

Grid map representations are particularly useful for the task of obstacle avoidance, due to the explicit representation of the free space. Besides, the occupancy grid repressentation is a powerful technique when we have sensory data in which reliable feature extraction is not practical. Something similar happens when we operate in non structured environments, where reliable feature extraction is highly difficult or not practical (Puente 1991).

Due to the context where this work is developped, the ESPRIT Panorama project aims to develop an autonomous system to operate in partially structured environments, that imply outdoor environments with some artificial and natural landmarks, the technique selected model the static world is an occupancy grid.

A more general model encodes multiple properties in a cell state. If the cell property is limited to occupancy, it used to be called occupancy grid.

## Static Fusion

The static fusion process in occupancy grids depends on the way we express the certainty about the occupancy state of the cell. If we use a classical probabilistic aproach to this problem, the occupancy certainty of the cell is determined by the occupancy probability $P[s(c_i) = occ|r_t]$ after the sequence $r_t$ of measurements. Depending on $P[s(c_i) = occ|r_t]$, that cell is supposed to be empty ($< 0.5$), occupied($> 0.5$) or unknown ($\approx 0.5$).

If we use a probabilistic aproach to indicate the uncertainty about the cell occupancy, the fusion of the new sensorial information provided by the sensors used to be made by the *Bayes's theorem*. Given a current estimate of the state of a cell $c_i$,

$$P[s(c_i) = occ|r_t] \qquad (1)$$

based on observations $r_t = r_1, ....r_t$ and given a new observation $r_{t+1}$, the improved estimate is given by:

$$P[s(c_i) = occ|r_{t+1}] =$$
$$\frac{p[r_{t+1}|s(c_i) = occ].P[s(c_i) = occ|r_t]}{\sum_{s(c_i)} p[r_{t+1}|s(c_i)].P[s(c_i)|r_t]} \qquad (2)$$

In this recursive formulation, the previous estimate of the cell state, $P[s(c_i) = occ|r_t]$, serves as the prior probability and is obtained directly from the occupancy grid and $p[r_{t+1}|s(c_i) = occ]$ is the measure provided by the sensors.

The new cell state estimated $P[s(c_i) = occ|r_{t+1}]$ is subsequently stored in the map. Since the cell states are exclusive and exhaustive, $P[s(c) = occ] + P[(scc) = emp] = 1$. In order to define the initial prior cell state probability, a $P[s(x) = occ](s) = 0.5$ is used.

More general models are possible by using a random vector and locating multiple properties in the cell state.

The multiple views acquired by the robot from different sensing positions, can be incorporated into the map updating process as a blurring or convolution operation performed on the occupancy grid.

- **World-based mapping.**

  In this case, the motion of the robot is related to an absolute coordinate frame, and the current robot view is blurred by the robots global positional uncertainty prior to composition with the global map. Since the global robot position uncertainly increases continuously , if there isn't

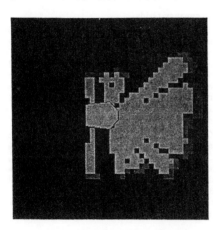

Figure 1: Example of occupancy grid

any other method to decrease the robot position uncertainty, this updating procedure has the effect that the news become progressively more blurred, adding less useful information to the global map.

- **Robot-based mapping.**

  In this case, it's estimated the uncertainty of the global map due to the recent moment of the robot, and the global map is blurred by this uncertainty prior to composition with the current robot view. A consequence of this method is that observations performed in the remote past become increasingly uncertain, while recent observations have suffered

little blurring. From the print of view of the robot, the immediate surroundings (of direct relevance to its current navigational tasks) are "sharp".

In the context of the Panorama project, a Robot-based mapping has been used.

## Dynamic Environment

The overall approach is depicted in Fig.2 where the mobile object state estimation method is based on a static map, management of hypothesis, and a Kalman filter. Several features have been incorporated into the algorithm to increase its usefulness for applications to real problems.

Observations are first compared with the occupancy grid to see if they correspond to the static environment. Actually, this step correspond to the building of the fix obstacle map. If the measurements do not correspond to the fix obstacle map, the measurements are considered as coming from possible mobile objects. The measurements are then matched with the mobile objects data base. If they correspond to an existing mobile object in the data base, the previous data of this mobile object are fused with the new ones using a Kalman filter. If the measurements do not match with any registered mobile object, a new hypothesis corresponding to a new target with low existence probability is created in the data base. The existence probability of a target increases with positive reports, and decreases when there are negative reports . The dynamic object database is maintained in order to eliminate hypothesis with very low existence probabilities and to join hypothesis which has been initially considered as belonging to different targets.

We model a moving obstacle by its geometric characteristics, its state vector, its estimation errors and its existence probability. Thus, a moving obstacle is characterized by $M = \{G, \mathbf{x}, \mathbf{P}, Pr\}$ where, $\mathbf{x} = [x, y, x', y']^T$ and $\mathbf{P}$ is the covariance matrix associate to $\mathbf{x}$, $G$ is in a simple aproach the object diameter estimation and $Pr$ is the probability that the moving target exists.

It is assumed that a mobile object evolves with time according to known laws of the form

$$\mathbf{x}_{k+1} = \Phi \mathbf{x}_k + \Gamma \omega_k \qquad (3)$$

where $\Phi$ is the state transition matrix, $\Gamma$ is the disturbance matrix , $\omega_k$ is a white noise sequence of $N(0, Q)$ normal random variables with zero mean $E[\omega] = 0$ and covariance $E[\omega\omega^T] = Q$.

These state variables are related to measurement $\mathbf{z}$ according to the following observation model

$$\mathbf{z}_k = H\mathbf{x}_k + \mathbf{v}_k \qquad (4)$$

where $\mathbf{H}$ is the measurement matrix and $\mathbf{v}$ is a white noise sequence of normal random variables $N(0, R)$ with zero mean $E[\mathbf{v}] = 0$ and covariance $E[\mathbf{vv}^T] = R$.

- **Mobile object detection.**

  In order to differenciate a sensor measurement from a mobile object, it is necessary to have information about the static environment around the mobile robot. In a first consideration, each sensor measurement can be originated by a mobile object, by a static object or it is simply a spurious measurement. If a measurement comes from a static object, it would be noticed and included in the static world model(certainty grid).

  If a sensor measure doesn't correspond to a static object, we can formulate the hypothesis that a mobile object has originated that measurement. The hypothesis will be rejected or not according posterior measures confirming or not that hypothesis.

- **Matching.**

  Each possible mobile object is matched with the mobile objects detected included in the mobile objects database. A possible mobile object matches the state prediction of a previous one if it is located inside the validation region around the predicted mobile object state. That is

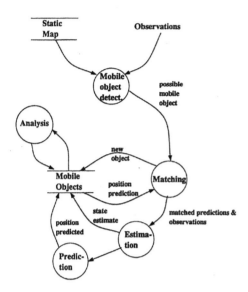

Figure 2: The Dynamic Environment Modelling

$$B = H \widehat{P}_{k+1|k} H^T + R \qquad (5)$$

$$(\mathbf{z} - H\widehat{\mathbf{x}}_{k+1|k}) B^{-1} (\mathbf{z} - H\widehat{\mathbf{x}}_{k+1|k}) \leq \eta^2 \qquad (6)$$

Measurements lying outside the validation are not incorporated with the Kalman filter. This accounts for the fact that the measurement used in the filter might have originated from a source different from the target of interest.

- **State Estimation.**

If a possible mobile object matches with a prediction in the mobile object database, this measurement is fused with the previous information about this mobile object located in the database by using the Kalman filter,

$$\mathbf{x}_{k+1|k+1} = \widehat{\mathbf{x}}_{k+1|k} + K[\mathbf{z} - H\widehat{\mathbf{x}}_{k+1|k}] \qquad (7)$$

$$P_{k+1|k+1} = \widehat{P}_{k+1|k} -$$
$$\widehat{P}_{k+1|k} H^T (H\widehat{P}_{k+1|k} H^T + R)^{-1} H\widehat{P}_{k+1|k} \qquad (8)$$

$$K = \widehat{P}_{k+1|k} H^T R^{-1} \qquad (9)$$

where K is the Kalman matrix gain.

- **State Prediction**

Once the measurement is fused with the previous prediction about the mobile object state, a prediction about the mobile object state in the next sensor cycle is done

$$\widehat{\mathbf{x}}_{k+1|k} = \Phi \mathbf{x}_{k|k} \qquad (10)$$

$$\widehat{P}_{k+1|k} = \Phi P_{k|k} \Phi^T + \Gamma Q \Gamma^T \qquad (11)$$

The process is repeated for every mobile object included in the mobile objects database.

- **Mobile object analysis.**

If a measurement marked as coming from a possible mobile object doesn't match with any of the mobile objects included in the database a new mobile object hypothesis is generated. Subsequent measurements matching with the new generated hypothesis will be used to increase the probability of newly generated hypothesis or to discard them if their probabilities lie below a threshold value.

All hypothesis have their corresponding validation regions. For each scan, every validation region (i.e., hypothesis) is evaluated to see if some measurements in the scan correspond to some hypothesis. The hypothesis supported by new measurements are reinforced in their probability. If there are hypothesis which are not supported by any measurement, the probability of the corresponding hypothesis is decreased according to the following Bayes formula.

If the probability associated with a certain hypothesis falls below a theshold, the mobile object hypothesis is discarded.

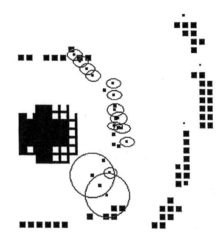

Figure 3: Tracking one target which is moving around the robot

## Implementation and Results

The world modelling method proposed in this paper has been simulated and implemented in our mobile robot laboratory yielding acceptable results. The experimental system existent at the laboratory of Automation of the UPM consists of one mobile robot with a belt of 15 ultrasonic sensors, a LAN with several SUN4 sparc 2 workstations, an infrared link to the robot at 9600 bps (for comand information only). The on-board hardware is based on a 68030 processor under the OS-9 operating system. The robot is connected to the workstation via Ethernet for downloading compiled programs and for human interface.

The sample time for each scan is 0.5 sec., the cell size is 20 cm, the number of cells is 50x50. All experiments were performed in an indoor environment composed of a large room of 5x9 m with walking researchers and students as targets. Fig. 3 depicts the fix obstacle map obtained on-line using the probabilistic Bayesian method mentioned above. We can see the floor plan superimposed to the sensor measurements. As expected, it is noted that the standard deviation of the state estimates decreases with additional observations. Thus, the algorithm works well even for maneuvering (i.e., not straight line) targets. It can be noted that the standard deviation of the Kalman filter estimates start with a relatively large value and quickly stabilizing to a smaller value. The standard deviation cannot be further decreased because of the directional uncertainty of the ultrasonic sensors.

## References

Asada, M. (1990). Map building for a mobile robot from sensory data. IEEE Trans. on Systems, Man, and Cybernetics, v:37, n:6, 1326-1336, November.

Ayache, N. and Faugeras, O.D. (1989). Maintaining representations of the environment of a mobile robot, IEEE Trans. on Robotics and Automation, 804-819.

Borenstein, J. and Koren, Y. (1989). Real-time obstacle avoidance for fast mobile robots, IEEE Trans. on <u>Systems Man and Cybernetics</u>, 19: 1179-1189, September.

Crowley, J.L. (1989). World modeling and position estimation for a mobile robot using ultrasonic ranging. Proc. IEEE Int. Conf. <u>Robotics and Automation</u>, 674-680.

Elfes, A. (1987). Sonar-based Real-world Mapping and Navigation, IEEE Journal of <u>Robotics and Automation</u>, Vol. 3, No. 3, June.

Leonard, J., Durrant-Whyte, H. and Cox, I.J. (1989). Dynamic map building for an autonomous mobile robot. Proc IEEE/RSJ Int. Conf on <u>Intelligent Robot Systems</u>.

Puente, E.A., Moreno, L., Salichs, M.A. and Gachet, D. (1991). Analysis of Data Fusion Methods in Certainty Grids Application to Collision Danger Monitoring, Proc. <u>IECON'91</u>, pp. 1133-1137, Kobe, Japan, Nov.

Richardson, J.M. and Marsh, K.A. (1988). Fusion of multisensor data. Int. J. of <u>Robotics Research</u>, v:7, n:6, December.

Smith, R.C. and Cheeseman, P. (1986). On the representation and estimation of spatial uncertainty. Int. J. of <u>Robotics Research</u>, v:5, n:4, Winter.

Thomopoulos, C.A. (1990). Sensor integration and data fusion. Jou. of <u>Robotics Systems</u>, 7(3), 337-372.

# A SYSTEM FOR AUTONOMOUS CROSS-COUNTRY
# NAVIGATION

**B. Brumitt, R.C. Coulter, A. Kelly and A. Stenz**

*Robotics Institute, Carnegie Mellon University, Pittsburgh, PA 15213, USA*

Abstract: *Autonomous Cross-Country Navigation
requires a system which can support a rapid traverse
across challenging terrain while maintaining vehicle
safety. This work describes a system for autonomous
cross country navigation as implemented on the
NavLab II, a computer-controlled off-road vehicle at
Carnegie Mellon. The navigation software is discussed. The perception subsystem constructs digital
maps of the terrain from range sensor data in real
time. The planning subsystem uses a generate-and-
test scheme to find safe trajectories through the map,
and validates them through simulated driving. The
planning process considers both kinematic and
dynamic constraints on vehicle motion. The system
was successful in achieving 300m autonomous runs
on moderate terrain at speeds up to 4.25 m/s.*

Keywords: *Autonomous Mobile Robots, Obstacle
Avoidance, Navigation, Image Processing.*

## INTRODUCTION

Autonomous Cross-Country Navigation (ACCN) can be
applied to tasks such as military reconnaissance and logistics, medical search and rescue, hazardous waste site study
and characterization, as well as industrial applications,
including automated excavation, construction and mining.
The need to safely travel at a reasonable velocity on terrain
with unknown obstacles is common to all these endeavors.

This paper addresses robot perception, planning and control required to support the fastest possible autonomous
navigation on rough terrain. This objective poses new
issues not encountered in slower navigation systems on flat
terrain.

- Uncertain environment: sensing must be done
  simultaneously with driving; a precomputed path is
  impossible.

- Vehicle safety: if computing fails, or a safe trajectory cannot be found, the vehicle must be brought
  to a stop to avoid collision.

- Computation time: as vehicle speed increases, perception and planning must be performed more rapidly.

- Rugged terrain: the geometry of the terrain is sufficiently complex that the assumption of flat terrain

with sparse obstacles does not suffice.

- Dynamics: the vehicle's speed is large enough that
  dynamics as well as kinematics must be modelled.

- Imaging geometry: the complexity of map generation increases in the context of rougher terrain.

- Sensor limitations: at higher speeds, technological
  limitations of the laser range finder are significant.

- Motion during digitization: given higher operating
  speeds, image distortion is an issue.

Early work in outdoor navigation includes the Stanford
Cart (Moravec, 1983). This system drove in a stop-and-go
manner at a slow speeds, modelling terrain as flat with
sparse obstacles.

Higher speeds have been acheived in system operating in a
constrained environment. Road following systems assume
benign and predictable terrain. Obstacles are assumed to
be sparse, so that bringing the vehicle to a stop rather than
driving around obstacles is a reasonable strategy. In this
way, speeds of up to 100 km/hr. have been achieved in
road following (Dickmanns, 1990; Pomerleau, 1991) using
cameras, and speeds up to 30 km/hr. have been achieved
on smooth terrain with sparse obstacles using a single
scanline lidar (Shin, 1991). While these systems consider
problems of dynamics and vehicle safety, they do not provide the capability to drive around obstacles.

Obstacle avoidance has been acheived on approximately
flat natural terrain with discrete obstacles (Feng, 1989;
Bhatt, 1987; Chang, 1986.) On a smooth gently sloping
streambed with sparse obstacles, the JPL rover has
achieved a performance of 100 meters in 8 hours (Wilcox,
1987)

When navigating over rough natural terrain, no assumptions about the shape of the terrain ahead can be made.
Obstacles may not only be discrete objects in the environment, but can correspond to unsafe configurations as well.
As a result, sophisticated planning is needed. Hughes'
ALV has reached speeds of 3.5 km/hr (Daily, 1988; Olin,
1991) in this manner. However, the speed of the ALV system is small enough that many concerns for vehicle safety
are diminished, and dynamics need not be modelled for the
planning process.

This paper presents an overview of the software and hardware system which addresses the eight problems introduced above, thus permitting ACCN in natural terrain at speeds higher than that of previous systems.

## SYSTEM OVERVIEW

The testbed for the system is the NavLab II, a computer-controlled HMMWV (High Mobility Multi-purpose Wheeled Vehicle). The equipment on the NavLab II can be divided into three categories: computation, sensing, and actuation. For computation, there are 3 Sparcstation II's connected by ethernet. For sensing, an ERIM, a 2-d raster scanning laser range finder, is used. The vehicle is totally self-contained, as there no provisions for off-vehicle communications or power supply while in the field. Power is supplied by two onboard generators.

The NavLab II has a central integrated motion controller; its main function is to coordinate control of the vehicles motion control axes to effectively control the vehicle as a whole. The NavLab II has three independent control axes: the throttle, the brake and the steering wheel. The controller determines the vehicle's state using a combination of encoders (mounted on the transmission output and the engine) and an inertial navigation system. Real time software, running on two processors, coordinates the control of the independent axes, keeps track of the vehicle's state, and handles motion commands (steering and speed) from the navigation computers.

In order to traverse unknown terrain, the navigation software must first sense the terrain in front of the vehicle, plan a trajectory across this terrain map, and then drive along the path. This process cycles as the vehicle drives.

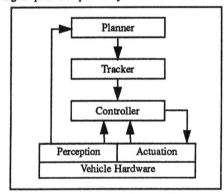

**Fig. 1. Software/hardware system interaction**

The sensing consists of taking a range image from the ERIM, and transforming the data into a 2.5D discretized terrain map. This operation requires approximately 0.5 seconds.

Path planning is comprised of a spatial planning phase, where a path is generated across the terrain map through heuristic modification to avoid obstacles, and a temporal planning phase, where each point along the path is assigned a speed by considering vehicle dynamics. The separation of dynamics and kinematics makes the planning problem tractable. Computation time for planning is about 0.5 seconds.

The planned path is passed to the path tracker, which

tracks the path by issuing commands to the vehicle controller. These components are illustrated in the software system design (Fig. 1). Previous work by Amidi (1990) thoroughly addresses issues involving the tracker and controller.

## PERCEPTION SUBSYSTEM

The goal of the perception system for the autonomous navigator is to generate a description of the terrain geometry that is convenient for the path planning subsystem to use, and will accommodate both higher vehicle speeds and rougher terrain than has been achieved in previous systems.

Fig. 2 presents a conceptual view of the software architecture of the perception system:

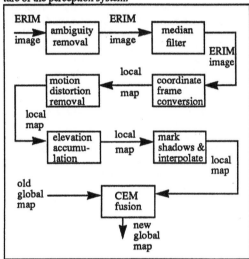

**Fig. 2. Perception software architecture**

The ambiguity removal and median filter modules remove noise and inaccuracies from the input image. Coordinate frame conversion converts the range information into elevation information. While this is being performed, the motion distortion removal module accounts for vehicle motion to correct the data. The elevation accumulation and shadow marking modules remove holes in the map and compute variation statistics when information overlaps. The Cartesian elevation map (CEM) fusion module is responsible for adding new elevation data to the global map, maintaining a scrolling window into the map which follows the vehicle, and estimating the vehicle's z coordinate by analyzing overlapping map data.

### Sensor Limitations

The data generated by the ERIM sensor is subject to limitations of ambiguity of the range measurement beyond the laser modulation wavelength, complete lack of information when the beam is reflected off a specular region of the terrain, and degraded accuracy as range is increased or in regions of high texture.

A special edge detector is used to find the ambiguity edge near the top of the range image and all information beyond it is ignored. A median filter is used to remove the outliers associated with regions of high terrain texture. No special treatment of specular regions (e.g. water puddles) was required since they are treated automatically as range shadows by the rest of the system.

## Vehicle Motion

Clearly, the transform of coordinates from the sensor frame to the world frame requires knowledge of the vehicle pose when each range pixel was measured. The sensor scanning mechanism requires about 1/2 second to complete a single scan. Hence, at higher vehicle speeds, the use of a single vehicle pose for the entire image will give rise to distortions in the map.

The motion distortion removal function uses a series of vehicle poses which were stored by a synchronous pose logger during image digitization. Currently, 8 poses are associated with each image, and, for each pixel, the function uses the pose that was measured closest in time to the instant when the pixel was measured. In this way, an obstacle that would have otherwise been elongated by 2 meters at current speeds is represented accurately in the map.

## Terrain Map Generation

The perception system generates a uniformly sampled grid data structure called a cartesian elevation map (Olin, 1991) which stores in each grid cell the elevation of the corresponding point in the environment. The generation of this data structure requires more than the trivial transformation of coordinates from the sensor frame to the world frame. Since one regularly sampled structure (the image) is converted into another (the map), some nontrivial issues rooted in the sampling theorem must be addressed.

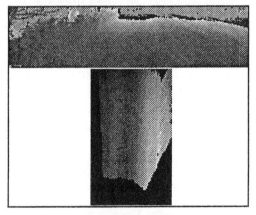

Fig. 3. Typical range image and terrain map

The terrain map generation problem is the inverse of the hidden surface removal problem of computer graphics. The sensor intrinsically removes hidden surfaces since the laser beam is reflected from the closest surface along its trajectory but cannot penetrate to be reflected from any surfaces beyond it. Hence, we must accept that some regions in the elevation map will contain no information. Figure 3 presents a single range image and the portion of the terrain map that results from it, as seen from above.

Clearly, this problem is caused by the sensor geometry. With the sensor attached to the vehicle, the ideal overhead viewpoint is not possible, yet the planner needs to see behind obstacles in order to plan optimally. The sensor configuration also causes regions to be oversampled close to the vehicle and undersampled far away. To some degree, the problem is overcome by fusing the results of several range images from different vantage points into a single map.

The elevation accumulation function computes the average and standard deviation of elevations encountered for oversampled grid points. Range shadows and undersampled regions are filled using a special interpolation algorithm that marks usually large unknown regions as range shadows and records an upper bound on elevation for the grids cells contained in them.

## PLANNING SUBSYSTEM

Path planning is defined as finding a path through space from a start point to a goal point while avoiding obstacles. Much of the previous work in this area (Lozano-Perez, 1979), addresses path generation for a polygonal body through a space containing polygonal obstacles. This approach assumes that all obstacles in the traversable space are known. This approach, and other approaches (Thompson, 1987; Feng, 1989) do not satisfy the requirements for ACCN, for the following reasons:

- Polygonal obstacle assumption is not appropriate for rough terrain, since the motion of an obstacle can include any pose (such a steep terrain) that is unsafe. This set of poses can be computationally prohibitive to precompute in it's entirety and may not assume a polygonal shape.

- A limited sensor horizon precludes knowledge of entire search space before planning

- The intrinsic limitations of the vehicle's ability to follow the given path are not considered (e.g. minimum turning radius.)

Work by Olin (1991) and Daily (1988), relaxes the flat world, polygonal obstacle assumption and introduces the notion of simulated traversal of paths, taking into consideration kinematic constraints at discrete points along these paths. Their system, however, limits the space of potential paths by only considering a static set of paths across the sensed area. This methodology does not allow small deviations around obstacles, and prevents more than a single change of direction within a planning cycle. This planner might find a region uncrossable simply because its static space of predetermined paths is insufficient to include a safe path. Furthermore, this system does not consider dynamics, as it operated in a speed region where dynamic effects can be neglected.

The trajectory planning software on the Navlab II considers both kinematic and dynamic constraints on the vehicle, plans paths through tight spaces, copes with a changing sensor horizon, and accounts for the intrinsic ability of the vehicle to track a given path.

The system first performs spatial searching based on kinematic constraints and then applies dynamic constraints to set the temporal portion of the path, that is, the speed at each spatial point along the planned path.

## Planning Configuration

The goal of the planner is to extend the known safe path

for the vehicle by planning a path across recently sensed terrain. Fig. 4 illustrates the geometric constraints on the planning problem.

The invariant in the planner is that the *next path* is planned while the *current path* is traversed. Each next path corresponds to a path across a map generated from an image taken at the first point on the current path.

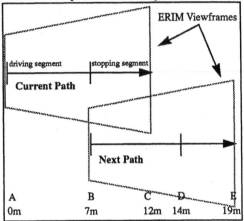

**Fig. 4. Geometry for planning**

Each path has a *driving segment* and a *stopping segment*. Consider the current path: at the beginning of the driving segment, the vehicle takes an image; as the vehicle drives along this segment, the system plans the next path; if an impasse is reached, and planning fails, the vehicle will drive along the stopping segment, and come to a stop, never having left known terrain. In Fig. 4, an image is taken at A, and path BE is planned while driving AB. If planning fails, BC is driven. Under normal operation, no stopping segment is ever driven, and the vehicle is in continuous motion.

This algorithm provides several guarantees. First, the vehicle can operate in an uncertain environment since the planned path can change radically when unknown obstacles are introduced into the terrain map. Secondly, the system ensures vehicle safety in case of a planning failure, which is essential for high speed driving in rough terrain.

Spatial Trajectory Generation

In the first stage of planning, a generate and test paradigm is used to choose and evaluate successive spatial paths until a safe path which traverses the known terrain is found. This method avoids the problem of preprocessing the terrain to find all unsafe regions. The generate and test method also permits the planner to compute any path needed without limiting the search to a precomputed set.

The planner begins initially with a *global path* describing a course route across a long stretch of terrain. During every sense-plan-drive cycle, the planner tries to move the vehicle along the global path, by choosing *control points*. Control points are points which define the ends of the local path, or which may lead the vehicle around an obstacle. A set of control points which will be used to generate a local path is called a *control path*. Initially, each planning cycle for the next path begins with a control path having two control points: a point from the current path for continuity,

and a point a fixed distance ahead to move the vehicle toward the global path.

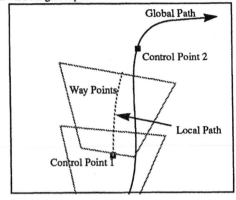

**Fig. 5. Global, control, and way points**

When a control path is selected for testing, a smooth spatial curve is fit to the control points, and closely-spaced *way points* are generated along the entire length of the curve. Fig. 5 illustrates these concepts. In this figure, Control Point 1 and Control Point 2 comprise the first control path.

The vehicle traverses the way points in simulation. When a way point is tested and found to be unsafe, up to two new control paths are generated. Each new control path will contain an additional control point found by the planner that is intended to modify the path so as to direct the vehicle around either side of the obstacle encountered. These new control paths are added to a list of control paths. The planner then selects a new control path from this list based on the estimated cost (as a function of path length and complexity) of traversing the path. This process repeats until a safe path is found. In theory, this algorithm should be capable of generating traversable trajectories through arbitrarily narrow admissible regions.

In the sections below, the test used to determine safety as well as the algorithm for identifying control points are described.

Kinematic admissibility. Kinematic admissability implies that the geometry of the path during vehicular motion will permit traversal. Traversal can be impeded by an obstacle, such as a tree or a rock, or by vehicle geometry, such as a minimum turning radius. Kinematic admissability may be defined as follows:

*A way point is kinematically admissable if there is no geometric cause to impede the proposed motion of the body and if the vehicle and its environment do not occupy common space.*

This definition is embodied in a set of four constraints, each of which is evaluated at a given way point to determine kinematic admissibility. A waypoint is admissible only if it satisfies all four constraints:

- Minimum turning radius; the curvature of the path at the way point cannot exceed bounds given by the vehicle's minimum turning radius.

- Locomotion support; the motion of the vehicle

must not be geometrically impeded by the presence of insurmountable obstacles in front of the wheels.

- Body collision; the vehicle cannot occupy the same space as its surrounding terrain.

- Unknown terrain; the vehicle must not be situated over unknown terrain, (off the map)

Vehicle statics are included with the of a kinematic admissability test. It is intuitive and straightforward to define static admissability as follows.

*A way point is Statically Admissable if the vehicle is in a state of static equilibrium at the point.*

Thus, the planner is conservative in that the vehicle is required to be statically stable at every point along the path, even though there may be some executable paths that do not meet this requirement.

Control point selection. If an inadmissable way point is found while traversing a local path in simulation, it can be assumed that an inadmissable region is located near the front of the vehicle at this way point. This inadmissable point is added as a candidate control point to the list of control points (the control path) used to generate the trajectory.

Once the inadmissable point is found, it is moved around the edge of the obstacle in successive sideways steps, on each step trying to proceed toward the next control point. When a step straight forward is found to be is admissable, it is assumed that the edge of the obstacle has been found. (See Fig. 6.)

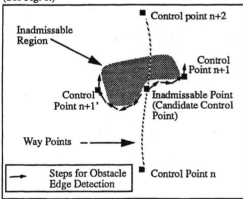

**Fig. 6. Example selection of new control points**

This procedure can be applied recursively between any two control points when an inadmissable way point is found. In this example (Fig. 7), two possible paths are found. In this way, an inadmissable point is moved around the obstacle, and becomes a control point which indicates the probable edge of an inadmissable region. Because the control point searching is done by continuously moving toward the edges of the known terrain during each search step, termination is guaranteed, as the point cannot be moved past the edge of the known map.

## Spatial Search Failure

There are two types of search failure. First, a traversable

path may not be calculated before the vehicle enters the stopping segment of the current path. In this case, the vehicle stops, drives backward to the last point in the intended region, and then drives over the next path. Second, the planner may not be able to find a path through the region, possibly due to an impasse. In this case, the current system will terminate the run after the vehicle comes to a stop on the current path.

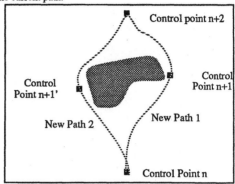

**Fig. 7. New paths with obstacle avoided**

## Temporal Path Generation

Once a spatial path is planned, it is necessary to specify vehicle velocities at each way point along that path. The high speed environment in which the vehicle operates necessitates the use of dynamic constraints on the vehicle's trajectory.

The velocity profile generator must calculate a set of maximum velocities along this path that ensures that the vehicle will not exceed dynamic limits, thus making the path dynamically admissable to execution. We define dynamic admissability in terms of a set of orthogonal accelerations called performance limits. These acceleration correspond to the vehicle's lateral, parallel and vertical directions.

A path is considered to be *dynamically admissable* if the accelerations generated during motion do not exceed the vehicular dynamic performance limits.

A *performance limit* describes a maximum permissible vehicle acceleration; it does not necessarily describe a vehicle's stability limit, but it is never greater. Any velocity profile that, at every way point, is exceeded by this generated profile is considered to be dynamically admissable.

Dynamic constraints are applied to determine the maximum permissible vehicle velocities along every point on the path. Given that the vehicle must be stopped at the last point on the stopping segment, the maximum permissible speed for the previous point can be calculated. Continuing this method of reverse state propagation gives maximum speeds to all points on the path. Permissible is described by the performance limits of the vehicle.

## RESULTS

Tests of the vehicle were performed in a large open area, characterized by gently rolling terrain, sparse vegetation, rock outcroppings, and water puddles. The system was implemented with a pure pursuit (Amidi, 1990) path tracker.

Obstacle avoidance was successful at speeds up to 2.0 m/s. Obstacles were detected, and paths planned at speeds in excess of 4.25 m/s. However, the ability of the vehicle's tracker and controller to accurately follow these paths prevented the vehicle from actually driving around obstacles while maintaining this speed. By all indications, an improved controller will enable us to achieve our theoretical maximum speed. Fig. 8 shows a map, vehicle position, and a path found around a clearly inadmissable region.

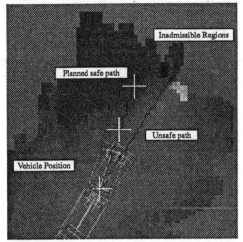

**Fig. 8. Example of rough terrain obstacle avoidance**

The rapid stop command is not yet implemented, thus testing of vehicle stop in the event of a planning impasse was precluded.

Most natural obstacles found in the testing environment required no more than 4 control points to define a path around them. A total of 4.5km of rough terrain was traversed during the latest series of test runs.

## CONCLUSIONS

A system for Autonomous Cross-Country Navigation has been developed and demonstrated successfully at speeds up to 4.25 m/s on moderate terrain. This perception and planning system are the first to support these speeds on natural barren terrain, to integrate a generate and test planning paradigm with simulated traversal of paths for ACCN, to include vehicle dynamics as part of the simulation process, and to account for the increased difficulty of perception at high speed on natural terrain.

The major difficulties encountered were limited sensor horizon and the limited response time of the vehicle. A longer sensor horizon would permit more time for planing, and would allow the system to make direction changes around obstacles more quickly. Better vehicle response would allow the system to successfully follow these planned paths. It should be noted that the vehicle is physically capable of following the planned paths, but that, in some cases, the controller lacks the needed response characteristics.

Future work will allow the system to travel along more tightly constrained paths at higher speeds. Sensor upgrades will permit the longer sensor horizon needed to achieve these goals.

Other future work will include feedback for active sensor pointing, planning for cluttered environments and more complete models of the vehicle's capabilities to permit more robust navigation

## ACKNOWLEDGEMENTS

As well as the authors, the software development team has included Omead Amidi, Mike Blackwell, William Burky, Martial Hebert, Jay Gowdy, Annibal Ollero, Dong Hun Shin and Jay West. The authors thank Dr. Behnam Motazed and the NavLab II design and retrofit team for producing a capable ACCN testbed.

This research was sponsored by the Defense Advance Research Projects Agency, through ARPA Order 7557 (Robot System Testing), and monitored by the Tank Automotive Command under contract DAAE07-90-C-R059.

## REFERENCES

1. Amidi, O. Integrated mobile robot control. Robotics Institute Technical Report CMU-RI-TR-90-17, Carnegie Mellon University, Pittsburgh.

2 Bhatt, R; Venetsky, L., Gaw, D., Lowing, and D., Meystel, A. (1987). A real-time pilot for an autonomous robot. Proc. of IEEE Conf. on Intelligent Control. 135-139.

3 Daily, M. (1988). Autonomous cross country navigation with the ALV. Proc. IEEE Internat. Conf. on Robotics & Automation, 4, 718-726.

4. Dickmanns, E.D. (1990). Dynamic computer vision for mobile robot control. Proc. of the 19th Internat. Symposium and Exposition on Robots, 314-27.

5. Feng, D. (1989). "Satisficing feedback strategies for local navigation of autonomous mobile robots." Doctoral Thesis, Electrical and Computer Eng. Dept., Carnegie Mellon University, Pittsburgh.

6. Lozano-Perez, T. and Wesley, M. A. (1979). An algorithm for planning collision free paths among polyhedral obstacles. Communications of the ACM, 22:10, 560-570.

7. Moravec, H. P. (1983). The Stanford cart and the CMU Rover. Proc. of the IEEE, 71, 872-884.

8. Olin, K. E. and Tseng, D. (1991). Autonomous cross country navigation. IEEE Expert, 8, 16-30.

9. Pomerleau, D.A. (1991). Efficient training of artificial neural networks for autonomous navigation. Neural Computation, 3:1, 88-97.

10. Chang, T.S., Qui, K., and Nitao, J. J. (1986). An obstacle avoidance algorithm for an autonomous land vehicle. SPIE Conf. Proc, 727, 117-123.

11. Shin, D. H., Singh, S., and Shi, W. (1991) A partitioned control scheme for mobile robot path planning. Proc. IEEE Conf. on Systems Eng.

12. Thompson, A. M. (1987) The navigation system of the JPL robot, IJCAI, 749-757.

13. Wilcox, B. and others. (1987). A vision system for a Mars rover. SPIE Mobile Robots II, 172-179.

# PATH GENERATION FOR A ROBOT VEHICLE USING COMPOSITE CLOTHOID SEGMENTS

### D.H. Shin, S. Singh and W. Whittaker

*Field Robotics Center, Carnegie Mellon University, Pittsburgh, USA*

**Abstract.** We present a method to generate fine grained, smooth paths for wheeled mobile robots. This method takes as input, a sequence of objective points that the robot must attain to avoid obstacles and to progress toward its goal. Segments of clothoid curves are used to join the objective points and hence provide a finer level of details for the robot controller. Previous path generation methods have sought to simplify a path by using arcs, superarcs, polynomial curves, and clothoid curves to round corners, which result from poly-line fits through a given sequence of points. The proposed method generates smooth and continuous curved paths directly from a given sequence of points without adhering to any straight line segments. By virtue of the property of clothoid curves, a generated path is continuous with respect to position, tangent direction, and curvature, and is linear in curvature. Aside from the properties innate to clothoid curves, the generated paths transition smoothly into turns, pass through all the way points, and sweep outside the corners.

**Keywords.** Robot vehicles; path generation; clothoid curve; linear curvature.

## INTRODUCTION

Mobile robot navigation is typically composed of three phases- perception, planning, and, control. A robot must perceive its state with respect to the world and use this information to plan its motion. Once a plan has been composed, it is necessary to enact the plan with requisite control actions. Typically, path planning occurs at two levels. First, path planning uses the information about the world to obtain an ordered sequence of objective points that the robot must attain to avoid obstacles and to progress towards its goal. These objective points constitute a coarse plan and a second step is necessary to generate a path in finer detail for the purposes of robot control.

This paper reports a method of path generation to develop a fine grained specification of a path from a sequence of objective points that immediately suggests a control law for the mobile robot. This problem has been studied by other researchers (Hongo, 1985; Kanayama, 1985; Nelson, 1989). Our work is an extension of work reported by Kanayama and accrues several advantages over the other reported schemes.

In the following section we first discuss common kinematic models for mobile robots. We have used a "bicycle" model to represent a robot vehicle. The advantage of such a model is that there exist simple geometric relationships between the curvature of the reference path and steering angle for the robot. Given such a model, we suggest criteria for "goodness" of candidate path segments that may be used to join the objective points. As with any control system, the response of a mobile robot in tracking a path is partly dependant on the nature of the reference path. We show that if the type of path generated is intrinsically easier for the robot to track, a prototypical robot following such a path experiences fewer errors in following it. We suggest that clothoid curves known to satisfy the criteria of goodness, be used to join the objective points. In Section 3 we present a method to join two arbitrary (within bounds) objective points that are each uniquely specified by position, orientation and instantaneous path curvature. In Section 4, we evaluate the performance of the presented method. Finally, Section 5 advances the conclusions and suggests the directions for future research.

## CONTINUOUS PATHS

### Guide Point

For our analysis a bicycle model is used as an archetype for modeling robot vehicles with two degrees of freedom, steering, and propulsion. The guide point is the point of the vehicle that is controlled to follow the given path. The choice of the guide point is an important decision—it affects the desired steering and propulsion functions required to follow the given path and speed. We have chosen the guide point to be at the midpoint of the rear axle (Fig. 1) resulting in the following advantages:

- The steering angle at any point on the path is determined geometrically, independent of speed, in the following manner:

$$\tan\phi = \frac{l}{r} \qquad ; \phi = \tan^{-1}cl \qquad (1)$$

where $l$ is the wheelbase of the vehicle, $r$ is the radius of the path, and $c$ is the curvature of the path. Also, the angular velocity of the driving rear wheel is determined only by the vehicle speed ($v$) and the radius of the wheel ($R_w$):

$$\omega = \frac{v}{R_w} \qquad (2)$$

If the guide point is placed elsewhere, expressions for $\phi$ and $\omega$ are more complex than shown in Eq. (1) and Eq. (2). The reference steering angle and angular velocity of the driving wheel must then be obtained by numerical integration.

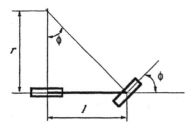

Fig. 1. Geometry of a bicycle model

- The vehicle is able to follow the minimum turning radius for the maximum steering angle (Nelson, 1989). In other words, the peak steering angle is smaller than that for any other guide point.

- The heading of the vehicle is aligned with the tangent direction of the path. This gives a more reasonable vantage point for a vision camera or a range scanner mounted at the front of the vehicle, as in Fig. 2, and a smaller area is swept by the vehicle.

Fig. 2. The heading alignment along a path

For the guide point at the center of the rear axle, paths with discontinuities of curvature will require infinite acceleration of the steering wheel. Consider the motion of the vehicle along a path which consists of two circular arcs. There is a discontinuity in curvature at the point where the two circular arcs meet. An infinite acceleration in steering is required for the vehicle to stay on the specified path provided that it does not come to a stop at the transition point, simply because it takes a finite amount of time to switch to the new curvature. In reality, moving through a transition point with non-zero velocity results in an offset error along the desired path. Likewise, rotary body inertia requires continuity of heading, and steering inertia requires continuity of curvature. This is not the case if the guide point is moved to the front wheel. However, this choice is at the cost of the advantages discussed above.

Selection of the guide point is important to the formulation of the path tracking problem. When the guide point is chosen on the center of the rear axle, steering and driving reference inputs are computed from the path parameters and vehicle speed, respectively, as in Eq. (1) and (2). This enables the planning and controlling of steering along a path to be independent of speed. It is referred to as *path* tracking, in contrast to trajectory tracking in which steering and speed are controlled along a time history of position (Shin, 1990).

### Vehicle Path Modeling

The configuration of a conventionally steered vehicle moving on a plane surface can be described completely by a set of coordinates $(x, y, \theta_V, \phi)$; vehicle position, heading, and steering angle, respectively. Continuity of $(x, y, \theta_V, \phi)$ is recommended, because the vehicle has inertia and finite control response, which preclude discontinuous motions. Otherwise, tracking errors will be greater. Since a path specifies an ideal motion of a conventionally steered vehicle, a path must guarantee the continuity of $(x, y, \theta_V, \phi)$ of a vehicle.

A path can be parameterized in terms of path length $s$ as $(x(s),$

$y(s))$. Tangent direction $(\theta(s))$ and curvature $(c(s))$ can be derived along a path:

$$\theta(s) = \frac{dy(s)}{dx(s)} \tag{3}$$

$$c(s) = \frac{d\theta(s)}{ds} \tag{4}$$

The continuity of $(x, y)$ is guaranteed if a path is continuous. But the effect of the continuity of vehicle heading and steering on the path depends on the guide point. Since the guide point is chosen at the center of the rear axle, the heading of the vehicle $\theta_V$ is aligned with the tangent direction of the path $\theta$, and the steering angle $\phi$ is determined by the curvature of the path, as in Eq. (1). Thus, the continuity of $(x, y, \theta_V, \phi)$ is tantamount to the continuity of $(x, y, \theta, c)$. If we define *posture* as the quadruple of parameters $(x, y, \theta, c)$, a posture describes the state of a conventionally steered vehicle, and a path is required to be posture-continuous for easy tracking.

Further, the rate of change of curvature (sharpness) of the path is important, too, since the linearity of curvature of the path dictates the linearity of steering motion along the path. If we assume that less control effort is required for an actuator to provide a linear velocity profile than an arbitrary nonlinear one, the extent to which steering motions are likely to keep a vehicle on a desired path can be correlated to the linearity of curvature of the path.

In the following subsection, clothoid curves are introduced and, are compared the performances of a vehicle tracking a clothoid path and an arc path.

### Clothoid Curves

Clothoid curves (Kanayama, 1985; Yates, 1952) are a family of curves that are posture-continuous, and are distinct in that their curvature varies linearly with the length of the curve:

$$c(s) = ks + c_i \tag{5}$$

where $k$ is the rate of change of curvature (sharpness) of the curve and $c_i$ is the initial curvature and $s$ is the curve length. Given an initial posture, sharpness, and the distance along the curve, the curve can be described by Eq. (5), (6), (7) and (8):

$$\theta(s) = \theta_i + \int_0^s c(\zeta)d\zeta = \frac{k}{2}s^2 + c_is + \theta_i \tag{6}$$

$$x(s) = x_i + \int_0^s \cos\left(\frac{k}{2}\zeta^2 + c_i\zeta + \theta_i\right)d\zeta \tag{7}$$

$$y(s) = y_i + \int_0^s \sin\left(\frac{k}{2}\zeta^2 + c_i\zeta + \theta_i\right)d\zeta \tag{8}$$

The response of a robot vehicle to different reference paths was compared. Open loop responses of a vehicle following a path composed of arcs and straight lines, and another composed of clothoid segments, are shown in Fig. 3. The vehicle is made to follow a set of steering commands, which are obtained from the path specification using Eq. (1), without trying to compensate for tracking errors. The dynamic responses of the vehicle are modeled as first order lag systems with hard limits on acceleration. In each case, the reference path is indicated by dashed lines, while the actual path is indicated by solid lines. Simulation results show that an arc path results in larger steady state tracking errors caused by discontinuities in curvature and support the claim that "good" paths are intrinsically easier to track.

### PATH GENERATION

This section presents a method to generate paths that are intrinsically easy to track. The problem is to produce a unique, easily trackable, continuous path from a given sequence of ob-

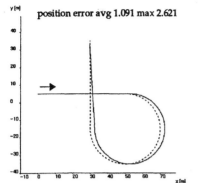

position error avg 1.091 max 2.621

(A) Along a path composed of arcs and straight lines

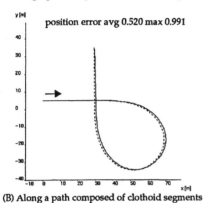

position error avg 0.520 max 0.991

(B) Along a path composed of clothoid segments

Fig. 3. Open-loop performance: time constant = 0.1 s, vehicle speed = 5 m/s

jective points. This problem is similar to those addressed by other researchers (Hongo, 1985; Kanayama, 1985; Nelson, 1989).

<u>Existing Path Generation Methods</u>

Hongo (1985) proposed a method to generate continuous paths composed of connected straight lines and circular arcs from a sequence of objective points. While paths composed solely of arcs and straight lines are easy to compute, such a scheme leaves curvature discontinuities at the transitions of the segments, as discussed previously.

Certain polynomial spline curves are candidates for path segments, because they guarantee continuity of posture. Nelson (1989) proposed quintic spline and polar spline curves obtained from two-point boundary conditions. However, these spline curves do not guarantee linear gradients of curvature. clothoid curves, by contrast, do vary linearly with the distance along the curve.

Kanayama (1985) proposed clothoid curves for transitions between straight line paths, where symmetric pairs of clothoid curves with zero curvature are used to round the corners of straight line junctions.

The above mentioned approaches have in general sought to simplify path representation by rounding corners of poly-line fits through a given sequence of points. However, in this paper, smooth and continuously curved paths are directly generated from a given sequence of objective points without adhering to any straight line segments.

<u>Path Generation from a Sequence of Points</u>

The following two-step method generates a unique posture-continuous path from a sequence of points. The first step is to derive a sequence of unique postures from the objective points; the second, is to interpolate between those postures with clothoid segments. Heading and curvature at the starting and ending positions are presumed from the configuration of the vehicle.

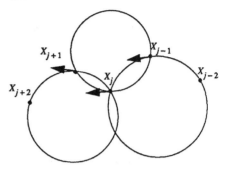

Fig. 4. Postures and associated circles from objective points

Let $[X = (X_0, ..., X_n) = \{ (x_0, y_0), ..., (x_n, y_n) \} ]$ be a sequence of objective points. The *associated circle* at $X_j$ is defined to be the circle which passes through points $X_{j-1}, X_j, X_{j+1}$ as in Fig. 4. Then the heading of the vehicle at $X_j$ is taken as the direction of the tangent to the associated circle at $X_j$, and the curvature is the reciprocal of the osculating radius of the associated circle, denoting the posture thus obtained as the *associated postures*. The next step is to connect neighboring associated postures with clothoid segments.

It is not always possible to connect two neighboring postures with one clothoid curve segment, because the four governing equations (5), (6), (7), and (8) cannot be satisfied simultaneously with only the two parameters (sharpness $k$ and length $s$) that a clothoid curve provides. To satisfy these four equations, at least two clothoid segments are needed. However, the general problem cannot be solved with only two clothoid segments. Fig. 5 shows two pairs of associated postures and their associated circles. Let $P_i$, $P_f$ denote the starting and the ending postures, respectively. $C_i$ and $C_f$ denote the circles corresponding to the curvatures at $P_i$, $P_f$. They are drawn by solid lines and dotted lines, respectively.

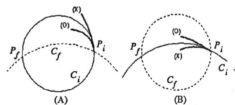

Fig. 5. Determination of the sign of the first clothoid segment

As in (A) of Fig. 5, if the orientation of $P_f$ is outward from $C_i$, the ending part of a solution curve should be inside $C_i$. Then, it is plausible that the starting part of a solution curve also lies inside $C_i$. Similarly, if the direction of the $P_f$ is inward into $C_i$, as in (B) of Fig. 5, it is plausible that the starting part of a solution curve lies outside $C_i$. Note that the sign of the sharpness determines the side of the associated circle on which the clothoid segment lies: (1) If the sharpness is zero, then the clothoid curve remains on the associated circle in question. (2) If the sharpness is positive, then the clothoid curve will be on the left side of the associated circle. (3) Otherwise, it will be on the right side of the associated circle. Hence, a solution curve should satisfy the following proposition:

If the direction of $P_f$ is outward from $C_i$, the sign of $k$ of the first clothoid segment must be chosen so that the curve lies inside $C_i$.

If the direction of $P_f$ is inward into $C_i$, the sign of $k$ of the first clothoid segment must be chosen so that the curve lies outside $C_i$

Otherwise, $k$ of the first clothoid segment is chosen so that the curve remains on $C_i$.[1]

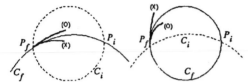

Fig. 6. Determination of the sign of the last clothoid segment

As a corollary to the above proposition, the sign of $k$ of the last clothoid segment can be determined (Fig. 6):

If the direction of $P_i$ is outward from $C_f$, the sign of $k$ of the last clothoid segment must be chosen so that the curve lies outside $C_f$.

If the direction of $P_i$ is inward into $C_f$, the sign of $k$ of the last clothoid segment must be chosen so that the curve lies inside $C_f$.

Otherwise, $k$ of the last clothoid segment must be chosen so that the curve lies on $C_f$.

Fig. 7 shows all possible cases of curvature variations between a pair of neighboring postures. Notice that the signs of $k$ for the first and the last segments are the same for each case. However, the sign is the opposite of that required for the curvature variation between the postures for all the cases except (C) and (D) of Fig. 7. Thus, the general problem to connect a pair of neighboring associated postures cannot be solved with two clothoid curve segments.

One adequate solution set of the clothoids is the set of three clothoid segments $(k, s_1)$, $(-k, s_2)$, $(k, s_3)$. The subscripts denote the order of the clothoid segments from $P_i$. This combination is plausible for the following reasons:

1. The signs of sharpness for the first and last clothoid segments are the same.

2. The sharpness for the second clothoid segment is equal in magnitude and opposite in sign to the first and last segments. This enables the curve of three clothoid segments to satisfy the curvature variation between the starting and the ending postures by varying $s_1, s_2, s_3$, even though the sign of the first and the last clothoid segments satisfies the curve location requirement.

3. There are four variables in the combination: $k, s_1, s_2, s_3$. It is possible to find a unique solution satisfying the following four equations which describe the mathematical relationship between the starting and ending postures:

$$c_f = c_i + k(s_1 - s_2 + s_3) \qquad (9)$$

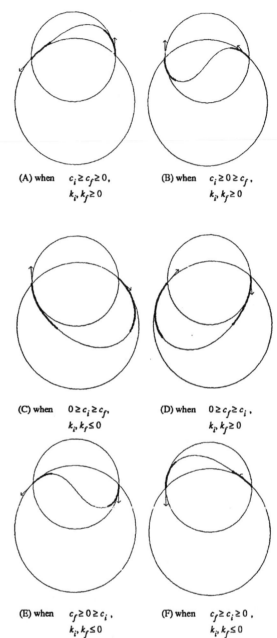

(A) when $\quad c_i \geq c_f \geq 0,$
$\qquad k_i, k_f \geq 0$

(B) when $\quad c_i \geq 0 \geq c_f,$
$\qquad k_i, k_f \geq 0$

(C) when $\quad 0 \geq c_i \geq c_f,$
$\qquad k_i, k_f \leq 0$

(D) when $\quad 0 \geq c_f \geq c_i,$
$\qquad k_i, k_f \geq 0$

(E) when $\quad c_f \geq 0 \geq c_i,$
$\qquad k_i, k_f \leq 0$

(F) when $\quad c_f \geq c_i \geq 0,$
$\qquad k_i, k_f \leq 0$

Fig. 7. Cases by curvature variation of $P_i$ and $P_f$

$$\theta_f = \theta_i + c_i(s_1 + s_2 + s_3) + k(s_1 s_2 - s_2 s_3 + s_3 s_1)$$
$$+ \frac{k}{2}(s_1^2 - s_2^2 + s_3^2) \qquad (10)$$

$$x_f = x_i + \int_0^{s_1} \cos\theta_1(\zeta)\, d\zeta + \int_{s_1}^{s_2} \cos\theta_2(\zeta)\, d\zeta$$
$$+ \int_{s_2}^{s_3} \cos\theta_3(\zeta)\, d\zeta \qquad (11)$$

---

[1]. In this case, $P_i$ and $P_f$ share the same associated circle, two postures are connected with the part of their associated circle, which is a clothoid curve of zero sharpness.

446

$$y_f = y_i + \int_0^{s_1} \sin \theta_1 (\zeta)\, d\zeta + \int_{s_1}^{s_2} \sin \theta_2 (\zeta)\, d\zeta$$
$$+ \int_{s_2}^{s_3} \sin \theta_3 (\zeta)\, d\zeta \tag{12}$$

where,

$$\theta_1 (\zeta) = \theta_i + c_i \zeta + \frac{k}{2} \zeta^2$$

$$\theta_2 (\zeta) = \theta_i + c_i s_1 + \frac{k}{2} s_1^2 + (c_i + k s_1) \zeta - \frac{k}{2} \zeta^2$$

$$\theta_3 (\zeta) = \theta_i + c_i (s_1 + s_2) + k (s_1 s_2) + \frac{k}{2} (s_1^2 - s_2^2)$$
$$+ (c_i + k (s_1 - s_2)) \zeta + \frac{k}{2} \zeta^2$$

Since Equations (11) and (12) contain Fresnel integrals, for which there is no closed form solution, the values of $k, s_1, s_2, s_3$ are computed using the numerical method (Press, 1986) outlined in Fig. 8. Initial values for $s_1$ and $s_2$ are chosen to be $1/3$ of the average of the lengths of two of the arcs that connect $P_i$ and $P_f$. Eq. (9) and (10) are used to compute $k$ and $s$. Then, $x_c, y_c$ can be computed for the quadruple $(k, s_1, s_2, s_3)$ using Simpson's approximation. Ideally $(x_c, y_c = x_f, y_f)$ and in fact values of $s_1$ and $s_2$ are adjusted until the difference is within a threshold.

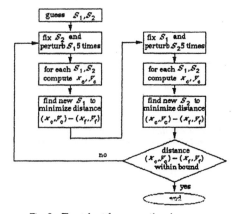

Fig. 8. Flow chart for computing $k, s_1, s_2, s_3$

### RESULTS

Connecting a pair of neighboring associated postures was accomplished successfully, as in Fig. 7, using three clothoid curves. Fig. 9 shows the graphic result of a posture-continuous path generated by clothoid curves through the given seven points. As the first step of the proposed method, a sequence of seven postures were generated. (Seven arrows in Fig. 9). Then, three clothoid segments are used to interpolate between neighboring postures.

Fig. 10 shows a comparison of the performance of the proposed method using redshank Kanayama's method (Kanayama, 1985). Parameters of curvature and sharpness were constrained equally for all methods. The maximum sharpness of Kanayama's method and the maximum curvature used in the arc method are set at the same levels as in the proposed method. Paths, curvatures and sharpness along the paths are compared. Paths resulting from the proposed method have the following advantages over other methods:

• The method proceeds from an arbitrary sequence of points. Generation of postures is essential to exploratory planning

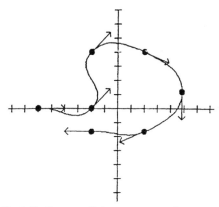

Fig. 9. Continuous path from a sequence of seven points

where goals are commonly posed as an evolving string of points. Paths generated by the proposed method pass through all the objective points, as in Fig. 10.

• The method guarantees continuity of position, heading, and curvature along the path. Further, sharpness is piecewise constant.

• Paths generated by the method always sweep outside the acute angles formed by straight line connection of the way points, as in Fig. 10. The resulting paths are especially useful for interpolating around obstacles that are commonly on the inside of angles. In contrast, the paths by other methods pass always inside the corners.

### CONCLUSION

We have presented two conditions for paths that are intrinsically easy to track: posture continuity and linearity of curvature along the path. Clothoid curves are good candidates to satisfy these conditions. Simulations show better performance when a vehicle tracks paths of clothoids versus when it tracks paths composed of arcs.

A method for generating a continuous path was developed. This method uses clothoid segments and consists of two steps: First, a sequence of the postures is obtained using the objective points. Then, each pair of neighboring postures is connected with three clothoid curve segments.

The method provides additional advantages in that preprocessing of the objective points is not necessary, as with arc and zero curvature clothoids. Further, the geometry of the paths generated always sweeps outside the acute angles formed by a straight-line connection of the way points. These are especially useful for interpolating around obstacles that are commonly on the inside of angles.

The method of obtaining postures, as in Fig. 4, requires that the circles formed by the radii of curvature of two postures intersect. Relaxing such a constraint would require heuristics to determine intermediate postures. Assuming intermediate postures could be found, such that the associated circles intersect, the method could then be used on the new set of postures. Thus far, the search for a completely general method that would generate a path between two completely arbitrary postures has not been fruitful.

Directions for future research include the following: (1) optimization of the lengths of more than three clothoid segments on the premise that some cost function can be used to find a

447

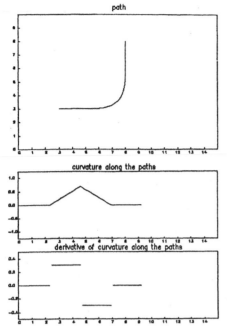

Fig. 10 - (A). Clothoid path with zero curvature transition

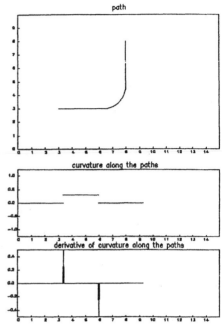

Fig. 10 - (B). Path with circular arcs and straight lines

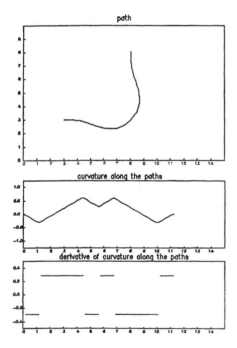

Fig. 10 - (C). Clothoid path with the proposed method

better solution than the results here; (2) improvement of the numerical method to connect postures through the Fresnel Integral to improve speed and accuracy.

REFERENCES

Hongo, T., H. Arakawa, G. Sugimoto, K. Tange and Y. Yamamoto (1985). An Automatic guidance system of a self-controlled vehicle—the command system and control algorithm. In *Proceedings IECON 1985.* 535-540.

Kanayama, Y. and N. Miyake (1985). Trajectory generation for mobile robots. In *Robotic Research: The Third International Symposium on Robotics Research,* Gouvieux, France. 333-340.

Kanayama, Y and B. Hartman (1989). Smooth local path planning for Autonomous Vehicles. In *Proceedings IEEE International Conference on Robotics and Automation.* 1265-1790.

Nelson, W. L. (1989). Continuous steering function control of robot cart. *IEEE Transactions on Industrial Electronics,* Vol. HFE-7. 330-337.

Press, W. H., B. P. Flannery, S. A. Teukolsky, W. T. Vetterling (1986). Parabolic interpolation and brent method. In *Numerical Recipes,* Cambridge University Press. 299-302.

Shin, D. H. and S. Singh (1990). Chapter 13: Vehicle and path models for autonomous navigation. In *Vision and Navigation,* Kluwer Academic Publisher. 283-307.

Yates, R. C. (1952). *Curves and Properties,* Classics Publishing Co.

# A REAL-TIME TRAJECTORY GENERATION ALGORITHM
# FOR MOBILE ROBOTS

**V. Feliu, J.A. Cerrada and C. Cerrada**

*Departamento de Ingenieria Electrica, Electronica y de Control, ETS Ingenieros Industriales UNED, Madrid, Spain*

Abstract. Trajectory generation for mobile robots is studied. Constraints in curvature
and sharpness are considered in order to get smooth trajectories that take into account
the physical limitations of the vehicle (centrifugal forces e.g.). An algorithm is proposed
to generate them in real time. The algorithm has two stages: in the first one, a basic
parabolic profile trajectory is generated; in the second stage, the discontinuity points in
the acceleration profile are smoothed by three-order polynomials. The whole algorithm
is based on searching and exploiting the symmetries of the trajectories, and has shown
to be extremely efficient from a computational point of view.

Keywords. Robots; vehicles; bang-bang control; navigation

## INTRODUCTION

Mobile robots represent an emerging area with
great potential applications, where many of the
most important problems in robotics can be
studied at the same time. One of them is the
efficient real-time generation of trajectories under
some mechanical constraints (Craig, 1986), like
maximum acceleration or maximum derivative of
the acceleration. Little literature exists about this
problem in mobile robots, in spite of being
critical in many real vehicles (Shafer and
Whittaker, 1988, e.g.). In fact, reference
trajectories generated without taking into account
these constraints may produce motions physically
not realizable for some vehicles (see Muir and
Neuman, 1986 or Kanayama and Miyake, 1986),
and such trajectories are inaccurately followed by
the vehicle.

Basic methods used to solve this problem have
been extended from joint trajectory generation for
standard manipulators to planar trajectory
generation for mobile robots, like spline curves
(Craig, 1986). Other methods have been defined
especially for mobile robots and take into account
the physical constraints of the vehicle: circular
arcs interconnected by straight lines - which
means constant curvatures and limited centrifugal
forces -(Latombe, 1991 e.g.), or arcs of clothoids
interconnected by circular arcs and straight lines -

which means limited and constant derivative of
the curvature - (Kanayama and Miyake, 1986).
Trajectories with a long straight line segment are
preferred in general because the vehicle can be
driven more easily. This condition makes difficult
the use of splines. Circular arc based trajectories
are extensively used in mobile robots, and are
typically assumed in collision avoidance problems
(Latombe, 1991, e.g.). But they exhibit
discontinuous changes in the curvature, such
trajectories being physically not realizable for
some vehicles. Clothoids based trajectories
(combined with straight line and circular arc
intervals) constitute the solution best suited for
point to point motion of mobile robots, because
they exhibit a limited and constant derivative of
the curvature (sharpness) in the transitions of the
curvature. But efficient algorithms have been
proposed only for cases in which the initial and
final curvatures are zero (Kanayama and Miyake,
1986). This solution is not utilizable when
complex trajectories with intermediate points
have to be followed.

This paper proposes a computationally efficient
algorithm for real-time trajectory generation for
mobile robots, that takes into account all the
requirements mentioned before. This algorithm
is based on a method previously developed for
generating continuous joint trajectories of
manipulators (Cerrada, 1987); and has been

modified and generalized to the mobile robot case. It combines polynomial trajectories of first order (straight lines), second and third order in order to move from an initial point to a final point, defined both by its cartesian position, orientation and curvature. Section 2 gives a general statement of the problem. Section 3 develops the algorithm for generating the above polynomials. Section 4 illustrates the method with some examples, and Section 5 states some conclusions.

## STATEMENT OF THE PROBLEM

Let us define a 2D Cartesian coordinate system as a medium of describing trajectories of a vehicle. Any curve can be defined using a parameter $s$:

$$(x, y) = (x(s), y(s)), \qquad (1)$$

where $s$ is assumed to be the length along the curve from a point $P_0$ on it. Let

$$\theta(s) = \tan^{-1}(y'/x') = \tan^{-1}((dy/ds)/(dx/ds)) \quad (2)$$

be the direction of the tangent at $(x(s), y(s))$.

The curvature $c(s)$ of a curve is the derivative of the direction of the tangent with respect to the length $s$

$$c(s) = d\theta(s)/ds . \qquad (3)$$

The curvature $c(s)$ is the reciprocal of the osculating radius $r(s)$. If a vehicle is travelling at a constant velocity $V$, $s = V \cdot t$ holds, where $t$ denotes time. Hence the centrifugal force at time $t$ applied to the vehicle at a constant velocity is

$$V^2/r(s) = V^2 \cdot c(V \cdot t) . \qquad (4)$$

Now, clearly the curve itself should be continuous. Furthermore, the direction of the tangent should also be continuous with respect to $s$. If the direction is not continuous at some $s$, the acceleration to be given to the vehicle at the point becomes infinite and the motion is physically impossible. Continuity in curvature is also highly recommended in any kind of vehicle because smooth change in centrifugal force given by Equation (4) is desirable. A quadruple

$$P = (x, y, \theta, c) \qquad (5)$$

of coordinates, direction and curvature is called a posture.

Then the problem may be stated as follows: given a posture pair $(P_i, P_f)$, a directed curve has to be found that starts at $P_i$, ends at $P_f$ and verifies the constraints

$$|c(s)| \le c_m , \ \forall s$$

$$|\frac{d\,c(s)}{d\,s}| \le d_m , \ \forall s . \qquad (6)$$

Solutions have been proposed by using splines, arcs of circumferences, and clothoids. Arcs of circumferences based trajectories exhibit very sharp changes in the curvature. Clothoids based trajectories are the best suited solutions for mobile robots, but they cannot be easily computed in the most general case of having non zero initial and final curvatures. Splines may be easily computed and provide with continuous second and third order derivatives of the trajectory, but limits for the absolute value of these derivatives can not be established.

We propose here a method based on the combination of polynomials of the general form

$$y = \sum_{i=0}^{3} \alpha_i \, x^i , \qquad \alpha_i \ \textit{are constants.} \quad (7)$$

The next lemma will prove that constraints in the x derivatives of y result in constraints in the s derivatives of the trajectory.

*Lemma*: Assume an y(x) trajectory that verifies limits in its derivatives:

$$|\frac{d^2y}{dx^2}| \le a_m , \qquad |\frac{d^3y}{dx^3}| \le b_m . \qquad (8)$$

Then curvature and derivative of the curvature are limited by the expressions

$$|c(s)| \le a_m \qquad (9)$$

and

$$|\frac{d\,c(s)}{d\,s}| \le b_m + 0.78 \cdot a_m^2. \qquad (10)$$

*Proof*: Differentiating equation (2) we get the curvature:

$$c(s) = \frac{d\theta/dx}{ds/dx} = \frac{d^2y/dx^2}{(1 + tg^2\theta)^{\frac{3}{2}}} \qquad (11)$$

and easily follows equation (9) from the first part of (8). Differentiating again we get

$$\frac{d\,c(s)}{d\,s} = \frac{dc/dx}{ds/dx} = \frac{(1 + tg^2\theta) \cdot \frac{d^3y}{dx^3} - 3 \cdot tg\theta \cdot (\frac{d^2y}{dx^2})^2}{(1 + tg^2\theta)^3} . \qquad (12)$$

It is easy to prove that

450

$$\left| \frac{3 \cdot tg\theta}{(1+tg^2\theta)^3} \right| < 0.78 \qquad (13)$$

and equation (10) follows.

Assume a vehicle that exhibits physical constraints represented by limits in the curvature and its derivative: $c_m$ and $d_m$. Then these constraints will be verified for a trajectory $y(x)$ whose limits in the second and third x derivatives are:

$$a_m = c_m , \qquad b_m = d_m - 0.78 \cdot c_m^2 . \qquad (14)$$

Based on that, the algorithm developed in the next sections will generate polynomial trajectories (7) whose x derivatives verify (14).

### TRAJECTORY GENERATION

Let $(P_0, \dots , P_n)$ be a sequence of postures in the world coordinate system, of a mobile robot trajectory as shown in Fig. 1. The problem is to find a curve that passes each posture in the established order, and verifying continuity constraints in all of them. It is well known that finding a directed curve for every consecutive posture pair satisfying continuity constraints will be sufficient. Two different approaches are considered that take into account different physical constraints.

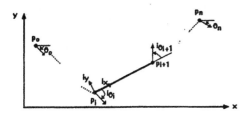

Fig. 1. Posture sequence of a mobile robot trajectory

#### Second order polynomic interpolation

In this approach only second order derivative of trajectory $y(x)$ is assumed limited to $a_m$. This means that only curvature constraint will be verified, but it states the basic formulation for the next approach. The kind of trajectory considered here between postures $P_i$ and $P_{i+1}$ will be composed in the general case of three segments: two of them with the second order derivative constant and equal to $\pm a_m$ (parabolic profiles) and the third one with second order derivative equal to zero (straight line profile). It can be easily proved that the case where the straight

segment is either in the starting or in the ending part of the trajectory is more restricted in general. Therefore, the algorithm described in this paper will only consider profiles starting with a parabolic segment, followed by a straight line segment and ending with another parabolic segment.

Let $({}^iX, {}^iY)$ be a local coordinate system whose origin is at the posture $P_i$. ${}^iX$ axis is in the connecting line of postures $P_i$ and $P_{i+1}$, pointing to the latter one, and ${}^iY$ axis is a normal vector in the counter clockwise direction. In order to improve the algorithm efficiency, a coordinate transformation to the local frame is firstly performed. Let $({}^iP_i, {}^iP_{i+1}) = ((0, 0, {}^i\Theta_i, {}^ic_i), ({}^ix_{i+1}, 0, {}^i\Theta_{i+1}, {}^ic_{i+1}))$ be the transformed posture pair. Let ${}^iy'_i = {}^idy/{}^idx|_i$ and ${}^iy'_{i+1} = {}^idy/{}^idx|_{i+1}$ be the tangent to the trajectory in the local coordinate system, at the starting and ending points respectively. It is assumed that absolute values of headings in the local frame are less than $\pi/2$. This assumption simplifies computations and it is usually considered by other authors (Amidi 1990). Trajectory will be generated in the local coordinate system, so a world coordinate system transformation is required afterwards.

Given these end point constraints, the problem is reduced to find the sign $S_s$ and $S_e$ of the constant second derivative at the starting and ending parabolic segments respectively, and the length $X_1$, $X_2$ and $X_3$ of each segment. Next expressions are verified:

$$\delta x = {}^ix_{i+1} - {}^ix_i = {}^ix_{i+1} = X_1 + X_2 + X_3 \qquad (15)$$

$$a_1 = d^2y/dx^2 = S_s \cdot a \qquad 0 \le {}^ix \le X_1 \qquad (16)$$

$$a_2 = d^2y/dx^2 = 0 \qquad X_1 < {}^ix < X_1 + X_2 \quad (17)$$

$$a_3 = d^2y/dx^2 = S_e \cdot a \qquad X_1 + X_2 \le {}^ix \le \delta x \quad (18)$$

A very simple solution can be achieved if a variable normalization based on the minimal criterion of Pontryagin (see Mc. Ausland, 1969) is carried out. Normalization considers ${}^nx$ as the independent variable which remains the same as ${}^ix$. Next expressions are used to normalize the other variables:

$${}^ny'({}^nx) = {}^iy'({}^nx) - {}^iy'_{i+1} \qquad (19)$$

$${}^ny({}^nx) = {}^iy({}^nx) - {}^iy_{i+1} + {}^iy'_{i+1} \cdot (\delta x - {}^nx)$$
$$= {}^iy'_{i+1} \cdot (\delta x - {}^nx) \qquad (20)$$

Normalization implies that, representing normalized variables ${}^ny$ and ${}^ny'$ in a phase plane, all possible trajectories will finish at the origin.

Equations (15) to (18) remain valid after normalization. Therefore, every trajectory can be solved in terms of the starting point conditions only. Resulting expressions are gathered in Table 1, where next definitions have been used:

$$y_{lim} = (^ny'(0))^2/(2.a_m) \tag{21}$$

$$C_1 = (^ny(0) < y_{lim}) \text{ .and. } (^ny'(0) < 0) \tag{22}$$

$$C_2 = (^ny(0) \leq -y_{lim}) \text{ .and. } (^ny'(0) > 0) \tag{23}$$

$$C_3 = (^ny(0) > -y_{lim}) \text{ .and. } (^ny'(0) > 0) \tag{24}$$

$$C_4 = (^ny(0) \geq y_{lim}) \text{ .and. } (^ny'(0) < 0) \tag{25}$$

$$V_M = -S_e.(\tfrac{1}{2}^ny'(0)^2 - a_3.^ny(0))^{\frac{1}{2}} \tag{26}$$

$$F_1 = (^ny'(0) - 2.V_M)/a_3 \tag{27}$$

$$F_2 = V_M^2/(a_3.^ny'(0)) \tag{28}$$

$$K_1 = \tfrac{1}{2}(^ny(0) - a_3.\delta x) \tag{29}$$

$$E_1 = V_2^2 - 2.K_1.V_2 + V_M^2 = 0 \tag{30}$$

$$K_2 = -^ny(0) + \tfrac{1}{2}(^ny'(0)^2/a_3) \tag{31}$$

$$E_2 = K_2/(a_m.\delta x + S_e.^ny'(0)) \tag{32}$$

where the variable $V_2$ represents the slope of the straight line segment of the trajectory in normalized coordinates. It is necessary to solve $V_2$ just for $X_1$ and $X_3$ computations purpose. $X_2$ is obtained from equation (15). Let us finally remark that validity condition of Table 1 states that $F_1$ is the minimum $\delta x$ required for a trajectory considered being possible.

TABLE 1    Solution of the second order polynomic interpolation

| Primary condition | Validity Condition | Ss | Se | V2 | X1 | X3 |
|---|---|---|---|---|---|---|
| C1 | F1 ≤ δx | +1 | -1 | E1 | | |
| C2 | F1 ≤ δx ≤ F2 | +1 | -1 | E1 | | |
| C2 | F2 ≤ δx | -1 | -1 | E2 | -(ny'(0)-V2)/a1 | -V2/a3 |
| C3 | F1 ≤ δx | -1 | +1 | E1 | | |
| C4 | F1 ≤ δx ≤ F2 | -1 | +1 | E1 | | |
| C4 | F2 ≤ δx | +1 | +1 | E2 | | |

Third order polynomic interpolation

This approach considers both physical constraints of equation (8). This fact implies that second derivative must change from zero to $\pm a_m$ with a controlled slope equal to $\pm b_m$, instead of stepwise. Therefore, trajectory $y(x)$ will be represented by a third order polynomial in those segments where their third order derivative equals to $\pm b_m$.

The algorithm developed in this case is an extension of the previous one. In practice, first and third segment of the previous approach are split up in three new subsegments each. Therefore, in this case trajectory is composed of seven segments. Let us call $X_1$ through $X_7$ to the length of each segment. There will be only two signs $S_s$ and $S_e$ as in the previous case, that will be determined from the same expressions. In the first segment, second derivative varies from 0 to $S_s.a_m$ with a $S_s.b_m$ slope. Second segment maintains constant the second derivative to the value $S_s.a_m$ (parabolic profile). In the third segment, second derivative goes to 0 with $-S_s.b_m$ slope. Fourth segment corresponds to a straight line profile. Finally, last three segments are similar to the first three ones.

Lengths $X_1$ through $X_7$ of every segment of this case are determined from $X_1$ through $X_3$ of the previous case. Computed lengths must be corrected for the new profiles, but these corrections are performed while real time trajectory generation. Equation (15) is now transformed to :

$$\delta x = X_1 + X_2 + X_3 + X_4 + X_5 + X_6 + X_7 \tag{33}$$

The algorithm is based on searching symmetries in the profiles. It was used before for industrial robot joint trajectories and its basic equations can be found in Cerrada, 1987. That algorithm has been extended here to mobile robot trajectories and following considerations must be taken into account for this specific case:

1.- First, third, fifth and seventh segments have constant lengths and they can be discarded before next computations. Following equation is verified:

$$X_1 = X_3 = X_5 = X_7 = a_m/b_m \tag{34}$$

2.- A symmetric point $P_s$ exists that could be located in the second segment or in the sixth one. Considering the first situation, let $X_s$ be the length from the beginning of second segment to point $P_s$. $X_s$ verifies next expression:

$$X_s = -^ny'(0)/a_m \tag{35}$$

which is valid for $S_s <> S_e$. analog expression can be obtained when $S_s = S_e$.

3.- From $P_s$ to the end of $X_6$ the trajectory is symmetric, which implies:

$$X_2 - X_s = X_6 = X_d \tag{36}$$

Notice that $X_6$ is the length of the shorter parabolic segment, but is not required to know in advance its value. The same is valid for $X_2$ which is related with $X_6$ by means of equation (36). Once $P_s$ is reached, difference $X_d$ is set to zero and a new error expression $e(^nx)$ is computed. Both $X_d$ and $e(^nx)$ are updated every sample length L. Next expression is used to update the error:

$$e(^nx) = e(^nx - L) + A.^nx + B \qquad (37)$$

where A and B are constants depending only on the initial conditions. $X_d$ coincides with $X_6$ when $e(^nx)$ changes its sign. Whenever this situation happens, $X_4$ is the only unknown of equation (33) and the trajectory is completely defined.

It is remarkable that this approach drives to the previous one when no constraint in derivative of curvature is taken into account. In such case equation (34) equals to zero and trajectory has only three segments. Validity conditions of Table 1 are also valid using $(\delta x - 4a_m/b_m)$ instead of only $\delta x$.

Computations in this algorithm are more shared than usual between the initial generation stage and the real time generation stage, due to the symmetries searching method. Other advantage is that continuity constraints are achieved using only third order polynomic interpolation. These features improve the efficiency of the algorithm compared to others methods using splines or clothoids.

## SIMULATION RESULTS

This section shows two different cases which have been simulated to test the algorithm. Trajectory between a posture pair is shown in both cases. Physical constraints $c_m = 0.01$ and $d_m = 0.000378$ have been considered. After applying equation (14) $a_m = 0.01$ and $b_m = 0.0003$ are obtained.

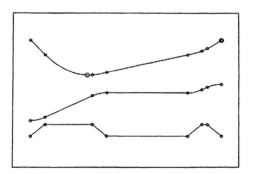

Fig. 2.   CASE 1: Trajectory with respect to x

The first case considers the posture pair in local coordinates $(^iP_i, \ ^iP_{i+1}) = ((0,0,-0.876, \ 0), (450,0,0.61,0))$. The solution obtained shows that sign $S_s$ is equal to sign $S_e$, and results are shown in Figures 2 and 3. The three curves of the Figure 2 represent, from top to bottom, the trajectory $^iy(^ix)$, its first order derivative $^iy'(^ix)$ and its second order derivative $^iy''(^ix)$ respectively. Small circles have been drawn in every segment change. A bigger circle shows the symmetric point $P_s$. The three curves of Figure 3, from top to bottom, represent the direction of the tangent $\Theta(s)$, the curvature $c(s)$ with its two boundaries $\pm c_m$, and the derivative of the curvature $dc(s)/ds$ with the boundaries $\pm d_m$.

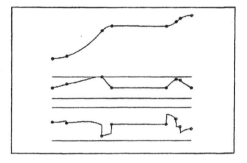

Fig. 3.   CASE 1: Trajectory with respect to s

The second case considers $(^iP_i, \ ^iP_{i+1}) = ((0,0,0.291,0), \ (450,0,0.915,0))$, and satisfies $S_s <> S_e$. Organization of Figures 4 and 5 is the same used with Figures 2 and 3.

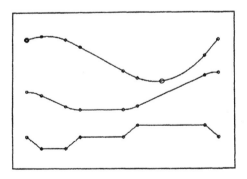

Fig. 4.   CASE 2: Trajectory with respect to x

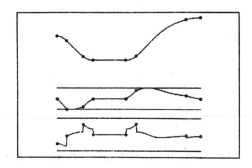

Fig 5.    CASE 2: Trajectory with respect to s

## CONCLUDING REMARKS

A new algorithm to generate real time trajectories for mobile robots has been presented. Its main features are:

1.- It takes into account all the physical constraints in order to calculate realizable trajectories

2.- It is appropriate to generate complex trajectories with intermediate points where the curvature is different of zero

3.- The method is very efficient computationally (more than the other methods mentioned) because of the simplicity of the curves used to generate the trajectory and the algorithms used to define the intervals of each curve. This allows the trajectory to be changed in real time which may be useful when following a mobile target e.g..

4.- This method gives a near minimum time trajectory. If we have constraints in the curvature and sharpness, the minimum time trajectory (assuming constant linear velocity) is given by a combination of straight lines, circle and clothoid arcs, which correspond to intervals of zero curvature, constant maximum curvature, or constant maximum sharpness, respectively. In our algorithm, these parameters are not kept constant, but they remain inside defined boundaries. Consequently our algorithm does not gives an optimal trajectory in the minimum time sense, but gives a close suboptimal solution.

## REFERENCES

Amidi, O., *Integrated Mobile Robot Control*, Technical Report CMU-RI-TR-90-17 Carnegie Mellon University, Pittsburgh (Pennsylvania-USA), 1990.

Cerrada, C., *Generación en Tiempo Real y Control de Trayectorias para Robots Industriales: Aplicación a un Robot de la Familia SCARA*, Ph.D. Thesis of the E.T.S. Ingenieros Industriales of the Universidad Politécnica of Madrid, 1987.

Cerrada, C., and Feliu, V., *An Efficient Algorithm for Generating Point to Point Trajectories in Industrial Robots*, 87 IMACS International Symposium on AI, Expert Systems and Languages in Modelling and Simulation, Barcelona (Spain), 1987.

Craig, J.J., *Introduction to Robotics. Mechanics & Control*, Addison-Wesley Publishing Co. 1986.

Kanayama, Y., and Miyake, N., *Trajectory Generation for Mobile Robots*, The Third International Symposium of Robotics Research. 1986.

Latombe, J.C., *Robot Motion Planning*, Kluwer Academic Publishers, 1991.

Mc. Ausland, I. *Introduction to optimal control*, John Wiley and Sons Inc., New York, 1969.

Muir, P.F., and Neuman, C.P., *Kinematic Modelling of Wheeled Mobile Robots*, Technical Report CMU-RI-TR-86-12 Carnegie Mellon University, Pittsburgh (Pennsylvania-USA), 1986.

Shafer, S., and Whittaker, W., *June 1987 Annual Report: Development of an Integrated Mobile Robot System at Carnegie Mellon*, Technical Report CMU-RI-TR-88-10 Carnegie Mellon University, Pittsburgh (Pennsylvania-USA), 1988.

# NEW CONTINUOUS-CURVATURE LOCAL PATH GENERATORS FOR MOBILE ROBOTS[1]

**V.F. Muñoz, J.L. Martinez and A. Ollero**

*Departamento de Ingenieria de Sistemas y Automatica, Universidad de Malaga, Plaza El Ejido S/N, 29013 Malaga, Spain*

Abstract. In this paper we propose two methods for wheeled mobile robots path generation based on B and β-Splines curves. Both methods provide continuity in position, orientation and curvature. Moreover, a smooth variation in curvature can be obtained. The efficiency of the methods is demonstrated for obstacle avoidance. Given the spatial localization of an obstacle on the global planned path to follow, the proposed methods generate a smooth local path to avoid the obstacle by providing the conditions to assure an efficient tracking of the path.

Keywords. Mobile Robots; Local Path-Planning; B-Splines; β-Splines, Autonomous Vehicles.

## INTRODUCTION.

The ability of a robotic wheeled vehicle to track a path depends on the vehicle kinematic and dynamic constraints, the navigation conditions, and the geometric properties of the path. If the path heading is not continuous at some points, the acceleration to be given to the vehicle becomes infinite and the motion is physically imposible. Moreover, several vehicles cannot be controlled to move along curves whose curvature is not continuous. Furthermore, continuity in curvature is necessary when looking for smooth changes in the centrifugal force.

Several path generation methods have been proposed in the mobile robot literature and implemented in navigation systems. The most simple technique is the concatenation of lines and circular arcs. However, it has been shown (Nelson, 1988) that these paths may give discontinuities in the steering for different configurations of the locomotion system.

Continuous-curvature paths avoid the problem providing good conditions to be followed by mobile robots and autonomous vehicles. Clothoids curve segments (Kanayama and Miyake, 1985) and cubic spirals (Kanayama and Hartman, 1989) give continuous curvature but have the disadvantage of lacking a closed-form expression to generate the

[1]This research has been partially supported by the CICYT Project ROB'89-0614-C03-02.

curve coordinates. Thus, the computer requirements may preclude its application particularly when avoiding obstacles without stopping. Quintic cartesian polynomials and polar polynomial curves (Nelson, 1989) have also been proposed. However, to avoid obstacles, different polynomial segments must be linked appropriately (boundary conditions) which complicate the path generation.

In this paper we present two methods for path generation based on parametric spline curves. These curves have closed forms and can be easily generated using efficient algorithms (Barsky, 1987; De Boor, 1978). In the Section 2 we state the local path planning and path definition problems, and the basic equations to generate the paths. Sections 3 and 4 present the proposed methods. Comparative experiments are shown in Section 5.

## LOCAL PATH SPECIFICATION/ GENERATION.

Consider a navigation system in which the global path planner generates a path to drive an autonomous vehicle accross a previously known area. The local navigator is concerned with the perception, planning and control operations to drive the vehicle across the path provided by the global path planner avoiding obstacles as they arise. In this paper we are concerned with the local path planner. Asume the start point (separation point from the global path), end point (meeting point with the local path), and obstacle-avoidance points (defined to avoid obstacles from the information provided by the perception system) are known. Each point is defined by the position (x,y coordinates in the two dimensional case), and heading ($\theta$). In some approaches, the curvature ($\kappa$) is also considered explicitly to represent the point. In this case we have the quadruple $\{x, y, \theta, \kappa\}$ . In this paper we consider the problem of fitting a continuous path through the start point, remaining the the the obstacle-avoidance points and the end point, preserving the continuity of the position, heading and curvature.

Let $V_1, V_2, ..., V_N$ be a control polygon of an order k B-Spline curve with parameter $\tau$. Let $\{\tau_i\}$ be a set of k+N+1 B-Spline knots with the first knot $\tau_0=0$ and the last knot $\tau_m=N-k+2$ each repeated k times. Then, the coordinates of the B-Spline point for a given value of the parameter $\tau \in [\tau_j, \tau_{j+1}]$ are given by (De Boor,1978):

$$f_k(\tau) = \sum_{i=j-k+1}^{j} V_i \cdot N_{i,k}(\tau) \qquad (1)$$

where $V_i$ represents the coordinates of the control points and $N_{i,k}(\tau)$ are B-Spline blending functions of order k recursively defined as:

$$N_{i,1}(\tau) = 1 \qquad \tau_i \le \tau \le \tau_{i+1}$$
$$N_{i,1}(\tau) = 0 \qquad \text{otherwise}$$
$$\text{and}$$
$$N_{i,k}(\tau) = \frac{(\tau - \tau_i) N_{i,k-1}(\tau)}{\tau_{i+k-1} - \tau_i} + \qquad (2)$$
$$\frac{(\tau_{i+k} - \tau) N_{i+1,k-1}(\tau)}{\tau_{i+k} - \tau_i}$$

Equation (1) applies for both x and y coordinates. This expression only involves the computation of the k numbers $N_{i,k}(\tau)$, i=j-k+1,...,j. The differentiation of the B-Spline curve can also be done by means of simple closed form expresions (De Boor,1978). Then, the curvature and heading for a value of the parameter $\tau$ are given by:

$$\theta(\tau) = \text{atan}\left(\frac{y'(\tau)}{x'(\tau)}\right) \qquad (3)$$

$$\kappa(\tau) = \frac{(x'(\tau) y''(\tau)) - (x''(\tau) y'(\tau))}{\sqrt{((x'(\tau))^2 + (y'(\tau))^2)^3}} \qquad (4)$$

where the apostrophes stands for the derivatives with respect to $\tau$. The parametrization in length of B-splines has been studied by several authors. It has been pointed out (De Boor, 1978) that it is usually not possible to explicitly obtain an arc length parametrization for a B-spline curve defined by a given control polygon. Numerical iterative techniques have been proposed (Sharpe, 1982; Guenter, 1990). In order to accelerate the path way point generation we have used a method based on binary search and linear interpolation of lookup table entries.

The $\beta$-Spline basis (Barsky,1984) can also be used for path representation and generation. The i-th segment of a $\beta$-Spline may be expressed as:

$$f_i(\tau) = \sum_{r=-2}^{1} V_{i+r} b_r(\beta_1, \beta_2, \tau) \qquad (5)$$

where $b_r$ (r=-2,1) is a cubic polynomial called the r-th basic $\beta$-spline function, and $\beta_1$, $\beta_2$ are two parametres called bias and tension. The bias $\beta_1$ provides control of the symmetry of the curve, and the tension $\beta_2$ controls the degree of adherance of the curve to the control polygon. When $\beta_1=1$ the curve is said unbiased and when $\beta_2=0$ untensed. For these values the corresponding $\beta$-spline is a cubic $\beta$-Spline. The parameter $\beta_2$ make possible to control the deviation of the generated path from the control points. When $\beta_2$ is increased, the curve flattens and uniformly approaches to the control polygon. However, the curvature tends to increase. Efficient methods for evaluating $\beta$-Spline using closed form expressions have been reported (Barsky, 1988).

## B-SPLINE CONTINUOUS PATH GENERATION.

Let $p_s$ and $p_e$ be the start and the end point of the local path defined by the postures $(x_s, y_s, \Theta_s, \kappa_s)$ and $(x_e, y_e, \Theta_e, \kappa_e)$ respectively. Furthermore, let $p_m$ be an obstacle avoidance point which coordinates $(x_m, y_m)$ are defined to have a safe security distance from the obstacle, and its orientation $\Theta_m$ takes the same value than the orientation $\Theta_q$ of the closest point $p_q$ in global path (see figure 1).

The problem is the computation of the coordinates of the B-Spline control points $(V_{xi}, V_{yi})$ in such a way that the generated curve starts in $p_s$, pass through $p_m$ avoiding the obstacle, and ends in $p_e$. In order to have smooth curvature transition we use quintic splines (k=6) and six control points with the relative positions

456

shown in figure 1.

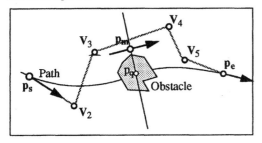

Fig. 1. Initial Data And Control Points.

The position and heading continuity in $p_s$ and $p_e$ provides the following conditions:

$$V_{x1} = x_s; V_{y1} = y_s$$
$$V_{x6} = x_e; V_{y6} = y_e$$
$$V_{y2} = \frac{V_{x2} - V_{x1}}{\tan(-\Theta_s)} + V_{y1} \qquad (6)$$
$$V_{y5} = \frac{V_{x5} - V_{x6}}{\tan(-\Theta_e)} + V_{y6}$$

It can be demonstrated (Martínez, 1991) that the curvature continuity in $p_s$ and $p_e$ imposes the following conditions on the coordinates of the vertices:

$$\kappa_s = \frac{(V_{x2} - V_{x1})(V_{y3} - V_{y2}) - (V_{x3} - V_{x2})(V_{y2} - V_{y1})}{\frac{8}{3}\sqrt{((V_{x2} - V_{x1})^2 + (V_{y2} - V_{y1})^2)^3}}$$

$$\kappa_e = \frac{(V_{x4} - V_{x5})(V_{y5} - V_{y6}) - (V_{x5} - V_{x6})(V_{y4} - V_{y5})}{\frac{8}{3}\sqrt{((V_{x6} - V_{x5})^2 + (V_{y6} - V_{y5})^2)^3}}$$

$$(7)$$

Furthermore, it can also be shown (Martínez, 1991) that the position and orientation in $p_m$ are given by the following expressions:

$$x_m = \frac{3(V_{x3} + V_{x4}) + V_{x2} + V_{x5}}{8}$$
$$y_m = \frac{3(V_{y3} + V_{y4}) + V_{y2} + V_{y5}}{8} \qquad (8)$$
$$\Theta_m = -\text{atan}\left(\frac{V_{x5} + V_{x4} - V_{x2} - V_{x3}}{V_{y5} + V_{y4} - V_{y3} - V_{y2}}\right)$$

Thus, we have 11 equations to determine the 12 coordinates of the control points. To solve this system, we use the relative position of the control

points and its relation with the maximum allowable curvature to provide an additional condition. By using these conditions and solving a second order equation, closed form expressions can be obtained and applied to determine all the vertices coordinates (Martinez, 1991).

## TANGENT CIRCLES β-SPLINE GENERATION METHOD

Consider an obstacle be modelled by a circle $C_b$ with radius $\rho_b$, and center in $c_b$; and let q be the point in the global path nearest from $c_b$. Let $C_m$ be an avoidance circle passing through a point a defined in the normal v to the path at a given security distance s from the obstacle (see figure 2).

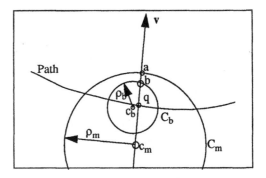

Fig. 2. The avoidance circle

Then, the center of such avoidance circle is given by

$$c_m = q + (d - \rho_m)v \qquad (9)$$

where $\rho_m$ is the radius of $C_m$ and d is the distance from q to a, which is also given by

$$d = \rho_b + s \qquad (10)$$

Let $C_s$ be a circle tangent to $C_m$ and to the global path in the starting point $p_s$ of the local obstacle avoidance path. Similarly, let $C_e$ be a circle tangent to $C_m$ and to the global path in the ending point $p_e$ of the local path (see figure 3)

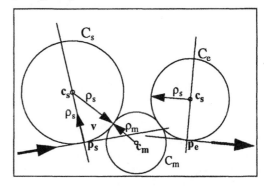

Fig. 3. Tangent circles and avoidance Arc.

457

The center of $C_s$ is given by

$$\mathbf{c_s} = \mathbf{p_s} + \rho \mathbf{v} \qquad (11)$$

and the radius is

$$\rho_s = \frac{2\,(x_m x_s + y_m y_s) + \rho_m - (x_s^2 + y_s^2 + x_m^2 + y_m^2)}{2\,(v_x\,(x_s - x_m) + v_y\,(y_s - y_m) - \rho_m)} \qquad (12)$$

Expressions (11) and (12) return the solution for a given avoidance direction $\mathbf{v}$ that corresponds to the avoidance in one of the two possible sides of the path (avoidance sense). The center and radius of $C_e$ can be computed in the same way by using $\mathbf{p_e}$ instead of $\mathbf{p_s}$.

The expresion (12) cannot be applied when the projection of the vector $\mathbf{c_s}$-$\mathbf{p_s}$ on $\mathbf{v}$ gives the radius of the avoidance circle. In this case the avoidance circle is tangent to the line defined by the heading $\theta_s$ at $\mathbf{p_s}$. Moreover, the expression (12) is valid if and only if the vector $\mathbf{c_s}$-$\mathbf{p_s}$ has the same direction and sense that $\mathbf{v}$. When any of these two conditions are found the obstacle must be avoided by the other side of the path.

The circles $C_s$, $C_m$, $C_e$ define a set of three arcs which compose a path avoiding the obstacle. This path has discontinuities of curvature at the junctures between two arcs. However, this path provides the basis for the definition of a $\beta$-Spline curve with smooth variations of curvature. The general idea to do that is the definition of a set of control points on the arcs. The $\beta$-Spline curve will tend to have the same curvature (constant) near the circle and will generate a smooth transition between arcs.

Let $\delta$ be the distance between control points defined on the arcs. Then, the torsions $\varepsilon_1$, $\varepsilon_2$ (variation of curvature with the lenght) when the path cross the tangent points (see figure 3) can be approximated by:

$$\varepsilon_1 = \frac{|1/\,(\rho_s - 1/\rho_m)|}{\delta} \qquad \varepsilon_2 = \frac{|1/\,(\rho_e - 1/\rho_m)|}{\delta} \qquad (13)$$

Additional control points are placed at the beginning and end of the local path to provide curvature continuity at $\mathbf{p_0}$ and $\mathbf{p_1}$ (Muñoz,1991). Using these control points, a $\beta$-Spline curve can be efficiently generated by means of (5).

Particularly, a $\beta$-Spline with tension $\beta_1$=1 and bias $\beta_2$=0 (cubic spline) has been used.

To guarantee the obstacle is avoided, the security distance s in (10) must be taken in such a way that the generated $\beta$-spline does not lie inside the circle $C_m$ modelling the obstacle. This distance can be computed by using the hull-convex property of the $\beta$-spline. Figure 4 shows the worst case in which the $\beta$-

Spline segment is inside the avoidance circle. Then, the security distance must be greater than the following value

$$s = \rho_m \left( 1 - \sqrt{\frac{1 - \cos\left(\frac{3\delta}{\rho_m}\right)}{2}} \right) \qquad (14)$$

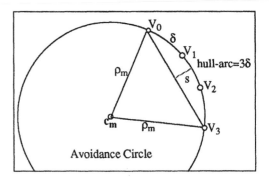

Fig. 4. Hull-convex property

## IMPLEMENTATION AND EXPERIMENTS

The above described methods have been implemented on a SUN Sparc 2 station on board of the mobile robot RAM-1 (Ollero and others,1992). Both methods can be used to generate paths avoiding obstacles without stopping. The $\beta$-Spline method is particularly efficient and can generate paths in about 10 milliseconds.

Figures 5 and 6 compare the results of both proposed methods. In Fig. 5 we show the original global path, the encountered obstacle and the local paths generated when using both methods. The start and end points are different in position, heading and curvature. The length of the avoidance-path generated by the $\beta$-Spline is lower. The curvature of both paths is compared in Fig. 6. Observe how both methods generate curvature continuous paths with maximum value of curvature lower than 0.5 (turning radius 2 meters). However, the curvature transition is smoother in the B-Spline method.

Figures 7 and 8 show the effect of the variation of the distance between control points in the $\beta$-Spline method. When the distance between control points is lower, the avoidance path is shorter, passing nearest the obstacle (see Fig. 7) and the torsions (13) are greater given a sharper transition of the curvature (see Fig. 8). The maximum value of the curvature can be controlled by means the radius of the avoidance circle, and the minimum radius of the tangent circles.

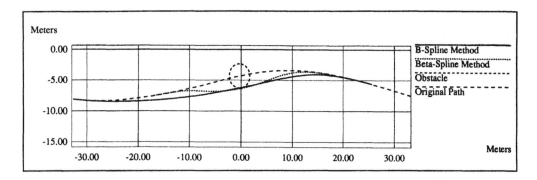

Fig 5. Comparation B-Spline and b-Spline paths

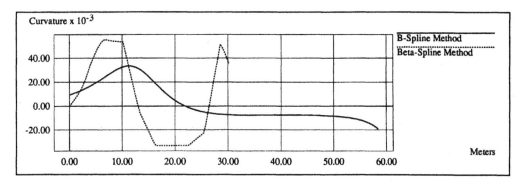

Fig 6. Curvature of paths shown in figure 5.

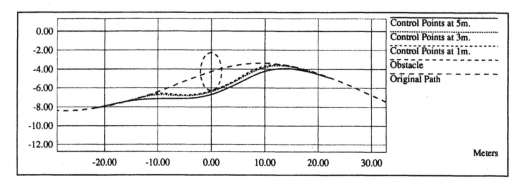

Fig 7. Path generated for several values of $\delta$.

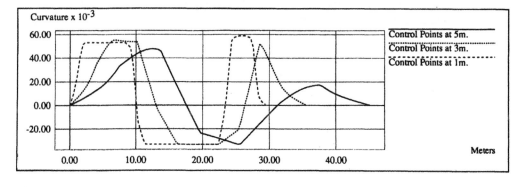

Fig 8. Curvature of paths shown in figure 7.

## CONCLUSIONS

In this paper we have presented two new methods for local obstacle-avoidance path generation methods for mobile robots. Both methods generate continuous curvature paths and assure the obstacle is avoided.

The first method uses the B-Spline basis. Six control points are computed to generate a quintic spline with continuity in position, heading and curvature. The method computes these control points and generates a path with 120 evenly spaced points in less than 30 milliseconds on a SUN Sparc 2 board in the vehicle's control system.

The second method generates a cubic β-Spline curve with control points placed on three tangent circles. The method takes about 10 msec for generation of a path with 100 evenly spaced points.

These methods can also be used to plan global paths with smooth curvature transitions.The methods can be applied to generate three dimensional path s without significant increment in the computational requierements. Furthermore, the methods are useful in conjunction with new path tracking methods based on the generation of smooth approach trajectories (Ollero and Amidi, 1991).

## REFERENCES

Barsky (1987), "Computer Graphics And Geometric Modeling Using Beta-Splines", Spriger-Verlag.

De Boor C. (1978), Applied Mathematical Sciences, volume 27, "A Practical Guide To Splines", Spriger-Verlag.

Guenter B. and R. Parent (1990), "Computing the Arc Length of Parametric curves", IEEE Computer Graphics and Applications, pp 72-78.

Kanayama Y. and Miyake (1985), "Trajectory Generation for Mobile Robots", Proc. 3rd Int'l Symp. on Robotic Research, O. D. Faugeras and G. Giralt, eds., Gouvieux, France, pp 333-340.

Kanayama Y. and B. Hartman (1989), "Smooth Local Planning for Autonomous Vehicles", Proc. IEEE Internatinal Conference on Robotics and Automation, pp 1265-1270.

Martinez J. L. (1991) "Evitación de obstáculos en Robots Móviles usando Basis-Splines". I.S.A. Labs. Internal report. RN191-UMA

Muñoz V. F. (1991) "Evitación de obstáculos en Robots Móviles usando β-Splines". I.S.A. Labs. Internal report RN291-UMA

Nelson W. (1988), "Continuous Steering-Function Control of Robots Carts". Bell Labs Internal Document.

Nelson W. (1989), "Continuous Curvature Paths for Autonomous Vehicles", Proc. IEEE Int Conference on Robotics and Automation, pp 1260-1264.

Ollero A. and O. Amidi (1991)"Predictive Path-Tracking of Mobile Robots. Aplications to the CMU Navlab". Proc. Fith International Conference on Avanced Robotics. pp 1081-1087 Pisa (Italy)

Ollero A. , A. Simón, F. García and V. Torres (1992) " Integrated Mechanical Design and Modelling of a New Mobile Robot". To appears in the proceedings of the IFAC Symposium on Intelligent Components and Instruments for Control Applications.

Sharpe R. J. and R. W. Thorpe (1982), "Numerical Method for Extracting an Arc Length Parametrization from parametric curves", Computer Aided Design, Volume 12, Number 2, pp 79-81.,

## ACKNOWLEDGMENTS

The basic ideas of the methods presented in the paper were developed during the stay of the third autor in the Field Robotics Center (Robotics Institute Carnegie Mellon University, Pittsburgh, USA). Comments and suggestions of William "Red" Whittaker, Director of this Center, B. Motazed, D. H. Shin, and T. Stenz are very appreciated.

# INTEGRATED MECHANICAL DESIGN AND MODELLING OF A NEW MOBILE ROBOT[1]

## A. Ollero*, A. Simon**, F. Garcia** and V.E. Torres*

*Departamento de Ingenieria de Sistemas y Automatica, Universidad de Malaga, Plaza El Ejido, 29013 Malaga, Spain
**Departamento de Ingenieria Mecanica, Universidad de Malaga, Plaza El Ejido, 29013 Malaga, Spain

Abstract: In this paper we present RAM-1 a new autonomous mobile robot designed as a testbed for the automatization of surveillance, manipulation and small part transportation. Particularly we describe VAM-1, the robotic vehicle designed for indoor and paved-floor outdoor navigation in unstructured environments. VAM-1 includes several software and hardware components for intelligent navigation. The mechanical design is the result of an integration approach by considering several criteria related with control, planning and perception issues in addition to structural design and other mechanical requirements. In this paper we also present the locomotion and control model of VAM-1. This model involves kinematic and dynamic relations and its parameters have been identified from experimental data.

Keywords: mobile robots, autonomous vehicles, kinematics, dynamics, position estimation.

## INTRODUCTION

The kinematic design of a wheeled robot greatly affects its ability to maneuver, navigate, following path and vehicles's positioning. Existing indoor autonomous wheeled vehicles offers a great variety of kinematic configurations. Wheeled arrangements include the classical four wheeled vehicles with Ackerman steering linkage (Moravec, 1980), vehicles with two diametrically opposed driven wheels with one castor (Hollis, 1977) or two castor (Nilsson, 1984) to maintain the stability, and tricycles with the front wheel steered and driven and the two rear wheels undriven (Cox, 1991). These tricycle-like solutions limit possible rolling motions. It is clear that such vehicles cannot be moved parallel to the fixed axis.

Full omnidirectional locomotion systems are the best solutions to navigate in cluttered environments. For horizontal motions omnidirectionality means independent and simultaneous translations in both axis of the horizontal plane and rotation with respect to the vertical axis. Existing solutions include four wheeled vehicles with omnidirectional roller wheels (Moravec, 1985), and vehicles with three steered and driven wheels with synchro-drive mechanism using concentring shafts (Holland, 1985) and belts (Fitzpartrick, 1989) to equalize all drive velocities and steering angles. However, the mechanical complexity and cost of these solutions, and its limitations to navigate in floors with irregularities, constraint yet its applications.

On the other hand, several outdoor autonomous wheeled vehicles are adaptations of conventional four wheeled vehicles with Ackerman steering.

These mechanical configurations respond to the different desing criteria related with the applications requirements. VAM-1 is a vehicle capable of carrying a robotic arm, small loads (parts and tools), power, electronic and computing components, and a variety of external sensors.

The robotic vehicle VAM-1 has been designed according to the following objectives: a)Indoor navigation in industrial environments; b)Outdoor navigation in paved floors; c)Autonomy to navigate in unstructured environments with on-board power, control, and computing; d)Precision enough to positionate an onboard arm for pick and place and simple manipulation operations with visual feedback; e)Flexibility to experiment different locomotion and control strategies.

The remaining of this paper is organized as follows. In section 2 we present the design methodology and the VAM-1 characteristics. Section 3 and 4 are related with the modelling. Finally, section 5 and 6 are for the conclusions and references.

## DESIGN ISSUES

The RAM-1 mechanical design is the result of several issues related with the objectives mentioned in the above section.

In the following we comment the main design criteria and the VAM-1 solutions.

### Structure

The dimensions of VAM-1 are constrained for the ability to navigate in industrial environments passing

---

[1]This research has been partially supported by the Spanish National Program PETRI for Technology Transfer, Project PTR´89-0150.

between machines and through corridors and doors. On the other hand, the robot needs space to accomodate the power system, standard electronic and computers enclosures, and a variety of sensors for navigation and operation. VAM-1 has a steel and aluminium octogonal structure with 1,100 mm in length and 830 mm high. A CAD system with finite elements methods has been used for structural design. The location of the onboard manipulator (see Fig. 1) has been decided to provide reach for a set of typical pick and place operations in industrial environments. A simulation system with graphical interfaces (Muñoz and others, 1991) has been used to resolve the traddeoffs between different operations.

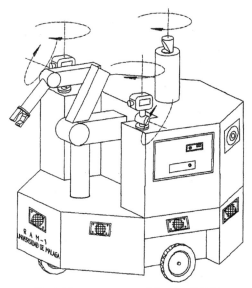

Fig. 1. The autonomous mobile robot RAM-1

According to the design objectives 1) and 2) the clearance of small obstacles such as electric wires, small terrain irregularities, and ramps is required. Thus, VAM-1 has independent suspensions in the driving wheels. Terrain irregularities also affect the locomotion system design because some omnidirectional systems require a perfectly flat floor (special wheels).

The lower platform of VAM-1 accomodates the power system (batteries, steering and driving motors, and power electronics). It should be noted that the power system also affects the structural design due to its dimensions and wheight. This system has been designed mainly from the locomotion requirements (clearance of 17% slope ramps, top vehicle velociy of 1.7 m/sec, and acceleration). We have also considered the power requirements for the onboard mechanisms, and for the electronic and computing. The VAM-1 power system has four batteries providing autonomy for two navigation hours in typical industrial operations. An automatic recharger system has also been designed.

The second floor accomodates the main electronic and computing elements including two VME standard buses one for the path tracking and lower level control system and the other with a Sparc-2 board and image processing boards for perception and planning.

The third floor provides a base for mounting sensor

devices and also has space to transport small loads (parts and tools).

Sensing and perception issues.

The VAM-1 position and orientation system involves the following functions:

• Low level position estimation by using dead reckoning techniques with a **north seeker** to measure the heading directly. The position is estimated from the locomotion model by using the measures of the wheels revolutions (**optical shaft encoders**) and the orientation.

• Line following: involves a **video camera** oriented to the floor, and image processing algorithms. Unexpensive optical devices under the VAM-1 lower platform can also be used.

• Landmark recognition techniques: beacons identification by means of oriented video cameras and pattern recognition techniques.

• Position estimation in unstructured environments. A **2-D laser range finder** scanning system is used for map building and precise position estimation. Both new iconic methods (Gonzalez and others, 1992) and feature based techniques can be used.

Furthermore, VAM-1 obstacle detection system uses a **ring of sonars**. The location of these sonars has been determined to scan all the space around the robot. A **special purpose VME board** has been designed to control the sonar ring and to implement security functions.

On the other hand, the surveillance, pick and place, and simple manipulation functions are also supported by the video cameras.

The placement of the sensors (see Fig. 1) has been determined to implement all the above mentioned functions. A simulation system with graphical interfaces representing the angular width of the vector field has been used to determine the optimal location of the video camaras (see Fig. 2). Furthermore, a mechanism with two stepping motors has been developed to orientate each camera. These mechanisms are controlled to perform the above mentioned functions, including the callibration and sinchronization of the cameras for stereo vision.

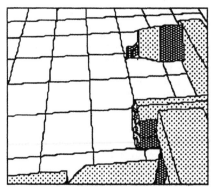

Fig. 2. Video camera virtual view

The perception system is also related with the design of the locomotion system because it is necessary that the frame frequency be sufficiently high that change of heading rates do not cause the vehicle to progress into terrain which has not been viewed. If $\dot{\theta}$ is the heading (yaw) rate, and $\gamma$ is the angular field of view about the yaw axis, the minimum frame frequency is $(\dot{\theta})/\gamma$. In practice it is necessary that the frame frequency be three or four times this value to have information enough for image processing, and the vehicle's top speed must be limited.

## Maneuverability and motion planning

The vehicle's maneuverabilty has a great impact in the complexity of the path planning problem. If the turned radius is not zero, several motions must be planned to locate the robot in a given position and orientation. For example, in the classical tricycle the maneuver needs to be decomposed in piecewise smooth arcs of pure rotations, and a prior analysis is required to construct the appropriate steering and driven rates. Furthermore, complex plans normally involves accumulation of errors that must be compensated through external sensing for position estimation which is in general a computing intensive task that takes a significant time to be carried out, affecting the efficiency of the navigation.

The VAM-1 locomotion system avoids these problems by providing, theoretically, zero turning radius. The vehicle has four wheels located in the vertices of a rhomb with a diagonal in the longitudinal axis (see Fig.3). The two parallel wheels are driven by DC motors with telescopic axis and double universal joints. The front and rear wheels are steered by a DC motor with a kinematic rigid link.

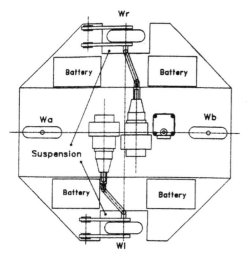

Fig. 3. The locomotion system

## Tracking and Control

The maneuverability is not the only issue for the design of the locomotion system. In fact, the ability of the vehicle to execute planned trajectories is also important. This problem also affects the kinematic design of the locomotion system.

The vehicle kinematic equations relate the wheels angular velocities and steering angles with the position and orientation of the vehicle. Thus, the kinematic control of the vehicle require the resolution of the inverse kinematic model. However this model is not resoluble in many kinematic configurations of current robotic vehicles (Muir and Newman, 1986). As we pointed out above, only some kinematic arrangements provide 3 degrees of freedom in the horizontal plane $(X, Y$ and orientation angle). In these vehicles three control variables (steering or driving) are required. However, most robotic vehicles, including like-cars vehicles with Ackerman steering, and conventional trycicles only have two degrees of freedom. In these cases, two control variables are required. Additional control variables may give actuator conflict producing slippage and shaky motion.

Consider now the problem of tracking simple planned paths like paths made up of line and circular arcs. Some kinematics design, including the simple like-cars vehicles and the tricycles, have discontinuities in the steering functions required to track the paths. In some cases (differential-drive) these discontinuities can be eliminated by shifting the guide point (Nelson, 1987).

To overcome these problems, VAM-1 (see Fig. 3) has a dual locomotion system in which any of the two possible locomotion modes is used as occasion demands. For maneuvering the front wheel is steered while in the path tracking mode only differential steering with the parallel wheels is used to avoid the actuator conflict and curvature discontinuities when following simple planned paths. The transition between modes is performed by means of a computer controlled clutch.

In the table 1 we compare VAM-1 with three different kinematic solutions from the point of view of its maneuverabilty, ability to be controlled, and ability to track simple planned paths (steering continuity).

TABLE 1 Comparison of kinematic configurations

|  | Maneuve-rability | Actuator conflict | Steering continuity |
|---|---|---|---|
| Ackerman | Low | No | No |
| T1 | Medium | Yes | No |
| T2 | Medium | No | Yes |
| VAM-1 | High | No/Yes | Yes |

T1: Steering & Driving front wheel
T2: Differential steering

## LOCOMOTION MODEL

Consider the differential steering navigation mode with the two independent driven and unsteered parallel wheels with velocities $\omega_l$ and $\omega_r$. For perfect rolling, the rolling axis of every wheel must pass through the Instantaneous Motion Center $(C_i)$. Thus, this center must be in the line defined by the common axis of both wheels at a distance $\rho$ from the center of mass $(S)$. We consider a coordinate frame $x, y, z$ attached to the vehicle at $S$, and the world coordinate frame $X, Y, Z$. Let V be the velocity of $S$. The vehicle's angular velocity $(\omega)$ is given by $\omega = V/\rho$. Then, the wheels velocities are:

$$V_l = \omega(\rho - R_t) \qquad (1)$$

$$V_r = \omega(\rho + R_t) \qquad (2)$$

where $R_t$ is the distance from $S$ to the wheels centers (see Fig. 4). The motion equations in world coodinates are the followings:

$$X = X_0 + \int_0^t V \sin\theta \, dt \qquad (3)$$

$$Y = Y_0 + \int_0^t V \cos\theta \, dt \qquad (4)$$

$$\theta = \theta_0 + \int_0^t \left(\frac{V_r - V_l}{2R_t}\right) dt \qquad (5)$$

Consider now the maneuvering mode. In this case, the front and rear pair of wheels $W_a$ - $W_b$ are steered simultaneously with angles $\alpha$ and $(-\alpha)$. If we consider the instantaneuous motion, the direction of the velocity of the center of the wheel $W_a$ is given by $\alpha$.

From the fig. 4 the following geometrical relations can be obtained:

$$\tan\alpha = \frac{CD}{C_i D} = \frac{R_g - a\cos\alpha}{\rho - a\sin\alpha} \qquad (6)$$

The value of the control variable $\alpha$ follows from the resolution of the equation:

$$\rho \sin\alpha - a\cos^2\alpha = 0 \qquad (7)$$

where $a$ is the distance from the rotation center to the center of wheels. In practice two or three Newton-Rapson iterations suffice to obtain the value of $\alpha$. The motion equations are (3), (4) and

$$\theta = \theta_0 + \int_0^t \left(\frac{V}{R_g} \cdot \tan\alpha\right) dt \qquad (8)$$

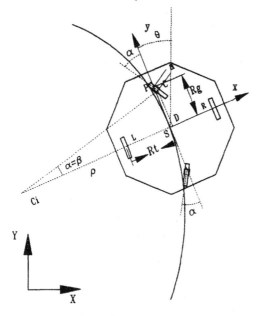

Fig. 4. Kinematic scheme

The values of $\omega_l$ and $\omega_r$ can be computed in such a way that the following relations holds:

$$\omega_r - \omega_l = \frac{1}{r} \cdot (V_r - V_l) = \frac{2VR_t}{rR_g} \cdot \tan\alpha \qquad (9)$$

where $r$ is the wheel radius. However, it should be noted that actuator conflict may arise because of the violation of the perfect rolling assumption, model unnacuracies, and accumulation of errors. Let us consider the force/torque equilibrium at the center of mass. If $T_l$ and $T_r$ are the motor torques of the lateral

wheels. Then, equilibrium equation is

$$(T_r - T_l)\frac{R_t}{r} - (Z_a + Z_b)\mu_r R_g \sin\alpha = 0 \qquad (10)$$

where $\mu_r$ is the friction coefficient, and $Z_a$, $Z_b$ are the normal reactions of the terrain on the front and rear wheels which can be determined from the dynamic analysis that follows.

If the left side of equation (10) is not zero, there is slippage. This slippage can be corrected by tuning the control velocities $\omega_l$ and $\omega_r$. Equations (5) and (8), and the measures of the current orientation and velocities, can be used to determine the control velocities $\omega_l$ and $\omega_r$ avoiding slippage.

Dynamic analysis

The main purpose of this analysis is to provide limit conditions which can be used in the trajectory planning and control. We are assuming navigation in paved floors with a top speed of 1.77 m/sec. Thus, we do not consider a full dynamic model involving all the vehicle-ground interactions.

Let $F_x$ and $F_y$ respectively be the centrifugal force in a curvilinear trajectory and be the braking force, both expressed in the frame attached to the vehicle. Consider a vehicle, such as VAM-1, with independent suspensions in the lateral wheels (see Fig. 5).

If the vehicle's roll angle at the equilibrium is $\beta$, the equations of this equilibrium can be written with the D'Alambert formulation as:

$$\sum M_a = 0; \ \sum M_x = 0; \ \sum F_z = 0 \qquad (11)$$

where $M_a$ is the moment at the supporting point of the front wheel, $M_x$ is the moment at the $x$ axis, and $F_z$ is the normal force at the origin of the coordinate frame attached to the vehicle.

Furthermore, the equilibrium equations at the suspensions are

$$Z_r \cdot b = R_r = k\delta_{cr} = k\delta_e + k(R_t \cdot \beta) \qquad (12)$$

$$Z_l \cdot b = R_l = k\delta_{cl} = k\delta_e - k(R_t \cdot \beta) \qquad (13)$$

where $Z_r$ and $Z_l$ are the normal reactions in the right and left wheels, $R_r$ and $R_l$ are the forces acting in the right and left springs, $k$ is the spring constant, $\delta_{cr}$ and $\delta_{cl}$ are the full deformation of the right and left springs, $\delta_e$ is the static deformation and $b$ is the eccentricity coefficient of the suspension. These equations can be easily solved to obtain the normal reactions in the front wheel $Z_a$, the rear wheel $Z_b$, the left wheel $Z_l$, and the rigth wheel $Z_r$:

$$Z_r = Ca_x + \frac{k\delta_e}{b}; \ Z_l = \frac{k\delta_e}{b} - Ca_x \qquad (14)$$

$$Z_a = \frac{Mh_g}{2R_g}a_y + A; \ Z_b = B - \frac{Mh_g}{2R_g}a_y \qquad (15)$$

In these expressions $A$, $B$, $C$ are functions of $k$, $\delta_e$, $b$, $R_g$, $R_{tt}$, and $h_g$, and $a_x$ and $a_y$ are the aceleration at the $x$ and $y$ axis, corresponding to the forces $F_x$ and $F_y$. These normal reactions can be used to assure the vehicle stability conditions. Assume the vehicle is turning left. When $F_x$ is increasing, $\beta$ increases until the reaction $Z_l$ is zero. In this case the equilibrium equation must be solved to determine the values of $a_x$ and $a_y$ for which the reaction in any other wheel is also zero. These values determine the vehicle stability margin.

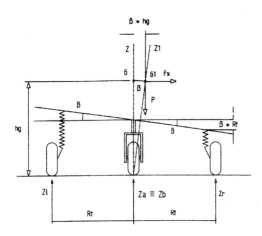

Fig. 5a Front view

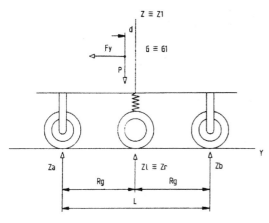

Fig. 5b. Lateral view

Fig. 5. The Suspension System

For a maximum braking acceleration $a_y = 2$ m/sec$^2$, the maximum allowable value of the centrifugal acceleration to avoid unstability is 10.5 m/sec$^2$. For the top velocity (1.77 m/sec) the minimum allowable value of the turning radius is 0.27 m.

The expressions (14) and (15) can also be used to study the lateral slippage. Assume that the frictional forces in the wheels $W_a$ and $W_b$ are perpendicular to the wheels plane, and the magnitude of these forces depends linearly of the normal reactions in the wheels (Kumar and others, 1989; Shiller and others 1991). Then, the equilibria equations can be formulated including the inertia torques. These equations can be solved to compute the values of $a_x$ and $a_y$ for which $F_r > \mu Z_n$ (Coulomb unequality), where $F_r$ is the friction force, $\mu$ is the static friction coeficient, and $Z_n$ is the normal force. In the fig. 6. we show the relations between the vehicle's velocity and the radius of curvature for several values of the braking acceleration $a_y$. These relations define the security margin to avoid the lateral slippage, and can be used as constraints for the velocity planning.

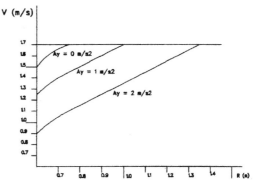

Fig. 6. Velocities to avoid lateral slippage

## CONTROL MODEL

The precise positioning of VAM-1 must be done by using detailed models of the vehicle's velocity and steering control loops. The identification of these models has been carried out by using the kinematic and dynamic equations presented above, and parameter estimation techniques from experimental data.

The steering control system for navigation in unstructured environments is diagrammed in Fig. 7. There are three low level control loops for the two lateral driven wheels (differential steering) and for the pair front-rear wheels (maneuvering steering). All the static and dynamic parameters of these control loops have been estimated. The value of the control voltage for the maximun angular velocity (17.73 rd/sec.) at the axis of the lateral wheel is 10.06 volts. The relation between the command voltage $V_c$ and the motor voltage $V_m$ provided by the amplifier is piecewise linear with a gain of 3.6 for $-0.58 \leq V_c \leq 0.58$, and a gain of 3.6 for $-10.06 \leq V_c < -0.58$ and $0.58 < V_c \leq 10.06$. The static relation between the command signal and the axis velocity is linear with a gain of 1.76.

The open loop dynamic of the positioning system can be represented by the transfer function

$$G(s) = \frac{2.004}{s(1 + 0.28s)} \qquad (16)$$

The consideration of this simple dynamic plays an important role in some steering methods (Shin, 1990; Ollero, 1991). The encoder resolution is 199 counts/rd.

The steering control loop use as reference signals the position $(X_k, Y_k)$, heading $(\theta_k)$ and curvature $(\kappa_k)$ provided by the local path planner. Both the pure pursuit method (Amidi, 1990) and predictive path tracking (Ollero, 1991) have been implemented. The position estimation and control period of this control loop is $T_s = 100$ msec. The position estimation use dead-reckoning equations assuming that the vehicle velocity and curvature are constant during $T_s$. In these conditions the position and heading are estimated by using the simplified differential model that results from the differentiation of equations (3), (4) and (5), by mesuring directly the heading $\theta$ with the north-seeker and applying the discretized equations to estimate the position.

Furthermore, to avoid error accumulation, the

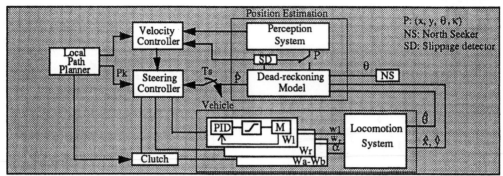
Fig. 7. Steering control system

perception system is requested, and the position estimation is reinitialized, when the distance travelled is large enough (see switch in Fig. 7).

The maneuvering control model is also activated by the local path planner. In this case the wheel $W_a$ is steered with $\kappa = \alpha$ and the references for wheels $W_l$ and $W_r$ are computed to maintain this steering. However, due to model imprecisions and unmodelled dynamics, some actuator conflict may araise, originating slippage. For slow motion this slippage can also be evaluated and corrected as described in the previous section. However, for fast motion, if the slippage is detected, the velocity is decreased and the perception system is requested to provide precise position estimation. This information is used for tuning the current relation between the velocities of wheels $W_l$ and $W_r$.

## CONCLUSIONS

We have summarized the mechanical design of VAM-1, a new wheeled robotic vehicle capable of navigating in indoor and paved floor unstructured environments. This design is the result of several issues related with control, planning and perception requirements for navigation and operation in several conditions. VAM-1 has a simple and unexpensive dual locomotion system for trajectory following and maneuvering in cluttered areas.

We have also presented the locomotion and control models of VAM-1 including both kinematic equations and simple dynamic relations. These dynamic relations are useful to detect and correct the slippage, and to provide limit conditions (constrains) to be used in trajectory planning and control. Furthermore, the dynamic of the actuation system is important for the steering and velocity control.

The presented models are currently being used for the design of new advanced motion planning and control systems.

## REFERENCES

Amidi O. (1990). *Integrated Mobile Robot Control*. The Robotics Institute Technical Report CMU-RI-TR-90-17, Carnegie Mellon University, Pittsburgh, Pensylvania, 15213.

Cox I. C. (1991). *Blanche - An Experiment in Guidance and Navigation of Autonomous Robot Vehicle*. IEEE Transaction on Robotics and Automation, Vol. 7, no. 2.

Holland J.M. (1985). *Rethinking Robot Mobility*, Robotics Age, Vol. 7, No 2, pp. 12-13.

Hollis R. (1977). *Newt: A Mobile, Cognitive Robot*, Byte, Vol. 2, No. 6, pp. 30-45.

Fitzpatrick K. W. and J. L. Ladd (1989). *Locomotion Emulator: A tesbed for navigation research*. World Conference on Robotics Research: The Next Five Years and Beyond

Gonzalez J., A. Stenz and A. Ollero (1992). *An Iconic Position Estimator for a 2D Laser Rangefinder*. To appears in the proceedings of the IEEE Conference on Robotics and Automation, Niza.

Kumar V. and J. Waldron (1989). *Actively coordinated vehicle system*. ASME J. Mechanisms, Trans. Automat. Design, Vol 111, pp. 233-231

Moravec H. P. (1980). *Obstacle Avoidance and Navigation in Real World by a Seeing Robot Rover*, PhD Thesis, Department of Computer Science, Standford University.

Moravec H. P. (1985). (editor) Autonomous Mobile Robots Annual Report. Robotics Institute Technical Report No. CMU-RI-TR-86-1, Carnegie Mellon University, Pittsburgh, PA.

Muir P. - Neuman Ch. (1986). *Kinematic Modeling of Wheeled Mobile Robots*. Carnegie Mellon University Pittsburgh, Pennsylvania 15213.

Muñoz V., J. Martinez, A. Ollero and J. González (1991). *Integrated Simulation System for Industrial Robots* . Second Congress of the Spanish Robotics Society, Zaragoza, pp. 341-348.

Nelson L. (1987). *Continous-steering function control of robots carts*. Internal Report AT&T Bell Laboratories.

Nilsson N.J. (1984). *Shakey the Robot*. Technical Note 323, Artificial Intelligence Center, Computer Science and Technology Division. SRI International, Menlo Park CA.

Ollero A. - Amidi O. (1991). *Predictive Path-traking of Mobile Robots. Aplications to the CMU Navlab*. Proc. of the Fifth International Conference on Advanced Robotics, Pisa, Vol. II, pp. 1081-1086.

Shiller Z. - Yu-Rwei (1991). *Dynamic Motion Planning of Automomous Vehicles*. IEEE Transactions on Robotics and Automation. Vol. 7, nº 2.

Shin V. (1990). *High performance tracking of explicit paths by roadworthy mobile robots*. Ph. D. Dissertation. Carnegie Mellon University.

## ACKNOWLEDGMENTS

Several design ideas was developed during the stay of the first author in the Field Robotics Center (Robotics Institute, Carnegie Mellon University, Pittsburgh USA). Comments and suggestions of William "Red" Whittaker, Director of this Center, and several researches are very appreciated.

# ON-LINE ESTIMATION OF SUBSTRATE AND BIOMASS IN ACTIVATED SLUDGE PROCESS

**R. Moreno\*, C. de Prada\*\*, I. Serra\*, M. Poch\*\*\* and J. Robuste\*\*\***

*\*Unidad de Ingenieria de Sistemas y Automatica, Universidad Autonoma de Barcelona,
08193 Bellaterra, Barcelona, Spain*
*\*\*Departamento de Ingenieria de Sistemas y Automatica, Universidad de Valladolid, 47011 Valladolid, Spain*
*\*\*\*Unidad de Ingenieria Quimica, Universidad Autonoma de Barcelona, 08193 Bellaterra, Barcelona, Spain*

## Abstract

One of the main problems to control in real-time the activated sludge process lies in the difficulty for getting reliable on-line measurements of two of the most important states: substrate and biomass concentrations. Therefore, it's crucial to provide the control system with a software sensor, in order to get on-line (estimated) measurements of these two states. This paper presents two different designs of a software sensor for substrate and biomass estimation. The first one is a two-steps on-line identification approach, based in RLS algorithm used to estimate the OUR variable, and non-linear Kalman filtering to estimate substrate and biomass, using the estimates values of OUR. The second one, based in a linear digital model of substrate and biomass dynamics, is a "black box" on-line identification approach. For both of them, specific developments and comparative results are presented and discussed. All results are compared with experimental data collected in a wastewater treatment plant, located near Barcelona.

## Keywords

Activated Sludge Process. Software sensosrs. Recursive on-line estimation. Kalman filtering. U-D algorithm.

## Introduction

The basis of the activated sludge process in waste water treatment lies in maintaining a microbial population (biomass) transforming, and reducing, the biodegradable pollution (substrate), with dissolved oxygen being supplied by aeratosrs. One scheme of this process is shown in Fig. 1.

Real-time control of the activated sludge process in waste water treatment constitutes a quite complex control problem, due to different factors as: the great variability of the process because on the different composition of the inlet water, the meteorological conditions or the complexity from a microbiological point of view.

However, the main difficulty to control in real-time this process lies in getting reliable on-line measurements of the variables to be controlled. In fact, main variables of the process state are measured by analysis in the laboratory. So, they are available to operation with delays of several hours.

Therefore, a software sensor algorithm have to be included in the real-time control system for getting on-line, but estimated, measurements of the variables to be controlled.

There are several variables of interest in the process state: the substrate S, concentration of soluble organic matter, the biomass X, concentration of microorganisms into the bioreactors, the OUR (Oxygen Uptake Rate), indicator of biological activity, or the concentration of dissolved oxygen C, into the reactors. In this paper, we will focus the attention on estimating substrate and biomass variables.

Fig. 2 shows a suitable control scheme to a wastewater treatment plant, where the delay block represents the time we must wait to get values of the state variables, because on the laboratory analysis. This characteristic marks the necessity of the estimation block, in order to get on-line values of the output variables and be able to control the plant in real-time.

This paper presents the design of two different estimation blocks suitable to be included in a scheme like that in Fig. 2, to the particular case of on-line estimation of S and X.

## Methodology and tools

All the work is related to the municipal wastewater treatment plant of Manresa, a town of 100,000 people situated 50 Km. from Barcelona.

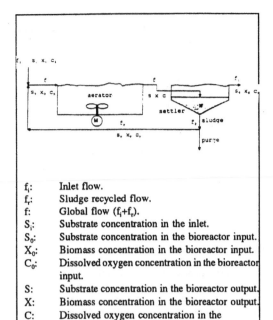

$$\frac{ds}{dt} = -\mu \frac{sx}{K_s+S} + f_{xd} k_d \frac{x^2}{S} + \frac{f}{V}(s_0-s) \quad (1)$$

$$\frac{dx}{dt} = y\mu \frac{sx}{K_s+S} - k_d \frac{x^2}{S} + \frac{f}{V}(x_0-x) \quad (2)$$

$$\frac{dc}{dt} = a(t) k_{1a}(cs-c) - OUR + \frac{f}{V}(c_0-c) \quad (3)$$

where,

| | | |
|---|---|---|
| V | : | Volume of the bioreactors. |
| a(t) | : | Air flow rate. |
| $k_{1a}$ | : | Global mass transfer of oxygen coefficient. |
| cs | : | Saturation level of the dissolved oxygen concentration. |

The model, completed with the settler dynamics, has been simulated with ACSL simulation language, applying experimental data sets, taken from the plant, for the inlet and recycled variables.

Through an optimization procedure, the parameters of the model have been adjusted, achieving an acceptable matching between the answers of equations (1) and (2) and the experimental measurements of S and X [12].

Thus, the model can be used as a substitute of the process to generate values of S and X, and be able to test the estimation algorithms in the "best situation". In addition, the model is a fundamental part of the estimation procedure based in Kalman filtering, as will be exposed later.

In general terms, all the work presented in this paper has been developed with off-line computation, but applying the mentioned experimetal data sets, and simulating the real-time operation of the plant.

**On-line estimation of substrate and biomass**

Method 1

The general strategy of this first method is shown in Fig. 3.

The first thing we can notice from this diagram is that this first method constitutes a two-steps indirect way to estimate S and X. First, OUR variable is estimated via Recursive Least Squares (RLS) identification algorithm. Then, S and X, are estimated via non-linear Kalman filter [3], where OUR estimates are given as a measured outputs.

OUR has been recognized long time ago as an important measure of the activated sludge process, from wich valuable information on the entire process can be inferred [6]. With the aid of a respirometer, it can be measured, but as a part of a real-time control system, it has to be estimated.

| | |
|---|---|
| $f_i$: | Inlet flow. |
| $f_r$: | Sludge recycled flow. |
| f: | Global flow ($f_i+f_r$). |
| $S_i$: | Substrate concentration in the inlet. |
| $S_0$: | Substrate concentration in the bioreactor input. |
| $X_0$: | Biomass concentration in the bioreactor input. |
| $C_0$: | Dissolved oxygen concentration in the bioreactor input. |
| S: | Substrate concentration in the bioreactor output. |
| X: | Biomass concentration in the bioreactor output. |
| C: | Dissolved oxygen concentration in the bioreactor output. |
| $f_1$: | Outlet flow. |
| $f_2$: | Sludge flow. |
| $f_3$: | Sludge purge flow. |
| $S_d$: | Substrate concentration in the outlet. |
| $X_d$: | Biomass concentration in the outlet. |
| $X_r$: | Biomass concentration in the recycled sludge. |

Fig. 1: Scheme of the activated sludge process.

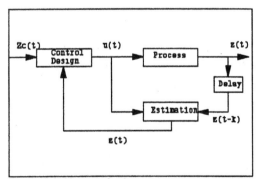

Fig. 2: Suitable control scheme to wastewater treatment plant.

Experimental data sets of substrate and biomass have been collected from the plant, in order to compare our estimated values with real measurements.

In addition, a dynamical model of the process, similar to the most extended in the literature, for control purposes, [1],[2], has been depeloped.

For the three main state variables of the process (s, x, and C, in the reactors), the equations of the model are:

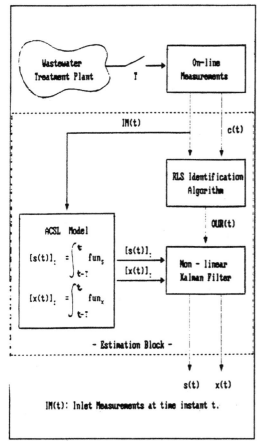

IM(t): Inlet Measurements at time instant t.

Fig. 3: On-line estimation of S and X via RLS and non-linear Kalman filter.

For the on-line identification of OUR, a digital model for the dissolved oxygen dynamics is used, obtained by linearization and discretization of equation (3). Considering this model as a digital CARMA model, we can estimate OUR as a parameter with RLS algorithm. In fact, we applied RLS, modified by U-D factorization for the covariance matrix P, and using directional forgetting factors [5]. This approach proved to be a powerful and numerically robust strategy, and good OUR estimates were obtained. Further details of this OUR estimation procedure can be found in [4].

Other important results about OUR estimation can be found in the literature, e.g. Goto and Andrews [7], Ko, McInnis and Goodwin [8], Rundqwist [9], Holmberg [10], Marsili-Libelli [11].

About the second step, the on-line estimation of S and X via non-linear Kalman filter, we can briefly expose what the situation is.

Asuming the process to be modelled by a non-linear equations system of the form:

$$x(k+1) = f_k[x(k),w(k)] \qquad (4)$$

$$y(k) = g_k[x(k),v(k)] \qquad (5)$$

the applied algorithm for non-linear Kalman filtering [4] has the form:

A. Propagation step.

$$x(k+1/k) = f_k[(x(k/k)] \qquad (6)$$

$$P(k+1/k) = A(k)P(k/k)A(k)^T + \Gamma(k)Q\Gamma(k)^T \qquad (7)$$

B. Actualization step.

$$x(k+1/k+1) = x(k+1/k) + L(k+1) \, [y(k+1) - C(k)x(k+1/k)] \qquad (8)$$

$$L(k+1) = P(k+1/k)C(k+1)^T \\ [C(k+1)P(k+1/k)C(k+1)^T + R]^{-1} \qquad (9)$$

$$P(k+1/k+1) = P(k+1/k) - L(k+1)C(k+1)P(k+1/k) \qquad (10)$$

being,

x, the process state array
P, the estimates covariance matrix
L, the Kalman gain matrix
Q, the covariance matrix of the state noise w.
R, the covariance matrix of the v output measurements noise.

$$A(k) = \frac{\delta f_k}{\delta x(k)} \Big|_{x(k/k)} \qquad (11)$$

$$\Gamma(k) = \frac{\delta f_k}{\delta w(k)} \Big|_{w(k/k)} \qquad (12)$$

$$C(k) = \frac{\delta g_k}{\delta x(k)} \Big|_{x(k/k)} \qquad (13)$$

In our application, we are considering OUR to be the output Y(k) of the process non-linear model (equation (4)), expressed as

$$OUR = K_o sx \qquad (14)$$

At each sampling instant t, equation (6) is computed via on-line simulation with the non-linear dynamical ACSL model, for getting the initial state predictions to the Kalman filter. This step constitutes an integration of equations (1) and (2) process (fun$_s$ and fun$_x$, in Fig. 3), over the time period (t-T,t), with initial conditions of this integration being done by s(t-T) and x(t-T), calculated, at the previous sampling instant, by the Kalman Filter, and needed information being completed by IM(T), as shown in Fig. 3.

Covariance matrix Q is obtained by off-line simulations, calculating the errors between the answers of the ACSL model and the experimental measurements of S and X.

Covariance matrix R is also obtained by off-line simulation calculating the errors between OUR estimated and measured values.

Considering a sampling period of 2 hours, the estimated values of S and X, compared with the experimental measurements are shown in Fig. 4.

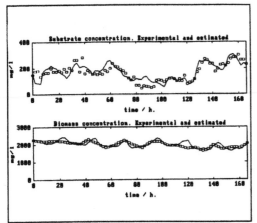

Fig. 4: On-line estimates of S and X via RLS and non-linear Kalman filter method.

In our opinion, these results can be considered acceptable for control purposes, realizing that errors in OUR estimation step are accumulated in the Kalman filtering procedure and that has to be reflected in the results.

We have also to notice the entire process is done without using any measurement of S and X into the reactors. Indeed, the only information needed about the process variables is:

-    Measurements of inlet and recycled variables.
-    Measurements of dissolved oxygen concentration into the reactors.

Te rest of involved variables are computed by the own method.

The next experiment we developed was to incorporate in the estimation procedure the available delayed information of S and X. This idea is shown in Fig. 5.

This delayed information is used in order to provide the initial conditions to the integration of equations (1) and (2) process. Thus, the integrations arise everytime from experimental values of S and X. In spite of that, we have to notice that now, the sampling interval for estimation has to be adjusted to the delay time we are receiving samples of S and X.

In the plant of Manresa, the time needed to analyze a sample of substrate is about 4 hours. Respect to the biomass, it is about 2 hours. Thus, we have to choose a sampling interval for estimation greater or equal to 4 hours.

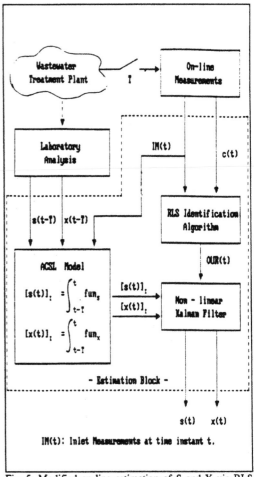

Fig. 5: Modified on-line estimation of S and X via RLS and non-linear Kalman Filter.

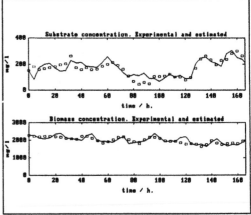

Fig. 6: On-line estimates of S and X via modified RLS and non-linear Kalman filter method.

Accordingly of that, we choose a sampling interval for estimation T=4 hours and the obtained results are shown in Fig. 6. Although the sampling interval is larger, these results are, as a minimum, as good as those shown in Fig. 4. The only thing we should to consider, in order to include this estimation approach in a real-time control system of the process, is if a sampling interval of 4 hours is acceptable in our aplication.

## Method 2

In this second method, we consider the substrate and biomass dynamics to be modelled by the following equations:

$$A_s s(t) = B_s x(t) + D_s f_i(t) + E_s f_r(t) + G_s s_i(t) + C_s (t) \quad (15)$$

$$A_x x(t) = B_x s(t) + D_x f_i(t) + E_x f_r(t) + C_x (t) \quad (16)$$

This is a discrete linear model in wich relationships between substrate and biomass, with the rest of the process variables, are expressed.

A, B, D, etc., are polynomials of the delay operator $z^{-1}$, whose orders have to be determined.

At each time instant t, since we only have values of $s(t-h_s)$ and $x(t-h_x)$, being $h_s$ and $h_x$ the delays due to the analysis of substrate and biomass, respectively, we can estimate only the parameters of the delayed model

$$A_s s(t-h_s) = B_s x(t-h_s) + D_s f_i(t-h_s) + E_s f_r(t-h_s) + \\ + G_s s_i(t-h_s) + C_s (t-h_s) \quad (17)$$

$$A_x x(t-h_x) = B_x s(t-h_x) + D_x f_i(t-h_x) + \\ + E_x f_r(t-h_x) + C_x (t-h_x) \quad (18)$$

and use this parameters in equations (15) and (16) to predict x(t) and s(t), under the assumption that the process dynamics is slow enough, because, for the predictions, we are using parameters delayed several hours.

The choosen time delays are: $h_s = 4$ hours and $h_x = 2$ hours, quite closed to the normal plant operation.

Notice that, in most cases (depending on the sampling period), in the application of (15) and (16) to predict S and X, we have to use previous estimated values for some of the involved variables. This feature is, of course, conditioning the behaviour of the estimation procedure.

Both discrete linear models describing the dynamics of substrate and biomass, can be seen as digital CARMA models and so, a couple of recursive identification algorithms can be used to estimate separetely the parameters of equations (17) and (18).

The identification algorithm applied is again, the RLS with U-D factorization and directional forgetting factors.

The entire identification scheme is graphically displayed in Fig. 7.

The first experiment we made to test the identification scheme was to generate values of S and X, by the non-linear ACSL model, using experimental data

sets for the inlet variables. Then, given the delays $h_x$ and $h_s$ and considering that we have a sample (but delayed) of s and x every 2 hours, we applied the exposed method to estimate S and X, with this sampling interval for identification. The results are shown in Fig. 8. Since the "experimental" values of S and X are obtained without any error, we can know what are the best results we can achieve with this estimation approach.

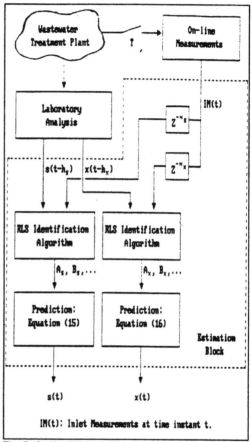

Fig. 7: On-line "black-box" estimation of S and X.

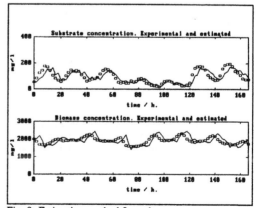

Fig. 8: Estimation method 2 results.

471

The following test to make is repeat the procedure, but now comparing the estimates with real experimental measurements of S and X, for the same sampling period of 2 hours. These results are shown in Fig. 9.

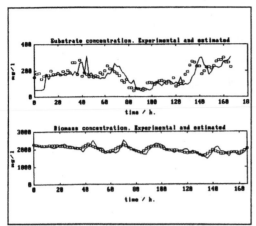

Fig. 9: Estimation method 2 results.

The final experiment was made considering we had a sample of substrate and biomass every 4 hours. For this sampling period, Fig. 10 shown the results of the identification procedure. We can see that the shape of experimental data is followed, but now, we had to use some previous estimated values of S and X in the prediction step, and this is reflected in the results.

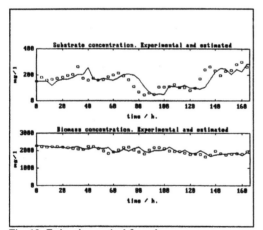

Fig. 10: Estimation method 2 results.

Respect to the orders of polynomials A, B, D, etc., arising from the continuous ACSL model, by linearization and discretization, the most appropriate order results to be 2 in all cases.

## Conclusions

In this paper, two different approaches to design an on-line recursive estimator of S and X in the activated

sludge process have been presented and discussed.

The first one, mainly based in non-linear Kalman filtering, can either works with or withoutout using the available delayed measurements of S and X, and good enough results for control purposes are achieved in both situations. The choice is referred to the sampling period we desire.

The second one, based in a digital model of S and X dynamics works using the delayed information of S and X and good estimation results are also achieved. In addition, because on the way we are modelling S and X dynamics, this second method is quite suitable to be included in a digital adaptive control system of the activated sludge process.

## References

[1]   Olsson, G. and Hansson, O.: Modelling and Identification of an Activated Sludge Process. Proc. of the 4th IFAC symposium on Identification and System Parameter Identification, Tiblisi, USSR. Pergamon Press, Oxford (1976).

[2]   Marsili-Libelli, S.: Modelling, Identification and Control of the Activated Sludge Process. A d v a n c e s   i n   B i o c h e m i c a l Engineering/Biotechnology, Springer-Verlag (1989).

[3]   A. Ollero. Control por Computador. 211-213. Marcombo, 1991.

[4]   C.de Prada, R.Moreno, M.Poch, J.Robusté. Recursive estimation of OUR in activated sludge process. ECC'91, Grenoble, France. July 1991.

[5]   Hagglund, T.: New Estimation Techniques for Adaptive Control, (Ph.D. 1983)

[6]   Stenstrom, M.: and Andrews, J. F.: Real Time Control of Activated Sludge Process. J. Environ. Eng. Div., EE2, 245-260 (1979).

[7]   Goto, M. and Andrews, J. F.: On-line Estimation of Oxygen Uptake Rate in the Activated Sludge Process in: Instrumentation and Control of Water and Wastewater Treatment and Transport System (ed. rake, R.A.R.), p. 867, Oxford, Pergamon Press (1985).

[8]   Ko, McInnis and Goodwin, Automática (1982).

[9]   Rundquist, ADCHEM'88.

[10]  U. Holmberg, Proc. IFAC Modelling and Control of Biotechnological process (1986).

[11]  S. Marsili-Libelli. Adaptive estimation of bioactivities in the activated sludge process. IEE Proceedings, Vol. 137, Pt.D, No. 6, November 1990.

[12]  J. Robusté. Modelització i identificació del procés de fangs activats. Ph. D. Thesis. Universitat Autónoma de Barcelona. October 1990.

## Acknowledgements

The authors wish to record their appreciation for the financial support received from the CICYT spanish program (ROB89-0479-C03-01).

# ADVANCED GOVERNOR FOR HYDROELECTRIC
# TURBINES[1]

## J. Riera and R. Cardoner

*Institut de Cibernetica (UPC-CSIC), Diagonal 647, 08028 Barcelona, Spain*

**Abstract.** This paper deals with a prototype of a speed and power control system for an hydroelectric power unit. Its structure and basic features are described. Main characteristics are the controller adaptive parameter algorithm with preprogrammed gains, the functions incorporated in sensors and actuators like autocalibration and self-diagnosis.

**Keywords.** : Hydrogenerator control; Adaptative Control; Reliability; Fault and error detection.

## INTRODUCTION

Continuously evolving conditions on power system networks require new and more demanding characteristics to generating units. In spite of patches, some old hydroelectric units are inappropiate and unefficient for use.

This is the case for a hydroelectric plant on the Ebre river. Its control system is composed of three concentric control loops (see figure 1). The inner, and oldest one, is the conventional hydro-mechanic speed governor with fixed structure and, despite certain adjustments can be made, its characteristics are also basically fixed. The speed is sensed by means of a fly ball pendulum and the regulating function is performed by a combination lever and a pilot valve actuating on sleeve valves. These control the oil flow to the servomotors that adjust the opening of the turbine runner and gate vanes.

The second loop is an electro-mechanic controller of the opening of the turbine gate. It gets the order from the outer loop and the actual opening value from a metering potentiometer. The control function is simply proportional. The dynamic behaviour of the system is strongly characterized by the dead band and saturations of the control elements.

The generated power is controlled by the outermost loop. The command signal is received from the Load-Frequency Control (LFC) of the Energy Management System (EMS) and compared with the output power. The controller is electronicaly implemented and performs a PI algorithm.

This control system has some serious drawbacks:
- The hydro-mechanic governor suffers from very limited flexibility: fixed controlling function, manually adjustable parameters with a reduced range values; a mechanic cam links turbine runner and gate controls; the maintenance has turned very difficult due mainly to its antiquity, no spare parts and the complexity of the precision mechanism.

- Neither of the most external control loops can compensate, in an effective way, the intrinsec defects of the internal levels: the opening control shows strong nonlinear behaviour; the power control presents large histeresi and limited adjusting capabilities.

All these disadvantages provided enough arguments to undertake the design of a new advanced controller compatible with the Integrated Control System (ICS) that was being developed for the plant.

## THE PROPOSED NEW CONTROL SYSTEM

### Objectives

A new controller for the plant had, evidently, to overcome the drawbacks of the existing one but had also to accomplish three other major objectives:

*Open architecture* for a not difficult incorporation of new features.
*Compatibility with the ICS* for a complete integration.
*Robustness and reproducibility.* Both necessary for an easy adaption to other turbines in the same and in other plants.

A brief description of the characteristics of the proposed system and its principal elements follows.

### Characteristics

The main characteristics of the control system are

- Speed and generated power control with local or LFC remote commands
- System dynamic response optimized in the whole operation range
- Control algorism allowing operation over small isolated or large interconnected networks
- Automatic unit synchronization with the external network
- Intelligent sensors and actuators with fault detection and self calibration
- Reduced cost of the interface with the hydraulic servomotors
- Supervision of the behaviour of the elements and whole system

### Structure

A schematic diagram of the structure of the proposed control system is shown in figure 2. The system is built around an Intel 286 based central controller which performs all functions related with speed and generated power as well as those functions related with the supervision and system security.

[1] This work has been supported by Empresa Nacional Hidroeléctrica del Ribagorzana - ENHER.

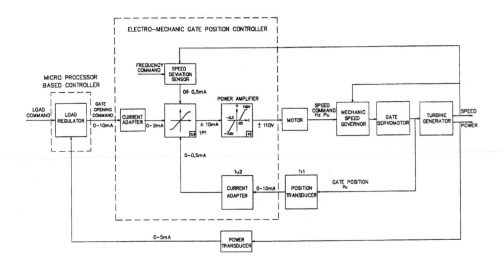

Figure 1: Original Control System.

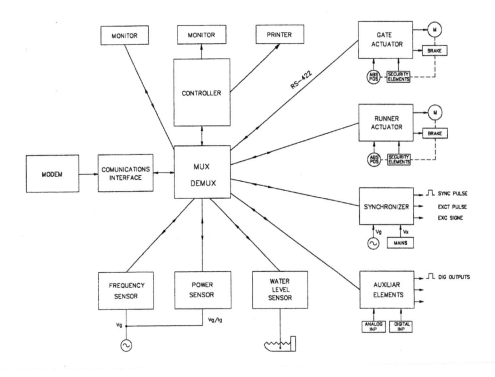

Figure 2: Structure of the Proposed Control System.

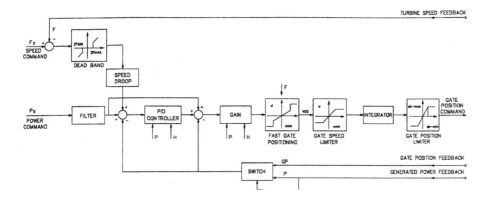

Figure 3: Control algorithm.

The information flux is coordinated by the Multiplexor De-multiplexor linking all the peripherals with the controller. The communications protocol allows reliable transmission of orders and data via a serial link.

The information required by the governor is of two kinds: the commands proceeding from the operator or from the EMS and the information proceeding from the process sensors (unit and grid frequencies, voltages and their relative phase; unit generated power; gate and runner openings; water head sensor) and information of auxiliary elements (oil and refrigerating systems....). These sensors send also information relative to its performance and can receive specific orders.

There are two sort of actuators: the synchronizer and the gate and runner actuators. The last ones allow to obtain the servomotors opening. The synchronizer allows the connection to the external grid. The orders and set points to the actuators are sent by the controller, which in turn receives complementary information about the actuators state or performance.

A more detailed description of these elements is given in following sections.

### The Controller

The controller has two main functions: first controlling the speed and the power of the unit and second supervising that the whole system and its elements work according to specifications.

The proposed control algorithm strategy permits adjust the response for the two set points commands (frequency and power) and operating in any of the two possible modes (group isolated - with or without load - and connected to the grid) without external information of the state of the network. This feature is attained by the feedforward stucture of the algorithm (see figure 3) and by automatic selecting the feedback between power and gate opening.

The dynamic behaviour of the turbine depends heavily on its water flow and on the water head of the dam. To optimize the control system response the digital controller has been provided with an adaptative parameter algorithm with preprogrammed gains: the best response is adjusted off-line for different working points and the corresponding parameters retrieved when the system is on-line. This agrees with Fasol (1987), Wozniak (1988), Orelind (1989) and Ye (1990). Interesting contributions concerning off-line tuning are proposed by Sanathanan (1988) and Wozniak (1990).

There are also emergency procedures that modify the algorithm structure, eg: when overspeed appears caused by sharp pertorbations.

Supervision is carried on in different ways: the response of each actuator is compared with that obtained from a reduced order model (this agrees with Jiang (1987); evaluating the coherency between the different sensors information and the behaviour of the whole system. The results of that supervision may be orders to peripherals like "change physical transducer 1 for physical transducer 2, or "repeat measure", or "reset", or "close servomotor"....

### Sensors and Actuators, Main Features.

The structure adopted for the control system, figure 2, requires that the peripherals present some degree of intelligence and realize automatically functions that, in a conventional controller, would be done by the central processor. Some of these functions are filtering, calibration, self-adjusting, fault detection...

Many advantatges are obtained: an important reduction of the noise existing in measured signals; an easier incorporation of new elements to the system; a rational task distribution implying more adapted and high level functions; facilitate comprehension of schemes and programs.

All peripherals are built around a Siemens 80515 microprocessor responsible of calculations, decisions and coordination of the different elements.

In the line followed by Ye (1990) and Ham (1988) redundancy is incorporated to the system. Each sensor has redundant measures from independent signal sources. After filtering and analysis of the sensed magnitudes, a comparison and selection is done depending on: amplitude with respect to the range of measurement; differences between sensors data; gradient of the data samples. Sensors also incorporate automatic fault detection in their hardware and software.

Main characteristics of actuators are: accurate position control; reliable operation; auto-calibration and automatic error detection; execution of specific complex commands e.g. controlled closure of the turbine.

The principal features of some of the most important peripherals are summarized in Table 1. Next sections describe more in detail two of the peripherals: the synchronizer and the servomotor actuators.

## TABLE 1 Main peripherals characteristics

| Element | Unit | N° Inp | Converter type | Analog Filter | Digit. Filter. | Accur. %/N.V. |
|---|---|---|---|---|---|---|
| Power Sensor | Mw | 4 | A/D 12bits Aprx/scc | 2nd ord. 2.5 Hz Butt. | 3nd ord. T=25ms | 0.2% |
| Freq. Sensor | Hz | 2 | Ck Count. | | Digital Fault detect. | .005% |
| Head Sensor | m | 2 | A/D 12 bits | | Digital | 0.2% |
| Syncro-nizer | V Hz Rad | 2 2 | A/D 12bits Aprx/scc Ck Count. | 2nd ord. 2.5 Hz Butt. | 3nd ord. T=25ms Digital Fault | 0.2% .005% .031% |
| Gate actuat. | mm | 2 | Absolut Encoder 12 bits. | no | Digital Fault Detect. | .024% |
| Runner actuat. | mm | 2 | Absolut Encoder 12 bits. | no | Digital Fault Detect. | .024% |

**The Synchronizer**. The synchronizer is essential for the correct coupling of the unit alternator to the network. Figure 4 shows the structure adopted for the synchronizer. Its basic functions are:

- Generation of control pulses to the synchronous machine excitation system in order to vary the unit output voltage level accordingly to a desired voltage. The reference set point is provided by the main system controller.

- Generation of the Closing pulse in order to connect the group to the external network. The closing order is also given by the central controller but the synchronizer verifies that instantaneous conditions ensure a good manoeuvre.

Variables needed by the synchronizer are:

- Vg Unit voltage
- Vx Network voltage
- Fg Unit frequency
- Fx Network frequency
- $\phi_{xg}$ Relative phase

Vg and Vx are obtained from the bus bars previous transformer reduction and isolation. Once rectified and analog filtered, signals are acquired with a A/D converter and digitally filtered to ensure the quality of the measure.

Determination of frequencies and relative phase is realized by a digital counter of internal clock pulses. The counter is governed by the zero crossed AC voltages signals. Since the calculation time is frequency dependent a dynamic error detector has been adopted: this takes into account the last measures, the physical value and rate of change limits.

The whole, measured and calculated, variables are sent to the central controller which determines the gate opening and the excitation pulses needed to carry on the syncronization.

The generation of the Closing pulse is done if a controller order exists and previous verification of the following conditions:

- The relative phase between Vg and Vx and its rate of change is in a predefined range
- The voltage difference between Vg and Vx doesn't exceeds a specified threshold
- The frequency difference between Fg and Fx doesn't overcomes a prescribed limit.

If any of the precedent conditions is not fulfilled the synchronization is aborted and a message of "failed manoeuvre" is sent to the Controller.

**The Actuators**. The role of actuators is to obtain the gate and running opening demanded by the controller in order to get the necessary mechanic intake power for the desired turbine speed and generated electric power. They have mechanic and electronic integrated functions and are the most complex elements in the system.

Figure 5 is a schematic description of the actuators structure. This structure has been previously designed and validated by simulation. Figure 6 shows the blocs diagram used.

Two loops define the system characteristics; the external one feedbacks the real opening (with information from an absolute optical position sensor) to the microprocessor; the inner loop (using data from an optical incremental encoder) fedbacks the sleeve valve position to the motor controller obtaining an accurate position control ( $\pm$ .015mm).

Since the actuator controls critical power elements, it has been designed to have the highest degree of reliability. So a series of actions have been taken in the electronic circuit and in the actuator controller. For example, a high degree of electric isolation exists between the microprocessor and the other elements. The actuator controller performs, in addition to the regulation algorithm, a succession of verifications of the whole system:

- The installation and calibration of the position sensor. If necessary an autocalibration is done.
- The mechanic installation of the motor-screw-sleeve valve.
- The motor-controller-encoder ensemble performance.
- The cyclic execution of the control algorithm.
- The mechanic interface with the power hydraulic system requires a previous centering of the sleeve valve position. This centering, and other tests concerning the performance of the hydraulic system, are done automatically at the start-up of the system. .

The results of all these actuations are communicated to the controller. Any irregularity detected causes a message to the controller which takes a decision and sends the appropiate orders, e.g. disconnect the unit from the grid, close the servomotor and shut down the unit.

**The Communications**.

The objective of the Multiplexor - Demultiplexor (M-D) and the communications protocol is to obtain a reliable communication system, with a compact communication message incorporating indistinctly data or orders and independently of the element (sensor or actuator) that sends or requires information.

So the basic tasks of the multiplexor are:

- Add to each information message proceeding from a sensor or actuator a header that identifies the emitting channel and transfer it to the controller. On the contrary, suppress in each information message proceeding from the controller, the header that identifies the destination, and send it by the corresponding channel.
- Notify the controller all the communication errors detected and supress all the defective messages.

The information protection is carried on in a double way: a geometric code that combines the horizontal with the vertical parity and a system that incorporates a two bits header to each byte ensuring the message synchronization.

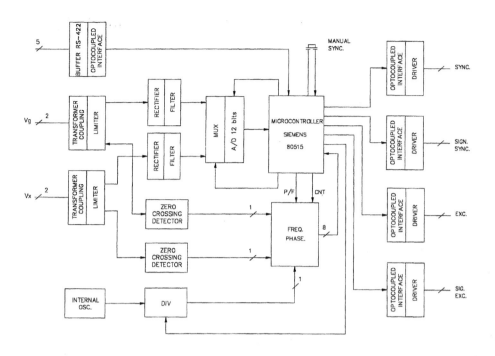

Figure 4: Synchronizer structure.

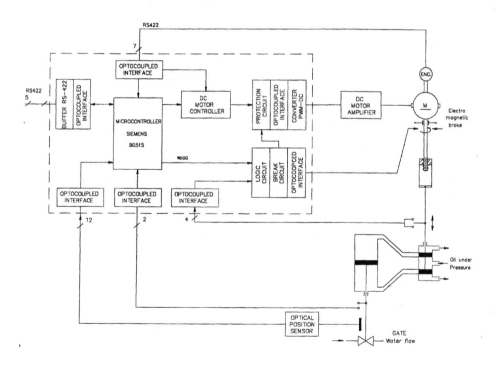

Figure 5: Schematic diagram of the actuators.

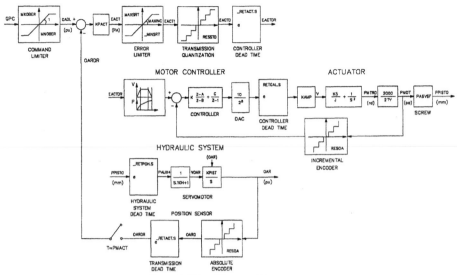

Figure 6: Blocs Diagram of the actuators.

The M-D is based on a CPU Z80-B and seven SIO/Z80-B peripherals allowing 12 communication channels with sensors or actuators.

All communications are realized using the RS-422 Full-Duplex communications standard with no Handshake hardware. This implies an additional software complexity, but allows a larger speed and a greater efficiency of the communications as well as a reduction of the economic cost of the system.

## CONCLUSIONS

The main objectives accomplished with the new control system are:

*Efficient control.* The dynamic response of the system fulfils all the specifications.
*Reliability.* The features incorporated in the controller and peripherals ensure high degree of reliability of the whole system.
*Robustness and reproducibility.* Both necessary for an easy adaption to other turbines, in the same and in other plants.
*Open architecture.* With easy incorporation of new features and with easy integration in the whole plant Integral Control System.
*Reduced cost.* The modifications necessaries in the oil-circuit that operates servomotors has been reduced to the minimum.

The system is yet implemented and has passed satisfactorily the first field tests.

## REFERENCES

Fasol,K.H., L. Huser and W. Westerthaler (1987). Digital Control of 180MW Francis Turbines: Controller Design, Simulation and Operational Results. International Conference on Hydropower: Waterpower'87.

Filbert, T. and L. Wozniak (1988). Speed Loop Cancellation governor for Hydrogenerators. Part II Application. IEEE Trans. on Energy Conversion, vol 3, No. 1, March 1988, pp 91-94.

Ham, P. and N. Green (1988). Developments and experience in Digital Turbine Control. IEEE Trans. on Energy Conversion, Vol. 3. No. 3, September 1988, pp 568-574.

Jiang, J. and R. Doraiswami (1987). Reliable Hydraulic Turbine Governor Based on Identification and Adaptive Filtering. IEEE Trans. on Energy Conversion, Vol. 2. No. 2, June 1987, pp 189-195.

Orelind, G., L. Wozniak and J. Medanic (1989). Optimal PID Gain Schedule for Hydrogenerators - Design and Application. IEEE Trans. on Energy Conversion, vol 4, No. 3, September 1989, pp 300-307.

Regalado, J. (1987). Frequency Control with Kaplan Turbines. Sulzer Technical Review, 3/1987, pp 33-37.

Sanathanan, C. (1988). A Frequency Domain Method for Tuning Hydro Governor. IEEE Trans. on Energy Conversion, vol 3, No. 1, March 1988, pp 14-17.

Ye,L., Wei S., Xu H., O. Malik and G. Hope (1990). Structure and Time-varying Parameter Control for Hydroelectric Generating Unit. IEEE Trans. on Energy Conversion, vol 4, No. 3, September 1990, pp 293-299.

Wozniak,L. and D. Bitz (1988). Load-Level-Sensitive Governor Gains for Speed Control of Hydrogovernors, IEEE Trans. on Energy Conversion, vol 3, No. 1, March 1988, pp 78-84.

Wozniak, L. (1990). A Graphical Approach to Hydrogenerator Governor Tuning, IEEE Trans. on Energy Conversion, vol 5, No. 3, September 1990, pp 417-421.

# PREDICTIVE CONTROL OF TURBO-ALTERNATOR TERMINAL VOLTAGE

## M. Saidy and F.M. Hughes

*Department of Electrical Engineering and Electronics, UMIST, Manchester, UK*

**Abstract.** A predictive form of generator excitation control is proposed which has the potential for consistent and significantly improved performance character-istics over conventional schemes. The basic principle of the proposed form of control are presented and its performance capabilities are demonstrated via simulation studies on a test system and compared with those of conventional Automatic Voltage Regulator (A.V.R).

Keywords . Control, excitation system, turbo-alternators, predictive control.

### INTRODUCTION

The increasing trend towards digital excitation control for turbo-alternators, the ever improving capabilities and reducing costs of digital processors, coupled with the availability of reliable and proven mathematical models of turbo-alternators promotes the opportunity to assess the capabilities of more sophisticated control schemes than those currently employed.

In this paper, an excitation control scheme based on one-step ahead prediction is proposed for the improved control of turbo-alternator terminal voltage.

The proposed controller employs measured values of terminal and field voltages, currents and rotor angle. Values direct and quadrature components derived from these are employed directly in the predictive model. By making use of current measurements, the non-linear coupling between the generator and the rest of the power system is essentially eliminated, enabling prediction to be based on a linear generator model. The controller consequently provides a performance which is consistent and varies little with changes in generator operating conditions. This is in sharp contrast to the situation when conventional automatic voltage regulators are employed.

The paper covers the basic principles of predictive controller and derives the digital algorithms which are employed. The discrete model used is based on a standard d-q axis model of a synchronous generator. Consideration is given to sampling period requirements, extrapolation methods for estimating future values of generator variables, and the effect of the excitation system dynamics on capabilities and performance.

The performance of the proposed control scheme is demonstrated through its application to a single generator feeding a large system. Generator responses are compared with those available from a conventional AVR controller for load changes on the generator busbar and following a three phase short circuit of duration 100 msecs.

### PREDICTIVE CONTROL CONCEPT

The performance of a synchronous generator varies significantly with operating conditions mainly due to the non-linear relationships which exist between power flow and current in a transmission line and the phase angle across it. Hence as a generator load angle varies, so do the dynamic characteristics between the field voltage and the terminal voltage.

The dynamics seen by an Automatic Voltage Regulator (which employs terminal voltage variations to automatically adjust field voltage) vary widely with operating conditions making it difficult to maintain good performance with fixed parameter controllers.

Adaptive controllers have been investigated widely to combat the situation as surveyed by Hughes (1990), but in this paper an alternative predictive control concept is proposed.

Observation of the tried and tested models of synchronous generator indicate that, if saturation is ignored, the equations relating generator voltages to disturbances in field voltage and stator currents are linear with constant coefficients. Hence if the generator stator currents, which are dictated by the system to which the generator is connected, are known, then a linear constant model can be used for the prediction of dynamic behaviour and can form the basis of a fixed parameter predictive controller.

Load variations on a power system produce changes in the stator currents of the synchronous generators feeding it. If these currents variations are known, then their influence on generator terminal voltage can be predicted and the variations needed in the generator excitation voltage to counteract the disturbance and maintain the terminal voltage at its desired level can be calculated.

In this paper it is shown how from measured values of stator terminal voltages and currents and field current a fixed parameter predictive controller can be developed which significantly improves generator voltage control performance.

479

## GENERATOR MODEL

A generator model, described by Hammons and Winning (1971) and widely used in power systems analysis is presented below. The generator rotor is modelled dynamically by a field winding plus damper windings on the d and q axes. The stator windings are assumed to have an instantaneous response and therefore are modelled by algebraic equations. The mechanical inertia effects of the rotor system are also modelled.

$$\frac{dE'_q}{dt} = \frac{1}{T'_{do}} \; [E_{fd}-(X_d-X'_d)I_d-E'_q] \qquad (1)$$

$$\frac{dE''_q}{dt} = \frac{1}{T''_{do}} \; [E'_q-(X'_d-X''_d)I_d-E''_q]+\frac{dE'_q}{dt} \qquad (2)$$

$$\frac{dE''_d}{dt} = \frac{1}{T''_{qo}} \; [(X_q-X''_q)I_q-E''_d] \qquad (3)$$

$$\frac{d\omega}{dt} = \pi f_o/H \; [P_m-P_e] \qquad (4)$$

$$\frac{d\delta}{dt} = \omega - 2 \, \pi \, f_o \qquad (5)$$

$$V_t = \sqrt{V_d^2+V_q^2} \qquad (6)$$

$$V_d = E''_d + X''_q*I_q \qquad (7)$$

$$V_q = E''_q - X''_d+I_d \qquad (8)$$

$$P_e = V_d*I_d + V_q*I_q \qquad (9)$$

It can be seen that the model has constant coefficients and can be expressed in state space form as

$$[\dot{E}] = [A][E] + [B][I] + [C][E_{fd}] \qquad (10)$$

where $E = [E'_q \; E''_q \; E''_d]^t$ - state vector;

$\quad\; I = [I_d \; I_q]^t$ - disturbance vector;

$\quad\quad E_{fd}$ - control input

In the studies carried out the following set of parameter values were employed which correspond to a salient pole generator typical of those use in hydro-stations.

Direct axis parameters :

Synchronous reactance $X_d$ = 1.445 p.u

Transient reactance $X_d'$ = 0.316 p.u

Sub-transient reactance $X_d''$ = 0.179 p.u

Transient time constant $T_{do}'$ = 5.26 s

Subtransient time constant = 0.026 s

Quadrature axis parameters :

Synchronous reactance $X_q$ = 0.959 p.u

Sub-transient reactance $X_q''$ = 0.162 p.u

Inertia constant H = 4.27 s

The application of a first order Euler approximation to the continuous equations enables these equations to be transformed to the following discrete form for predictive purposes.

$$\frac{E'_q(k+1)-E'_q(k)}{T}=\frac{1}{T'_{do}} \; [E_{fd}(k)-(X_d-X'_d)I_d(k)-E'_q(k)] \qquad (11)$$

$$\frac{E''_q(k+1)-E''_q(k)}{T} = \frac{1}{T''_{do}} \; [E'_q(k)-(X'_d-X''_d)I_d(k)-E''_q(k)]$$
$$+ \frac{E'_q(k+1)-E'_q(k)}{T} \qquad (12)$$

$$\frac{E''_d(k+1)-E''_d(k)}{T} = \frac{1}{T''_{qo}} \; [(X_q-X''_q)I_q(k)-E''_d(k)] \qquad (13)$$

$$V_d(k) = E''_d(k) + X''_q*I_q(k) \; ; \qquad (14)$$

$$V_q(k) = E''_q(k) + X''_d*I_d(k) \; ; \qquad (15)$$

$$V_t(k) = \sqrt{V_d^2(k) + V_q^2(k)} \qquad (16)$$

## PREDICTIVE ALGORITHM

Using the discrete model of the previous section, the generator responses to stator current disturbances can be determined.

From the quadrature axis damper equation

$$E''_d(k+1) = \frac{T}{T''_{qo}} \; (X_q-X''_q)I_q(k) - (\frac{T}{T''_{qo}}-1) \; E''_d(k) \qquad (17)$$

and since

$$E''_d(k) = V_d(k) - X''_q*I_q(k) \qquad (18)$$

$$E''_d(k+1) = (\frac{T}{T''_{qo}} X_q-X''_q)I_q(k) + (1-\frac{T}{T''_{qo}})V_d(k) \qquad (19)$$

The direct axis component of the terminal voltage at sample (k+1) is given as

$$\tilde{V}_d(k+1) = E''_d(k+1) + X''_q\hat{I}_q(k+1) \qquad (20)$$

where $\hat{I}_q(k+1)$ is an extrapolated value of current $I_q$ obtained from present and past measured values. Knowing this predicted value of $\tilde{V}_d(k+1)$, the value needed for $V_q(k+1)$ to maintain the terminal voltage at its required value $V_{t-set}$ can be calculated from

$$\tilde{V}_q(k+1) = \sqrt{V_{t-set}^2 - \tilde{V}_d^2(k+1)} \qquad (21)$$

where $V_{t-set}$ is the reference set point value for generator terminal voltage.

From $\tilde{V}_q(k+1)$, the required value for $E''_q(k+1)$ can be obtained as

$$\tilde{E}''_q(k+1) = \tilde{V}_q(k+1) + X''_d*\hat{I}_d(k+1) \qquad (22)$$

where $\hat{I}_d(k+1)$ is an extrapolated value for current $I_d$. Rearrangement of equations (11) and (12) enables the following expression to be derived

$$E_{fd}(k) = \frac{T'_{do}}{T} \tilde{E}''_q(k+1) + (\frac{T'_{do}}{T''_{do}} - \frac{T'_{do}}{T})\, E''_q(k) +$$

$$+ [1 - \frac{T'_{do}}{T''_{do}}]\, E'_q(k) + [\frac{T'_{do}}{T''_{do}}(X'_d - X''_d) + (X_d - X'_d)] I_d(k) \tag{23}$$

also since

$$I_{fd}(k) = E'_q(k) + (X_d - X'_d) * I_d(k) \tag{24}$$

and $\; E''_q(k) = V_q(k) + X''_d * I_d(k) \tag{25}$

equation (23) can be re-expressed as

$$E_{fd}(k) = \frac{T'_{do}}{T} \tilde{E}''_q(k+1) + [1 - \frac{T'_{do}}{T''_{do}}] I_{fd}(k) +$$

$$+ [\frac{T'_{do}}{T''_{do}} X_d - \frac{T'_{do}}{T} X''_d] I_d(k) + [\frac{T'_{do}}{T''_{do}} - \frac{T'_{do}}{T}] V_q(k) \tag{26}$$

$E_{fd}(k)$ is the field voltage required at sample (k), calculated from measured voltages and currents available at sample (k), to provide the desired terminal voltage at sample (k+1).

Values of direct and quadrature axes voltages and currents can be obtained from measured phase voltages and currents and rotor angle via Park's transformation, as described by Elgerd (1982).

## EXTRAPOLATION METHODS

Extrapolated values of current components $\hat{I}_q(k+1)$, $\hat{I}_d(k+1)$ are needed to estimate the values $\hat{V}_q(k+1)$ and $\tilde{E}''_q(k+1)$ respectively. Various methods for extrapolation exist.

### Zero Order Hold (Z.O.H)

The current components $I_d, I_q$ are considered constant during the sampling period so that

$$\hat{I}_d(k+1) = I_d(k) \tag{27}$$

$$\hat{I}_q(k+1) = I_q(k) \tag{28}$$

### Linear Extrapolation (L.E)

Knowing the previous values of $I_d$ and $I_q$ the sampling time (k), (k-1) the extrapolated values of $I_d(k+1)$ and $I_q(k+1)$ are evaluated as follows

$$\hat{I}_d(k+1) = 2 * I_d(k) - I_d(k-1) \; ; \tag{29}$$

$$\hat{I}_q(k+1) = 2 * I_q(k) - I_q(k-1) \; ; \tag{30}$$

### Quadradic Extrapolation (Q.E)

The values of current components $I_d$ and $I_q$ at the sampling time (k), (k-1), (k-2) are needed in order to evaluate $I_d(k+1)$ and $I_q(k+1)$. Using a second order Taylor's expansion $\hat{I}_d(k+1)$ and $\hat{I}_q(k+1)$ can be expressed as :

$$\hat{I}_d(k+1) = 2.5 * I_d(k) - 2 * I_d(k-1) + 0.5 * I_d(k-2) \; ; \tag{31}$$

$$\hat{I}_q(k+1) = 2.5 * I_q(k) - 2 * I_q(k-1) + 0.5 * I_q(k-2) \; ; \tag{32}$$

## TEST SYSTEM

The test system used to assess the capability of the predictive controller is shown schematically in Fig.1. It comprises a generator having a local load and connected to a large network via a reactive tie line. The large power network is represented as an infinite busbar connection. Local load disturbances can be applied by switching in specified load admittances.

The generator parameters are those stated in section 3. A static excitation scheme where the exciter response is very fast and can be considered instantaneous as shown in Fig.2.

The performance of the proposed predictive controller was compared with that of a conventional excitation control system (type 1S) as described in IEEE Committee Reports (1968, 1981) and shown in Fig.3.

## PERFORMANCE ASSESSMENT

### Extrapolation Method

In order to assess the influence of the type of extrapolation method, a series of studies was carried out involving a large disturbance in local load, in the form of a switched load admittance of Y = 0.5-0.3j.

The response obtained for sampling values of 10 msecs. and 60 msecs. are shown in Figs.4 and 5 respectively. With a 10 msecs. sample period the responses using linear and quadratic extrapolation are almost identical, and give improved performance over the case where currents are assumed constant between samples.

When the sample period is increased to 60 msecs. performance differences become more apparent with the quadratic extrapolation method proving best.

### Sampling Period

The effect on performance of varying the sampling period is shown in Fig.6, which presents responses following the local load disturbance for the linear extrapolation case. As expected, it can be seen that, the performance deteriorates as the sample period increases.

The best compromise between performance achieved and computational burden was considered to be provided by a sample period of 20 msecs. using linear extrapolation.

It can be seen that the predictive controller provides an excellent voltage response, with fast well damped recovery following the load disturbance.

### Comparison with Conventional A.V.R Type 1S

The performance of the predictive controller was compared with that of a standard static A.V.R (type 1S), for both a local load disturbance and for three phase short circuit at the generator terminal of 100 msecs. duration.

The responses of Figs.7 and 8 indicate that in both cases the predictive controller provides superior voltage control performance.

### Variation in Operating Condition

The consistency of the dynamic response and performance achieved with the proposed predictive control scheme is demonstrated by the responses shown in Figs.7,9 and 10. These correspond respectively to unity power factor (P = 1 p.u., Q = 0), lagging power factor (P = 0.8 p.u.,

Q = 0.6 p.u) and leading power factor (P = 0.8 p.u, Q = 0.2 p.u) generator operating condition for the same load disturbance.

The predictive controller provides an almost identical terminal voltage response characteristic which is fast and well damped for the three operating conditions considered.

## CONCLUSION

The proposed controller based on one step-ahead prediction has been shown capable of providing excellent terminal voltage control following load and short circuit disturbances.

With a conventional A.V.R where control is based essentially on a terminal voltage feedback signal, the dynamic performance changes significantly over the operating range. By making use of measured stator current values as a known disturbance in addition to voltage conditions, the proposed predictive controller eliminates a major source of dynamic performance variation and presents the capability of providing good consistent performance over the operating range of the generator.

## REFERENCES

Elgerd, O.I., (1982). Electric Energy Systems Theory : An Introduction. McGraw-Hill Book Company, New York, 2nd Ed.

Hughes, F.M., (1990). 8th Brazilian Automatic Control Congress, IFAC Sep 1990, Belem, Brazil, Vol.1, 18-26.

Hammons, T.J., Winning, D.J., (1971), Comparison of synchronous machine models in the study of the transient behaviour of electrical power systems. Proc.IEEE Vol.118, No.10, 1442-1458.

IEEE Committee Report (1968). Computer Representation of Excitation Systems. IEEE Trans.on Power Apparatus and Systems, Vol.PAS-87, 1460-1464.

IEEE Committee Report (1981). Excitation System Models for Power System Stability Studies IEEE Trans.on Power Apparatus and Systems, Vol.PAS-10, No.2, 494-509.

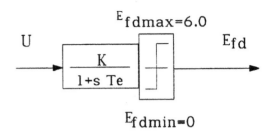

Fig. 2: Exciter System
K = 200 ; Tex.= 0 sec.

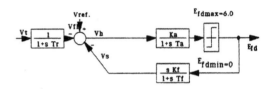

Tr = 0.015 sec.
Ta = 0.03 sec. ; Ka = 198
Tf = 0.371 sec.; Kf = 0.02

Fig. 3: AVR type 1S

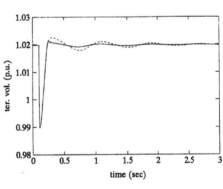

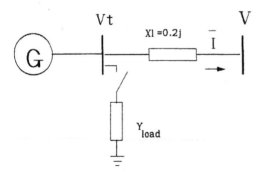

Fig. 1: Generator connected to an infinite busbar system

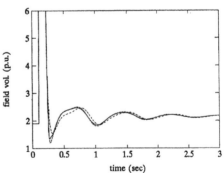

Fig. 4: Effect of Extrapolation Methods with T=10 msec. Tex.=0 sec.
-.-. Z.O.H. - - - L.E. —— Q.E.

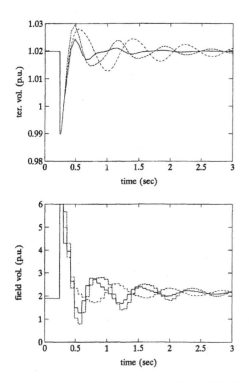

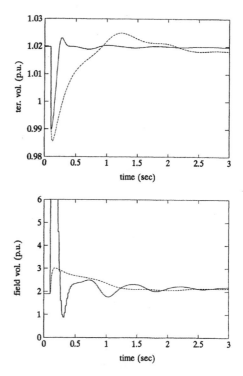

Fig. 5: Effect of Extrapolation Methods
with T=60 msec. Tex.=0 sec.
-.-. Z.O.H. - - - L.E. ⎯⎯ Q.E.

Fig. 7: Effect of A.V.R. type
(large load disturbance case)
- - - A.V.R. type 1S, Tex.=0 sec.
⎯⎯ Voltage Predictive Controller

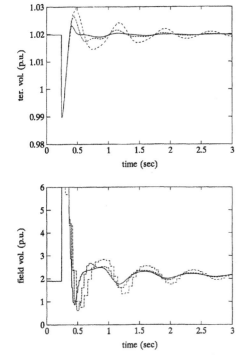

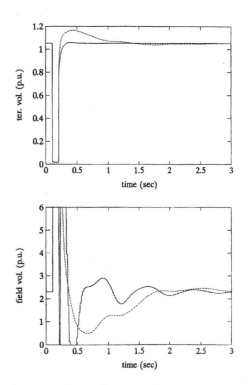

Fig. 6: Effect of Sampling Time T
with L.E. , Tex.=0 sec.
⎯⎯ T=20msec. - - - T=40msec.
-.-. T=60msec.

Fig. 8: Effect of A.V.R. type
(3-phase short circuit case)
- - - A.V.R. type 1S, Tex.=0 sec.
⎯⎯ Voltage Predictive Controller

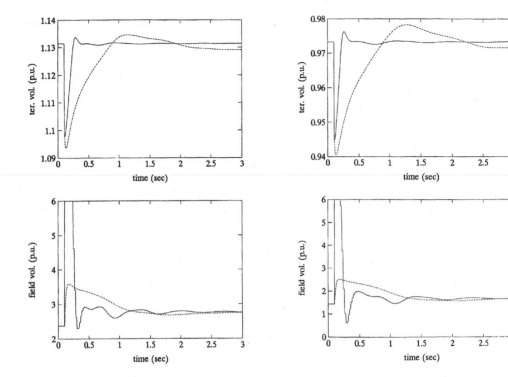

Fig. 9: Variation of Operating Condition
P=0.8 p.u. Q=-0.6 p.u
- - - A.V.R. type 1S
——— Voltage Prdictive Controller

Fig. 10: Variation of Operating Condition
P=0.8 p.u. Q=0.2 p.u.
- - - A.V.R. type 1S
——— Voltage Predictive Controller

It © IFAC Intelligent Components and Instruments
ɔl Applications, Malaga, Spain, 1992

# SURVIVABILITY IN SATELLITE INSTRUMENT CONTROL APPLICATIONS

## M.H. Jing and L.J.C. Woolliscroft

*Department of Automatic Control and System Engineering, University of Sheffield, Mapping Street, Sheffield S1 3JD, UK*

Abstract.A small fault tolerant system is described which has been designed as an instrument controller for space applications. It uses three processor modules and some limited shared resources. Central to the operation of this system is the manner in which the information on the performance of both hardware and software is updated and voted using adaptive voters. The system uses triplicated TMR to construct a dual level architecture and is capable of graceful degradation to duplex and even simplex configurations. A possible extension to five processors is outlined.

Keywords. Adaptive systems; aerospace control; data handling; fault handling; fault tolerance; redundancy; TMR; adaptive voter.

## INTRODUCTION

A satellite instrument control system is normally rather small having very limited replication of sub-systems and few resources for graceful degradation. This paper in concerned with the application of fault-tolerant ideas to such small systems and is intended to show how such techniques may be used in future space missions. It is based on the Cluster (ESA/NASA) and ELISMA (USSR Mars-94 mission) DWP instruments but takes fault-tolerance further than these two specific instruments (Woolliscroft, 1988). Three microprocessors are used in this system for data collection and handling, instrument control, scientific applications and internal task distribution.

In the space environment transient faults occur more often than with equivalent equipment in the terrestrial environment (Rasmussen, 1988; Siewiorek, 1982). This is mainly due to ionizing radiation causing single event upsets. Lack of a real-time closed loop control from the ground for many missions makes automatic onboard recovery from such transient faults valuable. Permanent faults are also more frequent in the space environment (due to radiation, thermal and launch vibration damage) yet the mission objectives usually call for extended survival. Thus it is important to consider both permanent and temporary faults when designing space instrumentation.

We have been studying the ways to improve the reliability of the DWP instrument through the fault-tolerant features. The system hardware architecture is shown in Fig. 1 (Jing, 1990).

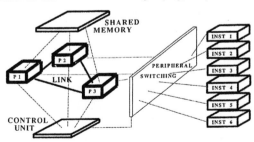

Fig. 1.    The system hardware architecture. The 3 processors may be connected to any instrument and the task allocation for each processor is related to the instruments actually connected. Three types of resource are the shared memory, control unit and links, are provided for the purposes of interprocessor communication and system maintenance data storage. These different resources also support the philosophy of hardware diversity (Gluch, 1986).

Limited redundancy is established to cope with hard (or permanent) failures at certain critical points and so increase the confidence in the fault-tolerant operation of the system. Due to the limited replication of hardware for redundancy, this system has multiple tasks whose functions overlap to obtain a certain functional redundancy (Abbott, 1990). System reconfiguration is then used to try to recover the lost functions. This work uses the concept of a function as the basic unit and it is not important whether a function is implemented in hardware, software or both.

Often, when in a deteriorated mode, a small system will not have sufficient high quality data about its own status to enable it to co-ordinate and maintain its own activities. The centralized and fault-tolerant maintenance data which are used to drive the system discussed in this paper are called System Information or SI. Each local processor uses redundant communication functions to construct a system triple TMR (triple modular redundancy) (Chen, 1978), resourceful design, adaptive voting and recovery mechanisms (Randell, 1975) can been used.

The design objectives for this system are to eliminate any single point failure, to make effective use of limited resources, to tolerate frequent transient faults by error masking and self-reconfiguration, and to increase the survivability (long life).

Fault avoidance is the prevention of faults from occurring and often results from conservative design practices such as the use of high-reliability components, component burn-in, and careful signal path routing. In this work particular fault avoidance techniques have included modular design, the functions have different design to avoid the possibility of common (or similar) errors, the processors are loosely coupled so that no processor idly waits due to a blockage or failure of the others, and each memory space of the critical data storage is isolated or partitioned to prevent the simultaneous update operations.

Fault-masking aims to maintain, by verification and evaluation, the system operation in the presence of errors. The most popular mechanism is majority voting to mask out an error (which is assumed to the minority). After voting a simple comparison of the input and the result will provide an error detection capability. Comparison after voting is used to provide better coverage for error detection within the system.

This system uses quality variables derived from the history of detected errors in the corresponding functions. The quality variables are used as the references for the operation of an adaptive voting mechanism which will described later. A benefit of using these quality variables is that the system may degrade gracefully from triplex down to duplex or simplex with increasing severity of faults.

This system has been built by using three ⓒIBM PC computers as the processor modules with a fourth PC to provide an overview of the system performance. This fourth PC also is used to inject errors into the system and to observe the consequences of these errors.

The paper will concentrate on the SI and the survivability achievement; other aspects of the design of this small system will be described in future papers.

TABLE, 1 The Content of System Information, SI.

| CATEGORY | CONTENT |
|---|---|
| References | Quality of functions, Inst. reconfiguration and Checking references. |
| Status | H/W check, Data stores, Failure of control buffers, Proc. ON/OFF. |
| Messages | Voting result, Proposals, Failure indication, Exception report. |
| Configuration | Current configurartion, Proposed configuration. |
| Control | Mode of operation, Timing, Synchronization signal. |

## THE SYSTEM INFORMATION, SI

The SI lies at the heart of this system since the system manager uses this to decide how to reconfigure the system in the event of a permanent failure and also to recover from a transient error. Physically, at any time, there are 23 copies of the SI in the hardware of the system. But for simplicity we do not discuss all of them here and concentrate on the conceptual number of copies which is 7 as shown in the system architecture (Fig. 2) (Jing, 1991). Once the system has started to function, and in the absence of faults, these are all the same. Some of the other physical locations are to be seen in Fig. 3 where the detail of the shared resources is slightly expanded. Actually in the communication links there are two local copies of the SI, the figure shows the logical connection after the selection by use of the quality variable as will be described later. An outline of the contents of the SI is given in Table 1.

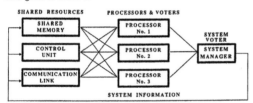

Fig. 2.    The system functional architecture — this is a triplicated TMR architecture with dual levels of processor and system level fault handling and management. The system information is the heart of this system. The system level SI is produced by mutual agreement and is feedback for system error checking, correction and for the establishment of system operational confidence.

To understand the way in which these SI are generated let us start with the assumption that all 7 are populated with a set of SI. How these are initially populated is an initialization task (which must itself be fault-tolerant) and will be described later. Each processor has within its scope four external copies of the SI as shown in Fig. 4. These are the control unit, shared memory and two of the dual-ported memory copies. It votes on three of these to produce its own internal local copy of the SI. The copy in the dual-ported memory with the better value for its quality variable is selected together with the copies in the shared memory and the control unit for the voting. This local voted result is then the SI which is used by that processor. The other two processors do the same to produce their own local voted SI.

Each hardware processor then selects the three local voted results i.e its own plus two from the other processor and the system manager in each processor (in turn) then votes to produce a global copy which is written back to the shared memory, control

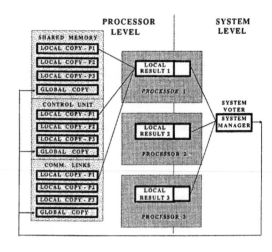

Fig. 3.    The SI distribution in the system is here expanded from Fig. 2. Each processor produces (by voting) a local result from the shared resources. These results support system level voting to produce the global copy. The feedback is performed by each processor for the comparison of local copies with the global copies. As long as any resource remains, the system should remain operation with confidence.

unit and the communication links (ie the dual-ported memories). Then, a little later, the next processor will use its system manager to do the same and thus there is effectively a distributed system manager. Clearly the timing and scheduling of all of these operations is important. Figure 4 shows the general timing.

What if the system manager in one processor is itself faulty? For most faults this would result in the processor quality variable being so low that the processor will be eliminated from the system by the two others. However, in case that does not happen there is a reasonableness check performed on the result of the system manager before the result is written back to the memories.
All of the voters, ie local and system level voters, are adaptive voters which use the quality variable to weight the inputs to the decision.

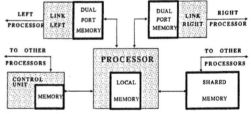

Fig. 4.    The physical location of SI.

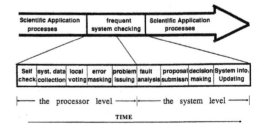

Fig. 5.    The system timing and scheduling: frequent checking is executed when a processor finds fault, the system level does the fault management.

486

## SYSTEM POWER-UP

When the system is started, there is a set-up operation using the initialization procedures. All the processors are not required to switch on at the same time. The first operated processor will set the mode to the simplex configuration. The system connection will start from default setting, and a delay time (e.g. 10 seconds.) is set for the other processors to power-up to eliminate unnecessary reconfiguration. The system's initial procedures do consider that the system may start with a non-healthy state from pre-existing failures.

When the second and third processors are logging in, the system does not need the same initial settings as for the first one, but health checking and fault(s) handling might be needed instead. With the varying number of processors in operation, the system may be reconfigured from simplex to duplex or triplex modes. The newly logged in processor will signal to the system and report its operational status. According to the new status, the system may rearrange the assignment of the system tasks. In this system any failed arrangement in any processor in any mode will activate the reconfiguration procedure.

### SYSTEM MAINTENANCE AND TIMING

In normal operation the system maintenance function is interleaved with the application tasks in each processor. There are frequent health checks and fault handling procedures which ensure that faults are detected.

Figure 5 shows the sequence for one processor. (Because the system is loosely coupled there is no need for the sequence to be tightly synchronized for the three processors; conversely there is a contention control mechanism to prevent a problem if two attempt to update simultaneously.) The figure shows the dual stages of the updating process, local processor level and system level.

When a change in a task is necessary, or a major fault has been found, the manager is activated by the local processor level. This provides the system with the capability of error masking and reconfiguration. This job has to be agreed by at least two of the managers as a minimum requirement to achieve a successful decision. The decision is sent to the operation control for execution and thus a closed loop feedback control of fault handling is completed (Fig. 2).

### FUNCTION CREDIBILITY

The quality variables of the fault-tolerant functions. All the fault-tolerant functions, for example the system manager, have quality variables associated with them. These variables are set by means of an error checking routine. This system has two means of error checking: the built-in self-checking and the checks after the majority voting (error masking). The occurrence of a minority vote for one function or a self-checking error will be counted as a fault, but there may not be a fault in the corresponding function due to intermittent or transient failures of this function. The performance of a particular function must be evaluated by the history of error occurrences over a certain period. This history defines the credibility of this function. Whenever a fault is detected, the credibility of this function is reduced. A quality variable is ascribed to each critical function to record the frequency (from the history) of transient faults. A function will be said to be failed either if it has permanent component faults or if the quality is poor.

The permanent failures of the circuits and interfaces can be detected by themselves or mutual checking. The one with lowest credit is determined from the comparison of the fault rate and the quality variable amongst the multi-versions of functions. The change of quality variables of different fault models has been simulated and shown in Fig. 6. In this figure the result of the single or transient faults had the result in a good condition and the intermittent and permanent failure will give a "BAD" quality.

This quality variable provides the weighting factor for the decision over which dual port memory copy is used for the SI voter.

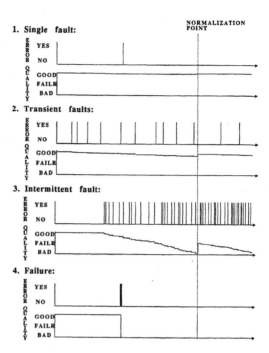

Fig. 6. The change of quality variables for different fault models; incidental or transient faults degrade the system functional performance. Intermittent faults are separated from transient faults and considered as permanent faults.

Updating the Quality Variables. Figure 7 shows the cyclic updating of the quality variable. A quality of a function may be represented as $Q_n(t)$ where $t$ is the discrete time, and the $n$ is the identifier of the different batch of normalization procedures. Figure 8 shows the simulated result of this algorithm which is as follows.

The initial value of a quality variable is

$$Q_0(0) = 100 \qquad (1)$$

The quality update of the first batch is

$$Q_0(k) = 100 + \sum_{t=0}^{k} b_0(t) \qquad (2)$$

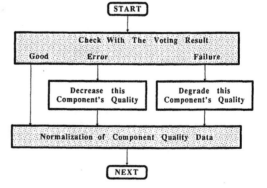

Fig. 7. The updating of the quality variables: following system checking, a failed function is immediately degraded and the quality variable of a function with an error is reduced. Normalization is periodically executed to increase the values of the variables.

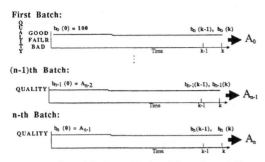

**First Batch:**

**(n-1)th Batch:**

**n-th Batch:**

Fig. 8.    The periodical normalization of the quality variables.

Where $b_0(t)$ is the update of each cycle as shown in Fig. 7. The value of each update is as follows

$$b_n(k) = \begin{cases} 0 & \text{if OK} \\ -1 & \text{if ERROR} \\ -b_n(k-1)+1 & \text{if FAILURE} \end{cases}$$

The updating equation always decreases the values of the variables. After a long run, assuming that most functions have experienced some transient faults, their quality variables will be reduced to a certain extent (they might have a quality variable close to zero). Because the quality variable is only for the comparison of the credibility of related functions, the functions might only have transient faults in some periods or have recovered from the error states, a normalization of the quality variables is needed. $A_n$ is the normalized value of each batch as is described below.

<u>Normalization.</u> The normalization of the variables has two purposes: to raise the value of the variables and to decrease the gap between the the values of the variables. The normalization can be described as

$$A_n = \sqrt{Q_{n-1}(p)} 10 \tag{3}$$

where $p$ is the item number of the last quality value of that batch.

TABLE, 2 Example of Recovery by Normalizations.

| F version(1) | F version(2) | F version(3) | |
|---|---|---|---|
| 1 | 25 | 81 | The initial value |
| 10 | 50 | 90 | 1st normalization |
| 10 | 50 | 81 | F(3) has faults |
| 32 | 71 | 90 | 2nd normalization |
| 56 | 86 | 95 | 3rd normalization |

An example is presented in the Table 2. Initially, the value of F(1) is 1 which is completely failed and the F(3) is in very good condition. We can see that the F(1) can be recovered if it has less fault rate and the F(2) gets closer to F(3). The generalized equation of the quality variable is then

$$Q_n(k) = A_{n-1} + \sum_{t=0}^{k} b_n(t) \tag{4}$$

The quality variables are used as the reference to give confidence in the function which has the lowest failure rate. They are also used as the weight value in the adaptive voter which is a mechanism used on the deteriorated system or the system with subsequent faults.

<u>Voter Design.</u> As this system is required to maintain the critical functions in an unmanned situation, the system should have the capability to degrade gracefully. The straight-forward majority voting for the full function does not meet this requirement, therefore a flexible system is provided which should operate reliably and survive in the deteriorated situation. A voter was designed which uses the quality variables so as to make the system more flexible.

The adaptive voter is shown in Fig. 9. There are three (or more) inputs, each of which also has two variables: the weighting factor and the data input. The weighting factor is directly from the quality of the function which produces the input. In full operation (without any failure), the system operates in multiple modular redundancy mode and the system may use majority voting mechanism. In the deteriorated operation mode, each module may not operate with the same quality and thus the majority may not always be correct (without good confidence). In some circumstances under which the inputs may not agree (total disagreement) or the system loses redundancy, the system is unable to get a majority from the vote, the system has to use comparison or get result from single inputs. In the total disagree or comparison mode, some references are needed.

As mentioned previously, the weighting factors are used to judge the inputs. They are derived from the quality variables which have been built up while the system operates. These references may be used to choose the result, solve conflicts or delete the inputs (actually the whole function). Normally the majority or the greater weighted data will be chosen. If the weight of one function has the greater weight than the sum of the other two, the lower weighted function can be considered as deleted.

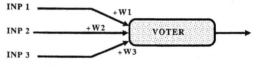

Fig. 9.    The adaptive voting.

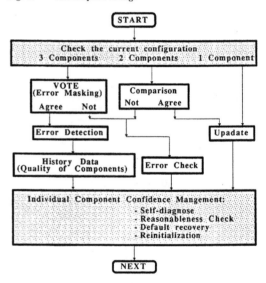

Fig. 10.    The system operation in degraded mode. Different modes of voting and comparison mechanisms are used in different modes. Also some confidence management procedures are used to double check the system information.

## THE GRACEFUL DEGRADATION

Majority voting in the TMR system is used for the single error masking. Due to the effect of the environment and the deterioration of the system, the majority voting mechanism is not absolutely fault free or without ambiguity. The originally redundant functions may lose their replicated members and thus the system will have to perform in a different configuration. In some cases the function may be completely lost and the system then must work in a degraded operational mode. Figure 10 shows the strategy that the system operates in different configurations.

In the TMR mode (triplex operation mode), the error detection function is used after the majority voting. The comparison of each individual data is performed to isolate any disagreement and to build the quality variable of this function (credibility). If there is no agreement in the result, the error check or a compromised result is needed. The compromised result is the result taken from the function with the highest quality or reasonable value.

The adaptive voter may have two or three inputs, depending on the current configuration. With two functions available, a pure comparison will be performed and the better quality function will have more chances to dominate the result. This is the normal way to produce a result from the pure comparison in a duplex mode. The chosen result will also be examined whether it is reasonable, and a wrong output is then eliminated if possible. Once again the unreasonable result will be consider as a failure of the related function.

If the differences is great or one of the results is always wrong, it is the time that a simplex mode should be introduced.
An autonomous system must consider all possible conditions, particularly in the situations under which the common error or the total disagreement take place. The "common error" is when two errors occur at the same time with the same outcome. This situation may result from a design error. This project has also found common errors caused by the operational environment, such as the loss of memory data when power surges occur due to the long wiring used to simulate radiation. The symptom of such errors is that they provide wrong, but uniform results. Another case is a total disagreement which may result in a risk of error contamination to the system. To eliminate this, the critical data have a quick domain check and, periodically, a reasonableness check is initiated. This is another step to find errors beside the normal voting as illustrated in Fig. 11 (Kanekawa, 1989). If any data are wrong (exceptional) (Cristian, 1991), they will be replaced by a default value and signal is sent to the system if necessary. In case of unrecovered error the system may reinitialize the system. This can be seen as a forward recovery procedure.

With a resource failure, the confidence of the system information maintenance is reduced but the system level is still in TMR configuration. If one or two processors fail, the system is then in different operational modes such as duplex or simplex. This is a so called Triplex, Duplex, Simplex (TDS) system, as shown in Fig. 12, and its reliability performance is shown in Fig. 13.

If there is only one processor left, the simplex mode is activated, and the reconfiguration function still remains and may permit a system recovery. If all processors are lost, a time out circuit can be designed and activated to attempt a cold restart which gives the system the last chance.

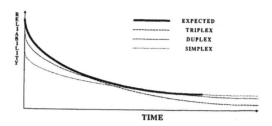

Fig. 13.    The reliability improvement of the proposed system.

CONCLUSION

Verification and Testing Results . There are aspects of the design of this system which have not been described here. It is, however, within the scope of the present work to discuss the results of testing the system. A fourth ©IBM PC has been used to display the status of the system and to permit the injection of faults. The result can be summarized as follows:

- The specific testing is the static testing on the designated (critical) points. The results for all of the individual errors (present as permanent faults) injected was that the system recovered successfully. This has demonstrated that this system has no "single point of failure".

- Transient faults did not cause any major system failure.

- This system may also tolerate certain subsequent and multiple faults. Figure 14(a) shows the modes of various performance, such as the full, minimum configuration, degraded and failure modes. This figure describes the system's capability for tolerating multiple faults after the system builds up its history. Here the adaptive voter can choose the one with the highest quality to serve the system and thus the multiple and transient faults may be injected. The system may also recover just after the errors have been released.

- The worst case of the system which may be recovered, is that one processor and one resource is left. Figure 14(b) shows the various operations from full to stand-alone mode.

- Design diversity eliminates the design errors and this increases the possibility of the system's fault-free design. This project was unable to check the real effect of this factor.

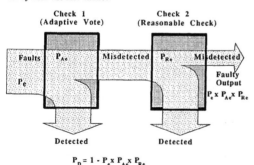

$$P_D = 1 - P_e \times P_{Ae} \times P_{Re}$$

Fig. 11.    The error elimination — from Fig. 10, two steps of error detection will reduce the overall errors.

**(a) Minimum Operation and Failure Mode:**

| MODE OF OPERATION \ BASIC MODULES | NUMBER OF PROC. | NUMBER OF RESOURCE | NUMBER OF PERIPHERAL | NUMBER OF COMMUNICATn |
|---|---|---|---|---|
| FULL OPERATn | 3 | 3 | 6 | 3 |
| MINIM. CONFIG. | 2 | 1 | 6 | 1 |
| DEGRADE MOD. | 1 | 1 - 3 | 1 - 5 | 1 - 3 |
| FAILURE  PROC. FAIL | 0 | 1 - 3 | 1 - 6 | 1 - 3 |
| FAILURE  RESOURCE F. | 2 - 3 | 0 | 1 - 6 | 1 - 3 |
| FAILURE  PERIPHL. F. | 1 - 3 | 1 - 3 | 0 | 1 - 3 |
| FAILURE  COMM. FAIL | 2 - 3 | 1 - 3 | 1 - 6 | 0 |

**(b) Various Operation Modes when Processors Failure:**

| MODE OF OPERATION \ BASIC MODULES | NUMBER OF PROCESSOR | NUMBER OF RESOURCES | NUMBER OF PERIPHERALS | NUMBER OF COMMUNICATn |
|---|---|---|---|---|
| FULL OPERATION | 3 | 3 | 6 | 3 |
| DUPLEX MODE | 2 | 1 - 3 | 2 - 6 | 1 - 3 |
| SIMPLEX MODE | 1 | 1 - 3 | 1 - 6 | 1 - 3 |
| STAND ALONE MODE | 1 | 0 - 3 | 0 - 6 | 0 - 3 |

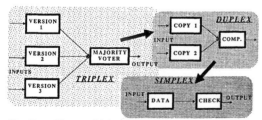

Fig. 12.    The graceful degradation of the TDS system.

Fig. 14.    The graceful degradation of the system.

Some comments on these results can be made as follow:

- The error masking redundancy is the most convenient mechanism to eliminate transient errors (with a good coverage),

- A small system may also adopt fault-tolerant technologies to increase its survivability,

- The testing support has to be planned at the very beginning, This reduces the activity at the integration and validation stages,

- A small system is easy to test, but it still needs a comprehensive testing task,

- The overall testing is very successful,

System Expansion. If more processing power is required in the application, more processors might be needed, for an example, a 5 processor system. With additional processors in this system, several problems might occur such as the complexity of the system and the problem to meet the physical constraints.

With more hardware modules, the system may have more reconfigurability. Figure 15 shows a 5-module system which has 5 of each of 5 processors and resources. As Fig. 16 shows, the adaptive voter in our design does increase the survivability with those extra inputs. Due to the shared resources design, it is necessary to know which of those circuits are the major additional components.

The hardware has used many buffers for switching and multiplexers for error detection purposes. Gate arrays and surface mounting technology are powerful techniques for implementation (Takano, 1988) with limited space and mass. The gate arrays, CPU, memories and other LSIs could be surface mounted on a ceramic base in chip form and sealed to constitute a hybrid IC.

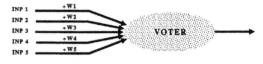

**5 INPUT ADAPTIVE VOTER**

5 INPUT ADAPTIVE VOTER WITH 2 INPUT DISABLED
(DEGRADED TO A TRIPLE MODULAR REDUNDANCY)

Fig. 16.    The adaptive voter of 5 module system.

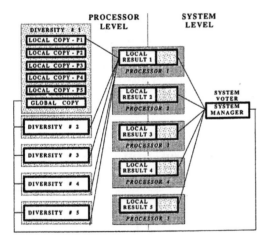

Fig. 15.    The system architecture for 5 module system.

## REFERENCES

Abbott, B.J., (1990), Resourceful System for Fault Tolerant, Reliability and Safety, ACM Computing Surveys., 22, No.1, pp.35-67.

Chen, L. and Avizienis, A. (1978), N-Version Programming: A Fault-Tolerant Approach to Reliability of Software Operation, IEEE Digest of FTCS-8, pp.3-9.

Cristian, F., (1991), Understanding Fault-Tolerant Distributed System, Communication of ACM, 34, No.2, pp.57-78.

Gluch, D.P. and Paul, M.J., (1986), Fault-Tolerance in Distributed Digital Fly-by-Wire Flight Control Systems,, Proc. AIAA/IEEE 7th Digital Avion. Syst. Conf., pp.507-514.

Jing, M.-H., Thompson, J.A., Alleyne, H.St.C., Woolliscroft, L.J.C. and Gough, M.P. (1990) On the Design of Fault-Tolerant instruments: Cluster and Elisma DWP Experience, 4th Intl. Seminar on Scientific Space Instrument Engineering, Frunze, USSR, Sept. 1989., Academy of Sciences of the USSR, IV, pp.1-11.

Jing, M.-H. and Woolliscroft, L.J.C., (1991), A Fault-Tolerant Multiple Processor System for Space Instrumentation, IEE, Intl. Conf. Control '91, 1, No.332, pp.411-6.

Kanekawa, N., Maejima, H., Kato, H. and Ihara, H., (1989), Dependable Onboard Computer Systems with a New Method - Stepwise Negotiating Voting, IEEE Digest of FTCS-19, pp.13-19.

Randell, B., (1975), System Structure for Software Fault Tolerance, IEEE Trans. Software Eng., SE-1, No.2, June 1975, pp.220-232.

Rasmussen, R.D., (1988), Spacecraft Electronics Design for Radiation tolerance, Proc. of IEEE, 76, 1527-1537.

Siewiorek, D.P. and Swarz, R.S., (1982), The Theory and Practice of Reliable System Design, Digital Press, Bedford, Massachusetts.

Takano, T. et al, (1988), Fault-Tolerant Onboard Computer, Proc. ISTS-16, pp.1097-1100.

Woolliscroft, L.J.C. et al, (1988), The Digital Wave Processing Experiment, ESA SP-1103, ISSN 0379-6566, pp.49-54.

# SIMULATION MODEL FOR CATALYTIC MONOPROPELLANT HYDRAZINE THRUSTERS

### A. Cavallo, G. de Maria and P. Marino

*Dept. di Informatica and Sistemistica, Universita Degli Studi di Napoli "Federico II", Via Claudio 21, 80125 Napoli, Italy*

**Abstract.** Hydrazine thrusters are often used as actuators for spacecraft attitude and orbit control systems, for example for the missions of METEOSAT, OTS, INTELSAT and GIOTTO. The reason is that hydrazine requires less hardware than other propellants, and thrusters with thrust levels down to 0.5 N with reproducible minimum impulse bit down to 0.005 N are available.

In order to design the spacecraft attitude and orbit system it is necessary to provide the control engineers with a simulation model of the thruster which allows to have informations at all possible system conditions, and especially during pulsed mode firings. As an analytical model, based on a complete description of the chemical and physical phenomena occurring during thruster firing is not easy to obtain, in this paper we will present a simulation model based on the test firings of the thruster. Simulation results will be provided for a 20 N catalytic hydrazine thruster.

**Keywords.** Actuators, Thruster Model, Thruster Performances, Hydrazine Thrusters

### Preliminaries.

A characteristic of hydrazine thruster time response is the presence, of oscillations and a positive slope, depending on the input pressure, that could dynamically interfere with the controlled system. In Fig. 1 a set of tests for a 20 N hydrazine thruster is reported [1].

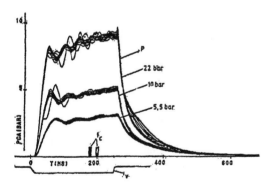

Fig. 1. Firing tests.

Usually systems that exhibit oscillations in the step response are modelers as second order systems, but this is not our case. A second order system would show overshoots also on the falling edge, and Fig. 1 shows that this behaviour is not realistic for a hydrazine thruster. For this reason a usual approach to modelling is to use two different curves, the one fitting the rising edge and the other the falling edge.

This framework is the basis of the work by Kollien and Weiss [2], where it is shown that the 2 N thruster response can be approximated by a linear combination of two exponential functions. But such a model, although adequate for 2 N thrusters, does not match the behaviour of 20 N ones, as one can deduce from Fig. 1. In fact the double exponential approximation, while preserves the mean value of the response, ignores the oscillations, hence its natural objective is to deal only with the energetic aspect of the phenomenon. In

the 20 N thruster case a spectral analysis of the oscillations reveals a meaningful contribution at low frequencies that could affect the structure form a mechanical point of view, and cannot be neglected.

In order to obtain a more accurate model for the 20 N thruster, some preliminary considerations are necessary. Although Fig. 1 shows that the real process is not deterministic, three considerations can be deduced:

*a)* the phenomenon shows an increasing response;

*b)* it is oscillating;

*c)* exhibits finite delays both on the raising and on the falling edges.

As the mathematical model of the process obtained from physical considerations is exceedingly complex, our strategy will be to describe an equivalent model showing the same input-output behaviour.

### Derivation of the model.

From a logical point of view, we can deduce three phases in the response of Fig. 1:
1) a quick rising front,
2) an oscillating phase with an increasing average value
3) a falling front slower than the rising one.
We will start from phase 2, which is the most difficult to describe.

Modelling phase 2 by a linear system would require a second order system at least, with complex conjugate poles. Such a system would show an inevitable link

Fig. 2. Bi-level modulator

among rising time, amplitude and frequency of the

491

oscillations. In order to avoid such a limitation it is possible to use a non linear system which is known in the Industrial Electronic area as *bi-level modulator* [3,4], whose scheme is reported in Fig. 2 and whose behaviour is depicted in Fig. 3.

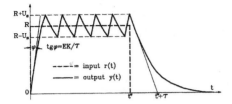

Fig. 3. Input and output of the bi-level modulator.

The response in Fig. 3 has a rising time determined by the variables E and τ in Fig. 2, while the frequency and the amplitude of the oscillation are settled by E, τ and the size of the tolerance band $2U_s$. Such a scheme still retain the interdependence among independent variables and parameters to adjust. Then a further step is required, namely to make the filter in Fig. 2 time-varying according to the differential equation

$$\dot{y}(t) = a(t)y(t) + b(t)u(t). \tag{1}$$

Finally, the increasing mean value of the response will be accomplished by using an input signal increasing with a suitable slope.
The tuning of the variables E, $U_s$ and of eqn. (1) completes the modelling of phase 2.

Now phases 1 and 3 must be described without altering the values previously set, so further manipulations are in order. More specifically, the most natural strategy is to use devices working only in phases 1 and 3 and disabled during phase 2. For the sake of simplicity, we will assume the input signal to be a rectangular pulse. As in phases 1 and 3 we are only interested in rising and falling fronts this assumption is consistent. Moreover we will neglect the time-variance of the filter (1) as the behaviour of the model in phases 1 and 3 is utterly imposed by the additional devices. This means that eqn (1) becomes

$$\dot{y}(t) = \hat{a}y(t) + \hat{b}u(t) \tag{2}$$

where

$$\hat{a} = a(t_1), \qquad \hat{b} = b(t_1) \tag{3}$$

during phase 1, and

$$\hat{a} = a(t_2), \qquad \hat{b} = b(t_2) \tag{4}$$

during phase 3.

A way to decrease the rising time during phase 1 is to increase the signal u(t) during this phase. This can be accomplished by adding to the scheme in Fig. 2 the non-linear device represented in Fig. 4 and labelled 2.

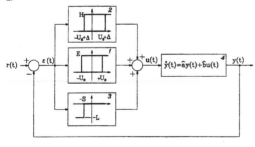

Fig. 4. Bi-level modulator with separately settings of rise-time and decay time.

Analogously the block 3 in Fig. 4 allows to force the desired behaviour in phase 3 without altering the remaining two phases.

Fig. 5 shows the behaviour of the "error" signal ε(t) resulting from a rectangular pulse input r(t).

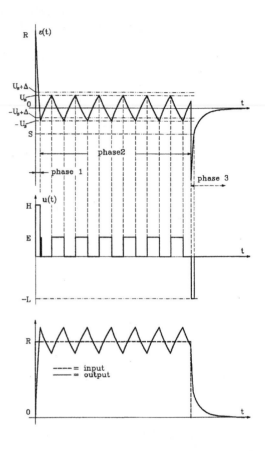

Fig. 5. Input, output and internal variables of the Fig. 4 modulator.

This picture shows that

a) block 2 operates only in phase 1, changing the signal u(t) from E to E+H.

b) block 3 works during phase 3 allowing a change of slope imposed by the values of the threshold S. The initial slope is imposed by the value -L, then, as soon as the signal ε(t) reaches the value -S, a free response phase starts.

A LTI first order filter is used to smooth the response of the model.

The delay is obtained by delaying the input r(t) by a suitable time.

Finally, the proposed model can be parameterized with respect to the inlet pressure. Actually each parameter of the thruster response depends on the inlet pressure, then at least the main parameters of the proposed simulator must be linked to the inlet pressure P. Specifically, for the parameters E, $U_s$, H, L and for the gain of the output filter a polynomial dependence on P has been selected.

The complete scheme is represented in Fig. 6.

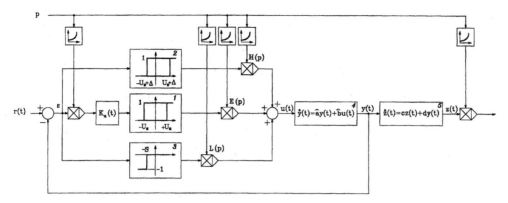

Fig. 6 — Complete model of Hydrazine Thruster

**Tuning the model.**

In this section the numerical values for the proposed model are reported.

To obtain decreasing amplitude oscillations the size of the tolerance band must be reduced. From Fig. 1 a reasonable rule to seems to be

$$U_s(t) = U_0 \exp(-t/\tau_U) \qquad (5)$$

Reducing $\Delta$ would increase the frequency of the oscillations. To avoid this effect the variable E must decrease with an analogous law

$$E(t) = E_0 \exp(-t/\tau_E) \qquad (6)$$

According to this structure, the first order linear time-varying filter must "reduce its settling time". More precisely the structure

$$a(t) = a_1(1 + \Delta_a t)$$
$$b(t) = b_0 a(t) \qquad (7)$$

can be selected for the filter (1).

Noting that the mean value of the thruster response is well approximated by the sum of a constant function and an exponential curve gives

$$r(t) = k_1 [1 + k_2(1-\exp(-t/\tau_R))] \qquad (8)$$

Finally the dependence on the inlet pressure P has been assumed to be quadratic.

With such a structure, the following expressions have been found:

$$a_1 = -0.8$$
$$a_2 = -11$$
$$\Delta_a = (a_2 - a_1)/(0.25\ a_1) \qquad (9)$$
$$b_0 = -1$$

$$U_0(P) = -0.0015P^2 + 0.0568P - 0.0167$$
$$U_1(P) = 0.0002P^2 + 0.0006P + 0.0014 \qquad (10)$$
$$\tau_U = \frac{0.25}{\log(U_0/U_1)}$$

$$E_0(P) = 0.002P^2 + 0.102P + 0.7778$$
$$E_1(P) = 0.0017P^2 + 0.0628P + 0.6037 \qquad (11)$$
$$\tau_E = \frac{0.25}{\log(E_0/E_1)}$$

$$S_e(P) = 0.1515P^2 + 2.6515P - 11.6667$$
$$H(P,t) = S_e(P)\ E(t) \qquad (12)$$

$$k_1(P) = 0.0017P^2 + 0.0295P + 0.537$$
$$k_2 = 0.3 \qquad (13)$$
$$\tau_R = 0.3$$

The transfer function of the output filter is

$$W(s) = \frac{K(P)}{0.1s+1} \qquad (14)$$

where K(P) is

$$K(P) = -0.0056P^2 + 0.1796P + 2.5154 \qquad (15)$$

In Fig. 7 the simulated responses for various values of pressure are plotted.

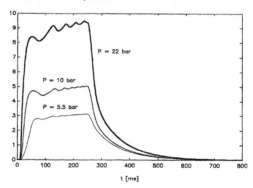

Fig. 7. Simulated system outputs.

In Fig. 8 the spectra of the real and simulated responses are plotted.

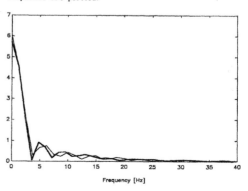

Fig. 8. Spectra of the real and simulated responses.

493

## Conclusions.

In this paper a simulation model for catalytic monopropellant hydrazine thrusters has been proposed. Such a model reproduces both the mean value and the low frequency oscillations of the thruster, and allows the designer to take into account the vibrations that affect the structure supporting the thruster. Moreover the model has been parameterized on the firing time and on the inlet pressure, so any operating condition can be simulated.

As an analytic model of the thruster based on physical considerations proves to be too complex, a "black box" approach has been used. The core of the model is the bi-level modulator, a classic device in the Industrial Electronic area. This suggests the possibility of an analogical realization of the simulator by using standard hardware devices.

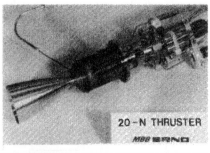

## References.

[1] MBB Deutsche Aerospace, Catalytic monopropellant hydrazine thrusters technical reference.

[2] J. Kollien and J. Weiss, "Interpolation of performance parameters for catalytic hydrazine thrusters", Proc. of the 2nd Int. Symp. on Spacecraft Flight Dynamics, Darmstadt, FR Germany, 1986.

[3] G. De Maria and L. Sciavicco, "Analysis and design of Pulse Ratio modulated feedback control systems", IEEE Trans. on Industrial Electronics and Control Instrumentation, vol. IECI no. 3, Aug. 1979.

[4] A. Izzo and P. Marino, "PRM modulator in DC drives", Int. Symp. on Electrical Drives, Cagliari, Italy, 1987.

# ITACA: AN INTELLIGENT URBAN TRAFFIC CONTROLLER[1]

A. Bahamonde, S. Lopez, P. Hernandez-Arauzo, A. Bilbao-Terol and C.R. Vela

*Artificial Intelligence Centre (AIC), Universidad de Oviedo en Gijon, Campus de Viesques, E-33271 Gijon, Spain*

**Abstract**. In this paper a new urban traffic controller called ITACA is presented. This system embodies a new architecture scheme to provide an integrated environment where both procedural and declarative knowledge can run cooperatively. To communicate those subsystems we have devised the so called virtual problem model. Throughout the paper we stress the advantage of using this methodology in order to get effectively running intelligent controllers.

Key Words. Advanced Control Strategies, Traffic Control, Knowledge Based Control, Artificial Intelligence, Virtual Problems.

## INTRODUCTION

The aim of this paper is to present the Urban Traffic Controller **ITACA** (InTelligent and Adaptive Control Area), an actually working system that has being developed in the Artificial Intelligence Centre (AIC) of the University of Oviedo at Gijón. The main goal of this system is to embody a new architecture scheme for knowledge based systems. This architecture provides an integrated environment where both procedural and declarative knowledge can run cooperatively.

The core point is the communication between algorithmic and knowledge-driven processes. To this end we have devised the so called **virtual problems**. Roughly speaking we have two process; the first one is the algorithmic problem solver that is working from a *virtual problem* provided by the knowledge driven one. This mechanism allow us to take advantage of procedural knowledge efficiency without penalizing the flexibility of declarative knowledge. Moreover, we can tailor the system to our needs just by editing the rules instead of

getting into cumbersome details of conventional software maintenance.

One important reward of using knowledge based systems is their possibility of providing explanations for the solutions that they compute We will spell out that *virtual problem* model can give higher lever explanations which go beyond the mere trace of partial results from procedural algorithms. These kind of comments are very important when dealing with traffic control because of the difficulty for humans to get a full understanding of other type of reports due to its numerical complexity

In general, we can say that the application field for this architecture is depicted by situations where some more flexible and declarative software is required in order to enhance conventional software systems.

Urban traffic controllers are systems that calculate and implement signal timings trying to match the latest traffic situation. Apart from the ability to calculate traffic parameters, there are another kind of knowledge that the system

---

1 This research has been supported by SAINCO TRAFICO S.A.E - under a CICYT grant number 61/85

should be able to deal with: local knowledge about concrete traffic nets, and control strategy.

The inclusion of local peculiarities into a procedural system is an acknowledged bottleneck. Moreover, we must take account that cities are living entities with continuous changes in their traffic topologies and behaviour; thus, this tunning problem is a very important one.

In procedural systems (like SCOOT (Hunt et al., 1981) for instance) the control strategy is encapsulated into a mass of computations and then is difficult to modify and update, and we must consider that it is possible that a control strategy which works well in one traffic situation can perform poorly in other situation. In fact, recent studies have shown that old fixed-time control gives better results than full algorithmic systems. Therefore, local experts must resolve manually critical (and typical) situations based on video images, radio reports, and other knowledge gathered around in previous problems.

Trying to avoid the gap of congestions some systems have been developed like SAGE (Forasté, 1986) and its outspread CLAIRE (Bell, 1991). Anyway, those systems have been designed on the basis of been computerized advisors instead of really responsible controllers as in ITACA.

## ITACA ARCHITECTURE

ITACA is an adaptive urban traffic control system enhanced with knowledge driven strategies. The system gets data from streets every five seconds; these data are measured by **loop detectors** placed either upstream or immediately after the stop-line for each traffic lane and each traffic lights group. According to the measured traffic flow and the knowledge added, ITACA provides slight improvements of the previous timing situations of traffic parameters: green/red splits, starting points for greens/reds (offsets), and cycle time. ITACA tries to implant on the streets its own recommendations as soon as possible.

Roughly speaking ITACA can be described as a set of concentric shells. The deepest one or **Kernel** is the procedural part and computes timings for each junction trying to optimize traffic flow from a local point of view; that is to say, the kernel takes decisions without considering the repercussions in the surrounding.

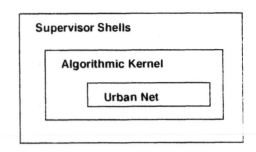

The **Supervisor** is the shallowest shell of ITACA and it is organized itself according to the traditional scheme rule/meta-rule/meta-meta-rule/ ... etc. The target of this subsystem is to provide a global point of view in the task of optimization of traffic flow. To this end the outputs of the kernel will not be directly implemented on the streets; in fact, they will be filtered or translated in order to accomplish a global strategy of optimization. Moreover the Supervisor will feed the Kernel also by means of the so called **virtual networks**.

The problem is that it is not always possible to get a global optimum just by adding local optima. Thus, some junctions must be sacrificed in their local pretensions. This means that local computations can be strongly constrainted or eventually skipped out. This unpleasant mission is confided to the Supervisor.

The conceptual architecture of ITACA can be sketched as follows (see Fig 1). The urban net is the source of data; these data plus the control knowledge are put together into the 'Knowledge Base'. The procedural part of the system ('Procedural Controller' in the graph) can ask for information through the 'Semantic Module'

## VIRTUAL PROBLEM MODEL

In this section we are going to describe the mechanism devised to communicate the knowledge and algorithmic modules: the virtual problem model. The idea is that we must be able to set problems to our algorithmic solver in such a way that its results can be translated to the real solution. Of course, depending of the case, the translation carried out by the Decoder may vary from trivial to a high complex process. Graphically we can present this model in Fig 2.

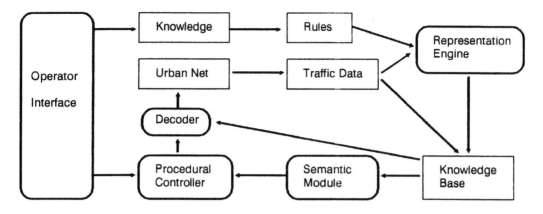

Fig 1 Conceptual Architecture

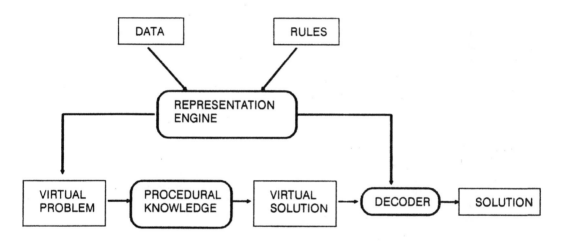

Fig 2 Virtual Problem Model

Besides virtual problems and decoders, we are stressing the specialized role of our **knowledge engine**. The aim here is to produce a solution path instead of providing the solution itself. Thus, our computation with knowledge is centrered around representation; that is to say we must set problems to a procedural module and keep track of the details of that representation because we will have to decoder procedural solutions to get the final answer. This is the reason why we are calling **representation engine** to our knowledge processor.

To accomplish this pattern we will use a highly parametrized algorithms (procedural knowledge) and rules about the deep knowledge on the solution process. We need to

know how to face problems supposing that we have at hand a very sophisticated calculator (the algorithm). Therefore we stress the relevance of expertise on how to deal with conventional resources availables and indeed how the algorithms are really built.

In ITACA the Supervisor starts by examining traffic situation and looking for a global strategy to face the optimization problem. A typical scenario where constraints should be imposed is offsets setting in a critical street. In order to implement global criteria for optimization we can choose between several strategies: the most often adequate one is to compute starting points for greens on a given junction of the street, and then we can spread the computations just by adding a given offset for the other junctions.

We will represent this case by transforming the real net

**junctions**

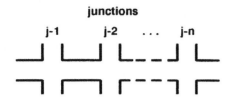

into the virtual one

**virtual-junction**

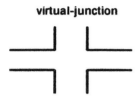

In this case the Kernel will be asked to just compute timings for "virtual-junction": this will be its virtual problem. The Kernel output should be translated to get a real solution for junctions 1 to n. The translating method and the virtual net will be the Supervisor goals in this framework.

Of course, there are other situations where it is convenient to "lie" to the Kernel. Note that to system eyes what it is going on in the street are the measures of loop detectors placed (usually) upstream. So, when the queue of vehicles reaches the loop, traffic simply cannot cross the loop because of the queue. Then we must not allow the Kernel to proceed normally since it is really blind and will tend to reduces green times.

The solution devised in these cases is to change the computing demand algorithm to another one which takes account that residual queues should be approached from the vehicles crossing loop detector when it is clear. In our model this assert for any lane can be told in a declarative style as follows:

```
IF
    Number.of.times.with.problems(lane) >
        Threshold-n,
    Unoccupied.periods(lane) > 2
THEN
    Update(residual.queue(lane),
        residual.queue(lane) + 3)
```

The basic classes of objects in ITACA are intersections and traffic lanes (like in the previous rule), and about their attributes we have several types of knowledge with different implementations. Let us review the most importants for this paper.

For instance, let us show how we can compute a problematic area in an urban traffic net. The idea is to merge saturated and connected areas in order to determine the congestion zone. Then we will be able to "gate" the area favouring outputs and penalizing inputs. If x and z are variables bound to junctions, and y takes values over saturation levels, we can say:

```
IF
    Saturation.level(x,y),
    Greater(y,'medium')
THEN
    Is.in.gating.area(x,'yes'),
    Gating.area.identifier(x,(gensym))

IF
    Is.in.gating.area(x,'yes'),
    Source.of(x,z),
    Is.in.gating.area(z,'yes'),
    Gating.area.identifier(x,N1),
    Gating.area.identifier(z,N2)
THEN
    Execute(merge.areas,(N1,N2))

IF
    Is.in.gating.area(x,'yes'),
    Source.of(x,z),
    Is.in.gating.area(z,'no'),
    Gating.area.identifier(x,N1)
THEN
    Make.split.using(z,N1)

IF
    Is.in.gating.area(x,'yes'),
    Source.of(z,x),
    Is.in.gating.area(z,'no').
THEN
    Favour.split.from.to(x,z)
```

Another kind of control strategy knowledge are those rules that remember the system for doing things. Operator mandatories can be expressed in this way with a very high priority.

The **explanatory ability** is probably one of the most important features of knowledge based systems. However, their outputs may be based purely on a backtrace of the heuristic rules that were needed to find a solution, what it is usually different from "a convincing rational argument why the solution is valuable, particularly if a lot of heuristic knowledge entered into the reasoning process" (Steels, 1985; p.213) Anyway, the problem is much more bigger if we are dealing with conventional algorithmic software: the trace of algorithms use to increase our puzzlement about the results. Urban traffic control is a good example of this situation.

The **virtual problem** model allows us to provide higher level explanations to the user because the knowledge process is about the solution path instead of the computations itself. Moreover, the user can easily choose the degree of details of the required explanations for solutions just by getting into deeper surfaces of the whole system. In fact, the model is based on an explanation principle; that is to say, we must effort to add knowledge about how to explain to an algorithmic module the way to face a problem (eventually a problem originally not in its area of competence) and then how to tell users how is the solution like in their own words.

## PROCEDURAL KERNEL

In this section we will review the most important features of the Procedural Kernel. The general idea, as has been said above, is to provide a powerful algorithmic calculator for dealing with the so called virtual problems built by the Supervisor and set through the communication module.

The Kernel is arranged by several concurrent processes sharing common data readable for all of them. In the figure below (Fig 3) those modules are drawn with some arrows to show the writing permissions in the common data structures.

The core of this subsystem is the **adaptive subsystem**. The aim of this module is to compute the timings for traffic parameters for each junction in the virtual net, and then try to implement them on the streets; this task is accomplished according to translations rules managed by the Decoder (see Fig 1) in order to recover the real net.

The **detector data** module takes data from loop detectors on the traffic lanes and builds up the traffic profiles of number of vehicles and occupation time. Some lanes do not have real detectors and thus we have to simulate them according to the sources and sinks of their traffic flow.

An important issue is the cycle time of each junction. In fact we have three kinds of time: real time, general cycle time, and cycle time for each junction.

The cycle time is a common data for a set of junctions and typically varies from 50 to 150 seconds. For all junctions the timings are calculated several times each cycle, and the implementation of them may change in fact this cycle time. For instance, if ITACA decides to increase the during of the last green for junction-n in 2 seconds, the next cycle will start two seconds later for junction-n; in other words, the delay of this junction will be modified in 2 seconds more with respect to the general cycle time.

This is a very usual situation, so we must take account of cycle delays accumulated for junctions whenever we are computing traffic demands (from profiles) and we are going to

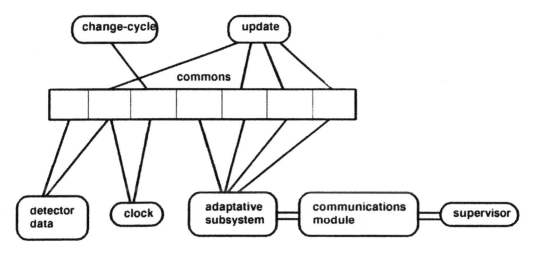

Fig 3 Adaptive Subsystem

implement the results of our computations. Notice that we can only act over future timings not over past.

The **clock** module is asked to keep track of all these details for our net. Additionally, the **updating** module is devised to compute, at the end of each cycle time, parameters like the residual queues, and to fix the split times for the next cycle.

## IMPLEMENTATION ISSUES

In the AIC we have developed an ITACA prototype that is actually running in Oviedo (Asturias, Spain). The testbed net is an arterial street connected to an important freeway.

The prototype is running on three computers: 2 MicroVAXes and 1 PC. The first VAX is devoted to communications tasks between the central control and the net. The second one executes the algorithmic part of ITACA and can also provide a simple operator interface mainly for informative and statistical issues. The Supervisor is placed on the PC in this preliminary version in order to have a fast prototyping platform for showing the feasibility of our ideas and to provide a friendly operator interface for control issues.

## BIBLIOGRAPHY

Bahamonde, A., and López-García, S.(1988). Diseño y Arquitectura de un Sistema Inteligente para el Control del Tráfico Urbano. Jornadas Técnicas Tecno-Traffic-88, Madrid, March, 1988.

Bell, M.; Scemama, G.; Ibbetson, L. (1991). CLAIRE an expert system for congestion management. Proceedings of the Drive Conference, Brussels, Feb. 1991.

Forasté, B.; Scemama, G. (1986). Surveillance and congestion traffic control in Paris by Expert System. Second International Conferencie on Road Traffic Control . London, April 1986

Hunt, P.B.;Robertson, D.I.;Betherton R.D., Winton R.I. (1981). SCOOT-a traffic responsive method of coordinating signals. TRRL Laboratory Report 1014.

Steels, Luc (1985). Second Generation Expert Systems. Future Generation Computer Systems, 1, No 4, 213-221.

# ADAPTIVE CONTROL OF A TRAFFIC INTERSECTION BY MEANS OF A NEURAL NETWORK

**C.J. Barnard\* and I.S. Shaw\*\***

*\*Tek Logic (Pty) Ltd., P.O. Box 420, Halfway House 1685, South Africa*
*\*\*Cybernetics Laboratory, Rand Afrikaans University, P.O. Box 524, Johannesburg 2000, South Africa*

Abstract. This study was undertaken to assess the feasibility of using a neural
network as a traffic controller. In particular, the optimization of the total traffic
flow rate through an isolated three-phase intersection is discussed. A
backpropagation-type neural network used samples of traffic density as inputs,
while the green times of the different phases constituted the outputs. The
network, trained with the desired target vectors, was found to provide good
estimates of the outputs on the basis of actual traffic flows. It is felt that a neural
network can successfully implement a dynamic, self-learning, adaptive control
system, however, subject to trade-offs between sensor cost constraints and control
performance.

Keywords. Adaptive control; neural nets; nonlinear control systems; self-adapting
systems; traffic control.

## INTRODUCTION

In traffic intersections the traffic lights must be
controlled so as to permit the vehicles approaching it
to move through the intersection in a reasonable time.
This paper deals with so-called isolated intersections,
where the traffic lights are not synchronized with those
of other intersections along either route. The control
of traffic lights is achieved by a sequential logic circuit
capable of cycling through a number of system states.
The control cycle consists of a number of phases (i.e.
possible directions of traffic movements) and within
each phase there are a number of states reflecting the
momentary status of the traffic lights. Thus the traffic
lights are in a certain state allowing or inhibiting traffic
to flow for a certain time during every cycle, after
which they are set into another state, etc. thus they go
through a number of states in a fixed sequence. The
so-called green phase is that particular set of states,
which provides a green light for a particular direction.
The green time, usually measured in seconds,
represents the time duration of the green phase. The
control strategy usually consists of varying the green
time as a function of the measured or assumed
average traffic demand. The control problem is thus
not trivial, since this relationship is non-linear, in as
much as vehicles are forced to wait while traffic is
flowing in another direction. The longer the waiting
time, the heavier the load that must pass through the

intersection at a later stage. The objective of the study
undertaken was to find an adaptive control strategy
whereby the total flow rate of traffic through the
intersection would be optimized on the basis of the
actual traffic load.

## STATE OF THE ART

Currently the problem is solved by the use of empirical
equations that describe the control of a set of traffic
lights, depending on actual traffic counts which
represent the traffic load at different times of the day
and in different seasons. Different levels of
sophistication are possible:

- the fixed control plan, consisting of a set of control
states computed from average measured traffic flows;
a number of these plans can be stored and at given
times-of-day, the controller is switched from one fixed
plan to another. This system depends on periodic
traffic counts (i.e. number of vehicles per unit time
period). Manual counts are costly and thus cannot be
repeated very often, hence the system is difficult to
keep up-to-date. On the other hand, permanent
electronic counting stations require expensive
hardware and maintenance; they can be read
periodically on site, or by means of long-distance
telephone lines, both of which are costly.

501

- real time systems, where the control strategy is computed according to actual traffic flows measured in real time and where the controller is periodically updated to implement the new strategy. The high cost of sensing instrumentation is again the limitation.

Mathematical models do exist for the description of the traffic flow through an intersection for a given traffic flow density, but the exact description is non-linear, and it is difficult to take into account the fluctuation of the traffic flow during the day from a measured average. Due to the time-varying traffic flow densities that approach the intersection from different directions, this approach is difficult and impractical to implement.

One of the limitations is that the system designer must get feedback information from time to time on system performance in order to cope with varying traffic requirements and adjust the control strategy accordingly. The costs of having to continually collect information in order to ensure an acceptable control strategy is an obvious disadvantage of the current approach.

## NEW TECHNOLOGY

With recent advances in the field of neural networks, a self-adapting alternative to the present traffic controllers has become possible. Neural nets have the ability to learn : they can change their behaviour in accordance with their inputs. When a neural network is given a set of inputs, together with a set of corresponding desired outputs, the network can adapt itself (i.e. learn) so as to generate the desirable response. This is the so-called training phase. In turn, when presented with a new set of inputs, it can calculate an output based on its previously learned experience.

When a neural net is trained, it is, to a reasonable extent, insensitive to small variations on its inputs. This ability to see though noise and distortion makes it highly desirable for pattern recognition in a real-world environment. This also explains why one of the major applications of neural networks is that of pattern recognition. One of the most significant differences between neural nets and conventional computers is that the neural net is not bound to clearly defined steps like the computer, but has the ability to interpolate between a set of data points.

A neural network has the additional ability to distinguish the most important feature of its different inputs. This means that it can recognize the most significant property of a given input pattern, on the basis of which the pattern may be classified.

These properties are eminently useful for traffic controller applications. The controller must continually learn the required outputs for a given set of inputs, even when these inputs vary in size and pattern. The

ability to interpolate is also essential, since no two days will have exactly the same traffic flow. Finally, the network must recognize certain properties of traffic flow patterns, so that a consistent optimal control strategy can be followed, independent of variations in flow inputs In other words, the neural network must function as a non-linear controller. Nguyen and Widrow (1990) have already shown that this is feasible.

## IMPLEMENTATION

The most important step of the whole design process is the choice of input/output requirements. As regards the implementation of a neural net, the first task was to determine the output variables to be used. The output was chosen to be the green time (in seconds), i.e. where traffic lights are green to allow through-traffic. The controller then cycles through the set of traffic light states in a predetermined sequence, changing the green time for each set according to the most recent inputs. When a set of traffic lights is green, the traffic lights in all other directions are assumed to be red, inhibiting the traffic flow.

An existing intersection in Randburg, Republic of South Africa, was selected for this study (refer to layout drawing in Fig. 1) for the following reasons:

a) It is not a case of two directly opposing directions of traffic flow (which would be simple to describe mathematically and control accordingly) but uses a three-phase operation.

b) Actual traffic flow density data to this intersection, as well as the current control plans, were available for the training of a neural net.

### Inputs

Traffic flow densities of the traffic approaching the intersection from different directions were chosen as inputs. This data consisted of counts of the number of vehicles that passed by each counting station (of which there was one on every approach to the intersection) during each 90-second sample interval. In addition, the moving average over the past ten samples of each of the counting stations was calculated. Traffic flow data was obtained for ten consecutive days. Since the purpose of this study was only to prove the feasibility of a neural net as controller, as opposed to a real implementation, ten days of flow data were deemed sufficient for our experiments. In general, a longer period would be necessary in order to take into account seasonal and weather-dependent fluctuations.

Pre-processing of data. An interesting question arises as to the training set that should be used for the initial training of the neural network. Should all the different days' data be combined to create a huge training set, with the same output training set for every day, or is there another alternative that can make training, as well as control, more efficient? To answer this question, one has to look at the behaviour of a neural net. When a neural network is presented with a varying set of inputs that must produce the same output, it tends to produce the same output for inputs that lie between the extremes of the inputs required to produce the same output. This can be seen as a kind of averaging of inputs. When the network is trained with a set of traffic flow data over a few days, it will tend to compute the desired strategy for the 'average day' or day in which the traffic flow is the average of the flows of the other days in the training set. This is undesirable, since the network doesn't learn how to handle days with high traffic densities, and can cause traffic to pile up. It was therefore decided to use the 'best day' for training the network, where the 'best day' was that with the maximum traffic throughput achieved. In other words, an artificial traffic flow data set was created, using the maximum value of all the available days at every sample point. Figure 2 shows the 'best day' in terms of the number of vehicles measured during each sample interval.

This approach has the additional advantage of a faster, more accurate training cycle for the network, since it has much fewer data points. In addition, there are fewer conflicting similar outputs for different inputs, whereby a more accurate approximation can be achieved.

It is acknowledged, however, that the number of vehicles waiting at the intersection from every approach during every cycle would be a more ideal input, in as much as this would provide feedback regarding the success of the control strategy followed. Yet this approach would have required a set of expensive detectors capable of reporting the number of vehicles waiting. In the absence of such an installation, this data was not available, and the training of the neural net was done only with the continually monitored traffic counts as described above. In practical traffic control situations, the installation of expensive sensors is usually difficult to justify economically. A traffic counter/timer with electronics, however, is often easily available. Using such counters, the 'best day' neural net training method, providing open-loop control, represents a good compromise without resorting to more sophisticated and expensive sensors capable of closed-loop control.

## Outputs

The existing control plan for the intersection, representing the desired outputs for the training of the neural net, was obtained from the municipal authority. This plan had been calculated using conventional traffic control techniques. The question should arise: in what sense would the neural net implementation improve upon the conventional system, when the conventional system is used as the training set for the neural net?

This can be explained as follows : the conventional control strategy is based upon traffic requirements at certain time periods. This strategy consists of a schedule whereby the time that each set of traffic lights is green during the control cycle is varied during the day. The plan implemented in Randburg switches between three different strategies during every day. It is important to note that these strategies are switched as a function of the time of the day, to cope with the varying load during peak times. It can be seen that this plan can be exactly accurate only three times per day (the exact times for which the applicable strategy was calculated).

The neural network implementation, however, does not use time as an input at all to decide on a control strategy. The only inputs that determine the strategy are the flow densities as counted at the counting stations. The network is therefore able to interpolate between the extreme points calculated for peak times, and find intermediate control parameters based on the traffic demand.

## RESULTS

Figure 3 shows the results that were obtained when training a neural network using Backpropagation. The straight lines show the current control strategy, which was used as the training set, and the fluctuating values show the output of the neural network for the input set shown in Fig. 2.

Figure 3 shows that the neural network output follows the desired plan quite well. It must be stressed that the network generates the output based on the actual traffic flow, whereas the target plan uses time as a basis. This is an important result, since it ensures that the network will be able to cope with variations in traffic flow in a way that corresponds to the ideal plan computed for a specific condition of traffic flow.

This explains the sharp peaks of fluctuations that can be seen on the neural network output. If this is undesirable, it can be reduced considerably by applying smoothing to the input data. This was already done to a certain extent through the use of moving data averages used as input to the network. Additional inputs using a greater number of points for data averaging can also be applied. At any rate, the rough, noisy input data available from the counting station should not be used directly.

The amount of data smoothing will be determined by both the sensitivity and stability required for the controller. In an intersection where busy routes meet, it may be better to avoid a too quick network response to variations in traffic flow. The reason is that one parameter that may be unrealistic due to one set of input values may cause a large traffic backlog during the next cycle. On the other hand, when a small route intersects a busy route, it is desirable to respond quickly to a change in flow density, so that the busy route is not unnecessarily inhibited by traffic along the smaller route.

The training of the neural net was done on an 12 MHz 80286 AT, and the training time was 314 minutes. The actual application software for network simulation is an algorithm called *opt*, which makes use of the so-called conjugate gradient descent training method as well as backward propagation, as described by Barnard and Cole (1989). Training proceeded to a point where the average squared error of the three normalised inputs was smaller than 3.8. In a real application, perhaps by using more neurons in the hidden layer, the network can be left to train to an arbitrarily small error, depending on the time available for training. The training time available in a practical implementation is discussed later.

## EVALUATION

From the training results it can be seen that the neural network has indeed the ability to track the traffic fluctuation, and dynamically produce control strategies according to the traffic flow data. Intuitively it can be seen from the network outputs, that an acceptable control strategy is produced which is at least as good as the current system. A crude simulation has also been written to simulate traffic flow through the intersection using a given control strategy, where the movement of vehicles is described by simple equations of motion, and the actual flow rates used as inputs. This simulation gave promising results, since it showed that the neural network yields a higher flow of vehicles through the intersection over a specified period.

It is recognized that the experimental system described is still not refined sufficiently to be actually implemented as a traffic controller, for the following reasons :

(a) Implementation : PC simulation
(b) Boundary conditions
(c) Input data type
(d) Continuous training
(e) Synchronization

These additional requirements will now be discussed.

### (a) Implementation : PC simulation

In an actual application, where the system will be optimized for both cosand effectiveness, a neural net should be implemented by using an inexpensive conventional microcontroller instead of the personal computer used in our experiments. The computing power of the PC is not required because of the relatively long time-constants involved. The feed-forward generation of a control strategy via a trained network involves relatively few multiply and summing operations, which can easily be handled by a conventional microcontroller.

The training operation, on the other hand, requires much more computing. Since, but however, the rate of change of traffic flow patterns (not daily or weekly cyclic fluctuations, but more permanent changes) is very slow, there is no pressure on the processor to train a network quickly. As many as two months may be available for training of a network, before it would be required to take control of the intersection.

Another alternative that may soon become feasible is the application of a hardware implemented neural net, as described by Hecht-Nielsen (1988). This may be both cost-effective and fast, apart from being an elegant solution, since it would implement the actual neural network-topology rather than a slow simulation.

### (b) Boundary conditions

It is necessary for certain constraints to be imposed on the output of the neural network before actual implementation in the intersection to ensure that the system remains stable under irregular conditions, such as, for example, a procession passing through or an accident occurring near the intersection. If this was not done, it could happen that traffic from a direction with a lower traffic density has to wait longer than is reasonable. It will be necessary to decide on a maximum waiting time for vehicles and to ensure that the actual red times does not exceed this. Another boundary condition is implied by synchronization, which will be discussed separately.

### (c) Input data type

It was stated earlier that using only traffic flow data to train the neural net was not ideal because of the lack of feedback from the intersection. In our experiments, the network was trained with data that only reflects the traffic demand flowing through the intersection. Since, however, the counting stations are at about a kilometre from the intersection, fluctuations within this distance would not show up in the traffic counts, and would not be taken into account.

As has been mentioned earlier, at the expense of more sophisticated instrumentation, better input data can be generated to reflect the success of the strategy, thus enabling the controller to continually adapt itself and improve on the current control strategy. One such data type would be counts of the number of vehicles waiting at the intersection from each approach during every cycle. More vehicles waiting would obviously imply that longer green times are required, and conversely, when no vehicles are waiting, it is no longer necessary to keep the traffic lights green, except on a main route, where traffic is more likely to arrive.

### (d) Continuous training

As traffic flow patterns gradually change with time, it is necessary for the controlling network to be continuously trained. Since a network being trained cannot yet produce reliable outputs, the same network cannot both be trained and be used to actively control the intersection. A solution would be to have two neural nets per controller: one to estimate the control outputs, to be used in feedforward mode, while another one being trained with current traffic data. When the network being trained produces an error smaller by a fixed margin than the estimator network on a given current data set, the roles of the networks would be switched around, so that the more recently trained network would take over the control of the intersection, while the other would go back to training with new traffic flow data. This cycle may be continued to ensure continued optimal control.

The question should arise: how long would the target output data (i.e. the traffic control plan) remain valid for the traffic flowing through the intersection? We have seen that the neural net can interpolate between the extremes given by the calculated plan, but can it extrapolate to cope with new traffic situations ? The answer is negative. Neural nets can only interpolate between points in the input space, but cannot extrapolate!

We come to the conclusion that it will be necessary to periodically supply the neural net with new output target vectors to enable it to interpolate between points. Thus a measure of designer interaction with the system will be required to ensure continued reliable operation of the system.

The process of the calculation of new target vectors according to the traffic demands can, however, be automated to a large extent. As standard algorithms are available for this, a conventional processor can also be implemented in the controller, and, making use of the available traffic flow data, new target vectors can be computed.

#### (e) Synchronization

This study investigated only so-called isolated intersections. That is, the intersection being controlled by the neural net was the only one present on the different intersecting routes. This means that there are no requirements for the intersection controller to be synchronized with other controllers along intersecting routes. Although this problem was not part of our study, it may be a fruitful area for future research.

## CONCLUSIONS

This study has shown a neural network is a good candidate for a traffic controller. It is well suited to the available input data, and there is good correlation between the input (varying traffic density) and the output. The method of input data collection is relatively simple and inexpensive. The neural network has the ability to follow certain requirements based on actual demand, and can interpolate between pre-calculated control strategies for the intersection.

The controller is self-structuring to a certain extent, but, because of the changing needs in terms of the output, it needs to be supplied with these changing requirements. It can therefore be seen that it is not an optimizing system: it only interpolates between data points. It must be noted, however, that this interpolation is greatly simplified by the fact that the neural net does not require a mathematical model of the function to be interpolated on, for it forms an internal representation of the function to be followed.

This study was only an initial look at the challenging field of traffic control. Nevertheless, it does provide encouraging results for the successful implementation of a neural network to give a dynamic, self-learning adaptive intersection traffic control system.

## REFERENCES

Barnard, E., Cole, R.A. (1989). A neural-net training program based on conjugate-gradient optimization. Technical Report No. CSE 89-014.

Hecht-Nielsen, R. (1988). Neurocomputing: picking the human brain. IEEE Spectrum. pp. 36-41.

Lippmann, R.P (1987). An Introduction to Computing with Neural Nets. IEEE ASSP Magazine. pp. 4-22.

Nguyen, D.H. and Widrow, B. (1990). Neural Networks for Self-Learning Control Systems. IEEE Control Systems Magazine.

Rumelhart, D.E., Hinton, G.E. and Williams, R.J. (1986). Learning internal representations by error propagation, Parallel distributed processing, vol.1. pp.318-62, Cambridge, MA: MIT Press.

Wasserman, P.D. (1989). Neural Computing : theory and practice. Van Nostrand Reinhold. New York.

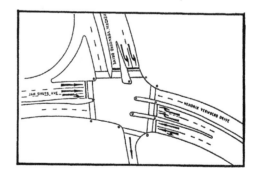

Fig. 1. Intersection in Randburg

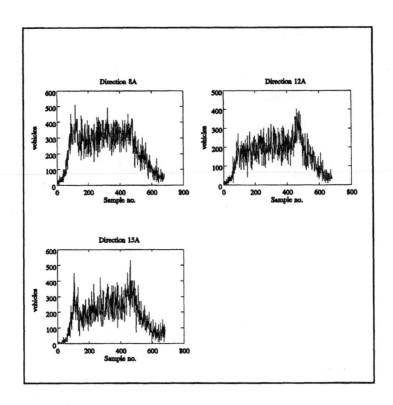

Fig. 2. Flow inputs of 'best day'.

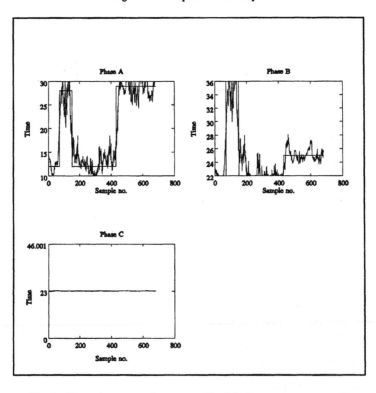

Fig. 3. Network target and response. (Straight lines represent target)

# UNI-TELWAY BUS MANAGEMENT IN A UNIX
# ENVIRONMENT

### J. Perez-Turiel, J.C. Fraile and J.R. Peran

*Department of Systems Engineering, Technical School of Industrial Engineering, C/Paseo del Cauce S/N, 47011*
*Valladolid, Spain*

**Abstract**. In this paper we describe a concrete implementation of
the local area network UNI-TELWAY for the interconnection of NUM
Numerical Controllers (NUM-760 and ROBONUM-800) with a process
management computer, in order to develop a Distributed Numerical
Control (DNC) system in a flexible machining cell. The software
modules to be executed at the computer for controlling and managing the cell elements has been developped under the UNIX Operating System, allowing different concurrent processes to access
the UNI-TELWAY bus without conflicts among them. We have also
implemented a polling strategy in the master computer that takes
into account the temporal requirements of the different messages
on the network, allowing both, synchronous and asynchronous traffic to share the use of the communication medium.

**Keywords**. Computer communication; Local area networks;
Multiaccess systems; Flexible manufacturing; CNC; NC Machine
tools.

## UNI-TELWAY COMMUNICATIONS BUS BASICS

UNI-TELWAY is a low cost, cell level
factory local area network (Daigle, Sede-
lman, Pimentel, 1988; Pimentel, 1990)
designed by Télémecanique to provide their
customers with a common interface for
their industrial devices (CNC's, Programable Logic Controllers, Robot Controllers,..). In this paper we describe the
use of this network to interconnect several CNC's (NUM-760 and ROBONUM-800) to
each other and with a cell master computer that manages and coordinates their
operation. Our goal is to implement a
Distributed Numerical Control (DNC)
(Decotignie, Grégorie, 1989) environment
that allows a flexible machining cell to
operate in an autonomous way.

In the hierarchy of industrial communication levels (Pimentel, 1990) cell networks
are placed between plant level networks
that support the supervisory and coordination activities required to integrate
diverse functional areas, and field buses
used to link sensors and actuators to its
controlling equipment by serial transmission.

Cell networks are usually fairly small (up
to 50 devices), low cost and property of
one vendor. The main candidate for standard at this level is the carrier band
mini-MAP network, but nowadays there is
not yet full consensus among the standari-
zation bodies.

UNI-TELWAY is a broadcast serial bus
for low level communication that allows
the interconnection of PLC's, Numerical
Controllers, robots, etc. to each
other and with field level devices such
as speed variators, I/O devices,
digital controllers, etc. Its functional

architecture, refered to the OSI model,
includes only the application, network,
link and physical layers ( Tejido,
Marinero, Turiel, 1991)

At physical layer it follows the EIA RS-
485 standard. The medium access
management is carried out by a fixed
master station that polls all the other
connected devices (secondary stations)
in a cyclic way, assuring their right to
send data.

A secondary station willing to send a data
frame must wait for a "polling" command
addressed to it. On the other hand, when
the master station must transmit a data
frame, it first sends a "select" frame,
designing a secondary station as receiver;
upon acknowledgement of this frame, the
master can start the transmission of data.
At this layer every data frame sent must
be acknowledged. The only exception is the
case of broadcast transmission: the master
station sends a data frame that is received by every secondary device. In order to
avoid conflicts no ack is required in this
case. Maximum size for data frames is 260
bytes (See fig. 1).

UNI-TE, the application protocol,
follows the client-server model and offers
a set of standard services (called
UNI-TE requests) that can be completed
with specific services for each concrete
product (NC, PLC, etc.).

Tipical services for CNC's are :

- Part programs downloading.

- Setting running modes in the machines.

- Reading status information.

- Reading and writing of memory objects.

- Starting and stopping program operation.

- Tool compensation tables downloading.

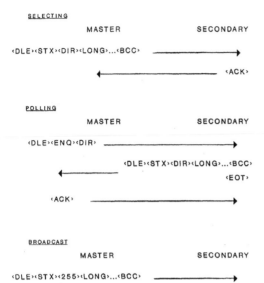

Fig. 1. Link layer frames interchange.

These services are similar to those provided by MMS (Manufacturing Message Specification, ISO DIS 9506), the ISO standard for Application layer in manufacturing networks (Lawrence, 1986). One of the goals of the european ESPRIT project CNMA (Communication Networks for Manufacturing Applications), is to provide facilities for the migration from proprietary protocols to standard ones such as MMS (Sanz, 1990).

Télémecanique neither provides a special communications card nor a software implementation of the UNI-TELWAY protocols for a external computer. Aimed to obtain a low cost and flexible solution we have decided to implement the communication protocols in the cell master station as software modules.

## MANAGEMENT OF CELL OPERATIONS

The CNC's that we use are provided with a special communications card, the LPC (Link Programable Controller) card, that includes the hardware and software items that allow the CNC to be attached to the UNI-TELWAY bus. This element, however, doesn't provide a standard electrical connection to the bus, but a serial RS-232 one. That's why a specific RS-232/RS-485 adapter must be used in order to garantee the electrical compatibility. The transmission rate can be programmed up to 19200 bauds.

In our network configuration we use an I-80386 based computer as master station. It must perform two different types of operations: control the access of secondary stations to the communications medium, and to manage the operations of the cell devices: A machining center, a robot for tool interchange, an automatic presetting machine and an I-80286 based PC that reads inputs from several sensors placed in the machining center to perform

tool wearing and breakdown monitoring. The presetting machine is linked to the master computer through a point-to-point line, while the other devices are secondary stations on the bus (See fig. 2).

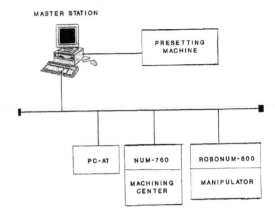

Fig. 2. UNI-TELWAY implementation.

The main supervision and control operations performed at the master computer are :

- Part programs editing, storage and transfer.

- Part programs execution control.

- Operation monitoring and gathering of statistics.

- Manufacturing operation synchronization.

- Tools management.

Because of the need of runing several concurrent processes in this computer and that our system doesn't have very strong real-time restrictions, we have decided to use the UNIX Operating System as platform for developping and running the control and communications software and the C language as programming tool (Poole, 1990; Ortiz, 1990). The advantages that this provides are :

- The posibility of adding terminals to access on-line the information concerning the cell operation.

- Transportability of the developed software.

- Availability of potent development tools.

- The posibility of linking the cell master computer to other UNIX systems in the factory, such as CAD/CAM stations, plant supervision computer and production management systems, by using an upper level network (factory network) such as Ethernet. This can be easily done through the networking utilities available under UNIX (TCP/IP, Streams, FTP, etc..)

The functions related to the UNI-TELWAY protocols have been implemented as a single software process that performs application and link layer tasks. The physical layer functions are performed by the communications card ( a serial one) and the network level functions are only necesary for inter-network operations (in our case, we are dealing with a single network).

This software module also acts as communications server at application level for the control and management processes that run on the computer. These require the UNI-TELWAY services to obtain the system information they need and send the corresponding commands. The available services are :

- Standard services: identification, test, status,..

- Non requested data: a secondary station is able to send data not previously requested (events notification)

- Access to elementary data: reading and writing of bits and bytes (process status)

- File transfer: part programs, PLC programs, machine parameter tables.

- CNC running mode : run, stop, step by step.

The control processes that want to use these services must provide a service identifier (request id.) and, optionally, the suitable parameters. These requests can be performed concurrently by several processes. The communications server procces is in charge of receiving the requests, building the data frames, polling the destination secondary station and transmiting the request. After receiving an answer it must send it to the process that made the request.

This involves an interprocess communication mechanism. Among the several IPC's that UNIX provides we have choosen that of message queues because of its good complexity/eficiency rate (Stevens, 1990).

The communications server process creates two message queues, one for requests reception and other for delivering of answers (if they exist). Before a message can be sent or received by a process, it must perform a system call to create :

- A message queue identifier that refers to the associated message queue and data structure.

- A data structure.

The message queues are used to store header information about each message that is being sent or received. This information includes the following for each message :

- Pointer to the next message on queue.

- Message type.

- Message text size.

- Message text address.

The data structure includes information related to the message queue : pointers to the first and last messages, number of messages currently on queue, queue's maximun size (in bytes), current number of bytes on queue, time of last arrival and departure of messages, etc..

In order to know the process that has sent a request, the associated message includes a type identifier related to this process. The answer will be delivered in a message of the same type.

BUS ACCESS MANAGEMENT

The communications server process checks in a cyclic way the requests queue status and generates a data frame for each message. In addition, it mus poll sequently all the secondary stations, even if it is not waiting for an answer, because the secondary stations must be able to send non requested data to report the occurrence of events (status changes in the manufacturing process or anormal events sucha as devices failures or malfunctions).

Once a request frame has been sent to a secondary station this must be polled to obtain the answer as soon as possible in order to be able to identify the answer's sender since the answer frame doesn't includes any information about the process that requested the incoming data.

The master station must poll the secondary ones according to a predefined strategy that depends on the temporal restrictions imposed by the traffic on the network. In our application four kinds of traffic exist :

a) File transfer: part programs, PLC programs, parameter tables. Size of these files varies from several hundreds of bytes to even Megabytes. Messages are not time critical and do not occur very often. Moreover, these transfers can be done off-line without disturbing the other transactions.

b) Command/responses: messages for the synchronization of cell operations, selection of running modes on the CNC's, activation of programs and utilities that allow the operator to command machining and peripheral operations. This class also includes messages for network operation management. These are mainly short messages (about 20/30 bytes) with delivery time constraints.

c) Status information : Gathering of data for monitoring the cell status and collecting statistics for preventive maintenance and for production reports. This traffic is periodic (synchronous), short size (about 20/40 bytes) and form the largest part of the complete workload.

d) Events notification: The host computer must know as soon as possible any event produced in the cell devices (state changes in the manufacturing process or anormal events such as device failure or malfunctions). The secondary stations send events notification by constructing a "non requested data" frame and sending it to the master upon receiving a polling from it. For this, secondary stations must be polled whenever possible.

Traffic of type c) is carried out polling the secondary stations in a cyclic way and in a periodic time basis. This operation can be interrupted by traffic of type b) that has higher priority. The remaining time is used for polling secondary stations giving them the opportunity to send event notification (See fig. 3).

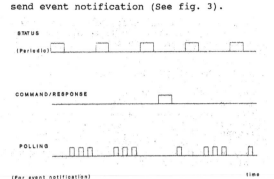

STATUS

(Periodic)

COMMAND/RESPONSE

POLLING

(For event notification)                                    time

Fig. 3. Traffic management over the bus.

The integrated PLC's at the numerical controllers are programmed to construct a status frame every 3 seconds. If the frame is not sent during this period data are lost because a new status frame is written over the previous one. Because of the soft real-time restrictions of the manufacturing process some status frames loss is tolerated.

A file transfer implies sending several consecutive frames. If this is made on line the channel is kept busy for a long period of time, without posibility of sending any other kind of traffic. In this case the master station manages the transmission allowing the secondary ones to send status and "non requested data" (if they exist) between every two consecutive file transfer frames.

Measures taken from an experimental implementation (See Table 1) show that at user layer every select/polling cycle takes on average 0.54 seconds for short frames (commands, status, events), and 0.68 seconds or 0.55 seconds (depending on data rate) for long ones (file transfer, writing and reading memory objects), allowing us to fulfil the requirements of periodic traffic.

The Polling/EOT row shows the time taken for giving a secondary station opportunity to transmit non requested data (event notification). Time available between two consecutive Select/Polling cycles is used for this action. As reflected in Table 1 the increasing of data rate only affects the performance of long frames transfer (file transfer) remaining unaffected the response time for short frames transmission. The critical factors in this case are the processing time in the stations attached to the network and the high overhead introduced for the polling mechanism that consumes bandwidth for control frames transmission reducing the effective utilization of the bus.

CONCLUSION

The implementation of the UNI-TE requests allows the Host computer to support full operation of the machine tool and, in adition, monitoring of the whole system operation.

The software modules to be executed at the master computer for controlling and managing the cell elements have been developed under the UNIX Operating System, allowing different concurrent

processes to access the UNI-TELWAY bus without conflicts among them.

The method implemented to allow the different processes to access the UNI-TELWAY services allows them not be concerned about the port access problems and the communications protocols management, simplifying the communications procedure.

We are presently developing an interface card for AT bus supporting the functionalities of the low layers UNI-TELWAY protocols. This will allow us to improve the response time of the network reducing the processing time at the master station. Further measures are to be taken in order to compare both implementations.

TABLE 1 User level medium response time

| TRANSACTION TYPE | RESPONSE TIME | DATA RATE |
|---|---|---|
| Select/Polling Cycle (Status) | 0.54 sgs. | 9600 |
|  | 0.53 sgs. | 19200 |
| Polling/EOT | 0.16 sgs. | 9600 |
|  | 0.16 sgs. | 19200 |
| File Transfer Frame / ACK | 0.68 sgs. | 9600 |
|  | 0.55 sgs. | 19200 |

REFERENCES

Daigle, J.N., A. Sedelmann, and J.R. Pimentel (1988). Communications for manufacturing: an overview. Network Mag., 2, 6-13.

Decotignie, J.D., and J.C. Grégorie (1989) Integrating the numerical controller and the FMS. Proc. IECON'89, pp. 675-680.

Gallardo, M. (1990). Tiempo real en el sistema operativo UNIX. Computer World Mag. (Spanish Ed.), 21-Nov., 17-25.

Lawrence, N. (1986). The use of MMS for remote CNC control. Proc. IECON'86, pp. 620-625.

Tejido, A.L., J.C. Fraile, and J.P. Turiel. (1991). Bus de campo UNI-TELWAY; interconexión de controles numéricos. Mundo Electrónico, nº 213, 57-62.

Pimentel, J.R. (1990). Communication Networks for Manufacturing, Prentice Hall, N.J. Chap. 14, 519-552.

Poole, G.A.(1990) Making UNIX predictable. UNIX World Mag., 7, nº 11, 93-96.

Sanz, J. (1990). Estrategias de migración desde protocolos propietarios a redes MAP/OSI: Gateways.VIII Congreso de Investigación, Diseño y Utilización de Máquina-Herramienta, San Sebastián (España).

Stevens, W. (1990). UNIX Network programming, Prentice Hall, N.J., Chap. 3.

# INTELLIGENT SUPERVISORY SYSTEM FOR MICROWAVE TELECOMMUNICATION NETWORKS[1]

## K. Tilly, I. Kerese, T. Zsemlye, B. Vadasz and Z. Szalay

*Technical University of Budapest, Department of Measurement and Instrument Engineering,
1521 Budapest, Muegyetem RKP 9, Hungary*

**Abstract.** To effectively supervise a telecommunication network an intelligent supervisory system is proposed consisting of a traditional process monitoring, a fault diagnostic, a communication handler and a database manager subsystem. The whole system is based on a generic expert system shell designed to operate in industrial environment. The diagnostic subsystem contains a two-level inference engine to operate on structural information and traditional if-then rules. To ensure easy mapping of the supervisory system to any telecommunication network a configuration environment consisting of several compilers integrated into a multi-window editor was also developed. The system was implemented on two interconnected IBM PC-s running MS WINDOWS 3.0.

**Keywords:** Supervisory control; expert systems; monitoring; communications control applications; software tools.

## INTRODUCTION

Telecommunication networks are complex, distributed systems of sophisticated components. Their reliable operation is crucial not only for TV channels or cellular telephone users, but also for air traffic controllers and ambulance cars, where a breakdown almost surely causes more trouble than mere inconvenience. A system of such complexity cannot easily be supervised, let alone diagnosed in case of malfunctions. To support these activities a distributed supervisory system was developed, which provides user friendly tools for the efficient maintenance of the network.

## THE MICROWAVE NETWORK

Microwave telecommunication networks are systems of high complexity consisting of several (sometimes hundreds of) radio relay stations, each of which receives information from other relays or primary input channels and transmits it to other relays or output channels.

A relay station consists of a receiver, a transmitter and a tree shaped multilevel multiplexer structure which is responsible for connecting the receiver and the transmitter to the input/output points which range from simple telex lines and telephone channels to radio and TV signals. Channels of different relay stations can be associated with one another thus establishing a network of communication lines.

A network like this is so sophisticated and the service it provides is so expensive that it cannot be operated fully automatically without human supervision.

The operator performs the following tasks:

- Controls some rather simple features of a station according to the current needs (e.g. to switch the lighting of the relay antenna on when it gets dark);

- Determines whether a given station is functioning properly and repairs it if necessary.

The relay stations are equipped with special remote control units which perform primary data acquisition and communication tasks and are responsible for maintaining connection between the relay station and the central monitoring station.

Services provided by the microwave network should be as reliable as possible. If despite of all efforts a failure in signal transmission occurs the reason should be detected and eliminated as soon as possible. The major difficulty is caused by the fact that a malfunctioning station sends alarm messages not only to the monitoring center which oversees the whole network but also to the relays which it is in connection with.

If a station receives an alarm message from one of its neighbours it also forwards it to the central monitoring station, therefore if an alarm message is received from a relay it does not necessarily mean that the station is faulty. Since unmanned relay stations are often located in remote places which are difficult to access (top of mountains, etc.) they should be monitored by systems providing powerful tools for quick and automatic error detection, localization and diagnosis to find the exact location and cause of the network malfunction before the service personnel set out for the site.

## GENERAL ARCHITECTURE OF THE SUPERVISORY SYSTEM

To match the requirements mentioned above we developed an on-line supervisory system providing the operator with explicit, high level and intelligent information about the current state of the network.

The architecture of the system was derived from the generic tasks of a supervisory system and from the special requirements

[1]This project was sponsored by the Hungarian National Committee for Technical Development.

of the monitored telecommunication network. The main tasks of a supervisory system are as follows:

- Data acquisition from the supervised object/process;

- Data evaluation and preprocessing to create output easy to interpret by the operator;

- Displaying preprocessed data and various system messages (e.g. current values of parameters, signal trends, special events, textual warning and error messages);

- Logging events and messages in log files;

- Receiving operator commands and transmitting them to the supervised object/process.

Some of these tasks require bidirectional communication between the supervisory system and the monitored object/process. In our case the communication media was the so called service channel of the network which in essence is a low speed serial bus connecting the relay stations with the central monitoring station.

When designing the communication protocol the following constraints had to be taken into consideration:

- The transmission speed of the service channel is merely 1200 bauds;

- The maximal number of stations is 256;

- The average number of parameters transmitted to a relay station is 128;

- The response time of relay stations for parameter requests in worst case is 90 seconds.

To avoid collisions on the service channel a flexible polling mechanism is used: although the central station is the exclusive master cyclically initiating polling of the relay stations, the polling order can vary accordingly to the current needs. To keep the amount of transmitted information on a minimal level we opted for a variable message length protocol with minimal overhead.

Resulting from the requirements and the above characteristics of the microwave network the proposed supervisory system consists of a monitoring, a diagnostic, a communication and a database management subsystem.

The supervisory system is implemented on two computers interconnected with an RS-232 serial line. One of them is charged with traditional monitoring functions while the other runs a fault diagnostic system. The operation of both computers is based on the network database storing information about the topology, actual parameter values and other characteristics of the supervised network.

Since our goal was to develop a supervisory system for a generic network and not for a particular one we had to assure easy mapping of the supervisory system to any network topology and station structure.

In order to achieve this goal we also created a configuration environment processing declarative descriptions of the targeted network, which makes it possible to define any network dependent feature in the supervisory system.

## THE MONITORING SUBSYSTEM

The monitoring subsystem - essentially a traditional supervisory station - performs communication with the network, generates data and trend displays, alarm and warning messages, and provides possibility for remote control. Besides these it is also responsible for providing data to the diagnostic subsystem.

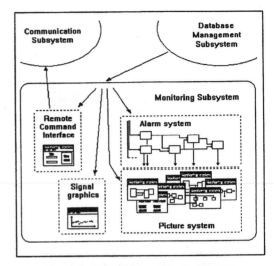

Fig. 1. Architecture of the monitoring subsystem

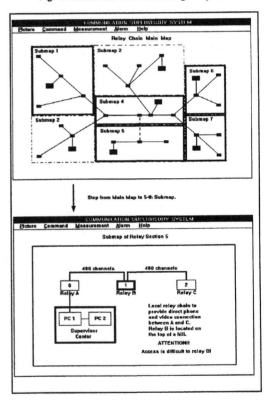

Fig. 2. The main map and a submap of the microwave network

The monitoring subsystem displays the supervised network or a part of it from different aspects. The main duties of this subsystem are as follows:

- Managing continuous updating of various pictures describing the microwave network;

- Alarming the operator when a relay station fails;

- Providing operator interface to issue remote commands.

The technological picture system helps to review the state of the microwave network. These pictures are organized in tree structure. The lower a picture is placed in this hierarchy the finer details of the system are shown on it.

The picture at the root of the tree describes a rough view of the supervised network representing the spatial and logical location of relay stations and supervisory systems. Pictures at the second level contain detailed representation of distinct parts of the network, usually comprising only a few relay stations.

Pictures at the next level of the hierarchy show the internal structure of the relay stations. These pictures describe the building blocks of the relays (i.e. the functional elements) along

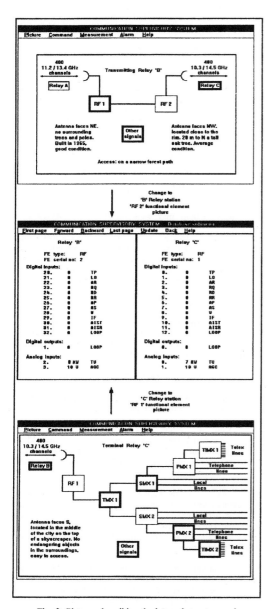

Fig. 3. Pictures describing the internal structure and signal values of relays

with their error state and connections to each other.

The most detailed information about a relay station is described on the fourth level pictures. These pictures show the values of analog and digital signals of the selected functional element and their counterparts on the connecting relay station. The point in displaying signal values of two functional elements placed on two distinct relays is that these two elements of the same type form a logical communication channel.

Displaying the pictures can be controlled with the help of menus or by clicking the mouse at "hot spots" on the screen.

The alarm system activates the display driver to indicate the error state of those faulty elements which are currently displayed. Depending on the seriousness of the detected failure two different error levels can be displayed: one for simple errors and one for emergency situations.

The alarm system can provide three different outputs:

- The computer beeps in case of an error;

- The symbol of the faulty element or relay station is colored differently from those operating normally. This feature is assured by the alarm propagation network, which is triggered by the changes of input signals;

- If the place of the error cannot be shown on the technological picture a message is sent to the operator.

The alarm system stores the information about the error in a buffer, so if an error is resulted from changes of several input signals the operator can check this buffer to see the changes of the system in details.

The monitoring subsystem provides a menu driven interface for the operator to issue remote commands which can be used to change the state and to perform remote measurements on the relay stations and to reconfigure the microwave channels. Results of measurements can be displayed graphically or in tables.

## THE DIAGNOSTIC SUBSYSTEM

The diagnostic subsystem - being an intelligent supervisory station - supplies the operator with high level information about the network, such as textual data logs, operational statistics (e.g. MTBF - Mean Time Between Failures - statistic) and intelligent diagnostic messages. Messages from operators of other monitoring stations are also displayed by this subsystem.

In alarm situations (e.g. when a station breaks down) it is very difficult for an operator to make fast and correct decisions, because he may receive too many and sometimes apparently contradictory alarm signals. However, the operator is urged to make his decision in a minimal time, although this increases the possibilities of mistakes.

A good method of decision support is to incorporate information about the structure and behaviour of the network (Davis, 1985) and of individual stations into the supervisory system. This information can be collected from network experts during the system definition phase, and the supervisory system can interpret this knowledge to provide explicit textual messages about the causes of errors. This method can assure accurate decisions within reasonable time, although enables the operator to override advices generated during diagnostics.

The operation of the diagnostic subsystem is based on the same communication and database management subsystems which are used by the monitoring subsystem, consequently the basic means of data exchange between the monitoring and the diagnostic subsystem is the network database. However, the diagnostic subsystem only reads the database while the monitoring subsystem is responsible for maintaining the network database consistency.

The diagnostic subsystem performs the following activities:

- Traditional textual data logging of significant events (e.g. remote commands, failures etc.) with a short description of the event; data logs can be watched on the display screen or hardcopied to a printer;

- Computing failure statistics on the operator's request for specified stations, communication lines and printed circuit board types.

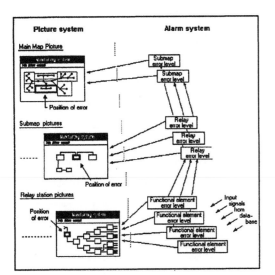

Fig. 4. Architecture of the alarm system

From the failure statistics the operator can acquire information about the relative reliability of different system components. After a longer period of operation the weak spots of the system (e.g. unreliable board types, etc.) can be pointed out, later supporting improvement of the network.

The system produces three types of statistics:

- The MTBF (Mean Time Between Failures) statistics shows the overall reliability of the system;

- The MTTR (Mean Time To Repair) statistics shows the seriousness of system faults; it also can be used to evaluate the abilities of the service personnel;

- Board failure statistics for every board with a short textual description of the nature of the fault entered by the operator. This feature supports detection of the weak construction points of the network system.

In a highly complicated network there may be more than one supervisory station; so called mobile supervisory stations can also be temporarily set up by service personnel during reparation of a faulty station. It's essential to provide the operator with the possibility to communicate with his colleagues in the form of textual messages. The main advantage of this method compared to e.g. telephone conversation is that this way the transmitted messages can be handled like events and can be automatically logged.

The diagnostic subsystem gives explicit textual advices to the operator regarding the current fault. Since it's the most interesting feature of the diagnostic subsystem we will discuss it in the rest of this section.

The advisory system is built on the knowledge-based approach with a multi-layered, modular knowledge representation scheme and a two level inference engine.

The advisory system has to meet several requirements, among which the most important are:

- The results of the advisory system have to be automatically generated from the actual network database, and as mentioned before this is a time critical operation. Although the time limit is not very strict (several minutes), this requires some kind of event driven real-time operation;

- Since the state of the network is changing continuously the system must be prepared for time variant operation,

i.e. some of the drawn conclusions must be withdrawn after a time; this feature requires nonmonotonic inference engine;

- The supervisory system must be able to efficiently handle different network configurations. Consequently, only methods which store information of the network in a very compact form are satisfactory, though this feature contradicts the high complexity of the network.

Considering the requirements above we chose a multi-layered solution: the two-level inference engine consists of a meta-forward level for handling structural information of the network and a forward level to perform rule-based inference in order to determine the cause of the occurred fault.

The forward inference engine (Papp and colleagues, 1989) performs event driven real-time operation. The incorporated expert rules are described in the traditional if-then form and make automatic search of the fault tree possible: current parameter values are matched against the conditional (if) parts of the rules and if the conditional part of a rule is satisfied the activities in the conclusion part are executed.

Systems with the complexity of a telecommunication network cannot be handled with one homogeneous rule-base: it would be too large, oversophisticated and slow to handle; therefore we had to divide the rule-base into modules. There are several possibilities to overcome this problem.

Since relay stations may have different configuration it is possible to associate a rule-base to every station. However, if we take a deeper look at the problem it turns out that even a single relay station is too complicated to create efficient and relatively small rule-bases with acceptable efforts, not to mention the possible number of different rule-bases which can be several hundred (Bagó and colleagues, 1986).

To overcome this problem we associated the lowest (rule-based) system level to the lowest (functional element) level of the network. Any station can be built using a very limited number (only six) of functional elements: receiver, transmitter and four types of multiplexers. Every functional element has its own, rather simple rule-base each of which holds less than 50 rules.

Rules can be categorized into two semantic classes:

- Advice generation rules are used to generate textual output for the operator about the actual failure causes;

- Fault propagation rules are the connections between individual rule-bases (i.e. functional elements). These provide information for the higher level inference engine to continue investigation on other functional elements by switching to their own rule-base.

When a fault occurs the meta-forward engine is started first. It automatically checks stations which have reported faults. First of all it determines the faulty station with a very simple algorithm: since the stations signal faults to each other, the station whose every neighbour reports fault is the cause of the network malfunction.

Every station has a tree shaped internal structure. After finding the faulty station the meta-forward engine determines the top level functional element of the station, activates the rule-base of it with the current set of fact-values and starts the forward inference engine.

If an advice generation rule is fired (i.e. the cause of the error is found), the inference procedure stops (we assume single faults).

If a fault propagation rule is fired, the meta-forward engine gets the control back and activates the rule-base of the next functional element which is determined using the so called meta-agenda. The meta-agenda is the main link between the meta-forward and the forward inference engine. Fault propagation rules put the identifier of the functional elements to be tested into the meta-agenda. The meta-forward engine takes the first item of it, switches to the actual rule-base and control is passed to the forward inference engine.

## THE COMMUNICATION SUBSYSTEM

The communication subsystem ensures information flow between the microwave network and the supervisory system. The communication subsystem periodically polls the relay stations requesting the values of various signals characterising the state of the relays. In order to reduce the polling period messages contain only the change in the state of the interrogated relay station.

If a relay station fails it is taken out of the regular polling order and is requested to send information about its current state in the form of various parameters. In order to collect these parameters the communication subsystem sends remote commands to the relay stations. The communication subsystem passes data arriving from the microwave network to the database management subsystem.

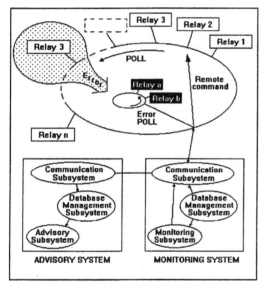

Fig. 5. Communication between the network and the supervisory system

## THE DATABASE MANAGEMENT SUBSYSTEM

The database management subsystem assures the database consistency of the supervisory system. The database is stored on both computers of the supervisory system, providing partial operation of the system even if one of them breaks down.

In order to maintain consistent databases their items are tagged with a time stamp which is refreshed when the item is updated.

The architecture of this subsystem is based on the object oriented approach. Every relay station is represented by an object which stores information about the inner structure of the relay and its connections to other stations. These objects are built up of local objects, which represent the functional elements of the relay, e.g. multiplexers and radio frequency transmitters. Digital and analog signals of the functional elements are stored in the database in groups.

The database management subsystem stores actual values of the input signals in the computer memory, while old values are stored on the hard disk. The values stored on the disk represent the time records of the signals. A data-driven object oriented processing chain prepares input data in order to generate statistics from the incoming data, and activates the alarm system if needed (Bagó and colleagues, 1986).

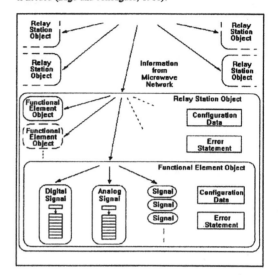

Fig. 6. The object oriented database

## THE CONFIGURATION ENVIRONMENT

The configuration environment of the supervisory system ensures easy mapping of the generic supervisory system to a dedicated communication network. It consists of a set of compilers integrated into a uniform multi-window editor which makes definition editing and compilation interactive, fast and simple. The output of the compilers is a set of pictures to be displayed by the monitoring subsystem, the network database and the preprocessed rule-base which can directly be used by the advisory system.

To increase reliable operation of the compilers they were developed using the LEX lexical analyzer generator and YACC parser generator programs usually used in UNIX systems. However, to avoid symbol collisions and the remarkable overhead caused by multiple inclusion in the final code of the same scanner and parser algorithms operating on different data certain modifications had to be introduced in the output of the generator programs. These changes made it possible to include the scanner and parser driver routines only once; switching between different compilers were carried out as a switch from one set of syntax descriptor tables to another. Other major components of the compiler set are the unified code generator and symbol table handler modules.

The environment contains four compilers: a main-map, a submap, a relay description and a rule compiler. The first three generates pictures and data for the network database, while the last one translates the high level rules of the advisory system to a "machine friendly" format.

To provide enough flexibility for future developments of the relays a configuration file containing the main features of possible functional elements is processed when the configuration environment is started.

Though the compilers stop translation when the first error is encountered, due to the usually small size of input files and the fast compilation it does not cause real inconvenience. On the other hand the compilers detect every possible error and provide detailed messages when necessary.

As an illustration of the pictures generated by the environment see Fig. 2 and 3.

## THE SOFTWARE BACKGROUND

As a base for the supervisory system a generic expert system shell called REALEX was chosen (Papp and colleagues, 1989; Papp, 1990), which is a high level programming environment designed to be a versatile toolkit of system designers to build industrial process supervisory systems. This toolkit provides support to perform common supervisory tasks such as event logging, data acquisition, alarming, etc.

The REALEX shell gives only a skeleton to create process supervisory programs. From the system designer's point of view it provides slots to be filled with descriptions of particular processes. These slots reflect the different aspects of the process to be supervised, e.g. events to log, transformations to be performed on measured data, etc. However, the shell also provides some AI originated tools: the supervisory systems generated from it are able to perform post-mortem diagnosis of the supervised system in case of a crash. Figure 7 shows the simplified architecture of the REALEX shell.

To provide adequate tools for system description the REALEX shell supports three different models of the supervised system.

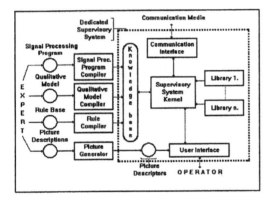

Fig. 7. Architecture of the REALEX shell

The mathematical model (which in essence is a signal processing program) is based on the data-flow concept and is implemented on object oriented basis (Tilly, 1992): the model of the supervised system can be constructed like chains of atomic transformations. The REALEX shell supplies a rich library of transformations, which can easily be extended to meet particular needs.

The heuristic (rule-based) model uses the well known rule-based paradigm. However, to ensure that the speed of inference does not depend on the number of rules a sophisticated preprocessing of them must be performed before the system starts operating (Vadász, 1992). This preprocessing is performed by the rule compiler. Rules are interpreted in data-driven (forward chaining) manner. This tool was used to implement the fault diagnosis part of the telecommunications network supervisory system.

The third (qualitative) modelling feature of the REALEX shell was not used in the supervisory system for telecommunication network.

The REALEX shell also provides additional facilities which were widely exploited. These are the multiprocessing (including remote processes) and interprocess communication (IPC) features. Relays of the microwave network can be accessed as remote processes via the standard IPC mechanism of the

REALEX shell. This IPC mechanism was designed to be capable to integrate various communication media to meet different application dependent needs.

The REALEX shell was designed and implemented at the Department of Measurement and Instrumentation Engineering of the Technical University of Budapest. It was written in C language, although the shell also provides a LISP interface for the users. This software architecture was chosen for and proved to be very efficient in integrating symbolic and numerical computation. The shell runs on IBM PC under MS-DOS, OS/2 and MS WINDOWS 3.0. operating system.

The dedicated software of the telecommunications network supervisory system was mostly written in LISP, except for some time-critical parts. In order to meet some special requirement of this application a communication loop as a new communication media was added to the module managing the inter-process communication. For the same reason the user interface was re-designed, although the original picture handling concept of REALEX was fully adopted. The dedicated microwave supervisory system and its configuration environment runs on two IBM PCs under MS WINDOWS 3.0.

## CONCLUSIONS

Supervision of telecommunication networks needs supervisory systems of high reliability and flexibility. A simple dedicated system is not acceptable because of the features of the application domain. The development work described above in detail resulted in a versatile supervisory system for telecommunication networks. It is easily configurable for different network structures and relay types. Exploiting possibilities of the underlaying REALEX shell the system successfully combines the traditional data acquisition with the knowledge based approach. This solution supports the operator with high level information preventing him or her being overloaded by the data in case of a fault in the supervised network. Moreover, it drastically reduces the overall cost of repairing.

This work has shown that the rule-based paradigm can be successfully used in the field of supervision and control of complex processes.

## REFERENCES

Davis, R. (1985). Diagnostic reasoning based on structure and behaviour. Qualitative Reasoning About Physical Systems, MIT Press, Cambridge, Mass. 347-410.

Papp, Z. and colleagues (1989). Expert system architecture for real-time process supervisor applications. In G. Rzevski (Ed.), Artificial Intelligence in Manufacturing, CMP & Springer-Verlag, Berlin. 223-240.

Bagó, B. and colleagues (1986). A multi-level signal processing system. Proc. of the 8th Ann. Conf. of the IEEE-EMBS, Vol. 2, 825-828.

Papp, Z. (1990). Programming tool for integrating numerical and knowledge based signal processing techniques. International Symposium Knowledge Based Measurement - Application, Research and Education. 1999, Karlsruhe, 181-188.

Tilly, K., Z. Papp, T Dobrowiecki (1992). A data-flow driven data processing and acquisition system for real-time industrial process monitoring applications. Proposed paper for the 18th IFAC/IFIP Workshop on Real-time Programming. Bruges, 23-25 June 1992.

Vadász, B. (1992). Implementational aspects of rule-based knowledge representation in real-time environment. Proposed paper for the 1992 IFAC/IFIP/IMACS Internat. Symp. on Artificial Intelligence in Real-Time Control. Delft, 16-18 June 1992.

# FLEXIBLE MULTIPLEXING OF CONTROL DATA

## R. Whalley and Z. Zeng

*Department of Mechanical and Manufacturing Engineering, University of Bradford, Bradford, West Yorks, BD7 1DP, UK*

Abstract. Variations in the control data sampling rates in digital control schemes where data transmission is via a data bus are investigated. Algorithms enabling the stability boundary and the range of acceptable data bus access times are proposed.

Keywords. Multiplexing, multi-dimensional systems, stability, computer control.

## INTRODUCTION

The employment of integrated monitoring, control and surveillance schemes was significantly advanced by the availability of low cost, digital electronics in recent years. Responding to the demand for micro-computer driven regulation schemes standard interfaces, data protocols, real time languages and communication links were proposed nationally and many manufacturers of digital electronic equipment now support European standards which resulted from these initiatives.

In fact substantial developments in digital technology can be traced back to military applications where the emphasis on high reliability, accuracy and robustness eclipsed all other considerations and where cost and flexibility remained low in importance. Nevertheless the combined command, control and communication requirements embedded in military specifications progressed the design, utilisation and development of digital regulation schemes while more general structures, suited to industrial applications, followed from this introduction.

A key element in the evolving architecture of industrial control systems in the last decade emerged as the data bus link between the process and computer. Several standard patterns of this communications channel now exist including the MODBUS, EUROBUS, FIELDBUS etc, all of which enable multiplexed data to be handled in a highly efficient manner. Access to the bus by way of real time interrupts with priority levels and protocols determining the queueing arrangements formed the principal feature of many real time operating systems for digital control systems wherein the execution of the high priority tasks associated with the transmission of control data normally absorbed minimal processor time. Thereafter a mass of "background" computation aimed at establishing system performance, tolerance validation, health monitoring, operational efficiency, start up and shut down requirements etc. is usually undertaken in the interval between the periodic, mandatory data processing intervals reserved for control data transmissions.

In applications where only one measured variable is included in the control scheme the simple structure invoked earlier would be quite acceptable. However, when more than one measurement is necessary to achieve the regulation required, this becomes less viable. Moreover on these occasions significant differences in time between the digital data measurements becoming available may be encountered. This is particularly so when, for example, the composition of mixtures, chemical analysis or flow rates are required whereas the digital sensing of voltage, pressure, temperature etc can be achieved very rapidly.

It is also true that in cases where the control data from various transducers has been synchronised, hardware failure or software errors may cause asynchronous data sampling by default, with all the consequences for system integrity which may follow from this condition.

Essentially the imposition of mono-rate data sampling for control purposes is quite artificial. Despite this self-inflicted constraint, texts on computer control system design and analysis rarely mention the possibility of doing otherwise. Moreover in the case of systems where fail-safe conditions must be maintained it is important to establish the extent to which asynchronous control action could be tolerated while remaining operational.

In this paper therefore the topic of asynchronous computer control will be addressed. Procedures for assessing the stability of two and multirate sample data control systems will be formulated and new algorithms for the determination of relative stability will be presented.

The results derived for stability assessment will be validated by simulation illustrating thereby that considerable variations in sample rates, gains etc. are often possible without incurring any significant change in transient performance or accuracy. This flexibility could be instrumental in improving the efficiency of communications via the multiplexing system and would relax the periodic requirement for interrupts at the highest priority level.

Otherwise the integrity of processes, where several transducers are employed and where the limits of system performance must be established, could be determined. This could now be predicted by solving the stability equation to give the maximum range of sampling rates for stable conditions, whereas extensive simulation and hardware trials are currently required to achieve this measure.

### Two Dimensional Models

With digital control schemes it is relatively easy to group the rapidly responding feedback control loops arranging access thereafter to the data bus at frequent intervals. Equally loops requiring a lower rate of servicing could be given access at some lower cycle time.

This is the simplest asynchronous, discrete feedback configuration where two different data sampling rates are incorporated and where all feedback measurements are sampled at either $T_1$ or at $T_2$ second intervals. Analytically however a major excursion in analysis - design is necessary in order to accommodate the transition from models in one dimension to descriptions in two.

In general input-output models in two dimensions in the transformed variables:

$$z_1 = e^{sT_1} \text{ and } z_2 = e^{sT_2}$$

Take the form:

$$(\sum_{j=0}^{n} \sum_{k=0}^{n} p_{jk} z_1{}^{-j} z_2{}^{-k})/$$

$$(\sum_{j=0}^{n} \sum_{k=0}^{n} q_{jk} z_1{}^{-j} z_2{}^{-k}) \qquad ----1$$

where the coefficients $p_{jk}$ and $q_{jk}$ $0 \leqslant j,k \leqslant n$ are real constants when the system model is Linear Shift Invariant (LSI). In two dimensional LSI schemes a principal objective is that of releasing the data multiplexing highway for monitoring, surveillance and logging traffic for as long as possible, without adversely affecting performance. Consequently when two different, independent rates of control data sampling are used techniques enabling relative stability to be assessed are an important part of the design - analysis exercise whereby the stable portion of the gain - sampling rate space can be investigated.

### Two Dimensional System Stability

A Nyquist like boundary image method of assessing two dimensional system stability is provided by De Carlo R.A. et al (1977) where in the absence of non-essential singularation of the second kind (NES), stability can be guaranteed providing:

$$\{ Q(z_1{}^{-1}, 0) \neq 0, \mathbin{|} z_2 z_2{}^{-1} \mathbin{|} \leqslant 1 \},$$

and $\{ Q(z_1{}^{-1}, z_2{}^{-1}) \neq 0, \mathbin{|} z_1 \mathbin{|} = 1, \mathbin{|} z_2{}^{-1} \mathbin{|} \leqslant 1 \},$

$$---2$$

Huang T.S (1981) comments on the second of these tests which he observes is difficult to execute accurately, is time consuming and tedius. In fact all that is necessary to establish stability is to test the minimum boundary image of:

$$\{ Q(e^{i\Phi_1}, z_2{}^{-1}), \mathbin{|} z_2{}^{-1} \mathbin{|} \leqslant 1 \},$$

ensuring that this minimum does not include the origin of C.

To calculate minimum boundary image for a two dimensional quadratic model the denominator of which, in accordance with equation 1 is:

$$Q(z_1{}^{-1}, z_2{}^{-1}) =$$

$$q_{00}+q_{01} z_2{}^{-1}+q_{02} z_2{}^{-2}+q_{10} z_1{}^{-1}+q_{20} z_1{}^{-2}+$$

$$q_{11} z_1{}^{-1}z_2{}^{-1}+q_{12} z_1{}^{-1}z_2{}^{-2}+q_{21} z_1{}^{-2}z_2{}^{-1}+$$

$$q_{22} z_1{}^{-2}z_2{}^{-2}$$

when $\mathbin{|} z_1{}^{-1} \mathbin{|} = 1$ and $\mathbin{|} z_2{}^{-1} \mathbin{|} \leqslant 1$ means that the minimum boundary image of:

$$\{Q(e^{i\Phi_1}, z_2{}^{-1}), \mathbin{|} z_2{}^{-1} \mathbin{|} \leqslant 1, 0 \leqslant \Phi_1 \leqslant 2\pi\}$$

is required. This becomes:

$$\{ \mathbin{|} q_{00}+q_{10}(Cos\Phi_1+i \sin \Phi_1)+q_{20}(Cos2\Phi_1+$$

$$i \sin2\Phi_1) \mathbin{|} > \mathbin{|} (q_{01}+q_{02} z_2{}^{-1}+q_{11}+$$

$$q_{12} z_2{}^{-1})(Cos\Phi_1+i \sin\Phi_1)+(q_{21}+q_{22} z_2{}^{-1})$$

$$(Cos2\Phi_1+i \sin2\Phi_1) z_2{}^{-1} \mathbin{|}, \mathbin{|} z_2{}^{-1} \mathbin{|} \leqslant 1)$$

The modulus expressions for the left and right sides of this equation are, respectively:

$$\sqrt{((q_{00}+q_{10}cos\Phi_1+q_{20}cos2\Phi_1)^2+(q_{10}sin\Phi_1+}$$

$$q_{20}sin2\Phi_1)^2) \text{ and }$$

$$\sqrt{((q_{01}+q_{02}+(q_{11}+q_{12})cos\Phi_1+}$$

$$(q_{21}+q_{22})cos2\Phi_1)^2+((q_{11}+q_{12})sin\Phi_1 +$$

$$(q_{21}+q_{22})sin2\Phi_1)^2)$$

Following routine manipulations and substitutions these expressions become:

$$\sqrt{(\alpha_1+2(\alpha_{12}+\alpha_{01})cos\Phi_1+2\alpha_{02}cos2\Phi_1)}$$

and

$$\sqrt{(\beta_1+2(\beta_{12}+\beta_{01})cos\Phi_1+2\beta_{02}cos2\Phi_1)},$$

respectively. Consequently $\{Q(e^{i\Phi_1}, z_2{}^{-1})$

$$\neq 0, \mathbin{|} z_2{}^{-1} \mathbin{|} \leqslant 1\}$$

providing:

$$(\beta_1+2(\beta_{12}+\beta_{10})cos\Phi_1+2\beta_{02}cos2\Phi_1)/$$

$$(\alpha_1+2(\alpha_{12}+\alpha_{01})cos\Phi_1+2\alpha_{02}cos2\Phi_1) \leqslant 1$$

where:

$$\alpha_1 = q_{00}{}^2 + q_{10}{}^2 + q_{20}{}^2$$

$$\alpha_{12} = q_{10}\,q_{20}$$

$$\alpha_{01} = q_{00}\,q_{10}$$

$$\alpha_{02} = q_{00}\,q_{20}$$

and

$$\beta_1 = (q_{01}+q_{02})^2 + (q_{11}+q_{12})^2 + (q_{21}+q_{22})^2$$

$$\beta_{01} = (q_{01}+q_{02})\,(q_{11}+q_{12})$$

$$\beta_{02} = (q_{01}+q_{02})\,(q_{21}+q_{22})$$

$$\beta_{12} = (q_{11}+q_{12})\,(q_{21}+q_{22})$$

Alternatively upon putting $x = \cos\varphi_1$, enables the last equation to be written as:

$$F_2(x), \quad -1 \leqslant x \leqslant 1$$

so that for stability:

$$Q(z_1{}^{-1}, 0) \neq 0, \; |z_1{}^{-1}| \leqslant 1$$

and $F_2(x) \leqslant 0, \; -1 \leqslant x \leqslant 1$  --- (3)

where:

$$F_2(x) = (\beta_1 - 2\beta_{02} - \alpha_1 + 2\alpha_{02}) + 2(\beta_{01}+\beta_{12} - \alpha_{12} - \alpha_{01})x + 4(\beta_{02} - \alpha_{02})x^2$$

Following exactly the same route the stability of a cubic two dimensional model can be established, in the absence of NES, if:

$$\{Q(z_1{}^{-1}, 0) \neq 0, \; |z_1{}^{-1}| \leqslant 1\}$$

and $F_3(x) \leqslant 0, \; -1 \leqslant x \leqslant 1$  ---(4)

where $F_3(x) =$

$$(\beta_1 - 2\beta_{02} - 2\beta_{13} - \alpha_1 + 2\alpha_{02} + 2\alpha_{13})$$

$$+ 2(\beta_{01}+\beta_{12}+\beta_{23} - 3\beta_{03} - \alpha_{01}$$

$$-\alpha_{12} - \alpha_{23} + 3\alpha_{03})x$$

$$+ 4(\beta_{02}+\beta_{13} - \alpha_{02} - \alpha_{13})x^2$$

$$+ 8(\beta_{03} - \alpha_{03})x^3 \leqslant 0, \; -1 \leqslant x \leqslant 1,$$

with:

$$\beta_1 = (q_{01}+q_{02}+q_{03})^2 + (q_{11}+q_{12}+q_{13})^2$$

$$+(q_{21}+q_{22}+q_{23})^2 + (q_{31}+q_{32}+q_{33})^2,$$

$$\beta_{01} = (q_{01}+q_{02}+q_{03})(q_{11}+q_{12}+q_{13}),$$

$$\beta_{02} = (q_{01}+q_{02}+q_{03})(q_{21}+q_{22}+q_{23}),$$

$$\beta_{03} = (q_{01}+q_{02}+q_{03})(q_{31}+q_{32}+q_{33}),$$

$$\beta_{12} = (q_{11}+q_{12}+q_{13})(q_{21}+q_{22}+q_{23}),$$

$$\beta_{13} = (q_{11}+q_{12}+q_{13})(q_{31}+q_{32}+q_{33}),$$

$$\beta_{23} = (q_{21}+q_{22}+q_{23})(q_{31}+q_{32}+q_{33}),$$

$$\alpha_1 = q_{00}{}^2 + q_{10}{}^2 + q_{20}{}^2 + q_{30}{}^3,$$

$$\alpha_{01} = q_{00}q_{10},$$

$$\alpha_{02} = q_{00}q_{20},$$

$$\alpha_{03} = q_{00}q_{30},$$

$$\alpha_{12} = q_{10}q_{20},$$

$$\alpha_{13} = q_{10}q_{30},$$

and

$$\alpha_{23} = q_{20}q_{30},$$

These stability tests are very easy to execute and similar expression exist for higher order models.

### Multiplexing Control Data

Usually multiplexing by means of a data bus is used to manage all data transmission on a priority basis apart from control data which is relayed at fixed intervals. Some priority conditions such as emergency shut down, alarms or failure trips could overide the control data interrupt facility. However, under normal working conditions absolute priority would be afforded to these transmissions.

In fact servicing the control loops at specific synchronised intervals is rarely necessary and efficient use of the data bus is inhibited by this lack of flexibility. To illustrate this condition a typical control scheme where digital data sampling is synchronised and confined to a standard rate of transmission will be considered.

In this system, which is shown in block form in Figure 1, both the inner and outer loop feedback is restricted to a sampling rate of 0.05 seconds. The response of the system following a step input of unity if for example, the sampling rate in the outer loop is now changed to 0.4 seconds and that for the inner loop is maintained at 0.05 seconds then the pulse transfer function for the closed loop system is as equation 1 with n = 3 and where the non-zero coefficients are:

$$p_{00} = 0, \; p_{01} = 0.70, \; p_{02} = 0.031,$$

$$p_{03} = -0.504, \; p_{11} = -0.666, \; p_{12} = -0.029,$$

$$p_{13} = 0.479$$

and

$$q_{00} = 1, \; q_{01} = -1.105, \; q_{02} = 0.926, \; q_{03} = -0.594$$

$$q_{10} = -0.902, \; q_{11} = 0.962, \; q_{12} = -0.836,$$

$$q_{13} = 0.56 \qquad \text{----(5)}$$

for $T_1 = 0.05$ and $T_2 = 0.4$ seconds.

Similarly if $T_1$ is maintained at 0.05 seconds and $T_2$ is increased 0.9 seconds the non-zero coefficients of equation 1 become

$$p_{00} = 0, \; p_{01} = 3.06, \; p_{02} = -0.184,$$

$$p_{03} = -1.82, \; p_{11} = -2.91, \; p_{12} = 0.175,$$

$$p_{13} = 1.731$$

and

$$q_{00} = 1, \; q_{01} = 1.643, \; q_{02} = 0.237, \; q_{03} = -1.824$$

$$q_{10} = -0.962, \; q_{11} = -1.632, \; q_{12} = -0.205,$$

$$q_{13} = 1.735 \qquad \text{----(6)}$$

Otherwise when $T_1 = T_2 = 0.05$ seconds the transfer function for this system is:

P/Q

where:

$$P = 0.012 (z^{-1} + 0.044z^{-2} - 0.956z^{-3})$$

and

$$Q = 1-2.669z^{-1} + 2.386z^{-2} - 0.7145z^{-3}$$

$$----(7)$$

## Relative Stability of Model

To establish the model's stability condition it is necessary to execute the relevant stability test given by equations 3 or 4 or for the case when $T_1 = T_2$ any of the standard tests given in Franklin G R et at (1980) for example will suffice. In fact when the inner and outer loop sampling rates are synchronised at 0.05 seconds the roots of the characteristic polynomial for the model given by equation 7 are:

$$z= 0.899, -0.885 \pm 0.104i$$

indicating that the response of the system is quite stable and monotonic. However, once the outer loop rate has been increased to 0.4 seconds per sample the coefficients of $Q(z_1^{-1}, z_2^{-1})$ given by equation 5 must be substituted into equation 4 to assess stability. This procedure must also be followed when the inner loop sampling rate has been further increased to 0.9 seconds.

In both of these cases only $q_{0j}$, $0 \leqslant j \leqslant 3$ and $q_{k0}$, $0 \leqslant k \leqslant 3$ have non-zero values so that in equation 4 only:

$$\beta_1 = (q_{01}+q_{02}+q_{03})^2+(q_{11}+q_{12}+q_{13})^2$$

$$\beta_{01} = (q_{01}+q_{02}+q_{03})(q_{11}+q_{12}+q_{13})$$

$$\alpha_1 = q_{00}^2+q_{10}^2$$

and $\alpha_{01} = q_{00}q_{10}$ $\quad ----8$

have values other than zero. Under these circumstances upon substituting for equation 8, equations 4 become:

$$Q(z_1^{-1},0) = 1+ q_{10}z_1^{-1}$$

and $F_3(x) = (\beta_1-\alpha_1) + 2(\beta_{01}-\alpha_{01})x$
$$----9$$

Substituting into equations 9 for the values of $q_{jk}$ $0 \leqslant j \leqslant 1$, $0 \leqslant k \leqslant 3$ from equation 5 gives for $T_1 = 0.05$, $T_2 = 0.4$ seconds

$$Q(z_1^{-1},0) = 1-0.902z_1^{-1}$$

$$F_3(x) = -0.746 + 0.744 x$$
$$----10$$

and from equation 6 for $T_1 = 0.05$, $T_2 = 0.9$ seconds equation 9 became:

$$Q_1(z_1^{-1},0) = 1 - 0.902z_1^{-1}$$

and $F_3(x) = -1.8 + 1.794 x$
$$----11$$

In equation 10 and 11 $Q(z_1^{-1},0)$ is none zero for $|z_1^{-1}| \leqslant 1$ so that stability depends upon $F_3(x)$ which must be negative definite for $-1 \leqslant x \leqslant 1$. The functions are shown in Figure 2 where the lines representing these functions are plotted in the vicinity of $x = 1.0$ which is the critical region for this model.

Though $F_3(x)$ for both $T_1 = 0.05$ and $T_2 = 0.4$ seconds and for $T_1 = 0.05$ and $T_2 = 0.9$ seconds indicate that stable operating conditions prevail, the acute angle of $F_3(x)$, $T_1 = 0.05$, $T_2 = 0.9$ seconds and close proximity of this line to $x = 1.0$ suggest sustained transient oscillations will occur. $F_2(x)$, $T_1 = 0.05$, $T_2 = 0.4$ seconds is less volotile owing to the shallower sloping characteristics and the slightly greater value of the interception with the x axis.

## Simulated Response of Model

To validate the predictions given earlier system simulation was employed using a reputable, commercially available digital system simulation package. Initially to demonstrate the response of the system with monotonic sampling with $T_1 = T_2 = 0.05$ seconds the output following a 1% step change on the reference input is shown in Figure 3 over an interval of 10.0 seconds. Subsequently an 8 fold increase in $T_2$ to 0.4 seconds while maintaining $T_1$ at 0.05 seconds is examined. The system output response change following a 1% change in the reference input is shown in Figure 4 over an interval of 10.0 seconds. Finally, to study the worst case conditions, $T_2$ is increased to 0.9 seconds while retaining $T_1$ at 0.05 seconds. The response for this configuration is shown in figure 5 following a 1% change in the reference input over an interval of 30 seconds when $T_2$ at 0.91 seconds results in instability.

## Conclusion

In this exposition a closed loop digital control system has been investigated in an attempt to establish the range of data sampling rates in the inner and outer loops which could be tolerated. Initially a single rapid rate of sampling of 0.05 seconds was employed in both loops. This gave an acceptable closed loop response, as shown in Figure 3 with no overshoot condition. However an equally acceptable response with a 30% overshoot can be demonstrated when the outer loop sampling rate has been lowered significantly to 0.4 seconds is available as Figure 4 shows. Moreover, instability only occurs after more than doubling the sampling time to 0.9 seconds as Figure 5 shows.

Essentially the tolerance of the system model to changes in the data sampling rates used is noteworthy. Certainly data sampling rates in the range:

$$0.05 \leqslant T_2 \leqslant 0.2$$

have virtually no effect on the performance of this model. Thus changes for this and other LSI system models can now be investigated directly by way of the stability algorithms used in Section 3. Thereafter, asynchronous data sampling within given, established limits may be employed enabling more efficient, effective use of the multiplexing limit to achieved without significant performance penalties.

## REFERENCES

De Carlo R.A., Murray J and Sacks R, "Multivariable Nyquist Theory". Int J. Control vol 25, No 5, pp 657-675, 1977.

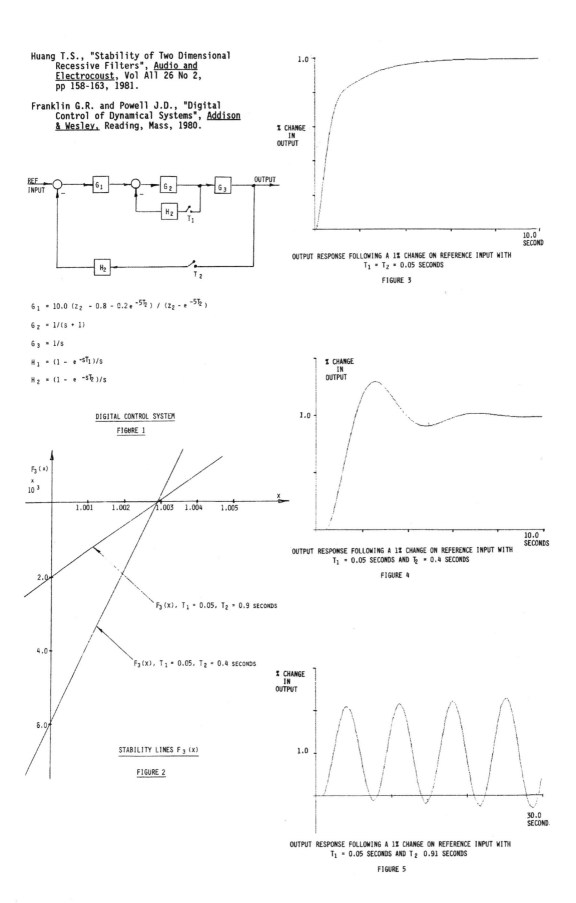

Huang T.S., "Stability of Two Dimensional
    Recessive Filters", <u>Audio and
    Electrocoust</u>, Vol All 26 No 2,
    pp 158-163, 1981.

Franklin G.R. and Powell J.D., "Digital
    Control of Dynamical Systems", <u>Addison
    & Wesley,</u> Reading, Mass, 1980.

REF INPUT — $G_1$ — $G_2$ — $G_3$ — OUTPUT

$H_2$    $T_1$

$H_2$    $T_2$

$G_1 = 10.0 \ (z_2 - 0.8 - 0.2 e^{-5T_2}) \ / \ (z_2 - e^{-5T_2})$

$G_2 = 1/(s + 1)$

$G_3 = 1/s$

$H_1 = (1 - e^{-sT_1})/s$

$H_2 = (1 - e^{-sT_2})/s$

DIGITAL CONTROL SYSTEM

FIGURE 1

% CHANGE IN OUTPUT

OUTPUT RESPONSE FOLLOWING A 1% CHANGE ON REFERENCE INPUT WITH
$T_1 = T_2 = 0.05$ SECONDS

FIGURE 3

$F_3(x)$
$x$
$10^3$

1.001  1.002  1.003  1.004  1.005      X

2.0

$F_3(x), T_1 = 0.05, T_2 = 0.9$ SECONDS

4.0

$F_3(x), T_1 = 0.05, T_2 = 0.4$ SECONDS

6.0

STABILITY LINES $F_3(x)$

FIGURE 2

% CHANGE IN OUTPUT

1.0

10.0 SECONDS

OUTPUT RESPONSE FOLLOWING A 1% CHANGE ON REFERENCE INPUT WITH
$T_1 = 0.05$ SECONDS AND $T_2 = 0.4$ SECONDS

FIGURE 4

% CHANGE IN OUTPUT

1.0

30.0 SECOND.

OUTPUT RESPONSE FOLLOWING A 1% CHANGE ON REFERENCE INPUT WITH
$T_1 = 0.05$ SECONDS AND $T_2$ 0.91 SECONDS

FIGURE 5

# MEDIUM LEVEL CONTROL OF PROCESSES:
# AN APPROACH TO THE CONEX DIRECT CONTROL

**F. Matia, R. Sanz, A. Jimenez and R. Galan**

*Department of "Automatica, Ingenieria Electronica e Informatica Industrial",*
*Universidad Politecnica de Madrid, J. Gutierrez Abascal, 2. 28006 Madrid, Spain*

Abstract. The Direct Control layer structure, inside the CONEX architecture, is presented. It is developed in C-Language over a real time operating system. Its level of working can be considered as medium, for it stays over the process interface layer but below the qualitative reasoning one. Its structure allows to include every control algorithm. In the commented present application, PID, fuzzy and state feedback control are employed. Dynamic modifications of the control algorithm parameters can be easily made either by the operator or by any other object of CONEX, and a set of controllers can be prepared for the same loop. Direct Control is independent of the machine in which the rest of the application is working. It is easy to configure from one process to another. These characteristics allow a wide application to the most different kind of processes. Finally an application to the cement industry is commented.

Keywords. Computer architecture; control applications; PID control; fuzzy control; state feedback; software development; real time computer systems; cement industry.

## INTRODUCTION

From the earliest controllers developed to the brand-new commercial control products, the level of their application has always been an object of discussion. Just above the plant, the PLCs allow to dispose a first control layer. This supplies lowest level control with simple classic loops, but the main difficult resides in its complex method of programming even for experimented users.

A first and commonly used step is to develop an upper control layer, usually implemented by high level programming languages which allow easier modifications of the control algorithms parameters and dispose of a friendly user interface. This level can include the classic control as PIDs, but also allow more sophisticated controllers such as the adaptive ones.

Expert systems have also been designed to resolve determinate control problems, usually employing rule based methods or even implementing fuzzy logic. In both cases we can catalogue them as a high level control layer.

## THE CONEX ARCHITECTURE

*Direct Control* (DC) is a software structure for real time control developed within the CONEX architecture (Sanz, 1991a). Basically, the CONEX architecture allows the integrated operation of several control layers and is organized following the principle of increasing precision with decreasing intelligence. It is composed of the following high level objects (HLOs) (see Fig. 1):

· Process Interface (CONEX-PI)
· Direct Control (CONEX-DC)
· Process Monitor (CONEX-PM)
· Model and Simulator (CONEX-MS)
· Intelligent Control (CONEX-IC)
· CONEX Monitor (CONEX-CM)
· Action Evaluator (CONEX-AE)

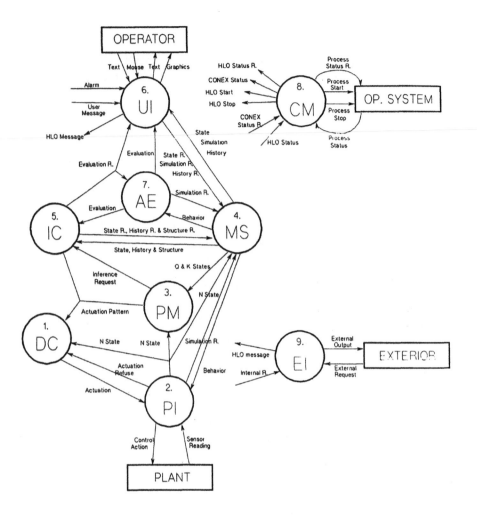

Fig. 1. The CONEX architecture.

- User Interface (CONEX-UI)
- External Interface (CONEX-EI).

DC is not in adversity with other control layers. In spite of this, it is in constant interaction with the other CONEX HLOs:

- DC sends direct control actuations through CONEX-PI to the plant, which returns either confirmation or refusal in answer to DC actuations.

- Allows the possibility of receiving orders (actuation patterns) from CONEX-IC and CONEX-PM which are translated into internal activities to attend, as actuations,

modification of control loop references and controller parameters.
- CONEX-CM carries the application giving the start and stop orders, and taking care of the correct execution of each HLO.

- CONEX-UI presents the direct control information as variable values and allows to modify control loop parameters.

## THE DIRECT CONTROL STRUCTURE

In order to support the last requirements, DC has been structured in eight internal objects (Matía, 1991), as it is shown in Fig. 2:

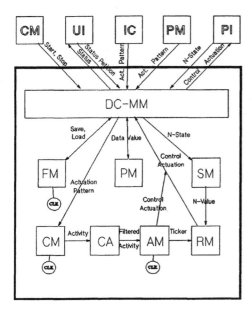

Fig. 2. Direct Control scheme

## The Message Manager

DC interacts with the other CONEX HLOs interchanging the following kind of messages:

Messages received:

- DC Start
- DC Stop
- Parameter Status Petition
- Parameter Configuration
- Actuation Pattern
- Numeric State of the variables
- Actuation Refuse/Confirmation

Messages sent:

- Stop Error
- User Message
- DC Status
- Control Action

The communication between DC and the other HLOs is done by using the CONEX Communication Layer functions developed over the TCP-IP transport services(Sanz, 1991b).

## The Parameters Manager

Its aim is to maintain the DC database information, so it may be consulted by the other HLOs. It also allows to modify the internal data values through configuration orders.

## The State Maintainer

It stores the numeric state of the plant variables received from CONEX-IP which are used by DC.

## The Command Manager

The command manager receives the actuation patterns (commands) coming form CONEX-IC and CONEX-PM, turning them into activities and sending these to the collision analyzer.

It is periodically executed by a clock.

Besides periodically, it takes patterns from the table in which they are achieved and reads the activities in them contained, sending to the collision analyzer those that must be immediately executed and deleting them from the pattern. As for the other activities, their remainder time for execution is refreshed.

There is an especial pattern, named *reset pattern* which deletes all activities (referring to modifications of actuations or loop references) achieved at that time. This is interesting when the user wants to actuate directly to plant without the CONEX help.

## The Collision Analyzer

The collision analyzer filters the new activities received from the command manager, and adds them to the table of activities.

It must determine if they strike with the present activities in the table of activities. It will be considered that a strike exists when an activity tries to make modifications over the same entities.

If the activity consists of a loop reference or a control actuation, we will have the following cases:

- if the activity is instantaneous it is directly sent to the activity manager to be immediately executed.

- if the activity is gradual (this means that the actuation must be increasingly done during a predeterminate period of time), the analyzer checks if it has also been fixed an initial value for the variable. In this case, this value is sent to the activity manager as an instantaneous activity. In this way, the information has been reduced in two new activities, where the second one is gradual.

If the activity consists of a controller, we may modify:

· the controller parameters

· the selected controller for a control loop.

## The Activity Manager

The activity manager is periodically executed by a clock and can be considered as the DC heart. It manages the temporal activities executing them. This may mean to send an actuation to CONEX-PI, to modify a control loop reference, or a ticker for the application of a control algorithm. In this last case, this object supplies to the regulation module the order to apply such algorithm.

It also supervises the temporal checking of the loop reference and actuation refreshments and those that must be immediately executed.

To allow this, it disposes of a table of activities that must be executed. This table is refreshed by the collision analyzer with the new filtered activities received. It executes those whose execution time has dropped off.

## The Regulation Module

The regulation module is the responsible of applying the control algorithms decided by the activity manager: PID, fuzzy or feedback state control (these are the control algorithms developed for this application, but any others can be used within this scheme).

This object may be considered the lowest level inside DC, because of the activity pattern filtering process, wich finally applies the control algorithm and supplies the control actuation.

It begins to run with a call from the activity manager and the corresponding control algorithm is immediately calculated. The control actuation is lastly sent to CONEX-PI.

The PID controllers have one input / one output, the fuzzy controllers allow three inputs / one output and the feedback state one allows three inputs / three outputs.

Another table achieves the existent relations between each control loop and its set of associated controllers, and the name of the controller selected at each time.

## TABLE 1 DC Internal Objects

| Object | TAG | Function |
|---|---|---|
| Message Manager | DC-MM | Interprets the network messages |
| Parameter Manager | DC-PM | Allows the access to the internal parameters |
| State Maintainer | DC-SM | Achieves the numeric state of the variables from the plant |
| Command Manager | DC-CM | Attends to the arrival of the commands from the other HLOs |
| Collision Analyzer | DC-CA | Analyzes the collisions among the commands |
| Activity Manager | DC-AM | Attends to the control activities |
| Regulation Module | DC-RM | Applies the control algorithms |
| File Manager | DC-FM | Manages the information between memory and the data files |

## The File Manager

The file manager resolves the equilibrium between the necessity to maintain the data files up-to-date, and the time employed in saving the data.

In case of failure, the system must be able to start over at the same point that it was.

The necessary information to be saved is the following:

Actuation patterns
Activities
Relations for each loop

Controller parameters
Control loop references
Numeric state
Control actuation
Global data

## CONTROL FEATURES

DC intends to support a medium level control and has been conceived to use classic PID control at the same time that other non-classic control algorithms as fuzzy techniques or feedback state control.

This flexibility can also be extended to the loop configuration, for the feedback state loops admit a three input/output controller, and the fuzzy controller accepts between three and seven linguistic terms, a variable number of inputs, different control ways as proportional-integral or proportional-derivative control (FPI, FPD), and independent design of the membership functions.

This converts the DC structure in a powerful tool of control for the most different kinds of processes.

## REAL TIME WORKING

An own characteristic of DC is its implementation based on a real time set of programs developed over a real time operating system as VMS in C-Language.

The communication between the DC internal processes is done by using shared memory functions and shared memory data techniques. Data access collisions are resolved using the System Event Flags (Burns, 1990).

Although DC needs an initial configuration of its loops (initial activities), its structure allows a dynamic configuration from external HLOs through actuation patterns.

Additionally, DC checks the correct execution of itself and its subprocesses, taking control and informing CONEX-CM in case of error. This is possible making an adequate use of the System functions (DEC, 1990).

The usual workings of DC are the following. In the most basic level, there are three objects working: the activity manager, the regulation module and the state maintainer. Only when a actuation pattern is received, the command manager starts its work.

Due to an operator request for some actuation from the user interface, or due to a CONEX actuation pattern, it begins a travel through the CONEX HLOs that is finally reflected in a direct control actuation on one or more PLC registers connected to the plant. The DC message manager receives the request in a actuation pattern form and achieves it as a set of activities to be interpreted. Parallelly, the command manager with a periodic timer fixed (1 minute in our application), reviews its pattern table, translating each pattern in a set of activities which are achieved in the activity table. This table is reviewed each 10 seconds by the Activity Manager (less is not necessary in cement processes) modifying the DC behaviour. This repercusses on a control actuation to CONEX-PI, that internally manages to send the new value to the plant. The following numeric state generated by CONEX-PI, will inform to CONEX (hence the operator) of the change, closing the loop.

Also, while all this is succeeding, the DC file manager is working, saving the main information of the database structure, and asynchronously other CONEX processes are attending their own tasks.

## THE CEMENT INDUSTRY APPLICATION

A software package application of this scheme has been developed for the Intelligent Control of a Cement Kiln, with ASLAND S.A. support. Initial software and control tests have been done in one of the ASLAND factories (Barcelona, Spain) to obtain guarantee of its correct working.

The software test is important to ensure the execution of the DC modules in parallel with the different processes from all the CONEX HLOs. The communication between each one and the rest of the objects, must reach the consideration of *real-time*. Without this condition would be impossible to handle the control we are searching for. This kind of test has demonstrated that parallel processing is not an utopia but a fact.

On the other hand, the control tests have been conceived to prove the reliability of conjugating the classic control with the fuzzy techniques and the state feedback control. Although the PID control does not need more test than the software one, for it is already implemented in the PLC, the most interesting case is to apply the fuzzy control to the traditional difficult loops as the control of the pressure under the clinker cooler grill.

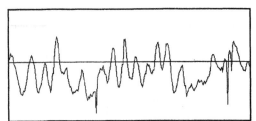

Fig. 3. Evolution of the pressure under the grill
(time = 25 minutes)

The material flowing (clinker) comes from the cement kiln and must be refrigerated in a cooler. This one has a grill with ventilators that blow air while the clinker goes over.

The clinker flow is measured with the pressure under the grill. In this way, the control of the clinker refrigeration is possible by modifying the grill speed and the air ventilators flow (Matía, 1990). The control loop should keep the pressure constant under the grill by modifying conveniently the grill speed (we know that an increase in the speed reference makes the grill speed grow up and decrease the pressure).

The typical evolution of the pressure signal is shown in Fig. 3. Notice there are continuous and persistent disturbances around the reference (horizontal line) due to the variability of the clinker flow, which is very unpredictable. There is also an asymmetric tendency to be under the reference, due to periodic unloadings of the grill slider. This clearly suggests the idea of employing a non-linear controller as the fuzzy one.

Before this work, only a PID controller was being used. Good results were obtained only for small deviations from the set point. But after applying fuzzy controllers, results show that they allow to implement easily the non-linearity required.

In other cases, a set of controllers can resolve a kind of problems in which the system to be controlled has a different behaviour at different workpoints. The CONEX-PM can detect this situations sending a pattern to DC that orders to change from one to another controller.

## CONCLUSION

As for contributions of this control structure we can say that:

- the controller parameters can be configured on-line both by CONEX and the operator

- DC allows to dispose of a set of controllers for the same loop.

- Changing from one kind of process to another it consists only in configurating conveniently data files, without touching the software.

Also besides a few limitations exist, because:

- its dependence from the other HLOs is only a function of the messages that they interchange.

- DC functioning is independent of the machine in which the other HLOs are working.

Finally, we want to consider as a significant result that one application is already running.

## ACKNOWLEDGES

We want to acknowledge the support of ASLAND S.A. and CDTI.

## REFERENCES

Burns, A. and Wellings, A. (1990). *Real-time systems and their programming languages*. Addison Wesley.

DEC (1990). *VMS programming reference manual*. Digital Equipment Corporation.

Matía, F. (1990). *Control y optimización de molinos en la industria del cemento*. Proyecto fin de carrera. Universidad Politécnica de Madrid.

Matía, F. and García, B. (1991). *Manual de referencia de Control Directo: estructura interna y algoritmos de control*. CONEX Project Report ASLAND/DISAM-04.91/02, Madrid.

Sanz, R., Jiménez, A., Galán, R., Matía, F. and Puente, E.A. (1991a). Intelligent process control: the CONEX architecture. *EURISCON congress*.

Sanz, R., Jiménez, A., Galán, R., Matía, F. and Puente, E.A. (1991b). Intelligent distributed process control. *P.D.COM congress*.

# AN INTEGRATED SYSTEM FOR SUPERVISORY PROCESS CONTROL USING ADA

**M. Marcos, F. Artaza and N. Iriondo**

*Departamento de Automatica, Electronica y Telecomunicaciones, Escuela de Ingenieros de Bilbao, Alameda de Urquijo S/N, 48013 Bilbao, Spain*

**Abstract.** The paper presents an integrated software package for supervisory process control using ADA language. This system performs process control using classical and heuristic controllers, supervision in order to detect abnormal situations and on-line monitoring.Using ADA language, the definition of the different processes(control, supervision and monitoring) can be easily implemented using the concurrent programming techniques inherent to the language. Specifically, a rule-based system which compensates for dead-zone phenomena in a position control servomechanism is developed. The supervisor module is designed to detect and avoid undesirable on-linear control phenomena such as steady state errors, limit cycling or excursions into unwanted zones. It constitutes an intelligent compensator for system input nonlinearities which may be generalized to other non-linearity types.

**Keywords.** Supervisory control; direct digital control; computer control; servomechanisms, real time computer systems.

## CONTENTS

In most practical applications of servomechanism control, the control system designed must contend with input nonlinearities,such as backlash, static friction, hysteresis, saturation and other generally undesirable characteristics that are usually unavoidable features of servoactuators. The present paper deals with one of these problems by addressing the control problem associated with dead-zone non-linearities due to static friction. Typical effects of dead-zone on the system response are steady state output errors and the possible appearance of limit cycling.

Dead-zone compensation using the injection of a high frequency periodic signal known as dither has been studied in great detail by several authors ( Mac Coll and Leroy, 1945 , Truxal 1955, Oldenburguer and Boyer 1962, Wagner 1972).

There are, however, situations in which dither compensation is inappropriate. Specifically, there are applications where mechanical components can be seriously worn or even damaged by continuous oscillatory dither action. For example, gear trains, valve membranes and piston seals are known to have their life cycles significantly reduced by dither.

It is also useful to note that digital computers technology has developed remarkably over the past thirty years. Such an improvement in both digital hardware and software makes it possible to use more complex control laws to mitigate the effect of non-linearities. In this context, Artificial Intelligence methods seem to have significant potential in the type of applications which involve a great amount of heuristic (Astrom and others 1986).

At present, there are numerous researchers using artificial intelligence tehniques to implement direct digital control such as fuzzy controllers (Francis and Leitch, 1984, Sripada and others, 1987, Ollero and García Cerezo, 1988, García Cerezo and others, 1986), or supervisory functions for detecting instalability trends ( Sanz and Ollero, 1986, Sanz and others, 1988, Ollero and Sanz, 1987).

This paper follows the ideas of Marcos and Wellstead (1990) who define a supervisory system for real-time position control of servomechanisms. This system detects abnormal situations and based on heuristic rules takes decisions in order to lead the process to a normal operation.

The goal is to present a software scheme which allows not only the definition of the different tasks: control, monitoring andsupervision, but assures a correct interaction between them.

The paper is laid out as follows: Section II describes the software system structure and main characteristics of the used software tool: the ADA language. Section III deals with the supervisory control structure which comprises the supervisor module and a set of specific-purpose controllers. Finally, Section IV shows an application example where thesupervisory structure is applied to a servomechanism with a time-varying dead-band provoked by static friction.

SOFTWARE SYSTEM STRUCTURE

The software structure to be designed must assure not only real-time process control but on-line supervision and monitoring. Because of that, the selected software tool must allow the definition of the different tasks: control, supervisor and monitor assuring synchronization between them.

The ADA language (Ichbiah 1983, Burns 1985) provides for the direct programming of parallel activities. Within an ADA program there may be a number of tasks, each of which has its own thread of control. It is thus possible to match the parallel nature of the application area with syntactical forms that reflect these structures.

Even more, a program written in ADA language will, if there are processes available, run each of these threads of control in parallel. Otherwise each of these theards will be interleaved.

Finally, it provides synchronization mechanisms, the rendezvous, which assures data integrity.

These characteristics as well as task priority assignment and pre-emptive scheduling make the ADA language a good choice for achieving our purposes.

Figure 1 shows the overall scheme which is proposed here. The "system state" represents all the information needed for controlling and supervising the process. This information is updated by the ACQUISITION TASK (reference and process output signal) and by the SUPERVISOR TASK (system history). At the same time, this information is consulted by the SUPERVISOR TASK, in order to decide if a controller switching is necessary and by the control algorithm active in order to compute the current control action.

The communication between CONTROLLER TASK and MONITOR TASK is implemented using the classical producer/consumer scheme.

The highest priority is assigned to the ACQUISITION TASK which uses preemptive runtime scheduling.This task constitutes the routine service clock signal interrupt.

The KEYBOARD and MONITOR tasks are executed by the scheduler using time slicing mechanism algorithm. They offer the user different possibilities for process monitoring such as setting scales, monitoring pauses, etc.

SUPERVISORY CONTROL STRUCTURE

The supervisory task is based on the possibility of switching control of the process between different controllers depending on the observed process behavior (Marcos and Wellstead, 1990). The switched mode action of Fig. 1 means that it is possible to use classical controllers during normal (linear) operation. However, when a non-desired operation which cannot be overcome by the classical controller is detected, the controller is switched to one specially designed to overcome the abnormal situation.

To achieve this goal, the SUPERVISOR TASK access to the "system state" which contains not only the acquired information (reference and process output signals) but the so called "system history".

This information is generated by a set of "detector modules" which establishes the system evolution. For instance, and recalling the input non-linearity under consideration is a dead-zone, the output detector establishes that:

(i)     steady output
(ii)    a change in output slope
(iii)   output in progress

Additionally, the control action detector establishes that:

(i)     control-action in progress
(ii)    control-action inside the dead-zone
(iii)   control-action too high or too low

Once the detector modules are evaluated, the SUPERVISORY TASK applies the set of implemented rules of type

if <premise> then <action>

Figure 2 shows the set of implemented rules for servomechanism position control.

The premises are logical combinations of detector outputs and previous history of the system.

A true-premise implies a selection of a controller (decision). When the premise also implies a controller switching (action), the SUPERVISOR TASK executes the selected controller algorithm. In summary, the SUPERVISOR TASK computes the premises sequentially until an action is carried out.

The correction controller must avoid and correct the errors provoked by the lack of precision of D/A and A/D converters as well as by the finite word length of the computer.

The Dead-Zone controller compensates for the actuator dead-zone and takes the system to the point it would have reached if it had been linear. In order to avoid oscillatory system response, two versions have been implemented, the fast and the slow dead-zone controllers. Their goal is to identify the dead-band increasing the control action until the dead-band is overcome.

## APPLICATION EXAMPLE

The structure defined in the present study has been applied to a servomechanism called "the ball & hoop apparatus" (Wellstead, 1983). This system consists of a hoop which is free to rotate under the action of a DC servomotor. The system involves a dead-zone non-linearity due to static friction. Moreover, the dead-band width varies in lubricant viscosity. The hoop position is taken as the control output and a discretized PID controller has been applied to the system. For the purposes of this paper we will focus upon the non-linearity compensation aspects of the ball & hoop control system. As mentioned above, a well known phenomenon associated with the dead-zone non-linearity is limit cycling in the system response. This can be observed in the ball & hoop system as illustrated in Fig. 3a under the action of a PID controller in which the control system was purely linear in nature, no non-linear compensation has been used.

Figure 3b shows the corresponding process response when the supervisory scheme is introduced. It can be seen that not only does the limit cycling disappear but the steady state error is also removed.

## CONCLUSIONS

A software system for real-time process control, supervision and monitoring, based on ADA language characteristics has been presented. This system defines the different tasks involve in a control system such as acquisition, supervision, control algorithms, process monitoring and user interface. In the same way, mechanisms for task interaction have been designed.

The system allows flexibility for different hardware environments as well as the integration of new controller algorithms and the inclusion of new rules depending on the characteristics of the process to be controlled.

The control structure presented here has been tested by applying it to a scale model of a servomechanism. Results of the real-time control have been shown and they proof the improvements on the response which are contributed by the supervisory level.

## REFERENCES

Astrom,K.J. and others. (1986) "Expert Control".
Automatica,22,pp.27-286.
Ichbiah, J. and others. (1983) "A.R.M. Reference Manual for the Ada Programming Language". ANSI/MIL-STD-1815A.
Burns,A. (1985) "Concurrent Programming in Ada". Cambridge University Press, Cambridge.
Francis,J.C. and R.R. Leitch (1984) "ARTIFACT: A Real Time Shell for Intelligent Feedback Control".Procs. BCS Conf. Expert Sys., pp.157-162.
García Cerezo,A. and others. (1988)."Expert Supervision and Auto Tuning of Digital Control Systems".IEEE Workshop on AI for Industrial Applications.
Mac Coll,L.A.and Leroy (1945)."Fundamental Theory of Servomechanisms".Van Nostrand, New York.
Marcos, M. and P.E. Wellstead (1990). "An Intelligent Rule-based Compensator for Control System Input Nonlinearities". Proc. of the 1990 American Control Conference",2,pp:1467-1473.
Oldenburger,R. and R.C. Boyer (1962) "Effects of Extra Sinusoidal Inputs to Nonlinear Systems". Trans. of ASME Journal of Basic Eng.,pp.559-570.
Ollero,A. and A. García Cerezo (1988)."Direct Digital Control, Auto Tuning and Supervision Using Fuzzy Logic". Fuzzy Set and Systems. North Holland.
Ollero,A and R. Sanz (1987). "Intelligent Workstation for Control and Supervision in Industrial Processes:Software Aspects". Procs. VLSI and Computers, 1st. Int. Conf. on Computers Tech., Systems and Applications. Hamburg
Sanz,R. and others. (1988)."Adaptive Control with a Supervisor Level using a Rule Based Inference System with Approximate Reasoning". 12th IMACS World Congress Paris.
Sanz,R. and A. Ollero (1986). "A Rule Based Inference Method for Supervision of Self-Tuning Controllers by Using Microcomputers". Preprints of IFAC Symp. Corp. Instr. and Tech. for Low Cost Aut. Valencia.
Spirada,N.R. and others. (1987). " A application for Process Regulation and Servo Control". IEEE Proc., 134, pp.251-259.
Truxal,J.G. (1955)."Automatic Feedback Control System Synthesis". Mc Graw-Hill, New York.
Wagner,J.A. (1972)."A Dither Technique Applied to a Constant Fuel-Rate Problem ". IEEE Trans. on Automatic Control, pp.162-164.
Wellstead,P.E. (1983)."The Ball and Hoop System". Automatica, 19, pp.401-406.

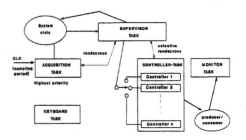

Fig. 1. Software system structure

| premise | conditions |
|---|---|
| 1. system inside dead-zone | previous decision not CC<br>previous decision not DZC<br>output is steady<br>error is nonzero<br>control-action inside dz<br>no change in set point |
| 2. limit cycling | previous decision not CC<br>previous decision not DZC<br>output in progress<br>settling time exceeded<br>transient state |
| 3. control degradation | previous decision not CC<br>previous decision not DZC<br>output is steady<br>error is zero<br>control-action in progress<br>steady state |
| 4. output degradation | previous decision not CC<br>previous decision not DZC<br>steady state<br>error is nonzero |
| 5. normal operation | previous decision not CC<br>previous decision not DZC |
| 6. system reaches steady state | previous decision DZC<br>output is steady<br>error is zero |
| 7. new transient state begins | previous decision DZC or CC<br>transient state |
| 8. process output is far from set-point | decision is DZC<br>error is big |
| 9. process output is near to set-point | decision is DZC<br>error is small |

| premise | decision | action |
|---|---|---|
| 1 | DZC | none |
| 2 | DZC | none |
| 3 | CC | switch on CC |
| 4 | DZC | none |
| 5 | LC | switch on LC |
| 6 | LC | switch on LC |
| 7 | LC | switch on LC |
| 8 | DZC | switch on FDZC |
| 9 | DZC | switch on SDZC |

Fig 2. Set of rules. CC stands for Correction Controller, DZC for Dead-Zone Controller, FDZC for Fast Dead-Zone Controller, SDZC for Slow Dead-Zone Controller and LC for Linear Controller

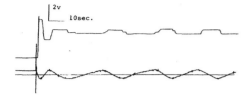

Fig 3.a. Discretized PID without supervision.
T=0.1 sec, Kp=0.2 Ki=0.2, Kd=0.1

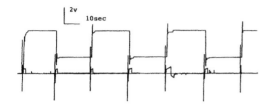

Fig 3.b. System response with supervision

# AUTHOR INDEX

# KEYWORD INDEX

Printed and bound by CPI Group (UK) Ltd, Croydon, CR0 4YY

03/10/2024

01040320-0014